The Mathematics of Oil Recovery

Based on the proceedings of a conference on the mathematics of oil recovery organized by the Institute of Mathematics and its Applications in association with the Society of Petroleum Engineers and held at Robinson College, Cambridge in July 1989.

Edited by

P. R. KING

Senior Petroleum Engineer
BP Research, Sunbury-on-Thames

CLARENDON PRESS · OXFORD · 1992

Oxford University Press, Walton Street, Oxford OX2 6DP

Oxford New York Toronto
Delhi Bombay Calcutta Madras Karachi
Petaling Jaya Singapore Hong Kong Tokyo
Nairobi Dar es Salaam Cape Town
Melbourne Auckland

and associated companies in
Berlin Ibadan

Oxford is a trade mark of Oxford University Press

Published in the United States
by Oxford University Press, New York

© Institute of Mathematics and its Applications, 1992

All rights reserved. No part of this publication may be reproduced,
stored in a retrieval system, or transmitted, in any form or by any means,
electronic, mechanical, photocopying, recording, or otherwise, without
the prior permission of Oxford University Press

A catalogue record for this book is available from the British Library

Library of Congress Cataloging in Publication Data
(Data applied for)

Printed in Great Britain by
St Edmundsbury Press, Bury St Edmunds, Suffolk

Preface

The First European Conference on the Mathematics of Oil Recovery was held at Robinson College, Cambridge in July 1989. It followed from a conference held two years earlier at the same place on the Mathematics of Oil Production. At that meeting it was decided that the time was right to hold a series of meetings, based in Europe, on the more mathematical aspects of oil recovery problems. The conference on which this book is based was the result, followed by a meeting in France in 1990 and further meetings every two years thereafter. It was organized by the Institute of Mathematics and its Applications in association with the Society of Petroleum Engineers.

There are, of course, many mathematical problems in oil recovery so the conference concentrated on three main themes:

The statistical description of the porous medium geological environment at various length scales.

The derivation of the appropriate spatially averaged equations of motion for multiphase displacement processes.

Solution of the resulting partial differential equations by numerical methods.

The need to solve these problems becomes economically more significant as oil production declines from established North Sea fields or is restricted to geologically more complicated fields. This is a very broad subject area with many active researchers from industry, government institutions and universities. The aim of the meeting was to bring together mathematicians and engineers from these groups in an informal atmosphere to discuss common problems.

These proceedings represent both the oral presentations made at the meeting and the poster presentations.

The editor would like to thank members of the international organising committee: Chairman, Prof. Sir Sam Edwards (Cambridge University), Prof. J Archer (Imperial College), Dr. I. Cheshire (ECL), Dr. C. Farmer (AEA Winfrith), Dr. J. Fayers (BP), Dr. D. Field (Hamilton/SPE), Dr. A. Grant (Shell), Dr. O. Guillon (ELFO), Dr. O. Jensen (Maersk), Dr. P. King (BP), Dr. B. Legait (IFP), Prof. D. Tehrani (Britoil) and Dr. P. Thomassen (Statoil). The authors of the papers and referees are also thanked for their help in making these proceedings possible. Finally, but by no means least, the IMA deserves many thanks for ensuring the smooth running of the conference and its great help in the production of this book.

P R King

BP Research, Sunbury

ACKNOWLEDGEMENTS

The Institute thanks the authors of the papers, the editor, Dr. P. King (BP Research) and also Mrs Karren Waller for preparing the papers.

CONTENTS

Contributors	xiii
A Review of Reservoir Simulation Techniques and Considerations on Future Developments by O.K. Jensen	1
Simulation of Waterflood Fracture Growth with Coupled Fluid Flow, Temperature and Rock Elasticity by P.J. Clifford	23
Corner Point Geometry in Reservoir Simulation by D.K. Ponting	45
A Boundary Element Solution to the Transient Pressure Response of Multiply Fractured Horizontal Wells by C.P.J.W. van Kruysdijk and G.M. Dullaert	67
A Convective Segregation Model for Predicting Reservoir Fluid Compositional Distribution by F. Montel, J.P. Caltagirone and L. Pebayle	85
Effects of Heterogeneities on Phase Behaviour in Enhanced Oil Recovery by F.J. Fayers, J.W. Barker and T.M.J. Newley	115
On the Strict Hyperbolicity of the Buckley-Leverett Equations for Three-Phase Flow by L. Holden	151
Application of Fractional-Flow Theory to 3-Phase, 1-Dimensional Surfactant Flooding by S.I. Aanonsen	161
Averaging of Relative Permeability in Heterogeneous Reservoirs by S. Ekrann and M. Dale	173
Equations for Two-Phase Flow in Porous Media Derived from Space Averaging by D. Pavone	199
Statistical Physics and Fractal Displacement Patterns in Porous Media by T. Jossang	211
Simulations of Viscous Fingering in a Random Network by M.J. Blunt, P.R. King and J.A. Goshawk	243
A Simple Analytical Model of the Growth of Viscous Fingers in Heterogeneous Porous Media by J.N.M. van Wunnik and K. Wit	263
Simple Renormalization Schemes for Calculating Effective Properties of Heterogeneous Reservoirs by J.K. Williams	281
Finite Element Principles for Evolutionary Problems, with Applications to Oil Reservoir Modelling by K.W. Morton and A.K. Parrott	299

Multigrid Methods in Modelling Porous Media Flow by F. Brakhagen and T. Fogwell	323
Local Grid Refinement by I.M. Cheshire and A. Henriquez	337
Analysis of a Model and Sequential Numerical Method for Thermal Reservoir Simulation by J.A. Trangenstein	359
Operator Splitting and Domain Decomposition Techniques for Reservoir Flow Problems by H.K. Dahle, M.S. Espedal, O. Pettersen and O. Saevareid	381
Simulation of Compositional Reservoir Phenomena on a Hypercube by J.E. Killough and R. Bhogeswara	393
The Parallelisation of Bosim, Shell's Black/Volatile Oil Reservoir Simulator by D.T. van Daalen, P.J. Hoogerbrugge, J.A. Meijerink and R.J.A. Zeestraten	411
Monte Carlo Simulation of Lithology from Seismic Data in a Channel-Sand Reservoir by P. Doyen, T. Guidish and M. de Buyl	423
Numerical Rocks by C.L. Farmer	437
Modelling Flow through Heterogeneous Porous Media using Effective Relative Permeabilities Generated from Detailed Simulations by A.H. Muggeridge	449
Viscous/Gravity Scaling of Pseudo Relative Permeabilities for the Simulation of Moderately Heterogeneous Reservoirs by J. Killough and Y. P. Fang	471
Microscopic Flow and Generalized Darcy's Equations by F. Kalaydjian and C-M. Marle	495
Prediction of Permeability from a Combination of Mercury Injection and Pore Image Analysis Data by D. MacGowan	515
The Inclusion of Molecular Diffusion Effects in the Network Modelling of Hydrodynamic Dispersion in Porous Media by K.S. Sorbie and P.J. Clifford	529
The Network Simulation of Displacement Processes in Fractured Reservoir Matrix Blocks by M.D. Santiago and R.A. Dawe	551
An Analytical Solution to a Multiple-Region Moving Boundary Problem: Nonisothermal Water Injection into Oil Reservoirs by R.B. Bratvold and R.N. Horne	573

Statistical and Experimental Calculation of Effective Permeabilities in the Presence of Oblique Shales by O.B. Abu-elbashar, T.S. Daltaban, C.G. Wall and J.S. Archer	597
Reservoir Simulation Using Mixed Methods, a Modified Method Characteristics, and Local Grid Refinement by M.S. Espedal, O. Saevareid, R.E. Ewing and T.F. Russell	617
Tracking an In-Situ Combustion Front Using the Thin Flame Technique by R. Davies	631
Averaging of Relative Permeability in Composite Cores by M. Dale	649
A Reservoir Simulator Based on Front-Tracking by F. Bratvedt, K. Bratvedt, C. Buchholz, D.T. Rian and N.H. Risebro	683
Effective Absolute Permeability in the Presence of Sub-Grid Heterogeneities: An Analytical Approach by H.N.J. Poulisse	699
Relative Permeabilities and Capillary Pressure Estimation Through Least Square Fitting by C. Chardaire, G. Chavent, J. Jaffré and J. Liu	721
Coning Simulation More Accurately by F.J.T. Floris	735
A Finite Difference Scheme for a Polymer Flooding Problem by A. Tveito and R. Winther	745
Two-Phase Flow in Heterogeneous Porous Media: Large-Scale Capillary Pressure and Permeability Determination by M. Quintard, H. Bertin and S. Whitaker	751
A Matrix Injection Simulator by E. Touboul	767
Effect of Fluid Properties on Convective Dispersion: Comparison of Analytical Model with Numerical Simulations by L.J.T.M. Kempers	775
Fundamental Mechanisms of Particle Deposition on a Porous Wall: Hydrodynamical Aspects by P. Schmitz, C. Gouverneur and D. Houi	785
Wetting Phenomena in Square-Sectional Capillaries by A. Winter	791
A Numerical Method for Simulation of Centrifuge Drainage by I. Aavatsmark	799
Gas-Coning by a Horizontal Well by P. Papatzacos	805

CONTRIBUTORS

S.I. AANONSEN; Norsk Hydro, P.O. Box 4313, N-5028 Bergen, Norway.

I. AAVATSMARK; Norsk Hydro, P.O. Box 4313, N-5028 Bergen, Norway.

O.B. ABU-ELBASHAR; Imperial College of Science and Technology, Department of Mineral Resources Engineering, Royal School of Mines, Prince Consort Road, London, SW7 2BP.

J.S ARCHER; Imperial College of Science and Technology, Department of Mineral Resources Engineering, Royal School of Mines, Prince Consort Road, London, SW7 2BP.

J.W. BARKER; BP Research Centre, Sunbury Research Centre, Sunbury-on-Thames, Middlesex, TW16 7LN.

H. BERTIN; Laboratory Energétique et Phénoménes de Transfert, EN-SAM, Esplanade des Arts et Metiers, 33405 Talence Cedex, France.

R. BHOGESWARA; 5000 Calhoon 116, Houston, Texas 77004, USA.

M.J. BLUNT; B.P. Research Centre, Sunbury Research Centre, Sunbury-on-Thames, Middlesex, TW16 7LN.

F. BRAKHAGEN; Gesellschaft Für Mathematik Und Datenverarbeitung (GMD), Postfach 1240, D-5205 Sankt Augustin, Germany.

F. BRATVEDT; Department of Mathematics, University of Oslo, P.O. Box 1053, Blindern, N-0316 Oslo 3, Norway.

K. BRATVEDT; Department of Mathematics, University of Oslo, P.O. Box 1053, Blindern, N-0316 Oslo 3, Norway.

R.B. BRATVOLD; IBM European Petroleum, Competency Group, Auglendsdalen 81, 4000 Stavanger, Norway.

C. BUCHHOLZ; Department of Mathematics, University of Oslo, P.O. Box 1053, Blindern, N-0316, Oslo 3, Norway.

J.P. CALTAGIRONE; M.A.S.T.E.R., Université de Bordeaux I, Talence, France.

C. CHARDAIRE; Institut Français du Petrole, 1-4 Air de Bois Preau, 92506 Rueil Malmaison, France.

G. CHAVENT, Ceremade, Université Paris-Dauphine and INRIA, Rocquencourt, France.

I.M. CHESHIRE; ECL Petroleum Technologies, Highlands Farm, Greys Road, Henley-on-Thames, Oxford, RG9 4PS.

P.J. CLIFFORD; B.P. Research Centre, Sunbury Research Centre, Chertsey Road, Sunbury-on-Thames, Middlesex, TW16 7LN.

H.K. DAHLE; Department of Applied Mathematics, University of Bergen, Allegt 55, N5007 Bergen, Norway.

M. DALE; Rogaland Research Institute/Rogaland Regional College, P.O. Box 2503, Ullandhaug, N-4001 Stavanger, Norway.

T.S. DALTABAN; Imperial College of Science and Technology, Department of Mineral Resources Engineering, Royal School of Mines, Prince Consort Road, London, SW7 2GP.

R. DAVIES; Department of Petroleum Engineering, Stanford University, California 94305, USA.

R.A. DAWE; Imperial College of Science and Technology, Department of Mineral Resources Engineering, Royal School of Mines, Prince Consort Road, London, SW7 2AZ.

M. DE BUYL; Western Geophysical, A Division of Western Atlas International, 3600 Briarpark Drive, Houston, Texas, 77042.

P. DOYEN; Western Geophysical, A Division of Western Atlas International, 3600 Briarpark Drive, Houston, Texas, 77042.

G.M. DULLAERT; University of Technology Delft, The Netherlands.

S. EKRANN; Rogaland Research Institute, P.O. Box 2503, Ullandhaug, N-4001 Stavanger, Norway.

M.S. ESPEDAL; Department of Applied Mathematics, University of Bergen, Allegt 55, N5007 Bergen, Norway.

R.E. EWING; University of Wyoming, Enhanced Oil Recovery Institute, P.O. Box 3036, Laramie, Wy 82071, USA.

Y.P. FANG; Department of Chemical Engineering, University of Houston, Houston, Texas, 77204-4792, USA.

C.L. FARMER; Winfrith Petroleum Technology, United Kingdom Atomic Energy Authority, Winfrith, Dorchester, Dorset, DT2 8DH.

F.J. FAYERS; BP Research Centre, Sunbury Research Centre, Sunbury-on-Thames, Middlesex, TW16 7LN.

F.J.T. FLORIS; TNO Delft, P.O. Box 285, 2600 AG Delft, Holland.

T.W. FOGWELL, Gesellschaft Für Mathematik und Datenverarbeitung (GMD), Postfach 1240, D-5205 Sankt Augustin, Germany.

J.P. GOSHAWK; B.P. Research Centre, Sunbury Research Centre, Sunbury-on-Thames, Middlesex, TW16 7LN.

C. GOUVERNEUR; Institut de Mécanique des Fluides de Toulouse, Avenue Camille Soula, 31400 Toulouse, France.

T. GUIDISH; Western Geophysical, A Division of Western Atlas International, 3600 Briarpark Drive, Houston, Texas, 77042.

A. HENRIQUEZ; Statoil, Den Norske Stats., Oljeselskapp AS, Postboks 300, 4001 Stavanger, Norway.

P.J. HOOGERBRUGGE; Koninklijke Shell Exploration and Production Laboratory, P.O. Box 60, 2200 AB, Rijswijk, The Netherlands.

L. HOLDEN; Norwegian Computing Centre, P.O. Box 1141, Blindern, 0314 Oslo 3, Norway.

R.N. HORNE; Department of Petroleum Engineering, Stanford University, California, USA.

D. HOUI; Institut de Mécanique des Fluides de Toulouse, Avenue Camille Soula, 31400 Toulouse, France.

J. JAFFRE; INRIA, Domaine de Voluceau, Rocquencourt, B.P. 105, 78153 Le Chesnaif Cedex, France.

O.K. JENSEN; Maersk Olie Og Gas A/s, Espalanaden 50, DK-1263 Copenhagen, Denmark.

T. JOSSANG; Department of Physics, University of Oslo, P.O. Box 1048 Blinden, 0316 Oslo 3, Norway.

F. KALAYDJIAN; Institut Français de Petrole, B.P. 311, 92 506 Rueil Malmaison Cedex, France.

L.J.T.M. KEMPERS; Koninklijke, Shell Exploratory en Produktie Laboratorium, Postbus 60, 2200 AB, Rijswijk, The Netherlands.

J. KILLOUGH; Department of Chemical Engineering, University of Houston, Texas, 77204-4792, USA.

P.R. KING; B.P. Research Centre, Sunbury Research Centre, Chertsey

Road, Sunbury-on-Thames, Middlesex, TW16 7LN.

C.P.J.W. VAN KRUYSDIJK; Koninklijke, Shell Exploratory en Produktie Laboratorium, Postbus 60, 2200 AB, Rijswijk, The Netherlands.

J. LIU; INRIA, Domaine de Voluceau, Rocquencourt B.P. 105, 78153 Le Chesnaif Cedex, France.

C.M. MARLE; Institut Français du Petrole, B.P. 311, 92 506 Rueil Malmaison Cedex, France.

J.A. MEIJERINK; Koninklijke Shell Exploration and Production Laboratory, P.O. Box 60, 2200 AB, Rijswijk, The Netherlands.

F. MONTEL; ELF Aquitaine, Centre J. Feger, Avenue Larribau, 64018 Pau Cedex, France.

K.W. MORTON; Numerical Analysis Group, Oxford University Computing Laboratory, 8-11 Keble Road, Oxford, OX1 3QD.

A.H. MUGGERIDGE; B.P. Research Centre, Sunbury Research Centre, Chertsey Road, Sunbury-on-Thames, Middlesex, TW16 7LN.

D. MACGOWAN; B.P. Research Centre, Sunbury Research Centre, Chertsey Road, Sunbury-on-Thames, Middlesex, TW16 7LN.

T.M.J. NEWLEY; B.P. Research Centre, Sunbury Research Centre, Chertsey Road, Sunbury-on-Thames, Middlesex, TW16 7LN.

P. PAPATZACOS; Rogaland University, Hogskolesenteret 1 Rogaland, Postbox 2557, Ullandhaug, 4004 Stavanger, Norway.

A.K. PARROTT; Numerical Analysis Group, Oxford University Computing Laboratory, 8-11 Keble Road, Oxford, OX1 3QD

D. PAVONE; Institut Francais de Petrole, B.P. 311, 92506, Rueil, Malmaison, Cedex, France.

L. PEBAYLE; M.A.S.T.E.R., Université de Bordeaux I, Talence, France.

O. PETTERSEN; Department of Applied Mathematics, University of Bergen, Allegt 55, N5007 Bergen, Norway.

H.N.J. POULISSE; Koninklijke, Shell Exploration and Production Laboratory, P.O. Box 60, 2280 AB Rijswijk, The Netherlands.

D. PONTING; ECL Petroleum Technologies, Highlands Farm, Greys Road, Henley-on-Thames, Oxon, RG9 4PS.

M. QUINTARD; Laboratory Energétique et Phénoménes de Transfert, EN-

SAM, Esplande des Arts et Metiers, 33405 Talence Cedex, France.

D.T. RIAN; Scandpower A.S., P.O. Box 3, N-2007 Kjeller, Norway.

N.H. RISEBRO; Department of Mathematics, University of Oslo, P.O. Box 1053, Blindern, N-0316 Oslo 3, Norway.

T.F. RUSSELL; University of Colorado at Denver, P.O. Box 170, Denver, CO 80204.

M.D. SANTIAGO; Imperial College of Science and Technology, Mineral Resources Engineering Department, London, SW7 2AZ.

O. SAEVAREID; Department of Applied Mathematics, University of Bergen, Allegt 55, N5007 Bergen, Norway.

P. SCHMITZ; Institut de Mécanique des Fluides de Toulouse, Avenue Camille Soula, 31400 Toulouse, France.

K.S. SORBIE; Department of Petroleum Engineering, Heriot-Watt University, Edinburgh, EH14 4AS.

E. TOUBOUL; Dowell Schlumberger, 21 Moling 1a Charcotte, 42003 St. Etienne, Cedex 1, France.

J. TRANGENSTEIN; Laurence Livermore National Laboratory, P.O. Box 808, Livermore, CA 94550 L-316, USA.

P. TVEITO; Department of Informatics, University of Oslo, P.O. Box 1080, Blindern, 0316 Oslo 3, Norway.

D.T. VAN DAALEN; Koninklijke Shell Exploration and Production Laboratory, P.O. Box 60, 2200 AB, Rijswijk, The Netherlands.

J.N.M. VAN WUNNIK; Koninklijke, Shell Exploratie en Produktie Laboratorium, Postbus 60, 2200 AB, Rijswijk, The Netherlands.

C.G. WALL; Imperial College of Science and Technology, Department of Mineral Resources Engineering, Royal School of Mines, Prince Consort Road, London, SW7 2BP.

S. WHITAKER; Department of Chemical Engineering, University of California, Davis, CA95616, USA.

J.K. WILLIAMS; B.P. Research Centre, Sunbury Research Centre, Sunbury-on-Thames, Middlesex, TW16 7LN.

A. WINTER; Geological Survey of Denmark, Thorvej 8, Denmark 2400, Copenhagen NV, Denmark.

R. WINTHER; Department of Informatics, University of Oslo, P.O. Box 1080, Blindern, 0316 Oslo 3, Norway.

K. WIT; Koninklijke, Shell Exploratie en Produktie Laboratorium, 2288GD Rijswijk, The Netherlands.

R.J.A. ZEESTRATEN; Koninklijke Shell Exploration and Production Laboratory, P.O. Box 60, 2200 AB, Rijswijk, The Netherlands.

A REVIEW OF RESERVOIR SIMULATION TECHNIQUES AND CONSIDERATIONS ON FUTURE DEVELOPMENTS

Ole Krogh Jensen
(Petroleum Engineering Department,
Mærsk Olie og Gas AS, Copenhagen)

ABSTRACT

Reservoir simulation technology has been developed ever since the first computers were built. By the late 1980'ies it is now a considerable industry in itself with a large userbase. The advancements of computing technology together with enhancements in numerical techniques have given the industry a sophisticated tool for prediction of reservoir behaviour.

Black oil and compositional reservoir simulation has to a large extent driven this development especially through the 1970'ies where also the engineering methodology was conceived to do reservoir simulation studies properly with pseudofunctions etc.

In the early 1980'ies the concept of integrated studies became to a larger extent a vital part of any reservoir simulation studies. The use of geological modelling with a computer has during the last years also become an important part of reservoir analysis for certain types of fields to give a better reservoir input definition.

The historical trends in the technological development are reviewed, and consideration of future technological developments is given. Reservoir geological modelling is foreseen to be a growth area. The standard black oil reservoir simulation will be continuously refined and compositional formulations will gradually be more and more used. The use of dimensionless analysis and numbers should be advanced.

1.0 INTRODUCTION

Reservoir simulation is today almost synonymous with the use of black oil reservoir simulation programs as they are used most widely in the industry. Before computer technology became available to any large extent, analysis of reservoir behaviour was done with mathematical techniques suitable for analytical (by hand) solutions. The work by Muskat (1) stands out as a pioneering achievement. With reservoir simulation being advanced as the prime tool for reservoir analysis for the oil industry and with the technological achievements through the 1960'ies, 70'ies and 80'ies, the development of classical type analysis and computer based methods separated into two distinct approaches to reservoir analysis. The main difference between computer based simulation and "classical" calculation methods are whether or not first principles of physics are integrated, geometric approximations, and the amount of data that can be handled.

The sophistication in reservoir analysis is ultimately dependent on the amount of reserves, oil production levels and data quality.

Nearly all of the reservoir simulation technology has been developed for medium-to-large size fields. Furthermore, the technology is today mostly used for these types of fields. For some types of fields, simulation software is, although still in its evolutionary stage, e.g. fractured fields.

A review of reservoir simulation technology can be done in many ways and several text books exist already. Although the author of this paper has been associated with reservoir simulation both as a developer and user, it still intrigues me how much you can get out of reservoir analysis with classical analysis - often sufficient to make major field development decisions. This review supports the idea that classical analysis and reservoir simulation should go hand-in-hand, and research efforts should be pursued in achieving generality.

2.0 CLASSICAL ANALYSIS

In 1937 M. Muskat (1) published the classical book "Flow Through Porous Media". In the introduction he specifically mentions that the book deals with homogeneous fluids - not heterogeneous systems like gas and oil. "Despite" this, his book is a classic in the oil industry and has (had) a wide applicability. Carslow and Jaeger (36) is a more "modern" extension of Muskat providing analytical solutions to many petroleum problems.

Mathematical modelling of heterogeneous rather than homogeneous fluids was initiated primarily by Schilthuis and Buckley-Leverett (74) with the concept of material balance and analytical solutions to immiscible flow problems.

Many text books (2, 3, 15, 19, 27, 51, 70, 74, 76) exist today summarizing the classical approach to reservoir analysis. Dake (19) gives probably the most allround fundamental background to the subject.

Analysis of the oil/gas behaviour using analytical methods had its most productive period in the 1950 and 60 (e.g. 5, 7, 8, 30, 57, 70, 71, 79, 80). A good example of how much insight can be gained by relatively simple analytical methods is the paper by Fetkovich (30). In this paper the normal gas well relation

$$q = C (P_e - P_w)^n, n = 2 \text{ gas wells}$$

is investigated for oil wells and the paper details through application to numerous field cases how field behaviour etc. can be deduced.

The classical analysis often leads to general solutions of classes of problems which have wide applicability (e.g. well test analysis). With computer technology becoming increasingly more sophisticated, analytical solutions to mathematical models will become even more important to provide verification of complex mathematical problems (77, 86).

3.0 RESERVOIR SIMULATION DEVELOPMENT

The late 1960s were the pioneering age of reservoir simulation. Coats, Peaceman, Thomas, Odeh were some of the prime researchers in developing the first black oil reservoir simulator (8, 9, 21, 32, 46, 52, 54, 55, 60, 64, 78); first in two dimensions and shortly after in three dimensions. After this initial period which also brought the first commercial programs, developers aimed at improving the solution algorithms, discretization methods, and others aimed at developing a reservoir simulator for other types of oil extraction where the assumptions regarding the black oil formulation would not suffice. Especially the first group of investigations has improved the technology rather steadily over the last 20 years (Figure 1).

HISTORICAL TRENDS

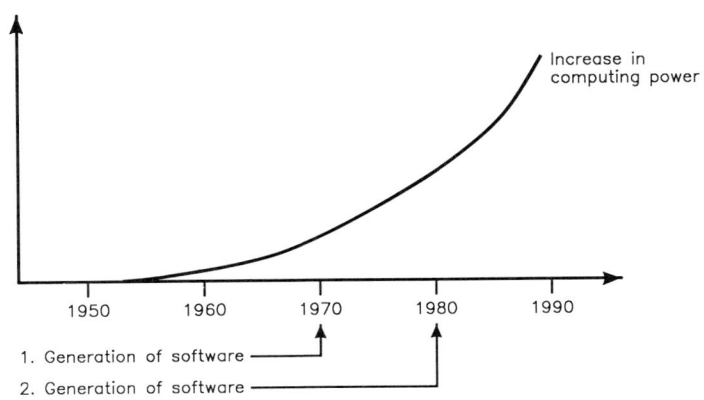

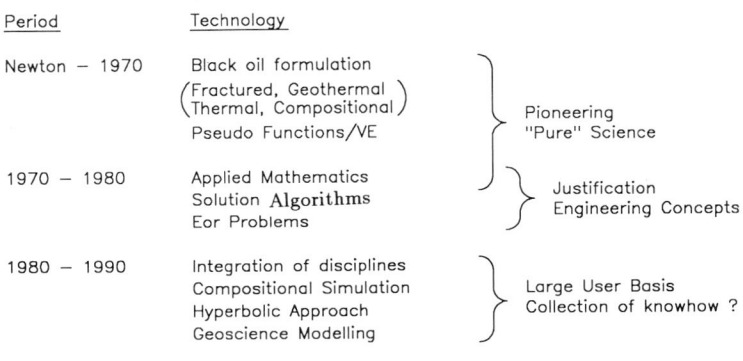

Figure. 1

3.1. Reservoir Simulation Technology Improvements

It was realized very early that considerable discretization errors in space and time were evident which basically were seen by oscillating solutions (Price et. al., 21), front smearing and grid orientation (Figure 2). Furthermore, most of the CPU time/storage was (and still is) spent on solving the linear equations (matrix inversion). In the next three sections are three aspects of the simulation technology developments reviewed.

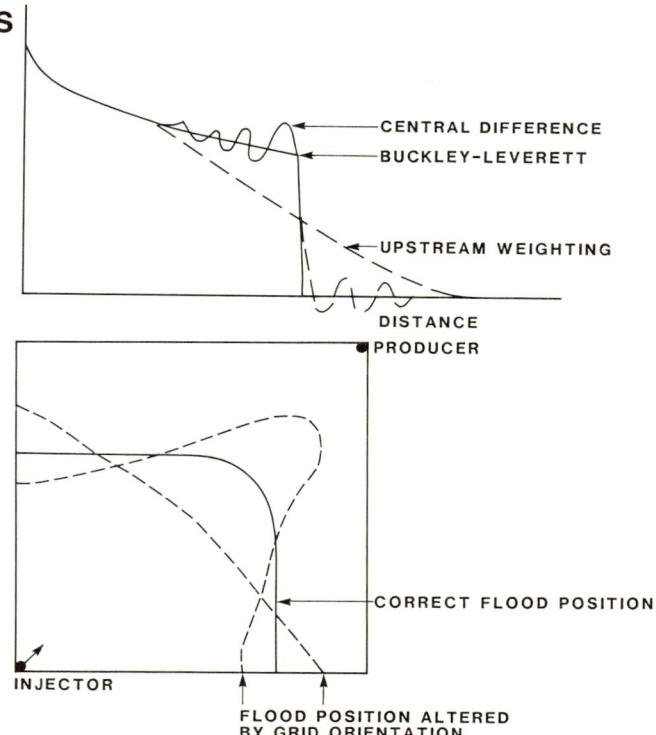

Figure. 2

3.1.1. Solution Methods

The first matrix solvers in the reservoir simulation were extensions of classical methods like the Thomas algorithms to two and three dimensions. Direct matrix solvers with ordering scheme of more and more complex type have been researched over the last 20 years. The D-4 ordering (Coats, 64) is considered one of the fastest. Sherman (83) gives a status of the application of direct methods in the industry -for small problems they still remain faster.

Iterative methods were recognised very early as having the greatest potential for building tailor made systems to achieve the fastest solutions, because of memory and number of equations to be made. The first work was done with line (altering direction) iterative methods (Thomas, Peaceman (4,54) give an overview), followed by SIP, SOR and other variations to the iterative solution technique (15). Most of the simulation software is now based on the conjugate gradient method with nested factorization (Vinsome, 26, Watts, 29). Improvements are still being investigated in relation to new matrix structures caused by local grid refinement etc.

The only other method which seems to hold some promise for further reducing matrix solution time is the multigrid method (84, 33).

With parallel computing technology being advanced, a series of technological developments for the best solution methods can be envisaged.

3.1.2. Space Discretization

The early work by Price, Varga and Warren (21) stands (almost) today as a "status" in this field dominated by finite difference methods. Central difference schemes, which have the lowest truncation error, give rise to oscillating solutions in the pressure or concentration profile when solving Buckley-Leverett or convection-diffusion problems. Forward (upwind) differencing solved this problem by smearing out the front, but it gives a higher numerical truncation error. This problem has given rise to a very large number of investigations (e.g. 20,21,31, 37,38,41,43,48,49). The finite difference method is ideal for grid systems of equal spacing, but irregular grid geometries such as faults and local grid refinements have given rise to interest in methods based on the variational method, Finlayson (39). The major problem in using methods based on variational principles is to find generalized methods for getting a stable solution and secondly getting matrix solution algorithms fast enough (38,41,44,46,47,50).

For nearly all practical reservoir engineering problems the black oil and compositional models will be a parabolic problem rather than a hyperbolic problem. The solution of the parabolic problem can, nevertheless, get infinitesimal close to the hyperbolic case.

This observation has led to interest in applying hyperbolic techniques and "adjusting them" to include a diffusive term (Peaceman, Concus, 31,35,40,42). It is especially important for many EOR simulations (44).

Comparison of finite difference methods and methods based on variational principles were examined by Jensen (41) in order to identify under which conditions the methods become unstable. Of all methods tried to solve this problem only 5 and 9 point methods and difference formulations (8,16,49) based on combinatory methods between finite difference and finite element methods has had any industrial acceptance for black oil simulation and mostly to solve grid orientation effects.

With computing power increasing renewed interest in methods based on variational principles can be expected since they provide the most rigorous mathematical background for any geometric configuration.

3.1.3.   Time Discretization

Two aspects are important - the number of equations to be solved and the order in which the equations are solved in any one iteration step i.e. the explicit/implicit nature of the time integration algorithm.

Considering both above aspects the so-called IMPES (implicit pressure, explicit saturation) algorithm was recognized early as the fastest and most stable algorithm (Peaceman, Coats, Thomas, 32,52,54,78). With the computer technology advances together with solution algorithm efficiency the ability to do fully implicit simulation (16) has to a large extent eliminated the use of IMPES even though it is perfectly suited for many reservoir analyses where differences in eigenvalues are small (rapid solution variations close to e.g. wells).

Because of the high discretization error in the spatial dimension, investigations related to higher order schemes (15,18,20,45,55) have not yet been considered for commercial simulators.

The control of discretization errors both with regard to the spatial dimension and time is still in its infancy -controls of timestep size are usually made with control of the solution variable is based on experience even though methods exist which are of a more general nature (45).

3.2. Conceptual Reservoir Simulation

In the 1970'ies, rapidly following the black oil model development conceptual modelling approaches were made:

- Compositional reservoir simulation (Coats et al., 11,47,55)
- Fractured reservoir simulation (Warren & Root, 57)
- Thermal reservoir simulation (Coats)
- Hydraulic fracture simulation (Settari, 63,65)
- Combustion reservoir simulation (Coats)
- Surfactant flooding simulation (Lake et al., 10,14,17)

From a scientific point of view the governing equations have been established early, but the computational use of these models remains a specialized discipline. One reason for this is the contraction between the sophisticated mathematical description and our inability to obtain field data and laboratory data to further explore these approaches. A good example of this contradiction is the technological developments for simulating fractured reservoirs. Firoozibadi (69) compares 10 different simulators with more than 5 different theoretical assumptions regarding the transfer functions between matrix and fractures but only few differences in the results are seen. For fractured reservoir simulation, there is a need to find methods to relate the fracture intensity and orientation to simulation parameters. This contradiction essentially makes the length scale of the matrix block size a history matching parameter. The question could be asked whether we need further enhancements in this simulation technology before we are able to relate geological observations and field data to the reservoir model (fracture distribution).

Similarly, compositional simulators are becoming increasingly popular even though field data are questionable and fundamental aspect of e.g. initialization and relative permeability relations are poorly understood. A considerable effort is required within the experimental area, field data calibration and reservoir simulation aspects before compositional reservoir simulation is a standard tool for reservoir analysis.

4.0 RESERVOIR DESCRIPTION MODELLING

Geoscientists and engineers have over recent years started to describe spatial variability of reservoir characteristics in formal statistical terms. Rock properties such as net to gross thickness ratios, permeability and porosity are treated as statistical or random variables rather than deterministic functions of space. Haldorsen et al. (71) and Lake et al. (22), give excellent overviews of the current status for this technology.

This type of technology has been applied to a number of large North Sea sand fields to generate better reservoir definition for reservoir simulation models. Although this technology provides a step in the direction of integration of geoscience and mathematics (24, 25, 71), the statistical formulation of the total problem including the continuous mechanics theory of flow through porous media has yet to be done (true stochastic reservoir simulation).

For fields where variations in permeability and porosity are caused by micro/macro fracturing the above approach does not help because of dual permeability effects. Jensen et al. (66) have proposed a methodology suggested by the geotechnical literature and newer petroleum exploration modelling. Backstripping of seismic profiles combined with depositional history models can assist in describing geological models of hydrocarbon accumulations. Matching of well data to depositional models can be made which thereby gives the geoscientist a method to check consistency in layering and fault patterns. Stress strain analysis of the depositional history can further aid in matching current pressure/strain patterns observed on a given field and indicate fracture intensity variations across a field.

One step further in the geoscience modelling compared to the approaches above is incorporation of the basin modelling approach used in exploration. This technology combines the above disciplines with fluid migration from the hydrocarbon source ("kitchen") to the accumulating reservoirs. Such models are currently not extended to production geology but in combination with depositional/stress modelling they could provide another method for doing true geological history matching of some rock properties across a given field.

5.0 APPLICATION OF RESERVOIR SIMULATION TECHNOLOGY

Most of the major oil companies have since the start of the technological development in the 1960s decided to pursue in-house development. The basis of nearly all reservoir analysis is now the black oil reservoir simulation. As the computer speed has increased, compositional aspects and refined geometry approximations have been the dominant enhancements.

Studying the literature in relation to the usage of reservoir simulation indicates that only reservoir analyses of medium-to-large reservoirs have used reservoir simulation but not smaller fields. Furthermore "difficult" reservoirs such as fractured fields must to a large extent use "classical analysis".

Figure 3 illustrates the difference between classical analysis and simulation technology. It is increasingly important for the industry to find methods to relate reservoir simulation to classical analysis.

5.1 Scale Ordered Structure for Reservoir Analysis

Fayers (71) gives an excellent overview of the specialist disciplines and models needed to do a full integrated reservoir analysis. The origin of the vertical equilibrium concept and the necessity to use pseudo functions (32,59,60,61,68) are probably the first investigations pointing out the scaling problems. Figure 4 shows the scale ordered analysis as given by Fayers. As was pointed out by Fayers a number of algorithms are being developed to fill out the total analysis procedure.

```
                    Engineering Tool box
           ─ Material balance Modelling
           ─ Decline Curve Analysis
           ─ Well Test Analysis
  Small    ─ PVT
  Fields
           ─ Displacement Calculations
           ─ Statistical analysis (field/laboratory data)
           ─ Well bore Hydraulics
                     First principles
```

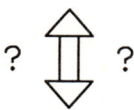

```
                    Simulation Tool box
           ─ Black Oil Simulation
           ─ Compositional Simulation
           ─ Fracture Reservoir Simulation
           ─ Thermal Simulation
  Large    ─ Reservoir Description Model        (R)
  Fields
           ─ Chemical Flood Simulation          (R)
           ─ Stress Strain Modelling            (R)
           ─ Reaction                           (R)
              Integration of first principles
```

5.2 Dimensionless Numbers

Figure. 3

In chemical engineering dimensionless analysis is classically used for nearly everything from flow in pipes, reaction kinetics, control theory and diffusion problems. This trend has not been carried over in petroleum engineering where nearly only well testing uses this concept.

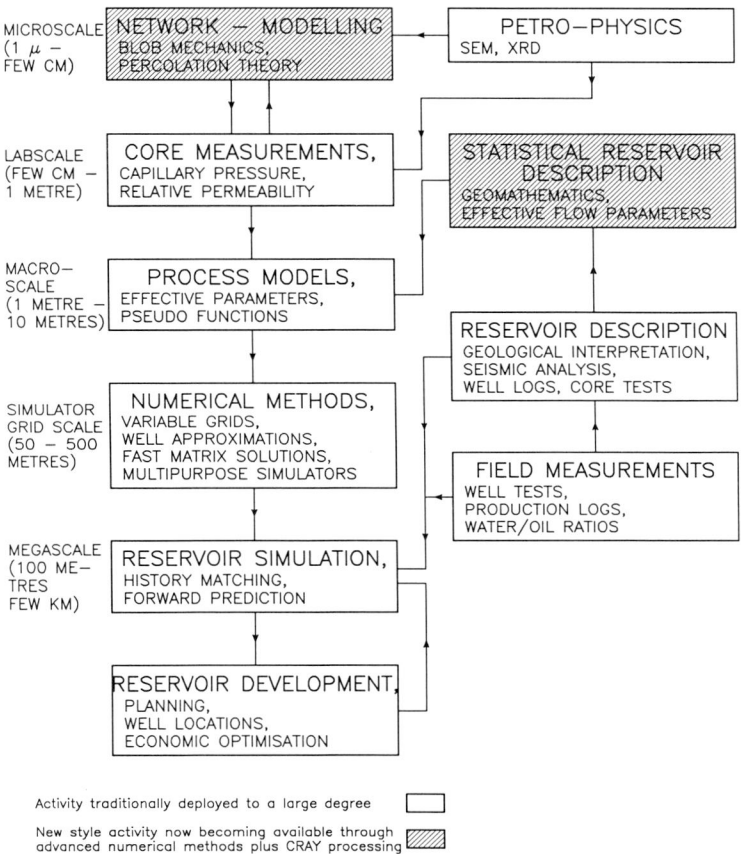

Figure. 4

An exception is parts of material balance modelling of production data. Dake (19) gives an excellent presentation of this. For a given reservoir a set of assumptions can usually be made which warrants use of the material balance concept although in some cases not solvable by hand (51,74,79) e.g. compartmental analysis (Stewart et al).

Combining material balance modelling with reservoir simulation can often be a cost effective tool and I believe this type of approach merits more attention from researchers and engineers.

In reservoir simulation the concept of pseudo-functions (Kyte and Berry) is very important for making three-dimensional modelling. Thomas (59) gives an overview of the concept and discusses limitations. Recent work by Kossack (68) has taken a renewed interest in pseudo-functions and uses viscous/gravity scaling (dimensionless grouping) to identify which types of pseudo functions are needed. This type of work should be extended to more general cases.

Another potential area for using dimensionless numbers is tailor made welltest analysis based on reservoir simulation. Several investigators (e.g. 77) are using black oil/compositional simulation to generate typecurves from reservoir simulation for cases where no analytical solution is possible.

Extension of Dietz classical analysis on gas displacement (Fayers, 86) is another good example of building bridges between classical analysis and reservoir simulation technology.

6.0 SUMMARY AND CONCLUSIONS

This review has attempted to cover over a thousand references in the petroleum literature. The detailed review of derivations and assumptions has, therefore, been omitted and it has been attempted to subdivide the technological developments into its major categories and, furthermore, suggested some new routes to pursue. The challenges of the oil industry have been and will continue to be enormous. Science advances our ability to do reservoir analysis and new approaches should rigorously be pursued with the reality of limited field data.

Petroleum engineering common sense approach using first principles "classical analysis" will continue to be important for reservoir engineers and (I believe) the impact of simulation technology in the oil industry could be even larger if results from reservoir simulation quickly and reliably can be related to classical analysis through the use of dimensionless analogs. Chemical engineers use them. Why not us. Having stated this a bit categorically a number of recent papers indicate a trend in this direction.

Reservoir description modelling and conceptual reservoir modelling for non-fractural and fractured fields will and should be a growth area in the 1990's.

For engineering purposes black oil reservoir simulation will probably continue to be enhanced technologically. Specifically well models, geometry and surface systems will be enhanced. Furthermore, the use of artificial intelligence methods could be useful in accumulating the experience from using these models, maybe particularly in the areas in PVT, pseudo functions and gridding (grid design). A trend towards the use of compositional models is thought to dominate the 1990s but in this case I believe we are stretching the limits of the reservoir simulation since field data and other basic data needed for this approach are lacking (at this point in time).

REFERENCES

1. Muskat, M., (1949) "Physical Principles of Oil Production", Mc-Graw-Hill, New York.

2. Bear, J., (1972) "Dynamics of Fluids in Porous Media", American Elsevier, New York.

3. Scheidegger, A.E., (1974) "Physics of Flow through Porous Media", Univ. of Toronto Press, Toronto.

4. Peaceman, D.W., (1977) "Fundamentals of Numerical Reservoir Simulation", Elsevier, Amsterdam.

5. Morel-Seytoux, H.J. (1965) "Unit Mobility Ratio Displacement Calculations for Pattern Floods in a Homogeneous Medium", Soc. Pet. Eng. J. pp 217-227.

6. Hirasaki, G.J. (1981), "Application of the Theory of Multicomponent, Multiphase Displacement to Three-component Twophase Surfactant Flooding", Soc. Pet. Eng. J., pp 191-203.

7. Perkins, T.H. and Johnson, O.C. (1963) "A Review of Diffusion and Dispersion in Porous Media", Soc. Pet. Eng. J., pp 70-84.

8. Koval, E.J. (1963) "A Method for Predicting Unstable Miscible Displacement in Heterogeneous Media", Trans. AIME 228, pp 145-154.

9. Todd, M.R. and Longstaff, W.J., (1972) "The Development, Testing and Application of a Numerical Simulator for Predicting Miscible Flood Performance", Trans. AIME, 253, pp. 874-882.

10. Pope, G.A., (1980) "The Application of Fractional Flow Theory to Enhanced Oil Recovery", Soc. Pet. Eng. J., pp 191-205.

11. Helfferich, F.G. (1981) "Theory of Multicomponent Multiphase Displacement in Porous Media", Soc. Pet. Eng. J., pp. 51-62.

12. Davis, H.T. and Scriven (1980), "The Origins of Low Interfacial Tensions for Enhanced Oil Recovery", SPE 9278.

13. Pope, G.A. and Nelson R.C. (1977) "A Chemical Flooding Compositional Simulator", SPE 6725.

14. Fleming, P.D., Thomas, C.P. and Winter, W.K. (1981) "Formulation of General Multiphase Multicomponent Chemical Flood Model", Soc. Pet. Eng. J., pp. 63-76.

15. Aziz, K. and Settari A. (1979), "Petroleum Reservoir Simulation", Applied Science Publishers, London.

16. Ponting, D.K., Foster, B.A., Naccache, P.F., Nicholas, M.O., Pollard, R.K., Rae, J., Banhs, D. and Walsh, S.K. (1983) "An Efficient Fully Implicit Simulator", Soc. Pet. Eng., 23, pp. 544-552.

17. Tood, M.R. and Chase, C.A. (1979) "A Numerical Simulator for Predicting Chemical Flood Performance", SPE 7689.

18. Sincovec, F.R. (1975) "Numerical Reservoir Simulation using an Ordinary Differential Equations Integrator", Soc. Pet. Eng., pp 225-264.

19. Dake, L.P. (1978) "Fundamentals of Reservoir Engineering", Elsevier.

20. Lantz, R.B. (1971) "Quantitative Evaluation of Numerical Diffusion (Truncation Error)", Soc. Pet. Eng. J., 11, pp 315-320.

21. Price, H.S., Varga, R.S. and Warren, J.R. (1966) "Application of Oscillation Matrices to Diffusion - Convection Equation", J. Math Phys, 45, pp 301-311.

22. Lake, L.W. and H.B. Caroll, Jr., Eds (1986) "Reservoir Characterisation", Academic Press.

23. Hewett, T.A. (1986) "Fractual Distributions of Reservoir Heterogeneity and their Influence of Fluid Transport", SPE 15386.

24. Begg, S.H. and King, P (1985) " Modelling of the Effects of Shales on Reservoir Performance Calculation of Effective Vertical Permeabilities", SPE 13529.

25. Angedal, H.O., Stanley, K.O., and Omre, H. (1986) "SISA BOSA - A program for Stochastic Modelling and Evaluation of Reservoir Geology" presented at Conference on Reservoir Description and Simulation with Emphasis on EOR, Oslo, Norway.

26. Vinsome, P.H.W. (1976)," Orthomin, An Iterative Method for Solving Sparse Banded Sets of Simultaneous Linear Equations", SPE 5729.

27. "Developments in Petroleum Engineering", Elsevier Applied Science Publishers, 1980.

28. Thomas, L.K., Dixon, T.N. and Pierson, R.G. (1983) "Fractured Reservoir Simulation" Soc. Pet. Eng. J., 23, p 42.

29. Wallis, J.R., (1983) "Incomplete Gaussian Elimination as a Preconditioning for Generalized Conjugate Gradient Acceleration", SPE 12265.

30. Fetkovich, M.J., "The Isochronal Testing of Oil Wells", SPE 4529.

31. Christie, M.A. and Bond, D.J. (1985) "Multidimensional Flux Corrected Transport for Reservoir Simulation" SPE 13505.

32. Coats, K.H., Nielsen, R.L., Terhune, M.H. and Weber A.G. (1969) "Simulation of Three-dimensional Two-phase Flow in Oil and Gas Reservoirs", SPEJ (377).

33. Behie, A. and Forsyth, P.A. (1985) "Multigrid Solution of the Pressure Equation in Reservoir Simulation", Soc. Pet. Eng. J., $\underline{23}$, p 623.

34. Wasserman, M.L. (1987) "Local Grid Refinement for Three-dimensional Simulators", SPE 16013.

35. Albright, N., P. Concus and W. Proskurowski (1979) "Numerical Simulation of the Multidimensional Buckley-Leverett Equation by a Sampling Method", SPE 7681.

36. Carslaw, H.S., J.C. Jaeger (1959) "Conduction of Heat in Solid", Oxford Univ. Press.

37. Chien, T.C. (1977) "A General Finite Difference Formulation of Navier-Stoke's Equation, Computer and Fluid", 5, pp 15-21.

38. Christie, I. and Mitchell, A.R., (1978) "Upwinding of High Order Galerkin Methods in Conduction-Convection Problems", Int. J. Num. Meth., Eng. 12, pp 1764-1771.

39. Finlayson, B.A. "The Method of Weighted Residuals and Variational Principles", Academic Press, New York, 1972.

40. Garder, A.O., Jr. (1964), Peaceman, D.W. and A.L. Pozzi, Jr., "Numerical Calculation of Multidimensional Miscible Displacement by Method of Characteristics", SPEJ.

41. Jensen, O.K. and B.A. Finlayson (1970) "Oscillation Limits for Weighted Residual Methods", Int. J. Num Meth. in Engineering, Vol. 15, pp 1681-1689.

42. Jensen, O.K. and Finlayson, B.A. (1980) "Solution of the Transport Equations Using a Moving Coordinate System", Advances in Water Resources, Vol. 3, pp 9-18.

43. Laumbach, D.D. (1975) " A High Accuracy Finite Difference Technique for Treating the Convection Diffusion Equation", Soc. Pet. Eng. J., pp 517-531.

44. Jensen, O.K. and Finlayson B.A. (1982) "A Numerical Technique for Tracking Sharp Fronts in Studies of Tertiary Oil Recovery Projects", SPE 12240.

45. Jensen, O.K. (1980), "Automatic Timestep Selection Scheme to Reservoir Simulation", SPE 9373.

46. McMichaels, C.L., Thomas, G.W. (1973) "Reservoir Simulation by Galerkins Method", Soc. Pet. Eng. J., pp 125-138.

47. Settari, A., Price, H.S., Dupont, T. (1976) "Development and Application of Variational Methods for Simulation of Miscible Displacement in Porous Media", SPE 5721.

48. Wheatley, M.J. (1979) " A version of Two Point Upstream for use in Implicit Numerical Reservoir Simulators", SPE 7677.

49. Yanosik, J.L., Macrachen, T.A. (1976) "A Nine-point Finite Difference Reservoir Simulator for Realistic Production of Unfavourable Mobility Ratio Displacement", SPEJ Num. Res. Conf.

50. Young, L.C. (1978) "An Efficient Finite Element Method for Reservoir Simulation", SPE 7413.

51. Van Golf Racht, T.D. (1982) "Fundamentals of Fractured Reservoir Engineering", Elsevier.

52. Peaceman, D.W. (1978) "Interpretation of Well Block Pressures in Numerical Reservoir Simulation", Soc. Pet. Eng. J., pp 183-194.

53. Young, L.C. (1987), "Equation of State Compositional Modelling on Vector Processors", SPE 16023.

54. Thomas, G.W. (1982) "Principles of Hydrocarbon Reservoir Simulation", IHRDC, Boston.

55. Coats, K.H. "Simulation of Gas Condensate Reservoir Performance", SPE 10511.

56. Rossen, R.H., "Simulation of Naturally Fractured Reservoirs with Semi-implicit Source Terms", Soc. Pet. Eng. J, pp 201-210.

57. Warren, J.F. and Root, P.J. (1963) "The Behaviour of Naturally Fractured Reservoirs", Soc. Pet. Eng. J., pp 245-255.

58. Hill, A.C., Thomas, G.W., Jensen, O.K. (1983) " A new Approach for Simulating Complex Fractured Reservoirs", SPE 13537.

59. Thomas, G.W. "An Extension of Pseudofunction Concepts", SPE 12274.

60. Coats, K.H., Dempsey, J.R., and Henderson, J.H. (1971) "The use of Vertical Equilibrium in Twodimensional Simulation of Threedimensional Reservoir Performance", Soc. Pet. Eng. J., pp 63.

61. Kyte, J.R. and Berry, D.W. (1975) "New Pseudo Functions to Control Numerical Dispersion", Soc. Pet. Eng. J. (1975) pp 269.

62. Heinemann, Z.E. (1988) "Using Local Grid Refinement in a Multiple - Application Reservoir Simulator", SPE 12255.

63. Settari, A. (1982) "Three-dimensional Simulation of Hydraulic Fracturing", SPE 10504, 10505.

64. Price, H.S. and Coats, H.H. (1974) "Direct Methods in Reservoir Simulation", Soc. Pet. Eng. J. 14, p 295.

65. Settari, A. (1989) "Physics and Modelling of Thermal Flow and Soil Mechanics", SPE 18420.

66. Jensen, O.K., Bresling, S., Christensen, O.W., Rasmussen, F.O. (1989), "Natural Fracture Distribution in Reservoirs by Backstripping and Finite Element Stress Analysis", SPE 18429.

67. Bech, N., Jensen, O.K., Nielsen, B. (1989), "Modelling of Gravity-Imbibition and Gravity-Drainage Processes: Analytical and Numerical Solutions", SPE 18428.

68. Kossack, C.A., Aasen, J.O., Opdal, S.T.(1989), "Scaling-up Laboratory Relative Permeabilities and Rock Heterogeneities with Pseudo Functions for Field Simulations", SPE 18436.

69. Firoozabadi, A., Thomas, L.K. "Sixth Comparative Solution Projects: A Comparison of Dual-porosity Simulators", SPE 18741.

70. Katz, D.L., "Natural Gas Engineering".

71. Edwards, S., King, P.R. (1987) "Mathematics in Oil Production", The Institute of Mathematics, New Series Number 18, Oxford Science Publications.

72. Whitson, C.H. (1983) "Characterising Hydrocarbon Plus Fractions", SPEJ, pp 685-696.

73. Katz, D.L. (1983) "Overview of Phase Behaviour in Oil and Gas Production", J. Pet. Tech., pp 1205-1214.

74. Amyx, J.W., Bass, D.M., Whiting, R.L. (1960), "Petroleum Reservoir Engineering", McGraw Hill, New York.

75. Stone, H.L. (1970), "Probability Model for Estimating Three-phase Relative Permeability", SPE 249, pp. 214-218.

76. Lee, SPE Monograph, 1988, Well Test Analysis.

77. Padmanabhan, L. (1979) "Welltest - a Program for Computer Aided Analysis of Pressure Transient Data from Welltest", SPE 8391.

78. Thomas, G.W., and Thurman, D.H. (1983), "Reservoir Simulation Using an Adaptive Implicit Method", Soc. Pet. Eng. J. 23 (1983), 759-768.

79. Aronofsky, J.S. (1957), "A Model for the Mechanism of Oil Recovery from the Porous Matrix due to Water Invasion", SPE 932.

80. Katz, D.L. (1962) "A Generalized Model for Predicting the Performance of Gas Reservoir Subject to Water Drive", SPE 428.

81. Coats, K.H. (1969), "Use and Misuse of Reservoir Simulation", SPE 2367.

82. Robertson, G.E. (1978), "Grid Orientation Effects and the Use of Curvilinear Coordinates in Reservoir Simulation", SPE 6100.

83. Sherman, A.H. (1985), "Sparse Gaussian Elimination for Complex Reservoir Models", SPE 13535.

84. Barman, J., Horne, R.N., "Improving the Performance of Parallel (and Serial) Reservoir Simulators", SPE 18408.

85. Anterion, F., Eymard, R., "Use of Parameter Gradients for Reservoir History Matching".

86. Fayers, F.J., "Muggeridge, A.H., "Extension of Dietz Theory and Behaviour of Gravity Tongues in Slightly Tilted Reservoirs", SPE 18438.

SIMULATION OF WATERFLOOD FRACTURE GROWTH WITH COUPLED FLUID FLOW, TEMPERATURE AND ROCK ELASTICITY

P.J. Clifford
(BP Research Centre, Sunbury)

ABSTRACT

Methods used to simulate thermal fracturing of water injection wells in three dimensions are described. The calculation combines a finite difference model of fluid and heat flow within the reservoir, with determination of the poro- and thermoelastic stress field, and with a boundary element fracture mechanics model to compute fracture growth. The components of the calculation are closely linked, but display behaviour over a range of timescales. Methods of coupling this complex problem with adequate efficiency and stability are discussed. Solutions are presented for a number of problems derived from field well behaviour. It is shown that behaviour can be governed by changes in the vertical extent of fracturing, and that different types of solution are derived for constant rate and constant pressure wells.

1. INTRODUCTION

Rock fractures are an important feature of most oil and gas reservoirs, either in the form of a naturally occurring fracture network, or as a form of rock failure induced around wells. The fractures act as channels for fluid flow, and are themselves created or influenced by the pressure and temperature changes associated with the fluids.

Waterflood thermal fractures show a particularly strong linkage between rock and fluid mechanics. They can be created around water injection wells, due to cooling and contraction of the surrounding rock, when the injected water is substantially colder than the reservoir. Under North Sea conditions, for example, a reservoir at 3000m depth will have a temperature of around 100°C, compared with a water injection temperature of less than 25°C. The cooling sets up tensile thermal stress, which partially counteracts the initial compressive earth stress. If the fluid pressure in the well bore exceeds the remaining minimum principal stress, a fracture can be propagated from the well. This is illustrated in Figure 1. The fracture propagates in a fixed vertical plane, which is perpendicular to the minimum principal stress direction.

Once fracturing has occurred, the fracture grows in a close
relationship with fluid and heat flow in the reservoir. Fracture
growth can have a positive feedback with the extension of the
cooled area, and hence of the area of stress reduction. In all
cases, the fracture will dominate the injectivity of the well (ie
the injection rate achievable for given bottom-hole pressure); if
it grows to achieve a length comparable to well spacings, or if
it is restricted in vertical extent, it can also influence the
fluid flow patterns and oil sweep from the reservoir as a whole.

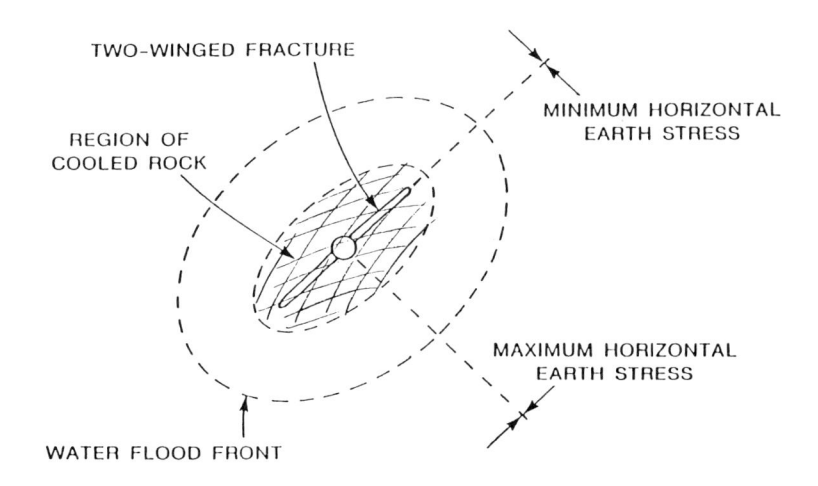

Figure 1. Diagram of Waterflood Fracturing in a Horizontal
Plane.

Models exist for calculating many of the individual physical
processes associated with waterflood fracturing, but have not
generally been brought together to treat the detailed and often
essentially 3-dimensional behaviour of real wells. A part of the
reason for this lies in the fact that the problem straddles areas
normally treated by finite difference (FD) and areas normally
treated by finite element (FE) techniques. Reservoir multi-phase
fluid flow simulation is almost invariably carried out with
finite differences, while stress and fracture calculations
generally use finite elements. Even if a single unified
computational method could be applied, the gridding needed for
different physical variables would not correspond. Stress, for
example, would be heavily concentrated at the fracture boundary,
while a temperature or saturation front might exist in quite a
different location. In general, models such as [7] which offer
the highest degree of coupling between thermal, mechanical and
hydraulic processes in the rock matrix suffer problems in dealing
with fractures and other forms of boundary condition.

Progress can be made towards a comprehensive solution when the rock approximates to a linear, elastic medium, as would be expected, for example, for a consolidated sandstone. In this case, the calculation of thermo- and poro-elastic stress, due to the distribution of fluid temperature and pressure, may be carried out independently of the fracture mechanics calculation, and on a separate grid. The calculation divides naturally into stages which can be solved, or at least iterated, in a much more efficient manner than if a unified model were used throughout.

The problem also becomes more straightforward when attention is given to the different timescales of the processes involved. In particular, the slow timescale of temperature and saturation change compared with pressure change, and of pressure change compared with fracture dynamics (sound speed), means that the fracture is always in a state of equilibrium with the stress due to the fluid pressure and temperature. Also, temperature can be treated explicitly, rather than being included within the pressure/stress iterative process at each timestep.

Previous oil industry work offers either models which incorporate a fracture as a source term into the fluid flow model, but lack a true rock mechanics calculation, or hydraulic fracturing codes which have much greater sophistication in rock mechanics, but do not couple into reservoir flow.

Models with some degree of coupling include those of Dikken and Niko [4] and Garon et al.[5]. Dikken and Niko have calculated 2-dimensional fracture growth using methods which have been in part adopted and extended to 3 dimensions in this study. Garon et al. report an implicit model based on highly simplified fracture mechanics with a fixed fluid pressure at the fracture boundary. They have also applied a conventional 3-D hydraulic fracturing code to a thermal stress distribution around an injector, but without coupling in a leak-off model appropriate to reservoir flows.

2. SIMULATION METHODS

2.1 FLUID AND HEAT FLOW CALCULATION

The coupling of waterflood fracture growth with reservoir pressure and flow over a long timescale requires a full treatment of reservoir behaviour. The approximations commonly adopted for hydraulic fracturing purposes, with their emphasis just on the well neighbourhood, would not be at all appropriate to waterflood fracturing. The fracture calculation is therefore set within a reservoir simulation code.

The host code chosen is an IMPES finite difference simulator, developed for a range of applications including chemical flooding [2]. The code incorporates a temperature calculation with heat conduction, convection and appropriate boundary conditions. It includes fluid compressibility, which is necessary for modelling of fracture transient behaviour, as during well step-rate and pressure fall-off (pfo) testing. Tracer and chemical options are of use for particular fracture applications. A fully implicit model, while ultimately desirable, would impose excessive overheads in coding and computing during a research phase. The flexibility of the IMPES formulation was essential to the testing of physical behaviour and numerical methods.

Modifications to the basic code to incorporate fracturing include:

(i) treatment of the fracture as a fluid source term;

(ii) selection of grids suitable for the full problem;

(iii) coupling of the finite difference solution with stress and fracture mechanics.

Formulation of the source term is a significant problem even for a fracture of constant dimensions. A fracture may be represented as a set of narrow blocks of high permeability [14] which, however, poses problems of stability, and requires repeated grid modification for a growing fracture. The main alternative (chosen here) is to treat the fracture as a separate entity, analogous to a well bore, which supplies fluid to the blocks according to a set of "connection factors", which are given as a ratio between the flow from the fracture into a given grid-block, and the difference in pressure between the fracture and the average pressure of that block.

Certain approximations are useful in arriving at appropriate connection factor formulae. The volume of a waterflood fracture (normally less than 1 cm width) is always negligible compared to the reservoir volume. It is frequently (but not always) also a good approximation to assume that the fracture is infinitely conductive to water. In that case, it behaves as a uniform pressure source, apart from gravity. Steady-state flow solutions exist for 2-D flow from such a line source either in an infinite reservoir or in a 5-spot well pattern. For the infinite reservoir, the potentials are ellipses whose foci are located at the fracture tips; there is a greater flow rate per unit fracture length close to the tips than near the centre.

The grid must be chosen to match the direction of fracturing. A grid diagonal to the fracture direction imposes complexity upon block-to-block transmissibilities as well as fluid input, and so has not been considered. Taking a Cartesian grid with fracture growth in the x-direction, and an arbitrary x- and y-coordinate of the well bore, it is only necessary to calculate connection factors, and not to change transmissibilities. It is reasonable to assume that flow from the fracture face is one-dimensional and perpendicular to its plane, if we are well away from its boundaries. The connection factor for blocks in this area, where the fracture slices through the block centre, must set the block pressure approximately equal to the local fluid pressure in the fracture, in order for the grid to calculate the 1-D flow correctly. (This excludes cases where there is a skin on the fracture surface). The greatest problem remains the treatment of blocks containing fracture tips or boundaries.

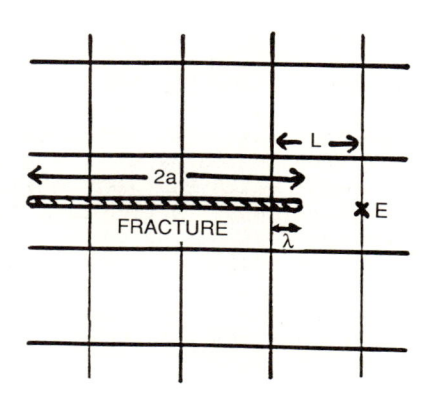

Figure 2. The Geometry of the Fracture-Grid Connection Formula.

For 2-D cases, the approach of Nghiem [11] has been adopted in modified form, in which it is assumed that the potentials approximate to an elliptical shape close to the tips. In this case, as adapted from [11], for a fracture of half-length a in an infinite, uniform medium, the pressure at the point E of Figure 2 is related to the fracture pressure p_f and rate Q_f by:

$$(p_f - p_E) = \mu Q_f \cdot \ln\{X + (X^2-1)^{1/2}\} / (2\pi k k_r \Delta z) \quad (2.1.1)$$

for $\quad X = 1 + (L-\lambda)/a$

where k, k_r, μ and Δz are the formation permeability, relative permeability, viscosity and thickness, and L is the block size in the x-direction. The flow through the end segment of the fracture, of length λ, is given by:

$$Q_1 = Q_f \cdot \cos^{-1}\{1-\lambda/a\} / \pi \qquad (2.1.2)$$

This allows a simple relationship between flow from the tip and pressure close to it, which can be carried forward as an approximation to less ideal cases. Note that the pressure assigned to the grid-block containing the tip is a suitably weighted mean of p_f and p_E, where the form of the weighting must be determined analytically, or from numerical experiment.

As Settari [13] has demonstrated, instability can result from non-smooth behaviour as a fracture tip grows through a block boundary, unless connection factors are carefully chosen. A weighting function which takes the tip grid-block pressure linearly from $(p_f + p_E)/2$ as the tip just penetrates the block, to p_f when the block is fully penetrated, was found to be adequately smooth in simulating fracture growth implicitly.

Inclusion of a 3-D or finite conductivity fracture cannot be based straightforwardly on analytic approaches. However, with sufficiently fine gridding, adequate connection factor formulations have been constructed based on the fraction of a block in the fracture plane which has been penetrated by the fracture. For some applications, particularly those with finite conductivity, pressure has been periodically calculated using a special grid refined in the region of the fracture, and including the fracture as a set of blocks, but with saturation and temperature updates only on the coarser reservoir grid. In such cases, connection factors for the coarse grid are obtained using the fine-gridded model.

A bordered matrix method has been used to incorporate the fracture source terms into the pressure matrix inversion. Some large off-band elements are generated, but a conjugate gradient method with incomplete Cholesky pre-conditioning (ICCG) has been found to be robust under these conditions, with minor modification.

Consideration has also been given to fine-gridding of temperature, given the importance of thermal stress, and to fracture face-plugging. Temperature fine-gridding perpendicular to the fracture growth direction is helpful when fracture extension is very rapid, and the cooled region becomes an eccentric ellipse, but is not important when the fracture lies well within the cooled zone.

The plugging of the fracture face by small solid particles (fines) in the injected water creates a "skin" which acts as a partial permeability barrier to the escape of water. It can cause pressure build-up and fracture extension. The level of plugging is a function of the quantity of water which has passed through unit area of fracture face. The use of the coarse reservoir grid for calculation of skin poses particular problems of stability when a fracture grows through a grid-block boundary to expose fresh surface. The problem can be overcome by the use of a fine gridding of skin build-up in fracture tip blocks.

2.2 STRESS CALCULATION

The assumption of linear elastic behaviour allows the rock mechanics calculation to be decoupled into two parts [1]:

(i) the calculation of the stress which would result from the fluid temperature and pressure distribution, in the absence of any fracturing;

(ii) the use of this stress field as a traction term in a separate fracture mechanics calculation.

Under (i) it is only necessary to calculate stress at the expected fracture surface (ie on a single known vertical plane), and it is adequate for most purposes to calculate a single stress component (normal to the fracture plane). It is not necessary to calculate explicitly the full stress field at any point under (ii), but rather just the stress intensity factors at the fracture boundary. This allows considerable economy in the calculation.

Thermo- and poro-elastic stress in a linear elastic medium are conveniently calculated using the Goodier displacement potential ϕ. This is defined in terms of strain [10] by:

$$\varepsilon_{ij} = \frac{\partial^2 \phi}{\partial x_i \partial x_j} \qquad (2.2.1)$$

and must (for the purely thermal case) be a solution of the Poisson equation:

$$\nabla^2 \phi = (1+\nu)/E \cdot A_T \Delta T \qquad (2.2.2)$$

where E is Young's modulus, ν is Poisson's ratio, and the thermoelastic constant A_T is defined as:

$$A_T = E\alpha_T/(1-\nu) \qquad (2.2.3)$$

The thermoelastic stress at any point is then given by:

$$\Delta\sigma_{ij} = E/(1+\nu) \cdot \frac{\partial^2 \phi}{\partial x_i \partial x_j} + A_T \Delta T \delta_{ij} \qquad (2.2.4)$$

The Poisson equation for ϕ could be solved using a variety of methods, for non-uniform rock mechanical properties. For the case of uniform properties, which has been used in current work, the most economical approach is a Green's function solution [8] which takes the form:

$$\Delta\sigma_{ij} = A_T \cdot \left\{ 1/4\pi \cdot \int d\underline{x}' \, \Delta T(\underline{x}') \frac{\partial^2(1/R)}{\partial x_i \partial x_j} + \Delta T \delta_{ij} \right\}$$

(2.2.5)

where: $R = |\underline{x} - \underline{x}'|$

At each point where stress is to be calculated, a singular quadrature over space must be performed. This is achieved by division of space into appropriate singular and non-singular domains. An interpolation of temperature from the finite difference grid is involved in the calculation.

It is important to note that poro-elastic stress obeys an identical formulation to thermoelastic stress, with replacement of the temperature change ΔT by the fluid pressure change Δp and of the thermoelastic constant A_T by the poro-elastic constant A_p, which is defined as:

$$A_p = E/(1-\nu) \cdot \alpha_p$$

(2.2.6)

where the poro-elastic expansion coefficient is:

$$\alpha_p = (1-2\nu)/E \cdot (1 - c_g/c_b)$$

(2.2.7)

in terms of grain and bulk matrix compressibilities c_g and c_b. This makes it possible, and generally desirable, to combine poro- and thermoelastic stress into a single quadrature. Typically, poro-elastic stress is compressive (due to the enhanced pressure around injection wells), and smaller (but by less than an order of magnitude) than the thermoelastic stress.

Thermal stress (in an effectively infinite medium) is a function of the shape, rather than the size, of the cooled region. The cooled region around an injection well characteristically consists of a stack of discs, corresponding to geological layers of different permeability and hence rate of convective cooling. The discs will be distorted by factors including heat conduction, non-radial flow patterns and fracturing, which can tend to extend them into ellipses whose major axis is the fracture propagation direction. In the limit of a cylinder of uniform temperature change, finite in radius but infinite in vertical extent, the internal hoop stress change is uniform and is given by:

$$\Delta\sigma_{yy} = 0.5 \, A_T \Delta T$$

(2.2.8)

A disc of small vertical extent and large radius has a larger internal stress change, equal to $A_T \Delta T$. This is a significant factor in deciding the location of thermal fractures, favouring thin high permeability streaks in the formation. A treatment of thermoelastic stress in relevant geometries is given by Perkins and Gonzalez [12], who illustrate also that a small compressive stress will exist outside the tensile cooled region, which helps to act as an effective barrier to fracturing beyond this zone.

2.3 2-D FRACTURE MECHANICS

Fracture growth can be approximated as a 2-dimensional areal problem if the fracture extends vertically over the full height of permeable rock in the reservoir, with plane strain boundary conditions, or if its vertical dimension much exceeds its horizontal dimension. These conditions do not often apply in practice, for reasons given below, but the 2-D case remains a useful semi-analytic comparison in examining the significance of certain physical factors.

For both 2-D and 3-D fracturing, the extremely long timescale of rock cooling compared to rock dynamical processes means that the fracture will evolve through a sequence of equilibrium configurations. The equilibrium is defined in terms of stress intensity factor K_i at the fracture tips, defined such that the stress in the rock at a small distance r from the tip limits as:

$$\sigma = K_i / r^{1/2} \qquad (2.3.1)$$

For simple materials, the fracture equilibrates when K_i becomes equal to a critical value K_{ic}, which is a constant material property.

The stress intensity factors for a crack in a 2-D plane may be calculated by the application of Cauchy integrals to analytic functions associated with stress in a linear, elastic medium. For a general, asymmetric, distribution of fluid pressure p and thermo- or poro-elastic stress σ along the line of the crack between its (unknown) tips at x=a and x=b (b>a), the equilibrium conditions are:

$$\int_a^b (p-\sigma)\{(b-x)/(x-a)\}^{1/2} dx = \{\pi(b-a)/2\}^{1/2} K_{ic}$$

$$\int_a^b (p-\sigma)\{(x-a)/(b-x)\}^{1/2} dx = \{\pi(b-a)/2\}^{1/2} K_{ic}$$

$$(2.3.2)$$

The length and location of the fracture are then determined by root-solving for a and b using Newton iteration. This procedure is fast enough to perform at each timestep of the reservoir calculation. However, unless the injector is constrained to operate at constant bottom-hole pressure, it is essential to include in the iteration terms expressing the dependence of pressure on a and b. This is achieved either by periodic additional reservoir pressure calculations with modified values of a and b, or (at less cost but greater risk of instability) by estimating from previous steps the derivatives of pressure with repect to a and b. The latter method relies on small amplitude numerical fluctuation to achieve accuracy, and is not robust.

The fluid flow problem in the reservoir will generally give a bottom-hole pressure which decreases as the fracture length increases, other factors being equal. A fractured well constrained to operate at constant injection rate will frequently evolve to a state in which the fracture lags behind the thermal front in a region of fairly uniform stress. Its equilibrium is then dominated by the requirement that it is just short enough to maintain a bottom-hole pressure equal to the confining stress.

2.4 3-D FRACTURE MECHANICS

Calculation of an equilibrium planar fracture in a complex 3-dimensional stress field is a significant problem, requiring the use of boundary element or finite element techniques. Here the fracture mechanics calculation must additionally be coupled with the computation of fluid pressure over a much larger domain.

The basis of the fracture calculation used in this code is a boundary element code developed by Gu [6] to solve the related hydraulic fracturing problem. In hydraulic fracturing, small quantities of high viscosity fluid are injected into a wellbore at high pressure, over a period of hours, to create and enlarge a fracture. It is normally assumed that only a modest proportion of the injected fluid leaks off from the fracture into the rock matrix. Fracture horizontal growth is limited by the rate at which the fluid can advance into the fracture, while vertical growth can be rapid, unless there are shale layers which provide a barrier of high compressive in-situ stress. The nature of the waterflood fracturing problem is different in certain important respects:

(i) the waterflood fracture size is normally determined through an equilibrium with the fluid pressure and rock stress field. Since water viscosity is low, and timescales are long, there is no restriction on fracture growth due to the water velocity within it;

(ii) the water in the waterflood fracture has a very high leak-off rate (ie almost 100% of the injected water leaves the fracture very rapidly), which means that pressure in the fracture cannot be derived without reference to the whole reservoir;

(iii) there is strong coupling between waterflood fracture growth, cooling and fluid pressurisation, and the stress field itself.

The principle of the calculation is to determine a fracture boundary such that the stress intensity factor K_i is equal to the critical stress intensity factor K_{ic}. We start with an initial guess for the fracture boundary within its vertical plane, and values of normal stress and fluid pressure in the fracture within this boundary. Values of stress at other points are not required. On the basis of the rock mechanical parameters (assumed uniform), the width w of the fracture (ie the small displacement of its faces perpendicular to the plane) is calculated at a set of nodes within the boundary, and from the calculated set of widths, the stress intensity factors at discrete nodes on the boundary are straightforwardly calculated.

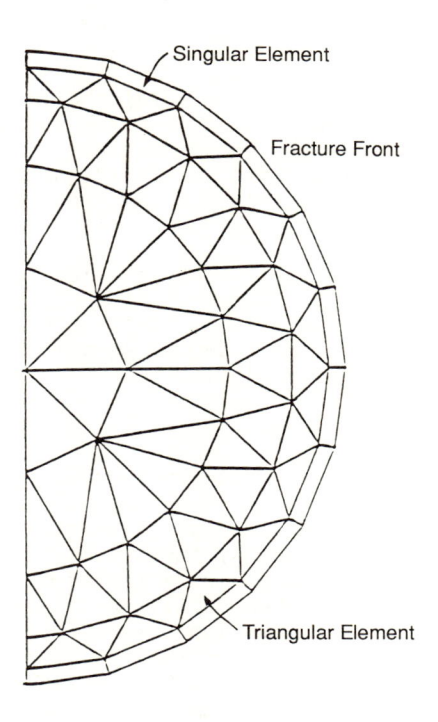

Figure 3. An example of the Fe Grid used in the 3-D Fracture Mechanics Calculation.

The widths are calculated at the nodes of a finite element grid in the (x,z) plane, of the type shown in Figure 3. They are obtained by inversion of the singular integral equation:

$$(p-\sigma)(x,z) = \frac{G}{4\pi(1-\nu)} \cdot \int \left(\frac{\partial}{\partial x}\left[\frac{1}{R}\right]\frac{\partial w}{\partial x'} + \frac{\partial}{\partial z}\left[\frac{1}{R}\right]\frac{\partial w}{\partial z'} \right) dx'dz'$$

where $R = [(x-x')^2 + (z-z')^2]^{1/2}$ \hfill (2.4.1)

and G is the shear modulus. The fracture surface constitutes the domain of integration. There is the boundary condition:

$$w(x,z) = 0 \hfill (2.4.2)$$

at the fracture boundary. This equation was derived by Kossecka [9] and Bui [3], using single-layer and double-layer elastic potentials, and by Weaver [15], using the boundary integral method.

The degree of singularity in the integrand is reduced by parts, and the integral equation is solved using the Galerkin method. The mesh has quadrilateral boundary elements which incorporate the requirement that the width varies as the square root of distance from the boundary, in the limit as the boundary is approached.

The fracture boundary is defined by the location of N points, at radius vectors $\underline{r}_j$, j=1,...,N from a fixed origin. For reasons of economy, N is a small number (N = 13 for a symmetric version of the code). Solution of the integral equation allows the calculation of K_i at each point. It is now necessary to iterate the values of r_j (normally choosing to keep the angles of the radius vectors constant) until $K_{ij} = K_{ic}$ for all points j.

In some circumstances (eg bottom-hole pressure too low) it is not possible to obtain a solution; this normally corresponds to fracture closure. If the fluid pressure exceeds the stress in some regions but not in others, then at most 2 solutions can generally be obtained. (A strict analysis of uniqueness has not been performed.) One of these solutions corresponds to a relatively large fracture (at least several metres in size) which is bounded by high stress, and where K_{ic} is not an important constraint. This is the physically acceptable case. The other solution corresponds to a very small fracture (centimetres in size) whose pressure just exceeds the minimum stress, and which is limited by the critical stress intensity. This is not a physical case, since rock cohesion will have been destroyed close to the well bore during the initial rupturing process. It is therefore important to choose starting radii r_{j0} with care.

The injection well constraints, the fracture conductivity, and the importance of poro-elasticity are the principal factors which determine how closely the iteration of the fracture boundary must be coupled with the fluid pressure solution. The timestep length is generally constrained by the Courant condition applied to saturation or temperature change in the explicit part of the BPOPE model, and is particularly limited by the use of fine grids for these variables. Thermoelastic stress changes slowly over a step, and so values from the end of the previous timestep are used throughout.

The simplest case is that of an infinite conductivity fracture under a constant bottom-hole pressure condition. This case essentially uncouples the fracture calculation from the reservoir over each timestep (since pressure at all points in the fracture is low), though in the longer term the reservoir and stress field do respond to the fracture. The fracture calculation is now a root-solving problem, where the number of simultaneous equations is equal to the number of boundary points.

The calculation is readily susceptible to Newton iteration, except that, since each alteration of the fracture contour requires calculation of a new FE stiffness matrix, it is computationally impracticable to calculate a full Jacobian:

$$J_{mn} = \partial K_{im}/\partial r_n \quad (2.4.3)$$

at each timestep, since this requires at least N contour changes, and may be required more than once per step. It is not generally robust or practicable to use Jacobians from previous steps, since the pressure does not, in all applications (eg well step-rate tests) vary slowly or smoothly. The practice has also been adopted (for many problems) of re-calculating the fracture size at intervals of several steps rather than at each step.

Rapid equilibration may however be achieved for this class of problem when it is noted that the influence of the position of boundary point n on the stress at point m has a strong dependence on the separation of m from n. When the three central bands of the Jacobian are calculated (using 3 contour changes) and the remaining elements are set to zero, accurate convergence is generally achieved for two Jacobian calculations, unless the initial conditions are particularly poor.

When the fractured injector is constrained by a constant water injection rate (but still has infinite conductivity), coupling of fracture and reservoir pressure calculations is essential. The coupling can be explicit when behaviour is smooth (using pressure and stress from the previous step in the fracture calculation). However, this degrades the original implicitness in pressure of the reservoir simulator and has not been found a robust or economical method. In particular, fracture size can be a very sensitive function of the fracture fluid pressure in (commonly encountered) cases where the stress field is fairly uniform in the region of fracturing, and is just balanced by the pressure. The code therefore uses implicit coupling of fracture dimensions with reservoir pressure, on those steps selected for a fracture mechanics calculation, and makes some sacrifice of accuracy by reducing the frequency of fracture steps in regions of smooth behaviour.

Since typically each iteration of the fracture mechanics routine is equivalent in expense to one reservoir pressure inversion, alternate evaluations of these processes take place several times within each fracture mechanics step. When it is a significant factor, poro-elastic stress must be re-evaluated corresponding to fluid pressure change. However, since it bears a close correspondence to the pressure field, approximate approaches may sometimes be used.

In this way, implicit 3-D coupled reservoir/fracture simulation can be achieved without exceeding the non-fractured simulator run-time by more than a factor of about 4, in a robust manner with a small sacrifice of accuracy. The assumption of uniform linear elasticity is undoubtedly a simplification for some reservoir materials. However, the model allows us to attempt, for the first time, prediction of fractured injector performance (and certain previously ill-understood physical processes) in typically complex layered reservoirs.

3. EXAMPLES OF WATERFLOOD FRACTURING IN LAYERED RESERVOIRS

The following examples are based on idealised reservoir cases.

3.1 PRESSURE-CONSTRAINED VERSUS RATE-CONSTRAINED FRACTURE GROWTH

A vertically layered reservoir model (Table 1) is considered, in which there is a modest permeability contrast of up to a factor of 5 between layers. The reservoir is subject to a uniform in-situ stress of 5700 psi, and the injection water is 70°C cooler than reservoir temperature. The rock mechanical parameters indicate a thermoelastic constant A_T of 10 psi/C. Poro-elasticity is neglected in these calculations, and the fracture has infinite conductivity to flow. The reservoir calculation places the injector at the centre of a 5-spot pattern, with producers at the corners of a square grid.

When the injection well is operated at a constant bottom-hole pressure of 5000 psi, fracture growth advances with the thermal front in the reservoir. The thermoelastic effect reduces stress in the cooled region around the well below the bottom-hole fluid pressure, almost instantaneously from the start of injection, and subsequently the fracture propagates such that its boundary extends into regions which are still hot, except for the

TABLE 1

RESERVOIR STRUCTURE FOR SIMULATED CASES

LAYER	THICKNESS (ft)	PERMEABILITY (mD)	PERFORATED
1	50	205	Y
2	43	203	Y
3	33	175	Y
4	11	150	N
5	33	136	N
6	46	279	N
7	59	282	N
8	74	326	N

immediate vicinity of the fracture face. Figure 4 illustrates a sequence of the successive boundaries of the advancing fracture, in the vertical plane of fracturing. It might be imagined that the fracture would propagate unstably, since as it advances, it cools all regions it comes into contact with. This is not in fact the case, since the thermoelastic stress change normal to the fracture plane is a function of the shape of the cooled region, and is negligible when the cooled region is thin normal to the plane. Nevertheless, the fracture grows rapidly, and substantially distorts the shape of the cooled region around the well from a circular to an elliptical shape. Over this period, the injection rate is found to increase corresponding to the growth of the fracture.

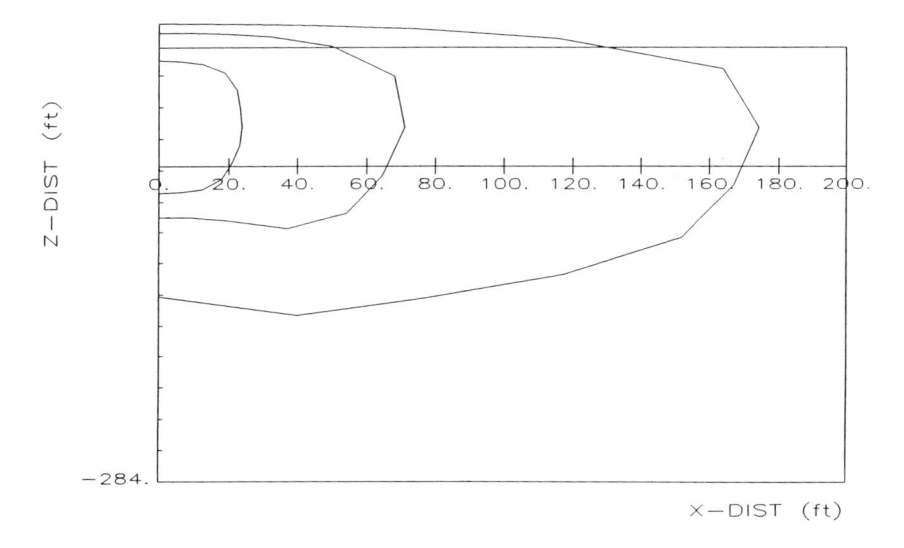

Figure 4. Successive Fracture Boundaries for a Pressure-Constrained Fracture.

When the injection well is set to a constant rate, and the production wells to a fixed pressure, behaviour is found to be considerably different. A rate is chosen such that, if the injector were not fractured, its pressure would exceed (slightly) the stress around it after cooling. After a few days, the fracture is found to lag significantly behind the temperature front, and to achieve a roughly constant surface area. Figure 5 shows the evolution of the fracture boundary, which displays just some changes in shape, as the stress distribution evolves. The fracture is restricted to the region of greatest thermal stress change, and its internal fluid pressure just balances the stress in this region. Its size is determined by the fact that, if it grows, its pressure must fall, on the basis of the reservoir fluid flow problem. A fall in pressure below the minimum stress will cause the fracture to close. The fracture is therefore maintained by a balance of its dimensions with its pressure, which can only be simulated by an implicit linkage between the fracture mechanics and pressure calculations.

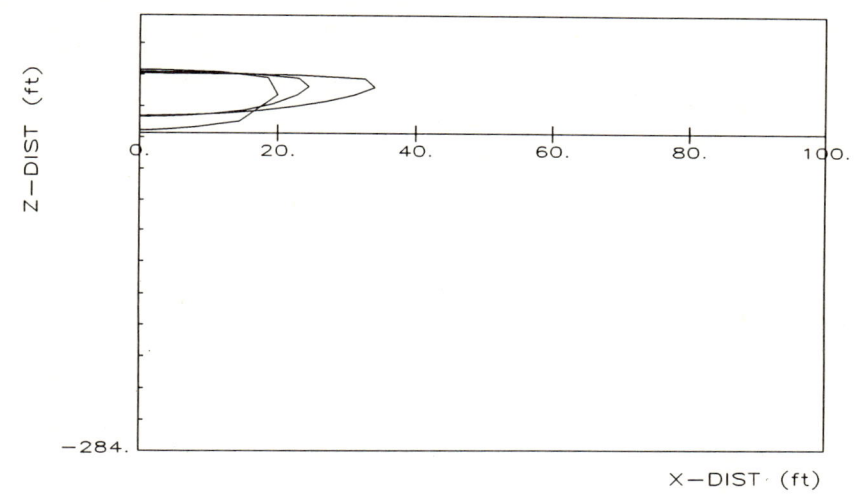

Figure 5. Successive Fracture Boundaries for a Rate-Constrained Fracture.

3.2 STEPPING UP OF WELL RATE OR PRESSURE

The model of Table 1 is used in the simulation of a step-rate test, in which the injection well is subjected to a series of increases in injection rate. This is a standard means to determine the performance of a fractured injection well in the field. As the rate is stepped up, the fracture size increases, which raises the well injectivity (rate of injection per unit increment in pressure).

The calculation here is performed in a fully-coupled manner, with both thermoelastic and poro-elastic stress. The poro-elastic stress ($A_p = 0.5$) is coupled into iterations of the fracture mechanics calculation, along with the reservoir pressure calculation. Rates are stepped up from very low values, at which the well is not fractured, to rates at which the fracture size is pushed towards its maximum limits, given the extent of cooling. A single implicit fracture mechanics evaluation is performed at each injection rate. This is permissible as a description of steady state fracture size at that rate, provided that pressure transients are damped on a timescale less than the step length, which is indeed the case, given the high permeability of the formation. It is then appropriate to exclude fluid compressibility from the calculation.

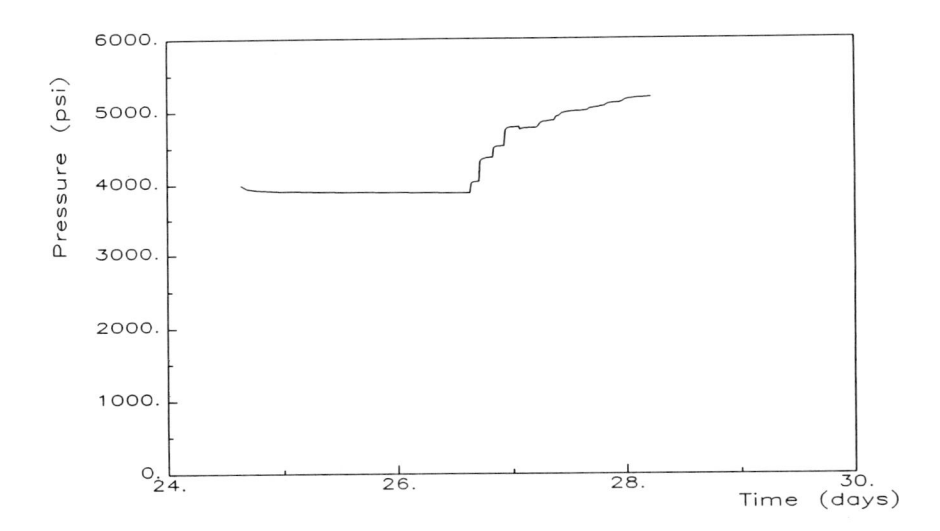

Figure 6. Development of pressure during a Step-Rate test in a Fractured well.

Figure 6 shows the calculated pressure behaviour of the fractured well as the rate is raised in steps from zero, starting at a time of 26 days. The initial steep section of the curve represents the low rate, non-fractured behaviour of the well, where an increase in well rate can only be achieved by an increase in its pressure. Subsequently, fracturing occurs, and the fracture grows, as illustrated by the successive fracture boundaries plotted in Figure 7. At this stage, the rate of increase of pressure goes down, since a large proportion of the rate increase is accommodated by the increase in the fracture size. The pressure increase is determined by the need for the fracture to expand into regions of somewhat greater stress. The simulation of a process of this type requires the code to deal robustly with a wide range of fracture behaviour.

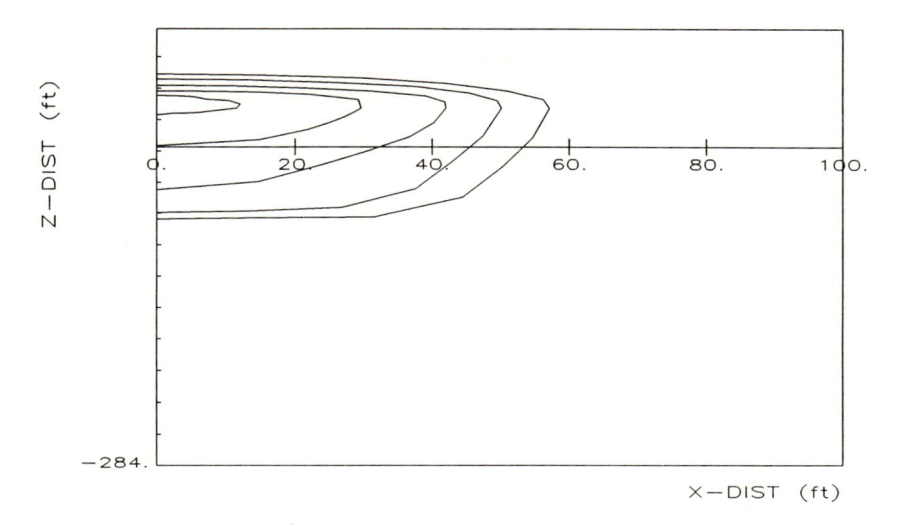

Figure 7. Successive Fracture Boundaries a Step-Rate test.

3.3 FACE-PLUGGING: FRACTURES AS A REGULATORY MECHANISM

A fracture which is rate-constrained, as in the example of Section 3.1, commonly exists in a region of fairly uniform stress, in the cooled region, well behind the thermal front. In these circumstances, it is capable of increasing its size substantially, without a significant change in the the well bottom-hole pressure, apart from an appropriate allowance for finite fracture conductivity, and the need for a pressure gradient to force the water through the fracture.

A fracture of this type will grow to compensate for factors which would otherwise (ie at constant fracture size) tend to lead to an increase in injection well pressure. Examples of such factors are mobility decreases in the reservoir, due for example to the advance of water of higher viscosity than the oil it displaces, or the blockage of the fracture faces by solid particles in the injected water. Behaviour in the latter case is illustrated by an example, in which a low permeability skin was allowed to form on the fracture face, with thickness proportional to the quantity of fluid injected per unit area. In between fracture mechanics steps, the pressure at the injector was observed to increase, corresponding to build-up of skin. However, on recalculation of the fracture size in an implicit manner (allowing here for non-infinite fracture conductivity) fracture growth was calculated, according to **Figure 8**, and pressure reverted to its original value.

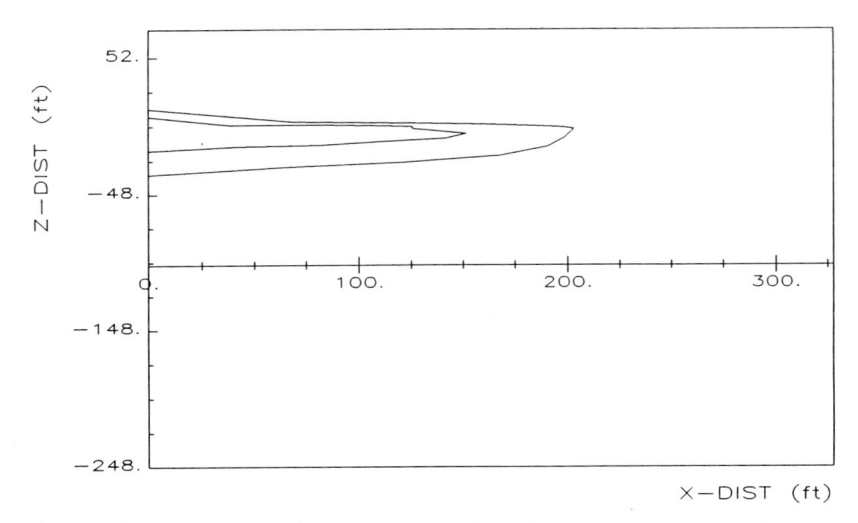

Figure 8. Increase in Fracture size due to Face-Plugging at Constant Rate.

Thus an apparently stable situation, in terms of measured well pressures and rates, is shown to be in fact accompanied by large and sometimes rapid changes in the injection well fracture system. This can be of particular significance and must be understood when vertical fracture growth is relevant to the pattern of fluid flow and ultimately oil sweep within the reservoir.

4. CONCLUSIONS

(i) Waterflood fracture growth has been calculated by coupling the simulation of fluid and heat flow, with thermo- and poro-elastic stress, and with a fracture mechanics calculation. Taking account of the varying timescales of the physical processes involved, the calculation may be divided into components which are performed independently and advanced explicitly in time. However, an implicit iteration between fracture mechanics and reservoir pressure is an essential component of the model.

(ii) The inclusion of an equilibrium 3-D fracture mechanics model, which follows the vertical and horizontal growth of the fracture between a succession of equilibrium states, poses significant problems of combining the fracture calculation efficiently and robustly with the reservoir fluid pressure. Characteristically, each iteration of the fracture code must be alternated with a complete inversion of the reservoir pressure matrix.

(iii) In the complex stress patterns which result from the cooling of a layered reservoir, waterflood fractures are commonly restricted in their vertical growth. Simulation is generally borne out by field measurement in this respect. It is rarely possible to treat the problem as 2-dimensional. On this basis, it is to be expected that field performance will be affected by vertical fracture behaviour.

(iv) We show that the nature of constraints applied to the injection well determines qualitative aspects of fracture behaviour. Wells maintained with a bottom-hole pressure above the stress in the cooled rock can show rapid fracture growth, with significant distortion of the flow field in the reservoir. Wells operated at a constant rate will evolve to have fractures of dimension less than the cooled zone around the well. Their behaviour then acts to conserve well bottom-hole pressure. Plugging of the fracture face, for example, will commonly lead to fracture extension, but without significant changes in well rate or pressure.

REFERENCES

[1] Barenblatt, G.I., "The Mathematical Theory of Equilibrium Cracks in Brittle Fracture", Advances in Applied Mechanics, Vol. 7, pp 55-129, 1962.

[2] Barker, J.W. and Fayers, F.J., "Factors Influencing Successful Numerical Simulation of Surfactant Displacement in North Sea Fields", In Situ, Vol. 12, No. 4, December 1988.

[3] Bui, H.D., "An Integral Equations Method for Solving the Problem of a Plane Crack of Arbitrary Shape", J. Mechanics Physics Solids, Vol. 25, pp 29-39, 1977.

[4] Dikken, B.J. and Niko, H., "Waterflood-Induced Fractures: a Simulation Study of their Propagation and Effects on Waterflood Sweep Efficiency", SPE 16551, presented at SPE Offshore Europe 87, Aberdeen, September 8-11 1987.

[5] Garon, A.M., Lin, C.Y. and Dunayevsky, V.A., "Simulation of Thermally Induced Waterflood Fracturing in Prudhoe Bay", presented at SPE California Regional Meeting, Long Beach, March 23-25 1988.

[6] Gu, H., "A Study of Propagation of Hydraulically Induced Fractures", Ph.D. Thesis, University of Texas, Austin, December 1987.

[7] Hart, R.D. and St. John, C.M., "Formulation of a Fully-Coupled Thermal-Mechanical-Fluid Flow Model for Non-linear Geologic Systems", Int. J. Rock Mech. Min. Sci. & Geomech. Abstr., Vol. 23, No. 3, pp 213-224, 1986.

[8] Koning, E.J.L., "Fractured Water Injection Wells - Analytical Modelling of Fracture Propagation", SPE 14684, 1985.

[9] Kossecka, E., "Defects as a Surface Distribution of Double Forces", Arch. Mech., Vol. 23, pp 481-494, 1971.

[10] Myklestad, N.O., "Two Problems of Thermal Stress in the Infinite Solid", J. Applied Mechanics, Vol. 9, pp A-136 to A-143, 1942.

[11] Nghiem, L.X., "Modelling Infinite-Conductivity Vertical Fractures with Source and Sink Terms", Soc. Pet. Eng. J., pp 633-644, August 1983.

[12] Perkins, T.K. and Gonzalez, J.A., "Changes in Earth Stresses around a Wellbore caused by Radially Symmetrical Pressure and Temperature Gradients", SPE 10080, presented at SPE Fall Conference, San Antonio, October 5-7 1981.

[13] Settari, A., "Simulation of Hydraulic Fracturing Processes", Soc. Pet. Eng. J., pp 487-500, December 1980.

[14] Settari, A. and Raisbeck, J.M., "Analysis and Numerical Modelling of Hydraulic Fracturing Processes during Cyclic Steam Stimulation in Oil Sands", J. Pet. Tech., pp 2201-2212, November 1981.

[15] Weaver, J., "Three-dimensional Crack Analysis", Int. J. Solids Structures, Vol. 13, pp 321-330, 1977.

Corner Point Geometry in Reservoir Simulation

D.K.Ponting

E. C. L., Highlands Farm, Greys Road, Henley, Oxon., U.K.

1 Abstract

Hydrocarbon reservoirs commonly possess a geometry which incorporates both dip and faulting. In order to distinguish depth variations due to these two effects, it is useful to specify a simulation cell through the positions of its eight corner points. The faces of such a cell may be bilinear surfaces, and may form part of a distorted grid. Such a grid may be chosen such that cell boundaries lie along faults, which may be vertical or sloping. Normal and fault connections may be treated on an equivalent basis. A system of 'coordinate lines' and corner point depths used to construct such a model is described. An exact analytic expression is obtained for the cell volume. Transmissibility values are calculated in terms of three-vector mutual interface areas, which automatically incorporate corrections for dip and inclined flow. The application of calculated transmissibilities to a simple five point model can lead to an inconsistent finite difference scheme, and to significant errors if the grid is highly distorted. A method of avoiding such errors is presented, which more accurately reflects the interaction between pressures and cell geometry to produce flows. This is derived from basic finite element type principles, and results are presented for test cases, comparing the corrected scheme with a simple five point model.

2 Cell geometry

The original motivation for treating reservoir simulation cells in terms of their corner points was the treatment of dip, and the problem of distinguishing this from faulting. This is difficult if only the cell centre depth difference is known, as this may be due to either effect. Traditionally, reservoir simulation cells have been represented as rectangular in section when calculating volumes, and as having a linear variation in cross sectional area between cell centres when calculating transmissibilities. The specification of the cell top and bottom corner depths suits data on horizon depths, and cell sections may be clearly represented graphically as quadrilaterals. Overlap areas between cells in different layers may be represented, and fault throws may be precisely defined.

In the case of a single dip angle in one direction, such a description reduces to the familiar one in terms of cell centre depths. Suppose, however, that two different dip angles exist in the x- and y- directions, a common physical situation. The question may then be asked what form the top and bottom surfaces take. Note that although simulation cells are usually thin, in comparison to their areal dimensions, depth variations across a cell may be large. If the cell is specified in terms of corner point depths it is natural to treat the top and bottom surfaces as bilinear.

Given the original motivation in terms of faults, the possibility of sloping cell edges may be considered. To do this we define a set of 'coordinate lines' for the grid, such that the corners of cells in a column lie on these lines. If one or more of these lines is sloping, the cell faces also become generally bilinear surfaces. For a grid with dimensions N_x by N_y, $(N_x + 1).(N_y + 1)$ such coordinate lines are required, each of which may be specified by two (x, y, z) points in space. A lower and upper corner point depth is specified on each of the four coordinate lines surrounding a cell.

The final assumed form for a grid block is shown in fig.1, being parameterised by the coordinates α, β and γ. This is as complex as one would wish, although has one slight irregularity : at a fault which is sloping by a changing amount, and along which the fault throw increases, the two bilinear surfaces of connecting grid blocks will not precisely match. The result is a 'fault volume', which is either omitted or double counted in the total cell volume. With this exception, the coordinate line and corner point depth system defines the grid system precisely. Cells may be separated from the layer above, or a layer of cells pinched out.

CORNER POINT GEOMETRY 47

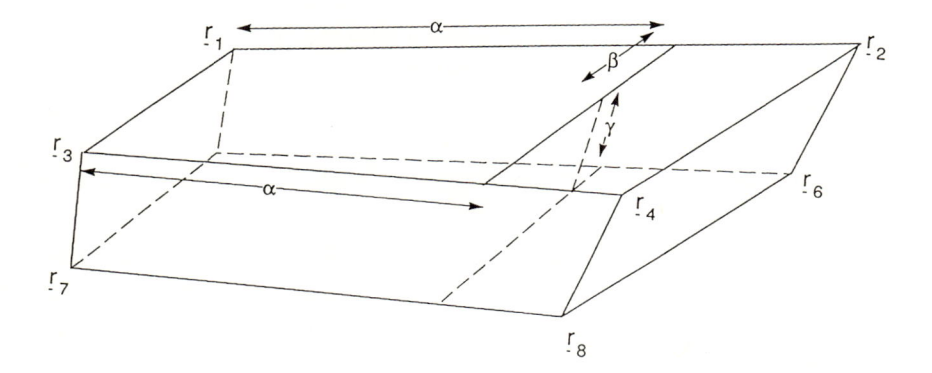

Fig.1. Cell geometry

A first essential is clearly to be able to define the cell volumes exactly. The volume of the cell may be obtained by integration over the domain $0 < \alpha < 1$, $0 < \beta < 1$, $0 < \gamma < 1$, transforming the integration measure $dx.dy.dz$ to $J.d\alpha.d\beta.d\gamma$. The cell volume is obtained in the appendix in terms of the corner point coordinates.

We now need to define transmissibilities between cells in such a way that fault transmissibilities (between different layers in the cartesian indexing grid) and non-fault transmissibilities are treated in the same manner. The mutual interface area between two cells should appear as a factor in the transmissibility, and the traditional form should be obtained in the case of a regular sloping layer.

A suitable expression for the transmissibility between a cell I and its neighbour J is

$$T_{I,J} = (1/T_I + 1/T_J)^{-1}$$

$$T_I = K_I |\mathbf{A}.\mathbf{e}_{\mathbf{d}I}|/d_I = K_I |\mathbf{A}.\mathbf{d}_I|/d_I^2$$

$$T_J = K_J |\mathbf{A}.\mathbf{e}_{\mathbf{d}J}|/d_J = K_J |\mathbf{A}.\mathbf{d}_J|/d_J^2$$

$\mathbf{A}$ is the mutual interface area between two cells, given by $\mathbf{A} = A_x\mathbf{e}_x + A_y\mathbf{e}_y + A_z\mathbf{e}_z$, and $\mathbf{d}_i$ and $\mathbf{d}_j$ are the vectors from the cell centres to the centre of the face which contributes to $\mathbf{A}$. It might be felt that the vectors $\mathbf{d}_i$ and $\mathbf{d}_j$ should go to the mutual interface area centre. This would, however, give incorrect results in the simple case of uniform one dimensional flow across a series of faulted or unfaulted columns, fig. 2. For horizontal flow the transmissibility from cell I to cell J in the un-faulted case should equal the total transmissibility from cell I to cell J and cell I to cell N in the faulted case. The above expression may also be used when the cells are not neighbours in the indexing grid, but have a mutual interface area due to faulting or a pinched-out layer.

This form preserves the traditional value for a uniformly converging layer of cells. For a uniformly sloping layer of cells, the usual $Cos^2\theta$ dip dependence is obtained, from the scalar product in the numerator and the increase in the distance between centres in the denominator.

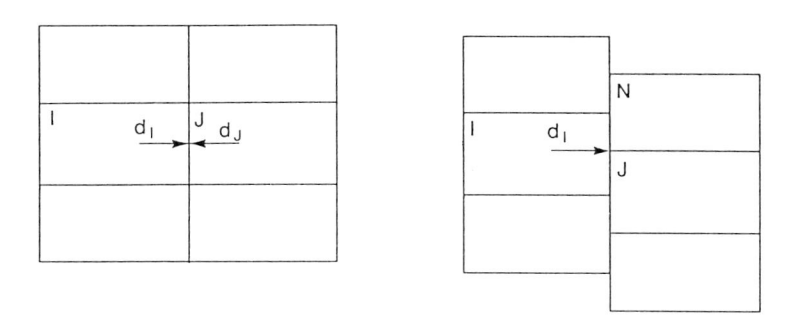

Fig.2. Column to column transmissibilities

The main difficulty in the transmissibility calculation is obtaining the projections of the surface areas A_x, A_y and A_z. The projection of any cell face is a quadrilateral, but at a fault interface the mutual area projections are of two overlapping quadrilaterals, with possibilities as shown in fig.3.

In figure 3 Case 1 would correspond to a scissors fault, and case 2 to a thin layer abutting a thicker layer, considering the x- or y-projections. Case 3 is the usual case of an interface between two matching cell faces. Case 4 might be obtained by as the z-component of the interface between two coordinate lines sloping in different directions. A general means of evaluating the overlap areas is as follows. For the case of the z-component of the interface area, project into the x-y plane, then :

1 Store the x-coordinates of all face corners
2 Store the x-coordinates of all edge intersections. These may be intersections between the two quadrilaterals, or self-intersections, as in the last example in fig.3.
3 Sort all the x-values. These then form a set of intervals. Within each interval the overlap area is a trapezoidal surface, defined by the innermost of the quadrilateral boundaries.
4 Sum the area contributions, allowing for the sign involved. In the last case of fig.3, for example, the two contributions are of opposite sign. (This case is obtained from the x-y projection of a surface between two sloped coordinate lines).

This calculation may also be done with respect to the y-axis, to form a useful null check on the evaluation. Calculations of the x- and y- projections of the interface area, into the y-z and x-z planes respectively, may be done in the same manner.

These transmissibility expressions are based on linear pressure gradients, although when layer thicknesses vary, (causing converging flow), the pressure will follow a logarithmic form.

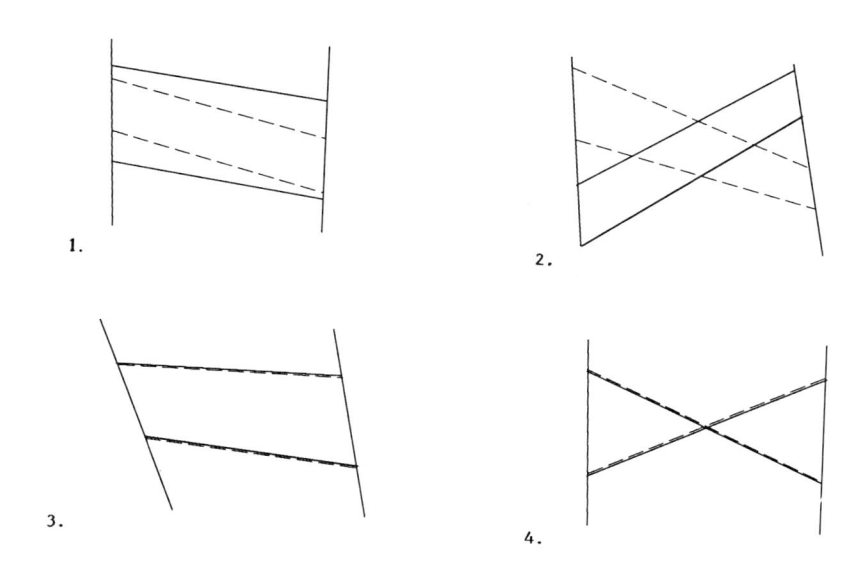

Fig.3. Possibilities for cell face overlap

3 Effects of grid non-regularity

Given that the volume and transmissibility expressions are in general three vector form, it is possible to have cells which are not rectangular in plan. Such a distorted grid may be used to follow geological boundaries or faults. The grid may be spread out to represent large aquifer cells at the boundary of the grid.

Such distortion can introduce rather larger angles than in the case of dip. This is particularly the case when interactive gridding packages are used. Errors are introduced in the simple five point transmissibility based picture. This occurs when the finite difference scheme does not calculate the correct flows for the current pressure values. These may be regarded as arising from two effects. Consider the grid shown in fig.4, assuming a constant pressure gradient in the y-direction.

The first effect is that in a five point scheme this would be interpreted as a pressure gradient in the j-direction, although in fact the component in this direction is $\Delta P.Cos\theta/\Delta d$. The second is that the y-direction pressure gradient then would cause flow across the interface from cell i to i+1, as

CORNER POINT GEOMETRY

the interface has a component in this direction.

This error in the five point scheme will occur in all simulation codes which model dip, unless a single dip angle is modelled by rotation of a regular grid, which has the disadvantage of introducing gravity effects in all three directions on the grid.

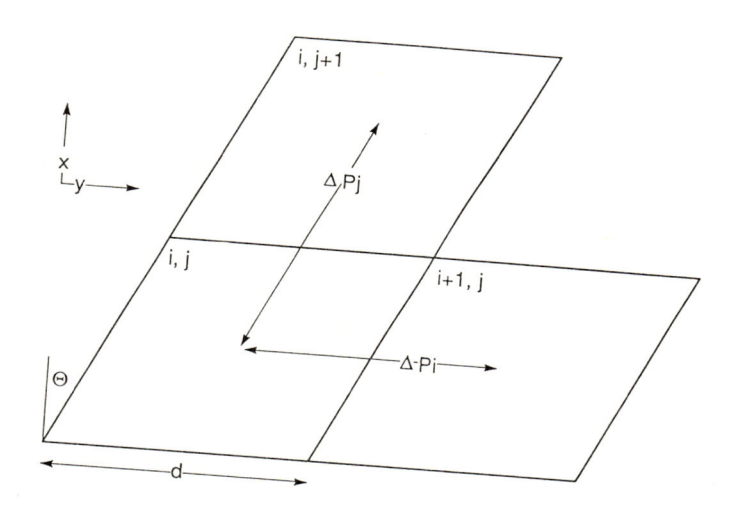

Fig.4. Simple case of sloping grid

The correct approach is to use the pressure values obtained from the grid (which will span the three dimensions) to construct pressure gradients in orthogonal directions, then construct the flows across each cell boundary as a function of these orthogonal gradients. This will yield tensor type behaviour(a dependence of the i-direction flow on the j-direction pressure difference), and set up a nine-point type of coupling for the flows in two dimensions, and thus also in the non-linear equation residuals. In the simple case of fig.4, the flows are given by :

$$F^{i,j \to i+1,j} = \frac{Kh}{Cos\theta}(\Delta P_i - Sin\theta \Delta P_j)$$

$$F^{i,j \to i,j+1} = \frac{Kh}{Cos\theta}(\Delta P_j - Sin\theta \Delta P_i)$$

The flows predicted by a five point scheme using transmissibilities would be

$$F^{i,j \to i+1,j} = KhCos\theta \Delta P_i$$

$$F^{i,j \to i,j+1} = KhCos\theta \Delta P_j$$

For a uniform pressure gradient in the j-direction, $\Delta P_i = \Delta P_j Sin\theta$, so the flows predicted the same. The five-point approach may thus still be used if the grid is aligned to the predominant local flow directions, as in Wadsley (1980).

4 The finite volume method

A suitable formalism for constructing a more accurate scheme is the finite volume method described by Rozon (1987). The method has been previously used in fields such as heat transfer calculations, as described by Baliga and Patankar (1980). In this technique, governing differential equations are integrated over cell volumes prior to discretisation. The resulting volume and surface integrals are estimated using local functions defined by discrete solution values. This type of scheme has the advantage of being locally conservative : mass balance may be applied to each grid block and the flows to and from it. This is vital in reservoir simulation, in which grid sizes are often large compared to the solution features. The fluid conservation equation for a grid block then becomes :

$$\int_V \partial/\partial t \phi b_f \, dV + \int_S \mathbf{F}.d\mathbf{S} + \int q_f \, dV = 0$$

where b_f is the density of fluid f (in surface volume per reservoir volume, or moles per reservoir volume). If flows are evaluated implicitly, for a time step of length Δt, the block conservation equation becomes :

$$R_f = M_f^{t+\Delta t} - M_f^t + \Delta t \int_S \mathbf{f}_f^{t+\Delta t}.d\mathbf{S} + \Delta t Q_f = 0$$

The mass accumulation term and the flows are expressed in terms of nodal values. The flow integral splits into a sum over interface flows into neighbouring cells.

On a regular grid, the finite volume method yields five- and nine-point schemes. To obtain a five-point scheme, the flow pressure gradient between cell (i,j) and (i+1,j) is assumed to be a linear function of the pressures of these cells only. Although the flow is obtained as an integral across a surface normal to the x-direction, transverse pressure effects are neglected. Finite difference equations have also been derived in this manner by Nghiem (1988). Non-uniform permeability values are simply included by obtaining the pressure at the cell interface as

$$P_{i+\frac{1}{2}} = \frac{(K_i P_i + K_{i+1} P_{i+1})}{(K_i + K_{i+1})}$$

and then using the pressure gradient in either cell to obtain the flow.

It is possible to take transverse pressure effects into account by using a bilinear pressure variation. The pressure dependence over a cell is split into four bilinear functions in two dimensions, or eight in three dimensions. In three dimensions, flow through a cell face is obtained by summing over four quadrants, each of which has its pressure variations specified by one of the bilinear patches. The result is a nine-point scheme in two dimensions, a twenty-seven point scheme in three. Note that in such a scheme, there are no diagonal flow terms : each of the flows to the adjoining cells is simply determined by more than two pressures. As shown by Rozon, the nine-point scheme obtained differs from the optimal Laplace form and that used for multiphase flow by Yanosik and McCracken (1979).

On a distorted grid, these two types of pressure variation may again be used. In the locally linear case, the pressure variation in the $\mathbf{e}_\alpha$, $\mathbf{e}_\beta$ and $\mathbf{e}_\gamma$ directions is assumed to be a linear function of the pressure differences in these directions. The pressure variation is thus known along three directions which span the space. In order to perform the flow surface integrals, values in three orthogonal directions are required.

The pressure gradients along the local grid directions $\mathbf{e}_\alpha$, $\mathbf{e}_\beta$ and $\mathbf{e}_\gamma$ are obtained as expressions of the form :

$$K_{\alpha i} \partial P / \partial \alpha = -2 K_{\alpha i} \cdot \frac{(K_{\alpha i+1}/d_{\alpha i+1})}{(K_{\alpha i+1}/d_{\alpha i+1} + K_{\alpha i}/d_{\alpha i})} (P_i - P_{i+1})$$

The permeability values in the directions e_α, e_β and e_γ for cell i are $K_{\alpha i}$, $K_{\beta i}$ and $K_{\gamma i}$ respectively, and equal flow conditions at the cell boundary have been used to eliminate the interface pressure value.

As the (x,y,z) coordinates are linear in α, β and γ, it is convenient to first obtain $\partial(x,y,z)/\partial(\alpha,\beta,\gamma)$ and invert this to obtain $\partial(\alpha,\beta,\gamma)/\partial(x,y,z)$. Then it is possible to obtain:

$$K\nabla P = T^{-1}.(K_\alpha \partial P/\partial \alpha, K_\beta \partial P/\partial \beta, K_\gamma \partial P/\partial \gamma)$$

where:

$$T = \begin{pmatrix} \partial x/\partial \alpha & \partial y/\partial \alpha & \partial z/\partial \alpha \\ \partial x/\partial \beta & \partial y/\partial \beta & \partial z/\partial \beta \\ \partial x/\partial \gamma & \partial y/\partial \gamma & \partial z/\partial \gamma \end{pmatrix}$$

This is a function of (α,β,γ), and may be obtained in terms of the coefficients $\{\mathbf{C}^{(0,0,0)},..,\mathbf{C}^{(1,1,1)}\}$ defined in the appendix. The flow terms are obtained as surface integrals:

$$F^{i\to i+1} = \int f_x dS_x + \int f_y dS_y + \int f_z dS_z$$

$$\int f_x dS_x = \int\int dydz (K\nabla P)_x M$$

where M is the mobility of fluid f.

In calculating the flow from cell (i,j,k) to (i+1,j,k) the pressure values in these two cells are used, and two other pressure values to obtain a complete set of pressure gradients. If these are the pressures in cells (i,j+1,k) and (i,j,k+1), these do not introduce additional nine point terms in the residual for cell (i,j,k), and it may at first appear that the Jacobian thus has a five-point (seven-point in three dimensions) form. However, the flow from cell (i,j,k) in the i-direction appears in the residual for cell (i+1,j,k) as well as (i,j,k), and the dependence upon the pressure in (i,j+1,k) is a nine-point term with respect to (i+1,j,k). This could be avoided if the out- and in-flow terms were calculated using different pressure patches, but this would introduce mass balance errors arising from discretisation.

The generalisation of a five-point scheme to a distorted grid thus yields a 'seven-point' type of scheme, in which each flow term depends on the usual two 'driving' pressures, and one or two 'steering' pressures, which may be the transverse neighbour of either of the two cells connected by the flow. This is not symmetrical in the neighbours, so that it is assumed that the gradient to the neighbour is the same in the positive and negative directions. This is essentially the locally linear assumption, that a single linear pressure variation exists in each direction.

To avoid this assumption, and produce a symmetrical pressure patch, each cell face is divided into four quadrants. The pressure over each of the quadrants is then determined by up to eight pressure values, as shown in fig.5. This results in a symmetrical scheme, which is a generalisation of the usual nine-point scheme, or a twenty-seven point scheme in three dimensions. The pressure across each cell face quadrant is defined by a pressure octant, containing contributions from up to 8 cells, each with different permeability values in each direction. By equating flows the pressure values at the interfaces may be obtained in terms of the cell centre pressure values. These pressures are not known, but if each pressure is expressed as a set of coefficients of the eight octant pressures, these coefficients may be manipulated like pressure values. The coefficients are obtained at the quadrant edges and used to determine the $\partial P/\partial \alpha$, $\partial P/\partial \beta$ and $\partial P/\partial \gamma$ values at the quadrant centre point.

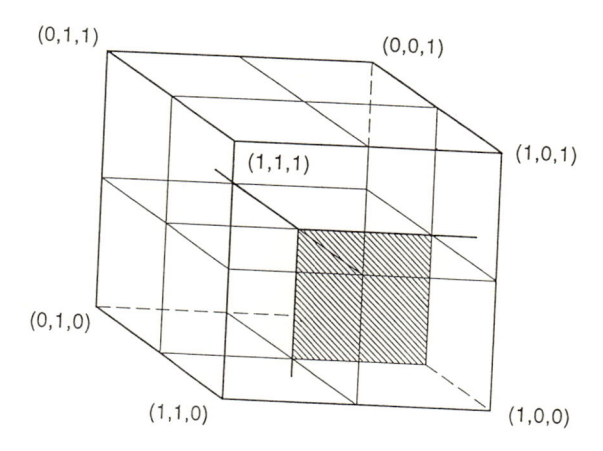

Fig.5. The eight pressure values used to define flow over a cell face quadrant. These are indexed by offset from the central pressure point, and all lie within the 27-point patch. The shaded area is the cell face quadrant over which flow is being obtained.

In all cases, the mass accumulation term is obtained by using the fluid density at the cell solution value, multiplied by the pore volume. The surface integrals defining flows are evaluated using a mid-point approximation. This is done at points such as $(\alpha, \beta, \gamma) = (\frac{1}{2}, \frac{1}{2}, 1)$, in the generalised seven-point case, or $(\alpha, \beta, \gamma) = (\frac{1}{4}, \frac{1}{4}, 1)$, in the generalised nine-point case.

On a distorted grid, therefore, the linear assumption, which would give rise to a five point scheme on a regular grid, gives rise to a nine point pattern , whilst a bi-linear local pressure variation yields a nine-point coupling on both regular and irregular grids.

We wish to generalise these results to the multi-phase or multi-component case. To obtain correct front heights and prevent fluid flowing from grid blocks in which it is not mobile, finite difference simulators conventionally upstream fluid mobilities. A simple initial option is thus to do this in the irregular grid case, so that the flow from cell I to cell J uses the mobility for cell I or J, depending on which is upstream. In a scheme with diagonal transmissibilities, such as that of Hegre et al (1986), upstreaming can be carried through to the flows to diagonal neighbours.

For the finite volume case, with simple upstreaming, and in the absence of depth variations and capillary pressures, the additional coupling terms are only functions of pressure. In a fully implicit scheme, therefore, the additional terms will be (N x 1) rather than (N x N). In any scheme in which depth variation occurs, and the reservoir density is a function of variables other than pressures (dissolved gas concentration, or mole fractions in the compositional case), the phase flows cannot be written as functions of a potential, but depend on the phase pressure gradients and a density value. The phase pressure includes the capillary pressures for all except the reference phase. These will introduce weak dependence of the cell to neighbour flows upon saturation and composition variables in the nine-point pattern, and the extra Jacobian terms will be generally full.

The major extra cost in an implicit simulation mode is in the linear solver. Using a nested factorisation type of linear solver (Appleyard and Cheshire (1983)) the extra terms are rather well placed. Each iteration involves :

OUTER SWEEP : Solve plane by plane tridiagonal, regarding planes as blocks

MIDDLE SWEEP : Solve line by line tridiagonal for a given plane, regarding lines as blocks.

INNER SWEEP : Solve tridigonal equations for a line of cells.

Each linear iteration involves one outer, two middle and four inner sweeps. The inner iteration is difficult to vectorise, as the tri-diagonal solution is recursive and the loop length short, although the middle and outer sweeps vectorise well. No extra nine-point terms occur in the inner sweep, eight in the middle and the remaining 16 in the outer.

5 Test runs

The three well problem originally described by Hirasaki and O'Dell (1970) was employed, using the data from Hegre et.al. (1986). This is a two phase oil-water areal study run using a regular and a slanting grid. Water is injected at a fixed rate by a centrally placed injector at one side of the grid and liquid produced by two wells at the other side. Results are shown in fig.6 for the pressure field at 500 days, using a slant grid with the simple five-point transmissibility scheme, and the corrected seven-point scheme. The problem, and the resulting pressures and saturations, should be left-right symmetrical.

The five-point results show a significant pressure error, in that the fluid flows preferentially to the right-hand well along the grid, resulting in a low pressure around the left-hand well. This pressure error is largely corrected when the seven-point scheme is used. The generalised nine-point scheme gives results very similar to those of the seven-point.

For a refined grid version of this problem, a good match is obtained in saturation and pressure using the corrected seven- and nine-point schemes, indicating that the flow correction results in a consistent scheme. The saturation profiles for refined grids are shown in fig.7. These are for a regular grid, a sloping grid using a five-point scheme and a sloping grid using the corrected seven-point scheme. In each case the refinement is by a factor of nine in each direction.

For the original, unrefined grid, however, the saturation profiles are not as symmetrical as those obtained by Herge et.al., even although the pressure field has been corrected. The injected fluid still moves preferentially along the grid block lines. This is due to the treatment of saturation dependence. The upstream mobilities used by Herge et. al. apply to the diagonal and normal transmissibilities. In the finite volume based scheme, however, there are no diagonal transmissibilities, and the fluid flow across the interface to the neighbour being corrected for dependence on the diagonal neighbour pressures, as the flow need not be along the grid lines. The upstreaming of the mobility to the cell centre, however, re-introduces this assumption. The error in this case vanishes as the grid size is reduced, but can still be serious for practical grids. The true solution is that the mobility should be upstreamed along the local flow direction. If this is done, a strong nine-point saturation coupling is introduced.

To obtain a simple generalisation of the five-point scheme, the upstream direction may be taken as that normal to the cell face. In this case a vector normal to the face is projected back until it intersects the cell centre line. This point will generally lie between the centre of the cell in question and its neighbour. The mobility is then taken as a linearly interpolated combination of the upstream neighbour mobility, and its transverse neighbour.

With this modification, the saturation distribution of fig.8 is obtained, yielding similar results to those of Herge et.al.

In the method described, the pressure modes have been taken at cell centres, so that the interpolation patches involve permeability values from eight cells. Another possibility would be to associate permeability values with the interpolation patches, and porosity values with the resulting control volumes. The volumes used for material balance calculations take a more complex shape, but each is generally a combination of eight sub-volumes, each of the form described above.

In this discussion, and that of Rozon (1989), a linear variation of pressure between nodal values has been assumed. For convergent grid cells, (or in the usual radial case) , a logarithmic form yields more accurate results.

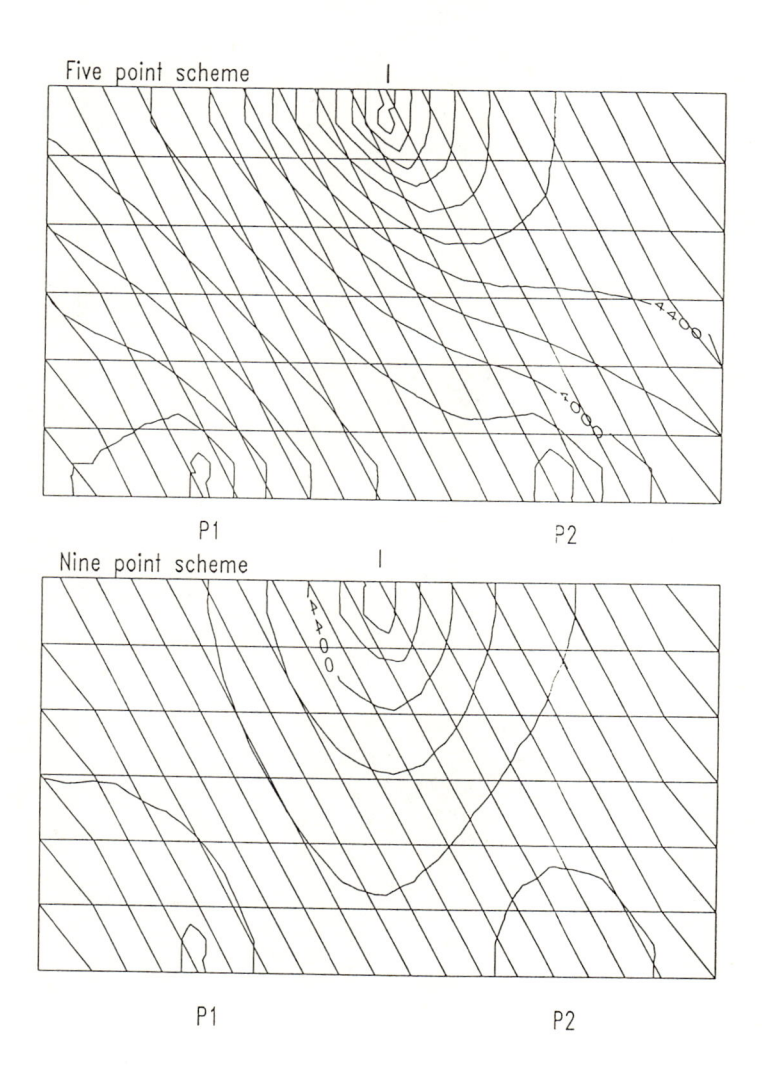

Fig.6. Pressures at 500 days, with 200 psi contour interval.

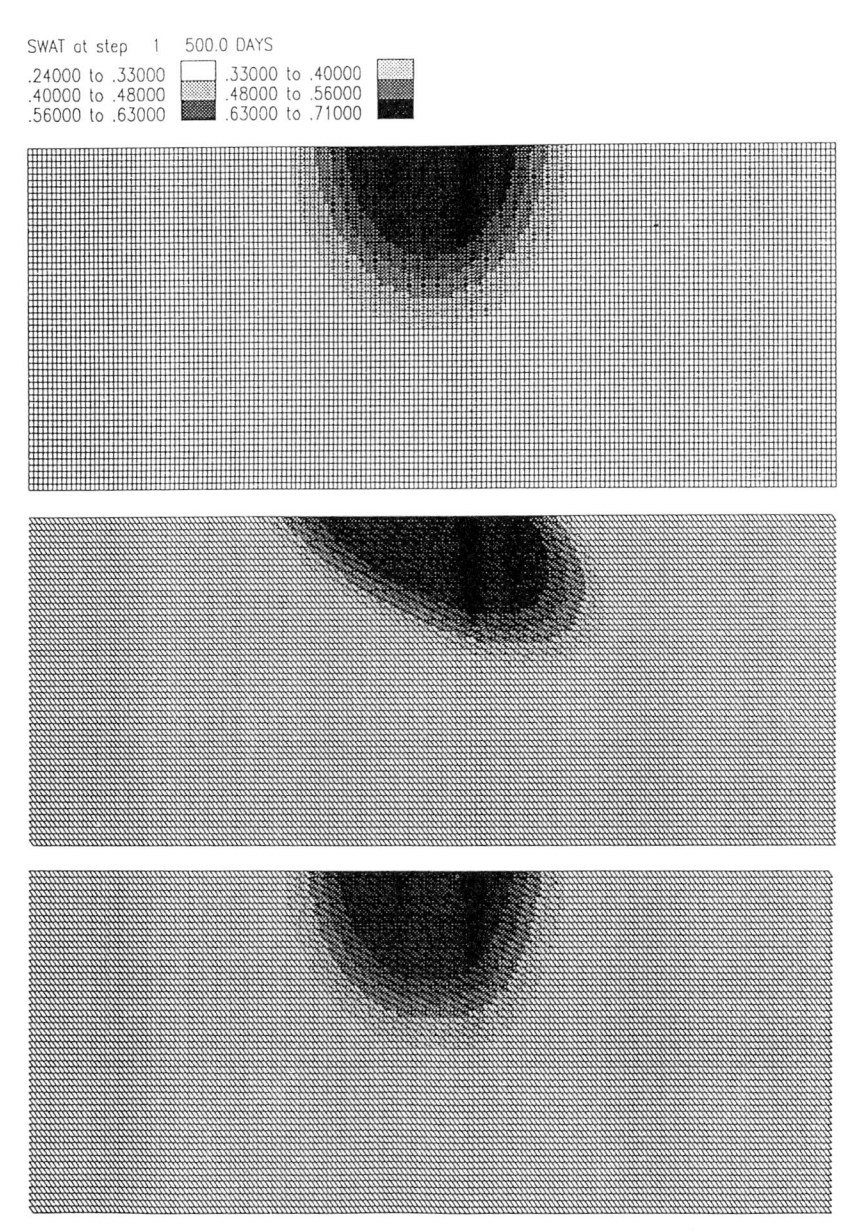

Fig.7 Saturations at 500 days for refined grids

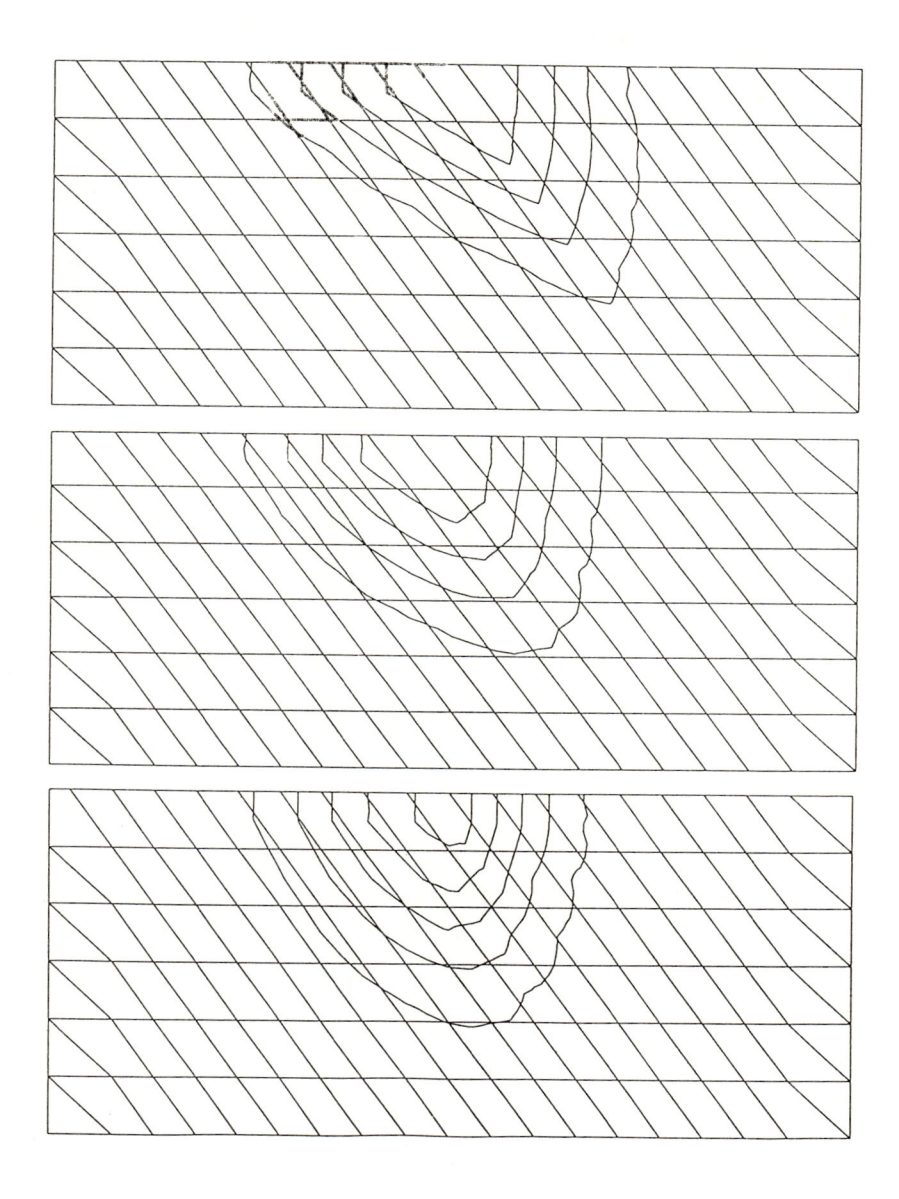

Fig.8 Water saturations at 500 days, with a 0.1 contour interval, using five-point, seven-point and seven-point with modified upstreaming schemes.

6 Conclusions

A method has been described of specifying simulation cells in terms of their corner point positions. Pore volumes and generalised transmissibilities may be defined which enable fault and normal connections to be treated symmetrically, and which reduce to familiar forms for simple situations.

For distorted grids in which flow is not predominantly along one grid direction, this introduces an error which does not vanish as the grid is refined. Using the finite volume method as a basis, flows may be expressed in a form which reflects the cross coupling between flows in one direction on the grid, and pressure variations in other directions.

The locally linear pressure assumption used to derive the usual five-point scheme will then give rise to a seven-point scheme. A bilinear pressure variation, which will give rise to a nine-point scheme on a regular grid, will also give rise to a nine-point scheme for a distorted grid. These schemes do not involve diagonal transmissibilities, the interface area appears as a factor, and fault flows may be obtained in the same manner.

For multi-phase flows, improved accuracy is obtained if the upstream direction is modified to allow for the direction of flow across the interface.

7 Acknowledgements

This investigation was prompted by suggestions and runs provided by C.L.Brown of Conoco UK. The highly refined runs were performed on an IBM 3090 at IBM UK, who also helped in vectorising and parallelising the code.

8 Nomenclature

N_x, N_y, N_z Number of cells in
              x-, y- and z-directions in a grid
N   Number of implicit solution variables per cell
f   Local flow/unit area vector
F   Flow of fluid through a cell interface
h   Thickness of cell in two-dimensional case

K   Rock permeability value
Q   Local well production rate/unit volume
Q   Well production rate from a grid block
x, y, z Coordinates in cartesian three-space
α, β, γ Coordinates used to parameterise
              grid blocks, in range 0 to 1
Δd Distance between cells
ΔP Pressure difference between cells
Δt Length of time step
ϕ Rock porosity

Sub- and superscripts

f   Fluid type ; oil, water or gas in black oil,
              component in compositional
i,j,k   Coordinates of a cell in the indexing grid
I,J,N   Cell numbers, in range 1 to $N_x.N_y.N_z$

9 References

Appleyard,J.R. and Cheshire,I.M., Nested factorisation, SPE 12264, Proceedings of 7th SPE Symposium on Reservoir Simulation, San Francisco, 1983.

Baliga,B.R. and Patankar,S.V. (1980) A new finite element formulation for convection diffusion problems., Numerical Heat Transfer, vol.3, pp 393-409,1980

Hegre,T.M., Dalen,V. and Henriquez,A.(1986) Generalized transmissibilities for distorted grids in reservoir simulation. SPE 15622, Proceedings of the 61st Annual Technical Conference and Exhibition, New Orleans.

Hirasaki,G.J. and O'Dell (1970),P.M.,Representation of reservoir geometry for numerical solution. Society of Petroleum Engineers Journal,1970,393-404.

Rozon,R (1989) A generalized finite volume discretisation method for reservoir simulation. SPE 18414, Proceedings of the Reservoir Simulation Symposium, Houston, Texas, Feb. 1989.

Wadsley,W.A.,(1980) Modelling reservoir geometry with non-rectanglar grids, SPE 9369, Proceedings of the 55th Annual Technical Conference and Exhibition, Dallas, Sept 1980.

Yanosik,J.L. and McCracken,T.A.,(1979) A nine-point, finite difference reservoir simulator for realistic prediction of adverse mobility ratio displacements. Trans. AIME, Aug 1979, 573-592.

10 Appendix : Cell volumes

The position of a point in the cell parameterised by (α, β, γ) is :

$$\begin{aligned}\mathbf{r} = &\quad \gamma(\beta(\alpha\mathbf{r}_8 + (1-\alpha)\mathbf{r}_7) + (1-\beta)(\alpha\mathbf{r}_6 + (1-\alpha)\mathbf{r}_5)) \\ &+ (1-\gamma)(\beta(\alpha\mathbf{r}_4 + (1-\alpha)\mathbf{r}_3) + (1-\beta)(\alpha\mathbf{r}_2 + (1-\alpha)\mathbf{r}_1))\end{aligned}$$

$$\mathbf{r} = \sum_{p_\alpha=0}^{1}\sum_{p_\beta=0}^{1}\sum_{p_\gamma=0}^{1} \mathbf{C}^{p_\alpha,p_\beta,p_\gamma} \alpha^{p_\alpha} \beta^{p_\beta} \gamma^{p_\gamma}$$

with

$$\mathbf{C}^{(0,0,0)} = \mathbf{r}_1$$
$$\mathbf{C}^{(1,0,0)} = \mathbf{r}_2 - \mathbf{r}_1$$
$$\mathbf{C}^{(0,1,0)} = \mathbf{r}_3 - \mathbf{r}_1$$
$$\mathbf{C}^{(0,0,1)} = \mathbf{r}_5 - \mathbf{r}_1$$
$$\mathbf{C}^{(1,1,0)} = \mathbf{r}_4 + \mathbf{r}_1 - \mathbf{r}_3 - \mathbf{r}_2$$
$$\mathbf{C}^{(0,1,1)} = \mathbf{r}_7 + \mathbf{r}_1 - \mathbf{r}_5 - \mathbf{r}_3$$
$$\mathbf{C}^{(1,0,1)} = \mathbf{r}_6 + \mathbf{r}_1 - \mathbf{r}_5 - \mathbf{r}_2$$
$$\mathbf{C}^{(1,1,1)} = \mathbf{r}_8 + \mathbf{r}_5 + \mathbf{r}_3 + \mathbf{r}_2 - \mathbf{r}_7 - \mathbf{r}_6 - \mathbf{r}_4 - \mathbf{r}_1$$

The expressions $\{\mathbf{C}^{(0,0,0)}, .., \mathbf{C}^{(1,1,1)}\}$ have a nice interpretation in terms of a type of multipole expansion - the last four terms are differences in the lengths of cell face diagonals and of the diagonals across the cell. For a cubical block only the first four terms would exist.
Then the cell volume is :

$$V = \int dx.dy.dz$$
$$= \int_0^1 d\alpha \int_0^1 d\beta \int_0^1 d\gamma J(\alpha,\beta,\gamma;x,y,z)$$

with the Jacobian :

$$J = \sum_{(a,b,c)=perm(1,2,3)} (-1)^{perm} \frac{\partial r_a}{\partial \alpha} \frac{\partial r_b}{\partial \beta} \frac{\partial r_c}{\partial \gamma}$$

where $r_1 = x, r_2 = y, r_3 = z$.

The $\partial/\partial\alpha$ operator projects out $p_\alpha = 1$ terms from the multipole expansion of $\mathbf{r}(\alpha,\beta,\gamma)$. With similar expressions in β and γ, obtain :

$$V = \sum_{(a,b,c)=perm(1,2,3)} (-1)^{perm} \sum_{p_\beta=0}^1 \sum_{p_\gamma=0}^1 \sum_{q_\alpha=0}^1 \sum_{q_\gamma=0}^1 \sum_{r_\alpha=0}^1 \sum_{r_\beta=0}^1$$
$$\alpha^{q_\alpha+r_\alpha} \beta^{p_\beta+r_\beta} \gamma^{p_\gamma+r_\gamma} C_a^{(1,p_\beta,p_\gamma)} C_b^{(q_\alpha,1,q_\gamma)} C_c^{(r_\alpha,r_\beta,1)}$$

Performing the integrals, the volume becomes :

$$V = \sum_{(a,b,c)=perm(1,2,3)} (-1)^{perm} \sum_{p_\beta=0}^1 \sum_{p_\gamma=0}^1 \sum_{q_\alpha=0}^1 \sum_{q_\gamma=0}^1 \sum_{r_\alpha=0}^1 \sum_{r_\beta=0}^1$$
$$\frac{C_a^{(1,p_\beta,p_\gamma)} C_b^{(q_\alpha,1,q_\gamma)} C_c^{(r_\alpha,r_\beta,1)}}{(q_\alpha+r_\alpha+1)(p_\beta+r_\beta+1)(p_\gamma+q_\gamma+1)}$$

A BOUNDARY ELEMENT SOLUTION TO THE TRANSIENT PRESSURE RESPONSE OF MULTIPLY FRACTURED HORIZONTAL WELLS

C.P.J.W. van Kruysdijk[+]
(Kon./Shell E&P Laboratorium, The Netherlands)

and

G.M. Dullaert[++]
(University of Technology Delft, The Netherlands)

ABSTRACT

One of the advantages of a horizontal well over a vertical one is that it can be fractured at a number of positions along its horizontal section. This technique is particularly useful in tight reservoirs where economic production can not be achieved by conventional means. As this type of reservoir will produce in transient state for a considerable part of its producing life, accurate transient inflow models are required, not just for well testing purposes, but also for production forecasting. We constructed a model that yields the transient pressure response of a horizontal well intersected by several fractures. The properties of the fracture, length, conductivity, position and orientation can vary from fracture to fracture. By using the Boundary Element Method in combination with the Laplace transform, we succeeded in minimising the discretisation errors for the early time transients without requiring massive amounts of CPU time.

[+] Currently at Shell Canada, Calgary Research Center, 3655 - 36th street N.W., P.O. Box 2506, Calgary, Alberta T2P 3S6, Canada
[++] Currently at Shell Nederland Chemie

1. INTRODUCTION

In most reservoirs, due to the overburden stresses, hydraulic fractures tend to be vertical. As a result, conventional wells will only support one or possibly two or three closely spaced fractures. Horizontal wells allow the induction of multiple fractures far enough apart to benefit production. Although this concept has been introduced as early as 1974 [18], only recently horizontal well drilling and fracture stimulation have matured sufficiently to allow full scale realisation. To date Maersk OG (Denmark) has three Multiply Fractured Horizontal Wells on production, with quoted Productivity Improvement Factors (with respect to a fracture stimulated conventional well) ranging from 2 to 6 [1]. This compares favourably with the increased well cost[*].

The outline of the paper is as follows; We start by reviewing the major developments in transient pressure modelling that predate the technique highlighted in this paper, followed by a discussion of the mathematical foundation of our work. Next we describe the Multiply Fractured Horizontal Well model in particular. Finally, the results of the MFHW model are presented as well as an evaluation of the technique followed by a summary of the main conclusions warranted by this study.

2. HISTORY

Over the years pressure transient analysis has become one of the major tools for characterising hydrocarbon reservoirs. Since the publication of the Horner analysis [10], which yields formation permeability and average reservoir pressure, more and more complicated models have been built to establish an increasingly detailed picture of the situation subsurface. Tracing the history of transient pressure analysis shows that, apart from numerical simulation, six distinct developments dominate the modelling

---

[*] 2 times cost conventional well. However, if 1 MFHW replaces 2 conventional wells total cost (including platform slot cost) is only 40% higher than conventional development.

effort.

Before the advent of the computer, mathematical techniques leading to closed analytical solutions were the primary, if not only, tools engineers had for analysing pressure transients. In 1949, van Everdingen and Hurst [7] introduced the Laplace Transform to the industry, thereby taking transient flow modelling a significant step forward. At the same time, however, it posed the inversion problem which generally could not be solved analytically. In 1970 Stehfest [17] published an inversion algorithm that was both simple and very well-suited for diffusion-type problems. Shortly after this, Gringarten and Ramey [9] rediscovered the possibilities of Green's function theory with respect to potential flow. Following these ideas, Cinco-Ley et al. [3] set out to solve the transient flow problem for a hydraulically fractured reservoir, and derived an implicit integral equation. After discretisation of the integrals, the resulting set of equations was solved numerically. In doing so, Cinco-Ley et al. introduced the Boundary Element Method (BEM) to the industry (for an extensive discussion of BEM see Banerjee and Butterfield [2]). Since then, several attempts have been made to combine BEM with the Laplace Transform [2][11]. It was not until 1988 that the inversion problem was overcome. Van Kruysdijk [13] showed that introduction of the Laplace Transform to the hydraulically fractured reservoir model simplified the modelling effort tremendously (both in mathematics and CPU time). At the same time Cinco-Ley et al. [4] published similar results for a slightly simpler problem. Moreover Kikani and Horne [12] showed the potential of BEM with respect to pressure transients in arbitrarily shaped reservoirs.

Incidentally in the same paper as referenced above [7], van Everdingen and Hurst present a relation for the water influx as a function of aquifer boundary pressure in the form of a convolution integral. Subsequently this concept was incorporated in reservoir simulators to model 'analytical' aquifers. This, essentially, combined the finite difference approach with a boundary element technique.

3. THEORY

The transient pressure response of a single phase, slightly compressible, fluid in a porous medium, assuming constant reservoir and fluid parameters and Darcy flow, is governed by the diffusivity equation (3.1):

$$\frac{k}{\mu} \nabla^2 p = \phi c \frac{\partial p}{\partial t} \tag{3.1}$$

The solution to this partial differential equation, following Green's function theory [9][8], can be given as

$$\Delta p(\underline{x},t) = (p(\underline{x},0) - p(\underline{x},t)) =$$

$$\frac{1}{\phi c} \int_0^t \{ \int q(t-\tau,\underline{x}') G(\underline{x},\underline{x}',\tau) d\underline{x}' \} d\tau + \int (p(\underline{x},0) - p(\underline{x}',0)) G(\underline{x},\underline{x}',t) d\underline{x}' \tag{3.2}$$

where $G(\underline{x},\underline{x}',t-\tau)$ is the response of the reservoir, at time t and place $\underline{x}$, to an instantaneous unit volume offtake at $\underline{x}'$ at time τ ($q = \delta(\tau,\underline{x}')$). The first integral reflects the contribution to the transient pressure response from fluid withdrawal / injection whereas the second one is due to the initial pressure distribution. An extensive list of Green's and source functions, for 1-, 2- and 3-dimensional problems is compiled by Gringarten and Ramey [9]. However, (eq 3.2) is considerably simplified when subjected to the Laplace Transform.

$$\Delta \bar{p}(\underline{x},s) = \frac{1}{\phi c} \int \bar{q}(s,\underline{x}') \bar{G}(\underline{x},\underline{x}',s) d\underline{x}' + \int (p(\underline{x},0) - p(\underline{x}',0)) \bar{G}(\underline{x},\underline{x}',s) d\underline{x}' \tag{3.3}$$

A comprehensive list of Green's functions in Laplace Space, pertaining to our problem, is presented by van

Kruysdijk [13]. Appendix A shows the derivation of the 1-dimensional bounded medium solution.

4. MFHW MODEL

4.1 *Physical model*

The model assumes the wellbore to be intersected by a number of fractures. The fractures penetrate the full formation height. Both fracture position (along the wellbore) and conductivity can be specified for each fracture individually. We assume the fluid to enter the wellbore through the fractures exclusively. The flow in this kind of geometry (Fig. 1) is predominantly horizontal.

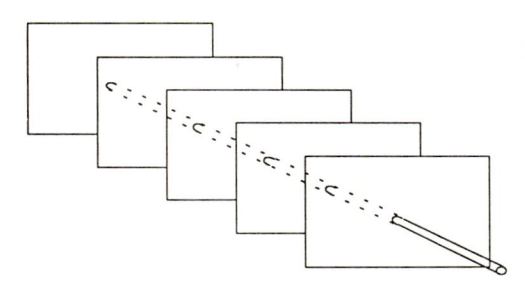

Fig. 1: Multiply fractured horizontal well geometry.

However the limited vertical communication between (finite conductivity) fractures and wellbore (Fig. 2a) introduces some vertical flow in the reservoir. The effect of the limited vertical communication is incorporated in the model by implementing a reduced apparent conductivity as suggested by Schulte [16]. In other words, the extra pressure drop incurred by the limited communication between fractures and wellbore is accounted for by reducing the conductivity of the fracture to a value that yields the same skin. This ensures that with the exception of very early times the transient is described properly. After the conductivity correction factors, based on a correlation derived by Mawer (Fig. 3) [15], have been applied to the model, we assume full communication resulting in horizontal flow only (Fig. 2b).

Limited communication is accounted for in a reduced fracture conductivity. The situation in 2a is replaced by 2b.

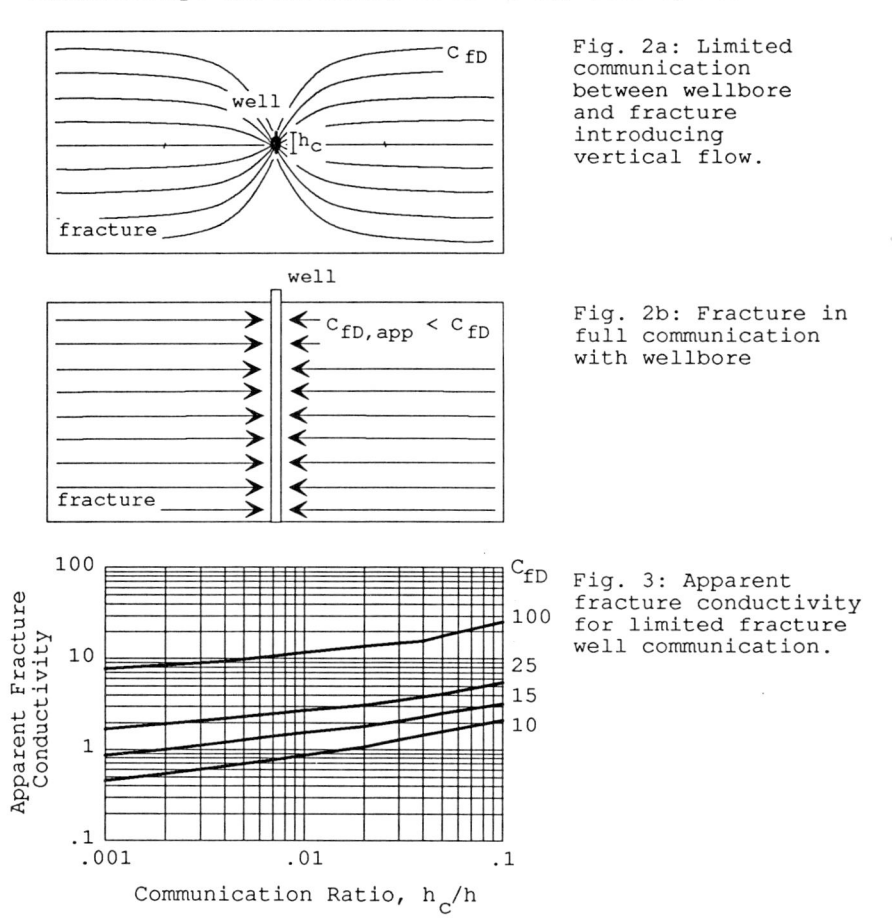

Fig. 2a: Limited communication between wellbore and fracture introducing vertical flow.

Fig. 2b: Fracture in full communication with wellbore

Fig. 3: Apparent fracture conductivity for limited fracture well communication.

This allows us to build a 2-dimensional model*.

---

* 3-Dimensional BEM models have been built successfully [14], but combined with the complexity of the multiple fractures this would result in prohibitive CPU usage.

4.2 Mathematical model

Following van Kruysdijk [13] we consider the system from two points of view, the reservoir and the fractures respectively. The pressure response (in Laplace space) at any given point in the reservoir, assuming no lateral boundaries and initial pressure equilibrium ($p(\underline{x},0) = p_i$), can be written as

$$\Delta \bar{p}_r(\underline{x},s) = \frac{1}{h (\phi c)_r} \sum_{i=1}^{n} \int_{\xi_{i0}}^{\xi_{i1}} \frac{\bar{q}_i(\xi_i',s) K_0(|\underline{x}-\xi_i'|\sqrt{s/\eta_r})}{2 \pi \eta_r} d\xi_i' \quad (4.1)$$

where n is the number of fractures, ξ_i is the space ordinate along the ith fracture, and ξ_{i0} and ξ_{i1} the respective fracture tip positions. The Green's function in Laplace space for a 2-dimensional medium is given by $K_0(|\underline{x}-\underline{x}'|\sqrt{s})$ (see e.g. [13]). Essentially, the reservoir 'sees' fluid disappearing at the fractures. If we now consider the fracture, we see fluid appearing all along the fracture face while at the wellbore intersection fluid is withdrawn. In mathematical terms, assuming 1-dimensional flow inside the fracture, this gives for the pressure response

$$\Delta \bar{p}_{fi}(\xi_i,s) = \frac{1}{hw_i (\phi c)_{fi}} \{ \bar{q}_{wi}(s) \frac{\bar{G}_{1B}(\xi_i,\xi_{iw},s/\eta_{fi};\xi_{i1}-\xi_{i0})}{\eta_{fi}}$$

$$- \int_{\xi_{i0}}^{\xi_{i1}} \bar{q}_i(\xi_i',s) \frac{\bar{G}_{1B}(\xi_i,\xi_i',s/\eta_{fi};\xi_{i1}-\xi_{i0})}{\eta_{fi}} d\xi_i' \} \quad (4.2)$$

with

$$\bar{G}_{1B}(x,x',s;x_e) = \frac{1}{2\sqrt{s}} \left\{ \frac{2\cosh((x-x')\sqrt{s})}{\exp(2x_e\sqrt{s}) - 1} + \exp(-|x-x'|\sqrt{s}) \right.$$

$$\left. + \frac{2\cosh((x+x')\sqrt{s})}{\exp(2x_e\sqrt{s}) - 1} + \exp(-|x+x'|\sqrt{s}) \right\}$$

For derivation of the 1-dimensional, bounded Green's function in Laplace space see Appendix A.

If we now impose pressure continuity along the fracture / reservoir interface,

$$p_r(\underline{x},t) = p_{fi}(\underline{x},t) \longrightarrow \bar{p}_r(\underline{x},s) = \bar{p}_{fi}(\underline{x},s) \qquad (4.3)$$

for all $\underline{x}$ coinciding with i^{th} fracture, for all fractures

we obtain an implicit integral equation for $q_i(\underline{x},s)$. This equation can be solved numerically by discretising the integrals. In other words, we approximate the fluid-transfer functions, $q_i(\underline{x},s)$, by step functions

$$\bar{q}_i(\xi_i,s) = \bar{q}_{i,j}(s) \quad \text{for} \quad \xi_{i,j-1} < \xi_i \leq \xi_{i,j} \qquad (4.4)$$

Subsequently we demand the pressure continuity constraint to hold for centres of the intervals. This provides us with as many equations as variables we introduced by discretising the fluid-transfer functions. However, n unknowns, q_{wi}, still remain. In order to complete the set of equations we assume incompressible flow in the wellbore which, together with the flux continuity constraint yields

$$q_w = \sum_{i=1}^{n} q_{wi}(t) \quad \longrightarrow \quad \bar{q}_w = \sum_{i=1}^{n} \bar{q}_{wi}(s) \tag{4.5}$$

where q_w is a boundary condition to our model. Furthermore, the wellbore is considered to be infinitely conductive. This results in the following set of equations

$$p(\underline{x}_{wi},t) = p(\underline{x}_{w1},t) \quad \longrightarrow \quad \bar{p}(\underline{x}_{wi},s) = \bar{p}(\underline{x}_{w1},s) \quad \text{for } i=2,n \tag{4.6}$$

thereby matching the number of unknowns with equations. Subsequently we employ a numerical linear equation solver to yield the values for the fluid-transfer function. Substituting these values back into either the reservoir (eq 4.1) or fracture (eq 4.2) relations renders the pressure response in Laplace space. Inversion to real time space is achieved by means of the Stehfest algorithm [17].

4.3 Validation

For a validation of the early time response (before interference between the individual fractures occurs) we compared our model against single fracture (Fig. 4) as well as multi-layer multi-fracture solutions [20] (Fig. 5).

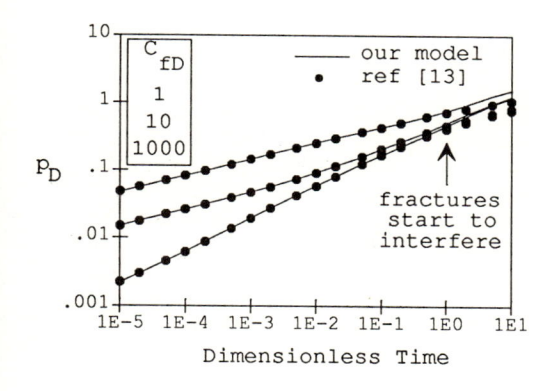

Fig. 4: Comparison of the early time response of our model with the single fracture model of van Kruysdijk [13].

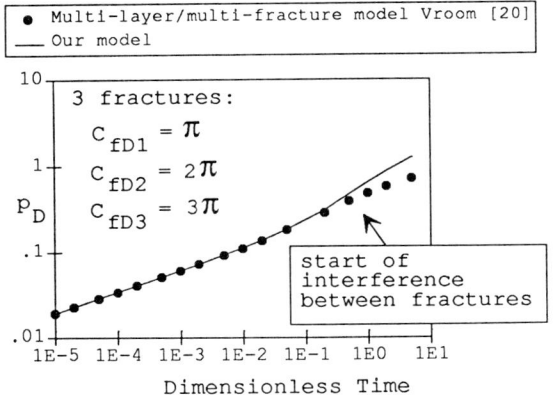

Fig. 5: Comparison of our early time results for three fractures of different conductivity with the multi-layer / multi-fracture model of Vroom [20].

3 fractures:
$C_{fD1} = \pi$
$C_{fD2} = 2\pi$
$C_{fD3} = 3\pi$

start of interference between fractures

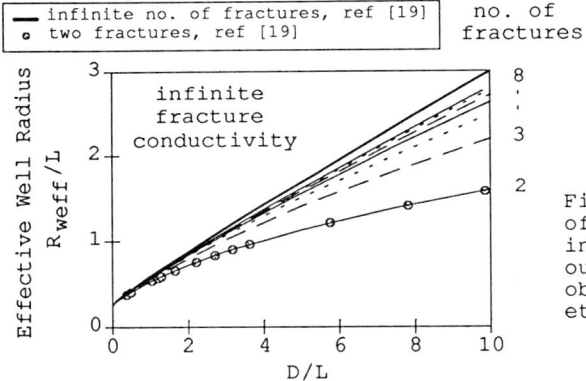

Dimensionless Time

Fig. 6: Comparison of the stabilised inflow results from our model with those obtained by v/d Vlis et al. [19]

Furthermore we compared the late time response (pseudo radial flow) we obtained against the results published by van der Vlis et al. [19] (Fig. 6). By using conformal mapping techniques on the steady state equation they derived effective wellbore radii for two parallel infinite conductivity fractures as a function of fracture spacing over fracture length, D/L. As demonstrated by (Fig. 4,5,6) excellent agreement was obtained. For a more extensive validation see Dullaert [6].

5. DISCUSSION OF THE RESULTS

As a complete discussion of the results would be prohibitively long as well as beyond the focus of the conference we limit ourselves to describing some of the results obtained for equidistantly spaced, identical fractures, perpendicular to the wellbore. Apart from validating our model for stabilised flow conditions, Fig. 6 shows the effective wellbore radii for MFHW's with up to 8 fractures. This diagram can be used to determine the optimum fracture spacing for a given horizontal well scenario. We now proceed to the analysis of the transient on a log-log plot of $\partial \Delta p / \partial \sqrt{t}$ versus t. This method of displaying the transient is instrumental in identifying the time frames dominated by linear flow*. Fig. 7 shows a number of curves for varying values of D/L.

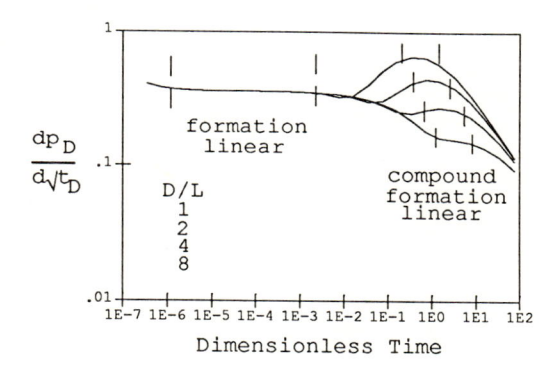

Fig. 7: Derivative plot of transient, highlighting the two linear flow periods.

Apart from an early horizontal line section, indicating formation linear flow [5], a second linear flow period is identified. We named this flow period, caused by linear flow in the formation towards the collection of fractures (Fig. 8), 'compound formation linear'.

---

* When the transient is dominated by linear flow the pressure response is proportional to $\sqrt{t}$ + constant. This results in a horizontal line-section on the plot described above.

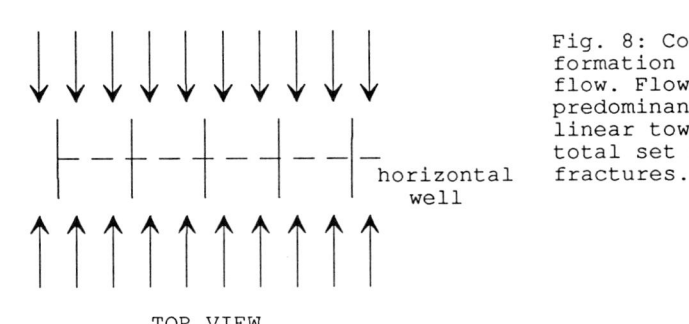

Fig. 8: Compound formation linear flow. Flow is predominantly linear towards total set of fractures.

The value of the $\sqrt{t}$-derivative at the horizontal line section* (i.e. the slope of the curve on a more conventional $\sqrt{t}$-plot) is approximately inversely proportional to $(\frac{n}{n-1}D\sqrt{k})$ (see Appendix B). Because the value of D should be known, analysis of this flow period yields an estimate for k. Conventional analysis utilises the pseudo-radial flow period to obtain a value for this parameter. However, pseudo radial flow generally does not develop until at least an order of magnitude in time later than compound formation linear flow. Introduction of our analysis therefore allows a significant reduction of the time needed for a well test.

6. DISCUSSION OF THE MODEL

The Boundary Element Method has two main advantages over the more conventional finite difference / element techniques. Firstly, it reduces the dimension of the problem by 1. A 2-dimensional problem is solved by modelling the 1-dimensional boundaries coinciding with the fractures. Secondly, although the coupling between reservoir and fractures is modelled numerically, the transients, both in the fracture and the reservoir, are handled analytically. This has an obvious advantage with respect to the early time response. Introduction of the Laplace

---

* For small values of D/L, when a horizontal line section does not develop, the tangent at the local maximum should be used.

Transform helps in three ways. First of all it simplifies the mathematics by eliminating one of the integral signs. This results in a faster algorithm* as well as a more stable one. Secondly it enables all the properties associated with the Laplace transform (see e.g. [13]), which allow simple addition of wellbore storage, calculation of time derivatives or integrals and multi-layer systems. Thirdly, a far larger number of Green's function, e.g. pertaining to double porosity or double permeability formations, have closed analytical expressions in Laplace space.

7. CONCLUSIONS

- We succeeded in building a fast* and stable model describing the transient pressure response of a multiply fractured horizontal well.
- By combining the Boundary Element Method with the Laplace Transform we obtained a flexible model capable of dealing efficiently with a large number of boundary conditions and formation characteristics.
- By introducing the concept of 'compound formation linear flow' we enabled comprehensive well test interpretation of multiply fractured horizontal wells from data sets far shorter than previously felt necessary.

ACKNOWLEDGEMENTS

The authors wish to thank Shell Internationale Research Maatschappij for permission to publish the paper. Furthermore professor de Haan is acknowledged for supervision of the second author.

---

* The computer implementation of our MFHW model takes about 100 [sec/log cycle/fracture] on a VAX 6220 assuming all fractures to have different dimensions. No significant effort has been spent to optimise the code.

REFERENCES

1. Andersen, S.A., Hansen, S.A. and Fjeldgaard, K., Horizontal Drilling and Completion: Denmark, SPE 18349, London, 1988.
2. Banerjee, P.K. and Butterfield, R., Boundary Element Methods in Engineering Science, McGraw-Hill Book Company (U.K.) Limited, Maidenhead, Berkshire, England, 1981.
3. Cinco-Ley, H., Samaniego, F. and Dominguez, N., Transient Pressure Behavior for a Well with a Finite Conductivity Vertical Fracture, SPE 6014, Society of Petroleum Engineers Journal, August 1978.
4. Cinco-Ley, H. and Hai-Zui Meng, Pressure Transient Analysis of Wells with Finite Conductivity Vertical Fractures in Double Porosity Reservoirs, SPE 18172, Houston 1988.
5. Cinco-Ley, H. and Samaniego, F.V., Transient Pressure Analysis for Fractured Wells, SPE 7490, 1978.
6. Dullaert, G.M., Development of a Model Describing the Inflow Performance of a Multiply Fractured Horizontal Well, MSc Thesis University of Technology Delft, Department of Mining and Petroleum Engineering, 1989.
7. van Everdingen, A.F. and Hurst, W., The Application of the Laplace Transformation to Flow Problems in Reservoirs, Petroleum Transactions, AIME, December 1949.
8. Greenberg, M.D., Application of Green's Function Theory in Science and Engineering, Prentice Hall, Englewood Cliffs.
9. Gringarten, A.C. and Ramey Jr., H.J., The Use of Source and Green's Functions in Solving Unsteady Flow Problems in Reservoirs, SPE 3818, Society of Petroleum Engineers Journal, October 1973.
10. Horner, D.R., Pressure Build Up in Wells, Proc., Third World Petroleum Congress, E.J. Brill, II, 503, Leiden, 1951.
11. Houze, O.P., Horne, R. and Ramey Jr., H.J., Infinite Conductivity Vertical Fracture in a Reservoir with Double Porosity Behavior, SPE 12778.
12. Kikani, J. and Horne, R.N., Pressure Transient Analysis of Arbitrary Shaped Reservoirs with the Boundary Element Method, SPE 18159, Houston, 1988.

13   van Kruysdijk, C.P.J.W., Semi-Analytical Modelling of Pressure Transients in Fractured Reservoirs, SPE 18169, Houston, 1988.
14   van Kruysdijk, C.P.J.W. and Niko, H., Alternatives for Draining Tight Naturally Fractured Gas Reservoirs: Horizontal Hole Drilling vs. Massive Hydraulic Fracturing, SPE 18339, London, 1988.
15   Mawer, A., Unpublished work.
16   Schulte, W.M., Production from a Fractured Well with Well Inflow Limited to Part of the Fracture Height, SPE 12882, 1988.
17   Stehfest, H., Numerical Inversion of Laplace Transforms, Communications of the ACM, Vol. 13, No. 1, 1970.
18   Strubhar, M.K., Fitch, J.L. and Glenn, E.E., Multiple Vertical Fractures from an Inclined Wellbore -- a Field Experiment, SPE 5115, Houston, 1974.
19   van der Vlis, A.C., Duns, H. and Fernandez Luque, R., Increasing Well Productivity in Tight Chalk Reservoirs, Tenth World Petroleum Congress, Vol. 3, Bucharest.
20   Vroom, J.K.A.B., A Composite Method for the Determination of Fracture Properties in Multiple Reservoir Layers, MSc Thesis University of Technology Delft, Department of Mining and Petroleum Engineering, 1986.

LIST OF SYMBOLS

η hydraulic diffusivity, $= \dfrac{k}{\phi \mu c}$

ϕ porosity

μ viscosity

ξ_i space ordinate along i^{th} fracture

c   compressibility

C_{fD} dimensionless fracture conductivity, $= 2 \dfrac{w k_f}{L k}$

D   distance between outer two fractures
h   formation height
h_c height of fracture in communication with wellbore
k   permeability

L   fracture length
n   number of fractures
p   pressure

p_D dimensionless pressure, $= \dfrac{2\pi kh}{q\mu} \Delta p$

q   flowrate
s   Laplace Space variable
t   time

t_D dimensionless time, $= \dfrac{\eta t}{x_f{}^2}$

w   fracture width
x   space coordinates
x_f fracture half-length, $= L/2$

subscripts

f   fracture
i   fracture number
r   reservoir
w   wellbore

APPENDIX   A

Derivation of 1-D bounded Green's function in Laplace space

differential equation : $\dfrac{\partial^2 \bar{G}}{\partial x^2} = \dfrac{s}{\eta} \bar{G}$ (A1)

let $\bar{G} = \begin{cases} \bar{G}_l & 0 \leq x \leq x_w \\ \\ \bar{G}_r & x_w \leq x \leq x_e \end{cases}$ (A2)

$\left.\dfrac{\partial \bar{G}_l}{\partial x}\right|_0 = 0 \; ; \; \left.\dfrac{\partial \bar{G}_r}{\partial x}\right|_{x_e} = 0$ no-flow boundaries at fracture tips (A3)

$$\frac{\partial \bar{G}_l}{\partial x}\bigg|_{x_w} - \frac{\partial \bar{G}_r}{\partial x}\bigg|_{x_w} = \frac{1}{\eta} \qquad \text{fluid offtake} \tag{A4}$$

(A1 + A2 + A3) --->

$$\bar{G}_l = c_l \left(\exp(-x\sqrt{s/\eta}) + \exp(x\sqrt{s/\eta}) \right) \tag{A5}$$

$$\bar{G}_r = c_r \left(\exp(-x\sqrt{s/\eta}) + \exp((x-2x_e)\sqrt{s/\eta}) \right) \tag{A6}$$

$$\bar{G}_l\big|_{x_w} = \bar{G}_r\big|_{x_w}$$

$$\text{----> } c_r = c_l \frac{\exp(-x_w\sqrt{s/\eta}) + \exp(x_w\sqrt{s/\eta})}{\exp(-x_w\sqrt{s/\eta}) + \exp((x_w-2x_e)\sqrt{s/\eta})} \tag{A7}$$

(A4 + A5 + A6 + A7) --->

$$\bar{G} = \frac{1}{2\sqrt{\eta s}} \left\{ \frac{2\cosh((x-x_w)\sqrt{s/\eta})}{\exp(2x_e\sqrt{s/\eta}) - 1} + \exp(-|x-x_w|\sqrt{s/\eta}) \right.$$

$$\left. + \frac{2\cosh((x+x_w)\sqrt{s/\eta})}{\exp(2x_e\sqrt{s/\eta}) - 1} + \exp(-|x+x_w|\sqrt{s/\eta}) \right\} \tag{A8}$$

APPENDIX B

For small values of D/L the geometry resembles a square (Fig. B1). and hence "compound formation linear" flow will never fully develop. Provided that D/L > 1, however, the $\sqrt{t}$ - tangent can be used to obtain an estimate for $\frac{n}{n-1}D\sqrt{k}$. For large values of D/L (> 6) the geometry better approximates a line (Fig. B2) and "compound formation linear" flow will develop. As can be observed from Figure 7, the straight line section is not quite horizontal. Fortunately this will hardly affect the analysis. Note that for a fracture stimulated vertical well the $\sqrt{t}$ - slope is only perfectly straight for high fracture conductivities (C_{fD} > 50). However, this hardly affects the $\sqrt{t}$ - analysis provided that C_{fD} > 5.

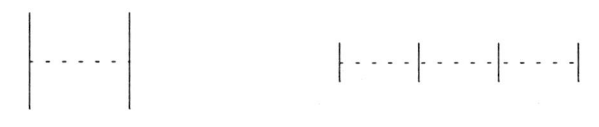

Fig. B1: D/L = 1   Fig. B2: D/L = 6

A CONVECTIVE SEGREGATION MODEL FOR PREDICTING RESERVOIR FLUID COMPOSITIONAL DISTRIBUTION

F.MONTEL

(ELF AQUITAINE, CSTJF, PAU, FRANCE)

and

J.P.CALTAGIRONE and L.PEBAYLE

(M.A.S.T.E.R., Université de Bordeaux I, TALENCE, FRANCE)

ABSTRACT

Convective rolls resulting from a geothermal gradient coupled with thermodiffusion and gravity effects may induce large spatial changes of fluid composition in hydrocarbon reservoirs.

This paper presents a theoretical modelling of these phenomena based on the equations of natural convection in porous media and of mass transfer.
The numerical model aims at predicting the fluid property distribution from a single representative sample.

The thermodynamical properties of the fluid are computed by the PENG-ROBINSON equation of state. Viscosity, diffusion and thermodiffusion coefficients of the various components are functions of temperature, pressure and total fluid composition. The porous medium can be anisotropic and heterogeneous.
The equations are solved by a control volume method and the numerical integration scheme is discussed.

An example is given for a typical reservoir. The distributions of the components and of the corresponding fluid properties are computed in different cases.

A simple criterion is proposed to define the application range of the model.

1 INTRODUCTION

The compositional grading in reservoirs has been widely studied [5], [8], [13], [14], [16], but as Jacqmin [6] has recently pointed out, consequences of the dynamical state of a reservoir on the distribution of the chemical species is an unexplored subject, although a large amount of work has been done on free convection and Soret effects in a multicomponent fluid [12] for chemical engineering purpose.

Compositional variations due to thermodiffusion and gravity effect within the reservoir can be very large, especially when the fluid is close to critical conditions. In reservoir simulation, neglecting these variations could lead to significant errors on liquid recovery.

In some cases, the steady state given by a static segregation model cannot explain the observed variations. These discrepancies can be induced by convective rolls resulting from the geothermal gradient.

In this paper, we describe the model that we have built in order to predict the properties distribution in a reservoir from a single local representative sample of fluid.

For reservoir engineers, the first step could be to calculate the time scale for the different phenomena to determine whether it is necessary to run such a model or not. For a reservoir considered as a rectangular box, a rough approximation of the time scales for convection and diffusion is given by :

$$t_c = \frac{\mu H}{\rho g \kappa (\theta \Delta T - \chi \Delta P) \sin(\beta)} \qquad (1-1)$$

$$t_D^i = \frac{H^2}{D_i^f} \qquad (1-2)$$

where H is the vertical thickness of the porous layer and κ its permeability.
β is the angle between gravity and temperature gradient, β is supposed to be small.
ρ, θ, χ, μ are respectively density, thermal expansion, compressibility and viscosity of the fluid.
D_i^f is the simple diffusion coefficient of component "i" in the fluid.

Comparing these values, 3 different situations can be found :

1 - $t_c \ll t_D^i$ for all components i

> Static compositional grading.
> One needs only a static model as we discussed previously [8].

2 - $t_c \approx t_D^i$ for at least one component i

> Complex composition distribution due to the interaction between gravity forces, thermodiffusion and convection.
> It is the typical application range of the dynamic model

3 - $t_c \gg t_D^i$ for all components i

> One could think he needs only a thermodynamical model for property calculation everywhere in the reservoir because of the convection overturning. However this case is also relevant of the dynamic model as we show later.

This time scale analysis applies only to an homogeneous composition of the fluid and this is a drastic assumption. We will discuss in the application section of this paper the validity of this criterion for real situations and how to improve it.

2 MODEL SYSTEM

2.1 BASIC EQUATIONS

The porous medium, which may be anisotropic and heterogeneous, is defined by the following fields which are functions of a space variable :

Φ porosity
$\underline{\underline{\lambda}}_m$ thermal conductivity tensor
$\underline{\underline{K}}$ permeability tensor

The subscript "m" refers to the reservoir properties (saturated porous medium), the subscript "f" refers to the fluid alone and "s" to the rock alone.

Reservoir properties are calculated when possible, by simple combination rules, for example, the volumetric heat capacity is :

$$(\rho c_p)_m = \Phi(\rho c_p)_f + (1-\Phi)(\rho c_p)_s$$

The governing equations for convection and diffusion within a multi-component reservoir fluid are Darcy's law, thermal and material balances :

DARCY LAW :

$$\underline{V} = -\frac{\underline{\underline{K}}}{\eta}(\nabla P + \rho \underline{g}) \qquad (2-1-1)$$

THERMAL BALANCE :

$$\frac{\partial H_m}{\partial t} = \nabla \cdot \underline{J}_T \qquad (2-1-2)$$

where H_m is the enthalpy of the impregnated rock. General expression for the thermal flux is :

$$\underline{J}_T = H_f \underline{V} + \underline{J}_q \qquad (2-1-3)$$

where H_f is the enthalpy of the fluid alone and J_q is the heat flux.

The thermal balance can be rewritten using the temperature instead of the enthalpy, assuming that the heat flux which results from the concentration gradient is negligible and that the variation of specific heat c_p is insignificant; we thus obtain the usual temperature equation :

$$(\rho c_p)_m \frac{\partial T}{\partial t} + (\rho c_p)_f \underline{V} \cdot \nabla T = \nabla \cdot \left(\underline{\underline{\lambda}}_m \cdot \nabla T\right) \qquad (2-1-6)$$

MATERIAL BALANCE :

$$\frac{\partial \rho_i}{\partial t} + \nabla \cdot \underline{J}_i = 0 \qquad (2-1-4)$$

where $\rho_i = \rho c_i$ is the product of density and mass fraction of component i.

General expression for the flux of component "i" is :

$$\underline{J}_i = \underline{J}_i^C + \underline{J}_i^T + \underline{J}_i^D \qquad (2-1-5)$$

$$\underline{J}_i^C = \rho_i \underline{V} \qquad \text{convection flux}$$

$$\underline{J}_i^T = \underline{\underline{\tau}} L_i^T \nabla T \quad \text{thermodiffusion flux}$$

$$\underline{J}_i^D = \underline{\underline{\tau}} \sum_{j=1}^n L_{ij}^D \nabla g_j \quad \text{mass diffusion flux}$$

where g_i is the chemical potential and $\underline{\underline{\tau}}$ is the dispersion tensor.

In the present work, we take into account gravitation forces and thermodiffusion effects.

The thermodynamic equations are discussed in section 3.

2.2 BOUNDARY CONDITIONS

The reservoir shown in figure 1 is represented by a simple inclined layer inscribed in a rectangular box. The dip angle is ϕ and β is the angle between gravity and temperature gradient.
The surrounding rock is considered as completely impermeable. "A" is the ratio between lateral extension and thickness of the box.
The edges of the box are used respectively as x and z axes.

With this geometric description, realistic boundary conditions are :

Mechanical conditions :

$$\underline{V} \cdot \underline{n} = 0$$

Thermal conditions :

Temperature is fixed by Dirichlet conditions on the boundary.

Mass conditions :

$$\underline{J}_i \cdot \underline{n} = 0$$

n is the unit vector normal to the boundary.

By using another mass boundary condition, it is possible to study the effect of fluxes or outfluxes of a fluid through the boundaries of the layer, like a diffusion through the sealing cap rock for example.

GEOMETRIC DESCRIPTION

.... Isobar   ------ Isotherm

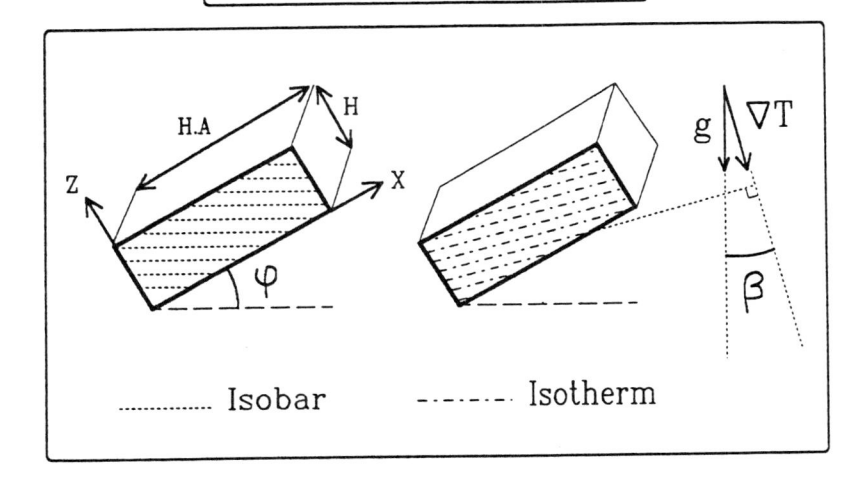

BOUNDARY CONDITIONS

Mechanical : $\underline{V} \cdot \underline{n} = 0$

Thermal : Dirichlet condition

Massic : $\underline{J}_i \cdot \underline{n} = 0$

Figure 1 : Geometric description and boundary conditions

2.3 REDUCED EQUATIONS

For the calculation, the set of equations is re-written in a dimensionless form, by using the following reduced variables:

$$\tilde{x} = \frac{x}{H}, \tilde{z} = \frac{z}{H} \qquad \tilde{t} = t\frac{\lambda_m^0}{(\rho c_p)_m^0 H^2} \qquad \tilde{V} = V\frac{(\rho c_p)_f^0 H}{\lambda_m^0}$$

$$\tilde{T} = \frac{T - T^0}{T_2 - T_1} \qquad \tilde{\rho} = \frac{\rho - \rho^0}{\rho_1 - \rho_2} \qquad \tilde{\lambda}_m = \frac{\lambda_m}{\lambda_m^0}$$

$$\tilde{\eta} = \frac{\mu}{\mu^0} \qquad \tilde{K} = \frac{\kappa}{\kappa^0} \qquad \tilde{P} = P\frac{(\rho c_p)_f^0 \kappa^0}{\mu^0 \lambda_m^0}$$

$$\tilde{g}_i = \frac{g_i}{g_i^0} \qquad \tilde{\rho}_i = \frac{\rho_i}{\rho_i^0} \qquad (2-3-1)$$

The reference value for the fluid properties, noted (^{0}), are evaluated in the centre of the box with the initial homogeneous composition, subscripts ($_1$) and ($_2$) refer to the bottom and the top of the box.

Then the set of equations from (2-1-1) to (2-1-6) can be re-written as follows :

$$\underline{\tilde{V}} = -\frac{\tilde{\underline{\underline{K}}}}{\tilde{\eta}}(\nabla\tilde{P} + Ra_m\tilde{\rho}\underline{k}) \qquad (2-3-2)$$

$$\frac{\partial \tilde{T}}{\partial \tilde{t}} + \underline{\tilde{V}} \cdot \nabla \tilde{T} = \nabla \cdot \left(\underline{\underline{\tilde{\lambda}}}_m \cdot \nabla \tilde{T}\right) \qquad (2-3-3)$$

$$M\frac{\partial \tilde{\rho}_i}{\partial \tilde{t}} = -\tilde{\nabla} \cdot \underline{\tilde{J}}_i \qquad (2-3-4)$$

$$\underline{\tilde{J}}_i = \tilde{\rho}_i \underline{\tilde{V}} + S_i^T \tilde{\nabla}\tilde{T} + \sum_{j=1}^{n} Le_{i,j}\tilde{\nabla}\tilde{\mu}_j \qquad (2-3-5)$$

with the following dimensionless numbers which represent the "coupling" between elementary processes.

Convective/Thermal
Rayleigh number
$$Ra_m^0 = \frac{(\rho_1 - \rho_2)g(\rho c_p)_f H \kappa^0}{\lambda_m^0 \mu^0} \qquad (2-3-6)$$

Diffusion/Thermal Lewis number	$$Le_{i,j} = \frac{\tau L_{i,j}^D g_i^0 (\rho c_p)_f^0}{\lambda_m^0 \rho^0}$$	(2-3-7)
Thermodiffusion/Thermal Soret number	$$S_i^T = \frac{\tau L_i^T (T_1 - T_2)(\rho c_p)_f^0}{\lambda_m^0 \rho^0}$$	(2-3-8)
Specific Heat Ratio	$$M = \frac{(\rho c_p)_f^0}{(\rho c_p)_m^0}$$	(2-3-8)

$\underline{k}$ is the unit vector along the direction of gravity.

2.4 DISPERSION

Within the pore space, the ratio between the time which is necessary for concentration mixing and the available time is the Peclet number :

$$P_e = \frac{Vd}{D_{if}}$$

where "d" is the pore length and "D_{if}" the diffusion coefficients (see chapter 3).
In our case, this number is very small (less than 0.01), and we can assume that the longitudinal dispersion is equal to transversal dispersion. and that there is no correction term in the following usual formula [10]:

$$\tau = \frac{1}{F\Phi}$$

where F is the formation factor.

3 THERMODYNAMIC PROPERTIES

3.1 FLUID CHARACTERIZATION

Starting from a detailed PVT analysis of the fluid, it is necessary to get a simplified representation of the fluid in several pseudocomponents. We suggest to apply our automatic clustering algorithm described elsewhere [7] in order to minimize computer time.

Each pseudo-component is characterized by its pseudo-critical properties which are obtained from the critical properties of the pure compounds which belong to the pseudo-component, according to the equation of state mixing rules. Heavy fraction properties are estimated by group contribution methods.

3.2 VOLUMETRIC AND EQUILIBRIUM PROPERTIES

The well-known Peng-Robinson Equation of State (PR-EOS) is used for the calculation of the fluid density and the chemical potential of each componentAppendix 1.

In the general case, we have to add the potentials of external forces to the chemical potential, as far as gravity forces are concerned :

$$g_i = g_i^c - M_i g h \qquad (3-2-1)$$

where h is the depth.

Chemical potentials are the driving forces for diffusion calculations and are used to check if the saturation pressure of the fluid remains below the static pressure everywhere in the reservoir.

The thermodynamical properties of the heavy fraction and the interaction coefficients are tuned in order to match experimental PVT data as shown in the application section.

3.3 DIFFUSION

The diffusion flux for component i is given by :

$$\underline{J}_i^D = \sum_{j=1}^{n} L_{ij}^D \nabla g_j \qquad (3-3-1)$$

$L_{i,j}^D$ are the generalized diffusion coefficients. Diagonal coefficients $L_{i,i}^D$ are inferred from classical diffusion coefficients :

$$L_{i,i} = D_i^f \frac{M_i}{M} \bigg/ \frac{\partial g_i}{\partial x_i} \qquad (3-3-2)$$

where the classical diffusion coefficients are calculated by WILKE and CHANG correlation

$$D_i^f = 7.4\,10^{-12}\,\frac{T\sqrt{M}}{\mu\,v_i^{0.6}} \qquad (3-3-3)$$

These coefficients are relative to a flux in mole per square metre per second and to a concentration gradient in mole per metre.
The molar volumes v_i and the derivative of the chemical potential are given by the EOS model.

The cross diffusion coefficients $L_{i,j}^D$ $(i \neq j)$ are linked by the ONSAGER reciprocity law :

$$L_{i,j}^D = L_{j,i}^D$$

We intend to define these coefficients to ensure the conservation law.
By summing the material balance equations, we obtain the classical continuity equation :

$$\frac{\partial \rho}{\partial t} + \nabla \cdot (\rho \underline{V}) = 0 \quad \left(\rho = \sum_{i=1}^{n} \rho_i \right)$$

provided that the following constraints are verified :

$$\sum_{i=1}^{n}\sum_{j=1}^{n} L_{i,j}^D \Delta g_j = 0 \quad \text{for any} \quad \Delta g_j$$

These constraints lead to a generalized expression for the cross diffusion coefficients :

$$L_{i,j,(i \neq j)}^D = -\frac{L_{i,i}^D}{n-1} - \frac{L_{j,j}^D}{n-1} + \sum_{k \neq i,j} \frac{L_{k,k}^D}{(n-1)(n-2)} \qquad (3-3-4)$$

3.4 THERMODIFFUSION

Thermodiffusion coefficients are estimated from the experimental data given by Costeseque [1],[2]. A composition dependency is assume in order to ensure nil value for pure compound limit and for nil concentration value :

$$L_i^T = \alpha_i^T x_i(1-x_i) \frac{1}{T} \cdot D_{if} \qquad (3-4-1)$$

The values obtained this way are modified to ensure a nil global flux.

α_i^T varies strongly with the size of the molecule : heavy compounds would accumulate at low temperature boundary (the upper part of the reservoir) in the absence of gravity. But, gravity effects are about one order of magnitude larger than thermodiffusion effect in static conditions.

3.5 VISCOSITY

Viscosity is calculated from composition, temperature and pressure with a corresponding state method first introduced by Pedersen and al modified by Ducoulombier and al [3] Appendix 2 .

All the thermodynamic calculations are interfaced with the main program through a single subroutine in which the static pressure and temperature fields are added to the dynamic fields and mass concentrations are converted into mole fractions.

4 NUMERICAL METHOD

The governing equations in their conservative form are discretized according to the control volume formulation described by Patankar [9]. To avoid wavy velocity and pressure fields, we use staggered grids for velocity and mass diffusion flux, the grid points correspond to control-volume faces ; pressure, temperature and density are calculated at control volume centre as it is shown on figure 2.

Transport properties (λ, κ) are evaluated on the staggered grid point by harmonic averaging.

CONTROL VOLUME   VARIABLES POSITION

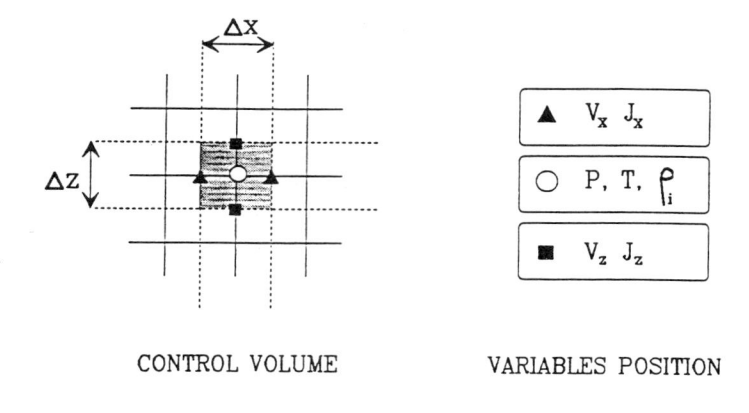

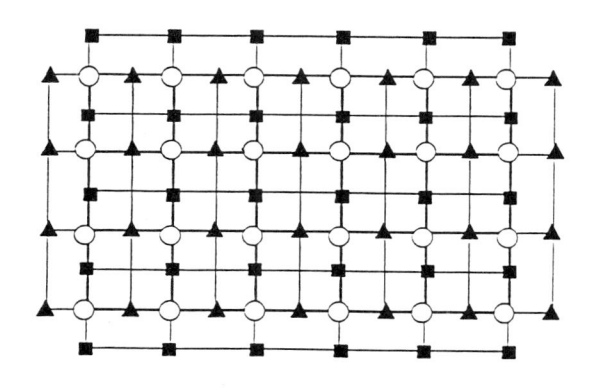

▲ V_x J_x

○ P, T, ρ_i

■ V_z J_z

STAGGERED GRIDS

Figure 2 : Numerical method

4.1 CONTROL VOLUME FORMULATION

FOR DARCY LAW

The generalized Darcy law coupled with an equation which ensures an incompressible flow is solved by the penalization method [11].

$$\left|\begin{array}{l} \dfrac{\partial \overline{\overline{V}}}{\partial \bar{t}} + \underline{V} = -\dfrac{\kappa}{\mu}(\nabla P + Ra_m^0 \rho \underline{g}) \\ \sigma \dfrac{\partial P}{\partial \bar{t}} + \nabla . \underline{V} = 0 \end{array}\right. \qquad (4-1-1)$$

where t is a fictitious time and σ a penalization parameter.

For a better conditioning, these equations are reformulated by using V/K as a new variable.

Steady state is reached by integrating the set of coupled equations by successive time integrations. This is justified by the large ratio between diffusion time scale and dynamic flow time scale.

FOR THERMAL BALANCE

The convective terms in the thermal balance equation are considerd in their conservative form. Time integration is made by the Alternating Directions Implicit (ADI) algorithm. We use a direct method to inverse the two corresponding tri-diagonal matrices.

FOR MATERIAL BALANCE

We rewrite the material balance equation in order to use the same ADI method :

$$M \dfrac{\partial \rho_i}{\partial t} + \nabla . (\rho_i \underline{V}) = -\nabla . \underline{J}_i$$
$$\underline{J}_i = \sum_{j=1}^{n} Le_{i,j} \nabla g_j + S_i^T \nabla T \qquad (4-1-2)$$

where $\underline{J}_i$ is now the mass diffusion flux.

The chemical potential is calculated at each grid point for current values of depth, temperature and pressure, by adding the dynamic and the static field of the state variables.

A second order Newton method is used to solve the cubic equation of state.

4.2 DISCUSSION OF NUMERICAL SCHEME

The order of magnitude required for the integration time is the inverse of the smallest Lewis number; on the contrary the reduced time step must be less than 0.01 : then, the total number of iterations is more than one hundred times the inverse of the smallest Lewis number. This value often reaches 10,000 !.
As a consequence of the thermodynamic calculations at each time step, the computing time needed for convergence could be very large.

The two-dimensionnal domain is covered with an array of (n x m) control volumes. We typically choose (16 x 16) for a satisfactory compromise between accuracy and computation time.

Trevisian and al. [15] have shown by scale analysis that the thickness of the mass stream is proportional to the square root of the Lewis number, which implies a finer grid for the computation of the distribution of a heavy compound. This is the reason why the study of asphaltenes distribution is out of the scope of this paper, although it has very important consequences on the fluid properties as Hirschberg [3] showed it.

The cross diffusion coefficients are chosen in order to respect the closing equation $\sum c_i = 1$. However, in transient state, renormalizing improves numerical stability.

5 APPLICATION

In the case of the schematic oil reservoir shown in figure 1 and with the mean properties of fluid and porous medium shown on table 1, we get the following characteristic times for convection and diffusion phenomena across the layer :

t_c = 800 years

t_D^i = 1500 years for methane

We are in the application range of the dynamic model, and it is the situation for most of the petroleum reservoirs.

Table 1 : Reservoir fluid and porous medium mean properties

$<\mu> = 0.45\, 10^{-3} Pa/s$	$<\rho> = 700\, Kg/m^3$
$<\chi> = 6.5\, 10^{-4}$	$<\theta> = 1.2\, 10^{-3}$
$<H> = 70\, m$	$<\kappa> = 10^{-12}\, m^2$
$<\phi> = 8$ degrees	$<\beta> = 4.5$ degrees
$<\Delta T> = 3.5\, K$	$<D_i> = 10^{-7} m^2/s$ for methane

Our application was built with a real fluid from a north sea oil reservoir. The composition of the fluid is assumed to be the composition of the most representative sample, from the middle part of the reservoir.

Ideally the reference fluid must be the average fluid over the present distribution in the reservoir. The use of a particular sample would imply to solve an inverse problem iteratively, but we do not consider this situation and we solve only the direct problem.

A detailed analysis of reservoir fluid composition was performed using Capillary Gas Chromatography up to undecane, almost all isomers were quantified as shown in table 2.
Gel Permeation Chromatography was applied to obtain the molar mass distribution of the heavy fraction.

Table 2 : Detailed fluid composition.

COMPOUNDS	MOLE %
Nitrogen	.303
Carbon dioxyde	.576
Methane	45.773
Ethane	5.397
Propane	4.376
Isobutane	.817
Normal butane	2.073
Isopentane	1.017
Normal pentane	1.296
2-2 dimethyl butane	.013
Cyclopentane	.095
2-3 dimethylbutane	.037
2 methyl pentane	.312
3 methylpentane	.192
Normal hexane	1.326
Methylcyclopentane	.373
2-2 dimethylpentane	.013
Benzene	.263
2-4 dimethylpentane	.032
3-3 dimethylpentane	.009
Cyclohexane	.640
2 methylhexane	.209
1-1 dimethylcyclopentane	.054
2-3 dimethylpentane	.069
3 methylhexane	.235
1 trans 3 dimethylcyclopentane	.092
3 ethylpentane	.010
1 cis 3 dimethylcyclopentane	.086
1 trans 2 dimethylcyclopentane	.150
Normal heptane	1.056
1 cis 2 dimethylcyclopentane	.025
2-2 dimethylhexane	.007
1-1-3 trimethylcyclopentane	.043
Methylcyclohexane	.989
2-5 dimethylhexane	.029
2-4 dimethylhexane	.038
Ethylcyclopentane	.068
1 t2 c4 trimethylcyclopentane	.048
3-3 dimethylhexane	.008
Toluene	.618
1 t2 c3 trimethylcyclopentane	.051
2-3-4 trimethylpentane	.011
2-3 dimethylhexane	.034
2 methyl 3 ethylpentane	.012
2 methylhexane	.199
1-2-2 trimethylcyclopentane	.051
4 methylheptane	.072
3 methylheptane	.151
3-4 dimethylhexane	.015
3 ethylhexane	.017
1 c2 c4 trimethylcyclopentane	.013
1 cis 3 dimethylcyclohexane	.123
1 trans 4 dimethylcyclohexane	.123
1-1 dimethylcyclohexane	.026
1 methyl trans 3 ethylcyclopentane	.032
1 methyl trans 2 ethylcyclopentane	.072
Cycloheptane	.011
Normal octane	.673
1 trans 2 dimethylcyclohexane	.122
1 cis 4 dimethylcyclohexane	.045
2-2 dimethylheptane	.008
2-4 dimethylheptane	.029
1 methyl 4 ethylcyclopentane	.008
2-6 dimethylheptane	.086
2-5 dimethyl heptane	.059
Propylocyclopentane	.036

COMPOUNDS	MOLE %
3-5 dimethylheptane	.011
Ethylbenzene	.087
Ethylcyclohexane	.275
Dimethylcyclohexane	.008
1-1-3 trimethylcyclohexane	.114
1-1-4 trimethylcyclohexane	.024
1 c3 c5 trimethylcyclohexane	.019
Various naphtenes in c9	.222
2-2-5 trimethylhexane	.009
Para-xylene	.097
Meta-xylene	.343
2-3 dimethylheptane	.069
3-4 dimethylheptane	.003
1 c3 t5 trimethylcyclohexane	.023
4 methyl octane	.083
2 methyl octane	.110
3 ethylheptane	.019
3 methyl octane	.108
Ortho-xylene	.155
1-1-2 trimethylcyclohexane	.040
Isopropylcyclohexane	.013
1 c2 c4 trimethylcyclohexane	.014
1 methyl t4 ethylcyclohexane	.018
1 methyl c3 ethylcyclohexane	.074
Normal nonane	.559
Cumene	.070
1 methyl c2 ethylcyclohexane	.019
1 methyl t2 ethylcyclohexane	.014
1 methyl 1 ethylcyclohexane	.023
4-4 dimethyl octane	.053
2-5 dimethyl octane	.060
Propylbenzene	.068
2-6 dimethyl octane	.150
2-3 dimethyl octane	.051
3-4 dimethyl octane	.021
4-5 dimethyl octane	.032
1-3 ethyl toluene	.122
1-4 ethyl toluene	.046
1-2 ethyl toluene	.060
4 methyl nonane	.072
2 methyl nonane	.083
1-3-5 trimethylbenzene	.067
3 methyl nonane	.062
1-2-4 trimethylbenzene	.162
Sec butylbenzene	.019
Isobutylbenzene	.032
Terbutylbenzene	.008
Normal decane	.484
1-2-3 trimethylbenzene	.057
Various naphtenes in c10	.458
1 methyl 3 isopropylbenzene	.033
1 methyl 4 isopropylbenzene	.051
C11 fraction	2.823
C12/13 fraction	4.408
C14/15 fraction	3.612
C16/17 fraction	2.711
C18/19 fraction	2.092
C20/24 fraction	3.207
C25/29 fraction	1.802
C30/39 fraction	1.945
C40/49 fraction	.818
C50/74 fraction	.619
C74/99 fraction	.180
C100/149 fraction	.039
C150/200 fraction	.007
C200/249 fraction	.003

In order to get a simplified representation of the fluid into 3 pseudocomponents, we applied our automatic clustering algorithm described elsewhere [7]. Results are shown in table 3.

Table 3 : Pseudo-composition of the fluid

COMPONENTS	MOLE (%)	M (g/mole)	Tc (°K)	Pc (bars)	ω
CP1	63.00	24.93	241.20	43.76	0.0502
CP2	12.72	110.43	572.71	28.19	0.3257
CP3	24.26	293.09	832.23	23.16	0.8708

The properties of the heavy fraction and the interaction coefficients were tuned to fit the PVT measurements.

The fluid is assumed to be initially homogeneous : this is not very realistic but it is impossible to get initial conditions from a geological approach because of the very large number of physical and chemical processes involved.

One unit reduced time in our simulation corresponds to 250 years.

5.1 STATIC SEGREGATION

In a first step we apply our model in the case of gravity segregation alone. The permeability of the layer is set to a very low value in order to prevent convection. The results for different integration times are shown on figure 3. Time step is 1 (250 years).

The quasi-steady state is obtained after approximately 25 million years (reduced time = 100000). As we can see on figure 3, the concentration segregation appears first along the bottom and the top edges of the layer, then the concentration profile becomes perpendicular to gravity.

Final profiles for density, saturation pressure and temperature are shown on figure 4.

Concentration gradients are : 0.037 % per metre for the light component, and - 0.036 % per metre for the heavy component, the medium component concentration remains almost constant with depth.

The bubble pressure of the oil varies from 27.54 MPa at the top to 24.82 MPa at the bottom (i.e. 11 kPa per metre), and the density varies from 690.8 kg/m^3 to 708.2 kg/m^3.

GERDYNAMIC

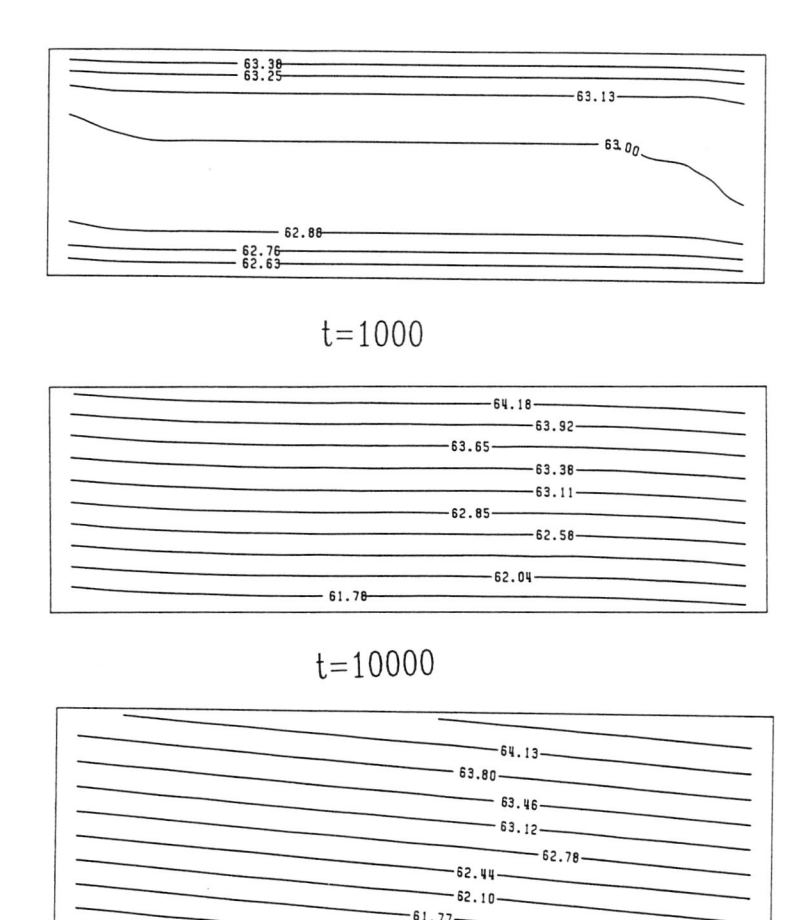

t=1000

t=10000

t=100000

Light component molar fraction

Figure 3 Static segregation, light component concentration field at different times.

GERDYNAMIC

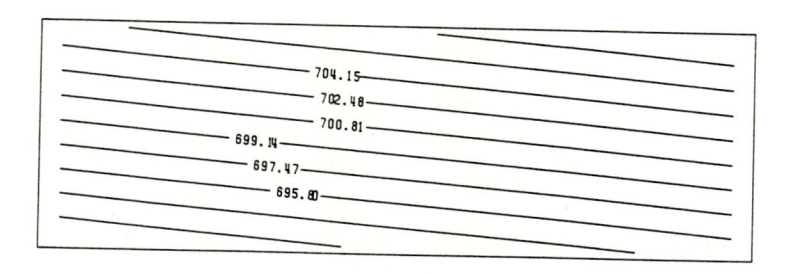

Fluid Density

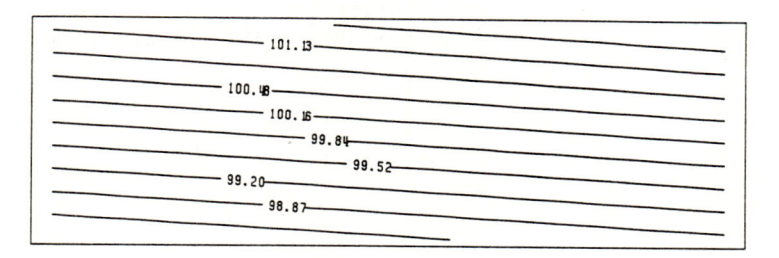

Temperature

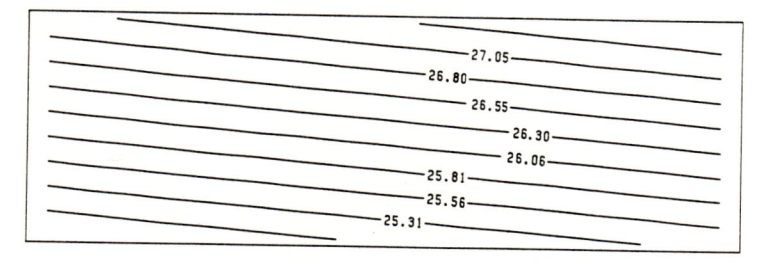

Sat. Pressure

Quasi Steady State

Figure 4 Static segregration, density, temperature and saturation pressure fields after segregation.

5.2 DYNAMIC SEGREGATION

With a real permeability value of 1 Darcy, natural convection occurs.

As shown on figure 5 and 6, the convective overturning is effective during a quarter million year, then it disappears. Time step is 0.01 (2.5 years).
During the first century, there is a single convective roll, then two secondary rolls appears at the top left and bottom right corners, finally they become parallel to the iso-concentration lines and vanish.

During the transient state, the concentration profiles are quite different from the static segregation profiles, but the final state is the same.

The static equilibrium concentration profile is reached in a very short time compared to static segregation : approximately one percent.

GERDYNAMIC

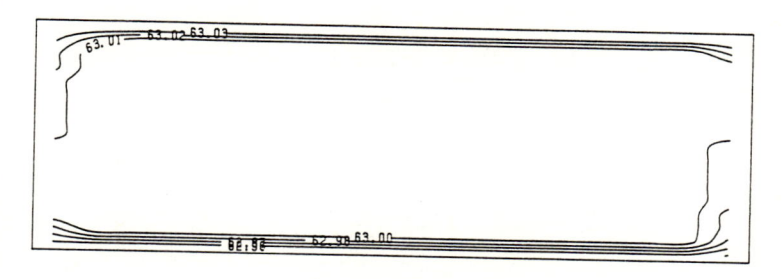

t=10

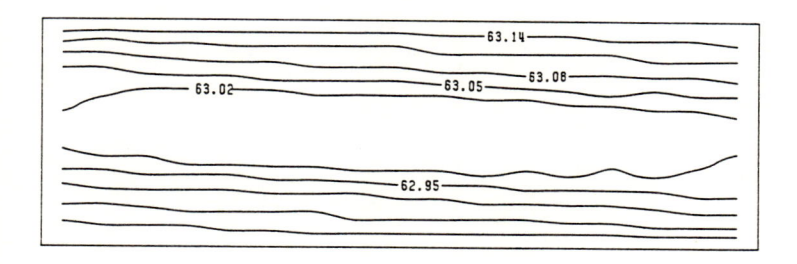

t=100

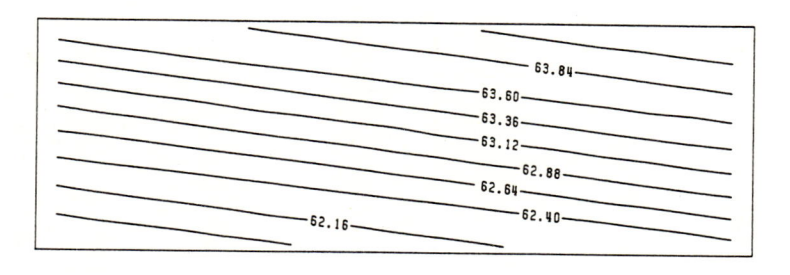

t=1000

Light component molar fraction

Figure 5 Dynamic segregation, light component concentration field at different times.

GERDYNAMIC

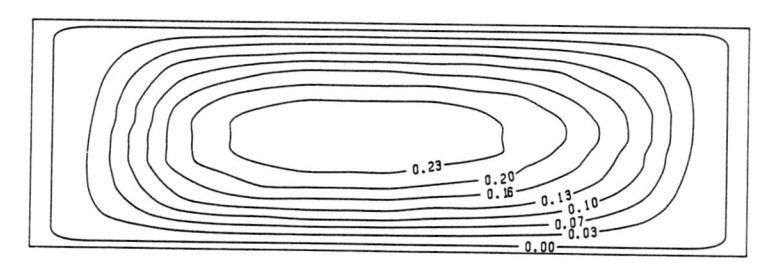

t=10

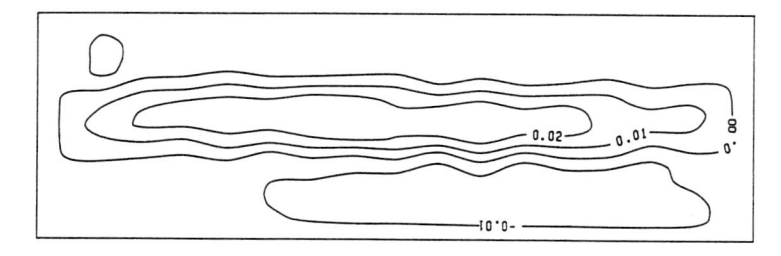

t=100

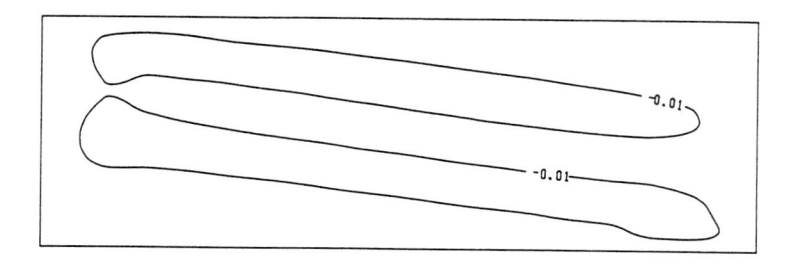

t=1000

Stream lines

Figure 6   Dynamic segregation, streamlines at different times.

5.3 DISCUSSION

Assuming a constant geological context, the simplified criteria outlined in the introduction do not apply to predict the steady state.

This is due to the fact that the convection time scale does not take into account the fluid density variations induced by gravity segregation.

The critical value for temperature variation across the layer corresponds to the equality between thermal expansion and gravitational contraction of the fluid.

In the case of an "ideal solution" we can estimate this critical value from :

$$\theta \Delta T = \sum_i \frac{x_i \dot{M}_i^2}{M} \cdot \frac{gH}{RT} + \chi \Delta P$$

In our case, thermal expansion is one order of magnitude below gravitational contraction.

For large β this simplified criterion does not apply and a convective movement can theoretically remain after the segregation process.

In our simulation we do not observe such a remaining flow. It probably happens for larger β values as in the Soret column describe in [1] where the thermal gradient is perpendicular to the gravity field. A lot of simulations have to be performed to define the critical β values for different fluids and reservoir properties.

Because of the very long time needed to reach the stationary state we can no longer assume constant pressure, temperature and global composition of the fluid. Although it is out of the scope of this paper to deal with such simulations, we can forecast that the dynamical process would restart if for example a lighter fluid comes in from below, or escapes through the cap rock.

Convection appears as the process responsible for creating the concentration grading within a realistic period.

6 CONCLUSION

Convection induced by geothermal gradient is rapidly softened by the segregation process in natural reservoirs.

The results show that the transient distribution of chemical components is strongly modified compared to the transient distribution obtained with gravity segregation alone.

For a typical reservoir, the steady state is the same in the two cases but the time needed to reach it is very long (million years). During this period, the geological parameters vary and they probably restart the dynamical process. This is an open new area of research.

Exceptionally, the static equilibrium concentration profile could be insufficient to prevent the convective overturning even if all the geological parameter remain constant. A lot of simulations have to be done if we want to define the critical values of the different parameters.

Anyway, the dynamic model is a very helpful tool for a better understanding of reservoir composition inhomogeneities, and this paper shows that it is not too difficult to carry out such a modelling.

ACKNOWLEDGEMENT

The authors thanks the management of ELF AQUITAINE, for permission to publish this paper, and acknowledge the assistance of Arthur PUCHEU in the preparation of the paper.

NOMENCLATURE

General

$(\underline{X})$: vector

$(\underline{\underline{X}})$: tensor

$(\check{X})$: dimensionless variable

(X_f) : fluid property

(X_s) : rock property

(X_m) : medium property

(X°) : initial or reference value

Variables

P : pressure

T : temperature

t : time

V : velocity

c_i : massic fraction of component i

x_i : mole fraction of component i

ρ_i : massic concentration of component i per unit of volume of fluid

Reservoir and fluid properties

$\underline{\underline{\kappa}}$: permeability

$\underline{\underline{\lambda}}$: conductivity

$\underline{\underline{\Phi}}$: porosity

$\underline{\underline{\tau}}$: dispersion

F : formation factor

H : layer thickness

ϕ : dip angle

β : angle between gravity and thermal gradient

ρ : density

Z : compressibility factor

θ : thermal expansion

χ : compressibility

M : molar mass

c_p : specific heat

H : enthalpy

J_T : thermal flux

μ : dynamic viscosity

Components properties

Pc_i : critical pressure

Tc_i : critical temperature

ω_i : acentric factor

v_i : molar volume

M_i : molecular weight

L_i^T : thermodiffusion coefficient

L_{ij}^D : generalized diffusion coefficient

D_i^f : simple diffusion coefficient

g_i : chemical potential

a_i : attraction parameter for PR EOS

b_i : covolume parameter for PR EOS

m_i : form factor

REFERENCES

[1]. COSTESEQUE P., HRIDABBA M. and SAHORES J. C.R. Acad. Sc. Paris, t. 304, Série II, n°17, p. 1069-1074, 1987.

[2]. COSTESEQUE P., ELMAATAOUI M. and SAHORES J. C.R. Acad. Sc. Paris, t. 305, Série II, p. 1531-1536, 1987.

[3]. DUCOULOMBIER D., ZHOU H., BONED C. PEYRELASSE J. SAINT-GUIRONS H. and XANS P. J. Phys. Chem. 1986, 90, 1692-1700.

[4]. HIRSCHBERG A. : "Role of Asphaltenes in Compositional Grading of a Reservoir's Fluid Column" JPT (Jan 1988) 40, 1 89-95.

[5]. HOLT T., LINDEBERG E. and RATKJE K. : "The Effect of Gravity and Temperature Gradients on Methane Distribution in Oil Reservoirs " SPE 11761 Usolicited paper (1983).

[6]. JACQMIN D. :"The Interaction of Natural Convection and Gravity Segregation in Oil/Gas Reservoir" paper SPE 16703 presented at the 1987 SPE Meeting, Dallas Sept 27-30.

[7]. MONTEL F. and GOUEL P.L. "A New Lumping Scheme of Analytical Data for Compositional Studies" paper SPE 13119 presented at the 1984 SPE meeting, Dallas Sept 16-19.

[8]. MONTEL F. and GOUEL P.L. : "Prediction of Compositional Grading in a Reservoir Fluid Column" paper SPE presented at the 1985 SPE Meeting, Las Vegas Sept 22-25.

[9]. PATANKAR S.V. : Numerical Heat Transfer and Fluid Flow, Hemisphere, Washinton D.C. 1980

[10]. PERKINS T.K. JOHNSTON O.C. : "A Review of Diffusion and Dispersion in Porous Media" S.P.E. J. March, 1963 70-84.

[11]. PEYKET R and TAYLOR T.D. : "Computional Method for Fluid Flow", Springer Series in Computational Physics, Spinger-Verlag 1983.

[12]. PLATTEN J.K. and LEGROS J.C. : Convection in Liquids, Springer-Verlag, Berlin Heidelberg New York Tokyo 1984.

[13]. RIEMENS W.G., SHULTE A.M. and DE JONG L.N.J. : "Birba Field PVT Variations Along the Hydrocarbon Column and Confirmatory Fiel Test" JPT (Jan 1988) 40, 1 83-89.

[14]. SHULTE A.M. : "Compositional Variations within a Column due to Gravity" paper SPE 9235 presented at the 1980 SPE Annual Meeting, Dallas, sept 21-24.

[15]. TREVISIAN O.V. and BEJAN A. Int. J. Heat Mass Transfer. Vol. 30, N°11 pp. 2341-2356, 1987.

[16]. WHEATON R.J. "Treatment of Variations of Composition With Depth in Gas Condensate Reservoirs" paper SPE 18267 presented at the 1988 SPE Meeting, Houston Oct 2-5.

APPENDIXES

Appendix 1 : Peng-Robinson Equation of State

Relationship between Pressure, Temperature and molar Volume is :

$$P = \frac{RT}{V-b} - \frac{a(T)}{V(V+b)+V(V-b)} \qquad (A-1-1)$$

The parameters a and b are defined by the following mixing rules :

$$b = \sum_i x_i b_i \qquad a(T) = \sum_i \sum_j x_i x_j \sqrt{a_i a_j}(1-k_{ij})$$

k_{ij} are the interaction parameters and the pure component parameters a_i and b_i are calculated from the critical properties :

$$b_i = 0.0778 \frac{RT_{ci}}{P_{ci}} \qquad a_i = 0.45724 \frac{R^2 T_{ci}^2}{P_{ci}} \left(1 + m_i \left(1 - \sqrt{\frac{T}{T_{ci}}}\right)\right)^2$$

P_{ci}, T_{ci}, and m_i are critical pressure, temperature and form factor of component i, the form factor is correlated with acentric factor ω_i by means of :

$$m_i = 0.37464 + 1.54226 \omega_i - 0.26992 \omega_i^2$$

The chemical potential is given by

$$g_i^c - g_i^{c*} = RT(Z-1)\frac{b_i}{b} - RT\log(Z-B) +$$

$$\frac{a}{2\sqrt{2}b}\left(\frac{2}{a}\sum_j x_j \sqrt{a_i a_j}(1-k_{ij}) - \frac{b_i}{b}\right)\log\left(\frac{Z+(1-\sqrt{2})B}{Z+(1+\sqrt{2})B}\right)$$

$$Z = \frac{PV}{RT} \qquad B = \frac{bP}{RT}$$

g_i^{c*} is the chemical potential in the ideal gas state of the component i in the mixture, and Z the fluid compressibility factor.
Density is given by :

$$\rho = \frac{MP}{ZRT}$$

where $M = \sum_{i=1}^{n} x_i M_i$ is the average molar mass of the fluid.

Appendix 2 : Viscosity

Corresponding state method for viscosity calculation, the n tetradecane is used as the reference compound (noted *) :

$$\mu(P,T) = \mu^*(P',T') \left(\frac{T_c^f}{T_c^*}\right)^A \left(\frac{P_c^f}{P_c^*}\right)^B \left(\frac{M^f}{M^*}\right)^C \qquad (A-2-1)$$

where

$$T' = T\left(\frac{T_c^*}{T_c^f}\right) \qquad P' = P\left(\frac{P_c^*}{P_c^f}\right)$$

and with :
    A = 1.385374   B = -0.756972   C = -0.532041

T_c^f, P_c^f are the pseudo-critical properties of the fluid.

EFFECTS OF HETEROGENEITIES ON PHASE BEHAVIOUR IN ENHANCED OIL RECOVERY

F.J. Fayers, J.W. Barker, T.M.J. Newley

(BP Research Centre, Sunbury)

ABSTRACT

The effects that heterogeneities can create on multi-phase flows and phase behaviour are discussed. A statistical indicator simulation method is used to generate two-dimensional heterogeneity descriptions for two hypothetical Representative Elements of Volume (REV's A and B). The REV's are used to investigate how the heterogeneities will influence the recovery of residual oil by several EOR processes. Finely gridded comparisons are made between homogeneous and heterogeneous cases in a compositional simulator, demonstrating the potential impact of heterogeneities.

A Dual Zone Mixing (DZM) procedure is proposed for representing the effects of heterogeneities within a compositional simulation. Each grid block is split into contacted and bypassed zones, whose sizes vary with time, and separate phase behaviour calculations are performed for the two zones. Correction factors are introduced in the transport terms in the conservation equations. An additional conservation equation is required in the general case, involving the rate of mass transfer between the dual zones. The form of this transfer term is discussed. The detailed solutions from the REV examples are used to determine the functional forms for the correction factors, and it is shown that the DZM method is capable of reflecting the effects of heterogeneity within a coarsely gridded (3 x 3) homogeneous reservoir problem.

1. INTRODUCTION

Large variations in rock properties, particularly permeability, occur at all length scales in petroleum reservoirs. Consequently, the effect of heterogeneities on average flow behaviour is the subject of a considerable amount of current research in the petroleum industry. The objective of much of this work is to determine effective flow parameters which can be used in the conventional equations governing fluid flow in a reservoir, without altering the form of these equations. Significant progress has been made for single phase flow, where the methods described by Begg and King[1] and Haldorsen[2] have become well known. For two-phase flow (eg water displacing oil), the use of pseudo or effective relative permeabilities is being extended to account for heterogeneity, in addition to large grid block size or reduced dimensionality for which pseudos were originally designed (see for example Lasseter[3], Kossack[4], and Muggeridge[5]). In compositional simulation, up to four flowing phases may be present, and a complicated phase behaviour calculation is necessary to determine the number, composition, and physical properties of the phases. Generation of pseudo relative permeabilities for these systems is a difficult problem. Additionally, the phase behaviour calculation is based on the assumption of complete mixing of all components present. This assumption may be valid at small length scales where diffusive effects ensure good mixing, but the presence of heterogeneities means that variations in fluid composition will still occur on scales smaller than the sizes of typical grid blocks used in reservoir simulation.

Assumption of uniform mixing can be in error even if the large grid block is uniform in permeability and porosity. Rapid changes in fluid composition over very short distances can occur as a result of viscous fingering in unstable displacement processes, gravitational segregation (particularly when gas is present), or the formation of fronts during the displacement process (a natural consequence of the hyperbolic character of the flow equations).

In this paper we use fine grid simulations to illustrate the effects that heterogeneities may have on compositional displacement processes, and show the nature of the errors that can occur if the heterogeneities are neglected. Vaporisation of residual oil by dry gas recycling is taken as the primary example process, but results are also presented for miscible displacement and surfactant flooding processes. Gravitational segregation effects and effects associated with the heterogeneities occur simultaneously for these processes. Results of displacement calculations on a 100 x 20 grid in which each block is assigned a different permeability and porosity are compared with those from simulations of the same displacements on the same grid but where the permeability and porosity in each block are set equal to the average values for the problem. For the heterogeneous cases, two realisations of the reservoir description are selected randomly from two distributions with similar variances but different correlation lengths. The homogeneous simulations are also run on a coarser 3 x 3 grid. The results presented illustrate the nature of the errors introduced by using a coarse grid, as distinct from replacement of the heterogeneous reservoir by a homogeneous one.

In the final sections of the paper a dual zone mixing method (DZM) is discussed which can be deployed within compositional simulators as a correction procedure to the uniform mixing assumption. This procedure has a simpler form for the oil vaporisation case and is shown to be a useful approximation for this type of problem.

2. ERRORS IN COMPLETE MIXING ASSUMPTIONS FOR HOMOGENEOUS CASES

Two different factors influence effective phase behaviour, one being numerical dispersion, which is an artefact of finite grid size often unavoidable due to practical computing limitations, and the other being physical and driven by the heterogeneity details of the porous medium which we cannot encompass. Thus, effective phase behaviour parameters may be required even in homogeneous problems.

To illustrate some numerical dispersion effects, consider a multi-contact miscible displacement using enriched gas. The compositional continuity equations for this system can be expressed in the form:

$$\nabla \left[C_{gi} \frac{Kk_{rg}}{\mu_g} (\nabla P + \rho_g g \nabla z) + C_{oi} \frac{Kk_{ro}}{\mu_o} (\nabla P + \rho_o g \nabla z) \right]$$
$$= \phi \frac{\partial}{\partial t} \left[C_{gi} S_g + C_{oi} (1-S_g) \right] \qquad (2.1)$$

The symbol definitions are given in the nomenclature.

Figure 1(a) illustrates a large grid block containing a partially waterflooded region with oil and water flow, a small developing miscible bank composed of a single phase liquid hydrocarbon, and an immiscible gas/oil zone where oil vaporisation is occurring. The three points A, B, C in this illustration are shown on a ternary phase behaviour diagram in Figure 1(c). A flash of the complete contents of the grid block, represented by point D, results in a two-phase hydrocarbon mixture with a larger oil saturation than point C, illustrated in figure 1(b). The distinction between the three zones is lost. Additionally, unless pseudo relative permeabilities are used, three-phase flow out of the block will be predicted instead of only oil and water flow. Even if this is corrected by a Kyte and Berry[6] approach, the composition of the flowing oil is not the same as the average oil composition computed from point D.

Prior to the inception of compositional simulators based on cubic equations of state, the phase behaviour was described through correlations for the partition coefficients defined by:

$$K_i = \frac{C_{gi}}{C_{oi}} \qquad (2.2)$$

Various correlative procedures were available for evaluating the K_i's as functions of C_i and reduced pressure, ($P_{ri} = P/P_{ci}$). Camy and Emanuel[7] proposed the use of pseudo K-values:

$$\overline{K}_i = \beta_i(C_i) K_i \qquad (2.3)$$

to approximate the true average phase composition in a large grid block, together with pseudo relative permeabilities (following Kyte and Berry) to approximate the true flow rate for each phase. Compositional correction factors $\alpha_{gi}(C_i)$ and $\alpha_{oi}(C_i)$ were also introduced to approximate the relation between the composition of each phase flowing out of the block and its average composition in the block. Thus, the left hand side of the equation (2.1) becomes:

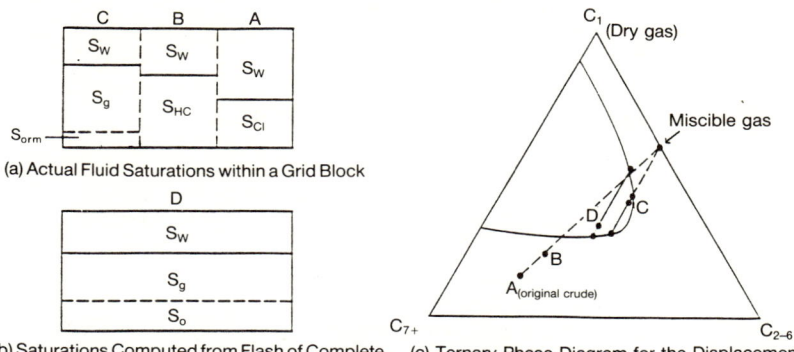

EFFECTS OF HETEROGENEITIES ON PHASE BEHAVIOUR 119

Fig. 1 Numerical dispersion effects in a multi-contact miscible displacement

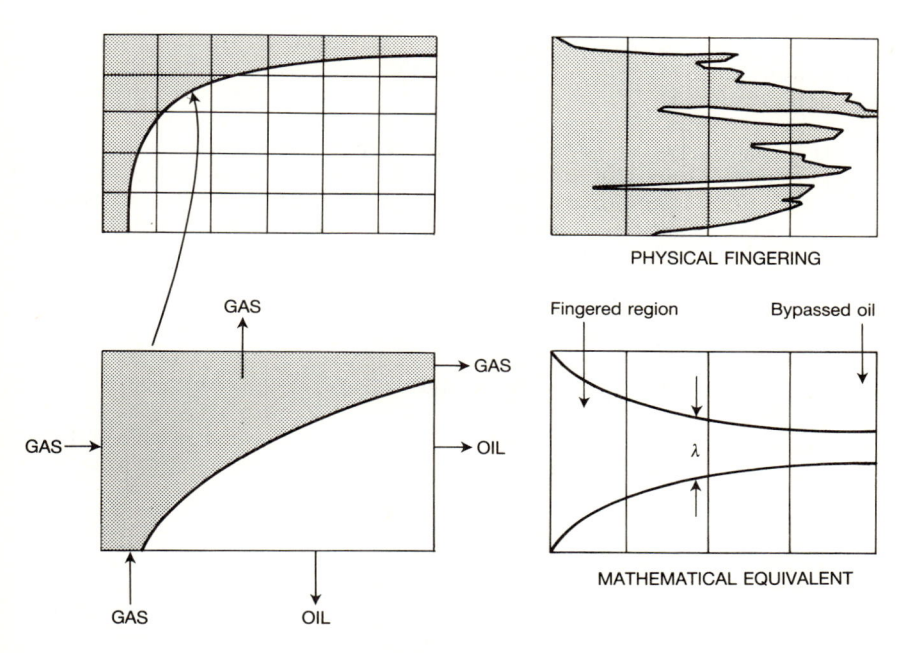

(a) Actual Fluid Saturations within a Grid Block

(b) Saturations Computed from Flash of Complete Contents of Block

(c) Ternary Phase Diagram for the Displacement

Fig. 2 Gravity effects on position dependence of pseudo relative permeabilities

Fig. 3 Representation of fingering flow in compositional simulation

$$\nabla \left[\alpha_{gi} \, C_{gi} \, \frac{Kk_{rg}}{\mu_g} \, (\nabla P + \rho_g g \nabla z) + \alpha_{oi} \, C_{oi} \, \frac{Kk_{ro}}{\mu_o} \, (\nabla P + \rho_o g \nabla z) \right]$$
(2.4)

The factors α_{gi}, α_{oi} and β_i were correlated with C_i by running finely gridded one-dimensional simulations, and examining average values over regions corresponding to blocks in the coarse grid, and flows between these regions. In principle, the α_{gi}, α_{oi} and β_i will depend on pressure, but Camy and Emanuel found this could be neglected for the CO_2 flood they studied.

In most reservoir applications gravity effects will be significant, as illustrated in Figure 2. Since the shape of the gas/oil contact depends on the displacement rate (i.e. the viscous to gravity ratio V_{gr}), the one dimensional fine grid model must be replaced by a representative finely gridded cross section which in some measure preserves the spatial distribution of V_{gr}'s which occur in a three-dimensional reservoir. This can be done by basing the cross section on the varying geometry of an average stream tube connecting the injection well to the production well. The pseudo relative permeabilities and compositional correction factors then become directional in nature and may need to be correlated with V_{gr}, or with position.

A further example of incomplete mixing in essentially homogeneous media is viscous fingering. In the Fayers model[8, 9] the fingering distribution in a coarse grid block is replaced by a mathematical finger (see Figure 3) containing a varying mixture of oil and gas, which bypasses unmodified oil. The finger fractional width is expressed as:

$$\lambda = a + (1-a) \, S_{gf}^{\alpha} \qquad (2.5)$$

where a, α are parameters, and S_{gf} is the gas saturation (or concentration) in the finger at any position. For miscible displacement this theory was shown to provide a satisfactory match to measurements for the choice:

$$a = 0.1 \text{ and } \alpha = 0.42 \left(\frac{\mu_o}{\mu_g} \right)^{0.4} \qquad (2.6)$$

These ideas can readily be extended to fingering in immiscible gas/oil displacement by replacing Eq (2.1) by its two-region equivalent:

$$\nabla \left[\lambda \left(C_{gif} \frac{Kk_{rgf}}{\mu_g} (\nabla P + \rho_g g \nabla z) + C_{oif} \frac{Kk_{rof}}{\mu_o} (\nabla P + \rho_o g \nabla z) \right) \right.$$
$$\left. + (1-\lambda) C_{oib} \frac{Kk_{rob}}{\mu_o} (\nabla P + \rho_o g \nabla z) \right]$$
$$= \phi \frac{\partial}{\partial t} \left[\lambda \{ C_{gif} S_{gf} + C_{oif} (1-S_{gf}) \} + (1-\lambda) C_{oib} \right] \quad (2.7)$$

The composition C_{oib} of the bypassed oil in the above equation is fixed, while the compositions in the fingered region C_{gif} and C_{oif} are determined from an equilibrium flash. The parameters defining λ given by Eq (2.6) may need modification for immiscible flows.

None of the above mentioned approximate methods of accounting for incomplete mixing within a homogeneous grid block are accomodated in existing compositional simulators. The modern preference for direct use of cubic equations of state has tended to preclude the use of pseudo K-values as defined in Eq (2.3); however some simulators do provide a K-value option which could be deployed in this manner. The λ-factors required in Eq (2.7), and the α_{oi} and α_{gi} required by Eq (2.4), could easily be added to existing simulators.

3. GENERATION OF HETEROGENEOUS TEST PROBLEMS

In order to illustrate some effects of heterogeneity on average phase behaviour in coarse grid blocks, it is necessary to generate a reservoir description in some statistically valid manner. There are a number of possible techniques available, but care must be taken to avoid a method which introduces spatial smoothing, which might obscure the effects being investigated. Journel and Alabert[10] have highlighted the need for statistical generation techniques which will preserve the spatial connectivity characteristics of extreme values; for example pathways of high permeability which may lead to bypassing of low permeability regions. We have therefore generated heterogeneous models by using sequential indicator simulation (SIS), as described in Ref 10.

It has been assumed for this purpose that the reservoir geology can be statistically described within a representative element of volume (REV), as defined by Haldorsen [11]. We have limited the REV to a 2D-vertical section of dimensions 100 x 10 metres. Our hypothetical problem is then to examine how phase behaviour effects associated with a 3 x 3 coarse grid (ie a grid block size of approximately 33 x 3.3 metres) might behave in relation to the more exact phase behaviour in a fine grid representation of the heterogeneous REV. Since the 3 x 3 coarse element is deemed to be embedded in a much larger reservoir problem, it will be necessary to examine both horizontal flow and vertical flow through the REV.

Permeability and porosity distributions for hypothetical reservoir descriptions have been developed through a three-stage sequential process similar to that deployed by Begg et al[12] for the Sherwood reservoir. The stages were as follows:

(i) Five permeable lithotypes (subsequently termed L1 to L5) representing a range from muddy silts to clean sands were selected. The lithofacies description is linked to hypothetical observed data at the left and right hand boundaries as shown in Figure 4a, and were generated to match the average boundary fractions in the proportions, L1 to L5 : 0.2, 0.25, 0.30, 0.15, 0.10. Distributions of the permeable lithotypes were then generated using the SIS technique on a 100 x 20 fine grid. Two different elements A and B have been constructed by using variograms which have two different sets of spatial correlation lengths as follows:

	Element A	Element B
Horizontal	20m	5m
Vertical	0.67m	1.67m

These two elements are shown in Fig 4b and c. REV A has longer and thinner heterogeneities than the somewhat square shaped heterogeneities of REV B, but the proportions of each type are correctly preserved, and they match the fixed distributions prescribed at the left and right boundaries.

(ii) Horizontal permeabilities were assigned randomly within each lithotype by using non-overlapping uniform distributions in appropriate ranges:

(a) INITIAL HYPOTHETICAL WELL DATA

(b) REV A-INTERPOLATED LITHOTYPES

Type 5
Type 4
Type 3
Type 2
Type 1
Shale Barrier

(c) REV B-INTERPOLATED LITHOTYPES

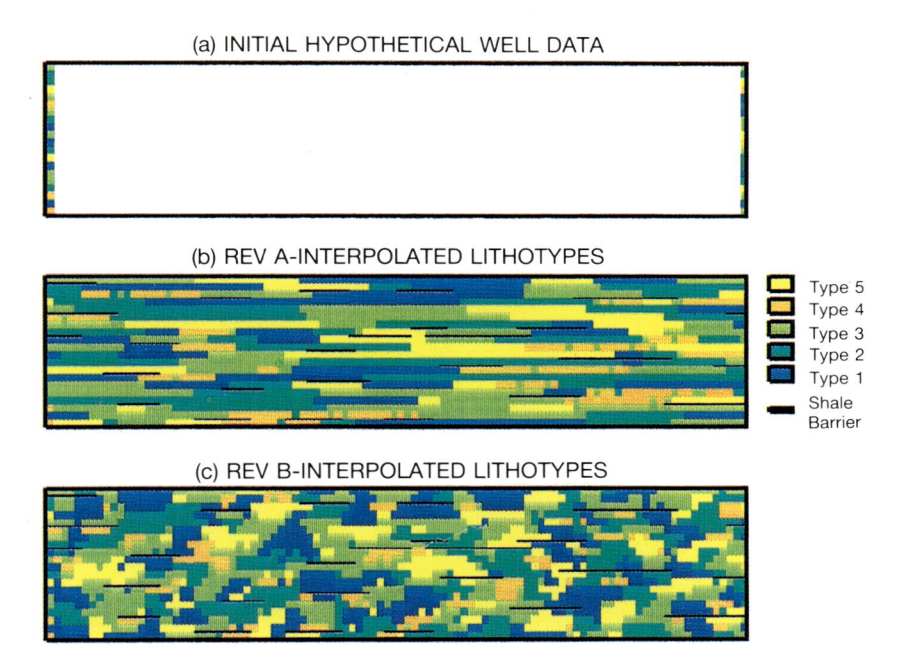

Fig. 4 Initial data and lithotype distributions

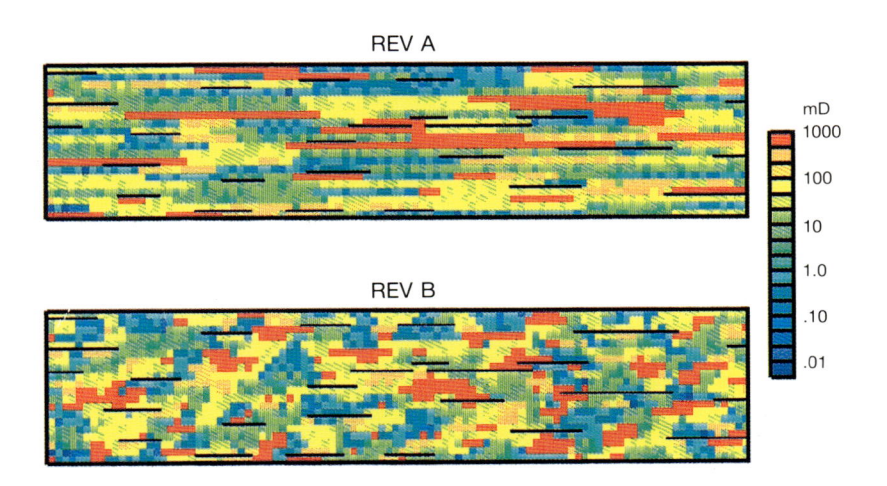

REV A

mD
1000
100
10
1.0
.10
.01

REV B

Fig. 5 Permeability distributions for Rev's A and B

Lithotype	Proportion	Permeability Range, mD
L1	0.2	$10^{-3}-1$
L2	0.25	1-20
L3	0.30	20-100
L4	0.15	100-500
L5	0.10	500-1000

The random variation of permeability within a lithotype is justified on the grounds that a lithotype is deposited by a single mechanism, and therefore its variations in permeability are likely to be random. Vertical permeabilities were correlated to the horizontal values by assuming that lithotypes L1 and L2 occur in muddy rocks (rock type 1) with $K_v/K_h = 0.4$, while L3, L4 and L5 belong to clean sands (rock type 2) with $K_v/K_h = 0.7$. Porosity was also assumed to be correlated with permeability using the relation; $\phi = 4 \log(K_h) + 13$, for all lithotypes. This generates porosities in the range $0.01 < \phi < 0.25$.

(iii) A fixed distribution of discontinuous impermeable thin shales was superimposed on both REV's A and B. The shale distribution was generated by selecting random positions within the element and by choosing the shale lengths from a triangular probability distribution (minimum 3 m, most likely 8 m, maximum 20 m). Shale assignment continued until the specified average frequency of 0.25 shales per metre depth had been achieved.

The resulting permeability distributions from steps (i) to (iii) are shown for REV's A and B in Figure 5. Every grid block within the 100 x 20 array is assigned its own permeability value. The shales in these models have negligble volume and are simulated in the displacement calculations by applying zero-transmissivity factors to the appropriate grid block faces.

Table 1 summarises various estimates of average permeabilities. The final column is obtained from a direct numerical solution of the pressure equation for single phase flow, following the procedure recommended by Begg and King[1]. These "effective" values will be used for the homogeneous simulations. Note that the other averages show considerable variations and do not include the shale effects because of their zero volume fraction. The standard deviations for K_h are 273 mD and 266 mD respectively for REV's A and B. REV A has a much lower K_v/K_h ratio than REV B which arises because of the long horizontal correlation length of its low permeability regions. Using a porosity cut-off of 12% to define non-pay, we obtain net to gross ratios of 0.89 and 0.86 for the two elements. The constant Dykstra-Parson's coefficient of variability of 0.86 implies fairly heterogeneous reservoir models, as was intended for this illustrative study.

TABLE 1

Average Properties Of Permeability Distributions For Rev's A and B

	Arithmetic Mean	Geometric Mean	Harmonic Mean	Effective Value
K_{hA}	163	19.7	0.53	73.5
K_{hB}	155	17.1	0.46	19.5
K_{vA}	113	10.5	0.21	1.4
K_{vB}	108	9.1	0.18	2.8
K_{vA}/K_{hA}	0.69	0.53	0.40	0.01
K_{vB}/K_{hB}	0.69	0.53	0.40	0.14
ϕ_A	0.182	–	–	0.182
ϕ_B	0.179	–	–	0.179

4. PHASE BEHAVIOUR EFFECTS IN OIL VAPORISATION BY LEAN GAS

The vaporisation problem arises in condensate fields where a dry hydrocarbon gas, or perhaps nitrogen, is injected to vaporise liquid hydrocarbon which has dropped out and become trapped during depletion of the gas condensate. The effect of reservoir heterogeneities on the efficiency of the vaporisation process is then a critical question.

For the vaporisation examples, hypothetical oil and gas compositions were set up in terms of a six pseudo-component description. These compositions are identified in Table 2. The injected lean gas consisted predominantly of component C_1. The phase behaviour was computed with the Peng-Robinson equation of state. The phase behaviour properties of the oil imply that up to 60% by volume might be vaporised after contact with a large throughput of the dry gas.

TABLE 2

Fluid Composition for Lean Gas Vaporization Study.

Component	Residual oil	Equilibrium Gas
C_1	0.4481	0.7782
C_2	0.1481	0.1756
C_3	0.1010	0.0345
C_4	0.1620	0.0142
C_5	0.0965	0.0004
C_6	0.0443	0.

The residual oil saturation S_{org}, which is also the initial oil saturation for this study, was assumed to be uniform at 0.25. The relative permeability to gas was chosen using Standing's formula [13]:

$$k_{rg} = \frac{S_g - S_{gr}}{1 - S_{gr} - S_{wc}} \left(1 - \frac{1 - S_g - S_{wc}}{1 - S_{wc}}\right)^{2/\nu + 1} \quad (4.1)$$

The connate water saturation S_{wc} was set at 0.2 and the critical gas saturation S_{gr} at 0.05. Since the oil saturation declines below S_{org} as a result of vaporisation, we have to use Eq (4.1) to predict shapes of gas relative permeabilities at saturations which are not available from conventional measurements. Furthermore, there are positive feedback effects associated with the fact that oil vaporisation will occur preferentially in high permeability paths, leading to consequent increases in the permeability to gas. The slopes of the gas relative permeability curves near S_{org} deduced from Eq (4.1) are quite large. The curves are shown in Figure 6, one for rock type 1 (shaly sands) with a pore size distribution factor $\nu = 0.5$, and the other for rock type 2 (clean sands) with a pore size distribution factor $\nu = 1.5$, which is appropriate to larger less evenly distributed grain sizes. Neither oil relative permeabilities nor capillary pressures are required for this problem, since the oil saturation will always be below the residual value of 0.25.

Both horizontal and vertical gas displacement were studied in the REV's, using equal gas injection and production rates of 0.365 pore volume (pv) per year. For the horizontal case gas injection was along the left boundary, with the total flow split between blocks in proportion to $K_h \Delta z$; for the vertical case injection was along the top boundary in proportion to K_v Δx. For the homogenised sections this implies uniform injection along the boundaries.

The BP-COMP code, which is a modified version of the SSI-COMP III simulator, has been used to solve the 6-component compositional problems associated with lean gas vaporisation in REV's A and B. The heterogeneous and finely gridded homogeneous cases have been solved using the same 20 x 100 grid used to define the permeability distributions described in Section 3. Mass diffusion effects are assumed to be of the same order of magnitude as the numerical diffusion aspects associated with the above grid. The complicated issues posed by spatially varying mass diffusion in a non-uniform multi-component situation are the subject of future work.

The volume fractions of residual oil recovered for horizontal displacement are illustrated in Figure 7a for REV A, and in Figure 7b for REV B. Recoveries are compared for detailed heterogeneous and homogeneous cases, and for the coarse grid (3 x 3) case. The results for vertical displacement are shown in Figures 8a and b. In all cases, the recoveries are becoming asymptotic at the large dry gas throughput of 5.0 pv. The final recoveries are summarised in Table 3. The worst recovery error for homogeneous flow is an overestimate of approximately 40%, which occurs for vertical flow for REV B. The shales exert a strong influence on the vertical flow; this is ameliorated to some extent in REV A by the larger horizontal correlation length which has produced paths more able to conduct flows around the shales. Thus REV B performs the most non-uniformly in vertical flow. The coarse grid results all show some minor compensation in errors, with the truncation error effects improving estimates of recovery slightly compared with the finely gridded homogeneous cases. It is often the case in reservoir engineering that we rely on the numerical dispersion errors associated with coarse grid blocks to give some measure of compensation for the omission of heterogeneities.

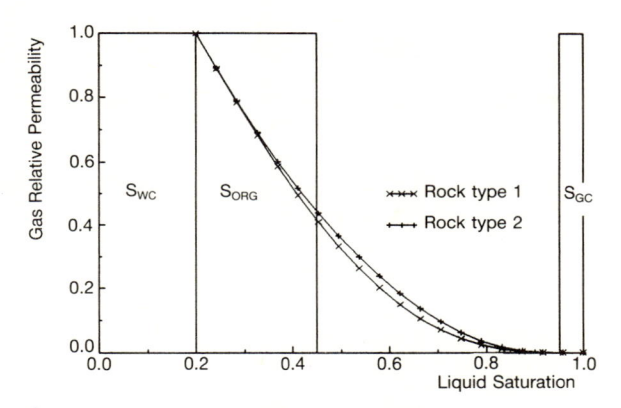

EFFECTS OF HETEROGENEITIES ON PHASE BEHAVIOUR 127

Fig. 6   Gas phase relative permeability for lean gas displacements

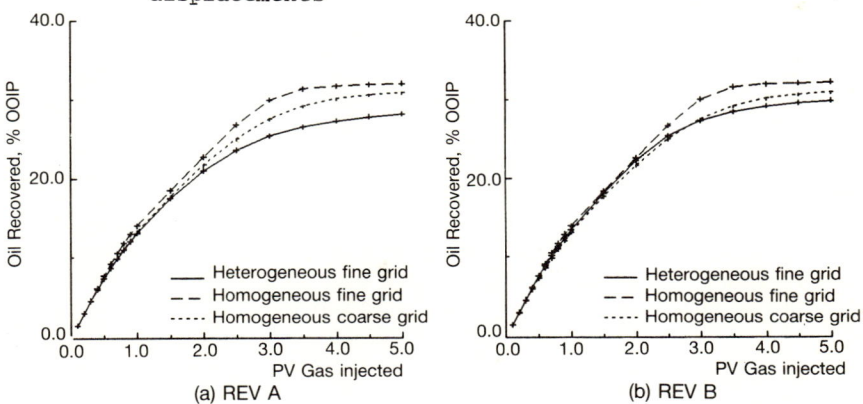

Fig. 7   Oil recovery for horizontal lean gas displacements

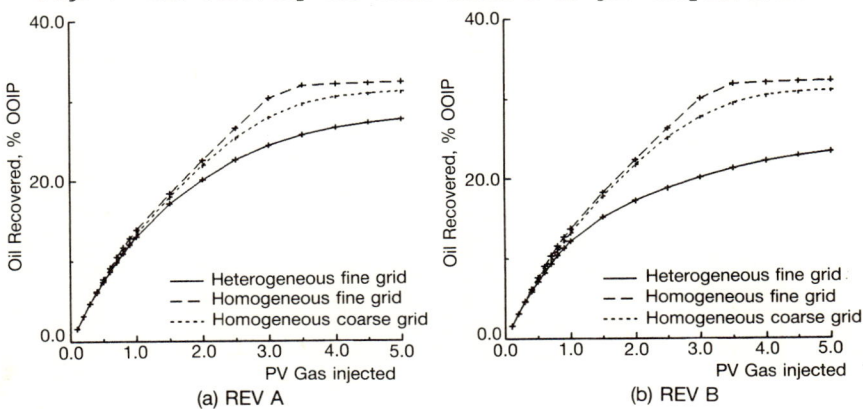

Fig. 8   Oil recovery for downward lean gas displacements

TABLE 3

Final Volume Fractions Of Residual Oil Recovered
by dry gas injection

	Horizontal Flow			Vertical Flow		
	(i)	(ii)	(iii)	(i)	(ii)	iii)
REV A	28.1%	32.1%	31.0%	27.7%	32.4%	31.0%
REV B	29.9%	32.3%	31.0%	23.4%	32.4%	31.0%

Case (i) fine grid heterogeneous
Case (ii) fine grid homogeneous
Case (iii) coarse grid homogeneous

Figure 9 compares the oil saturation distributions for the three calculations at 1 pv gas injection. For horizontal flow the longer and thinner heterogeneities of REV A are causing more non-uniformity of flow, than in REV B. The heterogeneous results are very different from their homogeneous equivalents. The homogeneous fine grid cases exhibit gravity override effects for horizontal flow, especially for REV Awhich has a larger effective K_h. The errors in predicted oil recovery resulting from omission of heterogeneities have thus been partially compensated by the segregation effects in this case.

5. MISCIBLE FLOOD EXAMPLES

In Section 2 we illustrated the kind of effect that failure to resolve different regions in a multi-contact miscible gas process might have on phase behaviour and miscible displacement efficiency. Even when the achievement of miscibility has been accomplished successfully in the heterogeneous geology, there are still problems of how complete the sweep efficiency will be, since the low viscosity miscible fluid preferentially enters the high permeability pathways and continues to flow in them, thus excluding low permeability routes. For the more extreme variations in correlated heterogeneities of the type considered in REV's A and B, the viscous fingering process will be governed by the physical distribution of the heterogeneities, rather than by any stochastic fingering aspects associated with unstable flow.

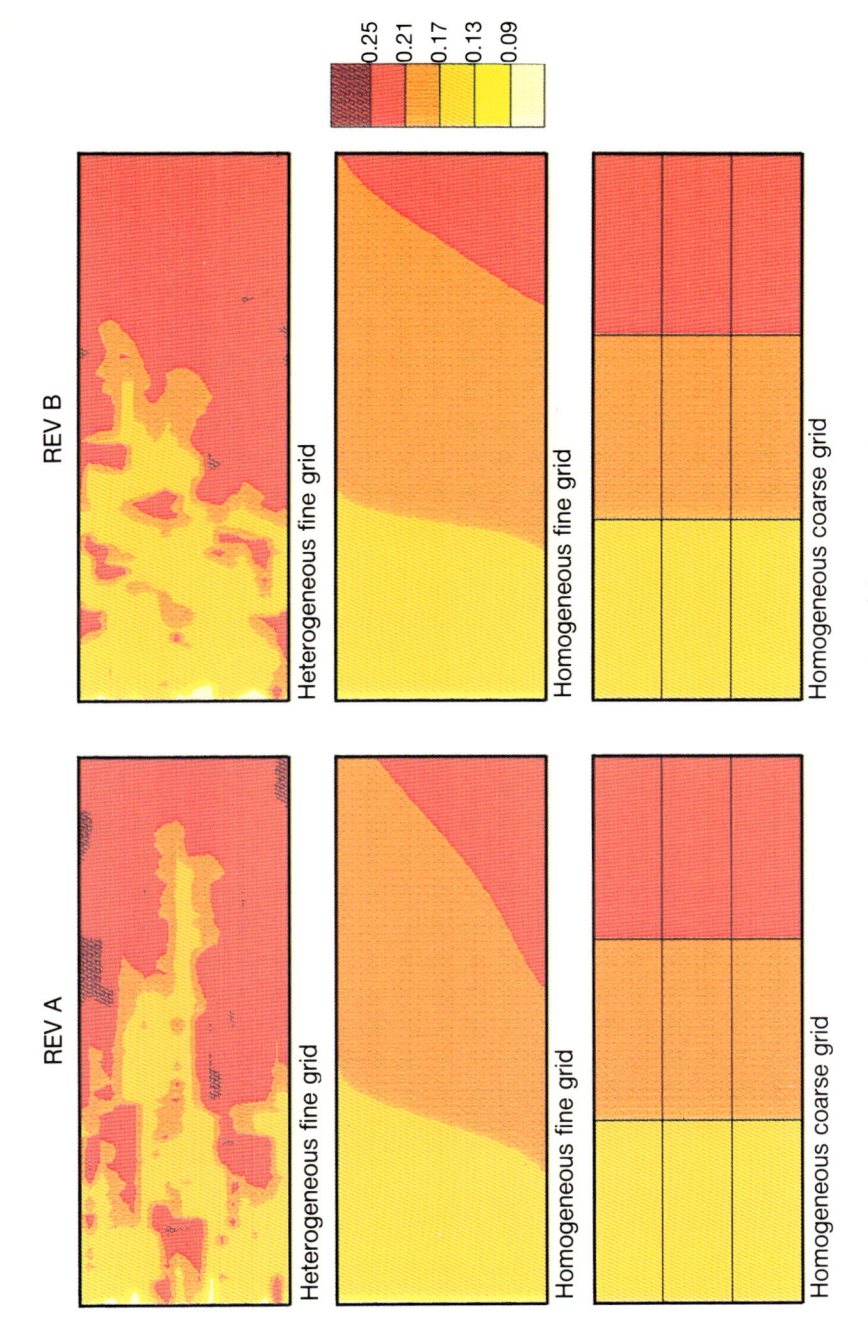

Fig. 9 Oil saturations at 1 pv for horizontal lean gas displacements

To illustrate these effects in miscible displacement we have chosen an example which involves tertiary recovery of residual oil from water flooding. Thus it is assumed that residual oil saturation from the waterflood has been established uniformly throughout the cross section with S_{orw} = 0.34. For the particular miscible injectant used, the system is first-contact miscible. Floods are conducted at an oil/gas viscosity ratio of 100 and an oil/water viscosity ratio of approximately 18. The densities of gas, oil and water at reservoir conditions are ρ_g = 421 kg/m^3, ρ_o = 907 kg/m^3, and ρ_w = 998 kg/m^3, so some gravity segregation is expected during the displacement. The Larson parameterization was used for relative permeabilities: $k_{ro} = (S_o - S_{orw})^2$ and $k_{rw} = 0.23 (S_w - 0.25)^6$.

Horizontal miscible displacements have been simulated using BP COMP for both REV A and REV B, at an injection rate of 0.365 pv/year. Oil recoveries for the heterogeneous fine grid, homogeneous fine grid and coarse grid are shown in Figures 10a and 10b. The miscible injectant concentration in the hydrocarbon phase is shown in Figure 11 at 0.2 pv injected.

Displacement in the heterogeneous fine grid is dominated by the effect of the heterogeneities: in REV A, gas channels through the central region, and there is minimal gravity segregation. By contrast, there is a clear tendency for gravity override in the simulations for the homogeneous fine grid, but the displacement is otherwise very uniform. Recoveries are significantly over-estimated.

The coarse grid homogeneous simulations give very poor concentration distributions. The coarse grid is less able to resolve gravity segregation, with the consequence that the predicted oil recoveries are even larger than for the fine grid homogeneous cases. The fine grid heterogeneous oil recovery is about 80% at 2.0 pv gas injection, whereas the coarse grid homogeneous case is already completed with close to 100% recovery.

6. SURFACTANT FLOOD EXAMPLES

The same two heterogeneous REV's were used for simulations of horizontal displacement for a surfactant flood process. The

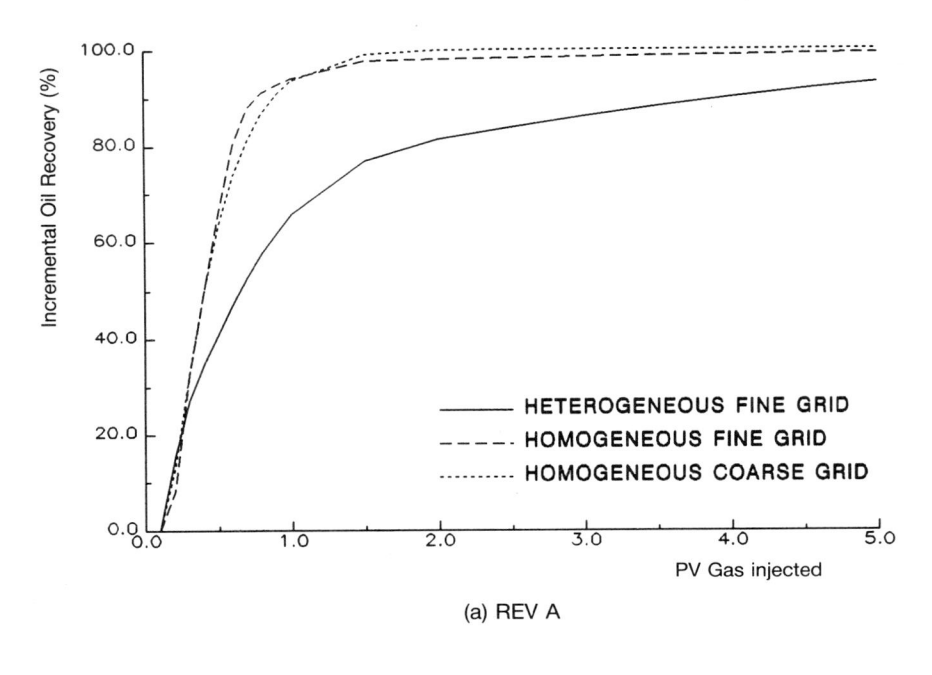

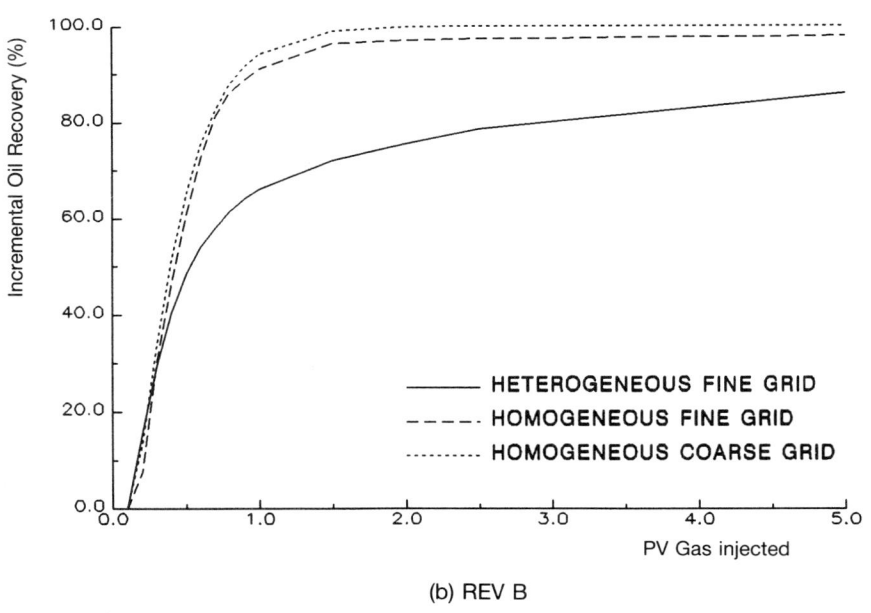

Fig. 10 Tertiary miscible flood recovery

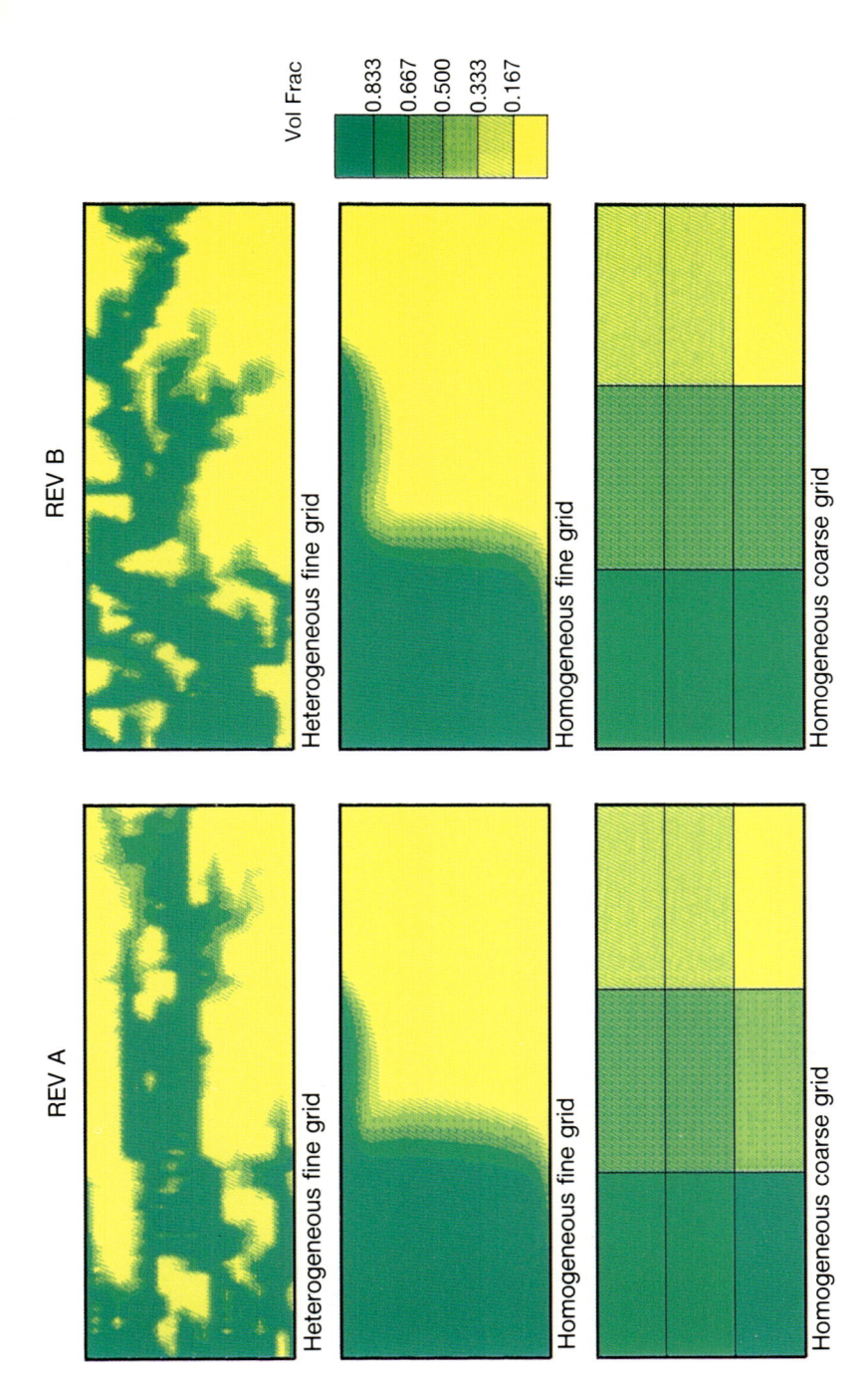

Fig. 11 Injection gas concentrations at 0.2 pv for tertiary miscible floods

BPOPE chemical flood simulator (briefly described in reference 14) was used for the simulations. The initial oil saturation was taken to be the waterflood residual value (0.29). An aqueous solution of surfactant at concentration 15,000 ppm was injected at a rate of 0.365 pv/year. Constant temperature and salinity were assumed throughout, and divalent ion effects neglected. The surfactant is therefore at optimal salinity[15] throughout. A type III phase diagram[15] was used with the middle phase containing 4,000 ppm surfactant, an interfacial tension of 2.4×10^{-3} dyn/cm between the middle and excess phases, and middle phase viscosity 0.75 cP. Oil and water viscosity were 1.2 cP and 0.3 cP respectively. A surfactant adsorption of 0.5 mg/g of rock was assumed, using a Langmuir isotherm reaching 50% of the asymptotic value at 1,000 ppm. High tension (oil/water) relative permeabilities were derived from the formulae of Standing[13], assuming water to be the wetting phase, with pore size distribution factors for the two lithotypes as in Section 4. At low tension (high capillary number), straight line relative permeabilities were used for each of the three phases. The capillary desaturation curve[26] was taken to be a linear function of $\log(N_c)$ between capillary numbers of $10^{-4.5}$ (residual saturations equal to oil/water values) and $10^{-2.5}$ (residual saturations identically zero). Finally, component densities were 998 kg/m³ for water and surfactant, and 752 kg/m³ for oil; ideal mixing was assumed.

Oil and surfactant production profiles for the heterogeneous, fine grid homogeneous, and coarse grid homogeneous runs are shown in Figure 12. Oil and surfactant concentrations at 0.75 pv after the start of surfactant injection are shown (for element B) in Figure 13. For both elements, oil recovery in the fine grid homogeneous case reaches 100% at about 2 pv injected, but in the heterogeneous case recovery is slower and tails off after about 3 pv, reaching only 83% (element A) or 91% (element B) by 5 pv injected. Oil recovery in the coarse grid homogeneous case matches the heterogenous case quite well up to about 2 pv, but recovery thereafter is more rapid, and reaches 100% by 3-4 pv injected. Thus, the numerical dispersion associated with the coarse grid compensates for the effects of heterogeneity on the oil recovery quite well over the first 2 pv throughput, but not thereafter. Additionally, surfactant production in the coarse grid case differs more from the heterogeneous case than it does in the fine grid homogeneous case (Figure 12).

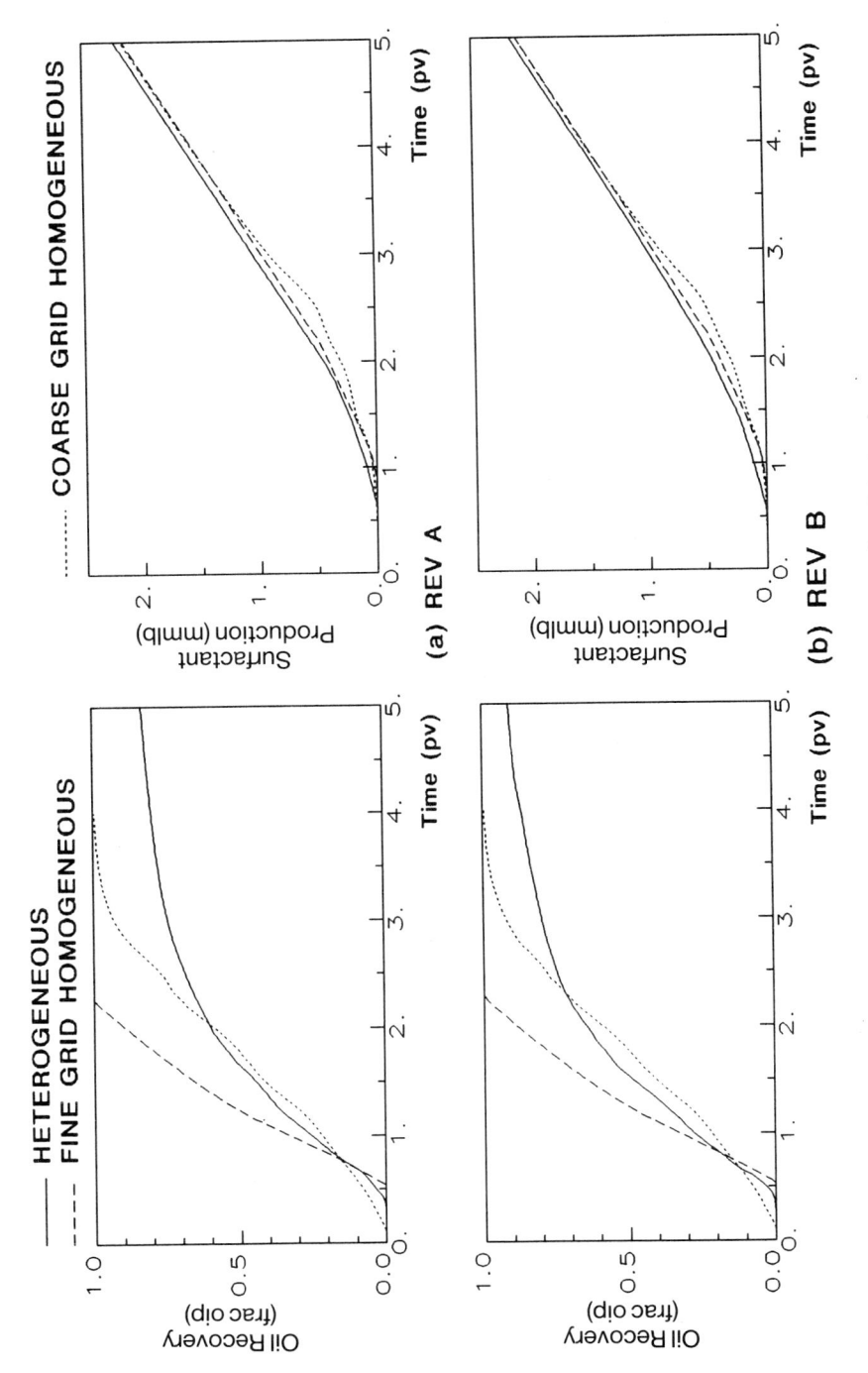

Fig. 12 Surfactant flood production histories

Figure 13 illustrates, for the fine grid homogeneous case, the formation of an oil bank (which moves towards the top of the reservoir under the influence of gravity) ahead of the surfactant front. A region of three-phase flow develops, where the surfactant, in the middle phase, also has a tendency to rise, and this is followed by a solubilisation front[16] behind which the oil concentration is zero. In the heterogeneous case, the same features are present, but they are broken up and intermingled by the heterogeneity. The heterogeneity has introduced a preferential flow path in the upper parts of the reservoir. In the coarse grid case, the various features are smoothed out almost beyond recognition; the reasonable match for the overall oil recovery between this case and the heterogeneous case is thus a result of separate errors cancelling out. Calculations were also run with a polymer added to the surfactant, increasing aqueous and middle phase viscosities by a factor of 8, but this had little effect on the sweep pattern caused by the heterogeneities in either REVs A or B.

7. DEVELOPMENT OF AN APPROXIMATE DUAL ZONE MIXING MODEL

The results of the previous sections have demonstrated that in a variety of applications the assumptions of uniform mixing of fluids on the scale of a typical reservoir simulation grid block can be seriously in error. In this section we develop a model that recognises that only part of the pore volume in a grid block may have been contacted by the injected fluid (or active component therein).

The idea of allowing two zones within each grid block has been used before in capacitance models for tracer flow[17], dual porosity models[18,19], and empirical fingering models[8,9,20]. The dual zone model developed here is an extension of these ideas which has been motivated by a desire to account for the effects of reservoir heterogeneity. It is formulated in a general manner so that small scale variations in fluid composition resulting from fingering, segregation, or the presence of sharp concentration fronts, can also be accounted for.

Dual Zone Mixing Model

(i) The coarse grid block is divided into two areas (figure 3); a fraction λ is contacted by the injected fluid (region c), and a fraction $1-\lambda$ of in-situ fluid is bypassed (region b). This is analogous to the fingering flow approach described in Section 2. λ is assumed to be a function of C_a, the average concentration of an active component (a) within the grid block. For the oil vaporisation problem, the active component would be the lightest hydrocarbon component, while for surfactant injection, it would be the surfactant itself. We determine λ by editing results from the fine grid simulation using some criterion (see Appendix) to discriminate between contacted and bypassed fine grid blocks. If C_a is the average concentration of species 'a' in all cells, then the function $\lambda(C_a)$ can be tabulated.

(ii) The total flow of each phase out of the coarse grid block is obtained by summation over the outlet face of the fine grid. This may be related to the average saturation and to the difference in inlet and outlet potentials ($\Delta\psi = \Delta P + g\bar{\rho}\Delta z$) to define pseudo-relative permeabilities according to the conventional Kyte and Berry rules. The pseudos will be directional. The average compositions C_{gip} and C_{oip} of the phases produced from the outlet face of the fine grid are also obtained, and are used to define compositional correction factors:

$$\alpha_{gi} = \frac{C_{gip}}{\bar{C}_{gi}}, \qquad \alpha_{oi} = \frac{C_{oip}}{\bar{C}_{oi}} \qquad (7.1)$$

which are tabulated against $\bar{C}_{gi}$ and $\bar{C}_{oi}$ respectively. Here $\bar{C}_{gi}$ and $\bar{C}_{oi}$, like $\bar{\rho}$ and the average saturations used to determine the pseudo relative permeabilities, are average values over the fine grid. The functions $\alpha_{gi}(C_{gi})$ are in principle dependent on component number i, direction and phase, but in practice may be less sensitive.

(iii) The overall mass balance equation for component i is now:

$$\nabla \cdot \left[\alpha_{gi} C_{gi} \frac{Kk_{rg}}{\mu_g} \nabla\psi_g + \alpha_{oi} C_{oi} \frac{Kk_{ro}}{\mu_o} \nabla\psi_o \right]$$

$$= \phi \frac{\partial}{\partial t} \left[C_{gi} S_g + C_{oi} (1-S_g) \right] \qquad (7.2)$$

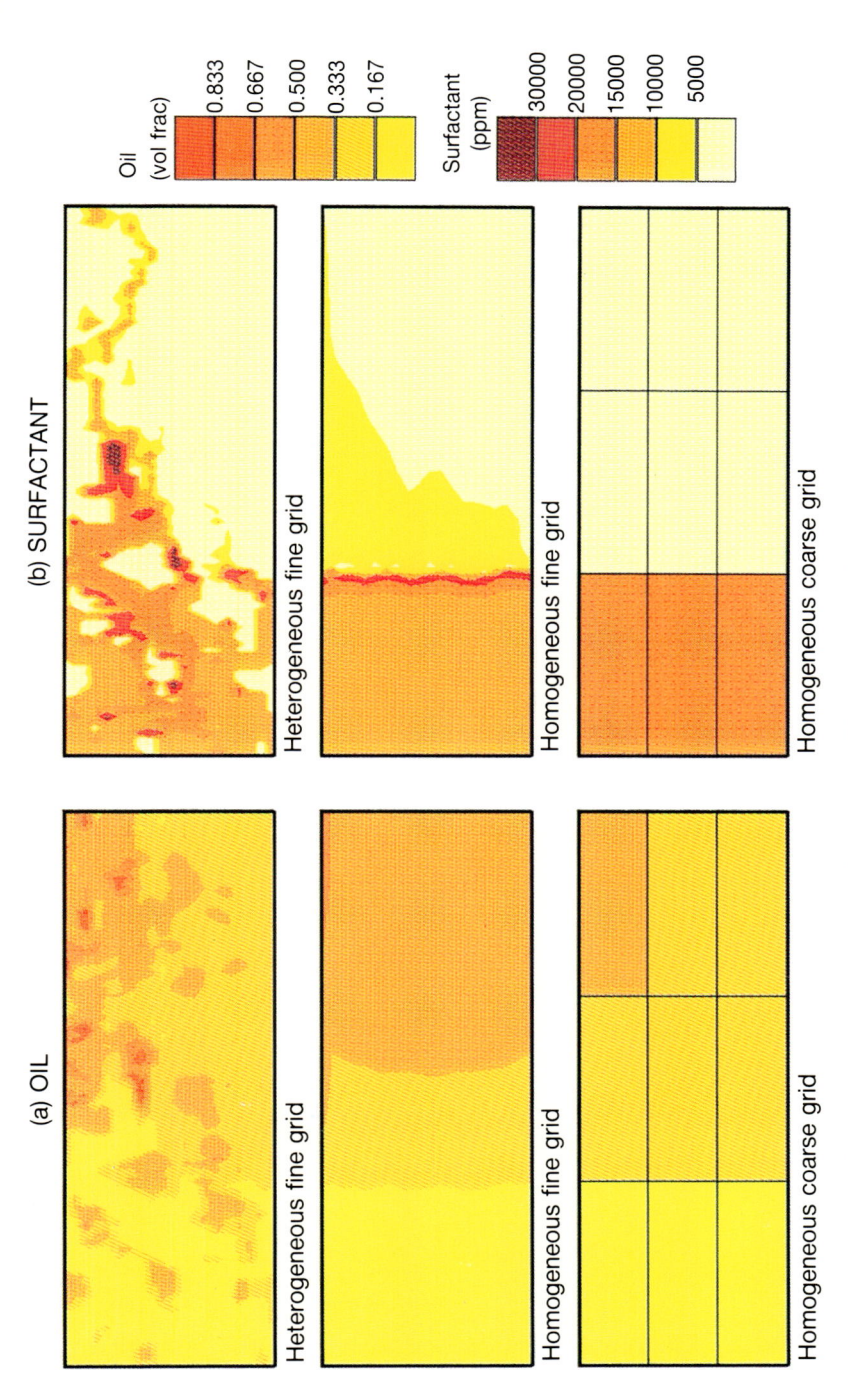

Fig. 13 Oil and surfactant concentrations at 0.75 pv for surfactant displacement in Rev B

for all i. This is the standard balance equation expressed in cell average quantities. The pressures derived from forming a pressure equation from Eq (7.2) define ψ_g and ψ_o, which we take to be common to regions c and b. The solution of the mass balance equations also give the total concentrations C_i for each component. The new value of λ is determined from $\lambda(C_a)$. Components i are divided between c and b according to rules to be specified to give concentrations C_{gic}, C_{oic}, C_{gib}, C_{oib}.

(iv) In some problems (including the oil vaporisation problem) the composition of the bypassed region will be known in advance. In general, however, a separate conservation equation is required to determine the concentrations in region b:

$$\nabla \left[C_{gib} \frac{Kk_{rgb}}{\mu_{gb}} \nabla\psi_g + C_{oib} \frac{Kk_{rob}}{\mu_{ob}} \nabla\psi_o \right] - \tau_i$$
$$= \phi \frac{\partial}{\partial t} \left[(1-\lambda) \{ C_{gib} S_{gb} + C_{oib} (1-S_{gb}) \} \right] \quad (7.3)$$

where the term τ_i represents the rate at which component i is transferring from the bypassed zone to the contacted zone. The form of this term is discussed in the appendix, where an example in which τ_i can be determined analytically is given. The direction of transfer (ie the sign of τ_i) may be different for different components. The effects of dispersion on τ_i, are also discussed in the appendix. Further research may reveal an adequate approximation for τ_i (see for example Ngheim et al[20]), but otherwise it appears that performing material balance on the results of the fine grid simulation must be used to tabulate τ_i as a function of λ. The fine grid results can also be used to determine the relative permeability k_{rgb} and k_{rob}; alternatively, it may be adequate to split the total relative permeability of a phase between the contacted and bypassed regions in proportion to the size of each region and the mobile phase saturation in that region, ie:

$$k_{rgb} = (1-\lambda) \frac{S_{gb} - S_{gc}}{S_g - S_{gc}} k_{rg} \quad (7.4)$$

and similarly for k_{rob}. Since λ is determined in step (iii), Eq. (7.3) can be used to update the total concentrations in the bypassed zone: they are used to determine S_{gb}, C_{gib}, C_{oib}, μ_{gb} and μ_{ob} from a flash calculation at pressure P.

(v) The total concentrations C_{ic} in region c are obtained by using the relation:

$$C_i = \lambda\, C_{ic} + (1-\lambda)\, C_{ib} \qquad (7.5)$$

in which C_i and C_{ib} are already determined in steps (iii) and (iv). A flash calculation now gives S_{gc}, C_{gic}, C_{oic}, μ_{gc} and μ_{oc}. The block average values of C_{gi}, C_{oi}, μ_g and μ_o required for the next time step in solving Eq (7.2) are then computed as weighted averages:

$$C_{gi} = \frac{\lambda\, S_{gc}\, C_{gic} + (1-\lambda)\, S_{gb}\, C_{gib}}{\lambda\, S_{gc} + (1-\lambda)\, S_{gb}} \qquad (7.6)$$

etc, where the denominator defines the average saturation S_g, which is required to compute pseudo relative permeabilities and k_{rgb} in Eq (7.4).

8. APPLICATION OF DUAL ZONE MIXING MODEL TO OIL VAPORISATION

In the oil vaporisation problem, the composition of the fluid in the bypassed zone is known; it does not change significantly from the initial composition because only equilibrium gas can flow in this zone. Thus, step (iv) of the full treatment can be omitted, and the compositions of the contacted zone can be inferred directly from Eq (7.5). The dual zone model is thus simplified in this case, as only one conservation equation is needed for the average concentration of each component, and the transfer term τ_i is not needed. It is not essential to calculate pseudo relative permeabilities for the coarse grid because only single phase flow of gas is involved with a slowly decreasing residual oil saturation.

The conservation equation (7.2) for the average saturations in the coarse grid must still retain the compositional transport correction factors α_{gi}. These factors have been introduced in the transport terms in the BP-COMP model. The pressure equation is obtained by an elimination procedure involving a flash of the overall grid block compositions. Solution of this equation provides the new values of the overall composition, including the methane (C_1) mol fraction, which is the quantity against which λ has been tabulated. The total compositions C_{ic} in the contacted region are then obtained from Eq (7.5), and a second flash calculation defines

C_{gic} and C_{oic}. Since the pressure is already determined, this flash may change the volume of the fluid in the contacted region, but in practice the resulting material balance errors are small. Having now defined the dual zone compositions, it is a comparatively simple matter to calculate the average compositions from Eq. (7.6) to be used in the transport terms in Eq. (7.2) for the next time step.

The values of $\lambda(C_i)$ and $\alpha_{gi}(C_i)$ to be used in this approach have to be derived by editing results of a representative fine grid heterogeneous solution. In an ideal situation the values obtained would be independent of the size of region over which the edits are taken. In practice this is unlikely to be achieved due to variability of the heterogeneity properties, to variations in the degree of gravity override, and to the influence of the type of boundary conditions assumed in the inlet and outlet sections of the fine grid.

We have used the detailed horizontal solutions for REV A obtained in Section 4 to derive the required parametric functions. The REV has been divided into three equal sections, referred to as inlet, middle and outlet. The average methane mol fraction is calculated for each section at appropriate time intervals, and the corresponding λ-values determined by examining the fraction of the section in the fine grid solution occupied by a methane mol fraction above a threshold value. The threshold value has to be chosen as a suitable average discriminator between contacted and bypassed zones. The functions α_{gi} are determined by integrating the mol flow rates of each of the six pseudo-components along the boundary at the outlet face of each section. The mol fraction of methane for the REV A detailed solution varies between 0.72 and 0.85. In Figure 14, we show values of λ for the three sections, generated using a threshold mol fraction of 0.76. The behaviour derived from the three sections is broadly similar with λ asymptotically increasing towards 1.0 at high throughputs of dry gas. We also show, in Figure 15, values of λ generated from the middle section using threshold methane mol fractions of 0.74, 0.76, and 0.78. The values of α_{gi} for the C_1 and C_2 pseudo-components are close to unity, but the C_3 to C_6 components show considerable variability between the three sections as illustrated in Figure 16.

We might expect that the middle section should be the most representative, being more remote from the inlet and outlet boundaries. This is confirmed in the plots of oil recoveries

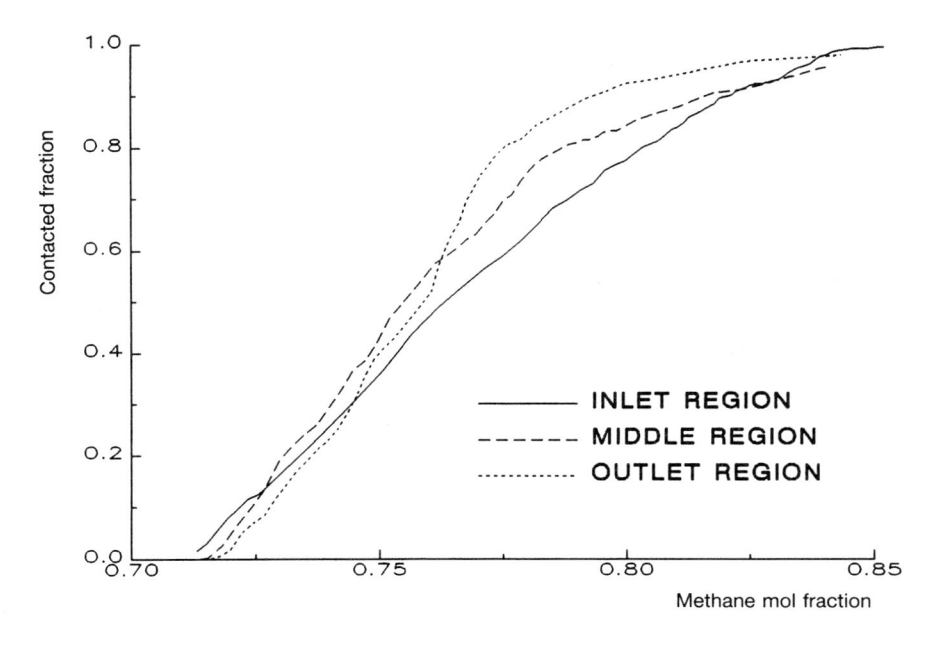

Fig. 14 Contacted fraction for different regions

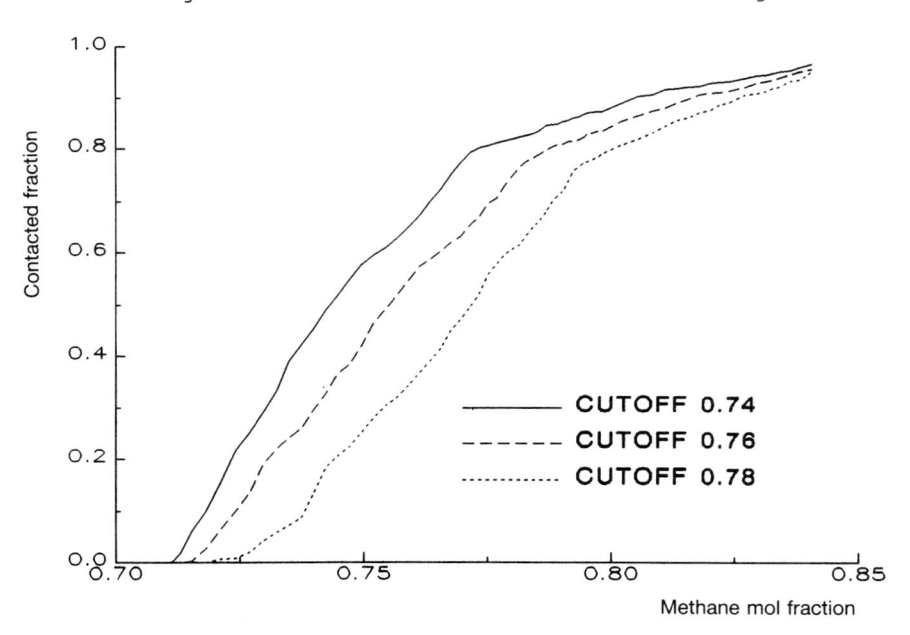

Fig. 15 Middle region contacted fraction for different cutoffs

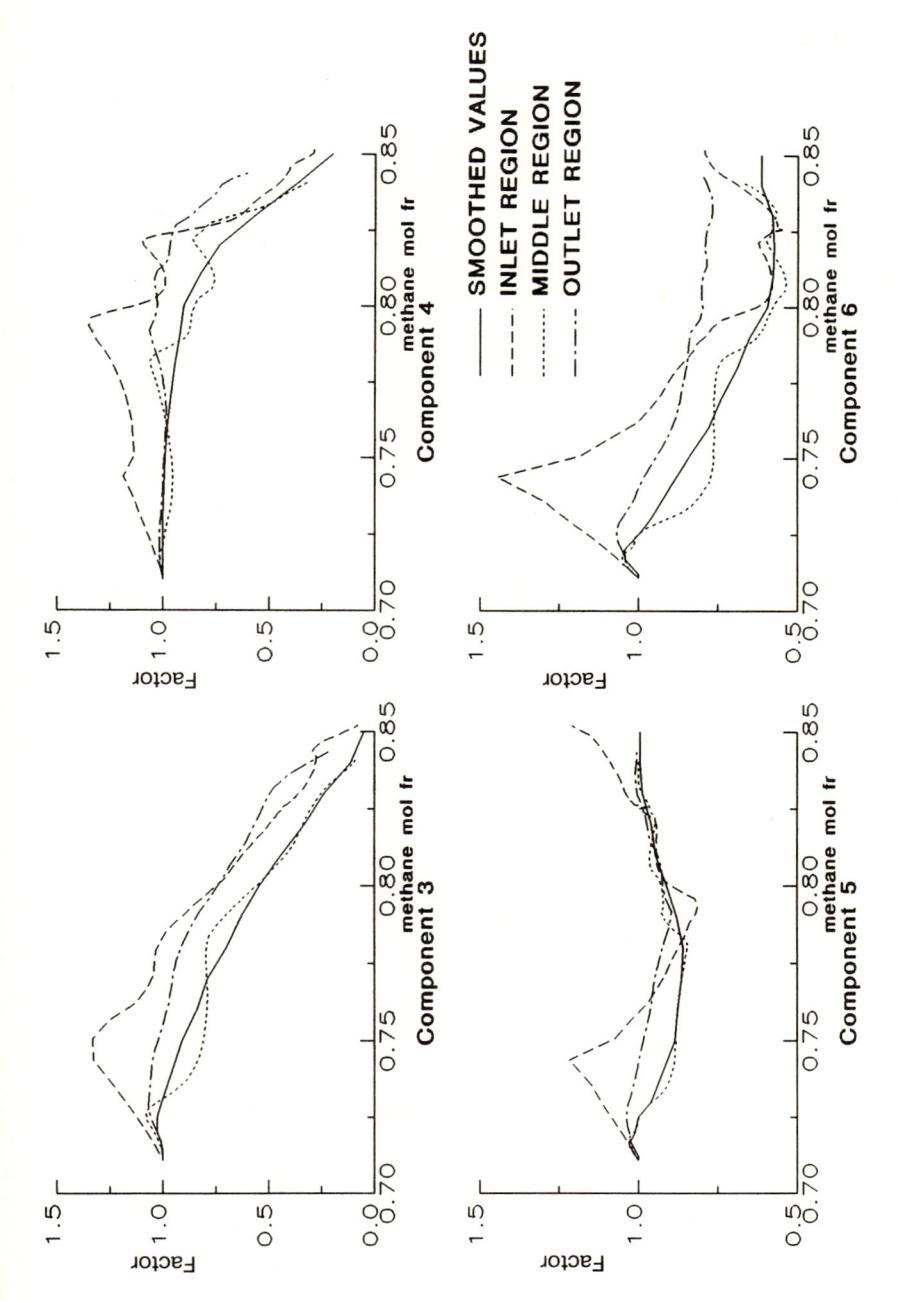

Fig. 16 Compositional correction factors for different regions

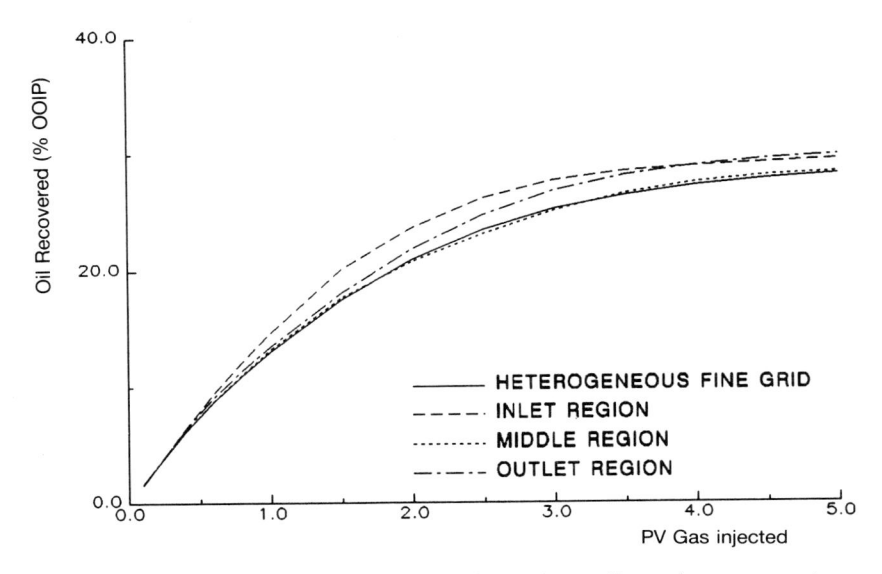

Fig. 17 Oil recovery as a function of region parameters

0.157	0.191	0.219
0.168	0.218	0.229
0.161	0.197	0.223
0.167	0.182	0.206
0.164	0.202	0.227
0.168	0.215	0.228

UPPER VALUES: DZM OIL SATURATIONS
LOWER VALUES: AVERAGE FROM HETEROGENEOUS FINE GRID

Fig. 18 Oil saturations from DZM

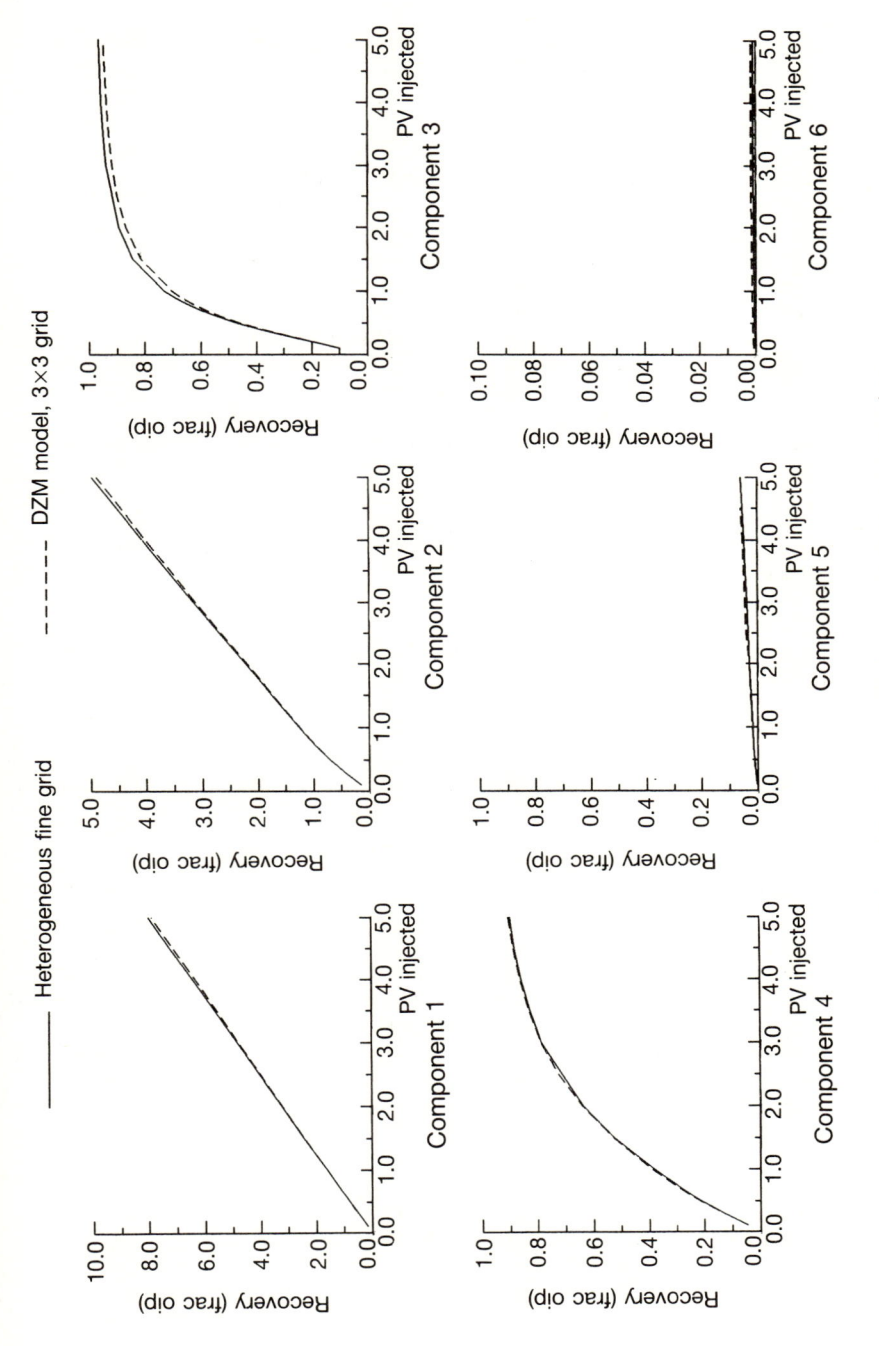

Fig. 19 Component by component recoveries

obtained from coarse grid (3 x 3) simulations using the DZM model with values of λ and α_{gi} derived from each of the three sections, which are shown in Figure 17. Use of the middle section values was also preferable to use of values derived from the whole REV. In practice it was not necessary to use the erratic variations of the original middle section values, and use of the smoothed values shown in Figure 16 was satisfactory. Figure 18 shows the average oil saturations in each block of the 3 x 3 DZM calculations after 1.0 pore volume of dry gas injection. The heterogeneous fine grid values shown in the figure were obtained by integration of the solution in each coarse grid block area. We also compare the total mol production of each pseudo-component from REV A for the two styles of calculation in Figure 19. Recoveries are expressed as a fraction of the number of mols of each component initially in place. Thus as components 1 and 2 are present in the injection gas, recoveries for these components are greater than one. The component-by-component recoveries obtained using the DZM model are in excellent agreement with the detailed simulation results for each component in the fluid model.

The α's and λ have also been determined for REV B, and recoveries predicted using DZM (results for this are not shown). Very similar behaviour is observed; λ's and α's calculated from the central section away from any boundary effects give rise to a corrected oil recovery for the coarse grid which is in good agreement with the heterogeneous fine grid simulations.

9. CONCLUDING REMARKS

(i) We have used the techniques of sequential indicator simulation (SIS) to generate two hypothetical cross section elements (REV's A and B) with statistically distributed heterogeneous properties on a 100 x 20 grid. These finely gridded representations have then been used to test how the consequent non-uniformities in flow and phase behaviour will influence recovery of residual oil.

(ii) The primary EOR process which has been studied is vaporisation by lean gas. For the horizontal displacements the recovery of residual oil is reduced by 10-15% by the presence of the assumed heterogeneities, when compared with a finely gridded calculation for a homogeneous problem. The latter is influenced by gravity segregation effects which causes some cancellation of errors in terms of oil recovery. The coarsely gridded homogeneous problem gives numerical dispersion errors which again cause cancelling error effects, but these are different due to the reduced gravity override.

(iii) Similar behaviour has been demonstrated for miscible gas displacement and surfactant displacement in REV's A and B. For the surfactant case the disposition of a middle phase dominates the recovery mechanism. For the worst miscible gas case omission of heterogeneities caused oil recovery to be overestimated by 20% after 2 pore volumes of gas injection.

(iv) A Dual Zone Mixing method (DZM) formalism has been developed in which each grid block is divided into two zones, whose extent varies with time, and for which separate phase behaviour calculations are performed. One zone represents the contacted region in which the invading fluid mixes with part of the in-situ fluid, while the dual zone represents bypassed fluid whose phase behaviour is essentially unmodified. It is shown that the transfer term between the two zones may be important and further work is required to establish whether the moving boundary between the zones can be chosen in a way which minimises transfer, and perhaps would allow the transfer to be approximated conveniently.

(v) The DZM method simplifies in the case of evaporation of residual oil in a lean gas cycling process. We have implemented this simplified method in the BP-COMP code and shown how to introduce appropriate correction factors into the equations of motion. This allows a problem computed on a 3 x 3 homogeneous grid to approximate satisfactorily previous results obtained from a 100 x 20 heterogeneous grid. Some sensitivities are caused by the need to choose the most appropriate region of the finely gridded problem as the basis for evaluating the correction factors. The transport correction factors α_{gi} vary significantly for the heavier pseudo-components and control the correction process. The dual zone boundary choice which defines λ has been found to be less sensitive. Bearing in mind that it is changes in convective flow which in turn shift average phase behaviour, it is not surprising that the α parameters will be important.

(vi) Heterogeneities occur at many length scales in petroleum reservoirs. We have assumed here that a Representative Element of Volume (REV) can be chosen for averaging in the DZM approximation. The scale length at which diffusive mixing becomes significant is not clearly defined (probably less than 1 metre), so a judgement is also involved in the level of detail required at the fine scale.

(vii) The heterogeneities used in this illustrative study are fairly severe, but not outside the range which may occur in some reservoirs. Careful study of actual reservoir properties would be needed to quantify these effects in a real case.

(viii) Further research is needed on the application of these techniques to three-dimensional grids of the size normally used for large compositional reservoir studies.

ACKNOWLEDGEMENTS

Thanks are due to Dr. S.H. Begg at BP Research for his contributions in preparing geological input for REV's A & B. Permission to publish this paper has been given by the British Petroleum Co plc.

NOMENCLATURE

C_i	overall concentration of component i
C_{gi}, C_{oi}	concentration of component i in gas and oil phases
g	acceleration due to gravity
K_i	partition coefficient for component i
K	absolute permeability
k_{rg}, k_{ro}	relative permeability of gas and oil phases
K_h, K_v	horizontal and vertical permeability
N_c	capillary number
P	pressure
q_i	rate of injection of component i
REV	representative element of volume
S_g	gas saturation
z	depth below datum
μ_g, μ_o	viscosity of gas and oil phases respectively
v	pore size distribution index [13]
ρ_g, ρ_o	density of gas and oil phases respectively
ϕ	porosity
ψ_g, ψ_o	potential of gas and oil phases respectively
Δx, Δz	grid block size.
b	bypassed zone
c	contacted zone
f	fingered zone

REFERENCES

(1) Begg, S.H. and King, P.R., (February 1985) "Modelling the effects of shales on reservoir performance: calculation of effective vertical permeability", SPE 13529, 8th SPE Symp. on Reservoir Simulation, Dallas.

(2) Haldorsen, H.H., Chang, D.M. and Begg, S.H., (1987) "Discontinuous vertical permeability barriers: a challenge to engineers and geologists", in North Sea Oil and Gas Reservoirs, Graham and Trotman.

(3) Lasseter, T.J., et al., (1986) "Reservoir Heterogeneities and their Influence on Ultimate Recovery", NIPER Reservoir Characterisation Tech. Conf.

(4) Kossack, C.A., et al., (February 1989) "Scaling Up Laboratory Relative Permeabilities and Rock Heterogeneities with Pseudo Functions for Field Simulations", SPE 18436, presented at SPE Reservoir Simulation Symposium, Houston.

(5) Muggeridge, A.H., (June 1989) "Generation of Pseudo Relative Permeabilities from Detailed Simulation of Flow in Heterogeneous Porous Media", presented at NIPER/DOE 2nd Int. Reservoir Characterisation Tech. Conf., Dallas.

(6) Kyte, J.R. and Berry, D.W., (August 1975) "New Pseudo Functions to Control Numerical Dispersion", SPE Journal.

(7) Camy, J.P. and Emanuel, A.S., (October 1977) "Effect of Grid Size in the Compositional Simulation of CO_2 Injection", SPE 6894, presented at 52nd Ann Fall Tech. Conf. and Exhib., Denver.

(8) Fayers, F.J., (May 1988) "An Approximate Model with Physically Interpretable Parameters for Representing Miscible Viscous Fingering", SPE Res. Eng.

(9) Fayers, F.J. and Newley, T.M.J., (May 1988) "Detailed Validation of an Empirical Model for Viscous Fingering with Gravity Effects", SPE Res. Eng.

(10) Journel, A. and Alabert, F.G., (October 1988) "Focussing on Spatial Connectivity of Extreme-Valued Attributes: Stochastic Indicator Models of Reservoir Heterogeneities", SPE 18324, presented at SPE Ann. Tech. Conf. & Exhib., Houston.

(11) Haldorsen, H.H., (1986) "Simulator Parameter Assignment and the Problems of Scale in Reservoir Engineering", in Reservoir Characterisation, L.W. Lake and H.B. Carroll, Jr., Eds., Academic Press.

(12) Begg, S.H., Carter, R.R. and Dranfield, P., (September 1987) "Assigning Effective Values to Simulator Grid Blocks in Heterogeneous Reservoirs", SPE 16754, presented at 62nd SPE Ann Fall. Conf. & Exhib., Dallas.

(13) Standing, M.B., (August 1974) "Notes on Relative Permeability Relationships", University of Trondheim.

(14) Barker, J.W. and Fayers, F.J., (December 1988) "Factors Influencing Successful Numerical Simulation of Surfactant Displacement in North Sea Fields", In Situ. Vol. 12, No. 4.

(15) Nelson, R.C. and Pope, G.A., (October 1978) "Phase Relationships in Chemical Flooding", SPE Journal.

(16) Pope, G.A., (June 1980) "The Application of Fractional Flow Theory to Enhanced Oil Recovery", SPE Journal.

(17) Coats, K.H. and Smith, B.D., (March 1964) "Dead End Pore Volume and Dispersion in Porous Media", SPE Journal.

(18) Warren, J.E. and Root, P.J., (September 1963) "The Behaviour of Naturally Fractured Reservoirs", SPE Journal.

(19) Kazemi, H., et al., (December 1976) "Numerical Simulation of Water-Oil Flow in Naturally Fractured Reservoirs", SPE Journal.

(20) Ngheim, L., et al., (February 1989) "A Method for Modelling Incomplete Mixing in Compositional Simulation of Unstable Displacements", SPE 18439, presented at SPE Reservoir Simulation Symp., Houston.

APPENDIX

DETERMINATION OF THE TRANSFER TERM τ BY THE METHOD OF CHARACTERISTICS FOR A LINEAR DISPLACEMENT PROBLEM

The oil vaporisation problem is dominated by the vaporisation of the intermediate components to leave a residual non-vaporisable heavy component. We therefore consider a three component model where component 1 (gas) exists only in the gas phase, component 2 (light oil) partitions between the gas and oil phases with fixed partition coefficient $K = C_{g2}/C_{o2}$, and component 3 (heavy oil) exists only in the oil phase. Assume the reservoir is linear and homogeneous with length L, area A, and porosity ϕ. Assume there is no water present, and neglect rock and fluid compressibilities. Assume that initially the reservoir contains fluid of uniform composition, with an immobile oil phase containing both light and heavy oil in equilibrium with a mobile gas phase containing gas and light oil. Fluid comprising only the gas component is injected at rate Q. Since the heavy oil exists only in the oil phase, which is immobile, the heavy oil concentration C_3 is constant throughout. The system can then be described by a single conservation law for the light oil concentration, C_2:

$$\frac{\partial C_2}{\partial t} + V_g \frac{\partial C_{g2}}{\partial x} = 0 \qquad (A.1)$$

where $V_g = Q/A\phi$, and by the constitutive relations:

$$C_{g2} = K C_{o2}, \qquad C_{g2} S_g + C_{o2}(1-S_g) = C_2 \qquad (A.2)$$

$$(1-C_{g2}) S_g = C_1, \qquad C_1 + C_2 + C_3 = 1$$

Equations (A.2) provide a relationship between C_{g2} and C_2, so that C_2 can be treated as the primary unknown.

The characteristic velocity associated with equation (A1) is:

$$v_c = V_g \frac{dC_{g2}}{dC_2} \qquad (A.3)$$

where from the constitutive relations (A.2):

$$\frac{dC_{g2}}{dC_2} = \frac{1}{2}\left(\frac{C_2 + b}{\{(C_2 + b)^2 + d\}^{1/2}}\right) \quad (A.4)$$

where $b = C_3 - K(1 + C_3)$, and $d = 4K C_3 (1-K)$

Hence:

$$\frac{d^2 C_{g2}}{dC_2^2} = \frac{2K C_3 (K-1)}{\{(C_2 + b)^2 + d\}^{3/2}} \quad (A.5)$$

so there are two cases to consider, according to whether K is less than or greater than one.

For $K < 1$, the characteristic velocity increases as the light oil concentration decreases. Thus the solution comprises a shock, moving with speed:

$$V_s = V_g \frac{C_{g2}^{(i)}}{C_2^{(i)}} \quad (A.6)$$

separating two regions of constant composition, namely the original composition ahead of the shock, and injected gas/residual heavy oil behind it (Figure A1). Note that the region ahead of the shock contains some injected gas: the injected gas front is a contact discontinuity propagating with speed $V_g/S_g^{(i)}$.

In applying the dual zone mixing model to this case, the contacted and bypassed zones are the regions behind and ahead of the shock respectively. The contacted zone is characterised by an increase in the overall concentration of the gas component (not by the presence of injected gas). Since component i can enter the contacted region only by injection or by transfer from the bypassed region, the transfer term r_i in equation (7.3) is the difference between the rate of change of the total amount of component i in the contacted region, and the rate of injection of component i into that region:

$$\tau_1 = (1-C_3) - \frac{V_g}{V_s} \frac{d\lambda}{dt}, \quad \tau_2 = 0, \quad \tau_3 = C_3 \frac{d\lambda}{dt} \qquad (A.7)$$

Note that the rate of transfer for the gas component (τ_1) is negative: injected gas is transferred from the contacted into the bypassed region.

For K > 1, the characteristic velocity decreases as the light oil concentration decreases, so there is a smooth transition between the initial and final (residual oil) states (Figure A1). The injected gas front lies ahead of the transition zone (as for the previous case).

Application of the dual zone mixing model to this case is very similar to the previous case, though there is some flexibility in the choice of the two zones. If the front of the transition zone is chosen as the boundary between the contacted and bypassed (non-contacted) zones, the contacted zone is again characterised by an increase in the overall gas component concentration. The presence of the smooth transition zone means that the overall composition is variable within the contacted zone, and performing a single flash calculation for this zone will introduce some errors. Some advantage might be gained by choosing some intermediate point in the transition zone as the boundary between the two zones: in this case, the overall gas concentration exceeding some threshold value can be used to characterise the contacted zone.

The transfer terms τ_i can be calculated for this case. Their exact form will depend on the definition used for the contacted zone, but will be qualitatively similar to the previous case. There will be a small amount of light oil transfer into the contacted zone, but the gas transfer will again be from the contacted to the bypassed zone.

In an actual oil vaporisation problem, most oil components will have K < 1, and one might expect a concentration shock to form. However, each component has a different value for K, implying a different velocity for the concentration shock. The solution would thus consist of a series of small shocks, and, in practice, the presence of dispersion will lead to a diffuse, continuously varying, profile.

The above example shows how the transfer term can be obtained from the exact solution to a particular linear displacement problem. In a heterogeneous reservoir, the contacted fraction λ increases as a result of diffusive, dispersive, gravitational and capillary pressure effects as well as by invasion of pathways of differing permeability. These other effects increase λ by means of an exchange of fluid between neighbouring contacted and non-contacted regions of the reservoir resulting in the entire original contents of the newly contacted region being transferred from the bypassed to the contacted zone. Note, however, that the increase in λ has occurred with no change in the overall fluid composition, merely with a redistribution of fluid within the REV. Thus, for processes involving only diffusive, dispersive, gravitational and capillary pressure effects, it is not possible to correlate λ with overall composition. Actual displacements will involve a combination of invasive and other processes. Proper accounting for the dependence of λ on overall composition and for the balance between invasive and other contributions to the transfer term can be achieved by means of fine grid simulations. The balance between invasive and other effects will depend on rate; if this dependence is to be neglected, the fine grid simulations must be performed at a representative displacement rate.

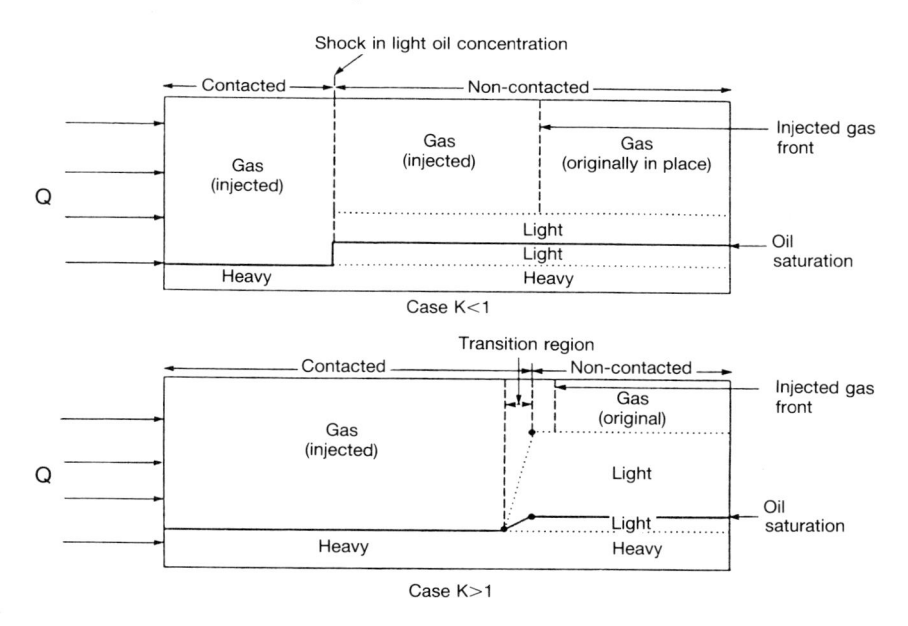

Fig. A1 Analytic solutions to three component oil oil vaporisation problem

ON THE STRICT HYPERBOLICITY OF THE BUCKLEY-LEVERETT EQUATIONS FOR THREE-PHASE FLOW

LARS HOLDEN †

Abstract. It is proved that the standard assumptions on the Buckley-Leverett equations for three-phase flow imply that the equation system is not strictly hyperbolic. Therefore, the solution of the Buckley-Leverett equations for three-phase flow is very complicated. We also discuss four different models for the relative permeability. It is stated that Stone's model almost always gives (an) elliptic region(s). Furthermore, it is proved that Marchesin's model is hyperbolic under very weak assumptions. The triangular model is hyperbolic and the solution is well-defined and depends L_1-continuously upon the initial values in the Riemann problem.

1. Introduction. The Buckley-Leverett equations describe the flow of three phases in a porous medium neglecting capillary effects. If the system is not strictly hyperbolic, the solution is much more complicated. If the system is not hyperbolic, then the solution is not stable. Therefore it is important to know whether the system is strictly hyperbolic, hyperbolic or not hyperbolic.

In the last years there have been several papers on the strict hyperbolicity of the Buckley-Leverett equations. Bell, Trangenstein and Shubin [1] showed that Stone's model, which is the most used model for three-phase relative permeabilities, may give an elliptic region. Shearer [9] proved that if two interaction conditions between the relative permeabilities are satisfied, then strict hyperbolicity fails. Shearer and Trangenstein [10] propose some other interaction conditions which also imply that strict hyperbolicity fails. They also discuss several alternatives

---

†Norwegian Computing Center, P.O.Box 114 Blindern, 0314 Oslo 3, Norway

to Stone's model. Fayers [2] examines the elliptic region(s) in Stone's model for different choices of the residual oil parameter. The properties of the solution near an elliptic region are not known. See [5] and [8].

In the following section the Buckley-Leverett equations are defined and the standard assumptions are listed. In the third section we will make an additional assumption called Stone's assumption. It is stated that with this assumption the Buckley-Leverett equations fail to be strictly hyperbolic. In the final section four different models are discussed.

2. The Model. Let u, v and $w = 1 - u - v$ be the water, gas and oil saturations. Let $f(u,v)$, $g(u,v)$ and $h(u,v)$ be the relative permeabilities divided by the viscosity for water, gas and oil respectively. Then the Buckley-Leverett equations for three-phase flow are

$$u_t + \left(\frac{f(u,v)}{f(u,v)+g(u,v)+h(u,v)}\right)_x = 0$$
$$v_t + \left(\frac{g(u,v)}{f(u,v)+g(u,v)+h(u,v)}\right)_x = 0. \tag{1}$$

The equations are defined in

$$\Omega = \{(u,v); 0 < u, v < u + v < 1\}. \tag{2}$$

We will assume that the functions f, g and h satisfy

$$\begin{aligned} f(0,v) &= 0, \\ g(u,0) &= 0, \\ h(u, 1-u) &= 0, \end{aligned} \tag{3}$$

$$\begin{aligned} f_u(0,v) &= 0, \\ g_v(u,0) &= 0, \\ -h_u(u, 1-u) - h_v(u, 1-u) &= 0, \end{aligned} \tag{4}$$

$$\begin{aligned} &f_u(u,v) > 0 \quad \text{and} \quad f_u(u,v) - f_v(u,v) > 0 \quad \text{for } (u,v) \in \Omega, \\ &g_v(u,v) > 0 \quad \text{and} \quad g_v(u,v) - g_u(u,v) > 0 \quad \text{for } (u,v) \in \Omega, \\ &h_u(u,v) < 0 \quad \text{and} \quad h_v(u,v) < 0 \quad \text{for } (u,v) \in \Omega, \end{aligned} \tag{5}$$

and

$$f_{uu}(u,v) > 0 \quad \text{and} \quad f_{uu}(u,v) - 2f_{uv}(u,v) + f_{vv}(u,v) > 0$$
$$\text{for } (u,v) \in \Omega,$$
$$g_{vv}(u,v) > 0 \quad \text{and} \quad g_{vv}(u,v) - 2g_{vu}(u,v) + g_{uu}(u,v) > 0$$
$$\text{for } (u,v) \in \Omega,$$
$$h_{uu}(u,v) < 0 \quad \text{and} \quad h_{vv}(u,v) < 0$$
$$\text{for } (u,v) \in \Omega. \tag{6}$$

(3) implies that the phase is not moving if it is not present. (4) states that the speed of the phase vanishes when the saturation vanishes. (5) and (6) state that the rate and the speed increase when the saturation increases and one of the two other phases has constant saturation. These assumptions are widely accepted. It is possible to give physical arguments for these properties. With an additional assumption in the corners of Ω, it is possible to prove that the system is not strictly hyperbolic, see [7].

3. Stone's assumption. It is also usual to assume that the relative permeability of water and gas only depends on the water and gas saturation, namely

$$f(u) = f(u,v) \quad \text{and} \quad g(v) = g(u,v) \tag{7}$$

This assumption is called Stone's assumption, see [11]. It uses the fact that water is usually wetting both in contact with oil and gas, and gas is usually not wetting neither in contact with water nor oil.

Some experiments indicate that the isoperms of the relative permeability of oil are concave, see [11]. See figure 1 for an illustration of concave isoperms. Then we may state the following theorem.

THEOREM 1. *Assume (3), (4), (5), (6) and (7) are satisfied. Then (1) is not strictly hyperbolic in Ω. If the three curves $f' + h_u = 0$, $g' + h_v = 0$ and $f' + g' = 0$ intersect at a point P in*

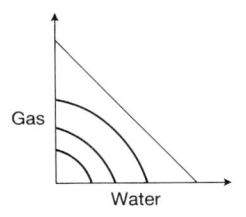

Figure. 1. Concave isoperms of the relative permeability of oil.

Ω, then (1) is not strictly hyperbolic at P. If the three curves intersect and the isoperms of h are concave, then the system is strictly hyperbolic except at P. If the three curves do not intersect in Ω, then there is at least one elliptic region in Ω.

This theorem is proved in [7]. See figure 2 for an illustration.

Since the gas viscosity is much smaller than the viscosity of oil and water, the intersection point between $g' + h_v = 0$ and $f' + g = 0$ has very small gas saturation. Therefore, the elliptic region which is caused by the non-intersection of the three curves, has very small gas saturation. The elliptic region reported in [1] is another elliptic region. In their example there is also another elliptic region which is much smaller with smaller gas saturation.

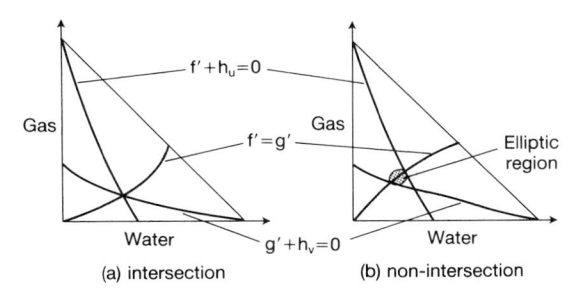

Figure. 2. The three curves $f' + h_u = 0$, $f' + g = 0$ and $g' h_v = 0$.

4. Four different models.
Since only two-phase relative permeabilities are measured, the formulas for three-phase relative permeabilities are usually only interpolation formulas in the three-phase region. In the following section we will discuss four different models for the relative permeabilities.

4.1 Stone's model.

This is the most commonly used expression for the three-phase relative permeabilities. See [11]. It assumes that the relative permeabilities satisfy (7) and in addition that

$$h(u,v) = \frac{(1-u-v)\, h(u,0)\, h(0,v)}{(1-u)(1-v)}. \tag{8}$$

$f(u)$, $g(v)$, $h(u,0)$ and $h(0,v)$ are usually found from experiments. With reasonable assumptions on the two-phase relative permeabilities, Stone's model satisfies assumptions (3), (5), (6). (4) is satisfied for the water and gas relative permeabilities and only in the corners for the relative permeability of oil. Therefore, it is not possible to use Theorem 1 directly. If we make the following minor modification on the model

$$h(u,v) = \frac{(1-u-v)^{1+\epsilon}\, h(u,0)\, h(0,v)}{(1-u)^{1+\epsilon}(1-v)^{1+\epsilon}} \tag{9}$$

for $\epsilon > 0$, then (4) is also satisfied for the relative permeability of oil. Then we may use Theorem 1 and state that the modified Stone's model is not strictly hyperbolic. Since the modified Stone's model is almost identical to Stone's model, we may state that except for very special two-phase data, Stone's model (8) is not strictly hyperbolic.

4.2 Marchesin's model.

In Marchesin's model [9] it is assumed that the relative permeability for each phase only depends on its own saturation, i.e.

$$h(1-u-v) = h(u,v) \tag{10}$$

in addition to (7). Therefore, we may use Theorem 1 making the additional assumptions (3), (4), (5) and (6). It is trivial to see that the three curves in Theorem 1 intersect at a common point. We will instead prove a theorem with much weaker assumptions on the relative permeabilities. Let us first define the following

$$f_D = \{y \in R; f'(u) = y \text{ for } 0 < u < 1\},$$
$$g_D = \{y \in R; g'(u) = y \text{ for } 0 < u < 1\},$$
$$h_D = \{y \in R; h'(u) = y \text{ for } 0 < u < 1\},$$
$$H = f_D \cap g_D \cap h_D \quad \text{and}$$
$$K = \{(u,v) \in \Omega; f'(u) \in H, \ g'(u) \in H, \ \text{and} \ h'(u) \in H\}.$$

We may then formulate the following theorem.

THEOREM 2. *Assume that (7) and (9) are satisfied. Then the system (1) is hyperbolic. If K is empty, then (1) is strictly hyperbolic. If K is not empty and (5) is satisfied, then (1) is strictly monotone except at a unique point P.*

Proof. It is straightforward to prove that (1) with the assumption (7) and (9) is strictly hyperbolic except if, and only if,
$$A = (B - C + E)^2 + 4BC > 0$$
where
$$B = g(f' + h_u), \ C = f(g' + h_v) \ \text{and} \ E = h(f' - g').$$

The system is hyperbolic if, and only if, $A \geq 0$. If $BC > 0$, then obviously $A > 0$.

Assume $B \geq 0$ and $C \leq 0$. Since $f > 0$, $g > 0$ and $h_u = h_v$, $E > 0$. This implies that
$$A = (B - C + E)^2 + 4BC >$$
$$(B - C)^2 + 4BC = (B + C)^2.$$

The argument is similar for $B \leq 0$ and $C \geq 0$. Thus we have proved that A is non-negative and vanishes if, and only if $f' = g' = h_u = h_v$. If K is empty, then obviously (1) is strict hyperbolic. If K is not empty and (5) is satisfied, it is easy to prove that (1) is strictly hyperbolic except at a unique point. □

4.3 The triangular model.

The fractional flow function of gas is

$$G(u,v) = \frac{g(u,v)}{f(u,v) + g(u,v) + h(u,v)}.$$

In [6] it is proposed to let the fractional flow depend only on the gas saturation i.e.

$$G(v) = G(u,v). \qquad (11)$$

Since the gas viscosity is much smaller than the oil and water viscosity, this is a good approximation. This results in a model

$$u_{i_t} + f_i(u_1,...,u_i)_x = 0 \quad \text{for} \quad i = 1,...,n. \qquad (12)$$

This model was studied in [6], [3] and [4] with Riemann initial data i.e.

$$u_i(x,0) = \begin{cases} u_{i_-} & \text{for } x < 0 \\ u_{i_+} & \text{for } x > 0 \end{cases} \quad \text{for} \quad i = 1,...,n. \qquad (13)$$

The following theorem was proved in [6].

THEOREM 3. *Assume that f_i is continuous, piecewise smooth and that $\frac{\partial f_i}{\partial u_i}$ increases or decreases faster than linear when u_i increases to ∞ or decreases to $-\infty$ for i=1,...,n. Then there is a solution of (12) and (13). The solution is unique except for $f \in M$ and for each f, $(u_{1_-},...,u_{n_-},u_{1_+},...,u_{n_+}) \in M_f$. M has measure zero in the supremum norm and M_f has Lebesgue measure zero for all f not in M. There is always uniqueness for $n < 3$.*

For the Buckley-Leverett equations we may state the following theorem.

THEOREM 4. *Assume (10) is satisfied. Then (1) is strictly hyperbolic except at a curve from one corner to another corner where the system is hyperbolic. The solution of the Riemann problem exists uniquely, is well-defined in Ω and depends L_1-continuously on the inital data.*

The first part of this theorem is obvious. Gimse proves the second part of the theorem in [3] and [4].

4.4 The hyperbolic model. It is possible to find a model which satisfies (3), (4), (5), (6), (7) and in addition ensures that the three curves in Theorem 1 intersect. In [7] is given an example of such a model which always is hyperbolic. The disadvantage with such models is that they become technically difficult in order to ensure that the three curves in Theorem 1 intersect.

Acknowledgement. The author wishes to thank Helge Holden for valuable discussions.

References

[1] Bell, J. B., Trangenstein, J. A. and Shubin, G. R., Conservation laws of mixed type describing three-phase flow in porous media, Siam J. of Appl. Math., 46 (1986), pp. 1000-1117.

[2] Fayers, F. J., Extensions of Stone's Method I and the Condition for real characteristics in three-phase flow, Presented at SPE conference in Dallas, September 1987.

[3] Gimse, T., A numerical method for a system of equations modelling one- dimensional three-phase flow in a porous medium, in Nonlinear hyperbolic equations - Theory, Computation Methods, and Applications, J. Ballmann and R. Jeltsch (Eds.), Vieweg, Braunschweig, Germany, 1989.

[4] Gimse, T., Thesis, University of Oslo, Norway, 1989.

[5] Holden, H., Holden, L. and Risebro, N. H., Some qualitative properties of 2x2 systems of conservation laws of mixed type, This proceeding.

[6] Holden, L. and Høegh-Krohn., A class of n nonlinear hyperbolic conservation laws, J. Diff. Eq. (to appear)

[7] Holden, L., On the strict hyperbolicity of the Buckley-Leverett Equations for Three-Phase Flow in a Porous Medium, Preprint, Norwegian Computing Center, 1988.

[8] Pego R. L. and Serre D., Instabilities in Glimm's scheme for two systems of mixed type, Siam J. Num. Anal., 25 (1988), pp. 965-988.

[9] Shearer, M., Loss of strict hyperbolicity of the Buckley-Leverett Equations for three-phase flow in a porous medium, in Numerical Simulation in Oil Recovery, Edited by M.F. Wheeler, Springer-Verlag, New York-Berlin- Heidelberg, 1988, pp.263-283.

[10] Shearer, M. and Trangenstein, J., Change of type in conservation laws for three-phase flow in porous media, Preprint.

[11] Stone, H. L., Estimation of Three-Phase Relative Permeability and Residual Oil Data, J. Cnd. Pet., 12 (1973), pp. 53-61.

Application of Fractional-Flow Theory to 3-Phase, 1-Dimensional Surfactant Flooding

S. I. Aanonsen

Norsk Hydro a.s

Abstract. Surfactants are used in enhanced oil recovery to decrease the interfacial tension (IFT) between oil and water, and thus the residual oil saturation. The lowest IFT values are obtained in a Type III phase environment, where a 3-component system of oil, water and surfactant may appear as 1, 2, or 3 phases depending on overall composition.

An analytical solution of the Riemann problem describing continuous injection of an aqueous surfactant solution into a core containing oil and water is presented for a constant salinity, Type III system. Special consideration is given to the problem of determining how the composition path crosses the phase boundaries.

1 Introduction

Applications to flow in porous media have recently caused an increase in the interest in systems of hyperbolic conservation laws.

In this paper, we consider a special form of a system of two conservation laws arising in surfactant flooding when a solution of surfactant in water is injected into a core containing oil and water. Surfactants are used in enhanced oil recovery to decrease the interfacial tension (IFT) between oil and water, and thus the residual oil saturation. Depending on the phase behaviour in the multiphase region, surfactant systems are usually classified in 3 groups: i) Type II(−) systems where the surfactant is preferably water soluble, and most of the surfactant stays in a lower-phase microemulsion in equilibrium with an excess oil phase; ii) Type II(+) systems where most of the surfactant stays in an upper-phase microemulsion in equilibrium with an excess brine phase; and iii) Type III systems which may separate into either 2 or 3 phases depending on overall composition. The phase behaviour of surfactant systems is usually described by ternary diagrams, and the diagrams corresponding to the 3 types are shown in Fig.1.

Fractional-flow theory has previously been applied to Type II(−) and II(+) systems [1][2][3], and several papers have adressed 3-phase systems without taking into account single- and two-phase regions [4][5][6][7]. In

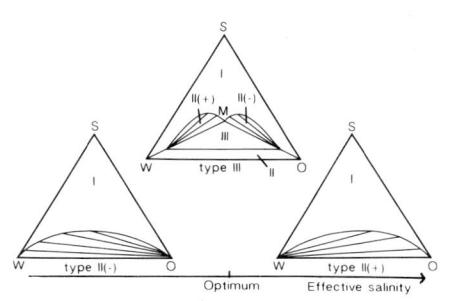

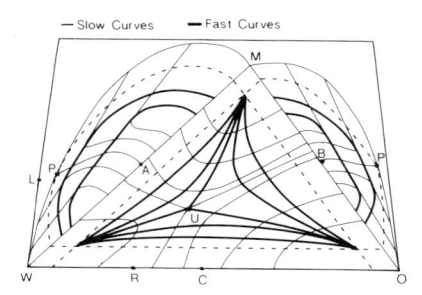

Fig.1. Ternary diagrams   **Fig.2.** Eigenvector curves

this paper we consider the full Type III phase diagram, but will limit the problem to cases where a fluid containing water and surfactant is injected into a core containing oil and water. The 3 components are assumed to be pure, implying—from Gibbs' phase rule—an invariant middle-phase composition.

In the next section, a precise formulation of the model is presented, while the rest of the paper concentrates on the solution of the Riemann problem.

2 Formulation of the model

Let C_i be the overall volume fraction of component i, and C_i^j the volume fraction of component i in phase j (throughout this paper subscripts are used to denote components and superscripts to denote phases). We then have $C_i = \sum_{j=1}^{M} C_i^j S^j$, $i = 1, 2, 3$, where M is the number of phases and S^j the saturation of phase j.

Similarly, the overall fractional flow of component i is given by $F_i = \sum_{j=1}^{M} C_i^j f^j$, where f^j is fractional flow of phase j.

We will neglect gravity, capillary pressure, and adsorption of surfactant on the rock surface. The phases are assumed to be incompressible, and the component densities are assumed to be the same in all phases. Since $\sum C_i = \sum F_i = 1$, conservation of mass may then be expressed as:

$$C_t + F(C)_x = 0, \qquad C \in [0,1] \times [0,1], \qquad (1)$$

where $C = (C_i, C_j)$, $F = (F_i, F_j)$, and i,j correspond to 2 independently flowing components. x and t are dimensionless distance and time, normalized by the length of the core and the time to inject one pore volume, respectively.

Initially the core is assumed to contain only oil and water, and at $t = 0$, a solution of surfactant in water is injected at a constant rate at the left end of the core. We also assume that all oil is solubilized and thus displaced from an infinitesimally small region at the inlet. Consequently, the physical situation may be represented by the Riemann problem for eqn. (1), i.e., a weak solution of the Cauchy problem with initial data:

$$C(x, t = 0) = \begin{cases} C^L, & x < 0; \\ C^R, & x > 0;. \end{cases} \quad (2)$$

Here the left state, C^L, is defined by the injected composition, and the right state, C^R, by the initial condition.

Weak solutions of eqn.(1) consist of rarefaction waves with concentration paths in the state space along eigenvector curves of dF/dC, and shock waves satisfying the Rankine-Hugoniot (RH) condition.† In addition, the shocks will have to satisfy certain entropy conditions, which to date are not fully understood.

Let $H(C^-)$ be the Hugoniot locus corresponding to a given left state C^- (set of right states for shocks from C^- satisfying the RH condition), and let $\sigma(C; C^-)$, $C \in H(C^-)$ be the shock velocity. In the following, we will examine the admissibility of shocks using the Liu entropy condition[10]:

A shock from C^- to C^+ is admissible if $\sigma(C; C^-) \geq \sigma(C^+; C^-)$ for all C on the Hugoniot locus between C^- and C^+; (3)

and the Lax entropy condition[11], which relates the shock velocity and the eigenvalues, λ_i, of dF/dC:

$$\begin{array}{ll} \sigma < \lambda_1(C^-) < \lambda_2(C^-), \quad \lambda_1(C^+) < \sigma < \lambda_2(C^+), & \text{slow shocks} \\ \lambda_1(C^-) < \sigma < \lambda_2(C^-), \quad \lambda_1(C^+) < \lambda_2(C^+) < \sigma, & \text{fast shocks} \end{array} \quad (4)$$

Following Keyfitz and Kranzer[12], we generalize Lax' condition to include shocks where the endpoints are limits of points on the Hugoniot locus satisfying condition (4).

Refering to the Type III phase diagram, we will impose the following limitations on phase behaviour:

i) The lower 2-phase region (II) is neglected, i.e., the base line of the 3-phase triangle coincides with the base line of the ternary diagram.

ii) The plait point of the II(+) lobe is close enough to the lower left corner of the diagram so that none of the tie lines may be extended to include the injected composition.

Note that the neglection of Region II introduces a jump in the residual saturation from the high value without surfactant to the low value in the 3-phase region. As will be shown, however, the solution of the Riemann problem consists of a shock from the base line into Region III, and Region

---

† These conditions correspond to the differential and integral coherence conditions as defined in Ref. [2]

II is not believed to be of importance for the problem considered in this paper. The second condition has been satisfied for most of the systems we have considered for applications. From a mathematical point of view, however, this assumption is more crucial, as it will be shown to result in a jump in concentrations directly from the injected composition to a value inside Region III.

We assume further that the 3-phase relative permeabilities are given by the Marchesin/Pope model[5][8], in which the relative permeabilities are functions of their respective saturations only. The system is then hyperbolic everywhere in the 3-phase triangle, and strictly hyperbolic except for umbilic points at the vertices of the triangle defined by the dashed lines on Fig.2 and a single umbilic point, U, in the interior[5].

Few relative permeability experiments have been reported for 3-phase surfactant systems. Some results[8] indicate that the Marchesin model may be appropriate for these systems, while other investigators have found a more complicated behaviour including strong hysteresis effects[9]. Changes in capillary number during the experiments may, however, have complicated the latter results.

Figure 2 shows the simplified multiphase region with the left and right states marked L and R, respectively. The dashed lines indicate residual saturations, and the solid lines the structure of the eigenvector curves. For details on the eigenvector curves, see Refs. [3], [5], or [7], e.g.

3 Solution of the Riemann problem

3.1 Solubilization front

In general, the solution of the Riemann problem for a Type III system may contain regions of single-, two- and three-phase flow. However, in this section we show that under the previously specified assumptions, the injected composition can be connected to a right state at the base line of the phase triangle only through a shock directly from L into the interior of the 3-phase region.

First, note that a wave with non-zero velocity crossing one of the phase boundaries has to be a shock or a contact discontinuity. This follows from the fact that along an eigenvector curve crossing a region where one of the phases does not flow, the corresponding eigenvalue is zero. Moreover, the only route by which a phase boundary can be crossed without passing such a region is through the plait points. It is also well known that if a tie line can be extended to include the injected composition, the route entry from the single-phase region into the 2-phase region will be along this tie line, and if the injected composition is outside the tie-line extensions, the route entry will be a contact discontinuity with velocity 1 through the plait point along the equivelocity curve (curve where $f^j = S^j$)[2]. (In our case, we have 2 plait points with corresponding equivelocity curves, one for each

2-phase lobe.)

Since $C_3 = 0$ at R, any solution must involve states where $F_3 = C_3 = 0$. Assume now that the composition path involves points in one of the 2-phase lobes. The solution will then contain a contact discontinuity with left state at L and right state at one of the equivelocity curves due to assumption ii) about the plait point position. At this right state, $F_3 = C_3$, and the velocity of the discontinuity is 1.

Consider now possible composition paths from L via this state on the equivelocity curve to a state where $F_3 = C_3 = 0$. If F_3 is plotted vs. C_3 along this path, the velocity of a given state given by dF_3/dC_3, and shocks are connected by straight lines with the velocity given by the slope.† Since all curves, except the straight line $F_3 = C_3$, must involve points where the slope is less than one, the requirement of increasing wave velocity from L to R implies that all the points must lie on the straight line $F_3 = C_3$. That is, states in the 2-phase lobes can only be included in the solution as a part of a discontinuity crossing the 2-phase lobe into the 3-phase triangle. Possible end-points of such a discontinuity on the border of the 3-phase region are points where $F_i = C_i$, and it will be shown below that none of these can be part of the solution. Consequently, the solution of the Riemann problem has to contain a shock from L into Region III. Upstream of this shock, all oil is displaced, and the shock is called the *solubilization front*.

The RH condition for the solubilization front may be written:

$$\sigma = \frac{F_3^+ - F_3^L}{C_3^+ - C_3^L} = \frac{F_2^+ - F_2^L}{C_2^+ - C_2^L} = \frac{F_1^+ - F_1^L}{C_1^+ - C_1^L}$$

$$= \frac{f^{m+} - \frac{C_3^L}{C_3^m}}{S^{m+} - \frac{C_3^L}{C_3^m}} = \frac{f^{m+} + \frac{f^{o+}}{C_2^m}}{S^{m+} + \frac{S^{o+}}{C_2^m}} = \frac{f^{m+} + \frac{f^{w+}}{C_1^m} - \frac{C_1^L}{C_1^m}}{S^{m+} + \frac{S^{w+}}{C_1^m} - \frac{C_1^L}{C_1^m}}. \quad (5)$$

(Here superscripts w, o, and m represent water, oil, and microemulsion, respectively. The $+$ sign indicates the right end-point of the shock.)

Figure 3 shows example fractional flow curves for oil and surfactant with the corresponding eigenvector curves in the region left of the eigenvector curve through M and the umbilic point U. It is assumed that the derivatives of the fractional flow curves approach zero at the end-points, but in general, the curves may have a more complicated form than indicated in Fig. 3. Consider for example the oil curve corresponding to the eigenvector curve 5. Depending on the composition of the middle phase, the oil concentration, C_2, may have a maximum value along the curve making $F_2(C_2)$ multivalued. However, the solution will not be influenced by a complicated shape of the fractional flow curves, and hence, the curves on Fig. 3 are drawn as simple as possible for illustrative purposes.

---

† These paths need not be connected for shock waves in the 3-phase region (see Ref. [6] or [7], e.g.).

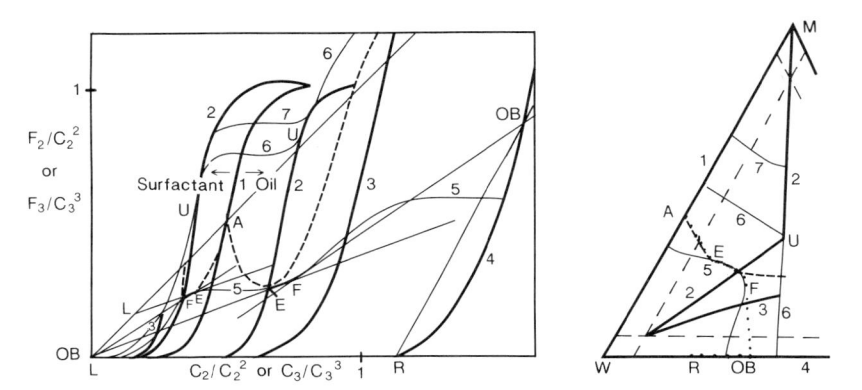

Fig.3. Eigenvector curves in 3-phase region with corresponding fractional flow curves for surfactant and oil.

Note that the slopes of the fractional-flow curves when plotted along the eigenvector curves correspond to the eigenvalues and thus determine the wave velocities. Note also that the oil and surfactant curves, when normalized by C_2^m and C_3^m, respectively, coincide along the border between the II(+) lobe and the 3-phase region (curve 1 on Fig. 3), and that inside the 3-phase region the surfactant curves lie to the left, and the oil curves lie to the right of this common curve.

Two points are marked L on the plot. The one at the origin corresponds to the left state for oil, and the other to the left state for surfactant ($F_3^L = C_3^L$ = injected concentration). From Eq. (5), it follows that a point on the surfactant curve is a possible end-point for a shock from L if the straight line from L through this point is parallel to the straight line from the origin throgh the corresponding point on the oil fractional-flow curve. Starting from point A (which is the end of the equivelocity curve in the II(+) lobe), the Hugoniot locus corresponding to the left state L may be constructed graphically and is indicated by the dashed lines on Fig.3. Similarly, the part of the Hugoniot locus lying in the right part of the triangle can be constructed from a plot of the surfactant and water fractional-flow curves. It follows that the only possible entrances into the 3-phase region are through points on the border where $f^j = S^j$. That is, points W, O, M, A, B, and C on Figs. 2 and 3. Note that the shock velocity always decreases from 1 along the curve through A and reaches a minimum value, while the shock curve starting at the point B may have a shock velocity increasing or decreasing depending on the injected composition, C_3^L. It may also be shown that shock velocities corresponding to the shock curves starting from point C always will increase from 1. According to Liu's condition, shocks with right states on the curve starting at A will then always be admissible, shocks with right states on the curve starting at C will never be admissible, while shocks with right states on the curve starting at B may

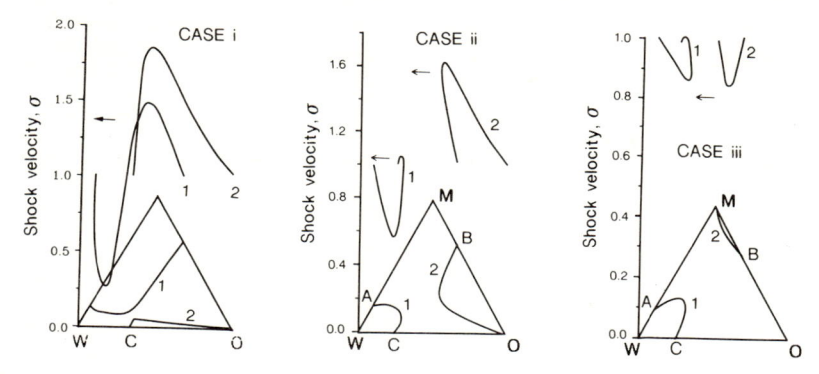

Fig.4. Shock curves and shock velocities for the solubilization front. Case i: $C_3^L = 0.005$, Case ii: $C_3^L = 0.02$, Case iii: $C_3^L = 0.1$.

be admissible in some cases. By a similar argument, it can be shown that of the shock curves entering at the vertices, only points on the curve starting from point M are acceptable right states for the solubilization front. We have here used that the Hugoniot locus corresponding to a left state at L includes the whole single-phase region in addition to the equivelocity curves in the 2-phase lobes. Hence, all the above mentioned entry points, except point C, may be seen as direct continuations of shock curves starting at point L.

The Hugoniot locus for a given system will always consist of two separate curves, and may be classified in 3 groups depending on the surfactant concentration in the injected fluid.

Typical examples of shock curves for the 3 cases are shown in Fig. 4 together with the corresponding shock velocities. Defining C_3^A as the surfactant concentration corresponding to the equivelocity point A on Fig. 3, the 3 cases correspond to $C_3^L < C_3^A$, $C_3^A/C_3^m < C_3^L/C_3^m < C_1^L/C_1^m$, and $C_3^L/C_3^m > C_1^L/C_1^m$. (It is easily shown that these 3 cases cover the whole range of injected compositions.) When $C_3^L \to C_3^A$ from below, curve 2 approaches curve 1 giving two curves crossing each other for $C_3^L = C_3^A$, while curve 2 merges with the line MO at the transition between case ii and case iii. The low-IFT relative permeability data used in the interior of the 3-phase triangle are listed in Table 1 together with the composition of the middle-phase microemulsion.

Considering Case i and ii in the figure, the only right states satisfying condition (3) are points on curve 1 left of the minimum shock velocity, including the minimum. For Case iii, we get points on curve 1 similarly, but for curve 2, both end-points of the curve may be connected to L, and depending on which direction is defined as positive, points both to the left and to the right of $\sigma = \sigma_{min}$ satisfy condition (3).

By examining the eigenvalues, the shocks satisfying Lax' condition may also be determined. This analysis shows that for curve 1 the two entropy

Residual saturations			End-point mobilities (1/cp)			Rel. perm. exponents			Composition of middle-phase microemulsion		
S_{wr}	S_{or}	S_{mr}	λ_w	λ_o	λ_m	E_w	E_o	E_m	C_w^m	C_o^m	C_s^m
0.1	0.1	0.0	0.3	4.0	0.5	2.0	2.5	1.5	0.55	0.40	0.05

Table 1. Example 3-phase region rock/fluid data.

conditions coincide in all 3 cases, and the shocks are all slow shocks. For curve 2, only points to the right of the minimum satisfy condition (4). The states A and B may thus be said to define the "natural" entry points into the 3-phase triangle.

3.2 Surfactant front

In the previous section, we have discussed possible endpoints for the solubilization front, and in this section we will show that only one of these endpoints may be connected further to the base line of the triangle without violating the entropy conditions.

Let $C(s)$ be a parametrization of the shock curve corresponding to a given left state, C^-. The tangent to $C(s)$ is then an eigenvector to dF/dC at a given point, C^*, if and only if the shock velocity, $\sigma(C^-; C(s))$ has a stationary point at C^*[13]. In addition, the corresponding eigenvalue is then equal to $\sigma(C^-; C^*)$. That is, a shock can be connected directly to a rarefaction wave on the right (without any intervening constant state) only at points where the shock velocity has a local minimum, and to the left only at points where the shock velocity has a local maximum.

Considering the shock curve starting at point A, it is seen from Fig. 3 that $\lambda_1 < \sigma < \lambda_2$ before the minimum (marked E on the plot). Hence, any wave continuing from a point on this curve will have to be a fast wave. Since the solution has to enter the base line through a shock, it follows that this shock will have $\sigma < \lambda_2^+$, which violates the entropy condition for a fast shock. The only possible exception seems to be the non-physical case where the fractional-flow curve along the base line is located to the left of that for the fast curve between W and O just above the base line. That is, the residual oil saturation with surfactant is higher than without. Consequently, for any "normal" form of the water-oil fractional-flow curve, it is impossible to connect R with points before the minimum. At the minimum, however, $\sigma = \lambda_1$, and the solution can be continued by a rarefaction wave along the slow eigenvector curve and further connected to the base line by a slow shock satisfying the generalized Lax entropy condition. In fact, from Fig.3 and the discussion of Fig.5.1 in Ref.[6], it follows that point E always can be connected to a state on the base line with a regular slow wave. This state corresponds to the oil bank marked OB on Fig. 3.

The velocity of the surfactant front is determined by the tangent from the origin to the surfactant fractional flow curve passing through E (tangent

point marked F), and the exact location of the oil bank is determined by
the intersection of the corresponding tangent to the oil curve with the pure
oil/water fractional-flow curve (curve 4).

For admissible shocks from L ending on curves through B, a similar
examination shows that $\lambda_1 < \sigma < \lambda_2$ before the minimum, and $\sigma = \lambda_2$ at
the minimum. That is, $\lambda_1^- < \sigma \leq \lambda_2^-$, $\sigma < \lambda_2^+$, and it follows that none
of the points on this curve can be connected to the base line. (The only
exception is also here the unphysical case mentioned previously, and in this
case it seems possible to construct a non-unique solution.) Note that most
of the difficulties reported in [6] when solving the general Riemann problem
for the Marchesin model—involving the so-called transitional shocks—are
avoided for our system.

3.3 Construction of the total solution

Disregarding unphysical fractional-flow curves, we have shown that there
is one and only one way of connecting the left state to the base line of the
triangle with a composite slow wave, and that it is not possible to connect
fast waves from the right state to this slow wave, except at the base line.
At this connection, which is the point where the composition path of the
surfactant front crosses the base line, we get a constant state—the oil bank.
The fast wave connecting the oil bank to the right state is then an ordinary
Buckley-Leverett type solution.

The solution of the Riemann problem thus consists of a composite
slow wave (the solubilization front and the surfactant front connected by
a rarefaction wave), and a fast wave along the base line of the triangle.
The solution is constructed by first drawing the Hugoniot locus for the
solubilization front starting at point A, and then localize the point E corresponding to minimum shock velocity. This velocity should coincide with
the tangents from L to the fractional flow curves along the slow eigenvector
curve through E (Fig.3). The surfactant front is given by the tangent to
the F_3 vs. C_3 curve through the origin, while the oil bank is determined by
the intersection between the corresponding tangent to the F_2 vs. C_2 curve
and the pure oil-water fractional-flow curve. Finally, the oil-bank front is
determined by a Buckley-Leverett type solution from the oil bank to R.

Usually, R will correspond to the residual oil saturation after waterflooding, S_{orw}. However, R may be located anywhere along the base line
of the phase diagram. For instance, if the chemicals are injected as a secondary process, R is located at irreducible water saturation, and the oil
bank front will be a shock from OB to S_{wi}. Thus, the surfactant delays
the water break through compared with a pure water flood. The fast wave
from OB to R may also be a composite rarefaction-shock, or even disappear
completely if the initial composition coincides with the right end-point of
the surfactant front.

If the fractional-flow curves are straight lines, the solubilization front

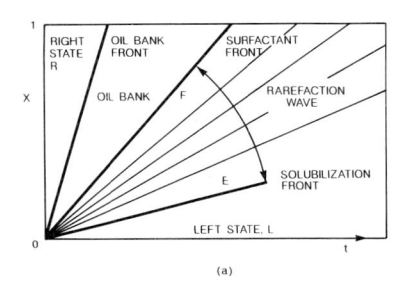

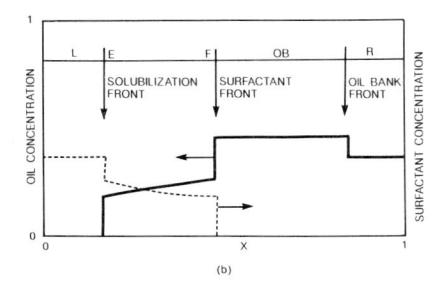

Fig.5. Example time/distance diagram (a) and concentration profiles (b) for a Type III displacement.

and the surfactant front will merge into a single front with velocity 1. This corresponds to a completely miscible displacement and does not seem likely, except for the extreme case where $C_3^m \to 0$, i.e., the multiphase regions disappear completely.

The solution path in the 3-phase region for R at S_{orw} is shown by the dotted line in Fig.3. The 4 simple waves making up the total solution corresponds to the line segments L–A–E, E–F, F–OB, and OB–R, respectively.

Figure 5 shows a time/distance diagram of the solution together with the oil and surfactant concentrations at a given time, and it is seen that the structure is very similar to the corresponding solutions for Type II(−) and Type II(+) flows [3].

4 Conclusions

If the injected fluid is a solution of surfactant in water lying outside the tie-line extensions of the II(+) lobe, the continuous injection problem for a Type III surfactant system has one and only one solution satisfying the Liu and generalized Lax entropy condition. However, it may be possible to construct non-unique solutions by using non-physical fractional flow curves.

The solution consists of 2 waves connected by a constant state—the oil bank. The upstream wave is a composite wave of 2 shocks connected by a rarefaction wave. The downstream wave is a standard Buckley-Leverett type solution connecting the oil bank with the initial reservoir state.

5 References

[1] Pope, G. A. (1980). The Application of Fractional Flow Theory to Enhanced Oil Recovery. *Society of Petroleum Engineers Journal*, 191–205; *Transactions* AIME, **269**.

[2] Helfferich, F.G. (1981). Theory of Multicomponent, Multiphase Displacement in Porous Media. *Society of Petroleum Engineers Journal*, **21**, 51–62; *Transactions* AIME, **271**.

[3] Hirasaki, G. J. (1981). Application of the Theory of Multicomponent, Multiphase Displacement to Three-Component, Two-Phase Surfactant Flooding. *Society of Petroleum Engineers Journal*, **21**, 191–204; *Transactions* AIME, **271**.

[4] Schaeffer, D. G. and Shearer, M (1987). The Classification of 2×2 Systems of Non-Strictly Hyperbolic Conservation Laws, With Application to Oil Recovery. *Communications on Pure and Applied Mathematics*, **40**, 141–178.

[5] Shearer, M. (1988). Loss of Strict Hyperbolicity of the Buckley-Leverett Equations for Three Phase Flow in a Porous Medium. *Numerical Simulation in Oil Recovery*, M.F. Wheeler (ed.), *IMA Volumes in Mathematics and Its Applications*, **11**. Springer-Verlag, New York.

[6] Isaacson, E. L., Marchesin, D., and Plohr, B. (1988). The Structure of the Riemann Solution for Non-Strictly Hyperbolic Conservation Laws. Preprint, Pontifícia Universidade Católica do Rio de Janeiro.

[7] Falls, A. H. and Schulte, W. M. (1989). Features of Three-Component, Three-Phase Displacement in Porous Media. Paper SPE 19678, presented at the 64th Annual Technical Conference and Exhibition of the SPE, San Antonio, Oct. 8–11, 1989.

[8] Delshad, Moj., Delshad, Moh., Pope, G. A., and Lake, L. W. (1987). Two- and Three-Phase Relative Permeabilities of Micellar Fluids. *SPE Formation Evaluation*, **2**, 327–337; *Transactions* AIME, **283**.

[9] Skauge A. and Matre, B. (1989) Three Phase Relative Permeabilities in Brine–Oil–Microemulsion Systems. *Proceedings of the 5th European Symposium on Improved Oil Recovery* (Budapest, April 25–27, 1989), 473–482.

[10] Liu, T-P. (1974). The Riemann Problem for General 2×2 Conservation Laws. *Transactions of the American Mathematical Society*, **199**, 89–112.

[11] Smoller (1974). Shock Waves and Reaction-Diffusion Equations. Springer-Verlag, New York.

[12] Keyfitz, B. L. and Kranzer, H. C. (1980). A System of Non-Strictly Hyperbolic Conservation Laws Arrising in Elastisity Theory. *Archive for Rational Mechanics and Analysis*, **72**, 219–241.

[13] Wendroff, B. (1972). The Riemann Problem for Materials with Non-convex Equations of State. II: General Flow. *Journal of Mathematical Analysis and Applications*, **38**, 640–658.

AVERAGING OF RELATIVE PERMEABILITY IN HETEROGENEOUS RESERVOIRS

by

Steinar Ekrann and Magnar Dale*
(Rogaland Research Institute)

ABSTRACT

In recent years, so-called <u>dynamic pseudo functions</u> have been increasingly used to capture the effects of small scale heterogeneities, in coarse grid reservoir simulation. This mode of use suggests a close relationship to the concept of <u>effective properties</u>, e.g. effective relative permeabilities. The two concepts are reviewed in some detail in the paper, and the relationship between them is explored. It is shown that fundamental differences exist.

Effective properties can be defined only on a much larger spatial scale than that of the heterogeneities themselves. Effective properties become universally valid in the sense that they appear as parameters in the ensuing partial differential equations for the large scale variables.

Dynamic pseudos are constructed from one single (approximate) solution of the governing equations. No scale restrictions apply to this "parent" solution. If properly constructed, dynamic pseudos allow exact reproduction, in a subsequent coarse grid simulation, of the parent solution. Dynamic pseudos generally become dependent on the particular parent solution chosen, and on the numerical grid (and numerical method) with which they are later to be employed.

* *Now with Rogaland Regional College*

1. INTRODUCTION

Full field reservoir simulation must, by economic necessity, proceed with numerical grids far too coarse to allow resolution of all detail of interest. It is generally believed that small scale behaviour may have important large scale consequences. Thus, the need has long been recognized for reliable methods to incorporate the large scale manifestations of sub grid scale effects into coarse grid simulations.

We consider available methods to fall into two broad categories, namely that of <u>effective properties</u> and that of <u>dynamic pseudo functions.</u> In the present paper, we explore these two classes of methods and the relationship between them.

2. GOVERNING EQUATIONS

We will consider only the comparatively simple black oil type mechanistic problem of incompressible two-phase flow in porous media. Our starting point will be the usual macroscopic equations

$$v_i^f = - \lambda_{ij}^f (p_{,j}^f - \rho^f g_j) \qquad (2.1\text{ a})$$

$$\Phi S_{,t}^f = - v_{i,i}^f \qquad (2.1\text{ b})$$

$$S^w + S^n = 1 \qquad (2.1\text{ c})$$

$$p^w - p^n = p^c \qquad (2.1\text{ d})$$

which we assume to be everywhere valid. We allow the mobility tensors $\bar{\bar{\lambda}}^f$ and the capillary pressure p^c to be functions of spatial position. At each point in space, these parameters are assumed to be functions of saturation only. In the equations, the superscript (f = w,n) indicates fluid phase. Subscripts (i,j) indicate Cartesian components of tensors. A subscript preceded by a comma denotes differentiation with respect to the corresponding spatial or temporal (t) coordinate. The summation convention is employed.

Bear et. al.[1] have shown that one should generally expect the ratio between two given components of $\bar{\bar{\lambda}}^f$ to be a function of the saturation S^w. Thus, the usual practice of factorizing two-phase permeability into a scalar saturation dependent relative permeability and a constant tensorial absolute permeability is suspect. Reinforcing the argument, Quintard and Whitaker [2] give a nice example where the effective (large scale) relative permeability is anisotropic even if the small scale relative permeability is not. In the present treatment, therefore, we retain as parameters unfactorized mobility tensors.

3. EFFECTIVE PROPERTIES

3.1 Introduction

Heterogeneous materials are of practical importance in a variety of technological fields. If the process of interest takes place on a spatial scale much larger than that of the heterogeneities, it becomes impractical to take individual heterogeneities into account. If the material is "statistically homogeneous" [3-6], it is possible to define an equivalent homogeneous material, the properties of which one refers to as the "effective properties" of the original heterogeneous material.

In non-linear processes, small scale variation of the dependent variables themselves may play a similar role to that of material heterogeneities, because of non-linear feedback. The geometry of process-induced small scale variation is unknown a priori, forming part of the solution which is sought. Thus, an extra complexity is introduced. In the present problem, such process-induced small scale variation is provided by e.g. viscous fingering.

Implicit in the previous discussion is that the concept of effective properties is meaningful only on a spatial scale much larger than that of the small scale variation. Large scale variables are most naturally defined via suitable volume averaging (to be discussed in the next section). Alternatively, a stochastic approach may be used [3,6], substituting ensemble averaging for volume averaging. Restrictions and assumptions are very similar in the two approaches.

3.2 Large scale volume averaging

Let f be a function of spatial position. Let $<\cdot>$ denote volume averaging, i.e.

$$<f> \stackrel{def}{=} \frac{1}{V_0} \iiint_{V_0} f \, dV \qquad (3.1)$$

Generally, $<f>$ will depend on the size, shape, orientation and position of the averaging volume V_0. For definiteness, let V_0 be a parallelepiped aligned with the coordinate axes, with fixed dimensions $L_1 \cdot L_2 \cdot L_3$. Associating $<f>$ with the centroid $\mathbf{r}_0$ of V_0, $<f>$ is then a function of $\mathbf{r}_0$ only.

Let $\wedge$ denote spatial deviation, i.e.

$$f(\mathbf{r}) \stackrel{def}{=} <f>(\mathbf{r}) + \hat{f}(\mathbf{r}) \qquad (3.2)$$

All variables in eq.(3.2) are taken at the same spatial position $\mathbf{r}$. In the sequel, for simplicity of notation, the dependence on spatial position will not be explicitly shown.

In eq.(3.1) we have tacitly assumed that f is defined everywhere within the averaging volume. In the present application, this may not always be true. In non-permeable parts of the porous medium, for instance, pressures and pressure gradients will not be defined. For any given medium, we circumvent this problem by considering only an approximate porous medium, where permeabilities and porosities are everywhere non-zero (but possibly small). In the two-phase case, in particular, we require both phases to be mobile everywhere. This approximation greatly simplifies notation (compare e.g. [2]), without, we believe, essentially altering the physics of the problem.

With the above approximation, it can be shown (see [2,7] and references therein) that

$$<p^f_{,j}> \; = \; <p^f>_{,j} \qquad (3.3)$$

$$\langle s^f_{,t}\rangle = \langle s^f\rangle_{,t} \qquad (3.4)$$

$$\langle v^f_{j,j}\rangle = \langle v^f_j\rangle_{,j} \qquad (3.5)$$

3.3 Effective absolute permeability

In the one-phase case, eq.s (2.1a) and (2.1b) take the forms (with a trivial transformation to remove the gravity term)

$$v_i = -\lambda_{ij}\, p_{,j} \stackrel{\text{def}}{=} -\frac{k_{ij}}{\mu}\, p_{,j} \qquad (3.6a)$$

$$v_{i,i} = 0 = (k_{ij} p_{,j})_{,i} \qquad (3.6b)$$

where k_{ij} is a component of the absolute permeability tensor and μ is the fluid viscosity. Decomposing the variables (see eq.(3.2)) and inserting the result into eq.(3.6a), one obtains after averaging

$$\mu\langle v_i\rangle = -\langle\langle k_{ij}\rangle \langle p_{,j}\rangle\rangle - \langle\langle k_{ij}\rangle \hat{p}_{,j}\rangle$$

$$\quad - \langle \hat{k}_{ij} \langle p_{,j}\rangle\rangle - \langle \hat{k}_{ij}\, \hat{p}_{,j}\rangle \qquad (3.7)$$

Now, we ask under what circumstances can the large scale variables $\langle v_i\rangle$ and $\langle p\rangle_{,j}$ satisfy an equation of the same form as eq.(3.6a), i.e.

$$\mu\langle v_i\rangle = -k^{\text{eff}}_{ij}\, \langle p\rangle_{,j} \qquad (3.8)$$

Recalling eq.(3.3), this requires the right hand side (r.h.s.) of eq. (3.7) to be proportional to $\langle p_{,j}\rangle$. If $\langle k_{ij}\rangle$ and $\langle p_{,j}\rangle$ can be regarded as constants on the scale L of averaging, then

$$\langle\langle k_{ij}\rangle \langle p_{,j}\rangle\rangle = \langle k_{ij}\rangle \langle p_{,j}\rangle \qquad (3.9)$$

Furthermore, it is easy to show under the same assumptions that $<\hat{p}_{,j}>$ and $<\hat{k}_{ij}>$ vanish, as do, therefore, the second and third r.h.s. terms of eq.(3.7). For the last term, rewrite eq.(3.6b) to read

$$(k_{ij} \hat{p}_{,j})_{,i} = - k_{ij,i} <p_{,j}> \qquad (3.10)$$

Eq.(3.10) is an equation for $\hat{p}$, with the r.h.s. acting as a source term. Imagine that the boundaries are sufficiently far removed for the (small scale) boundary conditions not to influence the solution, which is then determined exclusively by the source term. By the linearity of the equation, proportionality must then prevail between the pressure gradient deviation and the average pressure gradient. The proportionality constant will generally be a function of space:

$$\hat{p}_{,j} = \hat{\alpha}_{jl} <p_{,l}> \qquad (3.11)$$

(A more detailed derivation of eq.(3.11) was given by Quintard & Whitaker [2]). With eq.(3.11), eq.(3.8) is satisfied, with

$$k_{ij}^{eff} = <k_{ij}> + <\hat{k}_{il} \hat{\alpha}_{lj}> \qquad (3.12)$$

In these equations, $\hat{\alpha}_{jl}$ is a rapidly (hence the ^) varying function of space, determined by k_{ij}. k_{ij}^{eff} is the effective absolute permeability. It enters as a parameter in the equation (3.8) for the large scale variables $<v_i>$ and $<p>$, and depends only on the properties of the medium itself. Applicability is limited, however, as indicated by the the previous discussion:

Two basic assumptions were made, namely those that $<k_{il}>$ and $<p_{,l}>$ both could be regarded as constants. For the assumptions to be true (approximately), it is sufficient that the following "separation of scales" conditions are satisfied:

$$l_j \ll L_j \ll \ell_{kj}, \ell_{pj} \qquad (3.13)$$

Here, l is the linear scale of (small scale) permeability variations and ℓ_k and ℓ_p are the linear scale over which $<k_{il}>$ and $<p_{,l}>$, respectively, vary significantly. In (3.13), we allow scales to be different in different coordinate directions j.

3.4 Effective two-phase mobilities

Define porosity weighted average saturations via

$$<S^f>^* \stackrel{\text{def}}{=} \frac{<S^f \phi>}{<\phi>} \qquad (3.14)$$

At a given point in time, assume the small scale saturation distribution to be known. Then, spatial averaging of eq.(2.1) can be carried out essentially as in the one-phase case, producing

$$<v_i^f> = - \lambda_{ij}^{f\,\text{eff}} (<p^f>_{,j} - \rho^f g_j) \qquad (3.15a)$$

$$<\phi> <S^f>^*_{,t} = - <v_i^f>_{,i} \qquad (3.15b)$$

$$<S^w>^* + <S^n>^* = 1 \qquad (3.15c)$$

$$<p^w> - <p^n> = <p^c> \qquad (3.15d)$$

With the given saturation distribution, eq.(3.15a) relies on similar assumptions as in the one-phase case. In particular, $<\lambda_{ij}^f>$ must be constant on the scale of spatial averaging. Now, mobilities are saturation dependent. Thus, condition (3.13) must be generalized to read

$$l_j \ll L_j \ll \ell_{pj}, \ell_{kj}, \ell_{Sj} \qquad (3.16)$$

where ℓ_S is the linear scale over which the average saturations $<S^f>^*$ change significantly. In (3.16) we have assumed small scale saturation variations to occur on the same scale 1 as the permeability variations.

Given the statistical properties of the medium itself, $\lambda_{ij}^{f\,eff}$ and $<P^c>$ will be determined by the assumed small scale saturation distribution.

For eq.(3.15) to be a meaningful set of equations for the large scale variables $<S^f>^*$, $<p^f>$ and $<v_i^f>$, the parameters $\lambda_{ij}^{f\,eff}$ and $<p^c>$ must be known as functions of these variables ($<S^f>^*$, in particular), over the full range. With the remark above, this requires the small scale saturation distribution to be an essentially unique function of large large scale variables. We write this condition formally as

$$S^f = S^f(\,<S^f>^*,\ <p^f>_{,j},\ <v_i^f>\,) \qquad (3.17)$$

When conditions (3.16) and (3.17) are satisfied, $\lambda_{ij}^{f\,eff}$ and $<P^c>$ qualify as effective properties. Note that we must allow dependence on $<p^f>_{,j}$ and $<v_i^f>$, essentially because the small scale saturation distribution, for a given large scale saturation, will generally depend on the balance between viscous, capillary and gravity forces.

Eq.s (2.1) describe a dynamic process. The small scale saturation distribution corresponding to a given set of values for the large scale variables will generally change with space and time, even in a statistically homogeneous medium. Thus, according to eq.(3.17), effective properties $\lambda_{ij}^{f\,eff}$ and $<P^c>$ exist, at best, only asymptotically, when the small scale saturation distribution has stabilized. This has its parallel in the one-phase case, where a certain distance to boundaries was required, in order that small scale boundary conditions not influence the averages. In an evolutionary problem, "sufficient" distance in time from the initial conditions is generally required in addition. In a given practical problem, it may well be that asymptoticity is not reached in the time scale of interest. In such cases, effective properties can play no role in describing the process.

The development of effective properties does not in any way refer to the numerical grids or the numerical methods which may be used for the approximate solution of the resulting large scale partial differential equations. Thus, there is no direct relation between averaging volumes and grid blocks. In fact, due to the near constancy of large scale variables and parameters required by eq.(3.16), one might generally advocate the use of numerical grids with blocks far larger than the averaging volume V_o, if the process allows description via effective properties.

3.5 Previous work

Quintard & Whitaker [2] do a careful analysis, in a formal setting as outlined above. In their practical example, they neglect gravity and assume the small scale saturation distribution to be governed by capillary forces. This is a low rate approximation, which does not capture the effect of viscous forces.

Dale [8] has considered the 1D case, neglecting gravity and capillarity. His effective relative permeabilities are based on the small scale saturation distribution corresponding to steady state flow. Dale was able to stringently discuss convergence as the scale of the heterogeneities approaches zero. For an example set of effective relative permeabilities produced by Dale's technique, see fig. 1.

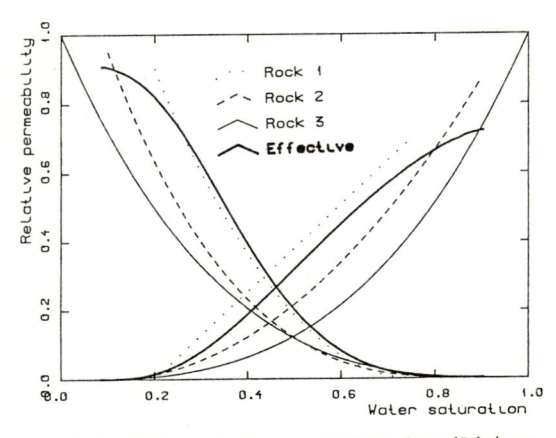

Fig.1. Effective relative permeabilities for a 1D heterogeneous medium made up of three different rock types.

The classical vertical equilibrium (VE) pseudos [9-11] are developed for stratified reservoirs. They are based on the assumption of a small scale (vertical) saturation distribution governed by gravity and capillarity. The associated neglect of vertical potential gradients allows analytic determination of pressure and velocity fields. With these assumptions, VE pseudos qualify as effective properties. Averaging is performed over the full reservoir height, implying trivial satisfaction of condition (3.16) for the vertical direction. With the assumption of stratification, we may set $l = 0$ for the horizontal directions. Even if horizontal averaging is not actually performed, we need to imagine the horizontal direction L's to be finite for satisfaction of the l.h.s. of eq.(3.16). Furthermore, the neglect of vertical pressure gradients is equivalent to assuming a constant average saturation, implying satisfaction of the r.h.s. of eq.(3.16).

4. DYNAMIC PSEUDO FUNCTIONS

4.1 Introduction

Dynamic pseudo functions [11,12-14] are constructed by back calculation from some single (normally numeric) parent solution of the governing equations. The general idea is for the pseudos to transfer, in some sense, the accuracy characteristics of the parent solution to subsequent coarse grid simulations. In the next sections we give a special development of dynamic pseudos, and discuss their properties.

4.2 Exact reproduction of given grid functions

Consider eq.(2.1). Let the computational domain be partitioned into a set of non-overlapping subdomains (grid blocks). A standard finite difference approximation to the equations could read

$$\tilde{F}_{s's}^{fm} = \tilde{\tilde{T}}_{s's}^{fm} (\tilde{P}_{s'}^{fm} - \tilde{P}_{s}^{fm} + G_{s's}^{f}) \qquad (4.1)$$

$$\tilde{V}_{s}^{fm+1} = \tilde{V}_{s}^{fm} + \Delta t^{m+1} \sum_{s'} \tilde{F}_{s's}^{fm'} \qquad (4.2)$$

$$\sum_{f} \sum_{s'} \tilde{F}_{s's}^{fm'} = 0 \qquad (4.3)$$

$$\tilde{P}_s^{wm} - \tilde{P}_s^{nm} = \tilde{\tilde{P}}_s^{cm} \qquad (4.4)$$

Here, a tilde signifies a finite difference (grid) variable, while a double tilde signifies a finite difference parameter. For reasons to become clear, we have let parameters be functions of time, when necessary, and not of saturations. Superscripts m and m' indicate the time level. With m' = m, the scheme above would be referred to as explicit, while m' = m+1 implies an implicit scheme. Subscript s is the grid block index. Summations above are taken over all grid blocks s' with a common boundary with block s. P denotes pressures, while V denotes (grid block) phase volumes. The F's are intermediate variables, whose physical interpretation is that of (grid block boundary) fluxes. G is a discretization of the gravity term.

Disregard for a moment the origin of eq.s (4.1)-(4.4). As they stand, with $\tilde{\tilde{T}}$, $\tilde{\tilde{P}}^c$, and G regarded as known, the equations form a complete machinery to advance in time any initial grid functions $\tilde{P}_s^{f0}$ and $\tilde{V}_s^{f0}$. The $\tilde{P}$'s are advanced via combination of eq.(4.4) and the equation obtained by substituting eq.(4.1) into eq.(4.3), while the $\tilde{V}$'s are advanced via eq.(4.2).

Now, it is easy to see that a given set of grid functions (P_s^{fm}, V_s^{fm}), regardless of its physical significance, can in principle (i.e. disregarding possible stability problems) be reproduced by this machinery, provided that parameters $\tilde{\tilde{T}}_{s',s}^{fm}$ and $\tilde{\tilde{P}}_s^{cm}$ are properly chosen:

$$\tilde{\tilde{P}}_s^{cm} = P_s^{wm} - P_s^{nm} \qquad (4.5)$$

$$\sum_{s'} \tilde{\tilde{T}}_{s',s}^{fm'}(P_{s'}^{fm'} - P_s^{fm'} + G_{s',s}^f) = (V_s^{fm+1} - V_s^{fm}) / \Delta t^{m+1} \qquad (4.6)$$

For satisfaction of the combined eq.s (4.1) and (4.3), eq.(4.6) introduces a restriction on the original grid functions:

$$\sum_f (V_s^{fm+1} - V_s^{fm}) = 0 \qquad (4.7)$$

which is equivalent to requiring them to satisfy a discretized version of eq.(2.1c).

Infinitely many sets of parameters $\tilde{T}_{s's}^{fm}$ satisfy eq.(4.6), since there there are two (f=w,n) equations per grid block, and 4, 8, or 12 unknowns in a regular grid (depending on the number of dimensions). Uniqueness can be obtained, and some physical significance attached to these parameters, if reproduction of fluxes is also required:

Let $FT_{s's}^{fm}$ be given grid functions interpreted as total volume of phase f flowing from block s' to block s during the time interval $t^{m-1} - t^m$. We require from these grid functions that

$$\sum_f \sum_{s'} FT_{s's}^{fm} = 0 \qquad (4.8)$$

and

$$V_s^{fm+1} - V_s^{fm} = \sum_{s'} FT_{s's}^{fm+1} \qquad (4.9)$$

Comparing with eq.s (4.3) and (4.2) we note that these restrictions force the given grid functions to have properties similar to those expected from an approximate solution of the original equations. Also note that the restrictions imply automatic satisfaction of (4.7).

With

$$\tilde{T}_{s's}^{fm'} = FT_{s's}^{fm+1} / (\Delta t^{m+1} (P_{s'}^{fm'} - P_s^{fm'} + G_{s's}^{f})) \qquad (4.10)$$

we have obtained a unique definition of $\tilde{T}^{fm}_{s's}$, which (together with $\tilde{P}^{cm}_s$ given in eq.(4.5)) implies exact reproduction of (V^{fm}_s, P^{fm}_s) as well as of (time averaged) fluxes $FT^{fm}_{s's} / \Delta t^m$. Note that $\tilde{T}^{fm}_{s's}$ will depend on the time stepping method (explicit, implicit) chosen in eq. (4.2). With $FT^{fm+1}_{s's} = - FT^{fm+1}_{ss'}$, $\tilde{T}$ becomes symmetric: $\tilde{T}^{fm'}_{s's} = \tilde{T}^{fm'}_{ss'}$.

4.3 Relation to classical dynamic pseudo functions

The concepts developed in the previous section may not be directly recognizable as similar to the dynamic pseudo functions developed by Jacks et. al. [30] and Kyte & Berry [31]. To see the connection, note first that our compounded transmissibilities $\tilde{T}^{fm}_{s's}$ can be factorized

$$\tilde{T}^{fm}_{s's} = (A_{s's} / \Delta_{s's}) \, \tilde{\lambda}^{fm}_{s's} \qquad (4.11)$$

where $A_{s's}$ is the area of the common boundary between blocks s and s', and $\Delta_{s's}$ is the distance between block midpoints. The resulting pseudo mobilities $\tilde{\lambda}^{fm}_{s's}$ could be further factorized, to produce pseudo relative permeabilities, for conformance with the classical treatment.

Furthermore, imagine the (coarse) grid functions (V^{fm}_s, P^{fm}_s, FT^{fm}_s) to have been produced from a prior (parent) solution to the governing eq.s (2.1). Normally, such a parent solution will be a fine-gridded numerical solution, but any (approximate or exact) solution will do. The only requirement is that the parent solution should allow definition of coarse grid functions. If eq.s (4.8) and (4.9) are satisfied, the coarse grid functions will be exactly reproduced using the finite difference scheme (4.1)-(4.4), provided that parameters (pseudos) are chosen according to eq.s (4.5) and (4.10). We will, somewhat loosely, refer to this as "exact reproduction of the parent solution".

Our pseudo capillarities $\tilde{P}^{cm}_s$ and pseudo mobilities $\tilde{\lambda}^{fm}_{s's}$ still do not resemble the classical concepts, because they are functions of (discrete) time, and not of saturations. If the parent solution is such that there is a one-to-one correspondence between time t^m and grid block saturations S_s^{wm} (essentially determined by V_s^{wm}), then the substitution

$$t^m \longrightarrow S_s^{wm} \qquad (4.12)$$

allows the pseudos to be written as functions of saturation;

$$\tilde{\tilde{\lambda}}_{s,s}^{f}(S_s^{wm'}) = \tilde{\tilde{\lambda}}_{s,s}^{fm} \qquad (4.13a)$$

$$\tilde{\tilde{p}}_s^c (S_s^{wm}) = \tilde{\tilde{p}}_s^{cm} \qquad (4.13b)$$

for a set of discrete saturation values. In eq.(4.13) the interblock mobility is related to the upstream (assumed to be s') block saturation. This is dictated by the use of standard one-point upstream weighting for relative permeabilities assumed for construction of $\tilde{\tilde{1}}_{s,s}^{fm}$. With other weighting schemes similar equations to eq.(4.13a) can be written down, but generally become more complicated. It is important to note that different weighting schemes produce different pseudo mobilities, even with the same coarse grid and the same grid functions, if exact reproduction is aimed for.

Two slight generalizations above classical treatments are inherent in the present derivations, both of them necessary to allow exact reproduction of the parent solution: The use of time averaged fluxes in eq.(4.10) might be termed "pseudoization in the temporal dimension". Traditionally, instantaneous fluxes are used. We also have to relate our pseudo mobilities to grid block boundaries (via subscripts s's), making them tensorial in nature, whereas one normally thinks of dynamic pseudo rel. perm.s as scalars. Directional pseudo rel. perm.s are occasionally referred to in the literature, however.

We shall consider the ability to exactly reproduce a parent sclution to be an inherent characteristic of dynamic pseudo functions, even if not necessarily present in any given implementation. This characteristic brings out the intimate relationship between a set of dynamic pseudos and the particular parent solution used to generate them.

4.4 Properties of dynamic pseudo functions

It is evident from the discussion in previous sections that dynamic pseudos will depend on the parent solution used to generate them. Also, dynamic pseudos can not be constructed without reference to the (coarse) spatial and temporal grid on which they are later to be employed. Finally, if exact reproduction is wanted, the intended numerical method itself also influences the pseudos. We shall give some examples illustrating the effect of these non-physical parameters.

Consider the simple problem of 1D Buckley-Leverett displacement in a homogeneous reservoir. Relative permeability curves are linear. End point saturations are zero and unity, for both phases. Note that effective relative permeabilities in this case are identical to their rock counterparts, since the reservoir is homogeneous, making comparison between effective properties and dynamic pseudos straightforward. The problem allows analytical solutions, which serve as parent solutions in the different cases. Discretization is generally with three blocks, the middle block pseudo relative permeabilities being displayed in the figures (curves are continous and smooth because of interpolation between discrete saturation values). We use grid block average pressures in the (coarse) grid functions on which pseudos are based. If not indicated otherwise, an explicit integration in time is assumed, as well as one-point upstream weighting of relative permeabilities. Pseudos are constructed to allow exact reproduction of the parent solution, as described previously.

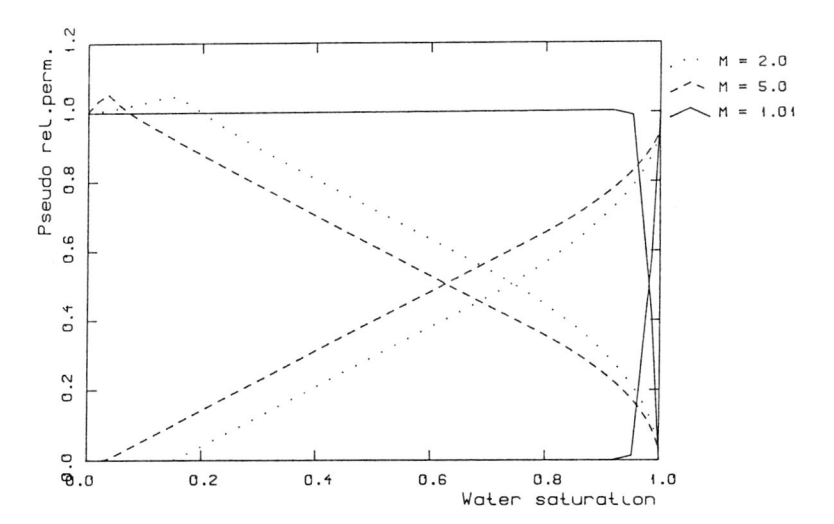

Fig.2. Dynamic pseudo relative permeabilities for a 1D homogeneous medium, for different viscosity ratios M.

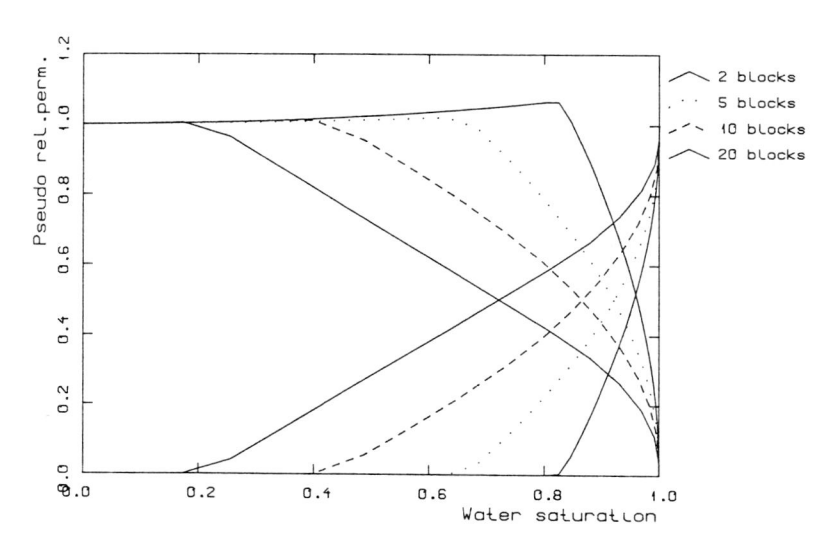

Fig.3. Dynamic pseudo relative permeabilities for a 1D homogeneous medium, for different discretizations.

AVERAGING OF RELATIVE PERMEABILITY   189

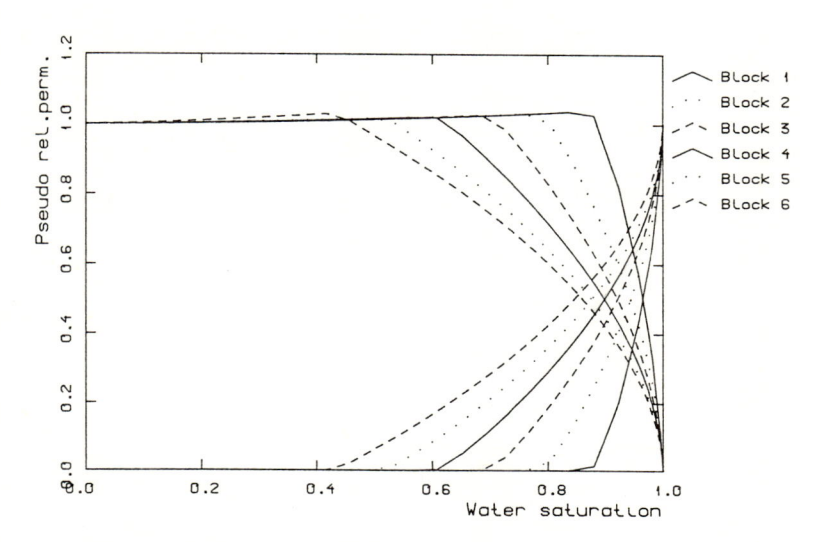

Fig.4. Dynamic pseudo relative permeabilities for a 1D homogeneous medium, in different grid blocks.

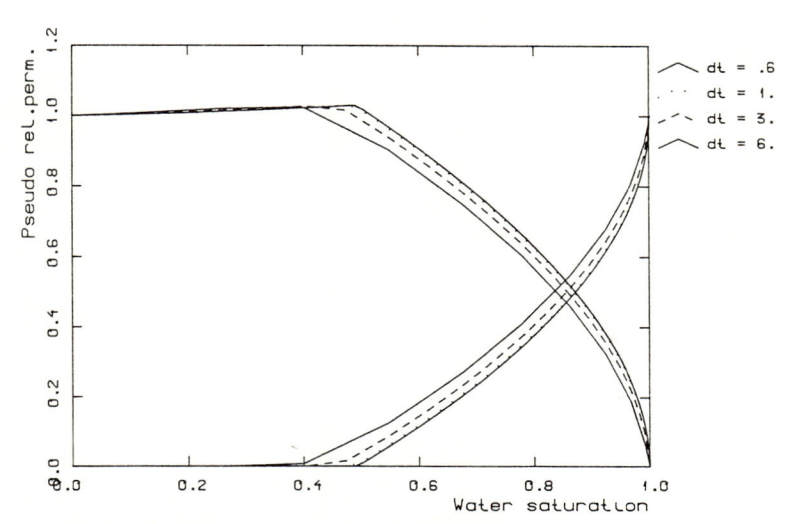

Fig.5. Dynamic pseudo relative permeabilities for a 1D homogeneous medium, for different time step lengths.

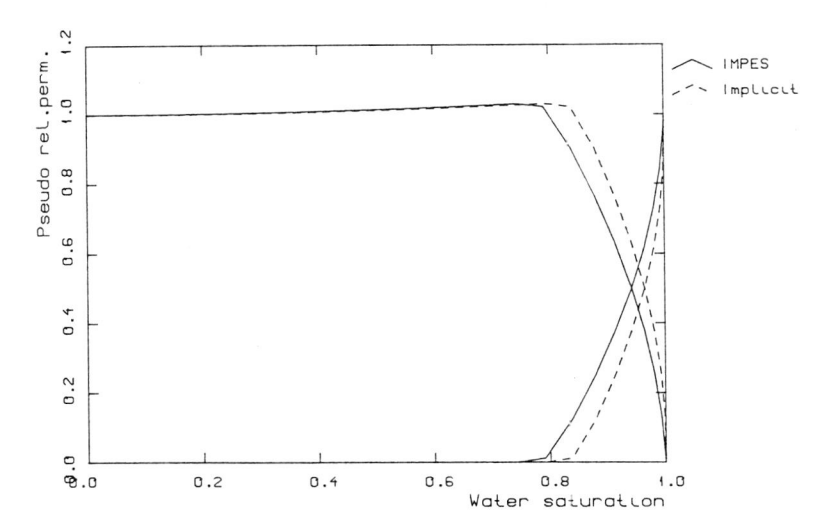

Fig.6. Dynamic pseudo relative permeabilities for a 1D homogeneous medium, for different time stepping procedures.

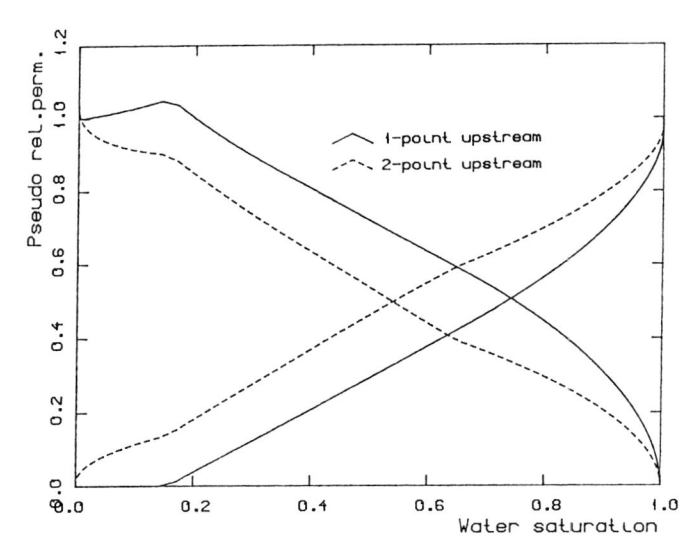

Fig.7. Dynamic pseudo relative permeabilities for a 1D homogeneous medium (block 2), with two mobility weighting schemes

In fig. 2, dynamic pseudo relative permeabilities for three different viscosity ratios M are displayed. As can be seen, these curves are markedly different from each other, and from the original straight line relative permeabilities.

In fig. 3, as in all subsequent cases except fig. 7, the viscosity ratio is 1.1. Pseudo relative permeabilities for four different discretizations are displayed, corresponding to approximately the same physical point. Again, dramatic differences exist, with fine discretization pseudos being closest to the rock curves.

In fig. 4, discretization is with 6 blocks, and dynamic pseudos for each block are displayed. The blocks closest to the injector have pseudos dramatically different from the rock curves.

In fig. 5, pseudos are shown for four different time step lengths. Differences are small, but significant.

In fig.s 6 and 7 (M=2), sensitivity to the coarse grid numerical method is examplified. Again, differences are significant.

By the previous discussion, there is no a priori reason to expect dynamic pseudos to have properties similar to those of their physical counterparts. Fig.s 2-7 all display apparently abnormal behaviour, in as much that pseudo relative permeabilities exceed unity. This is caused by the following effect: Displacement is always with an unfavourable viscosity ratio. Consider the period in time from water starts to enter the grid block in question until it starts to enter its downstream neighbour. The presence of water reduces the (grid function) pressure drop between the two blocks, without reducing the flux of oil (constant total flow rate). Therefore (compare eq. (4.10)), the oil pseudo relative permeability will exceed unity.

Also present in these figures, but not so clearly visible, is an effect connected to the substitution (4.12). If saturations remain constant for a period of time, the one-to-one correspondence between saturation and time is destroyed. If the pseudos change during such a period, multivalued pseudos result. In our cases, if the grid block in question has obtained full water saturation, only water flows out (at a constant rate). The (grid function) pressure drop between the block and its (downstream) neighbour will continue to fall, however, until the neighbour block has full water saturation, providing for a changing pseudo mobility for a fixed saturation. This translates into a vertical part in the pseudo relative permeability curves, at full water saturation.

5. RELATION BETWEEN DYNAMIC PSEUDOS AND EFFECTIVE PROPERTIES.

Traditionally [11,12-14], dynamic pseudos were regarded mainly as numerical devices for increased numerical accuracy, and for cheap pseudo 3D computation. In recent years [15-18], their use to capture the effects of heterogeneities suggests a close relationship to the concept of effective properties.

As discussed in previous sections, these two entities are conceptually very different. Effective properties can be defined only under relatively strict conditions (eq.s (3.16) and (3.17)). They become universally valid for cases where the conditions are satisfied, i.e. they appear as parameters in the ensuing partial differential equations for large scale variables.

Dynamic pseudos, on the other hand, are devices for reproduction of a given parent solution, in a subsequent coarse grid simulation. No restrictions apply to the parent solution, i.e. it can violate any and all of the conditions necessary to define effective properties. Dynamic pseudos can not be constructed without reference to the temporal and spatial numerical grid, and numerical method, with which they are later to be employed. Dynamic pseudos generally become dependent on the particular parent solution chosen, on the numerical grid, and also on the numerical method to be employed.

It is interesting to consider the dynamic pseudos resulting from a parent solution satisfying condition (3.16). Volume averaging will produce large scale variables $<S^f>$ and $<p^f>_{,j}$ which are constant (approximately) on the scale L of averaging. One would expect grid functions defined sensibly (i.e. via averaging) to behave similarily, if L now momentarily is taken to signify the linear (coarse) grid block scale. Then, one would expect

$$P_{s'}^{fm} - P_s^{fm} \sim \Delta_{s's} \tag{4.14}$$

Similarily, if Δt^m is not too large,

$$FT_{s's}^{fm} \sim \Delta t^m A_{s's} \tag{4.15}$$

Thus (compare eq.s (4.10) and (4.11)), one would expect dynamic pseudos to become independent of the coarse spatial and temporal grid. Likewise, we would expect the numerical method to have negligible influence on the result, essentially because any consistent numerical scheme will be able to approximate the large scale equations accurately, with the above assumptions.

We take the above as a strong indication that dynamic pseudos coincide with the corresponding effective properties, if constructed from a sufficiently accurate parent solution satisfying conditions (3.16) and (3.17), with L now being the (coarse) grid block scale. This observation provides a link between the two concepts. The insights provided by the study of effective properties may then be utilized in discussing the behaviour of dynamic pseudos.

It may be instructive to discuss some of our 1D examples in this light. Due to their simplicity (1D homogeneous), the l.h.s. of scale condition (3.16) is trivially satisfied. If the r.h.s. is satisfied, so is eq.(3.17). Thus, only the r.h.s. of (3.16) is relevant in these cases.

Dynamic pseudos are in all cases significantly removed from their effective counterparts. The two "abnormal" effects discussed previously can both be thought of as arising from violation of eq.(3.16), in that the (average) saturations change significantly over the averaging (grid block) scale.

In fig.s 2-4, the direction of change of the pseudo relative permeabilities is explainable by reference to (3.16). In fig. 2, as M increases, the saturation gradients decrease, providing for a better satisfaction of the r.h.s. of (3.16). This results in pseudo relative permeabilities becoming more rectilinear as M increases. Fig. 3 illustrates the same effect, but obtained by reducing the averaging (or grid block) volume. In fig. 4, saturation gradients decrease with the distance from the injector.

Fig. 8 is intended to illustrate the point made about the numerical method loosing its importance as (3.16) becomes better satisfied. Data are as in fig. 7, except that more blocks have been added to the reservoir. Block 9 pseudos are displayed. Comparing with fig. 7, one and two-point pseudos have become much more similar (and more close to the rock curves), again because of smaller saturation gradients.

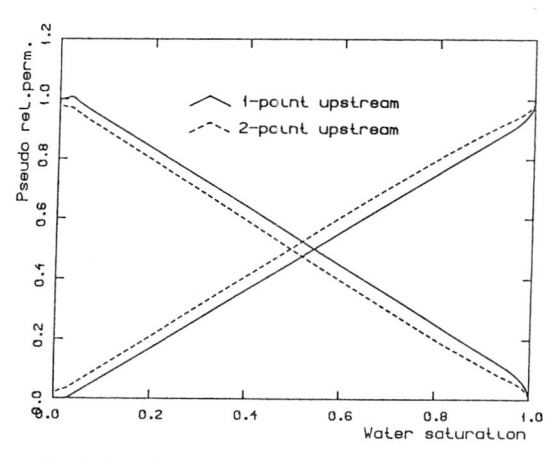

Fig.8. Dynamic pseudo relative permeabilities for a 1D homogeneous medium (block 9), with two mobility weighting schemes

6. ACKNOWLEDGEMENT

The present paper emerged from diverse work performed over the last few years. The work was funded by various organizations, mainly Statoil, through a British-Norwegian research program, BP Petroleum Development (Norway) Ltd. and the SPOR research program.

7. NOMENCLATURE

Latin symbols

A	-	Grid block boundary area
f	-	General function of spatial position
F	-	Flux (volume per unit of time)
FT	-	Time-integrated flux (volume)
g	-	Acceleration of gravity vector
G	-	Discretized gravity term
k	-	Permeability
k^{eff}	-	Effective permeability

l	-	Scale over which small scale variation takes place
L	-	Scale (linear dimension) of averaging volume
£	-	Scale over which variation of large scale parameters and variables take place. Subscripts k, p and S refer to permeability, pressure gradient and saturation, respectively.
M	-	Ratio between oil and water viscosities
p	-	Pressure
P	-	Grid block pressure.
p^c	-	Capillary pressure
r	-	Radius vector
$\mathbf{r}_0$	-	Radius vector to centroid of averaging volume
S	-	Saturation
t	-	Time
V_0	-	Averaging volume
V	-	Grid block phase volume.

Greek symbols

α	-	Proportionality constant for pressure deviations
Δ	-	(Representative) distance between grid blocks
Δt^m	-	$t^m - t^{m-1}$ - Time increment
λ	-	Mobility
λ^{eff}	-	Effective mobility
μ	-	Viscosity
ρ	-	Density
Φ	-	Porosity

Subscripts

$\left.\begin{array}{c}i\\j\\l\end{array}\right\}$ = 1,2,3 - Cartesian components (coordinates)

s,s'   - Spatial grid indices

s's   - Signifies quantity common to two neighbouring grid blocks

t   - Time

,   - Differentiation with respect to subsequent spatial or temporal coordinate.

Superscript

f   = n,w - Fluid phase

m   - Time level

Above symbol

     - Capital latin variables (P,V,FT) are grid functions used to construct pseudos

=   - Two-index tensor

~   - Finite difference variable

≈   - Pseudo function

^   - Spatial deviation

8. REFERENCES

[1]   Bear, J., Braester, C. and Menier, P.C.; Effective and Relative Permeabilities of Anistropic Porous Media, Transport in Porous Media, 2, 301-316, 1987.

[2]   Quintard, M. and Whitaker, S.; Two-Phase Flow in Heterogeneous Porous Media: The Method of Large-Scale Averaging, Transport in Porous Media, 3, 357-413, 1988.

[3]   Dagan, G.; Statistical Theory of Groundwater Flow and Transport: Pore to Laboratory, Laboratory to Formation, and Formation to Regional Scale, Water Resources Research, vol. 22, no. 9, 120S-134S, 1986.

[4]   Christensen, R.M.; "Mechanics of Composite Materials", John Wiley & Sons, 1979.

[5]   Batchelor, G.K.; Transport Properties of Two-Phase Materials with Random Structure, Ann. Rev. Fluid Dyn., 6, 227-255, 1974.

[6]   Beran, M.J.; "Statistical Continuum Theories", Interscience Publishers, 1968.

[7]   Crapiste, G.H., Rotstein, E. and Whitaker, S.; A General Closure Scheme for the Method of Volume Averaging, Chemical Engineering Science, vol. 41, no. 2, 227-235, 1986.

[8]   Dale, M.; Effective Relative Permeability for a 1D Heterogeneous Reservoir, paper presented at the NIPER/DOE Second International Reservoir Characterization Technical Conference, held in Dallas, 1989.

[9]   Coats, K.H., Nielsen, R.L., Terhune, M.H. and Weber, A.G.; Simulation of Three-Dimensional, Two-Phase Flow in Oil and Gas Reservoirs, SPEJ, 377-388, Dec. 1967.

[10]   Coats, K.H., Dempsey, J.R. and Henderson, J.H.; The Use of Vertical Equilibrium in Two-Dimensional Simulation of Three-Dimensional Reservoir Performance, SPEJ, 63-71, March 1971.

[11]   Thomas, G.W.; An Extension of Pseudofunction Concepts, SPE 12274, paper presented at the SPE Reservoir Simulation Symposioum, held in San Fransisco, 1983.

[12] Jacks, H.H., Smith, O.J.E. and Mattax, C.C.; The Modelling of a Three-Dimensional Reservoir with a Two-Dimensional Simulator - The Use of Dynamic Pseudo Functions, SPEJ, 175-185, June 1973.

[13] Kyte, J.R. and Berry, D.W.; New Pseudo Functions to Control Numerical Dispersion, SPEJ, 269-276, Aug. 1975.

[14] Starley, G.P.; A Material Balance Method for Deriving Interblock Water/Oil Pseudofunctions for Coarse Grid Reservoir Simulation, SPE 15621, paper presented at the 61st SPE Annual Technical Conference and Exhibition, held in New Orleans, 1986.

[15] Kortekaas, T.F.M.; Water/Oil Displacement Characteristics in Cross-Bedded Reservoir Zones, SPE 12112, paper presented at the 58th SPE Annual Technical Conference and Exhibition, held in San Fransisco, 1983.

[16] Davies, B.J. and Haldorsen, H.H.; Pseudofunctions in Formations Containing Discontinuous Shales; A Numerical Study, SPE 16012, paper presented at the 9th SPE Symposium on Reservoir Simulation, held in San Antonio, 1987.

[17] Lasseter, T.J., Waggoner, J.R. and Lake, L.W.; Reservoir Heterogeneities and Their Influence on Ultimate Recovery, in "Reservoir Characterization", edited by Lake, L.W. and Caroll, H.B., Academic Press, 1986.

[18] Kossack, C.A., Aasen J.O. and Opdal, S.T.; Scaling-up Laboratory Relative Permeability and Rock Heterogeneities with Pseudo Functions for Field Simulations, SPE 18436, paper presented at the 10th SPE Symposium on Reservoir Simulation, held in Houston, 1989.

EQUATIONS FOR TWO-PHASE FLOW IN POROUS MEDIA DERIVED FROM SPACE AVERAGING

D. Pavone

Institut Français du Pétrole

Abstract We start from mass and momentum balance equations which are valid at the pore level. Then, space averaging is used to define macroscopic variables and to derive equations that link these variables. That is the way a macroscopic mass balance equation and a macroscopic momentum balance equation are derived. But, the momentum balance equation fails to produce the generalised Darcy equation. Hence, thermodynamics of irreversible processes is used to generate two phenomenological laws. One is a generalised Darcy-like equation, the other is a capillary pressure equation. The generalised Darcy-like equation is standard except that it includes viscous and temperature couplings. The derived capillary pressure equation takes macroscopic fluid/fluid and fluid/solid interfacial areas into account plus a dynamic term. Finally, an explicit calculation of the derived capillary pressure equation in a conical capillary is given. It validates the derived equation and indicates a new way to compute capillary pressure in porous media.

1 Introduction

1.1 Background

Most oil recovery processes are based on the displacement of one fluid by another in porous media. The problem is to model fluid displacements without solving them in each pore. Of course, flow equations and boundary conditions are known in every pore of the porous medium, but the boundary locations are not.

Bundles of nonintersecting tubes are the simplest idealized and well-known mathematical structures that give boundary locations. Dullien [4] investigated more elaborate stuctures in which each tube radius may have axial variations. Since fluid retention is not explained by such models, Chatzis and Dullien [3] investigated two-dimensional networks.

As percolation theory does not need boundary locations, it was applied to capillary displacement (Larson et al. [7], Lenormand and Zarcone [8]). However, the percolation theory applied to porous media deals only with the minimum amount of porosity or fluid required for flow.

In fractal geometry, boundaries are not located exactly, but they are assumed to be self-similar. Jacquin and Adler [5] found that some geological structures are fractal. Adler [1] attempted to calculate the properties of such a medium from a continuous point of view. The longitudinal flow of a Newtonian fluid has been computed in a fractal network and single-phase permeability has been derived.

Space averaging was introduced to provide a theoretical justification of the Darcy-like equations for two-phase flow. Attempts made to justify these laws can be divided in two groups. The first uses volume averaging over a "representative elementary volume" (Marle [10], Slattery [13], Whitaker [14], Kalaydjian [6]). The second assumes that porous media are periodic, and space averaging is done over a period (Auriault and Sanchez-Palencia, [2]).

In 1982, Marle [10] formulated the basis of a volume averaging technique. The author treated the two-phase case [11]. He derived a generalised Darcy equation with couplings between phases and temperature and a capillary pressure equation which was the sum of a static and a dynamic term. The static part of the capillary pressure is proportional to the mean curvature of the interfaces.

2 Introduction

The following equations are generally required to model immiscible two-phase flow in porous media. Eq. 2.1 indicates that only fluid-a and fluid-b are present in the porous medium. Eq. 2.2 is the fluid-a mass balance equation. Eq. 2.3 is the fluid-a generalised Darcy equation and Eq. 2.4 links the fluid-a/fluid-b pressure difference to one saturation. Eq. 2.4 is called the capillary pressure equation. (Nomenclature in appendix.) The objective of this work is to derive, complete, or change such equations by theoretical means.

$$S_a + S_b = 1 \tag{2.1}$$

$$\frac{\partial(\Phi S_a R_a)}{\partial t} + \nabla.(R_a U_a) = 0 \tag{2.2}$$

$$U_a = -\frac{K k_{ra}}{\mu_a}(\nabla(P_a) - R_a g) \tag{2.3}$$

$$P_a - P_b = P_c(S_a) \tag{2.4}$$

3 Assumptions

The solid is at rest. It contains two immiscible phases, a and b, and interfaces. There is no flux of mass across the interfaces and no mass is adsorbed on the interfaces. There is no chemical reaction. We are in a Stokes regime and fluids are newtonian. The total area of interfaces depends only on the saturation value. The space distribution of the interfaces is isotropic. This last assumption is certainly the most restrictive. However, as our results hold true in conical capillaries, where this assumption is not verified, we may suppose that the presented results may be extended to the general case.

4 Volume Averaging

The work described in sections 2, 4 and 5 was first proposed by C.M. Marle in 1982 [10].

In porous media, variables related to one phase are defined only within this phase. So, at first, microscopic variables for one phase are extended to the whole medium. Values are set to zero everywhere but in the phase. Hence, microscopic equations can be written everywhere, provided derivatives are taken in the sense of distributions. As a consequence, Eqs. 4.1, 4.2 and 4.3 must be used as derivatives.

$$\operatorname{grad} f = \{\operatorname{grad} f\} + [f]_a^b n_a^b \delta_\Sigma \quad (4.1)$$

$$\operatorname{div} f = \{\operatorname{div} f\} + [f]_a^b \cdot n_a^b \delta_\Sigma \quad (4.2)$$

$$\frac{\partial(f)}{\partial t} = \{\frac{\partial(f)}{\partial t}\} - [f]_a^b v_a^b \cdot n_a^b \delta_\Sigma \quad (4.3)$$

Here, $\{\operatorname{grad} f\}$, $\{\operatorname{div} f\}$ and $\{\frac{\partial(f)}{\partial t}\}$ have the usual sense. $[f]_a^b$ is the jump of f when one crosses the interface Σ. n_a^b is the unit space vector normal to the interface. δ_Σ is the Dirac function, i.e. distribution associated to the interface $\sum$ and v_a^b is the velocity of $\sum$.

Then a space averaging of the microscopic equations is performed over a volume, called the Representative Elementary Volume (REV). If $f(x,t)$ is a locally defined function, the space averaging is performed by defining a convolution product, i.e.

$$F = f * m(x,t) = \int_{REV} f(x-y,t)m(y)d(y) \quad (4.4)$$

F is called the macroscopic function. m is introduced as a weight function because, if m is of class C^∞, $F = f * m$ is of class C^∞ too. m is zero everywhere except in the REV. In addition, $\int_{REV} m(y)d(y) = 1$.

5 Macroscopic Saturation Equation

Let 1_i be the function equal to 1 in phase-i and zero elsewhere. As phases do not overlap,

$$1_s + 1_a + 1_b = 1 \tag{5.1}$$

Hence, volume averaging of Eq. (5.1) gives

$$1_s * m + 1_a * m + 1_b * m = \int_{REV} m(y)d(y) \tag{5.2}$$

This defines three macroscopic variables, i.e. the porosity Φ and the two saturations S_i.

$$\Phi = 1 - 1_s * m \qquad \Phi S_i = 1_i * m \tag{5.3}$$

Hence, Eq. 2.1 can be proved at a macroscopic scale

$$S_a + S_b = 1 \tag{5.4}$$

6 Macroscopic Mass Balance Equation

Eq. 6.1 is the fluid-a mass balance equation valid at the pore level (ρ_a is density, u_a is velocity).

$$\frac{\partial(pa)}{\partial t} + \nabla.(p_a U_a) = 0 \tag{6.1}$$

First, we take the volume averaging of Eq. 6.1. Secondly, derivatives are performed according to Eq. 4.2 and 4.3. Nonetheless, we can see in this example that Dirac functions coming from the two derivatives cancelled. Thirdly, we commute the convolution product with derivatives. And finally, we obtain Eq. 6.2.

$$\frac{\partial(p_a * m)}{\partial t} + \nabla.((p_a U_a) * m) = 0 \tag{6.2}$$

Now, provided we define the macroscopic density R_a and the macroscopic filtration velocity U_a by

$$\Phi S_a R_a = p_a * m \qquad R_a U_a = (p_a u_a) * m \tag{6.3}$$

Eq. 2.2 is proved at a macroscopic scale, provided Eq. 6.3 are defined

$$\frac{\partial(\Phi S_a R_a)}{\partial t} + \nabla.(R_a U_a) = 0 \tag{6.4}$$

7 Generalised Darcy Equation Derived From Macroscopic Momentum Balance Equation

The same technique applied to the momentum balance equation leads to the macroscopic momentum balance equation

$$\nabla.(\mathcal{T}_a) - \mathcal{F}_{ab} = \mathcal{F}_{as} + \Phi S_a R_a g = 0 \qquad (7.1)$$

Here, $\mathcal{T}_a$ refers to the macroscopic fluid-a stress tensor, while $\mathcal{F}_{ab}$ and $\mathcal{F}_{as}$ are the macroscopic forces exerted by fluid-a on fluid-b and on the solid [10]. g is gravity.

Indeed, Eq. 7.1 should be used in place of the experimental Generalised Darcy's Law. However, it is not very useful because $\mathcal{T}_a$, F_{ab} and F_{as} cannot be estimated. Nevertheless, if we assume that fluids are newtonian and if we use analogy with the single-phase Darcy law, we can show [12] that

$$\mathcal{T}_a = \Phi S_a P_a I \qquad \mathcal{F}_{as} = \beta_a U_a + \beta_b U_b \qquad \mathcal{F}_{ab} = \alpha_a U_a + \alpha_b U_b - P_a \nabla(\Phi S_a) \qquad (7.2)$$

where I is the identity tensor and P_a is defined by $p_a * m = \Phi S_a P_a$. In [12], we show that this definition of P_a is consistent with the definition of P_a induced by the macroscopic Gibbs-Duhem equation. In fact, hydrodynamic definition and thermodynamic definition of P_a are equivalent. α_i and β_i are unknowns. Substituting Eqs. 7.2 in 7.1 and combining this equation for phase-a with its homologue for phase-b gives rise to a Darcy-like equation. This "quick" demonstration is not satisfactory. In addition, nothing can be derived concerning the capillary pressure.

8 Generalised Darcy Equation Derived From The Entropy Source

The Entropy Source

Thermodynamics of irreversible processes (TIP) is used to generate phenomenological laws as follow. The entropy source (σ_e) must be presented in a bilinear form ($\sigma_e = \sum_i J_i X_i$). Near a local equilibrium, if J_i and X_i are both zero at the equilibrium state, linear relationships exist ($J_i = \sum_j L_i^j X_j$). According to the Curie principle, J_i and X_j are identical rank tensors. L_i^j can depend on all variables but not on their derivatives. Onsager equations are: $L_i^j = L_j^i$ or $L_i^j = -L_j^i$, depending on the rank of the tensors.

Macroscopic entropy balance equations for phase-a, phase-b and interface were first derived by Marle [10]. The whole entropy source σ_e is computed by summing up the entropy source of the phases and of the interfaces. The whole entropy source (multiplied by T for simplification) can be simplified [12] to obtain

$$T\sigma_e = U_a\left[-\nabla(P_a) + R_a g\right] + U_b[-\nabla(P_b) + R_b g]$$
$$-\gamma_{ab}\frac{(A_{ab})}{\partial t} + \gamma_{ab} Z_{ab,s} + [P_a - P_b]\frac{\partial(\Phi S_a)}{\partial t} + T(\sum Q_i)_{i=a,b,s,ab}.\nabla(1/T) \quad (8.1)$$

T is the temperature, γ_{ab} the interfacial tension and A_{ab} is the macroscopic ab interfacial area. $Z_{ab,s}$ is connected to the area swept by the three-phase line. The three-phase line is the line where phase a, phase b and the solid phase are in contact. If $v_{\partial\Sigma}$ is the velocity of the three-phase line, $\delta_{\partial\Sigma}$ the Dirac distribution of the three-phase lines and t the tangent to the interface perpendicular to the three phase line, $Z_{ab,s}$ is defined by [10]

$$Z_{ab,s} = ((v_{\partial\Sigma}.t)\delta_{\partial\Sigma}) * m \quad (8.2)$$

If we assume that the wettability contact angle θ is constant, we obtain

$$Z_{ab,s} = -cos\theta\,(\mid v_{\partial\Sigma}\mid \delta_{\partial\Sigma}) * m \quad (8.3)$$

The minus sign appears because t is directed outwards. Now, using Eq. A5 page 600 of [10], we have

$$\frac{\partial(\delta_{\Sigma as})}{\partial t} = \mid v_{\delta\Sigma}\mid \delta_{\partial\Sigma} \quad (8.4)$$

As the macroscopic phase-a/solid interfacial area is defined by $\delta_{\Sigma_{as}} * m = A_{as}$,

$$Z_{ab,s} = -cos\theta\frac{\partial(\delta_{\Sigma_{as}})}{\partial t} * m = -cos\theta\frac{\partial(A_{as})}{\partial t} \quad (8.5)$$

If it is assumed that A_{as} depends only on ΦS_a,

$$Z_{ab,s} = -cos(\theta)\frac{\partial A_{as}}{\partial(\Phi S_a)}\frac{\partial(\Phi S_a)}{\partial t} \quad (8.6)$$

The entropy source becomes

$$T\sigma_e = U_a[-\nabla(P_a) + R_a g] + U_b[-\nabla(P_b) + R_b g]$$
$$+[P_a - P_b - \gamma_{ab}cos(\theta)\frac{\partial A_{as}}{\partial(\Phi S_a)} - \gamma_{ab}\frac{\partial A_{ab}}{\partial(\Phi S_a)}]\frac{\partial(\Phi S_a)}{\partial t}$$
$$+T(\Sigma \vec{Q}_i)_{i=a,b,s,ab}\cdot\nabla(1/T) \quad (8.7)$$

Generalized Darcy Equation

We can show [12] that each term of the bilinear form (Eq. 23) is zero at the equilibrium state. Hence, the TIP leads to the conclusion that there are three rank-two tensors K_{aa}, K_{ab} and K_{aq} such that:

$$U_a = K_{aa}[-\nabla(P_a) + R_a g] + K_{ab}[-\nabla(P_b) + R_b g] + K_{aq}\nabla(1/T) \quad (8.8)$$

Eq. 8.8 is a generalised Darcy-like equation for two-phase flow in porous media. It proves the validity of Eq. 2.3. However, it calls for two remarks. First, this equation involves couplings between the two phases and the temperature gradient. Such a result was already found [3,15,22,25]. Secondly, we can write K_{aa} as $K_{aa} = K k_{ra}/\mu_a$, where K is the rank-two tensor single-phase permeability and μ_a is the viscosity. It defines the relative permeability k_{ra} as a rank-two tensor, but nothing tells us i) that k_{ra} does not depend on μ_a and ii) that k_{ra} depends only on the saturation.

Capillary Pressure Equation

TIP concludes that there is a parameter ($\pi \geq 0$) such that:

$$P_a - P_b = \gamma_{ab}cos(\theta)\frac{\partial A_{as}}{\partial(\Phi S_a)} + \gamma_{ab}\frac{\partial A_{ab}}{\partial(\Phi S_a)} + \pi\frac{\partial(\Phi S_a)}{\partial t} \quad (8.9)$$

which becomes, with the Young-Dupré equation

$$P_a - P_b = (\gamma_{as} - \gamma_{bs})\frac{\partial A_{as}}{\partial(\Phi S_a)} + \gamma_{ab}\frac{\partial A_{ab}}{\partial(\Phi S_a)} + \pi\frac{\partial(\Phi S_a)}{\partial t} \quad (8.10)$$

Eq. 8.9 (or 8.10) can be connected with Eq. 2.4 because we assumed that interfacial areas only depend on the saturation. Nonetheless, it is different from the standard capillary pressure equation (Eq. 2.4) because it shows that the capillary pressure depends on the saturation through the interfacial areas. This capillary pressure equation (Eq. 8.9) is threefold and each term has a physical meaning. The first term takes into account the pressure required to substitute Phase-a/solid intermelecular forces by Phase-b/solid intermolecular forces, whenever interfaces are moving. The second term takes the pressure required to extend the interface between Phase-a and Phase-b into account.

The third term is dynamic and we will follow D. Lhuillier in its interpretation. At the equilibrium state, if one pressure is suddenly increased, the curvature of the interface would not change instantaneously to balance the new pressure difference. Actually, the fluids would move (inducing a $\frac{\partial(\Phi S_a)}{\partial t}$) to reach a new equilibrium state corresponding to a new curvature. π takes the inertia induced in the relaxation time needed to reach the new equilibrium into account.

9 Capillary Pressure Equation Applied to a Conical Capillary

This section computes the capillary pressure equation (Eq. 8.9) in a simple case, i.e. the conical capillary. The aim is to validate the equation and to improve physical knowledge about capillary pressure. The conical capillary is described in fig. 1. Note that a spherical bubble without contact with the solid corresponds to $\alpha = \pi$. $\alpha = 0$ gives rise to the simple cylindrical capillary and a bubble lying on a surface is given by $\alpha = \pi/2$.

A moving interface (at constant θ) induces interfacial area and saturation variations. We can compute the two static terms of the capillary pressure.

$$\gamma_{ab}\cos(\theta)\frac{\partial A_{as}}{\partial(\Phi S_a)} = \frac{2\gamma_{ab}}{R}\frac{\cos(\theta)^2\cos(\theta)\sin(\alpha)^{-1}}{2 - 3\sin(\theta - \alpha) + \sin(\theta - \alpha)^3 + \cos(\theta - \alpha)^3\cot(\alpha)} \quad (9.1)$$

$$\gamma_{ab}\frac{\partial A_{ab}}{\partial(\Phi S_a)} = \frac{2\gamma_{ab}}{R}\frac{2 - 2\sin(\theta - \alpha)}{2 - 3\sin(\theta - \alpha) + \sin(\theta - \alpha)^3 + \cos(\theta - \alpha)^3\cot(\alpha)} \quad (9.2)$$

Fortunately, the sum of these two terms is very simple. Moreover, it corresponds to the well known Laplace equation for a capillary and hence, confirms the validity of Eq. 8.9.

$$P_a - P_b = \frac{2\gamma_{ab}}{R} = \frac{2\gamma_{ab}\cos(\theta - \alpha)}{r} \quad (9.3)$$

In this particular case, we must remark that $\frac{P_a - P_b}{\gamma_{ab}\cos\theta}$ is not an invariable parameter, as is generally assumed in porous media.

Conclusions

Volume averaging of pore-level mass balance equations gives rise to i) the definitions of macroscopic variables and to ii) equations which link these

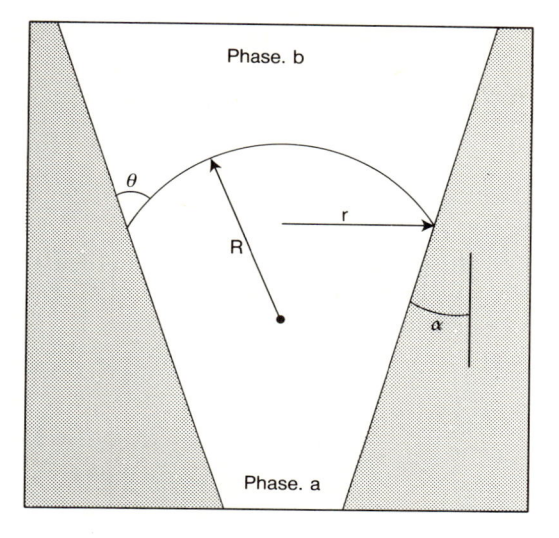

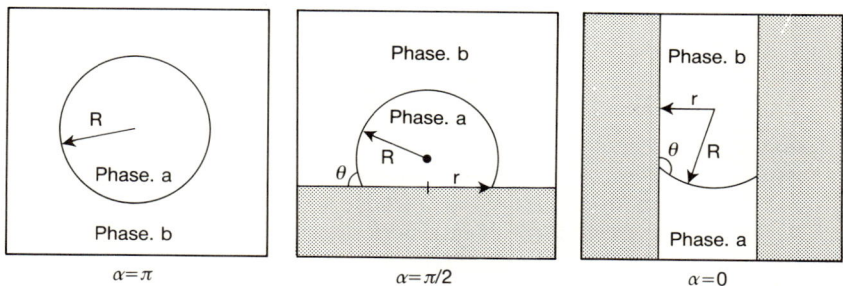

Figure 1. Description of a conical capillary for capillary pressure computation. Depending on α value, computation holds true for spherical bubble, bubble lying on a surface or cylindrical capillary.

variables. These macroscopic equations are the well known mass balance equations for porous media. The macroscopic entropy source is derived and thermodynamics of irreversible processes is used to generate two phenomenological equations.

The first equation is a generalised Darcy equation which contains viscous and temperature couplings. This law was already known for steady state flows [2]. Because we never need assumptions about steady state flows, this Darcy-like equation is still valid for unsteady state flows.

The second equation is a capillary pressure equation. This equation is different from the standard one because it involves macroscopic interfacial areas as well as saturation. However, each term that composes this equation has an obvious physical meaning. Moreover, when this equation is computed in a conical capillary, it gives rise to the well known Laplace equation. This capillary pressure equation may be a new way to improve porous media knowledge.

Acknowledgements

The author wishes to thank Prof. C.M. Marle of the Institut Français du Pétrole and Prof. D. Lhuillier of University Pierre et Marie Curie (Paris) for their advice and constant support.

Nomenclature

This nomenclature is consistent with nomenclature in [10] and [12].

a	Subscript for phase a.
b	Subscript for phase b.
i	Any subscript.
s	Subscript for solid.
A_{ab}	Macroscopic interfacial area between phase a and phase b.
A_{is}	Macroscopic interfacial area between phase i and solid.
$\mathcal{F}_{ab}$	Macroscopic force of interaction exerted by phase a upon phase b.
$\mathcal{F}_{is}$	Macroscopic force of interaction exerted by phase i upon the solid.
g	Acceleration due to gravity.
$J_i X_i$	Example of bilinear form.
K	Single-phase permeability.
k_{ri}	Relative permeability to phase i.
m	Function for convolution.
p_i	Microscopic pressure in phase i.
P_i	Macroscopic pressure in phase i.
P_c	Capillary pressure.
Q_i	Macroscopic heat flux in phase i.
ρ_i	Microscopic density in phase i.
R_i	Macroscopic density in phase i.
1_i	Presence function for phase i.
S_i	Saturation in phase i.
T	Macroscopic temperature.
T_i	Macroscopic stress tensor for phase i.
$\vec{t}$	Unit vector tangent to the interface, perpendicular to triple the line, directed outwards.
t	Time.

u_i Microscopic local velocity in phase i.
U_i Macroscopic filtration velocity in phase i.
$v_{\partial \Sigma}$ Three-phase line velocity.
$Z_{ab,s}$ Microscopic term taking the three-phase line into account.
μ_i Viscosity of the fluid in phase i.
γ_{ab} Fluid/fluid interfacial tension.
$\delta_{\Sigma_{is}}$ Distribution with support the interface is.
$\delta_{\partial \Sigma}$ Distribution of $\delta_{\Sigma_{ab}}$.
θ Wetting contact angle.
α_i Unknown multiplicative coefficient comming from TPI.
Φ Porosity.
σ_e Entropy source.
I Identity tensor.

References

1. Adler, P.M., "Transport Processes In Fractals", *Int. J. Multiphase Flow*, 11, 1, 91-108, 1985.

2. Auriault, J.L. and Sanchez-Palencia E., "Remarques Sur La Loi De Darcy Pour Les Ecoulements Biphasiques En Milieu Poreux", *J. of Theor. and Appl. Mech.*, special no, pp 141-156, 1986.

3. Chatzis, G. and Dullien, F.A.L., "Modelling Pore Stucture By 2-D And 3-D Networks With Application To Sandstones", *J. Can. Petrol. Tech.*, 97-108, 1977.

4. Dullien, F.A.L., "New Network Permeability Model Of Porous Media", *AIChE J.*, 21, 299-307, 1975.

5. Jacquin, C.G. and Adler, P.M., "Fractal Porous Media: Geometry Of Porous Geological Structures", *T.I.P.M.*, 2-6, 571-596, 1987.

6. Kalaydjian, F., "A Macroscopic Description Of Multiphase Flow In Porous Media Involving Spacetime Evolution Of Fluid/Fluid Interface", *T.I.P.M.*, 2, 6, 537-552, 1987.

7. Larson, R.G., Scriven, L.E. and Davis, H.T., "Percolation Theory Of Two-Phase Flow In Porous Media", *Chem. Eng. Sci.*, 36, 57-73, 1981.

8. Lenormand, R. and Zarcone, C., "Invasion Percolation In An Etched Network: Measurement Of A Fractal Dimension", *Phys. Rev. Letters*, 54, 20, 2226-2229, 1985.

9. Lhuillier, D., "Phenomenology Of Hydrodynamic Interactions In Suspensions Of Weakly Deformable Particles", *J. Physique*, 48, 1887-1893, 1987.

10. Marle, C.M., "On Macroscopic Equations Governing Multiphase Flow With Diffusion And Chemical Reactions In Porous Media", *Int. J. Engng. Sci.*, 20, 5, 643-662, 1982.

11. Marle, C.M., Personal communication, 1982.

12. Pavone, D., "Macroscopic Equations Derived From Space Averaging For Immiscible Two-Phase Flow In Porous Media", *Rev. Inst. Franc. du Pétrole*, 44, 29-41, 1989.

13. Slattery, J.C., "Multiphase Viscoelastic Flow Through Porous Media", *AIChE J.*, 13, 50-56, 1968.

14. Whitaker, S., "Flow In Porous Media: A Theoretical Derivation Of Darcy's Law", *T.I.P.M.*, 1, 1, 3-25, 1986.

STATISTICAL PHYSICS AND FRACTAL DISPLACEMENT PATTERNS IN POROUS MEDIA

Torstein Jøssang
(Dept. Physics, University of Oslo, Norway)

ABSTRACT

It is well known from early experiments that in the process of displacing a high viscosity by a low viscosity fluid in a porous medium, one creates irregular and highly ramified fluid-fluid interfaces which are fractals. Such processes and their resulting structures are the central and motivating theme for this commencing talk where stochastic modelling techniques from statistical physics and computer simulations will be extensively applied.

As an introduction the role of statistical physics in the modelling of disordered systems is commented on. Then the conceptual framework of critical phenomena, especially universality and (finite size) scaling, is briefly reviewed. The characteristic viscous fingering (VF) patterns obtained at high flow rates (where viscous forces dominate) in 2D porous Hele Shaw cells exhibit very similar branching structures to patterns obtained in very different experiments like dielectric breakdown, diffusion limited electrodeposition and chemical dissolution. In turn all of these show striking resemblance to the computer generated diffusion limited aggregates (DLA) for the same geometry. This common symmetry and the very close values of the fractal dimension indicates some kind of universality. In the invasion percolation (IP) regime at low flow rates (when capillary forces dominate) the interface bears close resemblance to well known percolation structures which have been extensively studied both by computer simulation and by statistical physics. Although different, VF (DLA) as well as IP structures exhibit scaling. Their self similar structures and recent results concerning their dynamic (scaling) properties will be reviewed.

In some respects such structures resemble snapshots of fluctuating equilibrium systems close to a critical point which suggests a deeper connection between the growth of fractal structures and the statistical physics of phase transitions. Employing methods of statistical physics, some encouraging results have been obtained. Our theoretical understanding is, however, still far from being complete even for the simplest nonequilibrium growth processes. While the value of present application of real space renormalization group methods to these problems is unclear, most of the activity in the field has been based on computer simulation, which has supplied us with valuable classification schemes and insight. Recent progress in both these fields will be discussed briefly.

1 INTRODUCTION

The observation of displacement patterns, like the interface between water vapour and air in moving clouds, is an everyday experience. The complicated geometry that mirrors the transport processes governed by the laws of statistical physics has fascinated artists and concerned scientists from ancient times. Similar phenomena are also observed in porous materials where the crumpled structure of moving fronts between fluids and also the dynamical changes within them, have become the new puzzle and challenge of *Statistical physics* and *Fractal geometry*. Even though its importance is recognized in many branches of science and technology, very little encouragement and support has been offered from industry and engineering projects.

This presentation aims at showing how Stochastic modelling based on the latest development in **Statistical physics** can describe complicated structure and dynamics of **Model experiments** and how new concepts in **Fractal geometry** bring some **Order** into the description of the apparent **Chaos** observed in fluid -- fluid displacement processes in **Porous media**.

Except for a few celebrated examples (harmonic oscillator and Ising model) statistical physics until recently has not been very successful in calculating the specific physical properties of systems where the elementary units (atoms ... grains) interact. *Statistical physics* as part of fundamental science, however, needs broad experimental verification if theoretical methods and results are to be trustworthy. Fortunately present strategy and methodology in the field are promising for many engineering applications. The task is, however, immense since *Statistical physics* and *Condensed matter physics* are also required both to have validity and to explain the physical properties of all systems of length scales which cover the whole range between atomic physics (10^{-7}cm) and astrophysics (10^{20}cm). But *Statistical physics* and *Condensed matter physics* also must explain, based on microscopic interactions, how physical systems suddenly change and display totally new properties, e.g. as happens in magnetic, superconducting and ferroelectric systems when they are, for instance, subject to changes in external influence such as temperature and pressure . This is the very profound theory of phase transitions and critical phenomena which we briefly review first in this talk. It is primarily the tools developed in the study of these phenomena that form the basis of the new strategy used in the statistical physics community today for stochastic modelling of disordered systems and processes taking place within them.

The application of the fundamental laws of statistical mechanics is above all concerned with obtaining the *correct averages* for the problem at hand. Due to the large number of particles, or alternatively the large number of collisions an individual particle participates in, one cannot keep track of all the microscopic information. One is thus forced to use *statistical averages* and keep only the relevant information. As the number of particles increases, the relevant *random distributions* become narrower, the relative fluctuations decrease and the average description becomes more and more appropriate. The noninteracting many particle systems (ideal gas etc.) are elementary and known to you from college physics.

It is, however, a discouraging fact that interacting many-body systems, which we treat in almost all engineering applications, constitute the toughest problems in theoretical physics. No general solutions exists. Until the renormalization group (RG) theory was invented --- for which Wilson was awarded the 1982 Nobel prize in physics --- we were not even able to give practical recipes for how to attack such problems by perturbation theory (techniques). The application of the new RG methods and use of the associated conceptual framework to transport problems in strongly disordered geometries, has only gained momentum in the last 5 years, and most of our present understanding of phenomena in porous media is even more recent.

The statistical physics community is convinced that a deeper understanding of the flow problems encountered in oil recovery and basin formation must be based on this novel approach. The enhanced knowledge to be accumulated in such a new approach will have an important impact on oil recovery activities and change the mathematical modelling used today in reservoir engineering.

In this lecture we shall review some on the background material and first give an introduction to the key concepts of universality, scaling, finite size scaling, crossover and finally fractals. Then we shall discuss stochastic modelling in percolation and viscous fingering. At the end we shall describe some recent results from computer and real model experiments done in our group in Oslo.

2 STATISTICAL PHYSICS AND PHASE TRANSITION

First a short tutorial with the purpose of presenting the three central concepts of *Universality*, *Scaling* and *Crossover* from modern statistical physics (Domb and Lebowitz (1970-1988)) on which we shall base our results for porous media. This will be done by suggestive arguments and examples from simple model systems also chosen from the field of porous media. Rather than beginning with the geometrical description (i.e., percolation type models) of porous media directly, I have chosen to start out from ordinary thermal phase transitions to emphasize the enormous and solid effort on which the qualitative new conceptual framework, now available to us, has been built. Following that treatment we show the relevance of those concepts to the physics of porous media.

2.1 The Ising Model and Phase Transitions

Among the models displaying phase transitions, the Ising types are the simplest, physically correct and realistic models known. These models consists of ions with magnetic moments μ_i, all of identical size ($|\mu_0|$), which sit on each lattice site of a (hyper-) cubic lattice having N sites along each of its edges in d-dimensions (square lattice in fig 1a). The $\mu_i's$ can only have two orientations, pointing up or down in the Ising models. The magnetic interaction energy between the ion at i and the ion at j is simply proportional to $-\mu_i \cdot \mu_j$ if they are in vacuum. Because they also interact with the rest of the ions in the lattice, the energy for a given configuration is

$H = -\sum_{pairs} a_{ij}(r) \times \boldsymbol{\mu}_i \cdot \boldsymbol{\mu}_j$

We treat here the *ferromagnetic Ising model* only, which in the simplest version considers nothing but the nearest neighbours interactions, and for those all $a_{ij} = a > 0$, (whereas all other $a'_{ij}s$ are zero). Since the two possible orientations of each $\boldsymbol{\mu}_i$ can just as well be described by $s_i = \pm 1$, the energy of any configuration can be written

$$H = -J \sum_{nearest pairs} s_i s_j \qquad (1)$$

The total magnetization for a specific configuration of magnetic moments would simply be:

$$M = \sum_i \boldsymbol{\mu}_i = N^d \times \boldsymbol{\mu} \qquad (2)$$

where the sum extends over all the lattice sites and $\boldsymbol{\mu}$ is the average magnetization per lattice site. Employing the correct statistical mechanical procedures for ensemble averaging the value of M can be shown to be zero at very high temperatures since thermal disturbances (entropy effects) make the orientation of the constantly fluctuating spins random. In the other limit of $T = 0$, M is equal to $M_0 = N^d \times \boldsymbol{\mu}_0$, reflecting the fact that when entropic effects are absent all the moments line up parallel due to the ferromagnetic interaction. Consider now the square lattice where we know the exact solution. At the critical temperature

$$T_c = \frac{2J}{k_B \ln(\sqrt{2}-1)} = 2.2692\ldots \frac{J}{k_B} \qquad (3)$$

where k_B is Boltzmann's constant, the average magnetization per spin, $\boldsymbol{\mu}$, changes from zero to a finite value (figure 1(b) solid curve). Below T_c

$$|\boldsymbol{\mu}| \sim (T_c - T)^\beta \qquad (4)$$

These results tell us that the system at T_c --- suddenly and as a result of cooperative interactions between the 'atomic magnets' $\boldsymbol{\mu}_i$ --- displays the totally new property of *macroscopic magnetization* when we cool.

The Norwegian physicist Lars Onsager (1942) demonstrated how to obtain the correct statistical mechanical solution to the

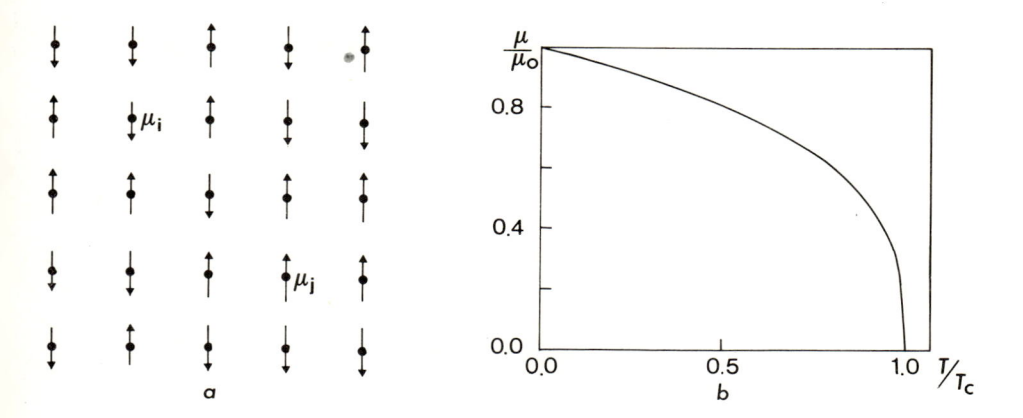

Figure 1: a). Two dimensional square lattice of magnetic ions μ_i b). The magnetization in a 3-dimensional cubic lattice of spins as function of temperature.

problem defined by eq.(1). Soon after the result $\beta = \frac{1}{8}$ for the two-dimensional ferromagnetic lattice, which is the only exactly solvable case, was obtained. Later, clever numerical methods have been developed and they give $\beta \sim 0.312$ in 3-dimensional systems, as illustrated in fig. 1b. The Ising model and Onsager's solution might seem academic only, at first. That is, however, not true:

- There exists 3-dimensional crystals in nature with regular mechanical properties, which due to their magnetic interactions, should behave as predicted in eq.(4) both for 2- and 3-dimensional types of interaction, *and they really do*.

- The universality principle, to which we will return below, clearly demonstrates the practical importance of the Ising model.

- The conceptual framework developed eventually led to the renormalization group theory and initiated a new way of thinking which later was applied to many areas in science, including the physics of porous materials.

We shall discuss some of the main strategies and results of this approach and try to illustrate its usefulness in treating strongly disordered materials like porous media and transport processes within them.

2.2 Phase Transitions — Universality and Scaling

The hallmark of every phase transition --- as indicated above --- is that a new property (structural and physical) suddenly arises. This can only happen in systems with "interactions" between the elementary units (like the a_{ij} in the Ising model).

This quantity (μ in ferromagnets) that suddenly comes into existence is generally called the order parameter, OP. Asymtotically its temperature dependence just below T_c is a power law like that found for the Ising model:

$$OP = (|\mu|) \sim \left(\frac{T_c - T}{T_c}\right)^\beta \equiv \varepsilon^\beta \tag{5}$$

for small $\varepsilon > 0$. For many systems the order parameter may have several components, being a vector or a tensor. In all cases there is necessarily a symmetry difference between the high and low temperature phases and, in the language of statistical physics, the order parameter OP, is said to break the symmetry at T_c.

Since Onsager's discovery, thousands of physicists and mathematicians all over the world have tried to find the corresponding solutions for 3-dimensional systems, but no one has succeeded so far. Wilson, however, found a way around the problem with his RG approach. His method involves a new averaging strategy and a selfconsistent technique to obtain the relevant observable physical quantities and calculate such exponents as β.

2.2.1 Universality

Today we know that it is very improbable for the simple 2-dimensional Ising model which Onsager solved to occur in nature. For most magnetic systems the interaction will most certainly extend out to second nearest neighbours and even beyond. They will in general have different strengths (a_{ij}) and equation (1) becomes so complicated, even in 2-dimensions, that an analytical solution cannot be found. Nevertheless, today we know from experiments as well as theory, that the power law form in eq. (5) is independent of the range of interactions --- as long as their range is finite --- with exactly the same exponent β. This value of the critical exponent $\beta = \frac{1}{8}$ is also the same for all different 2-dimensional lattice symmetries, indicating a universal feature. The specific value of T_c and the constant of proportionality in eq. (5) are, however, material constants that may change from one material to another.

According to the old numerical methods and present methods based on RG-theory, $\beta \sim 0.312$ for all 3-dimensional Ising ferromagnets. This result has --- analogously to those in two dimensions --- been confirmed experimentally (over many decades in ε) for many real systems. The general understanding of these results is that the behaviour close to T_c is determined only by the *dimensionality d* of the specimen (counted according to the nature of the interaction responsible for the transition) and the *symmetry* (number of components) of the OP (μ).

This is **universality** in the sense of statistical physics. All materials with those two features in common, are said to belong to the same **universality class**. This means that when we have determined the properties of one material, we also know the critical behaviour of all others in the same universality class.

The gas -- liquid phase transitions (for fluids with simple molecules like those of water) are probably the most frequently observed thermal phase change in nature. The order parameter in this case is the difference between the densities of the liquid and the vapour, $\Delta \rho_{LG} = \rho_L - \rho_G$. It can be shown that this OP exhibits the same symmetry as μ of the Ising ferromagnet and we can therefore, according to the universality principle, expect the phase transitions in liquids to be described by eq. (5) with the same $\beta \sim 0.312$ in 3 dimensions. This is proven beyond doubt experimentally, illustrating the general nature of statistical physics laws commented on in the introduction.

Universality thus allows us to study the simplest available system in a universality class, to predict the physical properties close to the critical point of more complicated real systems in nature. As we see next, the concept of scaling has even deeper consequences.

2.2.2 Scaling

In addition to the order parameter itself, the specific heat at constant magnetic field in a ferromagnet (which was the problem Onsager originally solved for in 1942) or equivalently at constant pressure in a fluid system is a relatively simple quantity to measure. It was observed to have anomalously high values in real materials close to T_c long before theory could explain it. Present theory predict its singular behaviour to be:

$$C_h \sim |T - T_c|^{-\alpha} \sim |\varepsilon|^{-\alpha} \qquad (6)$$

Next we shall, in addition to the exponents α and β, introduce three others, γ, ν and η, belonging to the same zoo of critical exponents, to illustrate the concept of scaling.
The (magnetic) susceptibility χ, measures how the order parameter (magnetization) responds to its conjugate field (a uniform external magnetic field) h and is given by:

$$\chi = \left.\frac{\partial(OP)}{\partial h}\right|_{h=0} \sim |\varepsilon|^{-\gamma} \qquad (7)$$

where γ is a positive quantity and therefore also χ diverges at T_c. For fluid systems the corresponding quantity to χ is the compressibility (κ).

Let us next consider the conditional probability, $C(r)$, of having the same spin orientations at $r'(=0)$ and some distance away r:

$$C(r) = \langle S(0)S(r) \rangle \qquad (8)$$

This is nothing but the two-spin correlation function. $C(r)$ is large within clusters of parallel spins, but falls rapidly as we enter a cluster of antiparallel spins. The statistical average eq. (8) will therefore give a function with a typical range --- called the correlation length and denoted by ξ --- defined by

$$\xi^2 = \frac{\int d^d\boldsymbol{r}\, r^2 C(r)}{\int d^d\boldsymbol{r}\, C(r)} \qquad (9)$$

where $d^d\boldsymbol{r}$ is the volume element in the d dimensional system. ξ is another important critical quantity and can be shown to diverge at T_c as

$$\xi \sim |T - T_c|^{-\nu} \sim |\varepsilon|^{-\nu} \qquad (10)$$

This means that the size of the correlated regions or equivalently size of the clusters of the same phase, becomes increasingly larger as the system approaches its critical point. $C(r)$ has a singular dependence on r at the critical point itself:

$$C(r) \sim r^{-d+2-\eta} \qquad (11)$$

introducing the dimensionality d into the exponent of r on equal footing with the critical exponent η.

A very large number of spin configurations (fig. 1) have energies in a very narrow band (infinitely) close to the thermodynamic equilibrium energy. Snapshots or a movie of the spin configurations (e.g. from a dynamical Monte Carlo simulation of the magnet in thermal equilibrium) would show very disordered and ever changing configurational structures of the spins. If we illustrate down spins as black discs for a given snapshot, its representation will look much like the patterns shown in fig. 2a. When we approach T_c, ''connected'' clusters grow in size and their black/white borderlines separating down spins from up spins becomes increasingly complex. To the human eye these snapshots all look the same when we consider areas smaller than ξ. Also different parts of one snapshot are indistinguishable and smaller parts look like a reduced version of the whole frame at T_c where $\xi \to \infty$. Structures which exhibit this property are said to be **selfsimilar** and obey the laws of fractal geometry (Mandelbrot (1982)).
This observation --- i.e., that the only characteristic length which enters the correlation function near T_c is ξ --- forms

the very basis for the new scaling theory and it has deep consequences for the system behaviour and physical properties close to T_c. One of these is that the five exponents introduced here and the dimensionality of the system are related by the following three independent *scaling equations*:

$$\alpha + 2\beta + \gamma = 2 \quad \bullet \quad (2-\eta)\nu = \gamma \quad \bullet \quad \nu d = 2 - \alpha \qquad (12)$$

If we therefore know two among the critical exponents, the rest of them are given by the scaling relations. Once again we see the crucial role played by the dimensionality of the system. We shall return to the subject of dimensionality later. From a practical point of view, scaling combined with universality tell us that it is sufficient to determine the critical behaviour of two properties (e.g. magnetization and specific heat) in the simplest system one can find in that universality class to know the all critical properties of all the materials in that class.

2.2.3 Percolation

The self similar patterns displayed when $T \sim T_c$ in the maps of the snapshots of the fluctuating magnetic moments (spins) has built in correlations because of the interaction between them. At $T = 0$ all the moments are parallel and could e.g. point up (correlations extend to infinity according to eq. 10) , so the map would be uniformly white with our convention, except for the grid which is present only to indicate the lattice structure.

By randomly removing ions from the lattice (at $T = 0$) with a probability p and changing the rules such that absent ions are represented in black, black ''clusters'' consisting of nearest neighbour black discs only, will appear as islands in the white sea of magnetic ions with spins pointing up. For very low p the average size of the black clusters is small, but as a *critical concentration* is reached, the black clusters will disconnect the spanning white sea so that no continuous "channel" of magnetic ions penetrates the system anymore. Since our ferromagnetic model has only nearest neighbour interactions there is clearly no long range magnetic order possible at any temperature any more, and ferromagnetism is destroyed, not because we changed the interaction, but because of the *changes in geometry*. This is the case of the strongly diluted ferromagnet of much current interest in statistical physics. Here we shall, however, not discuss any further the thermodynamics of such systems but rather turn to the consequences of the disordered geometry itself.

The geometrical disordering process leads to a long range order by establishing spanning black channels for transport when p is sufficiently high. For a finite sample, the probability of forming a spanning channel is obviously finite for any p except $p = 0$. However, and this is the fundamental theorem of percolation theory, *in the limit of an infinitely large system, the probability that a given site belongs to the spanning cluster, $P_\infty(p) = 0$, for any p less than a critical value p_c. When p exceeds p_c, however, $P_\infty(p)$ is finite and given by:*

$$P_\infty \sim (p - p_c)^\beta \tag{13}$$

The use of this notation suggests that we are drawing an analogy between percolation and an ordinary thermal phase transition, and indeed we are. $P_\infty(p)$ can be shown to be the order parameter, analogous to $\mu(T)$ in the ferromagnet.

3 STATISTICAL PHYSICS AND POROUS MEDIA

Porous media contain pores and pore necks of all sizes connecting them. In addition paths of higher permeabilities --- like plane layers or more irregular patterns of coarser sands or cracks --- may penetrate the material. Porous media may therefore have very complicated and disordered structures which also often are heterogeneous.

The simplest model one can think of to describe such a medium is the plain *percolation model* (Stauffer (1985), Aharony (1986)), which is illustrated in figure 2. Communication (transport) between all the pores within a cluster is possible through the connecting pore necks. There is, however, no channel open for transport all the way through the medium in figure 2a, which was generated for $p = \frac{1}{2} < p_c \sim 0.593$ on the square lattice. In figure 2b, which is generated on a much larger lattice for $p = p_c$ and where only the largest cluster is shown, we see that

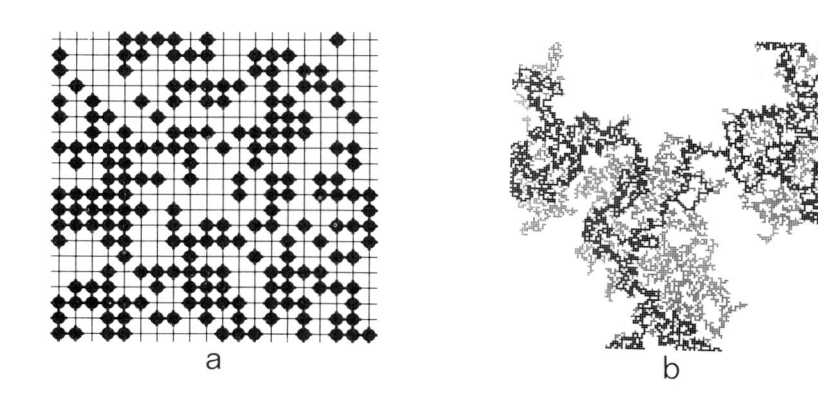

Figure 2: To illustrate the geometry of percolation. Part a) (Feder (1988)) is generated by stepping through all sites of a square lattice and tossing a coin at each point. Every time a head results, we mark that point with a black disc (occupation probability $p = \frac{1}{2}$). An n-cluster consists of n black discs which all have at least one nearest neighbour black disc. Observe the size distribution ($\langle s \rangle$, eq. 15) of clusters. (Here s is between 1 and 46). In part b) (Oxaal et.al.(1987)) which was generated in the same way, but for for $p = p_c \sim .593$ and on a much larger lattice, we have kept only the largest (percolating cluster), which connects all the outer boundaries. All the small ones have been removed, also those within the ''lakes'' in the percolating cluster itself. The part of this cluster which is represented in the darkest print, is called the **backbone** and it is the main channel for transport through the material.

all the outer boundaries are connected by the spanning channels of the percolating cluster. The main channel for e.g. fluid transport is however the **back bone** which is represented in the darkest printing in figure 2b. The back bone has the property that it breaks the rest of the cluster into subclusters (dangling ends), each of which is connected to the back bone itself with only one single pore neck (bond). An intuitive way to see that the back bone is the main transport channel is to regard the whole cluster initially filled with an incompressible fluid. Then we try to recover that fluid by injecting another fluid from the left. This invading fluid can than only move on the back bone since it would otherwise have to compress the defending incompressible fluid in the dangling ends.

When physical processes like this take place in disordered materials, methods based on valid results from statistical physics are needed to find the correct solutions. In the case of displacement flow in porous media most of the effort in the physics community has been concentrated on understanding 2-dimensional model experiments in the limits of fast (viscous fingering) and slow (invasion percolation) flow rates. We shall next describe the basic physics of these phenomena illustrated by the percolation model and thereafter study experiments in the two regimes.

3.1 Physics of Percolation Models

In the simplest version of the percolation model, one considers a d-dimensional lattice just as in the case of the diluted ferromagnet (at $T = 0$). But whereas the latter had built-in correlations because of the interacting spins, the percolation model is completely random as indicated above. In spite of its simplicity the resulting predictions may be very important since they are valid for the whole universality class to which it belongs. Instead of removing ferromagnetic ions we now take away material units (grains) or equivalently just state that a given site is occupied by a "crack" with probability p or left "blocked", with probability $q = 1 - p$. For small p, this normally creates only small clusters of connected "cracks". As p increases, the typical linear size of clusters ξ and the average number of "cracks" per cluster, $\langle s \rangle|$, increase (fig. 2a). For an infinitely large lattice, both diverge to infinity at the *critical threshold concentration*, p_c. Near p_c, these divergencies are described by

$$\xi \sim (p_c - p)^{-\nu} \qquad (14)$$

$$\langle s \rangle| \sim (p_c - p)^{-\gamma} \qquad (15)$$

The "*critical exponents*" β, ν and γ introduced here are quite analogous to those for thermal phase transitions and they also, in this purely geometric system, turn out to be *universal*, depending only on the dimensionality of the lattice, d, and not on other details (e.g. the symmetry of the lattice, the addition of ''cracks'' between next nearest neighbour vertices, etc.). Specifically, for this universality class, $\beta = 5/36 \simeq 0.14$ and 0.4, $\nu = 4/3 \simeq 1.33$ and 0.88 and $\gamma = 43/18 \simeq 2.39$ and 1.8 for $d = 2$ and 3 respectively. The same scaling relations eq.(12) as developed above also apply.

3.1.1 Finite systems — Scaling and Crossover Phenomena

Above p_c there exists a ''*spanning*'' cluster or equivalently an open channel, which connects the ends of the lattice. Thus, fluid will be able to "percolate" from one end of the sample to another. The probability of belonging to this cluster is given by eq.(13). If the "mass" of the spanning cluster was distributed uniformly over space, then the mass M in a volume of finite, linear size L would be

$$M = P_\infty L^d, \qquad (16)$$

with the average density P_∞. This turns out to be wrong on length scales L which are smaller than the typical linear size of a finite cluster, ξ. For $L < \xi$, measurements of M in boxes of size L show that (Aharony (1986)),

$$M \sim L^D, \; for L < \xi \qquad (17)$$

with the fractal dimensionality
$$D = d - \beta/\nu \qquad (18)$$
equal to $(\frac{91}{48})$ or about 1.9 and 2.5 at d=2 and 3, respectively:

Eqs. (16) and (17) may be combined into the *finite size scaling* form:
$$M(L) = L^D m(L/\xi) \qquad (19)$$
The scaling function $m(x)$ approaches a constant for $x < 1$, and becomes a power law, $m(x) \sim x^{\beta/\nu}$, for $x > 1$. This is a typical *crossover phenomenon*, which occurs whenever there is more than one length scale present in a system. It is absolutely necessary to observe this in the interpretation of laboratory experiments and computer simulations on finite size specimens if one is to find the correct answer for (larger) real systems

Equation (19) also implies that the *porosity*, i.e. the density of effective "cracks", which may participate in the transport of fluid through a sample, depends on the size of the sample. For sizes $L < \xi$ the average porosity depends on L, as
$$\frac{M(L)}{V} \sim L^{D-d} \sim L^{-\beta/\nu} \qquad (20)$$
where V is the volume (area) of the specimen and one cannot simply extrapolate from small samples to large ones! The necessity to observe the laws of finite size scaling in extrapolating from small to larger systems in real as well as computer experiments, cannot be over emphasized.

Similar crossover phenomena occur for all the physical properties of the spanning cluster. For example, the *permeability* k of a sample, i.e. the rate of flow for a unit pressure drop, behaves as
$$k \sim (p - p_c)^\mu \sim \xi^{-\mu/\nu} \qquad (21)$$
in the "homogeneous" regime, $L > \xi$, but depends on the size of the sample L, as
$$k(L) \sim L^{-\mu/\nu} \qquad (22)$$
in the "fractal" regime, $L < \xi$.

The fact that the *geometry* of the percolation clusters is *fractal* on length scales $L < \xi$ implies that there are very strong geometrical constraints on physical processes which occur within

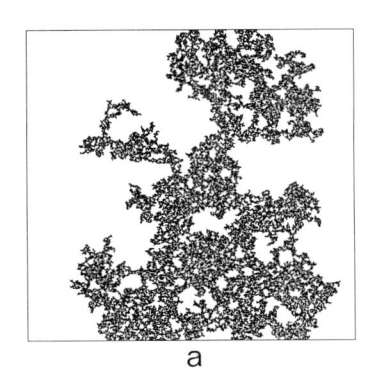

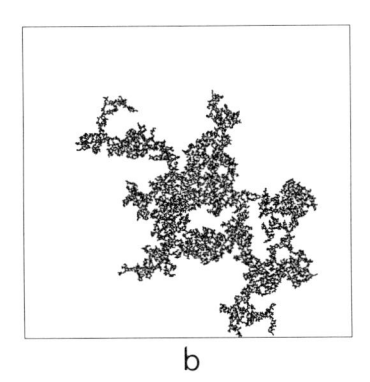

Figure 3: Ordinary percolation (a) and invasion percolation with trapping (b) on the same 400×400 lattice. The invader (black) enters from the centre and the defender (white) escapes through the square rim.

the "pores" or "cracks". The study of these effects has been one of the active fields in statistical physics in recent years.

3.2 Invasion Percolation

Is it possible to model the (time dependent) displacement process discussed briefly above? Which path would an invading fluid chose if it was injected into a porous material originally filled with another immiscible (defending) fluid which wets the solid material of the matrix. If the injection rate is very low, we expect the capillary forces to be dominant. To simulate such a process one could employ the rule that the invading fluid simply displaces the defending fluid through the widest pore-neck available at the perimeter of the invaded region. Of course, regions that are trapped cannot be invaded since the fluids are supposed incompressible. The process described by this algorithm has been baptised **invasion percolation**. We first describe the scaling properties of the structure it leaves behind and then proceed to discuss its dynamical scaling properties.

Imagine the pore size distribution to be given by random numbers assigned to the lattice points of a 2 dimensional lattice which is at p_c. After removing all the smaller clusters, we are left with the percolating cluster connecting the bottom and the top of the finite size specimen in figure 3a. Based on the very same pore size distribution we inject in the centre of this specimen an invading fluid. Guided by the rule to always invade the easiest pore available, the pattern in figure 3b results. We see that the main structural characteristics are common for the two patterns in figure 3, but there are also notable differences. Interestingly this 'greedy' algorithm where one always enters the 'best' pore results in the trapping of domains of all sizes. The structure of the invaded region is fractal so that the mass $M(L)$ increases with the size L of the invaded region in qualitatively the same way as for ordinary percolation (eq.17):

$$M(L) \sim L^D, \qquad (23)$$

where the fractal dimension now for invasion percolation with trapping is $D \simeq 1.82$, as compared to $D \simeq 1.89$ for ordinary percolation. It seems that the dynamics of the invasion percolation is driving the process towards a critical state where scaling ideas are necessary to obtain consistent results.

Only very preliminary experiments have been performed to study the *dynamics* of the invasion percolation process. We observed (Målø̸y et.al.(1985),Feder et.al.(1986)), however, that the front advanced in bursts and we have recently shown, by computer simulations, that this process leads to an interesting *dynamic scaling* which also describes the burstlike growth of the front (Furuberg et.al.(1988)) and a static structure in accordance with experiments. Qualitatively, this can be understood: The invading front first occupies all easily accessible pores. Gradually the front piles up against pores which are more difficult to invade. Then by forcing the invader through a difficult pore a new region, which may have new easily invaded areas, is made available to the front. The front then moves into this area until it gets stuck again having exhausted the new easily invaded pores.

To obtain a *quantitative* measurement of these time correlations, we considered the pair correlation function $N(r,t)$, giving the probability that a site a distance r from a reference site is invaded a time t later than the reference site. Thus $N(r,t)drdt$ is the conditional probability that if a site at the position r_0 was invaded at time t_0, another site at a distance between r and $r+dr$ away from r_0 is invaded in an interval dt around t_0+t. During the simulations the reference site is successively chosen to be the last invaded site. After some initial transient effects, the function $N(r,t)$ is found to be independent of t_0. This is expected since the invasion process is governed by local rules, and the surroundings of every new invaded site are statistically similar to the surroundings of any other invaded site independent of how far the process has developed.

We find that the correlation function obeys the *dynamic scaling* form
$$N(r,t) = r^{-1}f(r^z/t) , \qquad (24)$$
with the new dynamic exponent
$$z = D . \qquad (25)$$
We also find that the scaling function $f(u)$ is peaked at $u \sim 1$, and has an unusual limiting power law behaviour at both limits,
$$f(u) \sim \begin{cases} u^a, & u \ll 1 \\ u^{-b}, & u \gg 1 \end{cases} \quad \text{with} \quad a \simeq 1.4 , \ b \simeq 0.6 . \qquad (26)$$

This implies that the most probable growth occurs at $r_t \sim t^{1/D}$. At time t, most of the region within distance r_t has already been invaded, the rest of the region contains trapped defender fluid, and new growth there is rare. On the other hand, the probability to select a new growth site at $r \gg r_t$ decays rapidly with r. Roux and Guyon (1989) have recently shown that $b = 1/D$ consistent with our results. They also found deviations from the power law we suggested in the lower limit with $a = 1$, which also is in accordance with our numerical data, even though we conjectured scaling behaviour.

In figure 4 we show the correlation function $N(r,t)$ obtained in extensive simulations. First we note that for $t = 1$ growth

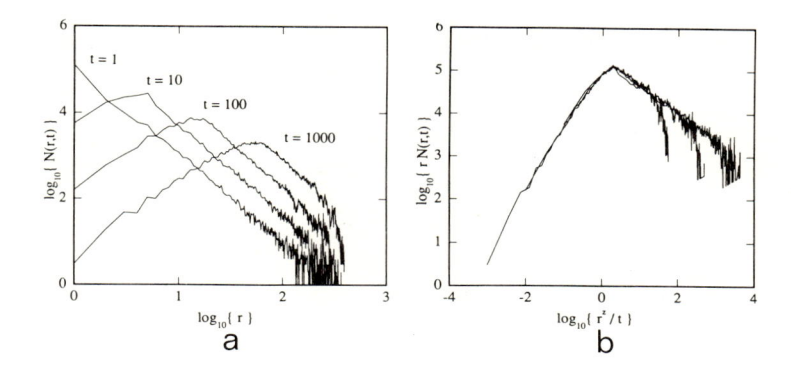

Figure 4: The data in the figure were generated in 8 simulations of the invasion percolation process on a lattice of size 300×600. (a) The correlation function $N(r,t)$ is a function of the separation r between sites invaded with a time separation t, and the figures shows results for time separations $t = 1$, 10, 100 and 1000. (b) Log-log plot of the scaling function $f(u) = r \cdot N(r,t)$ as a function of $u = r^z/t$ and with $z = 1.82$. Finite size effects result in a violation of the data collapse for $r \sim L$.

is most probable close to the previously invaded site, and that the growth decreases as a power law r^{-2} as r increases. For $t > 1$, we find that after a time t has passed, growth is most probable at some distance r_t away. We find that $r_t \sim t^{1/z}$ with $z \simeq 1.8$. These results suggest the scaling form given in (24), which leads to the satisfying data collapse shown in figure 4 with $z = 1.82$.

Experiments are in progress to measure the dynamics of the invasion percolation process in two-dimensional disordered models.

4 EXPERIMENTAL --- FLOW IN DISORDERED MODELS

Next we describe fluid-fluid displacement experiments for several geometries in two-dimensional systems and discuss the structural properties of the fronts and patterns that arise by displacement processes. All experiments were performed in 2-dimensional model cells, which were made porous by sandwiching a disordered, single layer of small glass beads between two stiff transparent plates as described in detail in Ref. Måløy et.al.(1987).

Depending on the displacement rates, viscosity ratios, miscibility, interfacial tensions and pore geometry a bewildering variety of displacement fronts arise which often are fractal (Mandelbrot (1982), Feder (1988)). For a systematic description of the many regimes observed under various conditions during two-fluid displacement processes in micromodels of porous media, see Lenormand (1988). In the typical experiments described here air (white) was used to displace coloured glycerol (black), which wets the model and initially is filling the pore space. The models had a porosity $\phi \simeq 0.7$.
To have a unique reference we introduce here the dimensionless *capillary number* Ca defined by

$$Ca = \frac{U\mu}{\sigma}, \qquad (27)$$

which measures the ratio of viscous (μ) to capillary forces (σ), and U is a typical front velocity.

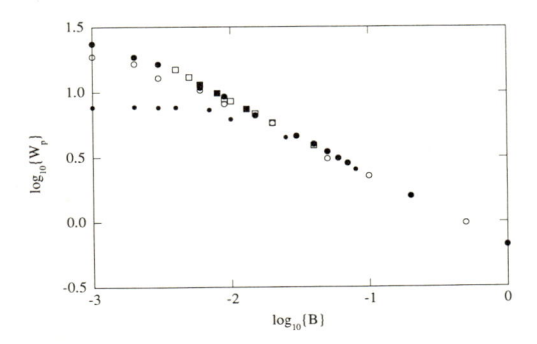

Figure 5: The width of the external perimeter as function of Bond number B.

4.1 Invasion Percolation with Gravity Effects

At very low displacement rates $Ca \ll 10^{-4}$ and with horizontal 2-dimensional cells like those described above, the experiments show (see fig. 7b) that the static structures observed are fractals as predicted for *invasion percolation* (Wilkinson and Willemsen (1983)). Here the capillary forces govern the process and the size of the system is the only relevant length scale. In 3-dimensional systems (Wilkinson (1984), Clément et.al.(1985)) or when our 2-dimensional cell is tilted, the process becomes influenced by buoyancy forces. The Bond number (B) gives the relative magnitude of gravitational and capillary forces. The buoyancy introduces a new length scale in the displacement front and in the structure of the residual oil. Here we (Birovljev et.al.(1989)) present preliminary results from slow displacement experiments in a tilted cell and corresponding computer simulations where the invasion percolation is modelled with a linear pressure gradient in the glycerol phase.

The external perimeter is the part of the coastline between air and glycerine that can be reached from the air-side by a finite size particle. This perimeter has a width W_p which, in a finite system of size L, is expected to scale as

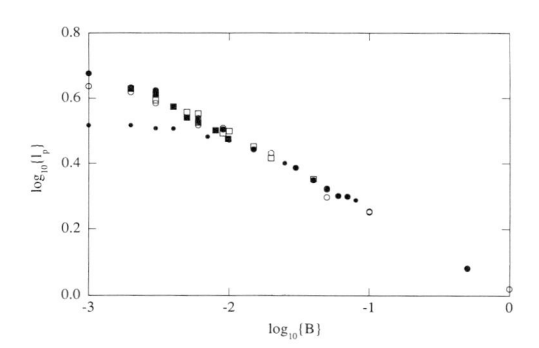

Figure 6: The length of the external perimeter as function of Bond number B.

(Biroljev et.al.(1989))(eq. 14):

$$W_p(B,L) \sim \left\{ \begin{array}{ll} L, & BL^{(1+\nu)/\nu} \ll 1 \\ (1/B)^{\nu/(1+\nu)}, & BL^{(1+\nu)/\nu} \gg 1 \end{array} \right\} \qquad (28)$$

The results of many measurements of W_p's from computer simulations on systems of different sizes L and Bond numbers B is shown in figure 5, where $\log(W_p)$ is plotted as a function of $\log(B)$. For finite systems and small B the width increases linearly the system size L and in the plot one clearly sees the crossover for small systems. The measured slope is: -0.57 ± 0.01, the same as predicted by theory: $\frac{\nu}{1+\nu} = 0.57$ for $\nu = 4/3$. Sapoval et al.(1985) obtain similar results for the diffusion front.

The first experiments have only been done on a small model giving only four points over one decade with the result $\sim -0.8 \pm 0.4$ for the measured slope.

When $BL^{(1+\nu)/\nu} \ll 1$, the external perimeter length scales with the system size: $l_p \sim L^{D_e}$. For large Bond numbers the external perimeter length scales linearly with the system size (L) and with a power expressed by D_e and the correlation length (ν) with the Bond number (B) (Biroljev et.al.(1989)):

$$l_p \sim \left\{ \begin{array}{ll} L^{D_e}, & BL^{(1+\nu)/\nu} \ll 1 \\ LB^{(D_e-1)\nu/(1+\nu)}, & BL^{(1+\nu)/\nu} \gg 1 \end{array} \right\} \qquad (29)$$

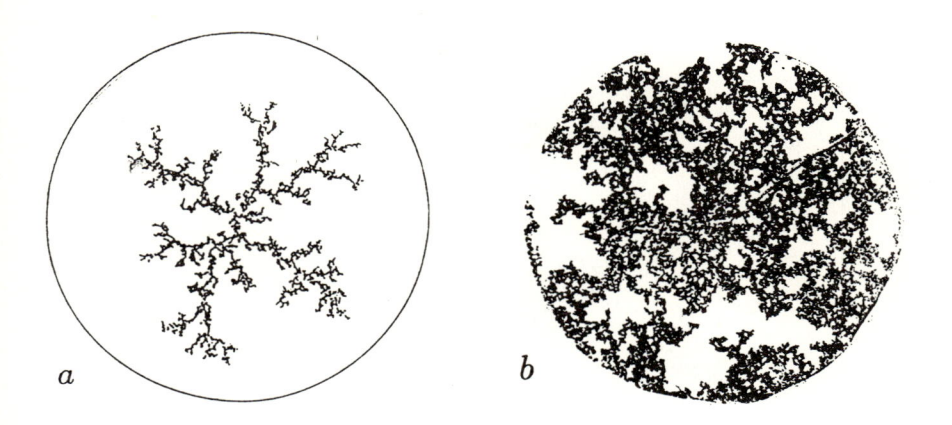

Figure 7: Displacement of glycerol (black) by air in a two-dimensional model consisting of a random layer of 1 mm spheres. (a) Fractal viscous fingering at high displacement rates, $D = 1.64$. (b) Invasion percolation at low displacement rates $D \simeq 1.8$;

Here the analysis of the computer simulations, figure 6, gives a measured slope of -0.23 ± 0.02, whereas theory predicts 0.19. More accurate experiments are in progress for this case.

4.2 Fractal Viscous Fingering

When a high viscosity fluid is displaced at high capillary numbers $Ca >> 10^{-4}$ by a low viscosity fluid in a porous medium, fractal viscous fingering (VF) results. This front structure can now be rationalized, as demonstrated by Chen and Wilkinson (1985) and Måløy et al.(1985). We analyzed fingering structures by digitizing pictures such as the one shown in figure 7a. The number $N(r) (\propto$ to the mass of displaced fluid $M(r))$ of black pixels was measured as a function of distance, r, from a point near the centre of injection. Fractal structures exhibit scaling and we expect $M(r)$ to have the form

$$M(r) \sim M(r/R_g)^D . \qquad (30)$$

Here D is the fractal dimension of the structure and R_g is its radius of gyration. M is the total number of black pixels, which correspond to the porespace from which glycerol has

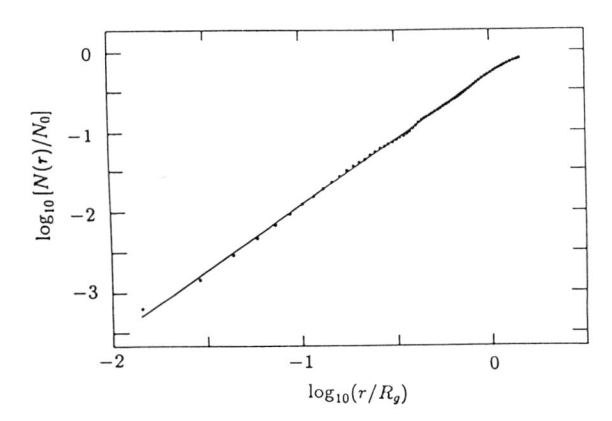

Figure 8: The normalized finger structure area $M(r)/M$ as function of the reduced radius r/R_g. The solid line obtained from a fit to the experimental data has a slope of 1.64

been displaced. The scaling power-law equation (30) is expected only for the range $a < r < R_0$, where a is a typical pore dimension, and R_0 is the radius of the model.

In figure 8 we plot $\log[M(r)/M]$ as function of $\log(r/R_g)$ for the structure shown in figure 7b. By fitting equation (30) in the range $2a < r < R_g$, we find a fractal dimension $D = 1.64 \pm 0.04$.

This fractal structure closely resembles that obtained from the diffusion limited aggregation (DLA) model of Witten and Sander (Witten and Sander (1981), Vicsek (1989)). Similar structures have also been obtained by fluid-fluid displacement in radial Hele-Shaw cells using non-Newtonian viscous fluids (Nittmann et.al.(1985), VanDamme et.al.(1986)). The relationship between fluid-fluid displacement in porous media and DLA was first discussed by Paterson (1984). The original DLA model of Witten and Sander specifies a time ordered sequence of events but is not explicitly time dependent. While the time dependent aspects of DLA and closely related processes have been extensively explored theoretically (Deutch and Meakin (1983)), (Hentschel et.al.(1984)) and by means of computer simulations (Meakin and Deutch (1984), Voss (1984)), experimentally this aspect of DLA has been explored only recently (Måløy et.al.(1985),

Måløy et.al.(1987)). The dynamics of the viscous fingering structures have also been discussed in terms of multifractal measures
(Måløy et.al.(1987),Meakin (1987)). Recently a crossover function has been shown to describe the structure of DLA and viscous fingers, which are not self-similar (Hinrichsen et.al.(1989)).

4.3 Fractals on Fractals

The fact that the *geometry* of the percolation clusters is *fractal* on length scales $L < \xi$ implies that there are very strong geometrical constraints on physical processes which occur within the "pores" or "cracks". The study of these effects has been one of the active fields in statistical physics in recent years. What happens when fractal processes take place on systems with fractal structures. This could shortly be referred to as *fractals on fractals*. There are many such examples, which are only at the beginning of their exploration. One example that is highly relevant to this conference concerns *viscous fingers on percolation* clusters. When a nonviscous fluid (air or water) is pushed into a viscous one (oil) in a porous medium, it develops an instability which results in a fractal fingering structure (Chen and Wilkinson (1985),Måløy et.al.(1985)) as we just discussed. In two dimensional models, the fractal dimension of these is about 1.7.

Different physical properties of the fractals turn out to require different subsets of the bonds on the fractals, which have different *fractal dimensionalities* (Aharony (1986)). For example, the flow of an incompressible fluid in the pores of the percolation model never enters "dangling ends", which are connected to the rest of the cluster only via a single "gate". The remaining biconnected part of the cluster, or the "*backbone*", has a fractal dimensionality D_B which is smaller than D, $D_B \simeq 1.6$ and 2.0 at d=2 and 3. Only bonds on the backbone contribute to the permeability, eq.(22). In practice, at most a fraction L^{D_B-D} of the fluid in the porous medium can be extracted. This fraction approaches zero for very large fractal samples!

We have recently (Oxaal et.al.(1987)) performed both computer and laboratory experiments of viscous fingering on a two dimensional dilute lattice model, with sites (nodes) at their percolation threshold. Clearly, the resulting fractal dimension of the fingers could not exceed that of the backbone, $D_B \simeq 1.6$. In practice, we found $D_{VF} \simeq 1.3$ and 1.5 for *fast* and *slow* flow rates. Ref. Oxaal et.al.(1987) contains a detailed discussion of these results. More recently, we have also studied the crossover from this fractal behaviour as one moves away from the extreme case of the percolation threshold, and found that the fully homogeneous result, $D_{VF} \simeq 1.7$, is approached very gradually (Meakin et.al(1989)).

4.4 Further Developments

On the experimental side physicists have concentrated most of their work on two-dimensional systems and even here the understanding of the geometry of porous media is only very preliminary. For example, using percolation concepts in modelling the structure of a porous material ignores *correlations* among cracks, which might modify many of the physical properties. The understanding of these correlations requires a study of the *development* of these cracks, which may be somewhat similar to that of the viscous fingers (both are governed, to some extent, by variants of the Laplace equation). However, we are now starting to obtain three-dimensional experimental results and further development of theory and these techniques have high priority in our research.

Another fruitful direction, which I had no time to review here, is to study physics on *regular fractals*, which imitate the geometry of the porous medium of interest. The advantage of such studies is that one can usually solve the physical problems *exactly and analytically* on regular fractals(Aharony (1986)).

In conclusion, I tried to give here a feeling for the rich variety of fractal phenomena one encounters when one studies statistical physics of and on porous media. The main message I hope to leave is that:

- The only safe way to obtain valid scientific results is to apply the correct strategies of statistical physics combined in both computer- and real- model experiments. In the oil industry, so far, we find a strong emphasis on computer experiments only.

- Extreme caution should be taken in extrapolating from small samples to large scales. Whenever there is more than one length scale present, *crossover phenomena* occur in all systems which are within the fractal regime. It is absolutely necessary to observe this in the interpretation of laboratory- and computer- experiments on finite size specimens if one is to find the correct answer for (larger) real systems.

ACKNOWLEDGEMENTS

Many interesting discussions with Amnon Aharony, Jens Feder, Liv Furuberg and Paul Meakin have been very helpful. I have used material in this presentation supplied by Aleksandar Birovljev, Finn Boger, Einar Hinrichsen, Knut Jørgen Måløy and Unni Oxaal who all are junior scientists in our research group on Cooperative Phenomena. This work has been supported by VISTA, a research cooperation between the Norwegian Academy of Science and Letters and Den norske stats oljeselskap a.s. (STATOIL), and by NAVF.

REFERENCES

Domb and Lebowitz, A coherent review of present accumulated knowledge in this field can be found in: In Domb C. and Lebowitz J. L., (1970--1988), editors, *Phase Transitions and Critical Phenomena*, Academic Press, New York.

Mandelbrot B. B., (1982), *The Fractal Geometry of Nature*. W. H. Freeman, New York.

Feder J., (1988), *Fractals*. Plenum, New York.

Oxaal U., Murat M., Boger F., Aharony A., Feder J., and Jøssang T., (1987), *Nature*, **329**, 32--37.

Stauffer D., (1985), *Introduction to Percolation Theory*. Taylor, Francis, London.

Aharony A., (1986), Percolation. In Grinstein G. and Mazenko G., editors, *Directions in Condensed Matter Physics*, pages 1--50, World Scientific, Singapore.

Måløy K. J., Feder J., and Jøssang T., (1985), Unpublished observations made during the preparation of reference [Feder et.al.(1986)].

Feder J., Jøssang T., Måløy K. J., and Oxaal U., (1986), In Englman R. and Jaeger Z., editors, *Fragmentation Form and Flow in Fractured Media*, pages 531--548. Ann. Isr. Phys. Soc. **8**.

Furuberg L., Feder J., Aharony A., and Jøssang T., (1988), *Phys. Rev. Lett*, **61**, 2117--2120.

Roux S. and Guyon E., (1989), *J. Phys. A: Math. Gen.*. in press.

Måløy K. J., Boger F., Feder J., Jøssang T., and Meakin P., (1987), *Phys. Rev. A*, **36**, 318--324.

Lenormand R., (1988), *J. Fluid Mech.*, **189**, 165.

Wilkinson D. and Willemsen J. F., (1983), *J. Phys. A*, **16**, 3365--3376.

Wilkinson D., (1984), *Phys. Rev. A*, **30**, 520 --531.

Clément E., Baudet C., and Hulin J. P., (1985), *J. Phys. Lett.*, **46**, L1163--L1171.

Birovljev A., Furuberg L., Aharony A., Feder J., and Jøssang T., (1989), *to be published*.

Sapoval B., Rosso M., and Gouyet J. F., (1985), *J. Phys. Lett.*, **46**, L149--L156.

Chen J. D. and Wilkinson D., (1985), *Phys. Rev. Lett.*, **55**, 1892--1895.

Måløy K. J., Feder J., and Jøssang T., (1985), *Phys. Rev. Lett.*, **55**, 2688--2691.

Witten T. A. and Sander L. M., (1981), *Phys. Rev. Lett*, **47**, 1400--1403.

Vicsek T., (1989), *Fractal Growth Phenomena.* World Scientific, Singapore.

Nittmann J., Daccord G., and Stanley H. E., (1985), *Nature*, **314**, 141--144.

VanDamme H., Obrecht F., Levitz P., Gatineau L., and Laroche C., (1986), *Nature*, **320**, 731--733.

Paterson L., (1984), *Phys. Rev. Lett.*, **52**, 1621--1624.

Deutch J. M. and Meakin P., (1983), *J. Chem. Phys.*, **78**, 2093.

Hentschel H. G. E., Deutch J. M., and Meakin P., (1984), *J. Chem. Phys.*, **81**, 2496.

Meakin P. and Deutch J. M., (1984), *J. Chem. Phys.*, **80**, 2115.

Voss R. F., (1984), *J. Phys. A*, **17**, L373--L377.

Måløy K. J., Boger F., Feder J., and Jøssang T., (1987), In Pynn R. and Riste T., editors, *Time-Dependent Effects in Disordered Materials*, pages 111--138, Plenum Press, New York.

Meakin P., (1987), Fractal aggregates and their fractal measures. In Domb C. and Lebowitz J. L., editors, *Phase Transitions and Critical Phenomena*, pages 336--489, Academic Press, New York.

Hinrichsen E. L., Måløy K. J., Feder J., and Jøssang T., (1989), *J. Phys. A: Math. Gen.*, **22**, L271--L277.

Meakin P., Murat M., Aharony A., Feder J., and Jøssang T., (1989), *Physica*, **A115**, 1--20.

Simulations of Viscous Fingering in a Random Network

M J Blunt, P R King and J A Goshawk

BP Research Centre, Sunbury-on-Thames, Middlesex. TW16 7LN. England.

We present a computer study of viscous fingering in an idealized porous medium. The porous medium is represented by an isotropic network of up to 80,000 nodes connected by thin tubes, which is modelled in two dimensions by the Delaunay triangulation of points placed at random in a circular region, and in three dimensions by a Voronoi tessellation in a sphere. We then simulate two-fluid displacements in this network and are able to demonstrate the effects of viscous and capillary forces. In the absence of capillary forces we show that the boundary between the two fluids is fractal. Furthermore we use the local average flow rates and pressures to calculate effective saturation dependent fractional flows. Using a radial Buckley–Leverett theory, the mean saturation profile can be inferred from the solution of the fractional flow equation, which is consistent with the measured saturation.

1 Introduction

We present simulations of immiscible pore–scale flow. We model the effects of viscous and capillary forces in a microscopic network of pore spaces. By generating large displacements we are able to describe the macroscopic properties of the flow by suitable averaged parameters.

Fluid displacements in porous media occur over a vast range of length scales. The fluids move through a disordered labyrinth of pores which are only a few microns in diameter, and yet displacements in oil reservoirs are typically several kilometres in extent. It is of interest to determine how the microscopic pore–scale physics of flow affect the overall macroscopic properties of the flood.

In our model, the physics at the pore scale is sufficiently realistic to mimic experimental displacements in networks of capillary tubes and yet simple enough to enable large–scale simulations to be performed. We are then able to measure empirical parameters describing the flood, such as fractional flow and capillary pressure functions, which in accordance with conventional theory are found to depend only on the local saturation.

Two and three dimensional isotropic random networks are used to represent a disordered matrix containing up to 80,000 pore spaces. Several simulations are performed where a non–wetting invading fluid is injected into the centre of a circular or spherical region which is initially full of the displaced fluid. This geometry is chosen to ensure that external boundaries do not influence the overall shape of the displacement. We present results for floods occupying up to 20,000 pores at a variety of adverse viscosity ratios and capillary numbers.

2 The Simulation

Network models have been used before to study fluid displacements. Recent work by Chen and Wilkinson [1] and Lenormand et al [2,3] simulated fluid flows in square grids of long thin tubes of varying radius. They were able to compare the numerical results with experimental patterns obtained from micromodels constructed from capillary tubes or etched networks. A porous medium was represented by a regular two dimensional network of interconnected pore spaces. Previous work by King [4], and Blunt and King [5] performed similar simulations in a hexagonal network.

Large displacements were often anisotropic and displayed the symmetry of the underlying lattice, whereas real porous media are irregular. Also, all the work on large lattices only studied two dimensional flows. Moreover, as yet there has been no consistent characterisation of the patterns produced in terms of conventional macroscopic parameters. This work will attempt to overcome all these difficulties. We perform simulations in both two and three dimensional isotropic networks and use Darcy's law and radial Buckley–Leverett theory to describe the displacements in terms of saturation dependent fractional flows.

2.1 Generating Random Networks

In this work we use large topologically disordered, isotropic networks. For our two dimensional network, points are placed at random (called Poisson points) in a circular region. The points are then connected to near neighbours by non–intersecting bonds. The bonds form a triangulation of the circle. A network where the triangles are as near equilateral as possible is called Delaunay [6] (Fig 1). We generate Delaunay triangulations containing up to 80,000 points. The methods we use will not be discussed here [7,8]. Essentially, the Delaunay triangulation is a consistent, straightforward way of connecting points to near neighbours. In a two dimensional Delaunay triangulation each point is connected to, on average, six others.

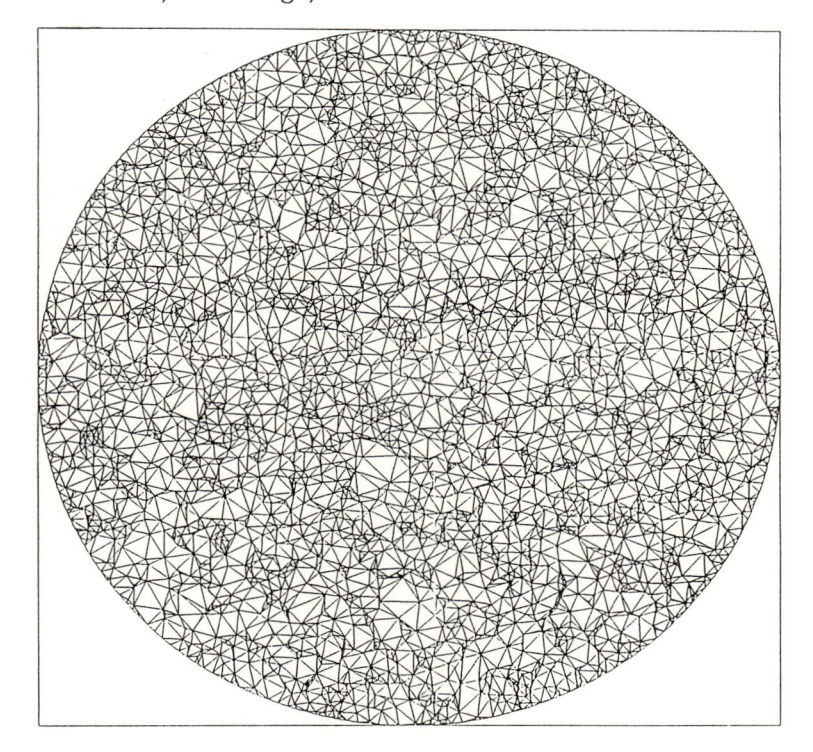

Fig. 1. A Delaunay triangulation of 4,000 points in a circle. In the simulations a network with 80,000 points was used

In the porous medium the Poisson points represent pore spaces of equal volume. The connections are thin tubes of an uncorrelated radius, r, chosen uniformly from the interval $[r_0(1 - \lambda), r_0(1 + \lambda)]$, where $1 \geq \lambda \geq 0$.

The three dimensional analogue of the triangulation above gives an average coordination number of over 14. There would only be sufficient computer memory to store all these connections for a relatively small three dimensional network. Instead we use a dual network, or Voronoi tessellation [6], which is formed from the Wigner–Seitz cells of each Poisson point: i.e; the territories nearer to each point than any other (Fig 2). An area is filled with polygons: a volume is filled with polyhedra. In the two dimensional Voronoi tessellation illustrated each node has three connections. Equivalent three dimensional networks, containing up to 50,000 vertices have been generated in a spherical region. The connectivity of these networks is exactly 4.

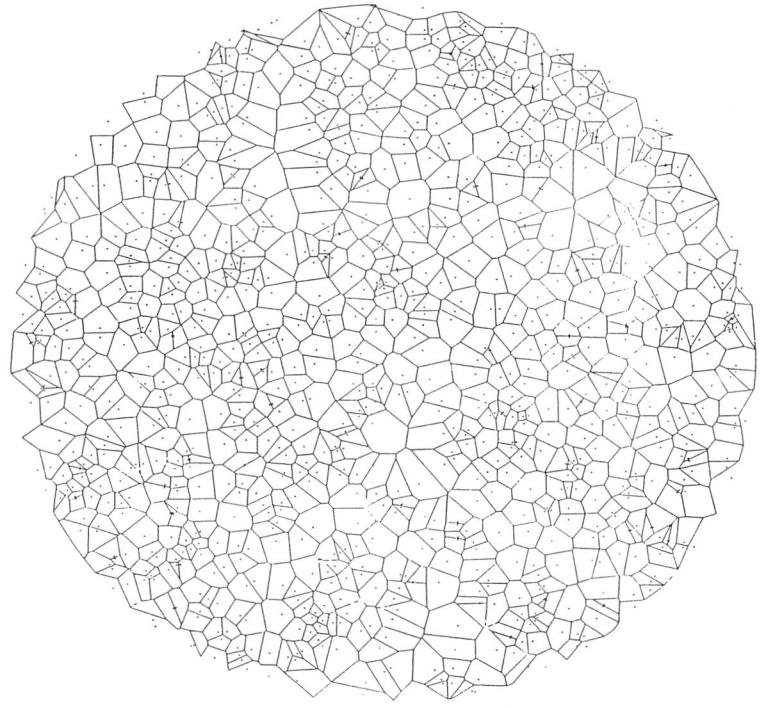

Fig. 2. A Voronoi tessellation of 500 points in a circle. Parts of the network which should lie outside the circle are placed around the circumference. The network contains approximately 3,000 nodes each connected to three neighbours. Three dimensional analogues with up to 50,000 nodes were generated.

The veritices of the Voronoi polyhedra represent the pore spaces and the edges between them are interconnecting tubes of variable radius. The radii are chosen in the same way as in two dimensions.

2.2 Equations of Fluid Flow

The fluid is considered to be contained entirely in the pores or nodes, which all have the same volume, but all the pressure drops occur in the bonds between them.

Invading fluid is injected through a central pore. There is a fixed pressure drop between the centre and the outside edge of the network. We assume that the flow rate Q_{ij} between the tube connecting nodes i and j is given by Poiseuille's law (with no fluid interface in the tube):

$$Q_{ij} = \frac{\pi(p_i - p_j)r_{ij}^4}{8l_{ij}\eta} = g_{ij}\Delta p_{ij} \qquad (1)$$

where p is a nodal pressure, r_{ij} and l_{ij} are the radius and length respectively of the tube and η is the viscosity of the fluid in the tube. This enables us to update the volumes of injected fluid in the nodes. In a time Δt, a nodal saturation $s_i(t)$ becomes:

$$s_i(t + \Delta t) = s_i(t) + \Delta t \sum_j Q_{ij} \qquad (2)$$

where the sum only includes bonds containing invaded fluid. Δt is chosen so that only one node in every time step becomes completely filled.

When a node is filled and s_i reaches 1, bonds connected to node i full of displaced fluid ('empty' bonds) may become full of invaded fluid. An empty bond of radius r_{ij} is filled with invaded fluid if:

$$p_i \geq p_j + p_c \qquad (3)$$

where p_c represents the capillary pressure jump across the fluid interface. p_c is given by the Young-Laplace equation: $p_c = -2\gamma \cos\theta/r_{ij}$. Notice that p_c is inversely proportional to the tube radius. γ is the interfacial tension and θ is the contact angle. In all the simulations $p_c \geq 0$: the invading fluid is non-wetting. If $p_j + p_c > p_i \geq p_j$ then the interface is frozen by capillary pressure and no flow occurs across it until p_i increases: i.e; the conductivity, g_{ij} is zero. We do not consider capillary pressure in the pores.

In some nodes s_i may decrease. If s_i reaches zero, a bond is filled with displaced fluid if $p_i \geq p_j$. Notice that if $p_c = 0$ the displaced and invaded fluids are treated symmetrically.

After each time step we solve for the pressure field. The fluids are assumed to be incompressible and hence $\sum_j Q_{ij} = 0$. We solve for p_i using successive over relaxation:

$$p_i = \beta \frac{\sum_j g_{ij} p_j}{\sum_j g_{ij}} + (1-\beta)p_i \qquad (4)$$

The relaxation parameter β is set to 1.7.

The scheme is then as follows:

(i) Solve for the pressure p_i, using eqn (4).

(ii) Calculate a time step Δt such that only one node is filled at a time.

(iii) Update the saturations using eqn (2).

(iv) If the saturation in a node reaches 1 or 0, alter the nature of the fluids in the bonds connected to that node. Then reset the corresponding conductivities g_{ij}, as described above.

(v) Repeat from step (i).

We simulate flow with a constant injection rate. The pressure at the outer boundary of the network is kept constant. At each timestep we modify the injection pressure to ensure that $\sum_j Q_{1j} = Q_0$ for constant Q_0, where the subscript 1 labels the central node and the sum runs over all tubes connected to the injector. The rest of the pressure field is then altered to make $\sum_j Q_{ij} = 0$ everywhere.

We are now able to characterise the flow with three dimensionless numbers:

(i) λ, which is the bond disorder parameter mentioned earlier. It is a measure of the heterogeneity of the network.

(ii) M, the ratio of displaced to injected fluid viscosities. We set the injected viscosity to unity.

(iii) N_c, which is a capillary number, defined by: $N_c = Q_0 l_0 \eta_i / \pi \gamma r_0^3$, where l_0 is the mean bond length and η_i is the viscosity of the injected fluid.

N_c represents a ratio of viscous to capillary pressures across a bond near the injection site. $N_c = 0$ gives capillary dominated floods, while $N_c = \infty$ is viscous dominated. The invading fluid is non-wetting.

This model is somewhat different from a previous simulation by the authors [4,5] which did not allow the nodal saturations to decrease and where all the fluid was considered to reside in the tubes and none in the nodes.

3 Viscous Fingering and Scaling Theory

Figure 3 shows a simulation for the case $N_c = 0$, $\lambda = 0.5$. Here capillary forces are dominant. This process is called invasion percolation and has been simulated directly. At the interface between the two fluids the flow only proceeds along the widest available empty bond (as long as it does not connect to a region entirely surrounded by invading fluid), where the capillary pressure is lowest, all other bonds are frozen. The advance of the front is independent of M as viscous forces may be neglected. Notice that we have loops enclosing pools of displaced fluid of many sizes.

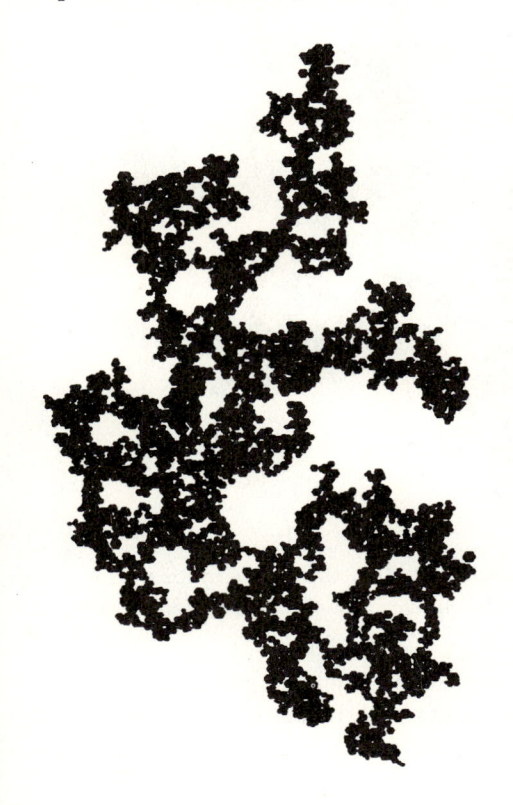

Fig. 3. Invasion percolation in a Delaunay triangulated lattice. The colours indicate the order in which nodes are invaded. The colour changes every 500 timesteps.

Figure 4 illustrates an opposite extreme: $N_c = \infty$, $M = \infty$ and $\lambda = 0.5$. Capillary forces are neglected and the invading fluid moves much more easily through the network than the fluid it displaces, creating a wispy pattern with many stages of branching. This case is mathematically analogous to a particle accretion process known as diffusion limited aggregation (DLA) [9].

In both these examples the displacements are fractal [10]. We measure the volume of invaded fluid, $V = \sum_i s_i$ and a root-mean-square radius r, where $r^2 = \sum_i s_i r_i^2 / \sum_i s_i$ and r_i is the radial distance of node i from the central injection site. As the displacement increases in size, we find:

$$V \sim r^D \tag{5}$$

Fig. 4. Fingering at infinite viscosity ration. Filled hexagons represent nodes full of invading fluid; white hexagons are partia filled nodes. The colour of the hexagons changes every 200 timesteps.

for some non–integer power D, called the fractal dimension. Alternatively we may cover the displacements with a square grid of size ϵ. If we count the number, $N(\epsilon)$, of grid squares containing some invading fluid, we find $N(\epsilon) \sim \epsilon^{-D}$ with the same exponent D as before, to within numerical error. Example results are given in figure 5, which shows logarithmic plots of V against r in the DLA limit, and figure 6, which shows $N(\epsilon)$ versus ϵ for both cases. Averaging our results, we find $D = 1.73 \pm 0.05$ for DLA and $D = 1.82 \pm 0.01$ for invasion percolation. In three dimensions our simulations give respective values of 2.6 ± 0.1 and 2.44 ± 0.05. A fractal dimension is a quantitative characterisation of the large scale geometry of the pattern. Moreover, D is dependent on the physical parameters of the flow (M and N_c) but not model–dependent properties (such as λ and the nature or connectivity of the networks). Our numerical values are consistent with previous simulations on different (usually regular) lattices [11,12,13]. The patterns are fractal since we have similar structure of all sizes from the scale of the individual pores upwards. There is no intermediate size characterising typical features.

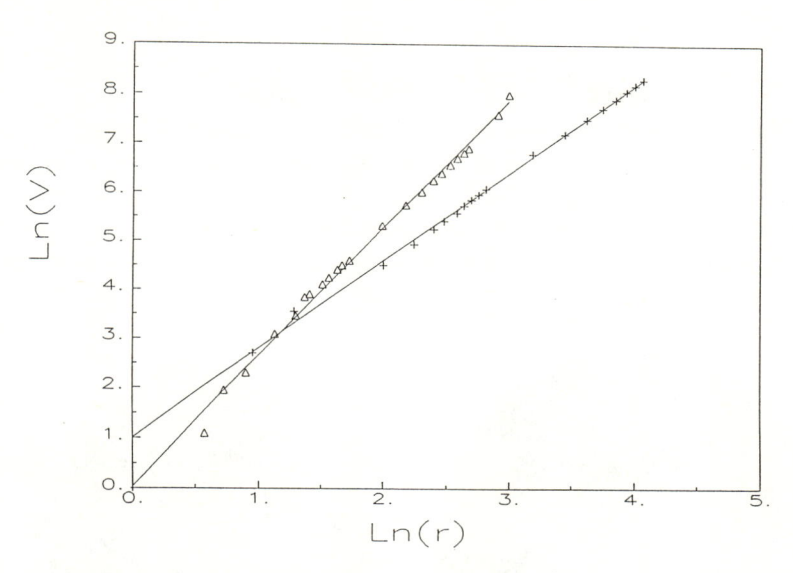

Fig. 5. A doubly logarithmic plot of fluid volume, V against radius r for viscous fingering patterns at infinite viscosity ration. The slope of the plots give an estimate of the mass dimension D. D is found from a least squares fit through the points, with each point weighted by a factor V. In two dimensions: crosses, $D = 1.79 \pm 0.06$. In three dimensions: triangles; $D = 2.62 \pm 0.10$.

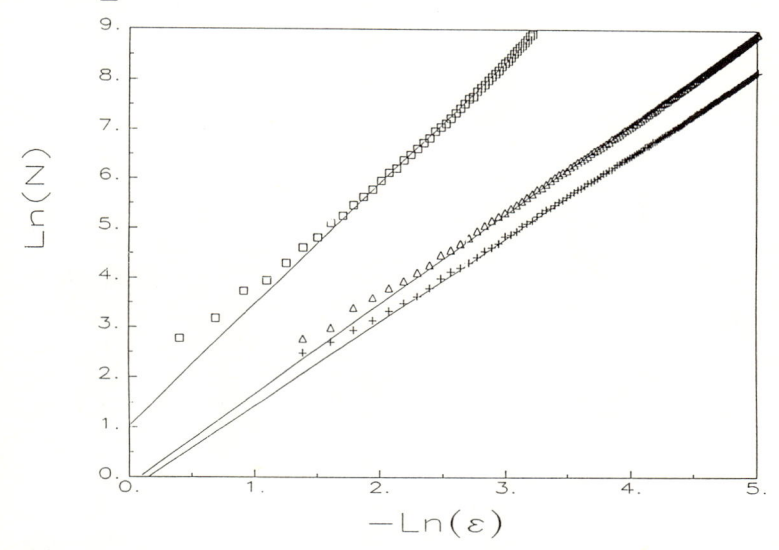

Fig. 6. Doubly logarithmic plot of number of fully occupied grid squares, $N(\in)$ against grid size $\in$. Slopes are found in the same way as in fig. 5. In two dimensions: crosses $M = \infty$, $D = 1.68 \pm 0.05$; triangles, invasion percolation, $D = 1.81 \pm 0.02$. In three dimensions squares, invasion percolation, $D = 2.44 \pm 0.05$.

3.1 Fingering at Finite Viscosity Ratio

Figure 7 illustrates two dimensional fingering patterns containing up to 20,000 filled nodes for $N_c = \infty$, $\lambda = 0.5$ and a variety of values of M. In these cases the displacement is unstable with many stages of branching. However, if the invading fluid has a finite viscosity, it cannot sustain the wispy patterns seen in the DLA limit, and the

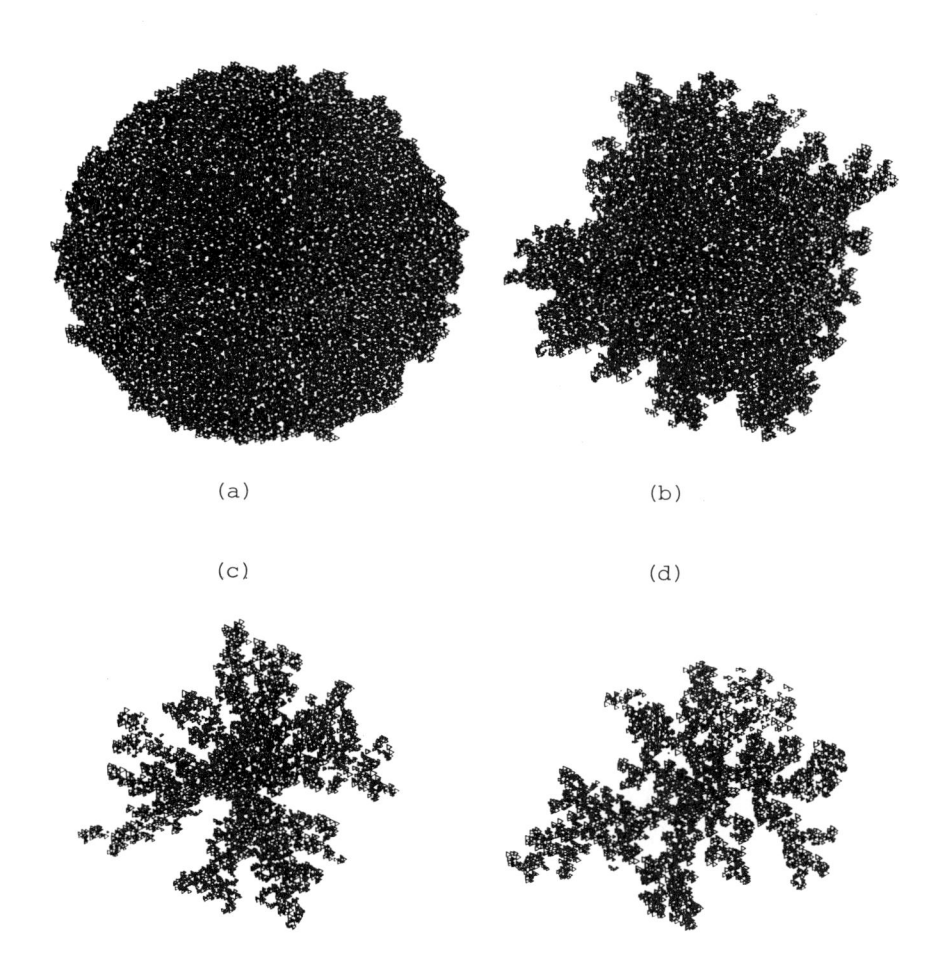

(a) (b)

(c) (d)

Fig. 7. Viscous fingering patterns for $N_c = \infty$, $\lambda = 0.5$. (a) $M = 1$. (b) $M = 10$. (c) $M = 100$. (d) $M = 1,000$. Open triangles represent nodes which have yet to become completely filled.

interior of the flood fills in and becomes compact. Figure 8 shows logarithmic plots of V against r [see eqn(5)] for $M = 1$ and $M = 10$. Notice that the results are consistent with a value of D equal to 2: the bulk of the patterns resemble a compact two-dimensional object. For finite viscosity ratio the interior of the displacement is not fractal. The anomalously high estimate of D in three dimensions cannot be sustained for very large r, since D must be less than or equal to the dimension of space [10].

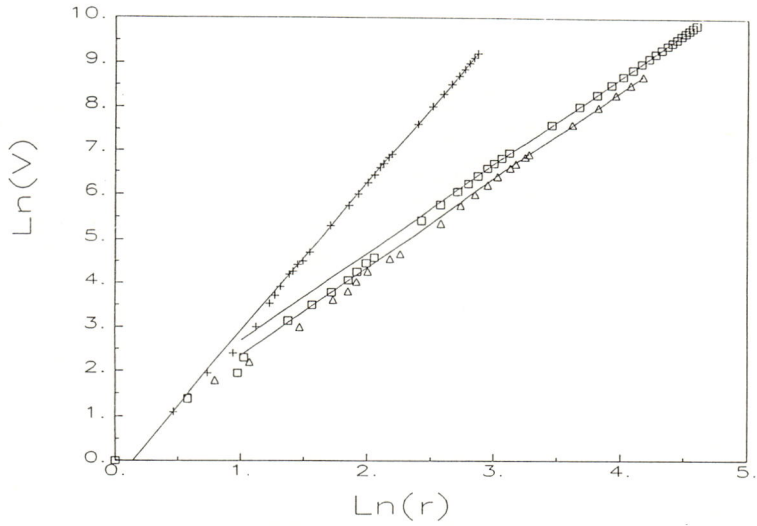

Fig. 8. Doubly logarithmic plots of fluid volume, V against radius, r for viscous fingering. In two dimensions: squares $M=1$, $D=2.00 \pm 0.01$; triangles $M=10$, $D=2.01 \pm 0.02$. In three dimensions: crosses $M=1$, $D=3.36 \pm 0.10$. Finite M the interior of the displacement is compact with D equal to the dimension of space.

However, the advancing fluid interface still appears to have structure of many sizes. We analyse the surface by a box-counting method similar to that described before. We count the number of empty grid squares adjacent to a square containing some invaded fluid. We find an approximate scaling of the form $N(\epsilon) \sim \epsilon^{-d_s}$ for small ϵ. See figure 9. The surface dimension, d_s increases from close to 1 for $M = 1$ to 1.5 for $M = 1,000$ and 1.7 (d_s and D are equal) for $M = \infty$. (Figure 10) These results are consistent, although slightly lower than previous work on hexagonal networks by the authors [4,5] and those obtained by real space renormalization theory [14]. d_s is a characterisation of the geometry of the displacement, which does not appear to depend on the microstructure of the network, but which is a function of visosity ratio, M.

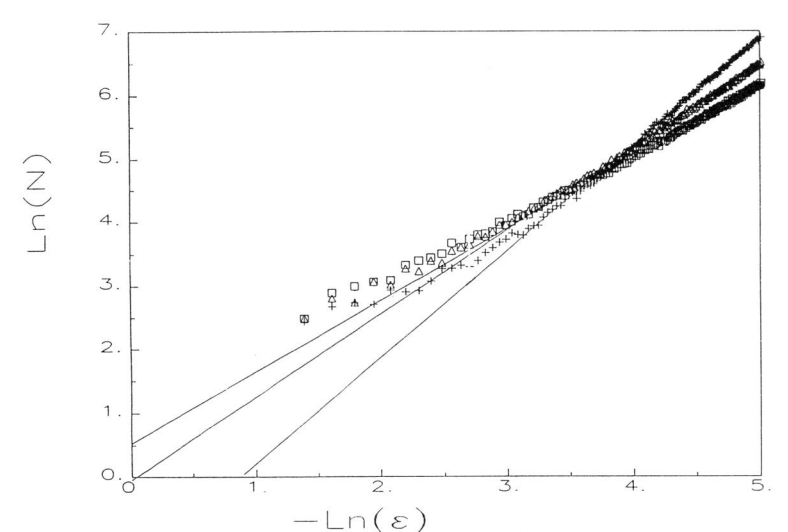

Fig. 9. Doubly logarithmic plots of $N(\epsilon)$ versus ϵ for the fluid interface. Crosses $M=\infty$, $d_s = 1.68 \pm 0.05$; triangles $M=10$, $d_s = 1.3 \pm 0.05$; squares $M=1$, $d_s = 1.1\overline{2} \pm 0.05$.

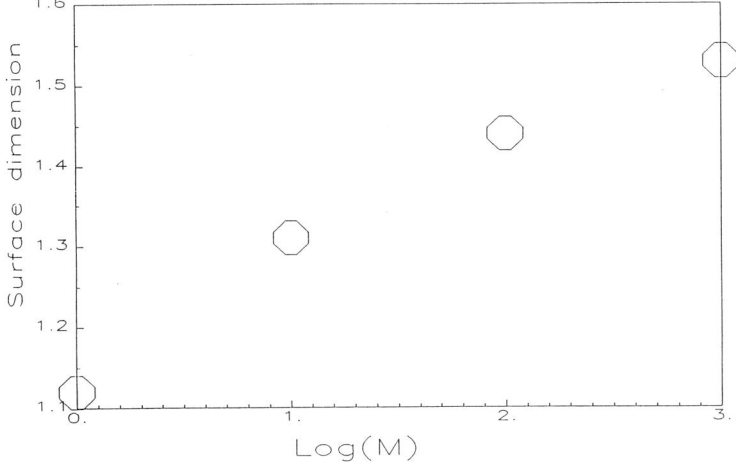

Fig. 10. The surface dimension d_s, as a function of viscosity ratio M. The size of the hexagons represent the spread of values over or more simulations. The results are consistent with previous simulations by the authors [4,5] and real space renormalisation calculations [14].

3.2 Fingering and Capillary Pressure

Figure 11 shows a series of fingering patterns for $M = 10$, $\lambda = 0.5$ as N_c is increased. When $N_c = \infty$ advance may occur at all points along the fluid boundary; on decreasing N_c more and more bonds at the fluid interface are frozen by capillary pressure, until, when $N_c = 0$

the flow is controlled entirely by the search for the easiest path of wide tubes through the network. In a real oil reservoir capillary forces are likely to dominate at the pore scale. A typical pressure drop along a single tube, Δp_v, may be one hundred to several thousand times smaller than the capillary pressure across the fluid interface. However, over large distances the viscous pressure drop will normally exceed the capillary force. We would expect capillary forces to control the flood on small scales and a viscous fingering like displacement when viewed over large distances. In a radial flood Δp_v decreases with distance from the injection node: we see viscous floods near the centre of the network with capillary forces becoming more important at larger radii.

4 A Conventional Description

We model pore–scale flow. However, simulations of reservoir flows describe the microscopic physics in large grid blocks by simple averaged parameters. A justification for this approach will be provided by this work.

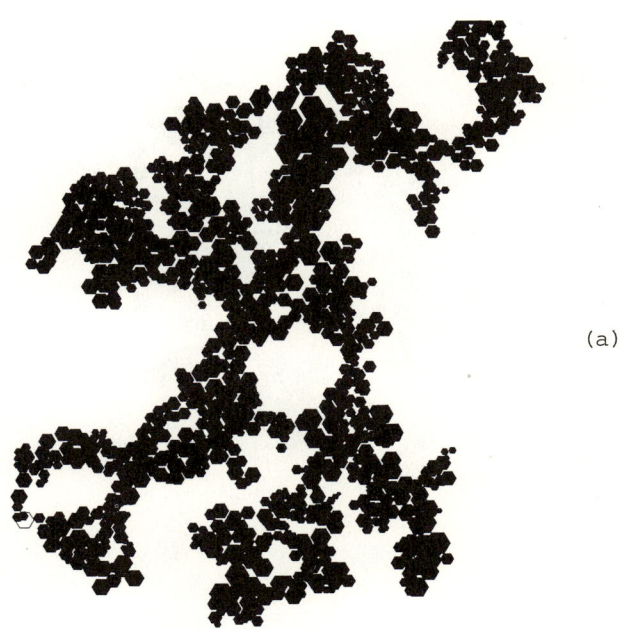

(a)

In the next section we shall attempt to use averaged properties of the flood to find a characterisation of the results, which confirms conventional theory.

256   BLUNT ET AL

(b)

(c)

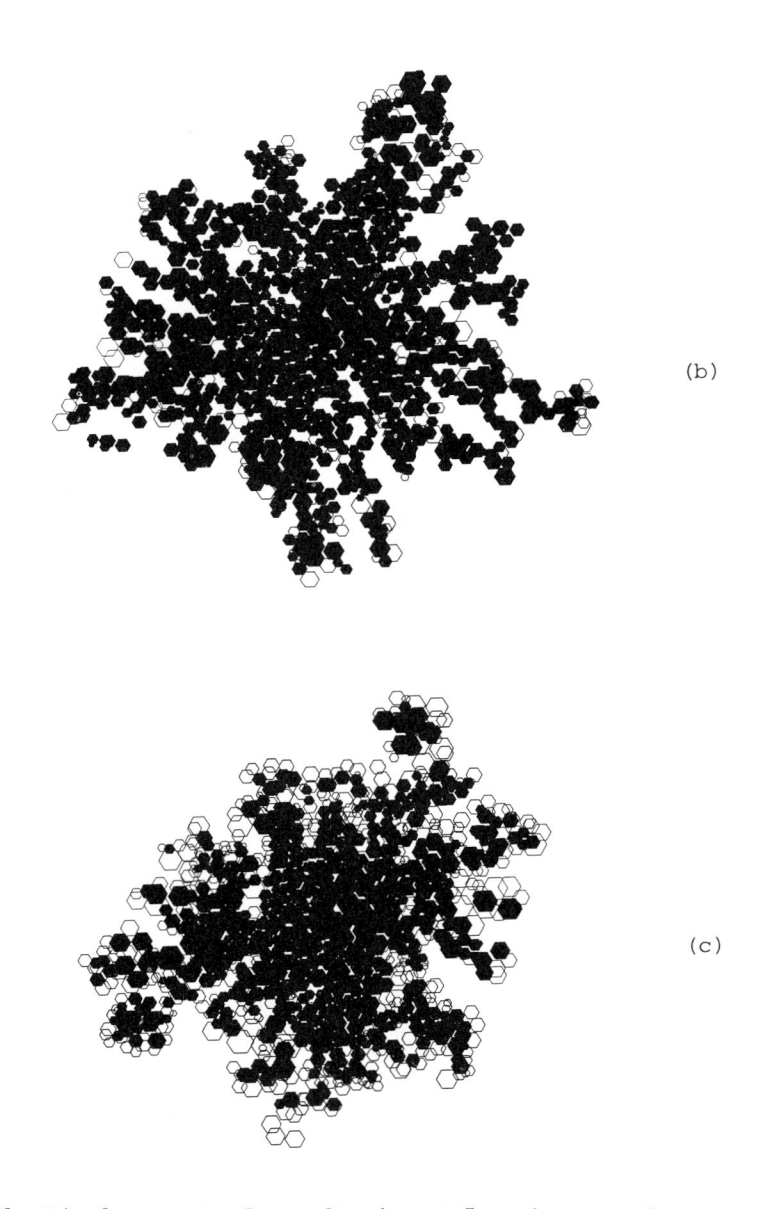

Fig. 11. Displacements for M=10, λ = 0.5 each occupying approximately 1,500 nodes. (a) N_C = 0. (b) N_C = 0.5. (c) N_C =5. Notice that the number of partially filled sites (shown by open hexagons increases as N_C decreases. For $N_C = \infty$ (fig 7) the displacements a completely surrounded by partially filled nodes. The exact microstructure of the network is different in all these cases.

4.1 Radial Buckley Leverett Theory

We shall attempt to describe the evolution of the displacement in terms of a radially averaged saturation profile for the invading fluid, $s(r,t)$, where r is the radius from the injection node and t is the time. $\int_0^1 2\pi r s(r,t) dr = Q_0 t$, where Q_0 is the (fixed) injection rate.

The equation of fluid conservation gives:

$$\frac{\partial s(r,t)}{\partial t} + \frac{1}{r}\frac{\partial}{\partial r}\Big(r Q(r,t)\Big) = 0 \qquad (6)$$

for non–zero r where $Q(r,t)$ is the averaged fluid flux of invading fluid at a radius r. Q depends on the pressure gradient. The functional form of Q may be deduced from the equations of two–phase fluid flow (Darcy's law), assuming that the averaged capillary pressures and relative permeabilities are functions only of the local saturation [15]. For a totally uniform flood with $M = 1$, $N_c = \infty$ and no matrix heterogeneity in a circular network, the radial pressure gradient falls as $1/r$ and hence Q would be $Q_0 s/2\pi r$. It can be shown that [15] $Q(r,t)$ may be written simply as $F(s)(Q_0/2\pi r - K\lambda_0 \nabla P_c(s))$, where the effects of viscous forces are accounted for by a fractional flow $F(s)$, which does not depend explicitly on time, but only on the local fluid saturation. The term $Q_0/2\pi r$ accounts for the macroscopic geometry of the flood and would be different for other shaped networks: i.e; it would be simply Q_0 for a linear flood. K is the absolute permeability of the medium, λ_0 is the displaced fluid mobility and P_c is a saturation dependent capillary pressure, which has a diffusive effect on the displacement front.

Equation (6) then becomes:

$$\frac{\partial s(r,t)}{\partial t} + \frac{1}{r}\frac{\partial}{\partial r}F(s)\Big(\frac{Q_0}{2\pi} - Kr\lambda_0(s)\frac{\partial P_c(s)}{\partial r}\Big) = 0 \qquad (7)$$

Eqn (7) may be written in terms of a single variable: $v = \pi^{1/2} r/(Q_0 t)^{1/2}$ to obtain (after a little algebra):

$$\frac{ds(v)}{dv}\left\{v^2 - \frac{d}{ds}\Big[F(s)\Big(1 - \frac{2\pi v K \lambda_0(s)}{Q_0}\frac{ds}{dv}\frac{dP_c}{ds}\Big)\Big]\right\} = 0 \qquad (8)$$

and thus a non–trivial solution is obtained when:

$$v^2 = \frac{d}{ds}\Big[F(s)\Big(1 - \frac{2\pi v K \lambda_0(s)}{Q_0}\frac{ds}{dv}\frac{dP_c}{ds}\Big)\Big] \qquad (9)$$

For a general $P_c(s)$ eqn (9) does not, unfortunately, have an analytic solution, although it may be obtained easily by numerical integration. However, it can be seen that the saturation profile is a function of only one variable, v and hence profiles taken at different times may be scaled on to each other. This is only possible for floods with non-zero P_c because we have a radial geometry. For $P_c = 0$ eqn (9) reduces to a radial Buckley–Leverett equation problem, whose solution is $s(v)$ where $dF/ds = v^2$.

4.2 Fractional Flow Curves

We need to test this analysis directly. For a variety of two-dimensional floods with different M and $N_c = \infty$, mean saturation profiles and fractional flows were measured at up to 5 different times. The network was divided into 40 annular sections. Within each annulus the mean saturation of invaded fluid is calculated as well as the fraction of the total flux across each annulus carried by the invading fluid. That this fractional flow is a unique function of the mean saturation is the central assumption of our analysis.

Example results are shown for $\lambda = 0.5$, $M = 10$ and $N_c = \infty$. Figure 12 shows the mean saturation profiles at different times successfully scaled on to a single curve. Figure 13 shows indeed that $F(s)$

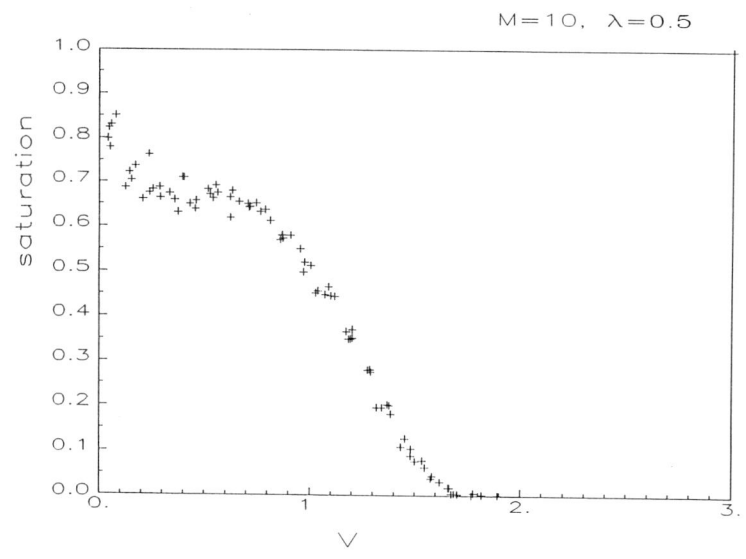

Fig. 12. The mean saturation profile s for $M=10$. $\lambda=0.5$ and $N_c=\infty$. The results of two simulations each taken at five equally spaced intervals during the flood have been scaled on to the same curve.

does appear to depend only on s and not independently on time or radius for given model parameters; Moreover, we can use $s(v)$ to calculate $F(s)$, since $dF/ds = v^2(s)$, then $F(s) = \int_0^s v^2(s)ds$. This is shown in figure 13: both direct and inferred fractional flows agree to within numerical error. Notice that the fingered front is very wide: the saturation profile does not develop a shock. The fractional flow function does not have a point of inflexion.

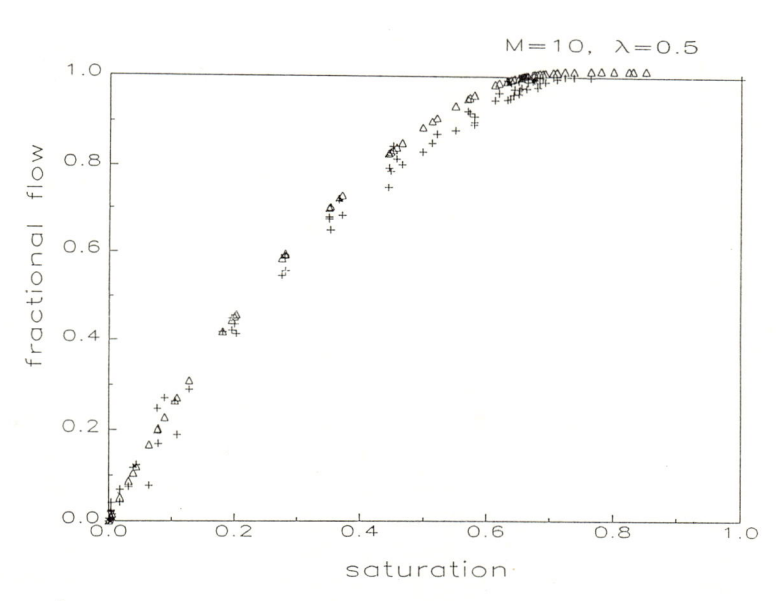

Fig. 13. The fractional flow curve $F(s)$ for $M=10$, $\lambda=0.5$ and $N_c=\infty$. F has been measured at five times during the growth for two different simulations. Crosses, direct measurement; triangles, calculated from the mean saturation profile.

Fractional flow curves for various M with $N_c = \infty$ and $\lambda = 0.5$ are illustrated in figure 14. Increasing M gives a more fingered front and lower overall recovery.

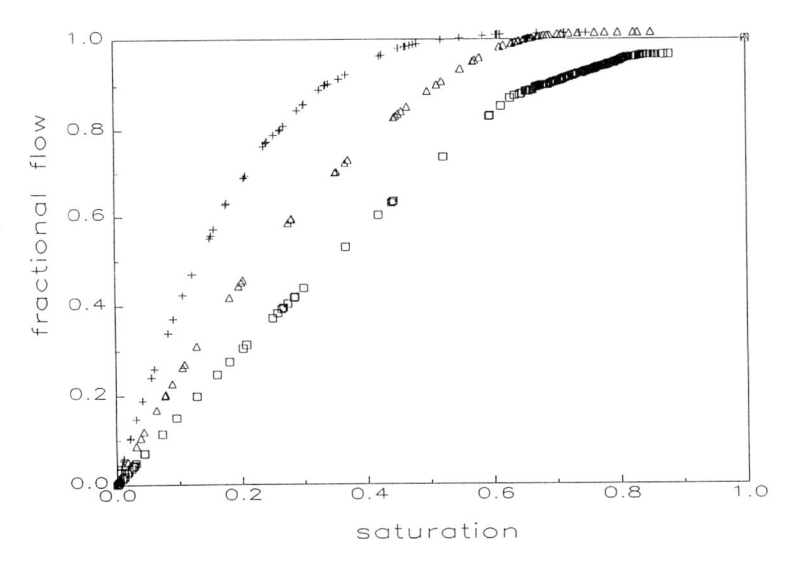

Fig. 14. Fractional flow curves calculated from mean saturation profiles for $N_c = \infty$ and $\lambda = 0.5$. Crosses $M=100$; triangles $M=10$; squares $M=1$.

Saturation profiles for $M = 10$, $\lambda = 0.5$ and various N_c are shown in figure 15. Notice again that s appears to be a function of only one variable, v. Viscous fingering is supressed on increasing the capillary pressure, but pools of displaced fluid become enclosed by the invader for low N_c, which causes the overall fraction of displaced fluid swept out to decrease. Overall, the capillary pressure has a diffusive effect on the flow: a comparison of figures 12 and 15 show that the profiles are smeared out for low N_c.

With a knowledge of the mean pressure field relative permeabilities and capillary pressure curves could be calculated. This will be the subject of further work [16,17].

On a microscopic scale the functions F and P_c are meaningless. However, we have demonstrated that the averaged properties of pore-scale flow may be described in terms of parameters which depend on the local saturation, and of course the properties of the network and the fluids, but not explicitly on time or radius.

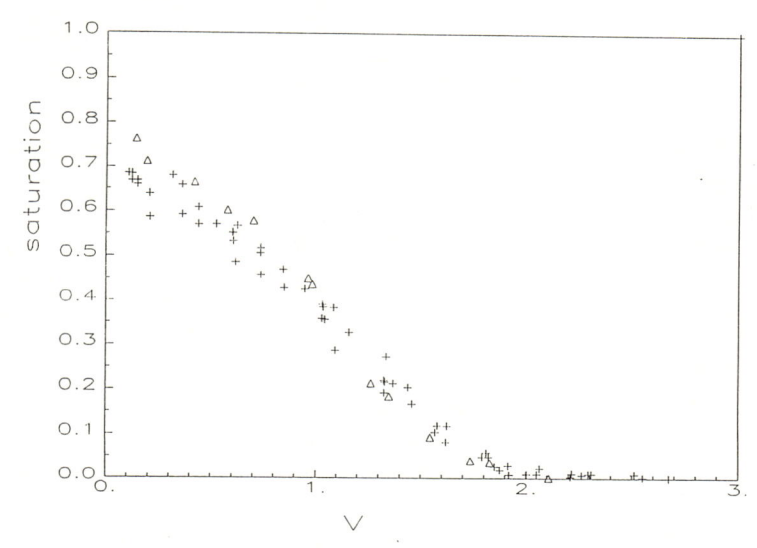

Fig. 15. Saturation profiles for $M=10$ and $\lambda=0.5$. Crosses $N_c=0.5$, triangles $N_c=5$.

5 Conclusions

We have described a network model of viscous fingering which can be used to study the effect of heterogeneity, viscosity ratio and capillary pressure on pore–scale displacements in both two and three dimensions. Further work could include a direct comparison with experimental random micromodels and bead packs, or a more thorough investigation of three dimensional displacements.

From this work we may make the following conclusions:

(i) In the absence of capillary forces, fingering at finite viscosity ratios produces a compact displacement with a fractal surface, whose dimension is a function of viscosity ratio, M.

(ii) In the limit $M = \infty$ the mass of the displacement is fractal with scaling properties similar to DLA.

(iii) The mean saturation profiles and fractional flow curves are consistent with a Buckley–Leverett theory. This enables the floods to be described consistently in terms of empirical parameters. These parameters may then be input into conventional simulations where each grid block encompasses many pores.

We should like to thank Dr R C Ball for useful discussions. We are also particularly grateful to Drs P K Sweby and J J Barley for providing us with programs for generating Voronoi meshes. We thank The British Petroleum Co plc for permission to publish this paper.

[1] J-D Chen and Wilkinson *Physical Review Letters* **55** 1892 (1985)

[2] R Lenormand, E Touboul and C Zarcone *J. Fluid Mechanics* **189** 165 (1988)

[3] R Lenormand *Proc R Soc London A* **423** 159 (1989)

[4] P R King *J Physics A* **20** L529 (1987)

[5] M J Blunt and P R King *Physical Review A* **37** 3935 (1988)

[6] B D Ripley *Spatial Statistics* John Wiley and Sons, New York 38-44 (1981)

[7] P K Sweby (unpublished)

[8] M J Blunt, P R King and J A Goshawk (to be published)

[9] L Paterson *Physical Review Letters* **52** 1621 (1984)

[10] B B Mandelbrot *The Fractal Geometry of Nature* Freeman, San Francisco (1982)

[11] For a recent review of a variety of fractal growth phenomena, including DLA, see P Meakin *Phase Transitions* **12** 335 (1988)

[12] D Wilkinson and J F Willemsen *J Physics A* **16** 3365 (1983)

[13] L Furuberg, J Feder, A Aharony and T Jøssang *Physical Review Letters* **61** 2117 (1988)

[14] T Nagatani *J Physics A* **21** 1109 (1988)

[15] D W Peaceman *Fundamentals of Numerical Reservoir Simulation* Elsevier, Amsterdam (1977)

[16] M J Blunt and P R King *Physical Review A* **42** (October 15, 1990)

[17] M J Blunt and P R King *Transport in Porous Media* (to be published)

A Simple Analytical Model Of The Growth Of Viscous Fingers In Heterogeneous Porous Media

J.N.M. van Wunnik and K. Wit

Koninklijke/Shell Exploratie en Produktie Laboratorium

1 Abstract

Fingering of unstable immiscible displacements in moderately heterogeneous reservoirs has been modelled analytically. The displacements are assumed to take place in the horizontal plane. Stabilising mechanisms such as capillary smear out or lateral dispersion are not incorporated.

An accurate expression for the viscous growth of an initially small disturbance (finger) in a displacement front is derived first. The porous medium is assumed homogeneous and the saturation of the invading liquid in the fingers constant. For a constant finger width, the derived expression shows exponential growth for small fingers and linear growth for large fingers. Next, this expression is merged with an expression which describes the development of a marginally stable displacement (mobility ratio M=1) in a heterogeneous porous medium. The results of this merging show that the more wide-spread and the more profound the permeability variations are, the faster the fingers initiate and the faster they show linear growth.

The results of the analytical model are compared with the results of a revised version of the numerical technique, diffusion-limited aggregation (DLA). The applied version of such a DLA simulator is summarised and its application justified. The results of the analytical model match the results of the numerical simulator, if in the latter only interface movements in the main flow direction (longitudinal) are included.

2 Introduction

In unstable displacements, bypassing can occur in the form of fingers. These fingering phenomena are determined not only by the viscous forces accompanying the flowing liquids, but also by the heterogeneity of the porous medium. Heterogeneities induce fingers but also influence the development of the fingers. For the extreme of a very heterogeneous medium by-passing is almost solely determined by the heterogeneities and not or just slightly by viscous forces. In contrast, in a homogeneous porous medium the fingers need to be initiated by external means after which they grow solely as a result of viscous forces. In this article the heterogeneity is taken as moderate which means that viscous forces as well as heterogeneities are important. The areal extension of the heterogeneities are assumed large with respect to the capillary smear out so that the smear out can be neglected.

In fingering processes in moderately heterogeneous reservoirs three phases can be distinguished:

i. The interface is becoming irregular because of heterogeneities in the porous medium. The irregularities are enhanced by viscous forces. However, the irregularities (fingers) are still so small that their mutual influence as a result of viscous forces can be neglected. Finger spacing is constant and determined solely by the porous medium.

ii. Finger length and the statistical variations in finger length are so large that mutual finger interactions become important. The fingers suppress each other and merge, the fingers become wider and their spacing increases. Ultimately one finger remains.

iii. The size of the remaining finger is so large that porous medium heterogeneities cannot effect finger growth anymore. The finger width is constant and its length grows as in a homogeneous porous medium.

Several analytical models for the description of the growth of identical fingers in homogeneous porous media have been proposed in literature. For small finger amplitudes perturbation type of approaches have been worked out [3,13,8]. These approaches result in exponential growth of the fingers. The models which are developed for large finger amplitudes [18,1], give linear growth. An expression which is valid over the total amplitude regime has been given by Mui and Miller [16]. However, their expression was valid for one finger only in an infinitely wide medium. In this article an expression for the growth rate of laterally periodic identical fingers is derived which is valid over the total amplitude regime.

For displacements with a mobility ratio $M = 1$ in heterogeneous porous media a number of analytical models have been proposed in the hydrology literature [7,21]. However, an extension of these models to displacements with an unfavourable mobility ratio ($M > 1$) has not yet been given. In this article the results of the $M = 1$ models for displacement in heterogeneous media are combined with the results of the viscous growth model in a homogeneous medium. The resulting expression is valid for those fingering phases where the finger width is constant, i.e. in phase i and iii. In phase ii the finger width increases. Because it is not known how the finger width increases, the analytical model cannot be applied.

The computer power presently available enables detailed numerical simulation of viscous fingering. In the simulation of viscous fingering in het-

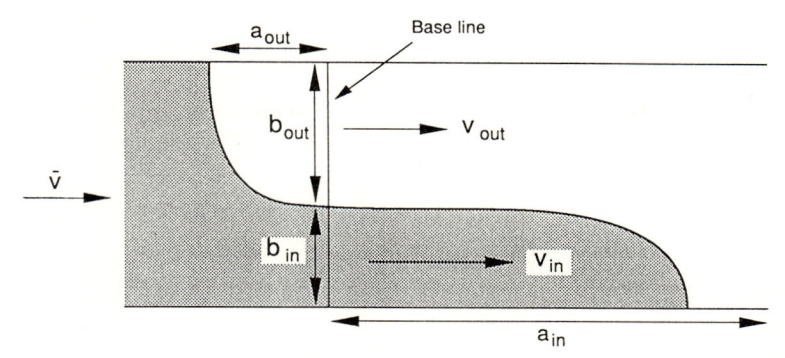

Figure 1. Symmetry element of a finger pattern which is composed of two half ellipses.

erogeneous media in general, two types of simulation methods are applied: the finite-difference method in which varying permeabilities are applied over the grid blocks [12], and the stochastic methods such as Monte Carlo and diffusion-limited aggregation (DLA). The last two methods inherently add perturbations to the displacement front [14,4,9]. The results of the analytical model are compared with a revised version of DLA. Next the numerical model will be applied to follow the fingering process into phase ii. However, before doing so, physical arguments will be presented that justify the application of the DLA method.

3 Growth by Viscous Forces of an Interface Perturbation ($M > 1$)

The growth rate (v_{vis}) of a perturbation in a displacement front (finger), developing in a homogeneous porous medium, was determined analytically by assuming that: 1) at any moment in time the finger structure can be modelled by forward and backward facing sets of half ellipses (see Fig.1) and 2) the liquid saturations are constant in the swept area. The geometry is taken rectangular. The expression for the growth rate (v_{vis}) could be formulated implicitly in terms of the length (a_{in}) and width (b_{in}) of the forward pointing ellipse, the length (a_{out}) and width (b_{out}) of the backward pointing ellipse and the liquid mobility ratio (M) of the displacing and the displaced liquids.

The growth rate is defined by the difference of the velocities at the tip of the forward and at the tip of the backward pointing ellipses $v_{vis} = v_{in} - v_{out}$. The equations determining the growth rate are derived in the appendix and are given by

$$b_{in}\alpha(a_{in}, b_{in}, B, P_{in}) + b_{out}\alpha(a_{out}, b_{out}, B, P_{out}) = b_{in} + b_{out} \quad (3.1)$$

in which
$$\alpha(a_i, b_i, B, P_i) = \frac{v_i}{\bar{v}} =$$
$$1 + C\left[\frac{a_i - b_i}{b_i} + \frac{P_i}{B + (P_i - 1)b_i}\left(B - \sqrt{B^2 + a_i^2 - b_i^2}\right)\right] \quad (3.2)$$
and
$$C = \frac{(P_i - 1)a_i b_i}{(a_i + b_i P_i)(a_i - b_i)} \quad (3.3)$$

in which $B = b_{in} + b_{out}$, the subscript i stands for respectively in and out, $P_{in} = M$ and $P_{out} = 1/M$ and $\bar{v}$ is the average front velocity. The average amplitude is defined by

$$\bar{a} = \frac{1}{2}(a_{in} + a_{out}) \quad (3.4)$$

For a given combination of $\bar{a}, b_{in}$ and b_{out} and M the values of a_{in} and a_{out} can be determined from the implicit set of equations 3.1 to 3.4. Via eq. 3.2 the value of $v_{vis}/\bar{v} = (v_{in} - v_{out})/\bar{v}$ can be calculated. Figure 2 shows the dependence of the finger growth rate as a function of $\bar{a}/b$; it is assumed that $b = b_{in} = b_{out}$. The mobility ratio is 2 and 20 respectively.

The data points in Figure 2 represent numerical results. The numerical technique applied to confirm the analytical results is the adaptive gridding method. This method enables refining the numerical grid in areas with large mobility gradients [19]. The required grid refinement is determined by calculating the flow rate in an elliptic patch of high mobility in a very large medium. An exact analytical expression exists for this geometry (see appendix). The grid refinement at the boundary of the highly mobile area is chosen such that the numerical results match the analytical results within 1%. This refinement is then applied to the fingering pattern build up out of ellipses. The analytical results closely resemble the numerical results. It has furthermore been shown that the results are not sensitive to the precise shape of the displacement front, substituting a sinusoidal shaped interface for the ellipses yielded, within a few percent, the same numerical results for the growth rates.

In the limit of small amplitudes ($\bar{a}/b \ll 1$) and $b = b_{in} = b_{out}$ the set of equations 3.1 to 3.4 gives

$$\frac{d\bar{a}}{dt} = 4\frac{M-1}{M+1}\frac{(1-\sqrt{3}-M)(1-\sqrt{3}-\frac{1}{M})}{\frac{1}{M}+M-2(1-\sqrt{3})}\bar{v}\frac{\bar{a}}{B} \quad (3.5)$$

which shows that small amplitudes grow exponentially with averaged swept distance; this behaviour is predicted by the perturbation approaches. For large amplitudes ($\bar{a}/b \gg 1$) and $b = b_{in} = b_{out}$, it follows that

$$\frac{d\bar{a}}{dt} = \frac{M-1}{M+1}\bar{v} \quad (3.6)$$

SIMPLE ANALYTICAL MODEL 267

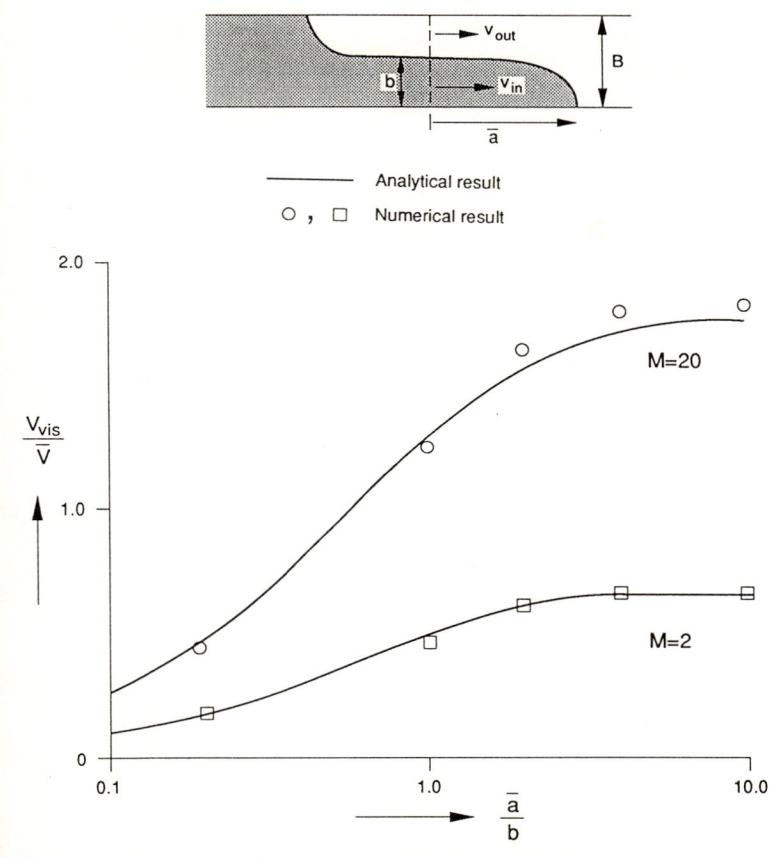

Figure 2. The difference of the fluid velocities inside and outside the finger (v_{vis}) over the average front velocity ($\bar{v}$), versus the ratio of finger amplitude ($\bar{a}$) over finger halfwidth (b); $b/B = 1/2$.

which shows the linear growth with swept distance. Figure 3 shows the results of the numerical integration of the set of equations 3.1 to 3.4 for the mobilities 2 and 20 and for the initial amplitudes $1b, 10^{-2}b$ and $10^{-4}b$ in which b is the half width of the initial disturbance; again $b = b_{in} = b_{out}$. Figure 3 shows that exponential growth in the initial phase of the displacement changes gradually to linear growth.

An alternative measure for the degree of fingering of the interface is the standard deviation. For an elliptically shaped interface with $b = b_{in} = b_{out}$ the relation between amplitude and standard deviation is given by

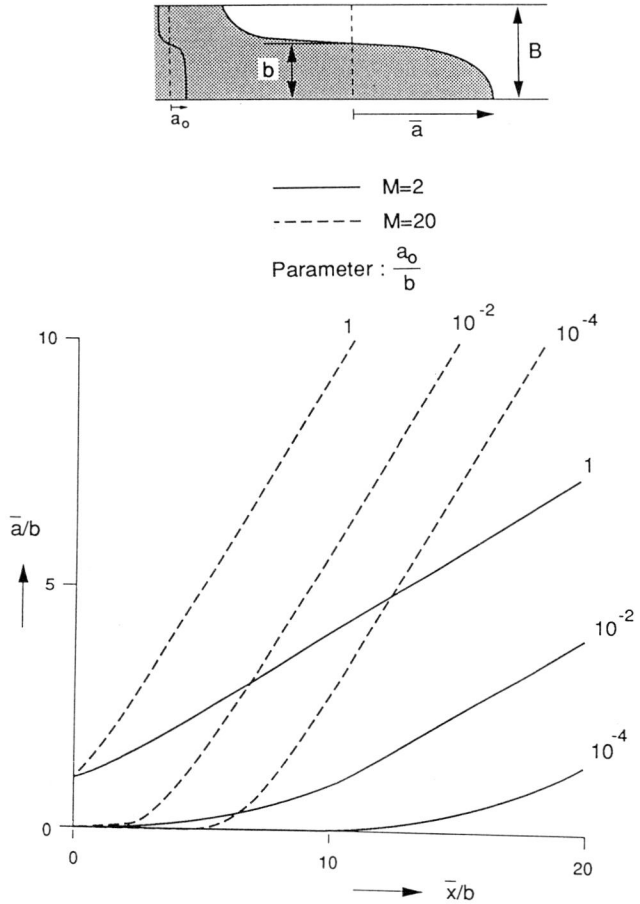

Figure 3. Growth of a finger in a homogeneous medium. $\bar{x}$ is the average position of the displacement front, b is the finger halfwidth, a_0 is the amplitude at $\bar{x} = 0$.

$$\sigma = \sqrt{\frac{2}{3}}\,\bar{a} \qquad (3.7)$$

4 Marginally Stable Displacement in a Heterogeneous Medium ($M = 1$)

The heterogeneous medium is assumed to consist of randomly distributed high and low permeability patches, which are small compared to the dimensions of the reservoir. The saturation is assumed constant in the swept zone.

Because of the varying permeability, the initially flat displacement front will be disturbed. The evolution of the disturbances can be described by a dispersion coefficient (D). The relation between the standard deviation of the displacement front (σ_{st}) and the position of the average displacement front ($\bar{x}$) is given by

$$\sigma_{st} = \sqrt{\bar{x}\frac{2D}{\bar{v}}} \tag{4.1}$$

in which $\bar{v}$ is the average velocity of the displacement front. We will only consider small variations in permeability k, i.e. we assume that $\sigma_{\ln k}^2 < 0.5$, in which k is assumed to be distributed log-normally ($\sigma_{\ln k}^2$ is the variance in $\ln k$). In this regime the dispersion coefficient is given by (see [6]) $\sigma_{\ln k}^2 \lambda \bar{v}$, and so

$$\sigma_{st} = \sqrt{\bar{x} 2 \sigma_{\ln k}^2 \lambda} \tag{4.2}$$

λ is the correlation length of the heterogeneities. In general, λ is of the order of 0.1 to 0.01 of the field scale [7]. The statistical evolution of the interface is thus characterised by $2\sigma_{\ln k}^2 \lambda$. From the definition of λ [6] it follows that the average width w of the heterogeneities is 2λ.

From statistical arguments it can be shown that the dominant wavelength of the interface disturbances is 8λ.

5 Finger Growth in a Heterogeneous Medium ($M > 1$)

The finger growth resulting from the combined effects of permeability variations and viscous forces is tentatively described by

$$\left(\frac{d\sigma}{dt}\right)^2 = \left(\frac{d\sigma_{st}}{dt}\right)^2 + \left(\frac{d\sigma_{vis}}{dt}\right)^2 \tag{5.1}$$

$d\sigma/dt$ is the growth rate of the standard deviation of the interface, $d\sigma_{st}/dt$ is the growth rate of the standard deviation caused by permeability variations and $d\sigma_{vis}/dt$ the growth rate caused by viscous forces. For $M = 1$, $d\sigma_{vis}/dt = 0$ and so $d\sigma/dt$ is determined solely by the permeability variations. For an extremely homogeneous medium, $d\sigma_{st}/dt$ is very small, and the growth rate of the standard deviation is determined purely by viscous forces $d\sigma/dt = d\sigma_{vis}/dt$. The squares are applied to account for the fact that it is assumed that the growth of the standard deviation of the disturbances in the time span t to $t + \Delta t$, owing to statistical fluctuations, is independent of the growth due to viscous forces.

The growth rate owing to statistical fluctuations can be determined from the eqs. 4.1 and 4.2 and is given by

$$\frac{d\sigma_{st}}{dt} = \frac{\sigma_{\ln k}^2 \lambda}{\sigma_{st}} \bar{v} \tag{5.2}$$

The actual value for σ at time t should now be substituted for σ_{st} in the denominator. The growth rate of σ_{vis} can be determined by differentiating eq. 3.7 and by applying $d\bar{a}/dt = \frac{1}{2}v_{vis}$ this yields

$$\frac{d\sigma_{vis}}{dt} = \frac{1}{2}\sqrt{\frac{2}{3}}\,v_{vis} \qquad (5.3)$$

v_{vis} can be determined if $\bar{a}$ and b_{in} and b_{out} are known. The value of $\bar{a}$ can be determined from σ by applying eq. 3.7. The value of b_{in} and b_{out} depends on which phase the fingering process is in. For phase i, as the porous medium determines the finger spacing $b = b_{in} = b_{out}$ approximately with wavelength 8λ, i.e. $b = 2\lambda = w$. For this constant value of b the numerical integration of eq. 5.1 in which eq. 5.2 and 5.3 are substituted, gives the results as depicted in Figure 4. The mobility ratios M are 1,2 and 20 and for the value of $\sigma^2_{\ln k}w$, which characterises the heterogeneity, is taken to be $0.1w$ and $0.001w$ respectively. Figure 4 shows that the fingers enter the linear growth regime more rapidly for larger mobility ratios and larger values of $\sigma^2_{\ln k}$. In the final stage, the growth rate is constant.

However, the results as depicted in Figure 4 are valid over a limited inflow distance only. After a certain distance, depending on heterogeneity and mobility ratio, the fingering process will enter phase ii, in which b starts to increase. As the behaviour of b with the average swept distance is not known, the viscous growth rate cannot be determined from eqs. 3.1 to 3.4. Consequently the growth in phase ii cannot be determined. Finally the fingering process will enter phase iii, in which only one finger is left. For this phase several studies, experimental [18, 17], analytical [15,11], as well as numerical [5,20] have shown that this finger has a tendency to occupy half the model width, so $b = b_{in} = b_{out} = B/2$. The viscous growth rate for a given value of $\bar{a}$ can now be determined from eq. 3.1 and 3.4 and so via eq. 5.1 the total growth rate can be determined.

6 The Numerical Simulator

The numerical technique applied is a revised version of the diffusion limited aggregation (DLA) technique. In DLA the displacement front advances via discrete saturation jumps; the position where a jump takes place is determined by a random procedure. The principle of the method is described in [14]. However in our simulator a few modifications are applied. Before giving the details of our revisions, a physical validation for the application of the DLA technique for displacements in heterogeneous media will be given first.

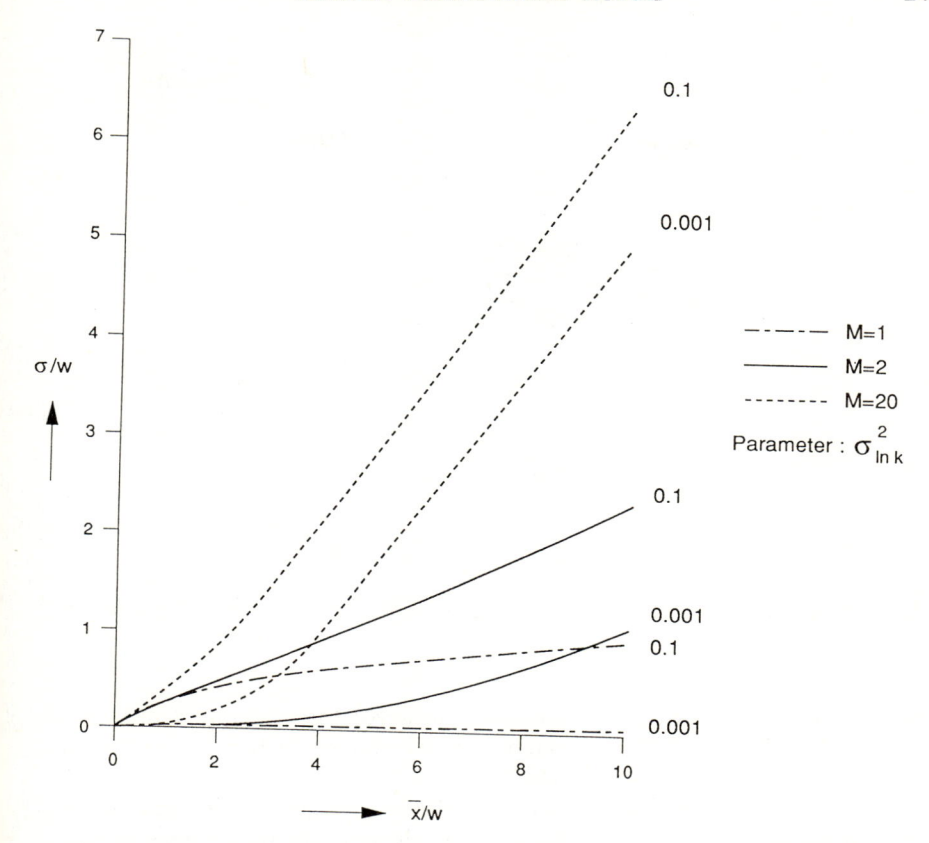

Figure 4. Development of the standard deviation σ of the displacement front as a function of the average position of the displacement front $\bar{x}$ is the width of the heterogeneities, which is equal to twice the correlation length. $\sigma^2_{\ln k}$ is the variance of the logarithm of the permeability.

6.1 Physical Validation

The high and low permeability regions ("patches") that the permeable medium consists of, will only disturb the flow pattern over a finite area. So patches far from the displacement front will not influence the actual growth of the displacement front. The front is only disturbed when it is close to a patch. After flowing over a patch the consequences of the patch are visible as humps in the displacement front (see Figure 5).

The influence of a patch on the displacement front is now modelled by a discontinuous jump of the displacement front forwards, shaped like a rectangular bulge. A jump occurs the moment the average position of the displacement front coincides with the centre of a patch. The dimensions of the jump are determined by the size and the permeability of the patch;

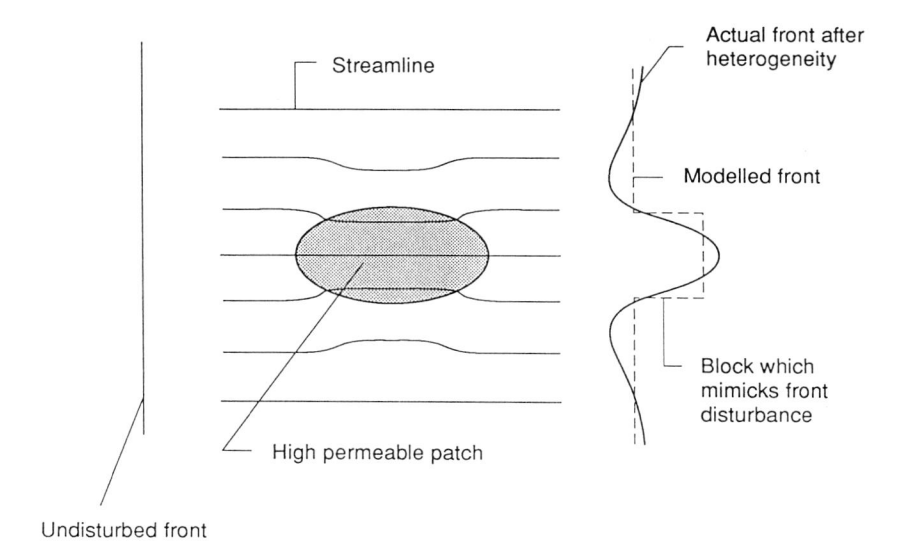

Figure 5. Schematic of the disturbance of a displacement front by a highly permeable patch.

the saturation in the jump is assumed equal to the (constant) upstream saturation. During the displacement, many patches will be encountered. In the following we will assume the density of the patches to be so large that the front effectively advances only in the form of jumps. These jumps occur randomly over the displacement front. In principle, the jumps have varying length and size, representing the size and permeability distribution of the patches. However, we will apply averaged properties; the average jump length is called l_{eff}. The average width is equal to twice the correlation length λ of the heterogeneous medium i.e. equal to w (see section 4). To relate l_{eff} to physical parameters we construct a model system of an $M = 1$ displacement in a heterogeneous medium. The model consists of N tubes of width w, which are positioned in the main flow direction. The swept area is built by subsequently allocating blocks, representing the jumps, to the N tubes. Each tube has a probability $1/N$ to receive a block. The distribution of the blocks over the tubes is given by the Poisson distribution and so the standard deviation σ_n of the number of blocks per tube is equal to the square root of the average number of blocks per tube. As the block length is l_{eff}, the standard deviation in length units is $\sigma = \sigma_n l_{eff} = \sqrt{(\bar{x} l_{eff})}$ in which $\bar{x}$ is the average position of the displacement front. Comparison with formula 4.2 yields for the block length l_{eff}

$$l_{eff} = 2\sigma_{\ln k}^2 \lambda = \sigma_{\ln k}^2 w \tag{6.1}$$

6.2 Simulator Description

The procedure as described above for building up the swept area can be extended to unstable displacements. To do this properly the random positioning of the blocks along the interface has to be weighted by the local interface velocity. The blocks are positioned in the downstream direction along the streamline which crosses the chosen interface position. A simulator based on this principle was built and the following provisions were taken:

- The block length l_{eff} can be a fraction of the spacing of the grid on which the displacement is simulated.

- During the displacement it may happen that displacing fluid is invaded by fluid to be displaced, for example if fingers are being pinched of, or if islands of fluid to be displaced are formed that subsequently move to the producers. To account for this kind of flow 'negative' blocks are introduced.

- The positioning of the blocks downstream of the chosen interface position may be longitudinal, transverse and diagonal across the grid.

- Numerical dispersion is suppressed.

A typical result of such a simulation run with $M = 20$ and $l_{eff}/w = \sigma_{\ln k}^2 = 0.1$ is shown in Figure 6.

7 Comparison of Analytical and Numerical Results

The analytical model can be worked out if the finger width is assumed to be constant during the fingering process, i.e. that the interface only moves in longitudinal direction. This assumption is appropriate for phase i. To compare the results of the analytical model with the numerical results, transverse and diagonal positioning of blocks must be excluded in the latter. Figure 7 shows the results of both models, for the mobility ratio $M = 2$ and 20 and the correlation length l_{eff} is $0.1w$ and $0.001w$, the numerical model is 20 blocks wide and 40 blocks long. The match of the analytical and numerical model with respect to the start of fingering and growth rate in the linear regime is satisfactory.

The deviations that occur for $l_{eff} = 0.001w$ can be attributed to a slight underestimation of the viscous forces by the numerical simulator if the average finger length is smaller than the grid spacing ($\sigma/w < 1$). Because initially the viscous forces are too small the start of the linear growth regime is delayed. The variations by chosing different random number sequences are small. The increased delay before the linear growth regime starts is apparent for decreasing values of l_{eff}.

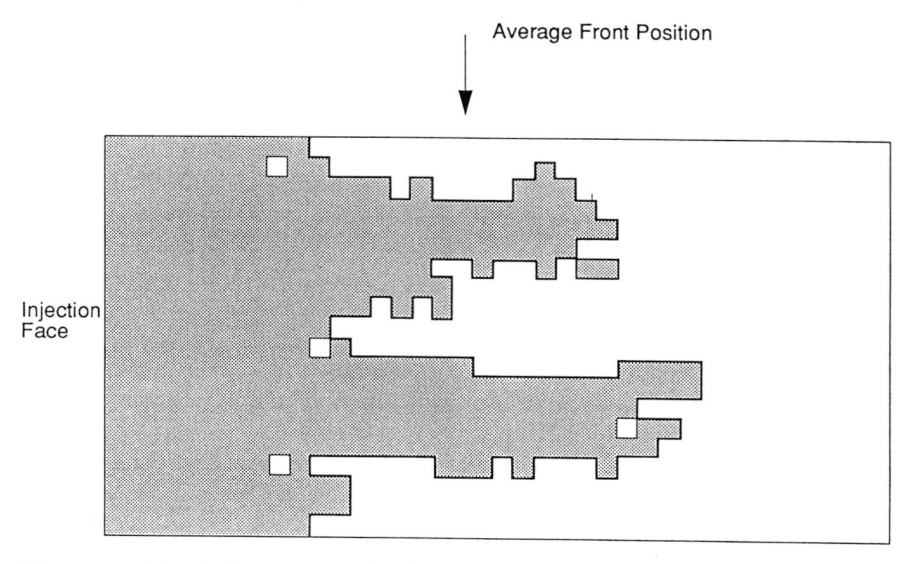

Figure 6. Typical example of a displacement pattern as calculated by the stochastic numerical simulator. The shaded area represents the flooded area. The 50 % saturation contour is shown. The mobility ratio is 20, $\sigma^2_{\ln k}$ is 0.1. The grid dimensions are 20x40. Lateral flow has been included.

The consequences of lateral interface movements can now be studied with the numerical simulator. Figure 8 shows the results of a numerical simulation in which positioning of blocks in transverse and diagonal directions is permitted, for comparison the results of the analytical model are shown too. The mobility ratios M are again 2 and 20 and $l_{eff} = 0.1w$.

The numerical simulations now show a reduction in the growth of the standard deviation by roughly a factor of two. The reason is that in the initial phase of the displacement, the size of the disturbances in the porous medium are still of the same order of magnitude as the size of the fingers, simply because the disturbances are also the initiation centre of the fingers. A heterogeneity can thus easily deviate the interface in a lateral direction thus stabilising longitudinal finger growth. In addition, the fingers are entering phase ii in which stabilisation is enhanced further by finger merging and suppression. This increases the finger width, as a consequence the ratio of finger length over finger width ($\bar{a}/b$) decreases and so the growth rate is reduced. It appears that the growth rate of the fingers as predicted by the analytical model, gives the upper limit of the growth rate. A single large finger ($b/w \gg 1$) in the numerical model (phase iii) showed the same growth rate as predicted by the analytical model.

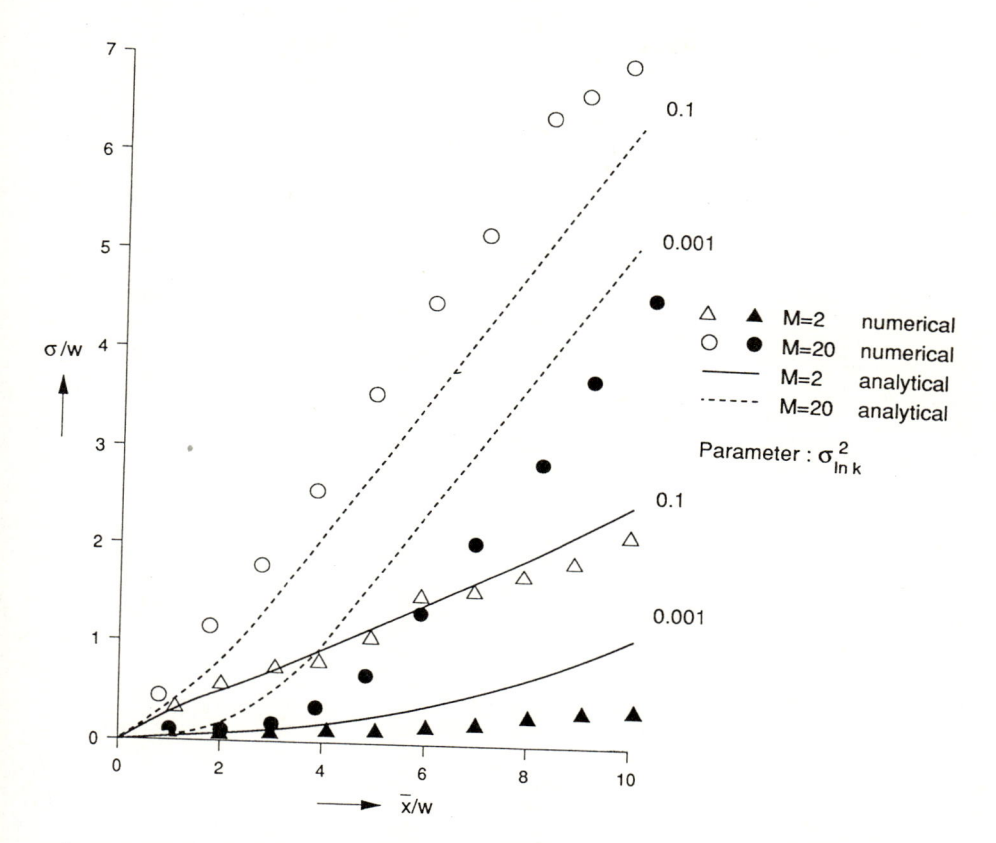

Figure 7. Comparison of the results of the stochastic numerical simulator with the analytical results. There is no lateral flow in the numerical simulator. See also the caption of Figure 4.

8 Conclusions

- An accurate expression describing immiscible finger growth in a homogeneous porous medium has been obtained.

- In a homogeneous medium a small finger starts growing exponentially but changes rapidly to linear growth.

- An analytical expression has been formulated which couples the growth of irregularities in a displacement front owing to porous medium heterogeneities, to the growth of these irregularities owing to viscous forces. Only the longitudinal effects of heterogeneities are considered, furthermore the heterogeneity must be moderate. ($\sigma_{\ln k} < 0.5$).

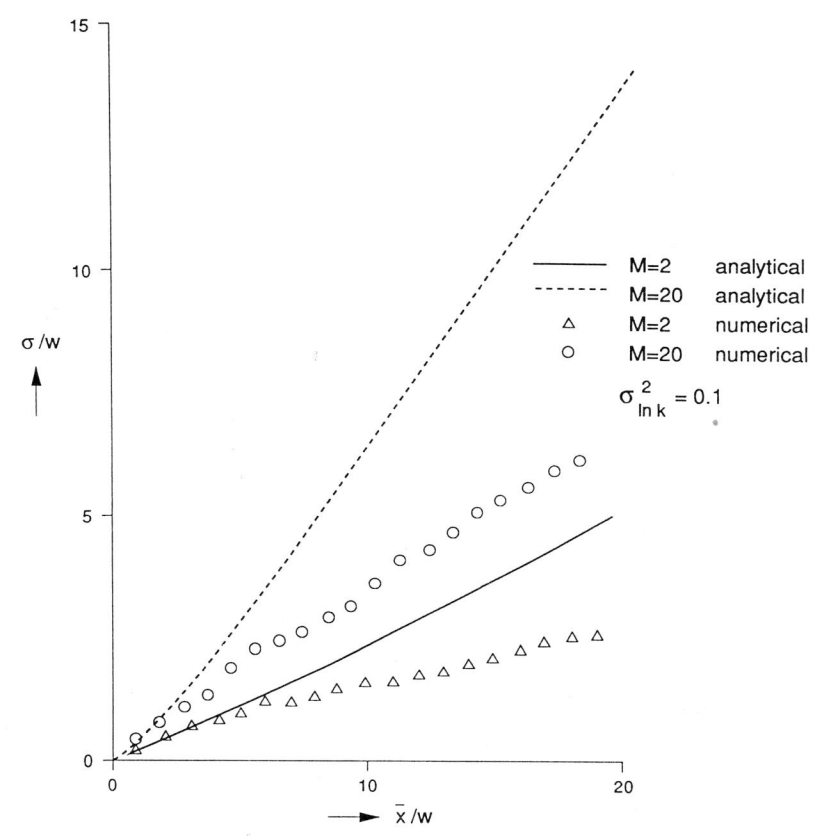

Figure 8. Comparison of the results of the stochastic numerical simulator with the analytical results. Lateral flow is included in the simulator. See also the caption of Figure 4.

- A larger heterogeneity gives rise to an earlier entering of the linear growth regime. The growth rate in the linear growth regime is sensitive to the mobility ratio but not to the degree of heterogeneity.

- To simulate fingering processes in a heterogeneous porous medium, a revised version of the diffusion limited aggregation technique has been applied. The results of the analytical model match the numerical results, if in the latter only interface movements in the main flow direction are included.

Acknowledgement

The authors are indebted to Shell Internationale Research Maatschappij B.V. for permission to publish this paper.

References

1. Bentsen, R.G., (1985), A new Approach to Instability Theory in Porous Media, *Soc. Pet. Eng. J.*, 765, October.

2. Carslaw, H.S. and Jaeger, H.C., (1959), Conduction of Heat in Solids, Clarendon Press, Oxford, 426-428.

3. Chuoke, R.L., von Meurs, P. and van der Poel, C., (1959), The Stability of Slow, Immiscible, Viscous liquid-liquid Displacement in Permeable, Media, Trans. AIME, 216, 188-194.

4. Degregoria, A.J., (1986), Monte Carlo Simulation of Two-fluid Flow through Porous Media at Finite Mobility Ratio - The Behaviour of Cumulative Recovery, *Phys. Fluids* 29 (11), November.

5. Degregoria, A.J. and Schwartz, L.W., (1986), A Boundary-integral Method for Two-phase Displacement in Hele-Shaw Cells, *J. Fluid Mech.*, 164, 383-400.

6. Gelhar, W. and Axness, C.L., (1983), Three-dimensional Stochastic Analysis of Macrodispersion in Aquifers, Water Resources Research, 19, no. 1, 161-180.

7. Gelhar, L.W., (1986), Stochastic Subsurface Hydrology, From Theory to Applications, Water Resources Research, no. 9, vol. 22, 135-145, August.

8. Huang, A.B., Chikhiliwala, E.D. and Yortsos, Y.C., (1984), Linear Stability of Immiscible Displacements including Continuously Changing Mobility and Capillary Effects: part 2 - General Basic Flow Profiles, SPE 13163.

9. Hughes, D.S. and Murphy, P., (1987), Use of Monte Carlo Method to Simulate Unstable Miscible and Immiscible Flow through Porous Media, SPE 17474.

10. Jackson, J.D., (1962), "Classical Electrodynamics", John Wiley and Sons Inc.

11. Jacquard, P. and Seguier, P., (1962), Mouvement de Deux Fluides en Contact dans un Milieu Poreux, Journal de Mecanique, no 4,1, December.

12. Jensen, J.L., Lake, L.W. and Hinkley, D.V. (1985), A Statistical Study of Reservoir Permeability: Distributions, Correlations and Averages, SPE 14270.

13. Jerauld, G.R., Davis, H.T., and Scriven, L.E., Stability Fronts of Permanent Form in Immiscible Displacement, SPE 13164.

14. King, M.J., (1987), Viscous fingering Utilising Probabilistic Simulation, SPE 16708.

15. Mclean, J.W. and Saffman, P.G. (1981), The Effect of Surface Tension on the Shape of Fingers in a Hele-Shaw Cell, *J. Fluid Mech.*, vol. 102, 455-469.

16. Mui, C. and Miller A., (1985), Stability of Displacement Fronts in Porous Media - Growth of Large Elliptical Fingers, *Soc. Pet. Eng. J.*, April.

17. Park, C.W. and Homsy, G.M., (1985), The Instability of Long Fingers in Hele-Shaw Flows, *Phys. Fluids* 28 (6), June.

18. Saffman, P.G. and Taylor, G.I., (1958), The Penetration of a Fluid into a Porous Medium or Hele-Shaw Cell Containing a more Viscous Liquid, *Proc. R. Soc. Lond.* A245, 312-329.

19. Schmidt, G.H. and Jacobs, F.J., (1988), Adaptive Local Mesh Refinement and Multi-grid Solution Methods in Numerical Reservoir Simulation, "Mathematics in oil production", Clarendon Press Oxford, 295-311.

20. Tryggvason, G. and Aref, H., (1985), Finger Interaction in Stratified Hele-Shaw Flow, J. Fluid Mech. 154, 287-301.

21. Yeh, T.C.J., Gelhar, L.W. and Gutjahr, A.L. (1985), Stochastic Analysis of Unsaturated Flow in Heterogeneous Soils, Part 1, Statistically Isotropic Media, Water Resources Research, no. 4, 21, 447-456, April.

APPENDIX

The Velocity of Finger Growth

To determine the growth velocity of fingers the following steps are taken: the velocity in an elliptic area of different mobility (M) in an infinitely large medium is determined. Next, an approximate expression is derived for an ellipse in a medium of finite width. Finally, by approximating the finger pattern by forward and backwards pointing ellipses a set of equations is derived, from which the growth rate of fingers can be determined.

The fluid velocity in an infinitely large porous medium, containing an elliptic area of different mobility (M), can be determined directly from potential theory [2]. A special feature of this elliptic shape is that the fluid velocity at all locations in the ellipse is constant. The flow direction is equal to the flow direction far from the ellipse. In the following we assume that one of the principal axes of the ellipse coincides with the flow direction far from the ellipse. Inside the ellipse

$$\frac{v_{in}}{v_\infty} = \frac{M}{1 + A_0(M-1)} \qquad (A1)$$

in which $A_0 = b/(a+b)$, a being the half length and b the half width of the ellipse, and outside the ellipse on the symmetry axes (y direction) perpendicular to the flow direction

$$\frac{v_{out}}{v_\infty} = 1 - \frac{M-1}{1 + A_0(M-1)} A_\lambda \qquad (A2)$$

with

$$A_\lambda = \frac{ab}{a^2 - b^2}\left(1 - \sqrt{\frac{y^2}{y^2 + a^2 - b^2}}\right) \qquad (A3)$$

M is the mobility in the ellipse and v_∞ the velocity far from the ellipse. The fluid velocity in a finite medium containing an elliptic area can be estimated by the method of mirror images (see for example [10]). By adding the potentials of the mirror image ellipses, the condition is fulfilled that perpendicular to the boundary the fluid velocities are zero. As the potential gradients outside the ellipse decrease only slowly we may assume that the sum of the mirror potential gradients in the area under consideration is approximately constant. As a result, the corresponding corrections in the fluid velocities inside and outside the high mobility structure due to the boundaries relate as

$$\delta v_{in} = M \delta v_{out} \qquad (A4)$$

Mass balance requires

$$\int_0^b (v_{in} - \delta v_{in})dy + \int_b^B (v_{out} - \delta v_{out})dy = \int_0^B v_\infty dy \qquad (A5)$$

in which B is the half model width or half the finger distance. Substitution and working out yields as the velocity in the elliptic area

$$\alpha(a, b, B, M) = \frac{v_{in} - \delta v_{in}}{v_\infty}$$

$$= 1 + C\left[\frac{a-b}{b} + \frac{M}{B+(M-1)b}\left(B - \sqrt{B^2 + a^2 - b^2}\right)\right] \quad (A6)$$

in which
$$C = \frac{(M-1)ab}{(a+bM)(a-b)} \quad (A7)$$

The expression for $\alpha(a, b, B, M)$ holds when inside the ellipse the mobility is M. For an ellipse surrounded by a medium with an M times larger mobility $\alpha(a, b, B, \frac{1}{M})$ holds. A consistent composition of the ellipses into a finger pattern requires the velocities at the base line (see Figure 1) to satisfy

$$b_{in}v_{in} + b_{out}v_{out} = (b_{in} + b_{out})\bar{v} \quad (A8)$$

as $v_\infty = \bar{v}$, there holds

$$b_{in}\alpha(a_{in}, b_{in}, B, M) + b_{out}\alpha(a_{out}, b_{out}, B, \frac{1}{M}) = b_{in} + b_{out} \quad (A9)$$

in which $B = b_{in} + b_{out}$. Furthermore we define the average amplitude by

$$\bar{a} = \frac{1}{2}(a_{in} + a_{out}) \quad (A10)$$

For a given value of $\bar{a}$ the equations A9 and A10 can be solved to give a_{in} and a_{out} and so also via A6 the velocities v_{in} and v_{out}. The growth velocity is then given by $v_{vis} = v_{in} - v_{out}$.

SIMPLE RENORMALIZATION SCHEMES FOR CALCULATING EFFECTIVE PROPERTIES OF HETEROGENEOUS RESERVOIRS

J.K. Williams

BP Research Sunbury Research Centre

Abstract This paper describes real space renormalization schemes for determining effective flow properties of heterogeneous reservoirs, focusing on effective permeabilities. Effective properties are calculated in a hierarchical fashion by dividing the system up into cells, consisting of a number of grid–blocks, and calculating the effective properties of each of these cells. Each cell is then treated as a single grid–block in a larger cell and so on. This paper discusses two types of scheme: small–cell methods, which are very simple to implement and computationally very efficient, and large–cell methods, which provide the greater accuracy required to deal with strongly anisotropic systems.
The large–cell method has advantages in determining the effective permeability of cross–sectional models, where discontinuous impermeable shales often cause difficulties for small–cell schemes with the locally imposed cell boundary conditions. Each of the large cells contains many grid–blocks and the effective permeability is calculated by replacing each cell by its equivalent resistor network. In 2D these resistor networks can be reduced to single equivalent resistors by a systematic sequence of exact resistor transformations mirroring the 'differential real space renormalization' technique developed in statistical physics. This large cell method appears more attractive for anisotropic systems with a finite fraction of impermeable material and can be extended to allow for mesh refinement. A test problem using outcrop data illustrates the capabilities of the method.

1 Introduction.

In many heterogeneous reservoir environments the flow geometry is primarily controlled by the spatial connectivity of the extreme permeability values and is rather less sensitive to the finer details of the spatial distribution of permeability. For example, low permeability shale inclusions often act as significant barriers to vertical flow, and high permeability streaks can act as preferential flow channels [1,2]. Techniques for determining the effective flow properties of these heterogeneous media must capture the underlying connectivity of the 'high' and 'low' permeability regions.

Renormalization schemes are powerful approximation methods ideally suited to this type of problem, where the permeability exhibits strong fluctuations on many length scales [3,4]. The essence of the renormalization methodology is to break a large problem down into a sequence of smaller and more manageable stages which are handled in a hierarchical fashion [5]. For example, the effective permeability of a large system is determined by dividing the system up into small cells, consisting of a small number of grid blocks, and calculating the effective permeability of each of these cells. The whole cell is then treated as an element, or single grid block, of a larger scale cell. The effective permeabilities of the larger scale cells are then determined; these cells are then used as elements in even larger scale cells, and so on. The scale dependent behaviour of the system is revealed by monitoring the evolution of the distribution of effective permeability values as the renormalization proceeds from one scale to the next. Typically, the average values of the cell effective permeabilities converge (approximately) towards the effective permeability of the whole system, and the width of the distribution narrows, as successive scales of heterogeneity are 'averaged' out.

There is no unique prescription for renormalizing in real space and a considerable number of approximate renormalization techniques have been developed in statistical physics [6,7]; many could be adapted for determining effective flow properties of heterogeneous media. In this paper we describe some simple renormalization schemes, focusing on the determination of effective permeabilities. We begin by outlining the analogy between a discretised porous medium and an equivalent resistor network. We exploit this analogy to determine the effective permeability of our renormalization cells. The choice of renormalization scheme is determined, in part, by how well the connectivity of flow paths is honoured. We consider two small cell schemes and a large cell scheme and apply them to some test problems; we note the need to treat anisotropic systems carefully. We then apply our large cell method to a shale problem which provides a severe test case. Finally, we present some concluding remarks and indicate future areas for research.

2 Renormalization Methodology: Illustrated in 2D.

For clarity, we focus on the problem of determining the effective permeability of a 2D heterogeneous region. The 3D case is similar in most respects.

2.1 Equivalent Resistor Networks and Small Cell Schemes.

Single phase incompressible flow through a permeable medium, $K(\mathbf{r})$, is described by the following equations, relating the Darcy flux $\mathbf{v}$ and the pressure P,

$$\mathbf{v}(\mathbf{r}) = -K(\mathbf{r})\nabla P(\mathbf{r}), \qquad \nabla \cdot \mathbf{v} = 0 \qquad (2.1)$$

(where, for clarity, we have set the viscosity = 1 and taken the permeability to be isotropic). These equations may be combined to yield the pressure equation and solved by using a standard five-point difference scheme. Introducing a regular grid $\{\mathbf{r}_i\}$, with unit spacing for convenience, we obtain a system of linear equations for the pressures $P_i = P(\mathbf{r}_i)$,

$$\sum_{j \in (i)} T_{ij}(P_j - P_i) = 0, \qquad T_{ij} = \frac{2K_i K_j}{K_i + K_j} \qquad (2.2)$$

for all $\mathbf{r}_i$ away from the boundaries, where $j \in (i)$ denotes all nearest neighbour blocks of i, and the 'average' transmissibility, T_{ij}, is given by the harmonic mean of the block centre permeability values. Equation (2.2) is formally equivalent to the Kirchoff current law for a regular network of conductances, T_{ij}, with P_i denoting the voltage at the ith node, this enables the effective permeability to be determined from an equivalent network resistance.

We now need to choose our renormalization cells. Small cells of grid-blocks are adequate for rapid approximate methods and the simplest suitable cell consists of 2 × 2 grid blocks, as shown in figure (1). In order to calculate the cell effective permeability in (say) the x-direction we set constant pressures over the inlet (left edge) and outlet (right edge) cell boundaries and no flow through the other boundaries (top and bottom edges), and then replace the 2 × 2 cell by an equivalent resistor network. Using simple resistor transformations we can find the equivalent network resistance and hence the corresponding cell effective permeability in the x-direction. We can then 'rotate' the boundary conditions and repeat the procedure to obtain the effective permeability in the y-direction. For isotropic grid-block permeabilities the x-direction cell permeability, K_{XE}, is given by,

$$\frac{4[A](K_1 + K_3)(K_2 + K_4)}{[A](K_1 + K_2 + K_3 + K_4) + 3(K_1 + K_2)(K_3 + K_4)(K_1 + K_3)(K_2 + K_4)} \qquad (2.3)$$

$$\text{where} \qquad [A] = K_1 K_2 (K_3 + K_4) + K_3 K_4 (K_1 + K_2) \qquad (2.4)$$

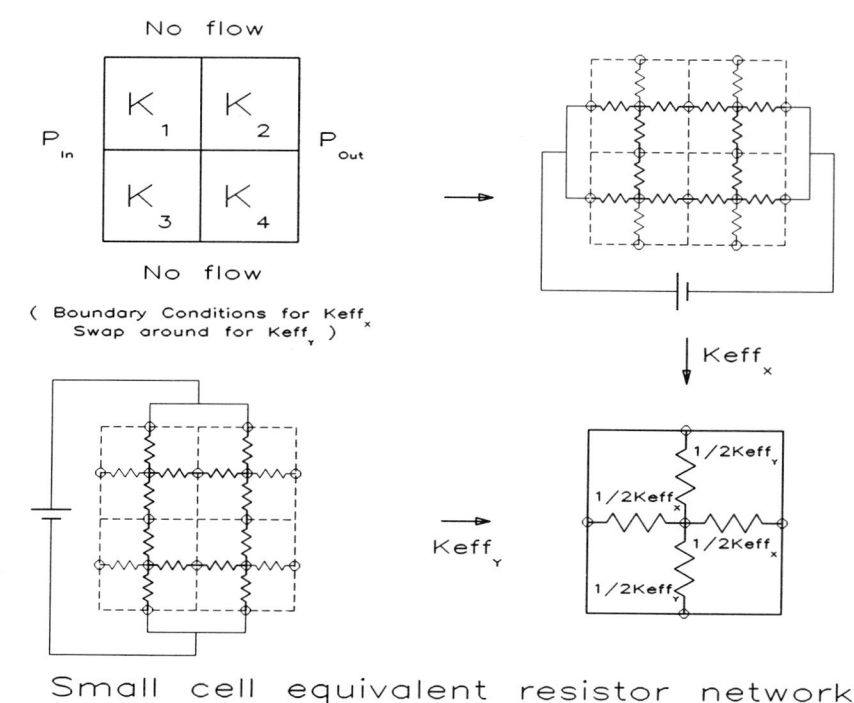

Figure 1.

A detailed derivation of this result appears in King [3,4]. An alternative scheme is shown in figure (2); it is essentially equivalent to a Migdal-Kadanoff (MK) renormalization transformation [7], and can be viewed as a two step process. The cell is broken up into two one–dimensional 2×1 sub–cells, figure (2b), which are then reduced to single effective permeabilities by series combination,

$$K_{12}^{(X)} = \frac{K_1 K_2}{(K_1 + K_2)}, \qquad K_{34}^{(X)} = \frac{K_3 K_4}{(K_3 + K_4)} \qquad (2.5)$$

These two sub–cell effective permeabilities are then combined in parallel to yield

$$K_{XE} = \frac{K_1 K_2 (K_3 + K_4) + K_3 K_4 (K_1 + K_2)}{(K_1 + K_2)(K_3 + K_4)} \qquad (2.6)$$

The resistor analog for this scheme is shown in figure (2c). Extensions of this type of scheme to 3D are relatively straightforward. Both schemes yield the correct effective permeability for perfectly layered configurations (eg.

SIMPLE RENORMALIZATION SCHEMES

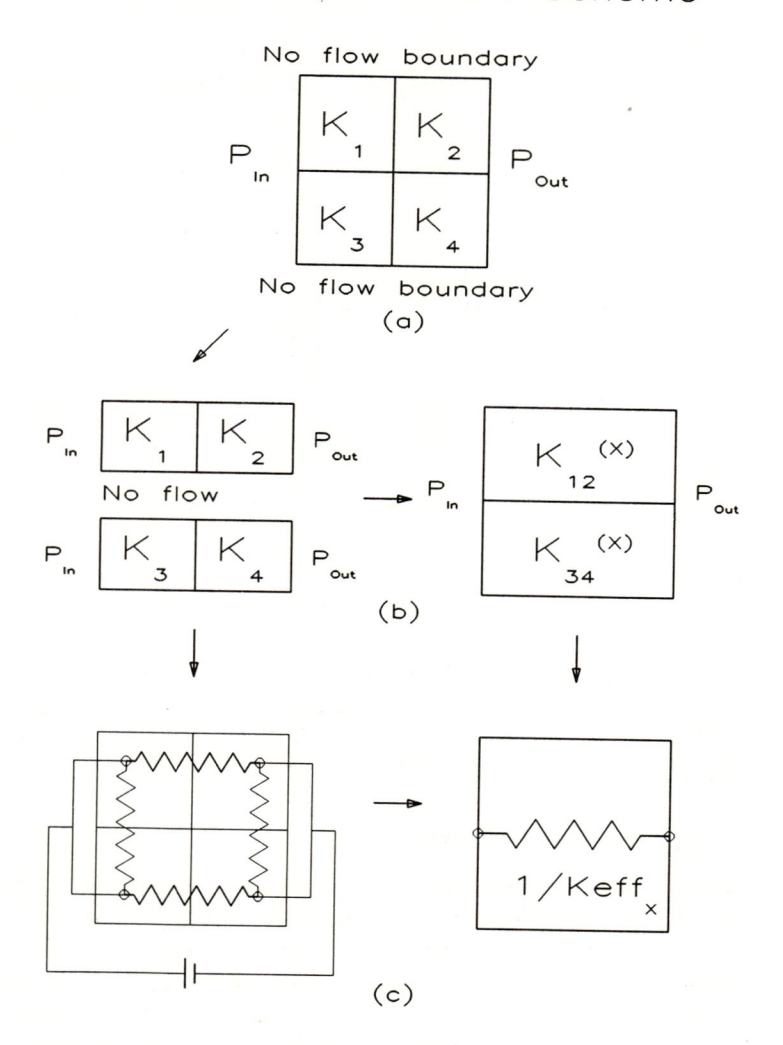

Figure 2.

Cell Configuration Classes

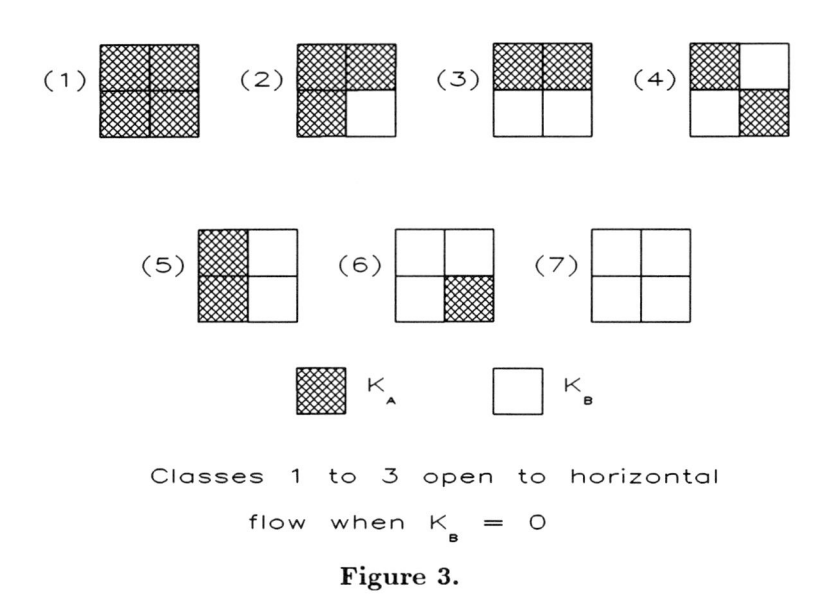

(1) (2) (3) (4)

(5) (6) (7)

K_A K_B

Classes 1 to 3 open to horizontal
flow when $K_B = 0$

Figure 3.

$K_1 = K_2$ and $K_3 = K_4$, or $K_1 = K_3$ and $K_2 = K_4$); giving the arithmetic average for flow parallel to the layering and the harmonic average for flow perpendicular to the layering. Differences between the schemes are seen for cell configurations which give rise to 'cross–flow'.

2.2 Capturing Connectivity.

We now indicate how these small cell approximations honour connectivity properties with the following simple example. Let G_0 denote the starting grid and suppose that each grid block is labelled either A_0 or B_0, with permeability $K_A^{(0)}$ or $K_B^{(0)}$; where A represents (say) a high permeability 'sand' and B a low permeability 'shale'. There are 2^4 possible cell configurations, which can be grouped together in 7 classes on the basis of their renormalized permeabilities, figure (3). Furthermore, for simplicity, we will assume that the shale is impermeable, i.e. $K_B^{(0)} = 0$. Then, according to the two schemes outlined above, only those cells with configurations belonging to classes 1, 2 and 3 have non–zero renormalized permeability for horizontal (x–direction) flow.

After one renormalization the renormalized grid, G_1, contains renormalized grid blocks (2×2 cells on G_0) some of which have zero permeability and others which have non–zero permeability. The permeabilities of the permeable blocks will depend, of course, on the underlying cell configuration (class) and the renormalization scheme employed. However, it is sufficient for our purposes to regard all the impermeable grid blocks as pseudo–shale, which we label B_1, and the permeable grid blocks as pseudo–sand, A_1. The pseudo–sands may be assigned some suitably weighted average effective permeability, $K_A^{(1)}$; the details need not concern us here. Cell configurations belonging to classes 1, 2 and 3 are renormalized to pseudo–sand, A_1, grid blocks; likewise cell configurations belonging to classes 4 to 7 yield pseudo–shale, B_1, grid blocks.

The procedure can be iterated, and after n iterations the grid, G_n, consists of permeable pseudo–sands, A_n, and impermeable pseudo–shales, B_n. After n+1 renormalizations the volume fraction, p_{n+1}, of A_{n+1} is the sum of volume fractions of cells on G_n associated with classes 1, 2 and 3 and this can be related to the volume fraction, p_n, of A_n. In the absence of spatial correlation we have,

$$p_{n+1} = R(p_n) = p_n^4 + 4p_n^3(1-p_n) + 2p_n^2(1-p_n)^2 = 2p_n^2 - p_n^4 \quad (2.7)$$

The transformation $R(p_n)$ relates the volume fractions of A_n and A_{n+1}; it has (trivial) stable fixed points at 0 and 1 and an unstable fixed point at $P^* = 0.618$. The behaviour of $R(p)$ is shown in figure (4). If we start with a volume fraction, $P_0^<$, lying between 0 and P^* then repeated iteration of $R(p)$ drives p away from P^* towards zero and the grid is eventually renormalized to a homogeneous grid containing pseudo–shale, B_m, only. Similarly, for a starting volume fraction, $P_0^>$, lying between P^* and 1 the grid is ultimately renormalized to a homogeneous grid containing pseudo–sand, A_m, only. We see that P^* is an unstable fixed point separating two regimes.

This divide in behaviour reflects (qualitatively) that of flow through sand/shale systems [8]; where for high volume fractions of sand the flow through connected sands controls the effective permeability, and at lower volume fractions, when the sand continuity is broken, the effective permeability is controlled by flow through the shales. Sharp decreases in effective permeability are seen near threshold sand volume fractions (percolation behaviour). The small cell schemes capture this behaviour with a remarkable degree of accuracy — at least for isotropic systems, as indicated in the above example where the fixed point obtained from renormalization of a 2×2 cell is close to the true threshold (0.593) for this simple example (essentially 2D site percolation).

Renormalized Connectivity

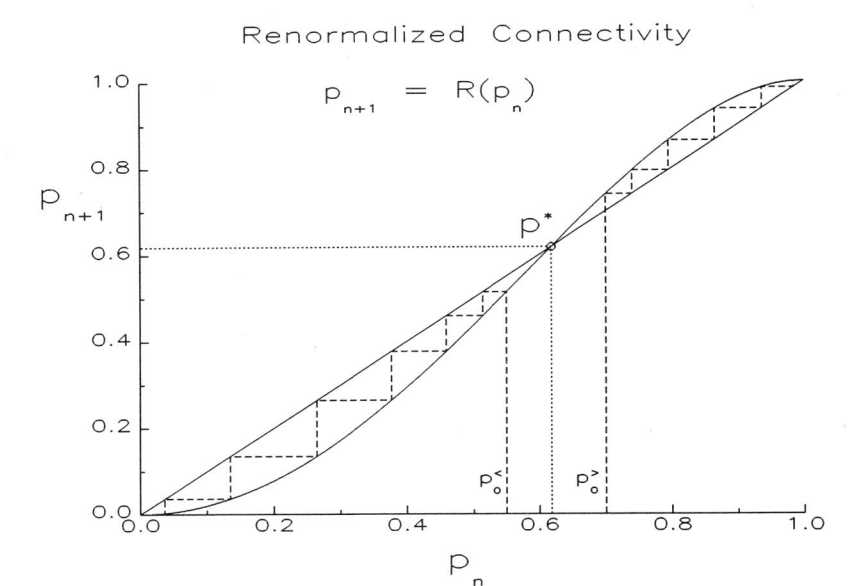

$$p_{n+1} = R(p_n)$$

Figure 4.

2.3 Large Cell Schemes.

A desired feature of a renormalization scheme is that the rescaled or renormalized system effectively mirrors the spatial connectivity of the high and low permeability regions of the original system. The controlling large scale features should be preserved by renormalization and the effects of the smaller scale features accumulated in the cell effective values as each length scale is probed and 'averaged over' in successive renormalizations.

In the small cell schemes errors result from the cell boundary conditions. On the one hand, the imposition of a constant pressure boundary condition on the inlet and outlet boundaries leads to an overestimate of the connectivity of high permeability paths over the length scale of several cells. On the other hand, with no–flow conditions imposed on the other boundaries small cells underestimate the connectivity of high permeability paths, as the cells fail to 'capture' the contributions of the more tortuous paths. Figure (5) shows a possible configuration for four adjacent 2 by 2 cells, where both types of error occur. In the original configuration there is a connected path from left to right via cells C, A and B, but in the renormalized configuration the path goes via cells C and D. In the original configuration cells A and B are connected, but are not in the renormalized configuration. Conversely, cells C and D are not connected in the original configuration, but are in the renormalized configuration. For isotropic

Small Cell Boundary Error

original configuration → renormalized configuration
(for horizontal flow)

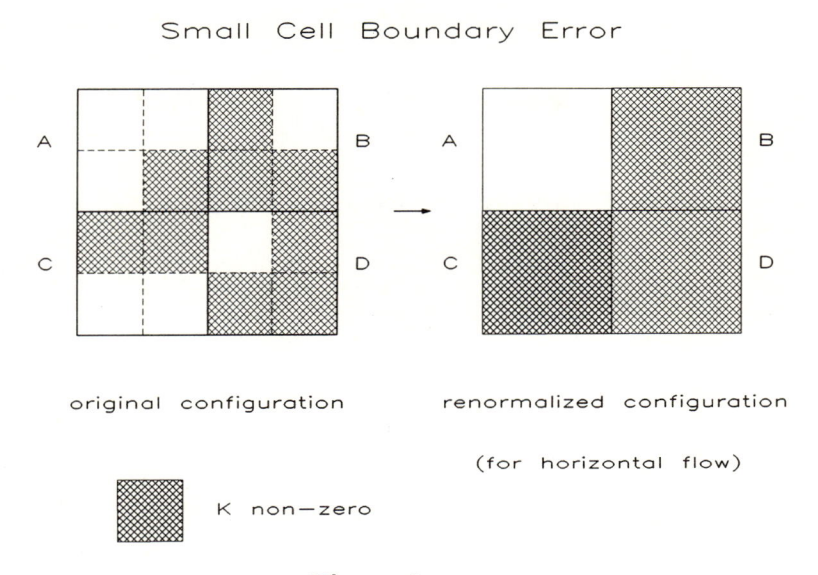

K non—zero

Figure 5.

systems a fair degree of cancellation appears to occur, as indicated by the example above; however, for anistropic systems the errors are more severe. These errors may be reduced by considering larger renormalization cells.

The effective permeability of a large cell of grid–blocks can again be determined by exploiting the analogy between discretised porous media and resistor networks, extending the 1st scheme outlined above to larger cell sizes. In 2D an efficient network reduction algorithm can be used to systematically reduce the analog network to a single equivalent resistor through a sequence of exact resistor transformations. This algorithm, based on the method of Frank and Lobb [9], is itself a form of differential real–space renormalization [7]. The sequence of resistor transformations is built up from elementary series and parallel reductions and 'star–triangle' transformations and is implemented numerically. The main elements of the reduction are outlined schematically in figure (6).

The network reduction (with boundary conditions appropriate for horizontal flow) is started by combining in series the two resistors attached to the top corner of the left–hand column of the network to generate a 'diagonal' resistor, as shown in figure (6c). This 'diagonal' resistor can then be propagated diagonally (down and to the right) through the network, using repeated star–triangle transformations, figure (6a), until it terminates at the boundary, figure (6b). The same procedure can now be followed with the next node down on the left–most edge, figure (6c). The remaining reduction can be carried out systematically column–by–column, and the final

Elements of 2D Network Reduction

(a) Propagation through an interior node

(b) Termination at boundary

(c) Column−wise reduction (schematic)

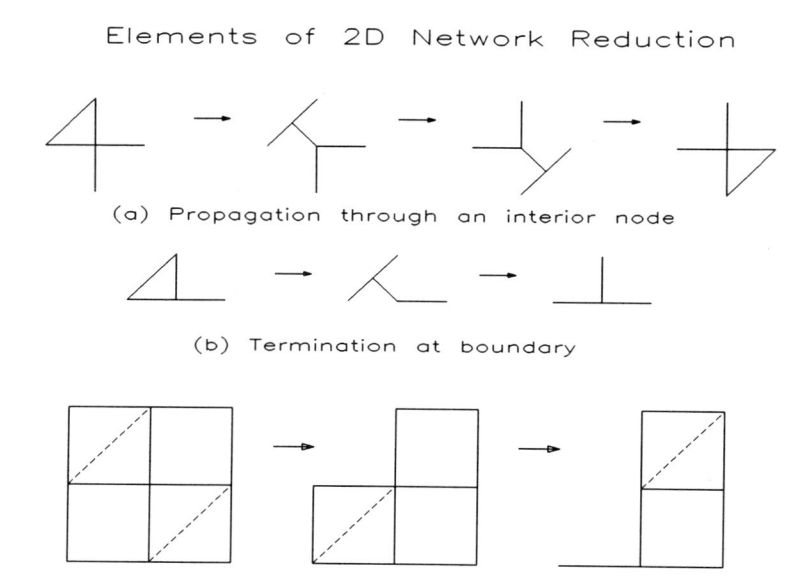

Figure 6.

string of remaining resistors combined in series to complete the reduction to a single resistor. The resistance of this single resistor is equal to the resistance of the original network and its reciprocal (network conductance) gives the cell effective permeability.

In 3D the 'star–triangle' transformations are not applicable, so standard alternatives for solving the cell pressure equation or analog resistor network must be used.

3 Test Results.

In this section we illustrate the use of the above schemes in determining the effective permeabilities for a number of 2D test cases, including both isotropic and anisotropic examples. In each case the fine scale grid–block permeabilities are drawn from a given distribution and the effective permeability determined for a number of realisations on 256 by 256 grids. For each realisation this is done in a hierarchical fashion using square cells of side b, where $b = 2,4,16$ or 256. In outline the procedure is as follows, using $b = 4$ as an example. We take a particular realisation (where permeability values k_x, k_y have been assigned to each grid–block) and divide it up into 4 by 4 cells yielding a coarser 64 by 64 grid. We then go through each of these cells in turn determining both the x and y direction cell effective permeabilities (Keff$_X$, Keff$_Y$), by replacing each grid–block by its equivalent resistor 'cross', then imposing the appropriate boundary conditions and performing the network reductions (as shown schematically in figures (1 & 6) — but for a larger cell). The procedure is then repeated on the coarser grid yielding an even coarser 16 by 16 grid, and so on. From the final 4 by 4 coarse grid we can determine the overall effective permeability (K_{XE}, K_{YE}) of the whole system.

Results obtained after averaging over a number of realisations for each case are presented in table (1) for isotropic examples and in table (2) for anisotropic examples. The numbers in parentheses after each entry give the statistical uncertainty in the last digit(s) as estimated from the standard deviation of the values obtained from different realisations. In each case the result of the 256 by 256 cell calculation provides a base figure against which the other calculations may be compared. For the isotropic examples results obtained from a simple 2 by 2 MK scheme (2.1.4) are also given.

In case (1) the isotropic grid–block permeabilities are selected independently from a log–normal distribution with a geometric mean $K_G = 1.0$ and variance of lnK given by $\sigma^2_{LN} = 1.0$. The corresponding sample estimates are shown in the table for the arithmetic mean, K_A, geometric mean, K_G, harmonic mean, K_H, and variance, σ^2. The results obtained for the effective permeability all yield reasonably accurate estimates with a small systematic error for the smaller cell sizes (about 6% for the 2 by 2 cell). The MK scheme gives a surprisingly good estimate; somewhat better than the alternative 2 by 2 scheme based on the 5–point discretisation.

Cases (2) and (3) are simple examples of systems with a finite fraction of impermeable material. The grid block permeabilities are independently drawn from a binary distribution with a fraction f being impermeable, K = 0, and a fraction $(1-f)$ having a permeability K. Once again the results are relatively insensitive to the cell size used with the smaller cells giving small but tolerable systematic errors. For this class of systems an effective medium approach provides accurate estimates of the effective permeability for a range of impermeable fractions away from the percolation threshold ($f_c \approx 0.41$); $K_E = 1.0 - \pi/2(2f - f^2)$, see [10], yielding effective permeabilities of 0.702 and 0.435 for impermeable fractions of 0.1 and 0.2 respectively. The renormalization calculations show that these are remarkably accurate estimates. The relative insensitivity of the results to the cell sizes used is generally the case for isotropic systems, see [3,4], and indicates that the two types of boundary error described above roughly cancel each other out. The advantage in using the smaller cell sizes resides in the smaller computational effort required.

Table (1) Isotropic Examples

1    $K_A = 1.62(1)$ $K_G = 1.00(1)$ $K_H = 0.62(1)$
    $\sigma^2 = 3.97(8)$ Log normal distribution, $\sigma_{LN} = 1.00$

Cell size	K_{XE}	K_{YE}
256	0.880(6)	0.880(5)
16	0.887(6)	0.891(6)
4	0.910(7)	0.912(6)
2	0.935(7)	0.935(6)
2 (MK)	0.884(6)	0.882(5)

2    $K_A = 0.900(1)$ $\sigma^2 = 0.090(1)$ Impermeable fraction 0.1

Cell size	K_{XE}	K_{YE}
256	0.697(4)	0.698(4)
16	0.697(5)	0.703(4)
4	0.711(6)	0.715(4)
2	0.733(4)	0.734(4)
2 (MK)	0.701(4)	0.702(5)

3    $K_A = 0.801(1)$ $\sigma^2 = 0.160(1)$ Impermeable fraction 0.2

Cell size	K_{XE}	K_{YE}
256	0.428(5)	0.427(5)
16	0.428(4)	0.431(7)
4	0.436(8)	0.438(8)
2	0.455(8)	0.454(8)
2 (MK)	0.413(9)	0.413(7)

Table (2) Anisotropic Examples

Impermeable fraction 0.1, microscopic anisotropy 0.1

Cell size	K_{XE}	K_{YE}	K_{YE}/K_{XE}
256	0.500(6)	0.0769(4)	0.15
16	0.529(5)	0.0766(5)	0.14
4	0.622(7)	0.0758(7)	0.12
2	0.708(3)	0.0760(6)	0.11

Impermeable fraction 0.2, microscopic anisotropy 0.1

Cell size	K_{XE}	K_{YE}	K_{YE}/K_{XE}
256	0.210(8)	0.0533(8)	0.25
16	0.233(9)	0.0523(9)	0.22
4	0.307(15)	0.0494(8)	0.16
2	0.422(18)	0.0497(9)	0.12

The remaining examples consider the effects of anisotropy in systems which are similar to cases (2 & 3). A microscopic anisotropy is introduced in these examples by simply assigning a fixed anisotropy ratio, K_y/K_x, to each grid–block keeping $K_x = 1$ and discretising appropriately. In case (4) we have an impermeable fraction of 0.1 and a microscopic anisotropy ratio of 0.1. On the macroscopic scale the effective anisotropy ratio is about 0.15, K_{XE} is reduced by about 40% from its value in the isotropic case whereas K_{YE} is about 10% greater than 0.1 of its corresponding isotropic value. Smaller renormalization cells systematically overestimate K_{XE}, with a 42% error for the 2 by 2 cell which corresponds to a K_{XE} value close to its value in the isotropic case. K_{YE} is much less sensitive to cell size. Case (5) shows the effect of increasing the impermeable fraction to 0.2 while keeping the microscopic anisotropy ratio fixed at 0.1. The effects of anisotropy are further reduced on the macroscopic scale with an effective anisotropy ratio of 0.25. Again K_{XE} is systematically overestimated by the smaller cells, with K_{YE} being much less sensitive to cell size.

This reduction in the macroscopic anisotropy can be understood as follows. In a system with no impermeable material the macroscopic effective permeability will directly reflect the microscopic anisotropy; flows in the x and y directions are essentially 'decoupled'. As the fraction of impermeable material increases, flow along a given direction becomes more difficult and is more dependent on 'flow path detours' in orthogonal directions. On the one hand, flow in the lower permeability direction is helped by the 'detours' along the higher permeability direction: on the other hand, flow in the high permeability direction is hindered by the 'detours'. The net effect is clearly to reduce the anisotropy exhibited at the macroscopic level. Ultimately, as the system approaches the percolation threshold the macroscopic anisotropy vanishes.

Such behaviour cannot be faithfully accounted for in small cell renormalization schemes. The limited number of paths 'captured' by the smaller cells lead to an asymmetry in the treatment of the x and y direction contributions to the cell effective values. In an x–directed b by b cell there are $2b^2$ 'resistors' pointing in the x–direction but only $2b^2(1 - 1/b)$ pointing in the y–direction which contribute to the effective x–direction resistance. To first order the fractional imbalance is given by $(1 - 1/b)/2$, and one might try to estimate K_{XE} (say) by plotting it against $1/b$ for a series of calculations using moderate cell sizes and then extrapolating to large b. The simple MK scheme is not applicable to these anisotropic systems as it totally decouples the x and y direction flows. These results for anisotropic systems illustrate the need to consider carefully the size of cell used. The usual method of 'stretching' the coordinates of a uniform anisotropic medium to turn it into an equivalent isotropic medium does not help much, as the aspect ratios of the heterogeneities/inclusions are altered by the change of coordinates.

Shale Map of Assakao Outcrop

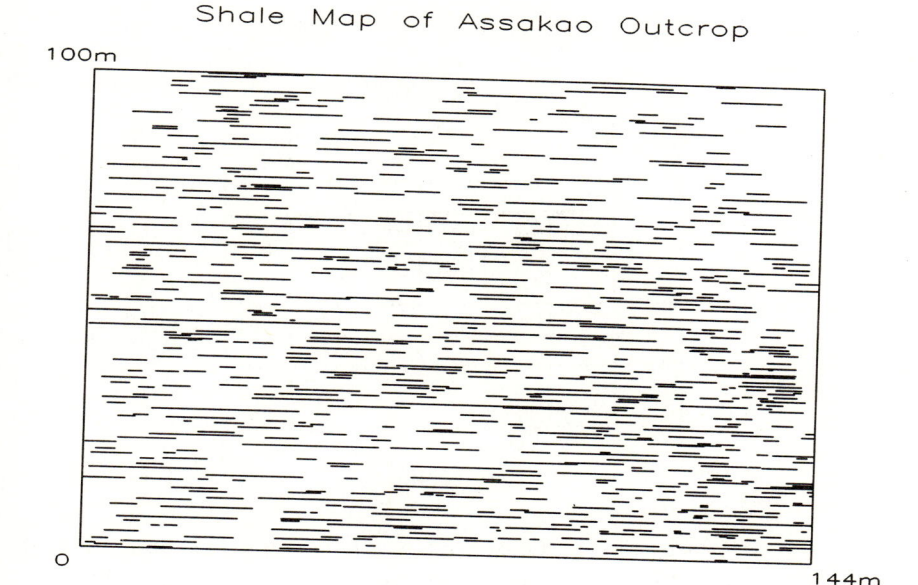

Figure 7.

4 A System with Shales: The Assakao Outcrop.

Many reservoirs contain shale inclusions which act as laterally extensive permeability barriers that strongly hinder vertical flow but have little effect on horizontal flow. A celebrated test problem is provided by data collected by Dupuy and LeFebvre from the Assakao Outcrop [11]. Figure (7) shows the shale configuration mapped from a 144 by 100 metre section of the cliff. A small cell renormalization approach encounters two sources of difficulty with this problem: firstly, the small cell size leads to poor resolution of the flow around the ends of the shales and secondly, the anisotropic character of the system requires an enormous number of grid blocks to provide adequate resolution to prevent the system from being spuriously driven to the 'shale fixed point' under renormalization [4].

We have tried using a large cell scheme on this problem by overlaying a fine grid, either 144 by 100 or 288 by 200, on the 'shale map'. A horizontal permeability value $k_h = 1$ is assigned to each grid block. The vertical transmissibility is set to zero in all grid blocks that are cut completely by shale(s); where grid blocks are only partially cut by shales the grid is locally refined (using a large cell scheme) to improve the flow resolution, figure (8). Cell sizes ranging from 2 by 2 to 32 by 32 are used for the local refinement. Extrapolating from the sequence of cell refinements we obtain estimates of 0.165 and 0.172 for the anisotropy ratio from the 144 by 100

Grid and Local Cell Refinements

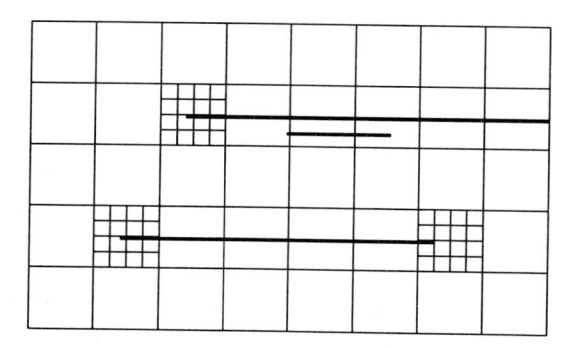

Figure 8.

grid and the 288 by 200 grid respectively. This compares tolerably well with stream–line and conducting paper estimates in the range 0.20–0.22, and is somewhat better than an estimate of 0.15 obtained from a direct pressure solve on a 432 by 300 grid (and computationally less demanding).

5 Concluding Remarks

In this paper we have described several renormalization schemes which can be used to determine the effective permeability of large heterogeneous systems. In common with other hierarchical procedures for tackling this type of problem (i.e. Multi–grid methods for solving the pressure equation) the renormalization approach is computationally efficient. Furthermore, unlike other hierarchical methods, only part of the problem (a single renormalization cell) need be held in memory thus reducing storage requirements and allowing very large systems to be handled. The cost of this 'decoupling' of parts of the problem is that the calculated effective permeabilities are only approximate; however, we have some control over these approximations through the choice of cell and renormalization scheme. The systematic improvement of results with cell size suggests that one should try to refine estimates of effective permeability by performing a series of calculations using small to moderate cell sizes and then extrapolating the results. This in turn could be combined with grid refinement studies (reduction of grid–spacing) to reduce 'discretisation' errors. Further studies should seek to determine optimum, and possibly 'adaptive extrapolation' strategies.

In many reservoir characterisation studies stochastic modelling techniques, such as conditional simulations, are used to provide realisations of the reservoir. Large numbers of realisations are required in order to obtain statistically meaningful results. As these realisations can be generated rapidly it is desirable to have quick and reliable methods for determining effective flow properties. The simple MK scheme and other small cell schemes are ideal for this, although it would be useful to find some way of handling the effects of anisotropy more accurately. Kasap and Lake [12] have recently proposed a method for calculating effective permeability tensors based on a similar series–parallel averaging approach, combining this with the full power of the renormalization framework is one promising avenue for future work.

The amalgamation of renormalization techniques with algorithms for characterising the spatial connectivity of good quality sands in heterogeneous, low net to gross systems should be another fruitful way of determining the flow behaviour in complex reservoir environments. In particular, connectivity algorithms which identify the 'backbone' [6] or 'flow–carrying part' of connected cluster(s) of good sands can be used in an adaptive strategy which honours the local and global topology of interconnections of the flow paths. For example, sparse matrix ordering schemes which honour tortuosity, 'bottle–necks' etc. of the flow paths could be used in conjunction with multi–level solution methods (renormalization or algebraic multigrid).

Heterogeneity on multiple length scales complicates many of the difficult problems in reservoir engineering. In our opinion, the most promising path to their solution, even if it is an arduous path, is the further refinement of renormalization and multi–level methods [13,14].

Acknowledgements

The author would like to thank the British Petroleum Company plc. for permission to publish this paper. I would also like to thank Dr P R King and Dr S H Begg for illuminating discussions on technical issues.

References

1. Begg S H and King P R, 'Modelling the effects of shales on reservoir performance: Calculation of effective vertical permeability', SPE 13529 (1985)

2. Haldorsen H H, Chang D M and Begg S H, 'Discontinuous vertical permeability barriers: a challenge to engineers and geologists', in 'North Sea Oil and Gas Reservoirs', 127-151, Graham and Trottman, (1987)

3. King P R, 'Effective values in averaging', in 'Mathematics in Oil Production', 217-234, Oxford, (1988)

4. King P R, 'The use of renormalization for calculating effective permeability', Transport in Porous Media 4 37-58 (1989)

5. Wilson K G, 'The Renormalization Group', Rev. Mod. Phys. 47, 773–840 (1975)

6. Kirkpatrick S, 'Models of disordered Materials', in 'Ill-condensed Matter', eds. R Balian, R Maynard and G Toulouse, 324, North Holland (1979)

7. various chapters in 'Real-Space Renormalization', Topics in current physics 30, eds T W Burkhardt and J M J van Leeuwen, Springer, (1982)

8. Desbarats A J, 'Numerical estimation of effective permeability in sand-shale formations', Water Resources Res, 23, 273–286 (1987)

9. Frank D J and Lobb C J, 'Highly efficient algorithm for percolative transport studies in two dimensions', Phys. Rev. B 37, 302 (1988)

10. Bernasconi J and Wiesmann H J, 'Effective-medium theory for site-disordered resistance networks', Phys. Rev. B 13, 1131 (1976)

11. Dupuy M and LeFebvre Du Prey E, Communication No. 34 Troisieme Colloque De L'Association De Recherche sur les Techniques de Forage et de Production, PAU, France (1968)

12. Kasap E and Lake L W, 'An analytical model to calculate the effective permeability tensor of a grid block and its application in an outcrop study', SPE 18434 (1989)

13. Brandt A, 'Multilevel computations: Review and recent developments' in Multigrid Methods, ed. S F McCormick, pp35–62 Marcel Dekker (1988)

14. Brower R C, Giles R, Moriarty K J M and Tomayo P, Combining renormalization group and multigrid methods', J. Computational Physics 80, 472–479 (1989)

Finite Element Principles for Evolutionary Problems, with Applications to Oil Reservoir Modelling

K. W. Morton A. K. Parrott

*Numerical Analysis Group, Oxford University Computing Laboratory,
11 Keble Road, Oxford OX1 3QD, ENGLAND.*

1 Introduction

In a finite element method the emphasis is on the approximation of the unknown quantities — rather than, say, the approximation of differential operators. Such a method provides flexible and powerful tools for this purpose, coping with distorted meshes and complicated geometries and allowing local mesh refinement or locally higher order approximation: and, when properly applied, the approximation it gives will be optimal in some sense. However, some mechanism is needed to determine this approximation. In an elliptic problem it is usually provided by a variational or extremal principle: but for evolutionary problems, especially hyperbolic problems, these are not available. Instead, we attempt to make use of: (a) the fact that information is carried along trajectories in space-time, such as along the characteristics; and (b) that integral conservation laws often hold.

During the last few years we have been applying these principles to the construction of finite element methods in various areas of computational fluid dynamics — see Morton and Sweby (1987), Childs and Morton (1989), Morton and Paisley (1989), Morton, Childs and Rudgyard (1988) and the references therein. Here we shall describe work done to apply these ideas to oil reservoir modelling. We shall see that the particular character of the equations suggests a choice of formulation which is less common in other areas.

2 Characteristic Galerkin methods

2.1 Outline of the methods

We begin with the methods which make most direct use of characteristics. Consider the scalar conservation law

$$w_t + f_x = 0 \qquad f = f(w). \tag{1}$$

Its solution is given by an evolution operator

$$E(\Delta t) : w(\cdot, t) \longrightarrow w(\cdot, t + \Delta t) \tag{2}$$

which carries the solution from time level t to level $t + \Delta t$. This can be approximated by the assumption that the solution is constant along the the characteristics — which is exact until shocks form — and leads to the evolution operator

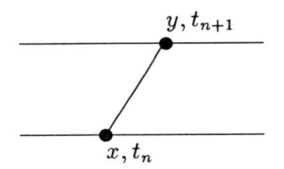

$$\widehat{E}(\Delta t) := \begin{cases} w(y, t + \Delta t) = w(x, t) \\ y = x + a(w(x))\Delta t. \end{cases} \tag{3}$$

Note the notation we shall use very commonly below, that (x, t) denotes the foot of the trajectory and $(y, t + \Delta t)$ the head.

Now suppose we approximate $w(x, t_n)$ at time level n by $W^n(x)$ given by the finite element expansion in basis functions $\phi_j(x)$, which we shall usually take to be piecewise constant or piecewise linear on some spatial mesh,

$$W^n(x) = \sum_{(j)} W_j^n \phi_j(x). \tag{4}$$

To obtain the approximation $W^{n+1}(x)$ at time level $n+1$ we first apply the evolution operator (3) and then use the Galerkin projection to obtain

$$\langle W^{n+1}, \phi_i \rangle = \langle \widehat{E}(\Delta t) W^n, \phi_i \rangle \qquad \forall \, \phi_i, \tag{5}$$

where $\langle \cdot, \cdot \rangle$ denotes the L^2 inner product over the spatial variable. Thus (5) can be written

$$\int W^{n+1}(y) \phi_i(y) dy = \int W^n(x) \phi_i(y) dy \qquad \forall \, \phi_i \tag{6}$$

and we call the result the Euler characteristic Galerkin, or ECG, method. For piecewise constant ϕ_i this is an explicit equation for W_i^{n+1}: but for piecewise linear basis functions we have a tridiagonal system of equations to solve, arising from the mass matrix with elements given by $\langle \phi_i, \phi_j \rangle$.

2.2 Properties of the ECG scheme

Key advantages of the ECG method include the following.

(i) Although it is essentially explicit, it is unconditionally stable. In principle, each value $W^n(x)$ is carried forward along a characteristic starting at (x, t_n) with speed $a(W^n(x))$ to contribute to the integrals for W^{n+1} : so any implicitness only occurs through the mass matrix. In practice, of course, only certain transition characteristic paths need to be considered — see Morton and Sweby (1987) for details of algorithms for the piecewise constant case, including the use of various recovery techniques of the type described below. Without recovery, and for a CFL number less than one, this algorithm is exactly equivalent to the Engquist–Osher first order upwind difference scheme. On a uniform mesh, if we mark the centre-points of the basis functions with dots as in the diagram and for simplicity consider the linear advection equation, all the ECG schemes can be associated with difference stencils: with a typical characteristic as shown, the piecewise constant ECG scheme involves the mesh points shown ringed; then the use of piecewise linear basis functions extends the stencil to include the points marked with a cross, and so on.

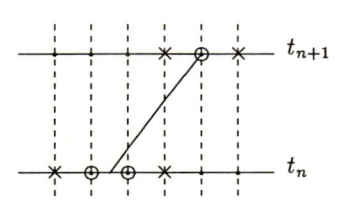

(ii) For smooth solutions and a uniform mesh, the piecewise constant basis gives a first order accurate scheme and the piecewise linear gives third order accuracy. In general, splines of order s give accuracy of order $2s - 1$: and this can be raised to order $2s$ if leapfrog time stepping is used — see Childs and Morton (1989) for detailed theoretical results.

(iii) Conservation is maintained exactly, for any basis that spans the unit constant. The piecewise constant basis gives the most compact conservation property, but in all cases the difference $\langle W^{n+1} - W^n, \phi_i \rangle$ can be expressed in terms of $(average\ flux\ in) - (average\ flux\ out)$.

(iv) The approximation (3) is still valid when shocks form. Values of W^n then overtake each other to form an "overturned manifold", but the projection embodied in (5) returns a smeared shock. Integrals are best considered as being carried out along the graph $[\widehat{E}W^n, x]$.

(v) The accuracy of the basic scheme can be enhanced through a *recovery procedure*. For example, suppose $\tilde{w}^n$ is an expansion in piecewise linears whose projection onto piecewise constants is the approximation W^n of (4), that is

$$\langle W^n - \tilde{w}^n, \phi_i \rangle = 0 \quad \forall \, \phi_i. \tag{7}$$

This gives a tridiagonal system to solve for the nodal values $\tilde{w}_j^n$. Now let $\tilde{w}^n$ be used in the next ECG step to get W^{n+1}:

$$\langle W^{n+1}, \phi_i \rangle = \langle \widehat{E}(\Delta t)\tilde{w}^n, \phi_i \rangle \quad \forall \, \phi_i. \tag{8}$$

Then the resulting scheme is second order accurate.

However, there is a major disadvantage to the ECG scheme when systems of conservation laws $\mathbf{w}_t + \mathbf{f}_x = 0$ are considered. For then one has to decompose the approximating vector $\mathbf{W}^n$ into waves corresponding to the eigenvectors of the Jacobian matrix $\mathcal{A} := \partial \mathbf{f}/\partial \mathbf{w}$, each travelling at the corresponding characteristic speed. This can be done readily enough for the equations of gas dynamics, especially with the use of Roe-type approximate Riemann solvers, and good results are given in Morton and Childs (1988). But for the more complicated and less well founded systems of equations used in oil reservoir models, this complexity is a significant deterrent to the use of these methods.

Moreover, characteristics for a system are no longer straight lines in general, and the dependent variables are not constant along them. Thus to achieve good accuracy it may be necessary to introduce some implicitness into the calculation of the characteristic paths, so departing from the simplicity of the ECG schemes. In the application below, in section (2.4), we shall use only the ECG algorithm but will need to limit the choice of the CFL number to lower values than can be used in the scalar case, in order to maintain accuracy.

2.3 Application to the Buckley–Leverett model

This familiar model for immiscible two-phase flow gives in one dimension the equation

$$\frac{\partial S_w}{\partial t} + (Q/A\varphi)\frac{\partial f_w(S_w)}{\partial x} = 0 \tag{9}$$

where $(Q/A\varphi)$ = flow rate/(cross-sectional area × porosity),

$$f_w(S_w) = \frac{(S_w - S_{wc})^2}{(S_w - S_{wc})^2 + (\mu_w/\mu_o)(1 - S_{or} - S_w)^2}$$

and (μ_w/μ_o) is the water/oil viscosity ratio. The ECG scheme using piecewise constant $S_w^n(x) = S_{w,i}^n$ for $x \epsilon (x_i - h, x_i)$, on a uniform mesh of size h, takes the form

$$S_{w,i}^{n+1} = \frac{1}{h} \int_{x_{i-1}}^{x_i} S_w^n(x) dy, \qquad y = x + a(S_w^n(x))\Delta t \tag{10}$$

where $a(S_w) = (Q/A\varphi)\frac{\partial f_w}{\partial S_w}$. If recovery with piecewise linears is used S_w^n is replaced by $\widetilde{S}_w^n$, where

$$\int_{x_{i-1}}^{x_i} \left(\widetilde{S}_w^n(x) - S_w^n(x)\right) dx = 0. \tag{11}$$

The results are shown in Fig.1 for a viscosity ratio of 3.0. Fig.1a gives results obtained with the explicit, first order, upwind difference scheme at a maximum CFL number 0.9 — so it is identical with a restricted use of the ECG method, using piecewise constants, without recovery. Fig.1b was obtained using piecewise linear recovery and a maximum CFL number of 5.5 — it shows that even sharper results are obtained here even though the time step is five times larger.

2.4 Application to a model surfactant system

A model for a low concentration, water soluble surfactant without absorption takes the form of the two equation system

$$\frac{\partial S_w}{\partial t} + (Q/A\varphi)\frac{\partial f_w(S_w, m_s)}{\partial x} = 0, \tag{12}$$

$$\frac{\partial m_s}{\partial t} + (\rho_w Q/A\varphi)\frac{\partial (C_{sw} f_w(S_w, m_s))}{\partial x} = 0, \tag{13}$$

where

$$f_w(S_w, m_s) = \frac{(S_w - S_{wc})^2}{(S_w - S_{wc})^2 + (\mu_w/\mu_o)(1 - S_{or}(m_s/S_w) - S_w)^2}. \tag{14}$$

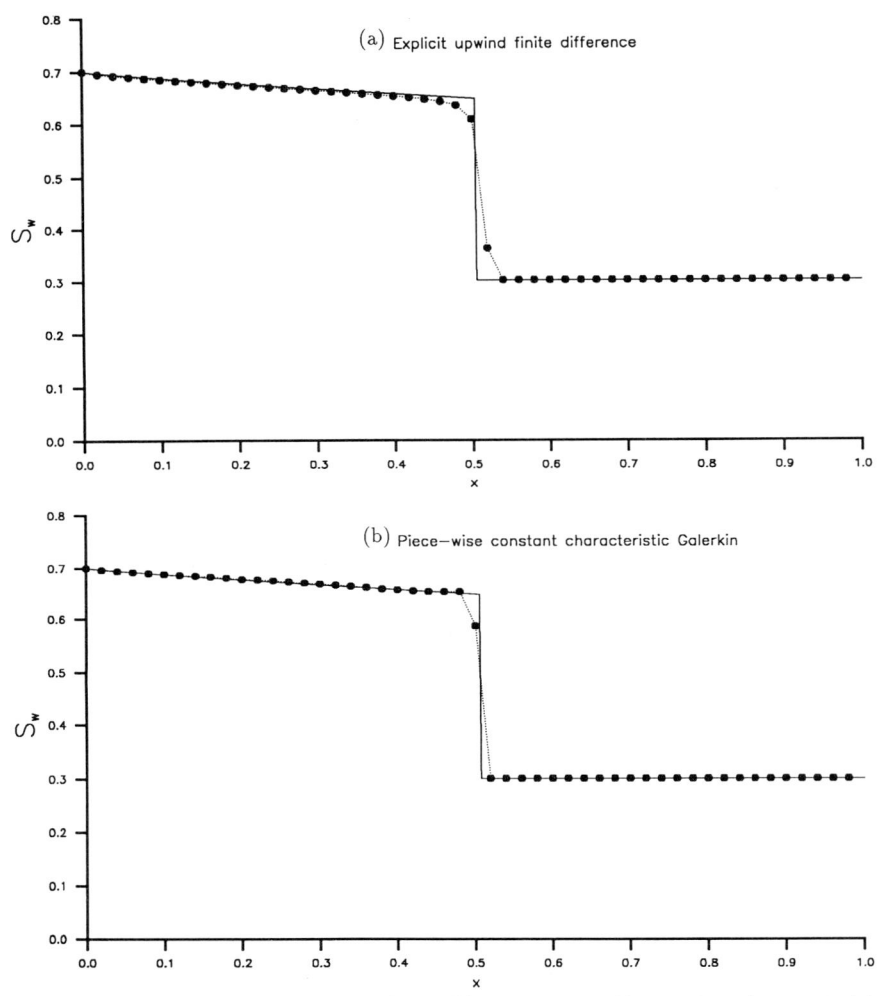

Figure 1: Solutions to the Buckley Leverett problem with $\mu_w/\mu_o = 3.0$: (a) explicit upwind finite difference at a maximum CFL number 0.9; (b) ECG with piecewise linear recovery at a maximum CFL number of 5.5.

Here $m_s = \rho_w S_w C_{sw}$ is the surfactant mass/unit pore volume and C_{sw} is the surfactant mass fraction in the water phase. The Jacobian matrix for the system is given by

$$\mathcal{A} = (Q/A\varphi) \begin{pmatrix} \dfrac{\partial f_w}{\partial S_w} & \dfrac{\partial f_w}{\partial m_s} \\ \dfrac{\partial(\rho_w C_{sw} f_w)}{\partial S_w} & \dfrac{\partial(\rho_w C_{sw} f_w)}{\partial m_s} \end{pmatrix} \tag{15}$$

which has eigenvalues

$$\lambda_1 = (Q/A\varphi)(\frac{\partial f_w}{\partial S_w} + \frac{m_s}{S_w}\frac{\partial f_w}{\partial m_s}), \ \lambda_2 = (Q/A\varphi)(\frac{f_w}{S_w}) \tag{16}$$

giving the characteristic speeds, and eigenmodes

$$\mathbf{e}_1 = \begin{pmatrix} S_w \\ m_s \end{pmatrix}, \quad \mathbf{e}_2 = \begin{pmatrix} \dfrac{\partial f_w}{\partial m_s} \\ \dfrac{f_w}{S_w} - \dfrac{\partial f_w}{\partial S_w} \end{pmatrix}. \tag{17}$$

Note that (16) are identical to the more familiar expressions found when derivatives are taken with respect to S_w and C_{sw} as the independent variables.

In applying the ECG scheme to this system, a Roe-type approximation to the Jacobian $\mathcal{A}$ is defined on each element: the approximation (S_w^n, M_s^n) at each time level is then resolved into the eigenmodes similar to those given by (17) before each is moved at the appropriate characteristic speed: the results are then recombined at the new time level. Note that the eigenmodes are parallel in the case of the exact Jacobian when $\lambda_1 = \lambda_2$, i.e. at the contact discontinuity. This could also occur with the eigenmodes of the approximate Jacobian, in which case the ECG algorithm is applied to the conserved variables since the wavespeeds are then equal. Fig.2b gives numerical results when this method is used with piecewise linear recovery at a maximum CFL number of 3.3: while Fig.2a gives the corresponding results obtained with the upwind finite difference scheme at a CFL number of 0.9. There is clearly a big improvement obtained by the ECG scheme: however, it is at the cost of considerable complexity.

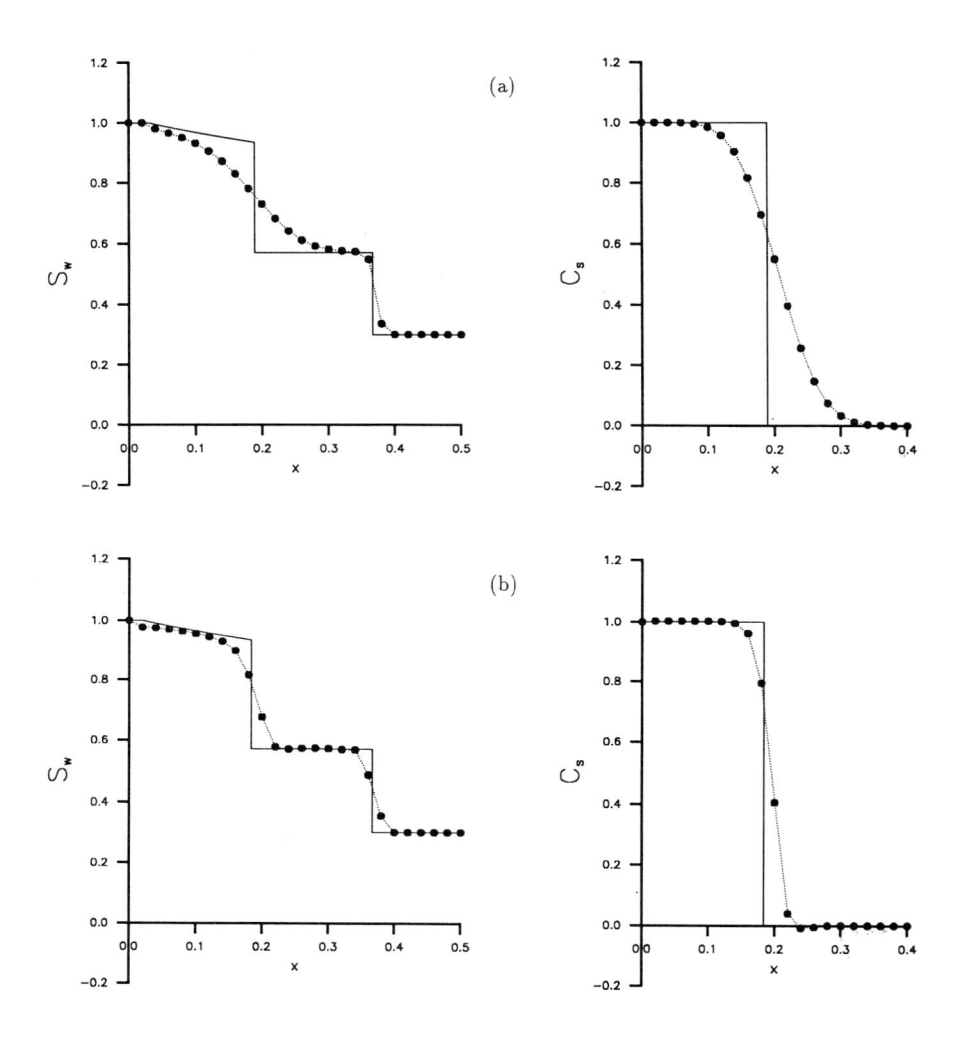

Figure 2: Solutions to the 1-D surfactant displacement problem: (a) explicit upwind finite difference scheme with a maximum CFL number of 0.9; (b) ECG with piecewise linear recovery at a maximum CFL number of 3.3.

3 Lagrange Galerkin methods

The complexity involved in working with the characteristic speeds and wave modes for a system of equations, and the even greater difficulties posed by generalising the algorithms to several space dimensions, has led to the development of simpler methods. Suppose, for example, the flow velocity **v** dominates the unsteady behaviour of a system of equations, as in the incompressible Navier–Stokes equations. Then we introduce the evolution operator corresponding to flow along this velocity field,

$$\tilde{E}(\Delta t) := \begin{cases} \mathbf{w}(\mathbf{y}, t + \Delta t) = \mathbf{w}(\mathbf{x}, t) \\ \mathbf{y} = \mathbf{x} + \int_t^{t+\Delta t} \mathbf{v} \, dt \end{cases}, \qquad (18)$$

and write the system of equations for the unknown vector **w** in the form

$$\mathbf{w}_t + (\mathbf{v} \cdot \nabla)\mathbf{w} + \mathbf{M}\mathbf{w} = 0. \qquad (19)$$

The first two differentials are combined to form a material or Lagrange derivative and the system then approximated in the form

$$\underbrace{\langle \mathbf{W}^{n+1} - \tilde{E}(\Delta t)\mathbf{W}^n, \phi_i \rangle}_{Lagrange\ Galerkin} + \underbrace{\langle \mathbf{M}[\theta \mathbf{W}^{n+1} + (1-\theta)\mathbf{W}^n], \phi_i \rangle}_{standard\ Galerkin} = 0. \qquad (20)$$

That is, only the advection along the flow lines is approximated as in the ECG schemes, and the remaining terms in the equations are approximated by a conventional Galerkin approach, with some choice of the time averaging for them. For the Navier–Stokes equations, when this is implicit (i.e. $\theta > 0$) a Stokes problem needs to be solved at each new time level for the pressure and the velocity field, the latter then being used to determine the evolution operator $\tilde{E}(\Delta t)$ for the next time step. Such schemes have been widely used and analysed in the literature — see Douglas and Russell (1982), Pironneau (1982), Süli (1988).

4 Weak formulation of the Lagrange Galerkin method

4.1 Derivation

There are at least three different ways of formulating the ECG or Lagrange Galerkin method — see Morton, Priestley and Süli (1988) for a discussion of these formulations and their implementation through quadrature. The weak form seems to be due to Benqué, Labadie and Ronat (1982), and begins with the introduction of a test function depending on both the space and time variables. We derive it in two dimensions for the scalar equation

$$w_t + div(f, g) = 0 \tag{21}$$

for which we use the test function $\psi(x_1, x_2, t)$. We begin by supposing that $\psi(\cdot, \cdot, \cdot)$ is differentiable in all its variables in a region of (x_1, x_2, t)-space bounded by the planes $t = t_n$, $t = t_{n+1}$ and some surface $\mathcal{S}$. Multiplying the equation (21) by $\psi(\cdot, \cdot, \cdot)$ and integrating over this region, we obtain

$$\int_{t_n}^{t_{n+1}} dt \int dx_1 dx_2 \, \psi[w_t + div(f,g)] = 0. \tag{22}$$

Then we integrate by parts with respect to all variables so that, if $\mathbf{d\mathcal{S}}$ denotes a surface element vector of $\mathcal{S}$ directed along the outward normal, we have

$$\int w^{n+1}\psi^{n+1} dx_1 dx_2 - \int w^n \psi^n dx_1 dx_2 + \int \psi\,(f,g) \cdot \mathbf{d\mathcal{S}}$$

$$= \int_{t_n}^{t_{n+1}} dt \int dx_1 dx_2 \, w[\psi_t + (\frac{f}{w}, \frac{g}{w}) \cdot \nabla \psi]. \tag{23}$$

We shall always choose the test function such that the right hand side is zero.

Clearly if $\psi \equiv 1$ over the whole domain in which (w, f, g) is non-zero, (23) merely expresses the global conservation property for w. If $\psi \equiv 1$ over a fixed cylinder, so that $\mathcal{S} = \mathcal{C} \times (t_n, t_{n+1})$ for some curve $\mathcal{C}$ in (x_1, x_2)-space, then (23) expresses conservation over the cylinder taking account of the flux through the sides. Such a choice is the basis for deriving the Godunov finite difference schemes, in which Riemann problems are solved to give approximations to the average flux through the cylinder sides.

In the weak form of the Lagrange Galerkin method we choose ψ^{n+1} to be one of the basis functions ϕ_i for the approximation W: it can then be seen that the right hand side of (23) is made zero by forcing $\psi(\cdot,\cdot,\cdot)$ to satisfy the advection equation

$$\psi_t + \mathbf{c} \cdot \nabla \psi = 0 \qquad \mathbf{c} = (f/w, g/w). \tag{24}$$

For example, if $(f,g) = w\mathbf{v}$ then $\psi(\cdot,\cdot,\cdot)$ satisfies the advection equation with the velocity field $\mathbf{v}$. The values of the basis function ϕ_i are carried unchanged along trajectories $d\mathbf{X}/dt = \mathbf{c}$ and, in particular, the surface $\mathcal{S}$ is formed by the trajectories from points on the boundary of its support, i.e. where $\phi_i = 0$. Hence, if the differential equation (21) holds in the whole of this region, the boundary term in (23) drops out and the weak LG method results from substituting the approximation W for w to give

$$\langle W^{n+1}, \phi_i \rangle = \int W^n(x_1, x_2) \phi_i(y_1, y_2) dx_1 dx_2, \tag{25}$$

where

$$\frac{d\mathbf{X}}{dt} = (\frac{f}{w}, \frac{g}{w})(W, t), \qquad \mathbf{X}(t_n) = (x_1, x_2), \qquad \mathbf{X}(t_{n+1}) = (y_1, y_2). \tag{26}$$

Boundary conditions for the differential equation will give values for the fluxes in the surface integral over $\mathcal{S}$ when this intersects with the boundary of the domain over which (21) holds.

So far our assumptions are valid for piecewise linear basis functions ϕ_i but not for piecewise constants. For the latter we must take a sequence of smooth test functions, with the same support as ϕ_i at $t = t_{n+1}$, whose limit in the L^2 sense is ϕ_i. Then clearly the integrals in (23) converge to those in (25) so long as w and W are in L^2, which we shall always assume.

4.2 Relationship between the ECG and weak LG methods

It is useful to pause here in our development to summarise how the ECG and weak LG schemes are related: we do so for the simplest case of the scalar conservation law in 1D, $w_t + f_x = 0$.

In the ECG scheme the trajectories, using the characteristic speed, carry the unknown function w or W forward in time:

4.4 Application of the weak LG method to a model polymer system

A model for a water soluble, absorbing polymer system consists of the two equations

$$\frac{\partial m_w}{\partial t} + (\rho_w Q/A\varphi)\frac{\partial(C_{ww}f_w(S_w, C_{pw}))}{\partial x} = 0 \tag{36}$$

$$\frac{\partial m_p}{\partial t} + (\rho_w Q/A\varphi)\frac{\partial(C_{pw}f_w(S_w, C_{pw}))}{\partial x} = 0 \tag{37}$$

where

$$f_w(S_w, C_{pw}) = \frac{(S_w - S_{wc})^2}{(S_w - S_{wc})^2 + (\mu_w(C_{pw})/\mu_o)(1 - S_{or} - S_w)^2},$$

and $m_w = \rho_w S_w C_{ww}$ is the water mass/unit pore volume, C_{ww} is the water mass fraction in the water phase, $m_p = \rho_w S_w (C_{pw} + \widehat{C}_p)$ is the polymer mass/unit pore volume, C_{pw} is the polymer mass fraction in water phase, $\widehat{C}_p$ is the polymer mass fraction adsorbed and $C_{ww} + C_{pw} = 1$. The approximate equations for the mass of each component, m_w and m_p, are then obtained from (34) with the trajectories for each case calculated from the two chord speeds.

Alternatively, one can modify the equations so that only one set of trajectories is used, namely that for the water component given by (36). To do so we rewrite the polymer component equation as

$$\frac{\partial(\rho_w C_{pw} S_w)}{\partial t} + (\rho_w Q/A\varphi)\frac{\partial(C_{pw}f_w(S_w, C_{pw}))}{\partial x} = -\frac{\partial(\rho_w \widehat{C}_p S_w)}{\partial t}. \tag{38}$$

The weak LG approximation can then be derived from

$$\langle \rho_w C_{pw}^{n+1} S_w^{n+1}, \phi_i \rangle - \int \rho_w C_{pw}^n(x) S_w^n(x) \phi_i(y) dx$$
$$= \int_{t_n}^{t_{n+1}} dt \int \psi_i(x,t) \left[-\frac{\partial(\rho_w \widehat{C}_p S_w)}{\partial t} \right] dx$$
$$\approx \int \phi_i(y) \left[-(\rho_w \widehat{C}_p^{n+1} S_w^{n+1} - \rho_w \widehat{C}_p^n S_w^n) \right] dy, \tag{39}$$

which, by recombination of the terms at t_{n+1} gives,

$$\langle M_p^{n+1}, \phi_i \rangle - \int \rho_w C_{pw}^n(x) S_w^n(x) \phi_i(y) dx = \int \rho_w \widehat{C}_p^n S_w^n \phi_i(y) dy. \tag{40}$$

$$\frac{dX}{dt} = \frac{\partial f}{\partial w} \qquad (x, t_n) \mapsto (y, t_{n+1}). \tag{27}$$

Then the integration is performed at the t_{n+1} level to give

$$\langle W^{n+1}, \phi_i \rangle = \int W^n(x) \phi_i(y) dy. \tag{28}$$

In the weak LG scheme the trajectories, using the "chord" velocity f/w, carry the test function back in time,

$$\frac{dX}{dt} = \frac{f}{w} \qquad (y, t_{n+1}) \mapsto (x, t_n), \tag{29}$$

and the integration is performed at the t_n level to give

$$\langle W^{n+1}, \phi_i \rangle = \int W^n(x) \phi_i(y) dx. \tag{30}$$

The two schemes are identical for advection by a given incompressible flow field, $\mathbf{f} = w\mathbf{v}$ with $div\mathbf{v} = 0$. The trajectories then define an area-preserving mapping for which $\mathbf{dy} = \mathbf{dx}$.

4.3 Trajectory calculations for the weak LG method in 1D

In the integration of (30) one needs the function $\phi_i(y(x))$. Clearly the piecewise constant case has a special advantage in its simplicity since it is independent of the mapping effected by the trajectories. We shall therefore assume that this is used and that second or higher order accuracy is obtained through a recovery procedure. That is on a fixed but non-uniform set of intervals given by the end points $\{x_i\}$ we assume

$$\begin{aligned} W^n(x) &= W_i^n \qquad x \in (x_{i-1}, x_i) \\ &= \frac{1}{x_i - x_{i-1}} \int_{x_{i-1}}^{x_i} \tilde{w}^n(x) dx, \end{aligned} \tag{31}$$

where $\tilde{w}^n(x)$ is the recovered (smoother) function. For each component W of a vector of unknowns $\mathbf{W}$, we need the trajectories $X_i(t)$ starting from the interval ends $\{x_i\}$ at time level t_{n+1} and ending at points we denote by $\{x^i\}$ at level t_n:

$$\begin{aligned} \frac{dX_i}{dt} &= c(\mathbf{W}, X_i) \qquad (c \equiv f/w) \\ X_i(t_{n+1}) &= x_i, \qquad X_i(t_n) = x^i. \end{aligned} \tag{32}$$

That is,

$$x_i - x^i = \Delta t \, [avge. \ c(\mathbf{W}, X) \ along \ X_i \ trajectory] \tag{33}$$

and then the integration can be performed to give

$$W_i^{n+1} = \frac{1}{x_i - x_{i-1}} \int_{x^{i-1}}^{x^i} \tilde{w}^n(x) dx. \tag{34}$$

The choice of the average used for $c(\mathbf{W}, X)$ is crucial and in all but trivial cases will depend on $\mathbf{W}^{n+1}$. Thus the scheme is implicit and we have used Picard iteration for the solution of the resulting system of equations.

The choice of the average for (33) depends on the assumed properties of the flux function. Let us consider the following typical situation as illustrated in the diagram: that is, we assume information flows from left to right and

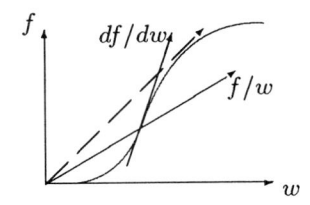

$$0 \leq c \equiv \frac{f}{w} \leq \frac{\partial f}{\partial w}.$$

The dashed line shows the critical speed where $f/w = \partial f/\partial w$.

Note that the ratio of interval lengths $(x^i - x^{i-1})/(x_i - x_{i-1})$ represents the average *amplification* from the function values $\tilde{w}^n$ to the values W^{n+1} over the $i^{\underline{th}}$ interval. Hence we use a weighted average of two values of c to obtain

$$x_i - x^i = \Delta t \left[(1-\theta)\bar{c}_i^n + \theta c_i^{n+1} \right], \qquad (35)$$

where

$$\bar{c}_i^n = c(\frac{1}{x^i - x^{i-1}} \int_{x^{i-1}}^{x^i} \tilde{w}^n dx, x^i)$$

and

$$c_i^{n+1} = c(\tilde{w}^{n+1}(x_i), x_i).$$

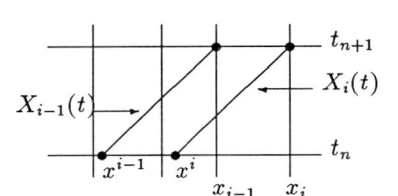

Thus $\bar{c}_i^n$ is calculated from a simple upwind average of $\tilde{w}^n$ using the same integral used in (34): while c_i^{n+1} is calculated from a recovered point value at (x_i, t_{n+1}). Note that even the choice $\theta = 0$ leads to an implicit scheme. As a first test of the algorithm Fig.3 shows results for the Buckley–Leverett problem considered earlier. The computation is with $\theta = 0$ at a CFL number of 1.7: it is compared with the familiar first order implicit upwind, finite difference scheme which gives strong damping.

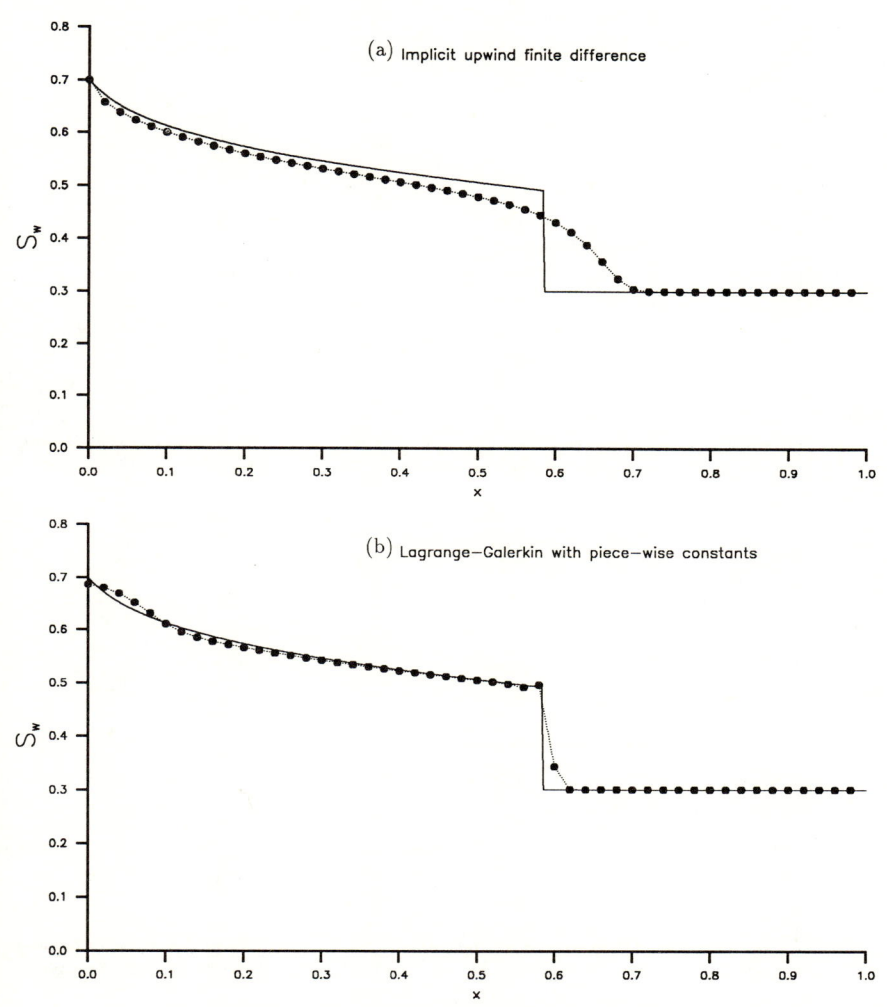

Figure 3: Solutions to the Buckley Leverett problem with $\mu_w/\mu_o = 0.3$: (a) implicit upwind finite difference at a maximum CFL number 1.7; (b) weak Lagrange Galerkin with piecewise constants at a maximum CFL number of 1.7.

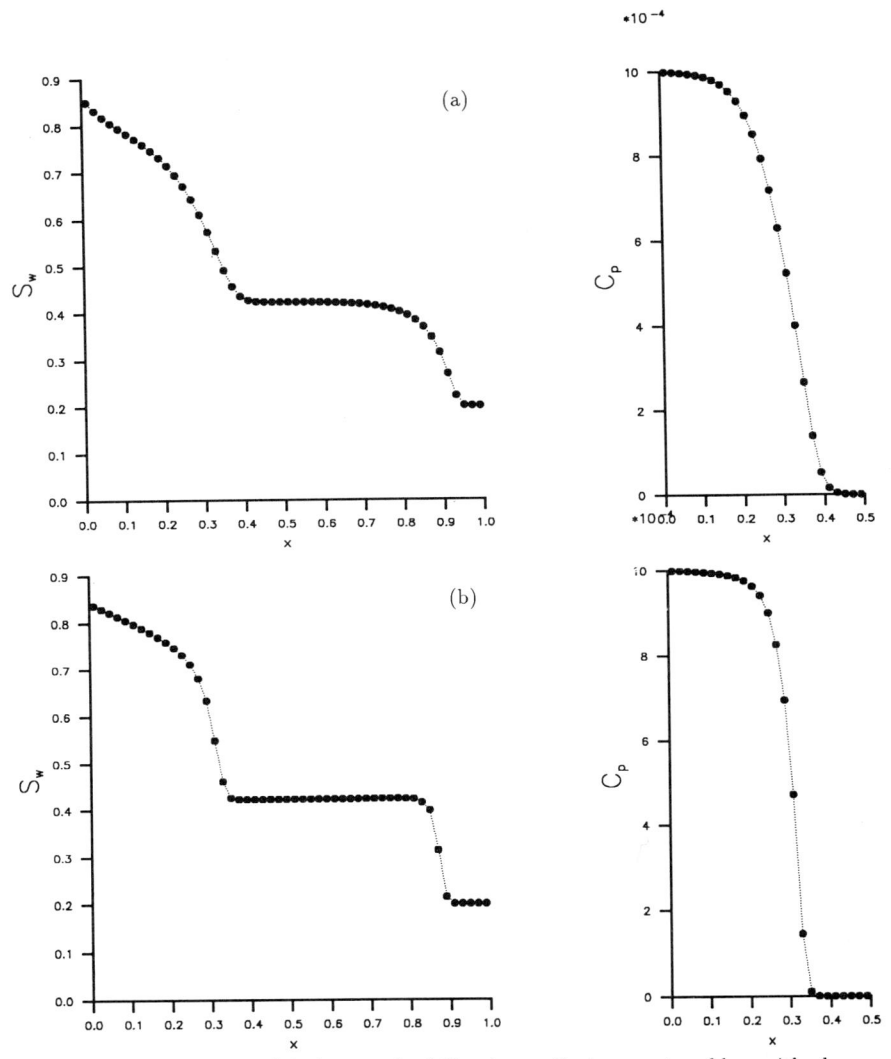

Figure 4: Solutions to the 1-D polymer displacement problem with absorption: (a) implicit upwind finite difference scheme with a maximum CFL number of 2.4; (b) weak Lagrange Galerkin with piecewise constants at a maximum CFL number of 2.4.

An alternative procedure, introduced by van Leer (1977, 1979), is to insert a single linear section in each interval. Then in $(x_{i-1} - x_i)$ we have an average W_i and a gradient G_i, say, and the problem is to choose the latter. The procedure appears to be more local, unless fairly sophisticated criteria are used in the choice of the G_i: unfortunately, without them there may be significant discontinuities remaining after the recovery and this leads to a lack of robustness in the trajectory calculations.

There are several alternative procedures which give completely local, continuous, linear sections: but they may do so at some cost in accuracy or smoothness. Consider the $i^{\underline{th}}$ interval of length h_i and with average W_i. To be completely local and to reproduce a linear function, the recovered function $\tilde{w}(x)$ should take the value $\tilde{w}_i = (h_i W_{i+1} + h_{i+1} W_i)/(h_i + h_{i+1})$ at x_i. Let us suppose that first we have a monotone sequence of values, i.e. that $(W_i - W_{i-1})(W_{i+1} - W_i) \geq 0$. Then we can introduce a broken linear function, taking the values W_i at a point splitting the interval into sections $\theta_i : 1 - \theta_i$, where $\theta_i = (\tilde{w}_{i+1} - W_i)/(\tilde{w}_{i+1} - \tilde{w}_i)$. It is easily verified that the average of the resultant function is just W_i. If, on the other hand, the sequence is not monotone, we can introduce three linear sections for $\tilde{w}_i$ of equal length and set $\tilde{w}(x_{i-1} + h_i/3) = \frac{3}{2} W_i - \frac{1}{12} \tilde{w}_i - \frac{5}{12} \tilde{w}_{i+1}$ and $\tilde{w}(x_{i-1} + 2h_i/3) = \frac{3}{2} W_i - \frac{5}{12} \tilde{w}_i - \frac{1}{12} \tilde{w}_{i+1}$, which will emphasise the extremum damped by the projection onto piecewise constants but also maintain the same average. An even simpler procedure which will cover both cases is to introduce a value at the mid-point of the interval equal to $2W_i - \frac{1}{2}(\tilde{w}_i + \tilde{w}_{i+1})$: that is, a mid-point deviation from W_i equal to minus half the sum of the deviations at the end-points. This also has the desirable property of depending continuously on the data.

All four procedures are illustrated in Fig.5. In every case it is easy to carry out the integrals over $\tilde{w}^n(x)$ to obtain W^{n+1}. Note that any short cuts or approximations used in carrying out these integrals should ensure that conservation is always maintained: this is readily accomplished in practice.

Ease of extension into two dimensions is one of the most important criteria in choosing between these procedures. Suppose that our basic approximation is piecewise constant over uniform squares, and suppose that for the continuous recovered function we take, as the value at each vertex, an average over the values in the four cells meeting at the vertex, and linearly interpolate along each edge. Then as for the last two procedures given above we can concentrate on recovery in each cell independently. The fourth and simplest procedure merely requires us to introduce a value at the centre which differs from the cell value W_i by minus half the sum of the deviations at the four vertices: $\tilde{w}(x_1, x_2)$ is then linear over each of the four

Note that the recovery process has to be implemented so that we can make a pointwise split of $\widetilde{m}_p^n$ into the two parts $\rho_w C_{pw} S_w$ and $\rho_w \widehat{C}_p S_w$.

Fig.4 shows results for the problem of polymer injection with an equilibrium absorption level of 30%, using two sets of trajectories; the use of a single set of trajectories gave virtually identical results (to three significant figures). The time step corresponds to a maximum CFL number of 3.4 and the comparison with an implicit upwind finite difference method shows how the typically high damping of the implicit scheme has been avoided. Further sharpening of the solution can be expected to result from the use of the linear recovery methods in the following section.

4.5 Procedures for recovery by piecewise linears

Although the use of piecewise constant basis functions simplifies the trajectory calculations necessary to compute W^{n+1}, the iteration for the trajectory may not converge unless the approximation at time level t_n has a bounded derivative. Hence it is important to have available a recovery procedure to produce a piecewise linear approximation $\tilde{w}^n$. We will describe briefly below two such procedures which have been published and widely tested and two simple alternatives which we have found to be generally adequate. We shall also discuss the question of extension into two dimensions.

The procedure with which we are most familiar is that described and used by Morton and Childs (1988) and Childs and Morton (1989). A parameter $\theta_i \in [0,1]$ is associated with each interval end x_i, and the discontinuity there between W_i and W_{i+1} is replaced by a linear section between $\tilde{w}_i$ at $x_i - \frac{1}{2}\theta_i h_i$ and $\tilde{w}_{i+1}$ at $x_i + \frac{1}{2}\theta_i h_{i+1}$, where $h_i = x_i - x_{i-1}$, $h_{i+1} = x_{i+1} - x_i$. Clearly if $\theta_i < 1$ there is a constant section in each interval either side of x_i. To fulfil the projection or conservation property (31), the new parameters $\{\tilde{w}_i\}$ have to satisfy a tridiagonal system of equations which for equal intervals reduces to

$$\frac{1}{8}\theta_{i-1}\tilde{w}_{i-1} + (1 - \frac{1}{8}\theta_{i-1} - \frac{1}{8}\theta_i)\tilde{w}_i + \frac{1}{8}\theta_i \tilde{w}_{i-1} = W_i. \qquad (41)$$

In the algorithms given in the above references (where the recovery is completely integrated into the ECG scheme), the parameters $\{\theta_i\}$ are chosen adaptively to be as large as possible without introducing oscillation: an effective criterion is to ensure that $\tilde{w}(x)$ has no more extrema than $W(x)$. The improvement in accuracy compared with the unrecovered procedure can be very large indeed if many time steps are carried out.

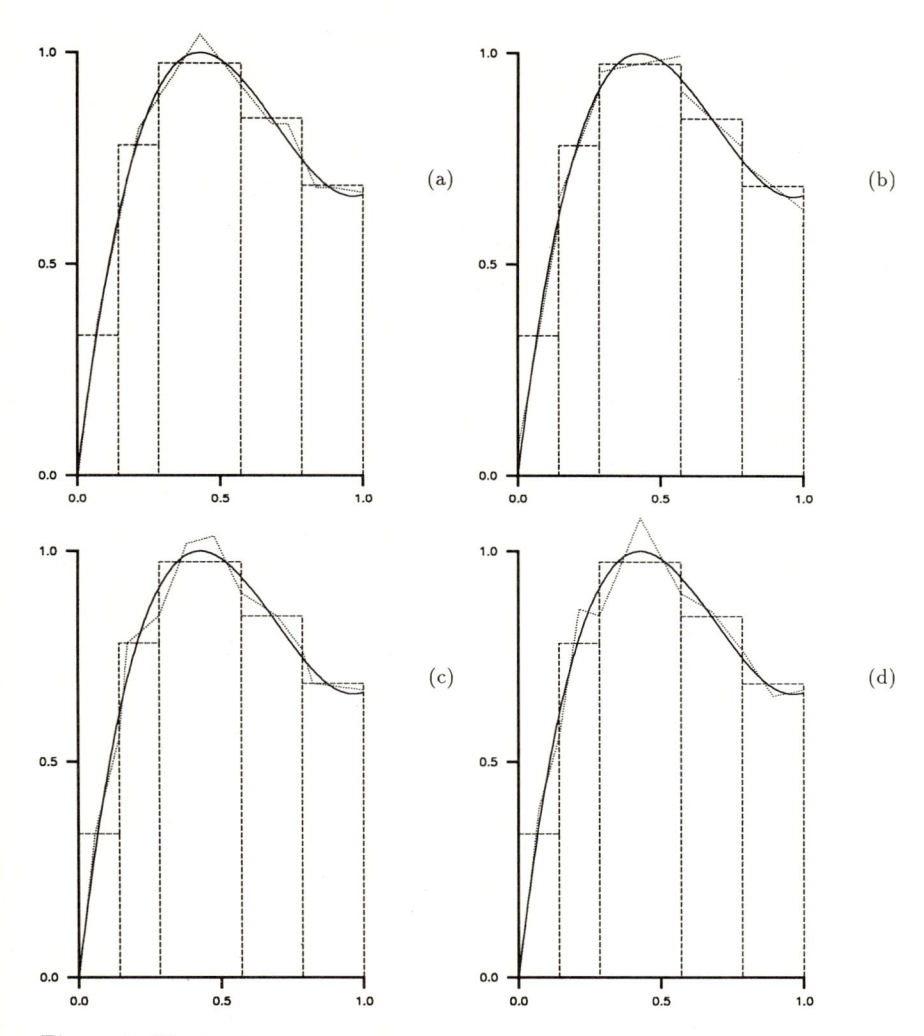

Figure 5: Illustrative recovery procedures in one dimension: (a) Morton and Childs; (b) van Leer; (c) local recovery with broken linear functions; (d) simplified local recovery.

triangles formed by drawing the diagonals of the cell. Clearly, by changing from the centre to another point in the square the central value can be chosen to take a different value: just as in the one dimensional case, it can be chosen to be zero if the average deviations on the left and the right edges have opposite signs, as well as those on the top and bottom edges – that is, if W is broadly monotone in both the x_1- and x_2- directions.

Because the calculations of the integrals are non-trivial in two dimensions, the details of the underlying recovered function is not of prime importance. Even without recovery one has to calculate the areas of polygons formed by the intersection of the fixed square grid and its transformed counterpart. This can be done exactly from the positions of the vertices (x_1, x_2): but then the effect of recovery on the computation of W^{n+1} is best carried out by an approximate transfer of contributions which maintains the conservation property – see Parrott and Keeping (1989) for the use of an FCT procedure to accomplish this.

5 Application of weak LG methods to the compositional equations

In this section we will outline a procedure for applying the techniques of the previous section to the full compositional equations in two dimensions. The development of the weak Lagrange Galerkin method has been directed towards and motivated by this objective and, although this scheme has yet to be coded and tested, it is therefore appropriate that we should end this paper with its description.

We use the notation

$m_\alpha(\mathbf{r},t) = \sum_{(l)} \rho_l C_{\alpha l} S_l,$ = mass of component α per unit pore volume
$\dot{m}_\alpha(\mathbf{r},t) = \sum_l \rho_l C_{\alpha l} \mathbf{u}_l$ = advective mass flow rate of component α per unit area
$C_{\alpha l}(\mathbf{r},t),$ = mass fraction of component α in phase l

where the phase velocity $\mathbf{u}_l$ is given by Darcy's law

$$\mathbf{u}_l = K \frac{k_l}{\mu_l} (\nabla p_{l_r} - \rho_l g \nabla h + \nabla P_{c,l}) \tag{42}$$

and the capillary pressure terms are with respect to some reference phase pressure p_{l_r}. The trajectories $\mathbf{X}_{\alpha i}$ are calculated using speeds $\mathbf{c}_\alpha = \dot{\mathbf{m}}_\alpha/(\varphi m_\alpha)$ which depend on the phase velocities $\mathbf{u}_l$. Then the equations to be approximated take the form

$$\frac{\partial}{\partial t}(\varphi m_\alpha) + \nabla \cdot \dot{\mathbf{m}}_\alpha = \nabla \cdot \mathbf{d}_\alpha + q_\alpha, \tag{43}$$

where q_α represents source terms and $\nabla \cdot \mathbf{d}_\alpha$ the effect of dispersion. These are approximated by

$$\langle M_\alpha^{n+1}, \phi_i \rangle - \int M_\alpha^n(\mathbf{r})\phi_i(\mathbf{s})d\mathbf{r} = \Delta t \int \phi_i(\mathbf{s})[\nabla \cdot \mathbf{d}_\alpha + q_\alpha]\,d\mathbf{s}, \tag{44}$$

where the trajectories define the mapping $(\mathbf{r}, t_n) \mapsto (\mathbf{s}, t_{n+1})$. It should be noted that the use of a piecewise constant basis $\phi_i(.)$ leads to a finite volume approximation for the dispersion term on the right here. For simplicity we assume that a rectangular mesh is used consisting of rectangles R_i with boundaries ∂R_i – so we can use the same numbering notation as in one dimension, with $\mathbf{r}_i$ being the top right vertex of R_i. Then the dispersion integral equals

$$\Delta t \int_{R_i} \nabla \cdot \mathbf{d}_\alpha \, ds = \Delta t \int_{\partial R_i} \mathbf{d}_\alpha \cdot \mathbf{dS} \qquad (45)$$

to evaluate which we need values of $\mathbf{d}_\alpha$ at the vertices of R_i, at the new time level.

The key issue is the evaluation of the velocities $\mathbf{u}_l$, or rather the reference phase pressure p_{l_r}, from which all phase pressures can be found. It is determined by the volume balance constraint described in Acs, Doleschall & Farkas(1985), using partial mass volumes V_α,

$$\langle \varphi \sum_{(\alpha)} V_\alpha M_\alpha, \phi_i \rangle = \text{const.}, \qquad (46)$$

which we shall satisfy by means of a pressure correction calculation. Suppose that the iteration for the trajectories and the values M_α^{n+1} at the new time level are started by assuming the pressures, associated with the centre of each cell, are those holding at the earlier time level. Now consider the effect of changes δp_{l_r} on the trajectories and the calculated values M_α^{n+1}. From $\nabla \delta p_{l_r}$ evaluated at each vertex, we obtain through $\delta \bar{\mathbf{c}}_\alpha$ a change $\delta \mathbf{r}$ to the transformed boundary B_i of ∂R_i. Using Gauss' theorem this becomes a volume integral

$$\langle \delta M_\alpha, \phi_i \rangle \approx \int_{B_i} M_\alpha^n \delta \mathbf{r} \cdot \mathbf{dS} = \int \nabla \cdot (M_\alpha^n \delta \mathbf{r}) \, \phi_i(\mathbf{s}) d\mathbf{r}. \qquad (47)$$

Imposing the constraint (46) on ∂M_α yields an elliptic equation for δp_{l_r} which is approximated by a nine-point finite volume scheme at the new time-level. A number of inner iterations are performed on this before new trajectories are calculated to continue the iteration for M_α^{n+1}.

6 Conclusions

From a review of characteristic Galerkin and Lagrange Galerkin methods for evolutionary problems, we have been led to advocating a weak Lagrange Galerkin formulation for oil recovery models. The basic approximation uses piecewise constant basis functions and so yields a first order approximation, but with no stability constraint on the time step. This is an important practical consideration where local mesh refinement is required, for example around injection and production wells.

Recovery procedures are recommended to obtain higher order approximation. New, simplified, linear recovery procedures are outlined which readily extend onto a two dimensional rectangular mesh. For the linear advection equation in one dimension they can be shown to lead to a second order accurate scheme, which for the special case of the CFL number taking half-integer values is identical with the third order scheme of Warming, Kutler and Lomax (1973). Since it is exact for integer CFL numbers, this scheme is clearly very accurate for linear advection and forms a very sound foundation for the weak LG methods.

Tests on model problems in one dimension confirm the effectiveness of the method and future work will concentrate on developments in two dimensions based on procedures outlined in section 5. Further work on the analysis of the schemes is required, particularly to confirm convergence to the entropy satisfying solution of the equations as in the work of Childs and Morton (1989).

References

Acs, G., Doleschall, S. and Farkas, E. (1985). General Purpose Compositional Model. *Society of Petroleum Engineers Journal*. 25(4), 543-553.

Benqué, J.P., Labadie, G. and Ronat, J. (1982). A new finite element method for the Navier Stokes equations coupled with a temperature equation. In *Proceedings of the Fourth International Symposium on finite element methods in flow problems, (ed. T. Kawai)*. 295-302. Amsterdam: North Holland 1982.

Childs, P.N. and Morton, K.W. (1989). Characteristic Galerkin methods for scalar conservation laws in one dimension. *Oxford University Computing Laboratory report NA86/5. (To appear in Society for Industrial and Applied Mathematics Journal on Numerical Analysis)*.

Douglas, J. Jr. and Russell, T.F. (1980). Numerical methods for convection-dominated diffusion problems based on combining the methods of characteristics with finite element or finite difference procedures. *Society for Industrial and Applied Mathematics Journal on Numerical Analysis Volume 19*. 5, 871-885.

Morton, K.W. and Childs, P.N. (1988). Characteristic Galerkin Methods for Hyperbolic Systems. *Proceedings of the Second International Conference on Hyperbolic Problems (eds. J. Ballman and R. Jeltisch). Notes on Numerical Fluid Dynamics*. 24, 435-455.

Morton, K.W., Childs, P.N. and Rudgyard, M.A. (1988). The cell vertex method for steady compressible flow. *Proceedings of the Conference on Numerical Fluid Dynamics III (eds. K.W. Morton and M.J. Baines).* Oxford University Press. 625-655.

Morton, K.W. and Paisley, M.F. (1989). A finite volume scheme with shock fitting for the steady Euler equations. *Oxford University Computing Laboratory report NA87/6. (To appear in Journal of Computational Physics).*

Morton, K.W., Priestley, A. and Süli, E.E. (1986). Stability analysis of the Lagrange-Galerkin method with non-exact integration. *Oxford University Computing Laboratory, Report No 86/14. (To appear in Mathematical Modelling and Numerical Analysis).*

Morton, K.W. and Sweby, P.K. (1987). A comparison of flux-limited difference methods and characteristic Galerkin methods for shock modelling. *Journal of Computational Physics.* 73(1), 203-230.

Parrott, A.K. and Keeping, B.R. (1989). The use of FCT as a multidimensional recovery procedure with monotonicity preserving properties. *(In preparation).*

Pironneau, O. (1982). On the transport-diffusion algorithm and its application to the Navier-Stokes equations. *Numerische Mathematik.* 38, 309-332.

Süli, E. (1988). Convergence and Nonlinear Stability of the Lagrange-Galerkin method for the Navier-Stokes equations. *Numerische Mathematik.* 53, 459–483.

Van Leer, B. (1977). Towards the ultimate conservative difference scheme. 4. A new approach to numerical convection. *Journal of Computational Physics.* 23(3), 276-299.

Van Leer, B. (1979). Towards the ultimate conservative difference scheme. 5. 2nd order sequel to Gudenov's method. *Journal of Computational Physics.* 31(1), 101-136.

Warming, R.F., Kutler, P. and Lomax, H. (1973). Second-order and third-order noncentered difference schemes for nonlinear hyperbolic equations. *American Institute of Aeronautics and Astronautics Journal.* 11(2), 189-196.

Acknowledgements

We are indebted to Ben Keeping for the practical implementation of the algorithms and all the computations reported in this paper. Support for the EOR applications is being provided by the Department of Energy through their funding of Mr Keeping.

Multigrid Methods in Modelling Porous Media Flow

F. Brakhagen   T. W. Fogwell
Gesellschaft für Mathematik und Datenverarbeitung
Postfach 1240, 5205 St. Augustin 1
Fed. Rep. Germany

Abstract

A new multigrid method is presented for the solution of the unsymmetric system of differential equations

$$\phi\frac{\partial S_w}{\partial t} = \nabla \cdot \lambda_w(\nabla P_o - \rho_w g \nabla D) - \nabla \cdot \lambda_w \nabla P_c - Q_w$$

$$-\phi\frac{\partial S_w}{\partial t} = \nabla \cdot \lambda_o(\nabla P_o - \rho_o g \nabla D) - Q_o$$

used to model incompressible multiphase flow in a porous medium. A semi-implicit discretization scheme in the unknowns S_w and P_o is used in order to allow larger stable time steps. Because of jump discontinuities in the coefficients, a special interpolation operator is defined using weights based on the discretization itself, but a restriction operator using full weighting is found to produce the best results. A Galerkin approximation is used for the coarse grid operators. The usual multigrid efficiency was observed, even for problems with large jump discontinuities in the anisotropic coefficients.

1 Introduction

The main purpose of this study is to evaluate the feasibility of using multigrid methods for the solution of the differential equations modelling multiphase flow in porous media, when they are discretized in the more standard manners. In a previous study [11] we demonstrated that multigrid was a satisfactory solver for the pressure equation in the IMPES (implicit pressure, explicit saturation) method and also for the symmetric system of difference equations arising from a formulation of the simultaneous solution method which has the phase pressures as unknowns and treats the relative permeabilities explicitly. The use of multigrid for the IMPES pressure equations was also investigated by Behie and Forsyth [4] with the ideas originally proposed in [1].

Here we show that the use of multigrid is feasible as a solver for a nonsymmetric system of difference equations which arises from a solution strategy requiring the solution of both pressure and saturation together in a semi-implicit formulation of the problem. This is one of the more standard formulations of the problem which allows for larger time steps [3, 12, 13, 14, 15, 16].

We have chosen physical problems which, in general, are difficult enough from a mathematical or computational point of view, but which tend to minimize modelling complexities. These problems, for example, can produce shocks and fingers. We have chosen incompressible, immiscible, two phase flow in a regular domain at constant temperature. The porous medium, however, can have any pattern of permeabilities which might be reasonable for an actual case. This problem then allows the porosity and the densities to be constant. We also assume that the viscosity μ in each phase does not change. The two phases are denoted by o and w. The equations which govern the flow arise from the conservation of mass in each phase and are given as

$$\phi \frac{\partial S_\ell}{\partial t} = -\nabla \cdot \mathbf{v}_\ell - Q_\ell, \qquad \ell = o, w. \qquad (1.1)$$

Here ϕ denotes the porosity, S_ℓ the saturation (volume fraction in phase ℓ), $\mathbf{v}_\ell$ the superficial velocity, and Q_ℓ the volumetric production at a point. If Darcy's law is assumed, we then have the superficial velocities for each phase determined by

$$\mathbf{v}_\ell = -\frac{K_\ell}{\mu_\ell}(\nabla P_\ell - \rho_\ell g \nabla D), \qquad (1.2)$$

where ∇P_ℓ is the gradient of the pressure, ρ_ℓ is the density, g is gravity acceleration, D is depth, and K_ℓ is the permeability, which, for several phases, is given as the product of relative permeability $k_{r\ell}$ and absolute permeability K. Thus, $K_\ell = k_{r\ell} K$. The absolute permeability K is a tensor, here assumed to be diagonal. We denote K_ℓ/μ_ℓ by λ_ℓ. On the boundary we have assumed no flow Neumann conditions given by $\mathbf{v}_\ell \cdot \mathbf{n} = 0$ for each phase, where $\mathbf{n}$ is the outward vector normal to the boundary. Combining (1.1) and (1.2) one obtains

$$\nabla \cdot [\frac{K k_{r\ell}}{\mu_\ell}(\nabla P_\ell - \rho_\ell g \nabla D)] = \phi \frac{\partial S_\ell}{\partial t} + Q_\ell. \qquad (1.3)$$

The following equations must also be satisfied:
$$S_o + S_w = 1, \quad (1.4)$$
$$P_c := P_o - P_w. \quad (1.5)$$

The capillary pressure P_c is assumed to be a known function of saturation. The unknowns of the problem are the phase pressures P_ℓ and the saturations S_ℓ.

2 Discretization by Finite Differences

The standard finite difference approximation [3, 14, 15] of the differential equations (1.3) in two dimensions is given by

$$\Delta_{xy}\left[C_{\ell xy}\, k_{r\ell}^m\left(\Delta_{xy} P_\ell^{n+1} - \rho_\ell g \Delta_{xy} D\right)\right] = \frac{V_p}{\Delta t}(S_\ell^{n+1} - S_\ell^n) + Q_\ell^m. \quad (2.1)$$

The spatial difference operator Δ_{xy} notation used here is such that

$$[\Delta_{xy}(T_{xy}\Delta_{xy}U)]_{i,j} = [T_x]_{i+1/2,j}(U_{i+1,j} - U_{i,j}) - [T_x]_{i-1/2,j}(U_{i,j} - U_{i-1,j})$$
$$+ [T_y]_{i,j+1/2}(U_{i,j+1} - U_{i,j}) - [T_y]_{i,j-1/2}(U_{i,j} - U_{i,j-1}).$$

The coefficients in (2.1) are defined by

$$[C_{\ell x}]_{i\pm 1/2,j} = \frac{\Delta y_j \Delta z}{\Delta x_{i\pm 1/2}} \frac{[K_x]_{i\pm 1/2,j}}{\mu_\ell}, \qquad [C_{\ell y}]_{i,j\pm 1/2} = \frac{\Delta x_i \Delta z}{\Delta y_{j\pm 1/2}} \frac{[K_y]_{i,j\pm 1/2}}{\mu_\ell},$$

$[V_p]_{i,j} = \phi \Delta x_i \Delta y_j \Delta z$, $[Q_\ell]_{i,j} = q_\ell \Delta x_i \Delta y_j \Delta z$, $\Delta x_{i+1/2} = x_{i+1} - x_i$, and $\Delta x_i = x_{i+1/2} - x_{i-1/2}$. For point-centred grids the block boundaries are defined by $x_{i+1/2} = 1/2(x_i + x_{i+1})$, whereas for a block-centred grid, $x_i = 1/2(x_{i-1/2} + x_{i+1/2})$. We use point-centred grids because such finite difference discretizations are consistent for irregular grids [3]. $\Delta y_{j+1/2}$ and Δy_j are similarly defined. Δz denotes the thickness of the reservoir. The absolute permeabilities $[K_x]_{i+1/2,j}$, $[K_y]_{i,j+1/2}$ at the block boundaries are defined using a harmonic mean. For the relative permeabilities we use upstream weighting. The superscripts n and $n+1$ in equation (2.1) refer to the time level, where m can either be n or $n+1$.

The equations (1.4) and (1.5) are used for the elimination of two of the unkowns, S_o and P_w, to obtain the following system of difference equations:

$$\Delta_{xy}\left[C_{oxy}\, k_{ro}^m\left(\Delta_{xy} P_o^{n+1} - \rho_o g \Delta_{xy} D\right)\right] = -\frac{V_p}{\Delta t}(S_w^{n+1} - S_w^n) + Q_o^m, \quad (2.2)$$

$$\Delta_{xy}\left[C_{wxy}\, k_{rw}^m\left(\Delta_{xy} P_o^{n+1} - \Delta_{xy} P_c^{n+1} - \rho_w g \Delta_{xy} D\right)\right]$$
$$= \frac{V_p}{\Delta t}(S_w^{n+1} - S_w^n) + Q_w^m. \quad (2.3)$$

Since the capillary pressure is a known function of saturation, (2.2) and (2.3) form a system in the two unkowns P_o and S_w.

Several methods, such as IMPES, use explicit relative permeabilities, i.e., $m = n$. It is well known [3, 15, 16] that the use of explicit relative permeabilities produces a conditional instability in the equations if the product of time step size and flow velocity is large. Setting $m = n + 1$ in (2.2) and (2.3) one obtains a fully implicit difference scheme. Although this method is unconditionally stable [3, 16], it has several disadvantages. The fully implicit system of equations is nonlinear and requires Newton iterations. This considerably increases the computational work per time step. Because the Newton iterations will not converge without a reasonable initial estimate, there is still a time step limitation, although the allowable time step is much larger than for the explicit permeability method [15]. Moreover, the time step size of the fully implicit method is limited by large truncation errors, which are always larger than those for the explicit permeability method [3].

The semi-implicit method has some advantages over the fully implicit method. It is more effective at overcoming the time step stability restriction without increasing the computational effort much beyond that required for explicit permeabilities, while producing a smaller truncation error than the fully implicit method [3]. In this method [12, 13, 14, 15] we set $m = n + 1$ in (2.2) and (2.3) but approximate the relative permeabilities $k_{r\ell}^{n+1}$ and the capillary pressure P_c^{n+1} as follows:

$$k_{r\ell}^{n+1} = k_{r\ell}^n + \overline{\frac{dk_{r\ell}}{dS_w}}(S_w^{n+1} - S_w^n), \tag{2.4}$$

$$P_c^{n+1} = P_c^n + \overline{\frac{dP_c}{dS_w}}(S_w^{n+1} - S_w^n). \tag{2.5}$$

The derivative terms in these expressions are given by chord slopes

$$\overline{\frac{dk_{r\ell}}{dS_w}} = \frac{k_{r\ell}(S_w + \delta S_w) - k_{r\ell}(S_w)}{\delta S_w}, \tag{2.6}$$

$$\overline{\frac{dP_c}{dS_w}} = \frac{P_c(S_w + \delta S_w) - P_c(S_w)}{\delta S_w}, \tag{2.7}$$

where δS_w is a specified increment in saturation. Nolen and Berry [14] suggest that δS_w be slightly larger in absolute value than the maximum saturation change expected over a time step.

With the substitution of (2.4) and (2.5), equations (2.2) and (2.3) become

$$\Delta_{xy}[C_{oxy}(k_{ro}^n + \overline{\frac{dk_{ro}}{dS_w}}(S_w^{n+1} - S_w^n))(\Delta_{xy}P_o^{n+1} - \rho_o g \Delta_{xy}D)] =$$

$$-\frac{V_p}{\Delta t}(S_w^{n+1} - S_w^n) + Q_o^{n+1}, \tag{2.8}$$

$$\Delta_{xy}[C_{wxy}(k_{rw}^n + \overline{\frac{dk_{rw}}{dS_w}}(S_w^{n+1} - S_w^n))$$

$$\cdot (\Delta_{xy}P_o^{n+1} - \Delta_{xy}(P_c^n + \overline{\frac{dP_c}{dS_w}}(S_w^{n+1} - S_w^n)) - \rho_w g \Delta_{xy}D)] =$$

$$\frac{V_p}{\Delta t}(S_w^{n+1} - S_w^n) + Q_w^{n+1}. \tag{2.9}$$

We use semi-implicit production rates [15], defined for $\ell = o, w$ by

$$Q_\ell^{n+1} = Q_t^n(f_\ell^n + \frac{df_\ell}{dS_w}(S_w^{n+1} - S_w^n)), \tag{2.10}$$

where $f_w = \lambda_w/(\lambda_o + \lambda_w)$, $f_o = 1 - f_w$ and $Q_t = Q_o + Q_w$.

The flow terms in (2.8) and (2.9) are nonlinear because they contain products of S_w^{n+1} and P_o^{n+1}. The semi-implicit system can be solved by Newton's method, as proposed by Nolen and Berry [14], or linearized, as proposed by Letkeman and Ridings [12] and by MacDonald and Coats [13]. We linearize the nonlinear system by assuming

$$(S_w^{n+1} - S_w^n)\Delta_{xy}P_\ell^{n+1} \approx (S_w^{n+1} - S_w^n)\Delta_{xy}P_\ell^n, \qquad \ell = o, w. \tag{2.11}$$

In doing so, we obtain the following linear system:

$$\Delta_{xy}[C_{oxy}k_{ro}^n(\Delta_{xy}P_o^{n+1} - \rho_o g \Delta_{xy}D)] +$$

$$\Delta_{xy}[C_{oxy}\overline{\frac{dk_{ro}}{dS_w}}(S_w^{n+1} - S_w^n))(\Delta_{xy}P_o^n - \rho_o g \Delta_{xy}D)] =$$

$$-\frac{V_p}{\Delta t}(S_w^{n+1} - S_w^n) + Q_o^{n+1}, \tag{2.12}$$

$$\Delta_{xy}[C_{wxy}k_{rw}^n(\Delta_{xy}P_o^{n+1} - \Delta_{xy}(P_c^n + \overline{\frac{dP_c}{dS_w}}(S_w^{n+1} - S_w^n)) - \rho_w g \Delta_{xy}D)] +$$

$$\Delta_{xy}[C_{wxy}\overline{\frac{dk_{rw}}{dS_w}}(S_w^{n+1} - S_w^n))(\Delta_{xy}P_o^n - \Delta_{xy}P_c^n - \rho_w g \Delta_{xy}D)] =$$

$$\frac{V_p}{\Delta t}(S_w^{n+1} - S_w^n) + Q_w^{n+1}. \tag{2.13}$$

The linearization necessary for a Newton iteration of the nonlinear semi-implicit equations (2.8) and (2.9) produces the same equations as the linearized semi-implicit difference scheme (2.12) and (2.13). Therefore, for a single iteration, the nonlinear semi-implicit method is identical with the linearized semi-implicit method. Nolen and Berry [14] found that in many situations the linear method performs nearly as well as the nonlinear method, except in case of very large time steps. Both semi-implicit methods allow considerably larger time steps than the explicit permeability method does [12, 13, 14].

If we order the unknowns in a vector

$$U^h = [(P_o)_1, (S_w)_1, \ldots, (P_o)_k, (S_w)_k, \ldots, (P_o)_N, (S_w)_N]^T,$$

we can write the difference equations (2.12) and (2.13) in matrix form

$$L^h U^h = F^h. \tag{2.14}$$

L^h is a five point operator. At each grid point (x_i, y_j) it can be written as

$$L_{i,j}^h U^h = W_{i,j}^h U_{i-1,j}^h + S_{i,j}^h U_{i,j-1}^h + C_{i,j}^h U_{i,j}^h + E_{i,j}^h U_{i+1,j}^h + N_{i,j}^h U_{i,j+1}^h, \tag{2.15}$$

where $W_{i,j}^h, S_{i,j}^h, C_{i,j}^h, E_{i,j}^h$ and $N_{i,j}^h$ are 2×2-matrices and $U_{i,j}^h$ are the values of U^h at (x_i, y_j). The form of these matrices depends on the direction of flow because the terms $\frac{dk_{r\ell}}{dS_w}(S_w^{n+1} - S_w^n)$ have to be calculated at the upstream points. In the following we assume that the flow is in the direction of increasing i and j. In this case the matrices $W_{i,j}^h$ and $E_{i,j}^h$ are given by

$$W_{i,j}^h = \begin{pmatrix} [T_{ox}]_{i-1/2,j} & -[T'_{ox}]_{i-1/2,j} \\ [T_{wx}]_{i-1/2,j} & -[T_{wx}]_{i-1/2,j} \overline{[\frac{dP_c}{dS_w}]}_{i-1,j} - [T'_{wx}]_{i-1/2,j} \end{pmatrix} \tag{2.16}$$

and

$$E_{i,j}^h = \begin{pmatrix} [T_{ox}]_{i+1/2,j} & 0 \\ [T_{wx}]_{i+1/2,j} & -[T_{wx}]_{i+1/2,j} \overline{[\frac{dP_c}{dS_w}]}_{i+1,j} \end{pmatrix}, \tag{2.17}$$

where $\ell = o, w$,

$$[T_{\ell x}]_{i+1/2,j} = [C_{\ell x} k_{r\ell}]_{i+1/2,j}, \tag{2.18}$$

$$[T'_{\ell x}]_{i+1/2,j} = [C_{\ell x} \overline{\frac{dk_{r\ell}}{dS_w}} \Delta \Phi_\ell^n]_{i+1/2,j}, \tag{2.19}$$

$$[\Delta \Phi_\ell^n]_{i+1/2,j} = (P_\ell^n)_{i+1,j} - (P_\ell^n)_{i,j} - \rho_\ell g (D_{i+1,j} - D_{i,j}). \tag{2.20}$$

Similar expressions are obtained for $S_{i,j}^h$ and $N_{i,j}^h$. The main diagonal block is given by

$$C_{i,j}^h = \begin{pmatrix} -\sum[T_o]_{i,j} & \sum[T_o']_{i,j} + \frac{[V_p]_{i,j}}{\Delta t} \\ -\sum[T_w]_{i,j} & \sum[T_{wx}]_{i,j}[\overline{\frac{dP_c}{dS_w}}]_{i,j} + \sum[T_w']_{i,j} - \frac{[V_p]_{i,j}}{\Delta t} \end{pmatrix},$$
(2.21)

where

$$\sum[T_\ell]_{i,j} = [T_{\ell x}]_{i+1/2,j} + [T_{\ell x}]_{i-1/2,j} + [T_{\ell y}]_{i,j+1/2} + [T_{\ell y}]_{i,j-1/2} \quad (2.22)$$

$$\sum[T_\ell']_{i,j} = [T_{\ell x}']_{i+1/2,j} + [T_{\ell y}']_{i,j+1/2}. \quad (2.23)$$

At the production wells $C_{i,j}^h$ has an additional term arising from (2.10). Obviously the matrix L^h is highly unsymmetric.

3 Multigrid Methods

We use a multigrid algorithm for the solution of the linearized semi-implicit system (2.12) and (2.13). Simulation of large reservoirs requires the solution of very large sets of equations, with the number of unknowns running up to several thousands. Several solution methods have been used for oil reservoir simulation [2]. The advantage of multigrid methods is that they provide convergence in $O(N)$ operations, where N is the number of unknowns [6, 7]. This means that for large problems, these techniques are superior to other methods.

The basic idea of the multigrid methods is to take advantage of the fact that iterative relaxation methods are very efficient at the elimination of high frequency error components. By performing a few relaxation steps the error is made smooth. This means the low frequency error components become dominant, so that the error can be accurately represented on a coarser grid. The residual equations are transferred to a coarser grid, where the corrections are solved for. The resulting system is handled in the same way as the original one. By performing a few relaxations the error and the residual are made smooth, and the residual equations are transferred to the next coarser grid. Proceeding in this way, one finally arrives at the coarsest grid, on which the system of equations can easily be solved exactly. The corrections are then interpolated back up the chain of grids to the finest grid, with some additional relaxations performed along the way.

In order to describe the multigrid algorithm in more detail it is necessary to define certain terms and designate certain operations. Suppose we have a sequence of grids G^k ($k = 1(1)M$) defined with $h_1 > h_2 > \ldots > h_M$, where

h_k is the mesh size of G^k. Let L^M be a finite difference approximation to the differential operator L on G^M and let L^k be an approximation to L^M on G^k for $k < M$. By X^k we denote the linear space of real valued grid functions defined on G^k:

$$X^k: \quad G^k \to \mathbf{R}.$$

We denote a restriction operator by

$$I_k^{k-1}: \quad X^k \to X^{k-1}$$

and an interpolation operator by

$$I_{k-1}^k: \quad X^{k-1} \to X^k.$$

For convenience we introduce the notation

$$\bar{u}^k := S_k^\sigma(u^k, L^k, f^k)$$

to denote that $\bar{u}^k$ is the result of σ relaxation steps applied to $L^k u^k = f^k$ starting with u^k as first approximation, where $u^k, \bar{u}^k, f^k \in X^k$. Similarly we denote

$$\bar{w}^k := MG_k^\gamma(u^k, L^k, f^k),$$

so that $\bar{w}^k$ is the result of γ successive applications of the following algorithm MG_k to $L^k u^k = f^k$ starting with u^k as the first approximation:

Algorithm MG_k

Step 1: If $k > 1$ go to step 2, else solve exactly

$$L^1 \bar{u}^1 = f^1, \tag{3.1}$$

$$MG_1 := \bar{u}^1. \tag{3.2}$$

Step 2: (Smoothing)
$$\bar{u}^k := S_k^{\sigma_1}(u^k, L^k, f^k). \tag{3.3}$$

Step 3: (Transfer to the next coarser grid)

$$\bar{f}^{k-1} := I_k^{k-1}[f^k - L^k \bar{u}^k], \tag{3.4}$$

$$v^{k-1} := 0. \tag{3.5}$$

Step 4:
$$\bar{v}^{k-1} := MG_{k-1}^\gamma(v^{k-1}, L^{k-1}, \bar{f}^{k-1}). \tag{3.6}$$

Step 5: (Transfer to the next finer grid)

$$w^k := \bar{u}^k + I_{k-1}^k \bar{v}^{k-1}. \tag{3.7}$$

Step 6: (Smoothing)

$$\bar{w}^k := S_k^{\sigma_2}(w^k, L^k, f^k), \qquad (3.8)$$

$$MG_k := \bar{w}^k. \qquad (3.9)$$

The multigrid method then consists of the following iteration:

Initialize: $u^{(0)} \in X^M$,

$$u^{(j+1)} := MG_M(u^{(j)}, L^M, f^M). \qquad (3.10)$$

γ in step 4 is either equal to 1 or 2. With $\gamma = 1$ we obtain a V-Cycle and with $\gamma = 2$ a W-Cycle. If the coefficients of the differential operator L are continuous, it is usually very efficient to use bilinear interpolation for I_{k-1}^k and the transpose of this interpolation operator for the restriction I_k^{k-1}. Moreover, in this case very good results are generally obtained if L^k for $k < M$ is the same finite difference approximation of L on G^k as L^M on G^M [6].

The coefficients of the oil reservoir differential equations may have jump discontinuities of several orders in magnitude, as well as strong anisotropies. Discontinuities in absolute permeablity occur between different layers having different geological structures. Moreover, the relative permeabilities are dependent on the saturations, which have large differences at oil-water interfaces. If the coefficients of the differential equations jump by orders in magnitude, the multigrid method exhibits poor convergence when bilinear interpolation is used [1]. In this case the use of a more appropriate interpolation is necessary to obtain the usual multigrid efficiency. Several authors [1, 4, 5, 8, 9, 10] suggest the use of the difference operator itself for the interpolation. It must also be guaranteed that the difference equations on the next coarser grid approximate those on the given grid [1]. The coarse grid difference operator L^{k-1} then is defined by the Galerkin approach

$$L^{k-1} = I_k^{k-1} L^k I_{k-1}^k, \qquad k = M(-1)2. \qquad (3.11)$$

L^{k-1} ($k = M(-1)2$) are nine point operators, although L^M may be a five point operator.

The interpolation operator is defined in the following way. Let L^k be defined at point $(x_i, y_j) \in G^k$ by

$$\begin{aligned} L_{i,j}^k u^k &= NW_{i,j}^k \, u_{i-1,j+1}^k + N_{i,j}^k \, u_{i,j+1}^k + NE_{i,j}^k \, u_{i+1,j+1}^k \\ &+ W_{i,j}^k \, u_{i-1,j}^k + C_{i,j}^k \, u_{i,j}^k + E_{i,j}^k \, u_{i+1,j}^k \\ &+ SW_{i,j}^k \, u_{i-1,j-1}^k + S_{i,j}^k \, u_{i,j-1}^k + SE_{i,j}^k \, u_{i+1,j-1}^k. \end{aligned} \qquad (3.12)$$

All the coefficients in this equation are 2 × 2-matrices and the $u_{\nu,\mu}^k$ ($\nu = i - 1(1)i + 1$, $\mu = j - 1(1)j + 1$) are two-dimensional vectors.

For horizontal lines embedded in the coarse grid the interpolation is defined as follows. Suppose $(x_{i+1}, y_j) \in G^k \setminus G^{k-1}$ and $(x_i, y_j), (x_{i+2}, y_j) \in G^k \cap G^{k-1}$, where $k = M(-1)2$. On G^{k-1} the latter two grid points are denoted by $(x_I, y_J), (x_{I+1}, y_J)$ respectively. The interpolation is defined by solving the matrix equation

$$(C_{i+1,j}^k + S_{i+1,j}^k + N_{i+1,j}^k)(I_{k-1}^k u^{k-1})_{i+1,j} = \quad (3.13)$$

$$-(NW_{i+1,j}^k + W_{i+1,j}^k + SW_{i+1,j}^k)u_{I,J}^{k-1} - (NE_{i+1,j}^k + E_{i+1,j}^k + SE_{i+1,j}^k)u_{I+1,J}^{k-1}$$

for $(I_{k-1}^k u^{k-1})_{i+1,j}$. A similar definition is given for the interpolation on vertical lines embedded in the coarse grid. At fine grid points, which do not lie on coarse grid lines, for example, (x_{i+1}, y_{i+1}), the interpolation is defined by solving the equation

$$L_{i+1,j+1}^k u^k = 0 \quad (3.14)$$

for $u_{i+1,j+1}^k$.

For symmetric problems several authors [1, 4, 5, 8, 9, 10] propose to define the restriction operator I_k^{k-1} by the transpose of the interpolation operator. In our computations [11] this choice of the restriction operator was very successful, as long as the discrete system of equations was symmetric and positive definite. But the linearized semi-implicit system (2.12) and (2.13) is highly unsymmetric. We applied the multigrid method with this restriction operator to this nonsymmetric system and observed poor convergence. Sometimes the algorithm was divergent. Alcouffe, et al. [1], point out that the transpose of the interpolation is not always the best choice for the restriction. In some cases they found better convergence using a restriction operator with fixed weights. The restriction operator is applied to the residuals only. We use the same weights for the residuals of both the oil and the water equations. If $(x_i, y_j) \in G^k \cap G^{k-1}$, then on G^{k-1} this grid point is denoted by (x_I, y_J). We then define the restriction operator by

$$[I_k^{k-1} r^k]_{I,J} = \sum_{\nu=-1}^{1} \sum_{\mu=-1}^{1} \rho_{\nu\mu} r_{i+\nu,j+\mu}^k, \quad (3.15)$$

where $r_{\alpha,\beta}^k$ is the residual at point (x_α, y_β). The weights are defined by

$$\begin{aligned} \rho_{0,0} &= 1, \\ \rho_{0,1} &= \rho_{1,0} = \rho_{0,-1} = \rho_{-1,0} = 1/2, \\ \rho_{1,1} &= \rho_{1,-1} = \rho_{-1,1} = \rho_{-1,-1} = 1/4. \end{aligned} \quad (3.16)$$

Using this restriction operator our multigrid method exhibits the usual multigrid efficiency.

A good smoothing procedure is crucial for the efficiency of the multigrid method. Because the oil reservoir differential equations have a hyperbolic part, we employ the collective Gauss-Seidel method. In this method the grid points are scanned one by one and at each point k the two unknowns $(P_o)_k$ and $(S_w)_k$ are changed simultaneously. The collective point Gauss-Seidel method gives acceptable smoothing rates, unless there are severe anisotropies. To handle anisotropic problems, our program provides three options, x-line Gauss-Seidel for $C_{\ell x} \gg C_{\ell y}$, y-line Gauss-Seidel for $C_{\ell x} \ll C_{\ell y}$, and alternating line Gauss-Seidel method if there are anisotropies of both kinds in the reservoir. One relaxation step of this last mentioned method consists of one x-line Gauss-Seidel step followed by one y-line Gauss-Seidel step.

4 Numerical Results

We tested the multigrid method described above for a reservoir Ω represented by the horizontal square $\{(x,y)|0 \leq x,y \leq L\}$ of constant thickness. Initially the entire reservoir is almost fully saturated with oil. Water is injected in the lower left corner at a constant rate and oil is produced in the upper right corner. The data are taken from a test problem in [3]: $\mu_o = \mu_w = 1$ cp, $K_x = K_y = 0.3$ Darcy, $\phi = 0.2$ and $L = 1000$ ft. We also considered nonhomogenous reservoirs with absolute permeabilities having jump discontinuities of several orders in magnitude. The multigrid algorithm was tested for the case

$$K_x = K_y = \begin{cases} 300.0 \text{ Darcy} & \text{for } (x,y) \in Q_2 \setminus Q_1, \\ 0.3 \text{ Darcy} & \text{else,} \end{cases}$$

where $Q_2 = \{(x,y)|0 \leq x,y \leq 0.2L\}$ and $Q_1 = \{(x,y)|0 \leq x,y \leq 0.1L\}$. We assumed a connate water saturation of 0.16 and a residual oil saturation of 0.2. The relative permeabilities were given by the smooth functions $k_{rw}(S_w) = (S_w - 0.16)^2$ and $k_{ro}(S_w) = (0.8 - S_w)^2$. We made tests with zero capillary pressure and with a nonzero capillary pressure, which was given by the function $P_c = (0.8 - S_w)/S_w$.

For the discretization we used a uniform point-centred grid with 65×65 = 4225 points, so that the total number of unknowns was 8450. It should be noted, however, that it is not necessary to use a power of two plus one as the number of grid points in one direction. The program can handle any number in each direction, while the data structure in our multigrid IMPES code allows for quite irregular domains. All problems were solved to a tolerance of $\|r^M\|_\infty < 10^{-6}$, where $\|r^M\|_\infty$ denotes the maximum norm of the residual vector r^M on the finest grid G^M. In the following we denote by NC the number of multigrid cycles per time step which are necessary to make $\|r^M\|_\infty < 10^{-6}$. To measure the speed of convergence, we define one G^k-work-unit as the work required for one relaxation on G^k. One G^k-work-unit is approximately equal to $4^{-(m-k)} G^M$-work-units. In the following, WU denotes the total work in G^M-work-units per time step required for relaxations.

For the problems described above the collective point Gauss-Seidel method proved to be a good smoother. We obtained the best convergence of the multigrid method using V-cycles ($\gamma = 1$) and performing just one relaxation on each grid ($\sigma_1 = \sigma_2 = 1$). In most cases we obtained $NC = 4$ or 5 and WU ranged from 10.70 to 14.40. The execution times on the IBM 3090 computer without vectorization lay between 2.50 sec and 3.30 sec per time step.

Solving the same problem with our multigrid code for the IMPES method [11] we found practically the same convergence speed. The numbers NC and WU lay between the same bounds. The execution times were, of course, smaller, ranging from 1.50 sec. to 1.90 sec., because in the IMPES method only a single equation per grid point is solved implicitly. But the IMPES method is only conditionally stable because of the explicit treatment of the relative permeabilties. Experimentally we found that the time steps of the linearized semi-implicit simultaneous solution method can be chosen up to eight times as large as for IMPES. This means that the overall execution times can be considerably reduced. If the time steps are chosen eight times as large as for IMPES, then the linearized semi-implicit simultaneous solution method requires about a quarter of the execution time needed for IMPES. Anisotropic problems could equally well be handled by the use of line relaxation.

References

[1] Alcouffe, R. E., A. Brandt, J. E. Dendy, and J. W. Painter. "The Multi-Grid Method for the Diffusion Equation with Strongly Discontinuous Coefficients." *SIAM J. Sci. Stat. Comput.* 2 (1981): 430 - 454.

[2] Allen, M. B. III, G. A. Behie, and J. A. Trangenstein. *Multiphase Flow in Porous Media—Mechanics, Mathematics, and Numerics.* Lecture Notes in Engineering, No. 34. Springer-Verlag, 1988.

[3] Aziz, K. and A. Settari. *Petroleum Reservoir Simulation.* London: Elsevier Applied Science Publishers, 1979.

[4] Behie, A. and P. A. Forsyth. "Multi-Grid Solution of the Pressure Equation in Reservoir Simulation." In *Proceedings of the Sixth SPE Symposium on Reservoir Simulation, New Orleans, Louisiana*, 1982.

[5] Behie, A and P. A. Forsyth. "Comparison of Fast Iterative Methods for Symmetric Systems." *IMA Journal of Numerical Analysis* 3 (1983): 41 - 63.

[6] Brandt, A. "Multi-Level Adaptive Solution to Boundary-Value Problems." *Math. Comp.* 31 (1977): 333 - 390.

[7] Brandt, A. "Guide to Multigrid Development." In *Multigrid Methods: Proceedings of the Conference Held at Köln-Porz, 23–27 Nov. 1981*, edited by W. Hackbusch and U. Trottenberg, 220-312. Berlin, Heidelberg, and New York: Springer-Verlag, 1982.

[8] Dendy, J. E. "Black Box Multigrid." *Journal of Computational Physics* 48 (1982): 366 - 386.

[9] Dendy, J. E. "Black Box Multigrid for Nonsymmetric Problems." *Applied Mathematics and Computation* 13 (1983): 261 - 283.

[10] Dendy, J. E. "Black Box Multigrid for Systems." *Applied Mathematics and Computation* 19 (1986): 57 - 74.

[11] Fogwell, T. W. and F. Brakhagen. "Multigrid Methods for the Solution of Porous Media Multiphase Flow Equations." *Nonlinear Hyperbolic Equations—Theory, Computation Methods, and Applications*. Vol. 24, *Notes on Numerical Fluid Mechanics*, ed. Josef Ballmann and Rolf Jeltsch, 139 - 148. Braunschweig: Vieweg, 1989.

[12] Letkeman, J. P. and R. L. Ridings. "A Numerical Coning Model." *Soc. Petrol. Eng. J.* (Dec. 1970): 418 - 424.

[13] MacDonald, R. C. and K. H. Coats. "Methods for Numerical Simulation of Water and Gas Coning." *Soc. Petrol. Eng. J.* (Dec. 1970): 425 - 436.

[14] Nolen, J. S. and D. W. Berry. "Test of Stability and Time-Step Sensitivity of Semi-Implicit Reservoir Simulation Techniques." *Trans. SPE of AIME* (1972): 253 - 266.

[15] Peaceman, D. W. *Fundamentals of Numerical Reservoir Simulation*. Amsterdam: Elsevier Scientific Publishing Company, 1977.

[16] Peaceman, D. W. "A Nonlinear Stability Analysis for Difference Equations Using Semi-Implicit Mobility." *Soc. Petrol. Eng. J.* 17 (1977): 79 - 91.

LOCAL GRID REFINEMENT

I.M. Cheshire
(Exploration Consultants Ltd)

and

A. Henriquez
(Statoil)

1. INTRODUCTION

Local grid refinement is a technique which aims to provide improved accuracy in sub-regions of a reservoir simulation study. HEINEMANN et al (1983) discuss dynamic grid refinement to improve the resolution at flood fronts. However, most papers on the subject published since MROSOVSKY et al (1973) focus on enhanced accuracy near wells. The objective is to develop efficient techniques for field scale studies with full resolution of awkward coning effects near troublesome production wells. Figure 1 illustrates typical radial and Cartesian refinements required by reservoir engineers. Growing interest in the application of horizontal wells gives added impetus to the technology. Recent work in local grid refinement has been reviewed by EWING et al (1989).

The conventional method for enhanced accuracy in sub-regions is to specify small cells in the region of interest. Because grid lines must extend to the reservoir boundary this leads to a large number of cells (Figure 2), particularly in three dimensions, and computing costs are prohibitive. The introduction of efficient fully implicit simulators with non-neighbour connection facilities (CHESHIRE et al (1980)) made it practical to incorporate radial refinements in a cartesian global grid (Figure 3). The use of non-neighbour connections for cartesian refinements (Figure 4) is less satisfactory, because, as pointed out by QUANDALLE et al (1985) it may be necessary to interpolate pressures (and depths) within the global grid in order to compute accurate flows between the global and local grid systems (Figure 5).

PEDROSA et al (1985) described a hybrid method which combines IMPES in the global Cartesian grid with the fully implicit method in the local grids surrounding production wells. This technique has considerable merit and is investigated further in the present paper. Because the global and local systems are decoupled at a high level they can be solved independently with obvious potential for parallelization. Computing effort can be concentrated on the difficult local problems without holding up the progress of the overall simulation.

To increase the global timesteps we have extended the hybrid method by applying the fully implicit method to the global grid and by the use of smaller local timesteps in the local grids. A material balance correction is introduced at the end of each global timestep to aid stability. The method permits a straightforward application of the pressure and depth interpolation of QUANDALLE et al (1985). Vertical equilibrium may be used in the global grid with dispersed flows in the local grids.

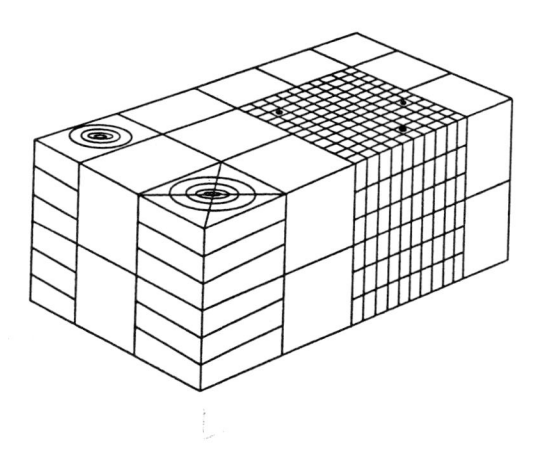

Figure 1: Typical local grid refinements

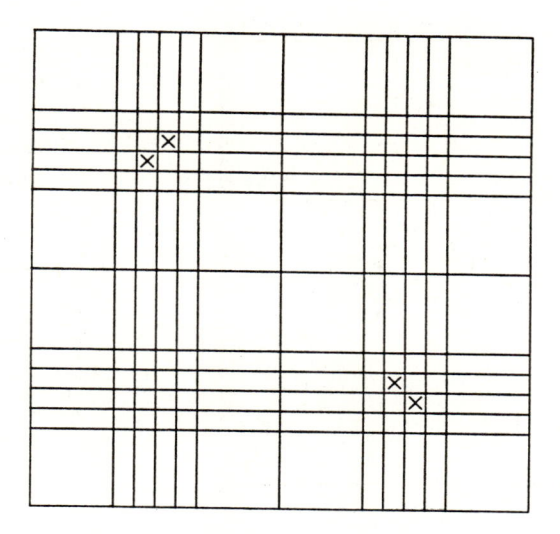

Figure 2:   The conventional method for generating local grids near wells

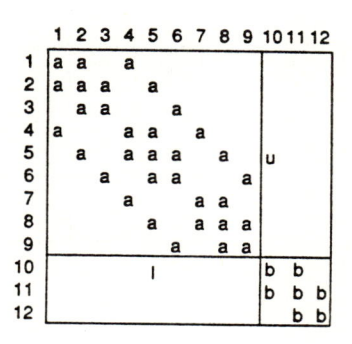

Figure 3:   A simple radial grid in a cartesian global grid and the corresponding matrix

340 CHESHIRE AND HENRIQUEZ

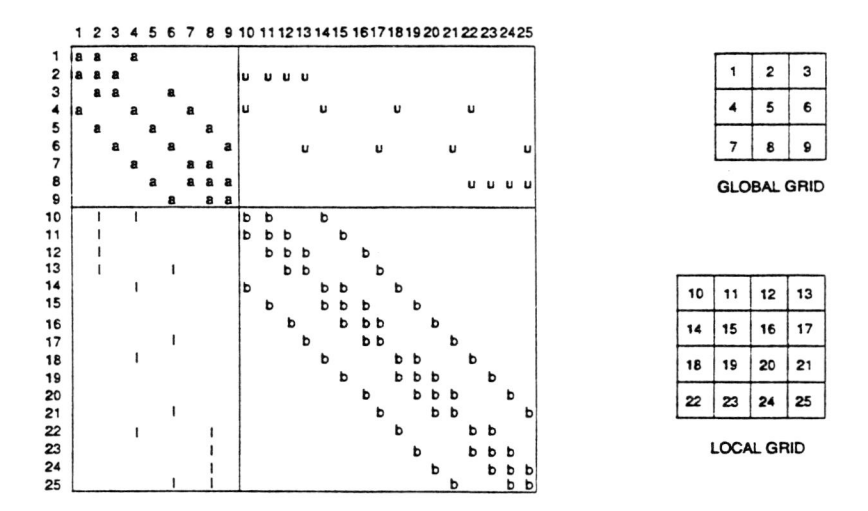

Figure 4: Global cell number 5 is replaced by the local grid. The non-neighbour connections are represented by I and u

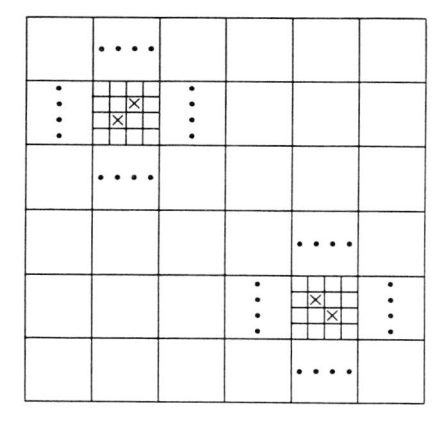

Figure 5: The local grid equivalent of Figure 2. Flows between the local and global grids should use interpolated pressures and depths at the points indicated in Figure 2.

2. THE LINEAR EQUATIONS

With the exception of PEDROSA et al (1985), papers on local grid refinement treat the solution of the global and local systems simultaneously. The linear equations take the form;

$$A.x + U.y = Rg \quad\quad\quad\quad\quad (1)$$
$$L.x + B.y = Rl \quad\quad\quad\quad\quad (2)$$

where A is the usual banded matrix for the global system, B is the set of banded matrices for the local systems and L and U are the coupling terms connecting the local and global systems. The elements of A corresponding to refined cells are either eliminated, since they are inactive, or retained to enhance vectorization but decoupled from other cells in the global system. The latter option is usually chosen since it is then possible to switch the local refinement off dynamically by reactivating the coupling between global neighbouring cells. Rg and Rl are the residuals in the global and local cells respectively and x and y are the solution increments (pressures and saturations) in the global and local systems. The matrix B may be regarded as a set of uncoupled matrices B1, B2, B3...one for each local grid system. Equations 1 and 2 may be solved in a variety of ways with various levels of efficiency for parallel and vector processors.

3. THE SIMULTANEOUS METHOD

The most natural solution procedure is to apply a powerful iterative method, such as nested factorization (APPLEYARD et al (1983)), to the combined set of equations 1 and 2. Here the L and U terms are simply non-neighbour connections which can be handled by well established techniques. Because the local grids are coupled only to the global grid and not to each other, the solution procedure can be parallelized if the cells are numbered with the global cells first followed by the cells in the first local grid followed by the cells in the second local grid etc. This method converges slowly because the off band coupling terms, L and U, are far from the diagonal.

To reduce the number of iterations required to solve the combined equations we must bring L and U closer to the diagonal. This can be done by numbering the cells with the local grids in place. We begin by numbering the cells in the global grid in the usual way until we reach a refined global cell. The next cell is then the first cell in the local grid. The numbering continues until all the local cells within the refined global cell are numbered. The numbering then reverts to the global grid and continues until the next refined global cell is reached etc. This method is probably the most powerful procedure on conventional machines. There is some loss of efficiency in vector processors because the preconditioned vector runs terminate at globally refined cells but this effect is small in commercial simulators. The main defect of the method is its loss of potential parallelization. The cell numbering system is illustrated in Figure 6.

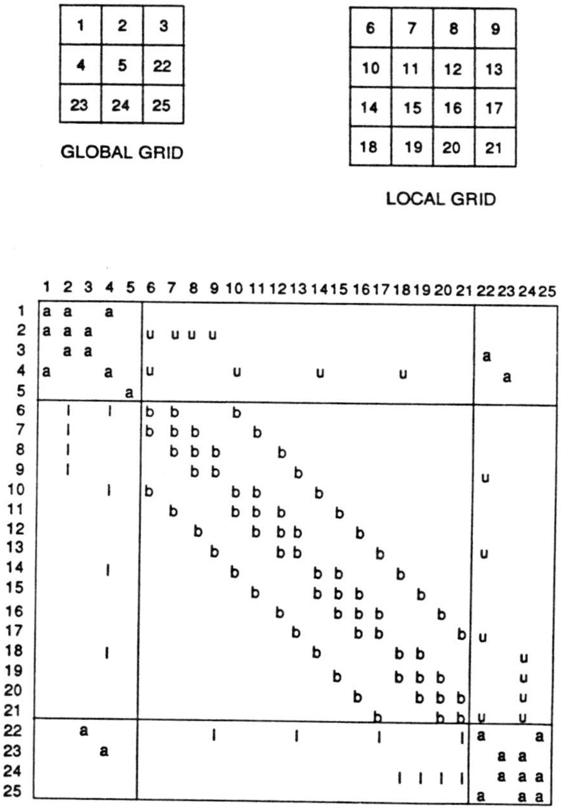

Figure 6: Local cells numbered in place

4. SEQUENTIAL METHODS

In a sequential method we first solve for x in the global grid and then solve equation 2 for y in the local grids. If the local solution variables, y, are formally eliminated from equations 1 and 2 we obtain

$$(A - U.B^{-1}.L).x = Rg - U.B^{-1}.Rl = Rt \dots\dots\dots(3)$$

To formulate an iterative method for the solution of equation 3, we must construct a good approximation to $A - U.B^{-1}.L$ to serve as the pre-conditioning matrix. If the column sum of the pre-conditioning matrix is the same as the column sum of $A - U.B^{-1}.L$ then material will be conserved exactly at each iteration and the number of iterations required to solve the equation will be halved. This suggests that we replace $U.B^{-1}.L$ by the diagonal matrix $colsum(U.B^{-1}.L)$ and then compute a mass conserving pre-conditioning matrix based on an approximation to $A - colsum(U.B^{-1}.L)$.

At each iteration we must update the residual (error) by calculating

$$r = Rt - (A - U.B^{-1}.L).x \dots\dots\dots\dots(4)$$

This calculation must be performed precisely. If an approximation to B is used at this step the wrong equation will be solved. It follows therefore that a precise solution of each local problem must be computed at each iteration of the global problem and the method becomes unattractive. If the local grids are sufficiently small to permit a direct solution method to be used then the inverse of B need only be computed once and the calculation of r becomes less expensive.

A more practical iterative procedure is based on the decomposition

$$\begin{vmatrix} C & L \\ U & D \end{vmatrix} = \begin{vmatrix} C & \\ U & E \end{vmatrix} \times \begin{vmatrix} I & C.L \\ & I \end{vmatrix} \dots\dots\dots\dots(5)$$

where C and D are easily inverted approximations to B and A respectively. Material is conserved at each iteration if the column sum of C is equal to the column sum of B, the column sum of D is equal to the column sum of A and

$$E = D - colsum(U.C.L) \dots\dots\dots\dots(6)$$

At each iteration, the solution sequence is

$$C.z = Rl \dots\dots\dots\dots\dots\dots\dots\dots\dots(7)$$
$$E.x = Rg - U.z \dots\dots\dots\dots\dots\dots\dots(8)$$
$$C.y = R - L.x \dots\dots\dots\dots\dots\dots\dots(9)$$

and the residuals are

$$rl = Rl - B.y - L.x \dots \dots \dots \dots \dots (10)$$
$$rg = Rg - U.y - A.x \dots \dots \dots \dots \dots (11)$$

A method analogous to that outlined above has been proposed by EWING et al (1989). The technique requires two inversions of the approximate local grid matrix C at each iteration. However, since the local grids are not coupled to each other this calculation may be parallelized.

5. THE HYBRID METHOD

PEDROSA et al (1985) proposed a hybrid method in which, at each timestep, a solution is first obtained in the global grid using IMPES, as if the local grid did not exist. Wells in the local grids are replaced by pseudo wells in the global grid. The pressures and saturations in the global grid at the end of each timestep are then used as boundary conditions for the fully implicit solution within the local grids.

This technique is highly efficient and clearly maintains full vectorization and parallelization potential. It has a strong appeal to common sense. IMPES is normally applicable to field scale studies with large grid cells and is unsuitable for the study of detailed coning effects near wells. By using the fully implicit method in the local grids we appear to have the best of both worlds. The local grids may be either radial or cartesian.

The automatic dimensioning feature of ECLIPSE makes it relatively easy for us to implement the hybrid method. The simulator is effectively converted to N+1 simulators where N is the number of local grids. In an attempt to overcome the IMPES timestep limitations in the global grid we have also applied the fully implicit method to the solution of the global system. In practice however we have also found that it is necessary to restrict the timestep in the global grid due to the fundamentally explicit nature of the method.

To retain material balance and to prevent the development of explosive instabilities we have found it necessary to apply careful material balance corrections at the end of each global timestep. Material in the local cells is summed and substituted for the erroneous material in the corresponding pseudo global cell. Material in global cells adjacent to refined cells is corrected by accounting for the difference between the flows from global to local cells and the flows from global to pseudo global cells.

A distinct advantage of the hybrid method is that smaller timesteps may be used in the local grids than those used in the global grid. This possibility was first investigated by QUANDALLE et al (1985). To obtain boundary conditions for the local grid problems the pressures and saturations in the global grid are assumed to vary linearly with time over a global timestep. Thus detailed coning effects near troublesome wells can be resolved without holding up the progress of the full field simulation.

Switching refinements on and off at various times during a simulation is a useful option which is relatively simple to implement in the hybrid method. When the refinement is switched off the pseudo cells in the global system are treated as normal global cells. In the loop over refined grids we simply skip any calculations for the refinements which have been switched off. When a refinement is switched on then the material in the global pseudo cell is initially distributed uniformly within the local grid. It is therefore advisable to switch the refinement on several timesteps before the local wells are activated to obtain a better local fluid distribution.

Another advantage of the hybrid method is that vertical equilibrium may be used in the global grid with dispersed flow in the local grids. This option is particularly effective in ECLIPSE which contains a collapsed VE option in which a 3D global grid is treated as a 2D areal system while retaining full vertical variation of rock properties.

The main defect of the hybrid method is that the pseudo cells in the global grid do not fully represent the local grids. The material balance corrections referred to above certainly help to maintain stability but it may also be necessary to develop special pseudo functions for the pseudo global cells to help mimic the behaviour of the local systems they represent. Pseudo wells in the global pseudo cells must be used to represent real wells in the local grids. At the end of each global timestep we compute phase dependent mobility multipliers for each pseudo well to align production rates between the pseudo and local wells. These multipliers are used at the next global timestep. Phase dependent transmissibility multipliers may also be required for the pseudo cells before the hybrid method becomes sufficiently reliable for routine use in a commercial simulator. This requirement is the focus of our current research and will be reported more fully in future.

6. THE USER INTERFACE

Static local grid refinements can be modelled in any standard implicit simulator with a non-neighbour connection facility. The user sets up separate grid systems for the global and local models together with a set of non-neighbour transmissibilities connecting the local grids to the global grid. The refined cells of the global grid must be disconnected from other cells in the global grid or simply blanked out.

The disadvantages of this approach are that it is extremely cumbersome for the user and there is some loss of accuracy because pressures are not interpolated in the global grid when calculating flows between the global and local systems.

A practical local grid refinement option should make its application as easy as possible for the user. The calculation of transmissibilities between the global and local grids must be computed automatically by the simulator. Data preparation should be as simple as possible. Output reports of pressures and saturations etc., should be reported separately for global and local grids.

7. PRACTICAL APPLICATIONS

Some field scale applications of local grid refinement are discussed below. These illustrate horizontal well modelling, gas and water coning, condensate dropout and well interference. All CPU times quoted are for an IBM 3090 with vector facility.

Horizontal wells may be represented in a simulation by modifying Peaceman's formula or by the inclusion of very small blocks of unit porosity and high horizontal permeability (KOSSAK et al (1987)). To test the local grid refinement option in ECLIPSE we used the high permeability method adding an additional 60 blocks to represent the well in a 1440 block 3-phase simulation. The problem was solved first using non-neighbour connections (NNCs) to attach the local well model to the global grid and secondly using the hybrid method with small local timesteps. The CPU times for the three cases were;

Modified Peaceman in the global grid	510 seconds
High permeability local grid with NNCs	372 seconds
Hybrid method with small local timesteps	213 seconds

This test convinced us of the clear advantages of small local timesteps and that the hybrid method was the only practical way of including large numbers of local refinements within a full field study.

Gas and water coning are characterized by large saturation changes in short times in the vicinity of production wells. Vertical spatial resolution of under one $_{metre}$ is often required. Large timesteps may be used once a stable cone has developed but short timesteps are required during the development of the cone and at gas or water breakthrough. Timesteps as short as one minute are not uncommon and these may occur at different times in different local grids. Spatial refinement alone is therefore insufficient to avoid prohibitive computing costs and the small timesteps required for coning calculations must be isolated to the local grid regions. The hybrid method with local timesteps was therefore used in the remaining examples.

Figure 7 shows an areal view of the simulation grid with 25*23*19 cells covering an area of 6km by 8km. This grid illustrates the need for cell amalgamation in the peripheral regions.

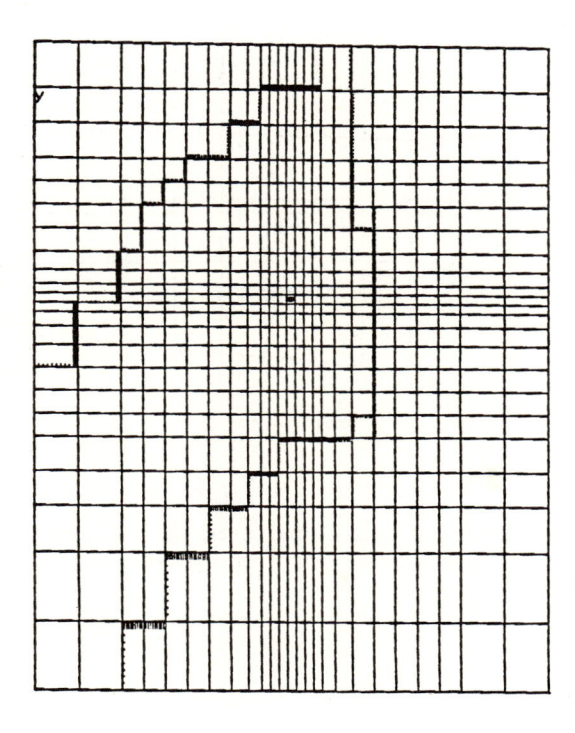

Figure 7: Horizontal well and coning, full field simulation grid. The well is located in the middle of the finer grid section.

Figure 8 shows an XZ cross section with the fluid contacts indicated by dotted lines. Figure 9 shows the refinement needed to capture the coning behaviour giving an additional 3780 cells. Figures 10 and 11 show water and gas coning in a cross section perpendicular to the horizontal well and for a cross section along the length of the well. Figure 12 shows the water cut with and without local grid refinement. As expected, breakthrough occurs earlier when local grid refinement is used. CPU time for the unrefined simulation with 10925 cells was 853 seconds compared with 1723 seconds for the local grid refinement run with the additional 3780 cells.

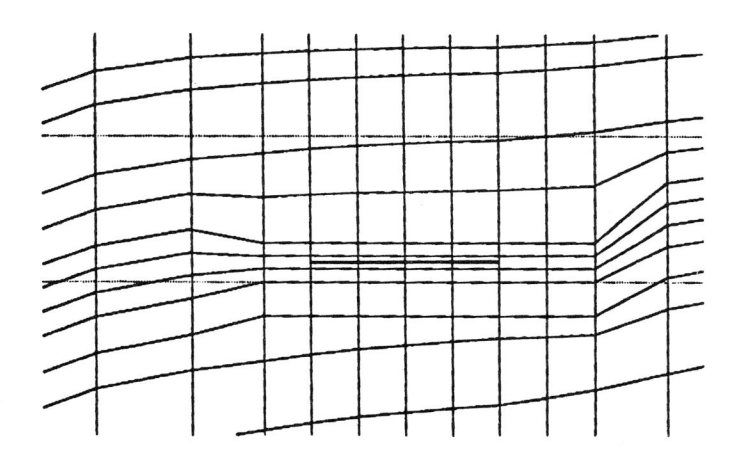

Figure 8: Horizontal well and coning, XZ cross section of the simulation grid in the well area.

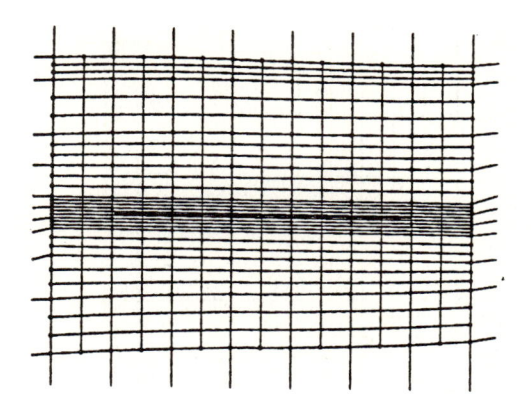

Figure 9: Horizontal well and coning, grid refinement for simulating the horizontal well.

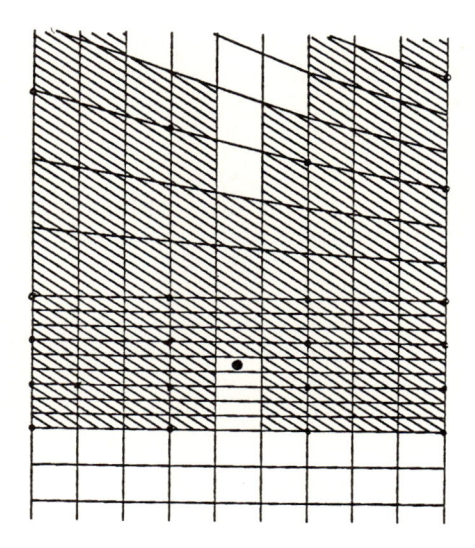

Figure 10: Horizontal well and coning, YZ cross section of cone: cross section perpendicular to the well showing the hatched blocks where there is no reduction in the oil saturation.

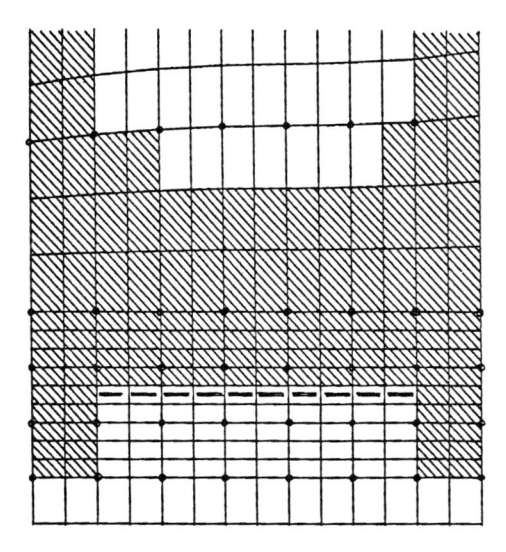

Figure 11: Horizontal well and coning, XZ cross section of cone: cross section along the well, with hatched blocks where there is no reduction in the oil saturation.

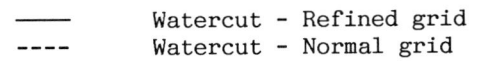

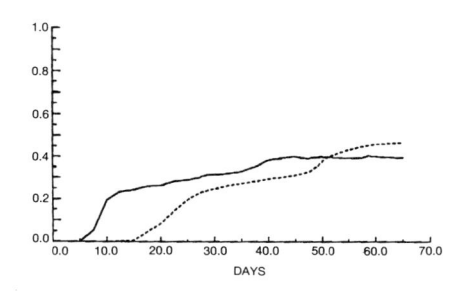

―――― Watercut - Refined grid
- - - - Watercut - Normal grid

Figure 12: Horizontal well and coning, water cut: water cut for simulations with coarse and refined grids.

Figure 13 shows a 3D full field grid of a gas condensate reservoir with the location of several 5*5 refinements. Figure 14 shows a cross section with two of the refinements. Figure 15 shows the oil saturation against distance from a well for runs with and without local grid refinement. The case for local grid refinement is clearly illustrated by the improved resolution of condensate dropout near the well.

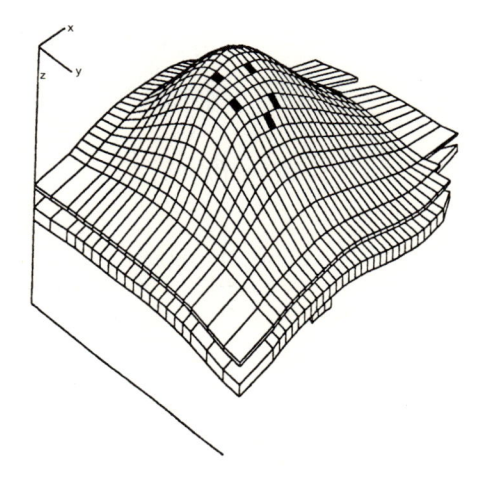

Figure 13:   Condensate dropout, 3D view of simulation grid: the location of the wells are marked with black grid cells.

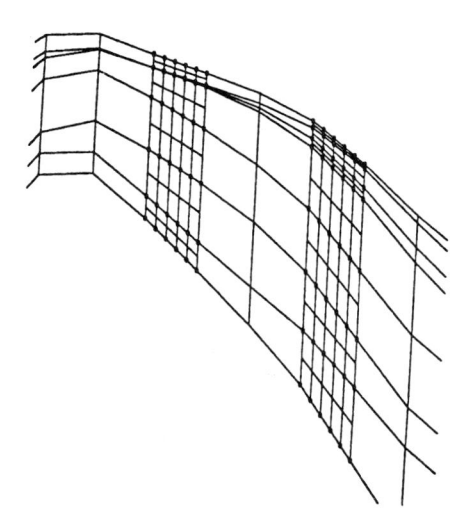

Figure 14: Condensate dropout, grid cross section: the refinement at two well locations is shown in a vertical cross section.

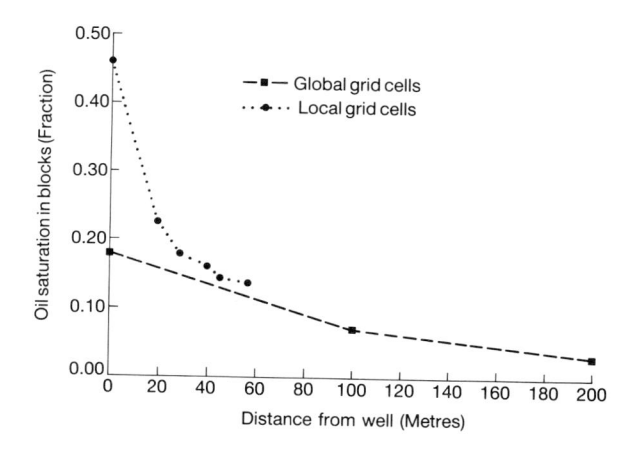

Figure 15: Condensate dropout, resolution of dropout: liquid saturation for a coarse well block simulation and for a refined well block simulation as a function of the distance to the well.

The global grid has 2699 active cells and 1750 local cells. The study was run without refinement, with refinement using NNCs and with refinement using the hybrid method. CPU times were 262, 638 and 377 seconds respectively which once again illustrates the relative efficiency of the hybrid method with short local timesteps.

In our final example we test the validity of PEACEMAN (1988) approximate method for representing the interference effect when several wells are included in the same grid cell. Our case had 24 wells in one block of 1600 by 1600m. Using local grid refinement the wells were resolved areally within an 11 by 11 local grid. Five new layers were also introduced to accurately represent the depths at which the wells are perforated. The global grid is shown in Figure 16. The areal and vertical refinements are shown in Figures 17 and 18 respectively. Figure 19 shows the production profile and Figure 20 compares the pressure in the refined block using Peaceman's correction with that obtained using local grid refinement. The results show the validity of Peaceman's approximation in this case. For cases where the aspect ratio of the block does not warrant the use of Peaceman's approximation, or where several phases are present it is expected that grid refinement will give results which are numerically more correct.

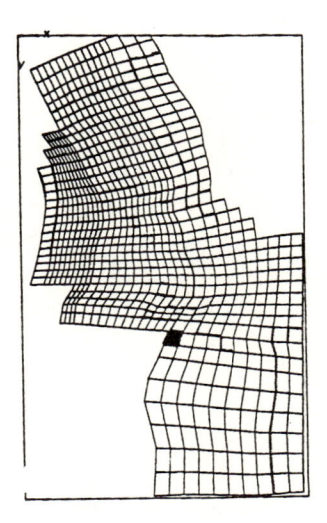

Figure 16: Well interference, simulation grid for full field: the block with 5 effective wells in the gas region is marked.

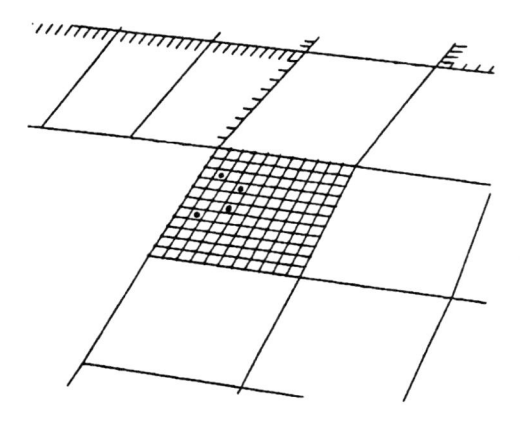

Figure 17:   Well interference, XY cross section of the grid with refined blocks: the refinement has 11x11x5 blocks.

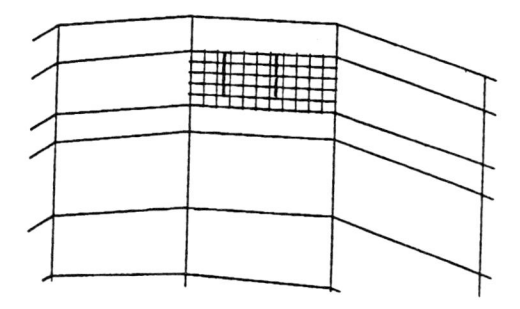

Figure 18:   Well interference, YZ cross section of the grid with refined blocks: the refinement has 5 extra layers in the second layer of the global grid. The wells are perforated in the refined layers 1 to 4.

—— Locally refined grid
---- Normal grid and well interference correction from Peaceman

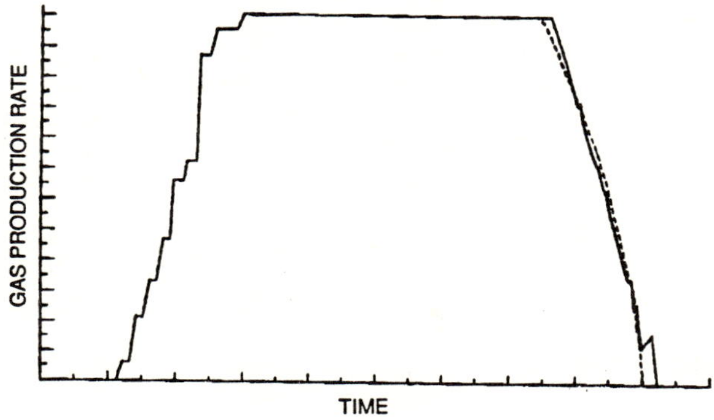

Figure 19: Well interference, gas production for the 5 well centre.

—— Locally refined grid
---- Normal grid and well interference correction from Peaceman

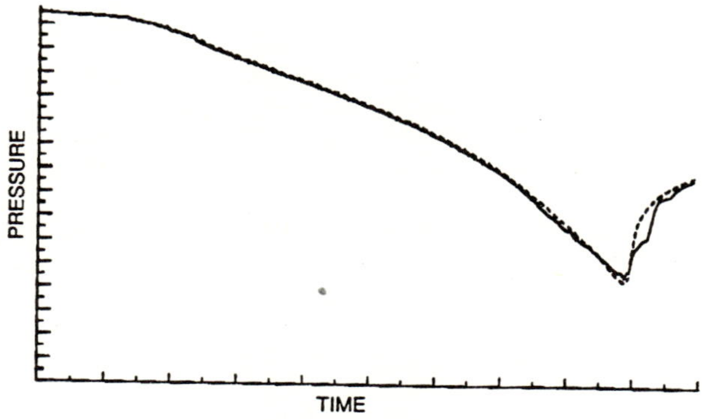

Figure 20: Well interference, pressure in the refined block.

8. CONCLUSIONS

1. The hybrid method offers a practical solution to the local grid refinement problem in field scale applications.
2. Local time stepping is easily incorporated into the hybrid method and has proved to be more efficient than alternative methods.
3. Material balance corrections are necessary in the practical application of the hybrid method.
4. Global timesteps are smaller using the hybrid method than for comparable runs without local grid refinement.
5. Well pseudos have proved to be a useful extension to the hybrid method.
6. The automatic dynamic generation of pseudos for globally refined cells may improve the technique further and remove the timestep restriction on the global solution.

9. ACKNOWLEDGEMENTS

The authors would like to thank Oystein Lie of Statoil who performed the simulation runs and Jon Holmes of ECL who coded the well model in the ECLIPSE local grid option. This work has been supported financially by Statoil, Norsk Hydro and Saga and the permission of these companies to publish the results is gratefully acknowledged.

10. REFERENCES

1. HEINEMANN Z.E., GERKIN G., VON HANDELMAN G., (November 1983), Using Local Grid Refinement in a Multiple Application Reservoir Simulator, SPE12255, presented at the Seventh SPE Symposium on Reservoir Simulation, San Francisco California.

2. MROSOVSKY I. and RIDINGS R.L., (January 1973), Two-Dimensional Radial Treatment of Wells within a Three-Dimensional Reservoir Model, SPE4286, presented at the Third Numerical Simulation Conference in Houston Texas

3. EWING R.E., BOYETT B.A., BABU D.K., HEINEMANN R.F., (February 1989). Efficient Use of Locally Refined Grids for Multipurpose Reservoir Simulation, SPE18413, presented at the Tenth SPE Symposium on Reservoir Simulation, Houston, Texas.

4. CHESHIRE I.M., APPLEYARD J.R., BANKS D., CROZIER R.J., and HOLMES J.H., (October 1980). An Efficient Fully Implicit Simulator, EUR179, presented at the European Offshore Petroleum Conference, London.

5. QUANDALLE P., and BESSET P., (February 1985). Reduction of Grid Effects due to Local Sub-Gridding in Simulations using a Composite Grid, SPE13527, presented at the Eighth SPE Symposium on Reservoir Simulation, Dallas Texas.

6. PEDROSA O.A., AZIZ K., (February 1985). Use of Hybrid Grid In Reservoir Simulation, SPE13507, presented at the Eighth SPE Symposium on Reservoir Simulation, Dallas Texas.

7. APPLEYARD J.R., CHESHIRE I.M., (November 1983). Nested Factorization, SPE12264 presented at the Seventh SPE Symposium on Reservoir Simulation, San Francisco California.

8. QUANDALLE P., BESSET P., (November 1983). The Use of Flexible Gridding for Improved Reservoir Modeling, SPE12239, presented at the Seventh SPE Symposium on Reservoir Simulation, San Francisco California.

9. KOSSAK C.A. and KLEPPE J. Oil Production from the Troll Field: A Comparison of Horizontal and Vertical Wells, SPE16869

10. PEACEMAN D.W., (September 1988). Near Singularities of Pressure and Concentration of the Wellbore in Reservoir Simulation, presented at the First International Forum on Reservoir Simulation, Alpbach, Austria.

ANALYSIS OF A MODEL AND SEQUENTIAL NUMERICAL METHOD
FOR THERMAL RESERVOIR SIMULATION

John A. Trangenstein
(Lawrence Livermore National Laboratory, U.S.A.)

ABSTRACT

In this paper we present and analyze a model for two-component, three-phase, non-isothermal fluid flow in a one-dimensional fluid reservoir. We discuss the thermodynamic principles that constrain the model functions, and analyze the effect of these thermodynamic principles on the flow equations. This analysis allows us to formulate a sequential approach to steam flooding: first a parabolic equation is solved to find the pressure and total fluid velocity, then a system of hyperbolic conservation laws is solved to update the fluid composition and energy. The thermodynamic principles allow us to compute the characteristic speeds and directions in the component/energy conservation equations, and to use these in a second-order Godunov method.

1. INTRODUCTION

Steam injection is the most widely-used and most successful tertiary oil recovery technique in the United States. A large number of the American proven oil reserves involve moderate to heavy oils, in which the mobility is greatly increased as heat is transferred from steam and hot water injection. The reasons for the engineering success of this technique also contribute to the difficulties in simulating the thermal recovery process. First of all, the mobility of steam is much greater than that of oil, leading to large variations in the characteristic speeds of the flow and making it difficult to choose stable explicit timesteps. The greater compressibility of steam leads to difficulty in determining the fluid pressure near phase changes. Third, sharp fronts lead to convergence problems in

---

[0]This work was performed under the auspices of the U.S. Department of Energy by the Lawrence Livermore National Laboratory under contract No. W-7405-Eng-48.

phase behaviour computations. Finally, the problems with heat
losses to the surrounding rock matrix create problems for both
the engineering and the simulation: heat losses reduce the heat
transferred to the resident oil, and require complicated energy
sink terms in the simulator that serve to spread the fronts.

The standard numerical approach for numerical simulation of hot
water and steam injection is to use a fully implicit treatment
of temporal differences, coupled with upstream weighting for
spatial differences [2,7,16]. However, the use of fully
implicit differences can lead to a larger amount of numerical
viscosity than corresponding explicit differences, and requires
the solution of large linear systems as part of the iterative
process in solving the nonlinear system for the updated flow
variables. These linear systems typically are very poorly
conditioned, partly because the unknowns are of radically
different types, and because the flow equations are partly
parabolic and partly hyperbolic. Often, fully implicit
simulation begs the use of fewer timesteps than desired for
accuracy considerations, in order to keep the cost of the total
simulation within acceptable limits. Recently, adaptive
implicit methods have been designed to reduce the computational
cost (see, for example, [11]), but these techniques typically
are implicit at the fronts, with the attendant smearing due to
fully implicit discretization, and involve a discontinuous
transition from implicit to explicit differencing.

Our goal in this paper is to take an initial step in developing
new numerical methods for the simulation of thermal recovery.
We begin by analyzing the thermodynamic and mathematical
structure of the flow equations, in order to use the analysis
to develop higher-order difference schemes appropriate to the
different types of partial differential equations in the
system. For the parabolic pressure equation we use a
lowest-order mixed finite element method, equivalent to a
block-centred finite difference scheme. For the
component/energy conservation equations, we use the
characteristic structure of the flow equations to
upstream-centre the difference scheme via a second-order
Godunov method. The characteristic analysis can also be used to
develop an adaptive implicit finite difference method, similar
to the techniques in [12].

The outline of this paper is as follows. In §2, we present a
simple two-component, three-phase model for the phase behaviour
of the reservoir fluid, and discuss the solution to phase
equilibrium. This gives us the basis for a discussion of the

thermodynamics of multi-phase mixtures and first-degree
homogeneity principles in §3. Afterward, we present the flow
equations in §4, as well as our sequential approach. In §5, we
show that the pressure equation is parabolic, and in §6 we use
the results of §3 to show that the component/energy
conservation equations are hyperbolic, by finding the
characteristic speeds and directions. This allows us to discuss
our numerical methods briefly in §7.

2. TWO-COMPONENT, THREE-PHASE FLUID MODEL

We shall consider a reservoir containing three components (oil,
water and rock) in as many as four phases (liquid, vapor, aqua
and solid). Only the liquid, vapor and aqueous phases are
allowed to be mobile. Furthermore, oil is the only component in
the liquid phase, water is the only component in the vapour and
aqueous phases, and rock is the only component in the solid
phase. Obviously, this is a gross simplification of the complex
chemistry and phase transfer that occurs in real reservoirs.
However, this model does contain a number of interesting
complications due to Gibbs phase rule. Also, we will be able to
present succinct thermodynamic principles that underly the
correct specification of this model.

Because we allow water to exist in two phases, Gibbs phase rule
allows that these phases coexist only if pressure and
temperature are not independent. Thus, there must be a
boiling-point curve, which we take in the form $T = T_b(p)$. Next,
we take the phase mass densities ρ_l, ρ_v and ρ_a, as well as the
enthalpies h_l, h_v, h_a and h_s, to be functions of pressure and
temperature. In order to simplify the model, we shall assume
that rock is incompressible. This means that the rock density
ρ_s is constant, and that porosity ϕ is constant.

In order to write the phase equilibrium problem in a concise
mathematical form, we will introduce the following notation.
The vector h of phase enthalpies and diagonal matrix R of phase
densities are defined by

$$h(p,T) = \begin{bmatrix} h_l(p,T) \\ h_v(p,T) \\ h_a(p,T) \end{bmatrix}, \; R(p,T) = \begin{bmatrix} \rho_l(p,T) & 0 & 0 \\ 0 & \rho_v(p,T) & 0 \\ 0 & 0 & \rho_a(p,T) \end{bmatrix}.$$

We shall write e for a generic vector of ones. From these, we
define the specific internal energies of the fluid phases and

of the solid phase by

$$u(p,T) = h(p,T) - R^{-1}(p,T)ep \;,\;\; u_s(p,T) = h_s(p,T) - \frac{p}{\rho_s}\;.$$

The vector n of component masses / pore volume, and matrix N of phase component masses / pore volume are defined by

$$n = \begin{bmatrix} n_o \\ n_w \end{bmatrix}, \; N = \begin{bmatrix} n_{ol} & 0 & 0 \\ 0 & n_{wv} & n_{wa} \end{bmatrix}.$$

We shall also have use for the constant matrices

$$X = \begin{bmatrix} 1 & 0 & 0 \\ 0 & 1 & 1 \end{bmatrix}, \; Y = \begin{bmatrix} 1 & 0 & 0 \\ 0 & 1 & 0 \end{bmatrix}, \; Z = \begin{bmatrix} 0 & 0 & 0 \\ 0 & 1 & -1 \end{bmatrix},$$

and the diagonal matrix

$$D_n = \begin{bmatrix} n_o & 0 \\ 0 & n_w \end{bmatrix}.$$

With these definitions, we can state the phase equilibrium problem for our model. Given the internal energy / bulk volume v, the component masses / pore volume n and the pressure p, we want to compute the temperature T and phase compositions N so that

1. mass is balanced:
$$n = Ne\,,$$

2. internal energy is balanced:
$$v = e^T N u(p,T)\phi + \rho_s u_s(p,T)(1-\phi)\,, \qquad (1)$$

3. the phase compositions are nonnegative:
$$0 \le N^T e\,,$$

4. and finally, vapour and aqua coexist only at the bubble-point:

   if $0 < n_{wa} < n_w$ then $T = T_b(p)$.

The steps in the solution to the phase equilibrium problem are straightforward. Since oil is found only in the liquid phase, $n_{ol} = n_o$. If the fluid were saturated, then we would have $T = T_b(p)$ and
$$N = D_n Y - Z n_{wa}^*\,,$$

where n_{wa}^* is defined implicitly by (1). Since p is given, we can solve this linear equation for n_{wa}^* to obtain

$$n_{wa}^*(n,p,v) = -\frac{v - n^T Y u(p,T_b(p))\phi - \rho_s u_s(p,T_b(p))(1-\phi)}{e^T Z u(p,T_b(p))\phi}. \quad (2)$$

If $0 \leq n_{wa}^* \leq n_w$ then the fluid is saturated, and $T = T_b(p)$, $n_{wa} = n_{wa}^*$ and $n_{wv} = n_w - n_{wa}^*$. On the other hand, if $n_{wa}^* > n_w$ then there is no vapour phase. In this case we define

$$\bar{R}(p,T) = \begin{bmatrix} \rho_l(p,T) & 0 \\ 0 & \rho_a(p,T) \end{bmatrix}, \bar{h}(p,T) = \begin{bmatrix} h_l(p,T) \\ h_a(p,T) \end{bmatrix},$$

and $\bar{u}(p,T) = \bar{h}(p,T) - \bar{R}^{-1}(p,T)ep$. It is easy to see that the phase compositions for this case are given by $n_{wv} = 0$, $n_{wa} = n_w$. It is also easy to see that the temperature is defined implicitly by the equation

$$v = n^T \bar{u}(p,T)\phi + \rho_s u_s(p,T)(1-\phi). \quad (3)$$

Inequalities (5) and (7) below will guarantee that this equation has a unique solution for $T \ll T_b(p)$. Finally, if $n_{wa}^* < 0$ then there is no aqua. In this case we define

$$\bar{R}(p,T) = \begin{bmatrix} \rho_l(p,T) & 0 \\ 0 & \rho_v(p,T) \end{bmatrix}, \bar{h}(p,T) = \begin{bmatrix} h_l(p,T) \\ h_v(p,T) \end{bmatrix}.$$

Then we find that $n_{wv} = n_w$, $n_{wa} = 0$ and that the temperature is again defined implicitly by (3).

3. THERMODYNAMICS AND FIRST-DEGREE HOMOGENEITY

In the previous section we discussed the relations between the pressure, energy and fluid component masses on one hand, and the temperature and phase component masses on the other. The solutions we found are not yet guaranteed to exist and be unique. In order to pose a realistic phase equilibrium problem, we must impose several physical constraints on the model functions. Other constraints will be needed to guarantee that fluid flow is realistic.

Obviously, all phase densities should be positive, as should porosity. Note that at the bubble point steam should be less dense than condensate; thus we require

$$\rho_v(p,T_b(p)) < \rho_a(p,T_b(p)). \quad (4)$$

Also, because of the latent heat of vaporization, the internal energy of steam should be greater than that of condensate; as a result we also require

$$u_a(p,T_b(p)) < u_v(p,T_b(p)). \quad (5)$$

This implies that the denominator in (2) is positive.

Since $\frac{\partial h}{\partial T}$ and $\frac{\partial h_s}{\partial T}$ are the heat capacities at constant pressure, we must have
$$\frac{\partial h}{\partial T} \geq 0 \,,\quad \frac{\partial h_s}{\partial T} \geq 0 \,. \tag{6}$$
Since the specific internal energy of each phase in our model is a function of p and T only, the specific internal energy must increase with temperature and decrease with pressure [8, p.136]:
$$\frac{\partial u}{\partial T} \geq 0 \,,\, \frac{\partial u_s}{\partial T} \geq 0 \,, \tag{7}$$
$$\frac{\partial u}{\partial p} \leq 0 \,,\, \frac{\partial u_s}{\partial p} \leq 0 \,. \tag{8}$$
These inequalities are sufficient to guarantee that the saturated and undersaturated phase equilibrium problems are mutually exclusive, and that the undersaturated phase equilibrium problem has a unique solution.

Next, we consider the more difficult issue of mechanical stability, which requires that the total fluid volume must decrease as the pressure increases, while n and T are fixed. For either case of undersaturation (no vapour or no aqua), the vector of existing phase volumes / pore volume is given by
$$\bar{w}(n,p,T) = \bar{R}^{-1}(p,T)\,n \,. \tag{9}$$
Then the pressure derivative of the total fluid volume / pore volume, for fixed p and T, is
$$\frac{\partial e^\mathsf{T} \bar{w}}{\partial p} = e^\mathsf{T} \frac{\partial \bar{R}^{-1}}{\partial p} n \,.$$
By varying n, we see that we must require $\frac{\partial \bar{R}^{-1}}{\partial p}e < 0$. This is equivalent to the following set of inequalities:
$$0 \leq \frac{\partial \rho_l}{\partial p} \,,\text{ for all } T\,;$$
$$0 \leq \frac{\partial \rho_v}{\partial p} \,,\text{ for } T > T_b(p)\,; \tag{10}$$
$$0 \leq \frac{\partial \rho_a}{\partial p} \,,\text{ for } T < T_b(p) \,.$$

In the saturated case, a different approach to mechanical stability is necessary. The difficulty is that Gibbs phase rule forces T to be a function of p, so the usual notion of mechanical stability is of no good to us. We have not been successful in finding references in the thermodynamic literature to mechanical stability (or the equivalent notion) for single-component, two-phase fluids. Instead, we have experimentally found that along the bubble-point curve the total fluid volume decreases as pressure increases, while the component masses/pore volume n and *internal energy* v are held fixed. As we will show, this reduces to a collection of inequalities, analogous to (10).

Recall that phase equilibrium has already determined $T(p)$ and $N(n, p, v)$. We define the array of phase volumes / pore volume by

$$w(n, p, v) = R^{-1}(p, T(p))N^T(n, p, v)e .\tag{11}$$

In order to determine the pressure derivatives of the phase volumes, we must first compute the pressure derivative of n_{wa}, using (2):

$$\frac{\partial n_{wa}}{\partial p} = \frac{1}{e^T Z u \phi}\left\{\phi e^T N\left[\frac{\partial u}{\partial p} + \frac{\partial u}{\partial T}\frac{\partial T_b}{\partial p}\right] + (1-\phi)\rho_s\left[\frac{\partial u_s}{\partial p} + \frac{\partial u_s}{\partial T}\frac{\partial T_b}{\partial p}\right]\right\} . \tag{12}$$

Next we can use the identity $N = D_n Y - Z n_{wa}$ to find that

$$-e^T Z u \phi \frac{\partial e^T w}{\partial p} = e^T Z u \phi \left\{ e^T \left[\frac{\partial R}{\partial p} + \frac{\partial R}{\partial T}\frac{\partial T_b}{\partial p}\right] R^{-2} N^T e + e^T R^{-1} Z^T e \frac{\partial n_{wa}}{\partial p}\right\}$$

$$= \phi w^T \left\{R^{-1}\left[\frac{\partial R}{\partial p} + \frac{\partial R}{\partial T}\frac{\partial T_b}{\partial p}\right]ee^T Z u + R\left[\frac{\partial u}{\partial p} + \frac{\partial u}{\partial T}\frac{\partial T_b}{\partial p}\right]e^T R^{-1} Z^T e\right\}$$

$$+ \rho_s(1-\phi)\left[\frac{\partial u_s}{\partial p} + \frac{\partial u_s}{\partial T}\frac{\partial T_b}{\partial p}\right]e^T R^{-1} Z^T e .$$

Since (5) guarantees that $e^T Z u = u_v - u_a > 0$, we want to guarantee that the right-hand side of this equation is positive. Note that we can vary parameters to obtain arbitrary values of w and $\rho_s(1-\phi)$. Thus, in order to guarantee that the total fluid volume decreases as pressure increases (for fixed n and v, and for $T = T_b(p)$), we must require that for all p,

$$0 \le R^{-1}\left[\frac{\partial R}{\partial p} + \frac{\partial R}{\partial T}\frac{\partial T_b}{\partial p}\right]ee^T Z u + R\left[\frac{\partial u}{\partial p} + \frac{\partial u}{\partial T}\frac{\partial T_b}{\partial p}\right]e^T R^{-1} Z^T e , \tag{13}$$

and

$$0 \le \frac{\partial u_s}{\partial p} + \frac{\partial u_s}{\partial T}\frac{\partial T_b}{\partial p} . \tag{14}$$

Again, we mention that these inequalities have been verified experimentally for the model functions used in this work. We would expect that in a more realistic thermal model, the vapor phase would contain water and hydrocarbons, so Gibbs phase rule would not cause these complications. We also note that although the vapor density increases as pressure increases along the bubble-point curve, the aqueous density *decreases* and the liquid density first decreases and then increases. Thus we could not have asked that $\frac{\partial e_v^T w}{\partial p} < 0$ for constant n and n_{wa}.

Another important thermodynamic principle involves first-degree homogeneity, which can be expressed through the partial derivatives of the phase volumes and temperature. Let us consider the partial derivatives of w for a saturated fluid at fixed pressure. Thus $T = T_b(p)$ is fixed. We can invert the definition (11) of w and internal energy balance (1) to write

$$n(w, p) = XR(p, T_b(p))w,$$

$$v(w, p) = w^T R(p, T_b(p)) u(p, T_b(p)) \phi + \rho_s u_s(p, T_b(p))(1 - \phi).$$

The chain rule for partial derivatives of w in terms of itself gives us

$$I = \begin{bmatrix} \frac{\partial w}{\partial n}, & \frac{\partial w}{\partial v} \end{bmatrix} \begin{bmatrix} \frac{\partial n}{\partial w} \\ \frac{\partial v}{\partial w} \end{bmatrix} = \begin{bmatrix} \frac{\partial w}{\partial n\phi}, & \frac{\partial w}{\partial v} \end{bmatrix} \begin{bmatrix} XR \\ u^T R \end{bmatrix} \phi. \quad (15)$$

Note that the matrix of partial derivatives of w is 3 x 3, so this expression tells us the inverse of the matrix of partial derivatives of the phase volumes.

Next, we recall that the solution to undersaturated phase equilibrium gave us $N(n)$ and $T(n, p, v)$, after which we computed $\bar{w}(n, p, v)$. Again, we invert to find

$$n(\bar{u}, p, T) = \bar{R}(p, T)\bar{w},$$

$$v(\bar{w}, p, T) = \bar{w}^T \bar{R}(p, T)\bar{u}(p, T)\phi + \rho_s u_s(p, T)(1 - \phi).$$

Then the chain rule for partial derivatives of $\bar{u}$ and T as functions of themselves gives us

$$\begin{bmatrix} I & 0 \\ 0 & 1 \end{bmatrix} = \begin{bmatrix} \frac{\partial \bar{w}}{\partial \bar{u}} & \frac{\partial \bar{w}}{\partial v} \\ \frac{\partial T}{\partial n} & \frac{\partial T}{\partial v} \end{bmatrix} \begin{bmatrix} \frac{\partial n}{\partial \bar{w}} & \frac{\partial n}{\partial T} \\ \frac{\partial v}{\partial \bar{w}} & \frac{\partial v}{\partial T} \end{bmatrix}$$

$$= \begin{bmatrix} \frac{\partial \bar{w}}{\partial n\phi} & \frac{\partial \bar{w}}{\partial v} \\ \frac{\partial T}{\partial n\phi} & \frac{\partial T}{\partial v} \end{bmatrix} \begin{bmatrix} \bar{R}\phi & \frac{\partial \bar{R}}{\partial T}\bar{w}\phi \\ \phi\bar{u}^T\bar{R} & \phi\bar{w}^T\frac{\partial \bar{R}\bar{u}}{\partial T} + \rho_s \frac{\partial u_s}{\partial T}(1 - \phi) \end{bmatrix}. \quad (16)$$

We will use these relations (15) and (16) in §6 to determine the characteristic structure of the component/energy conservation laws.

4. FLOW EQUATIONS AND THE SEQUENTIAL APPROACH

At this point in the paper, we have described the phase equilibrium problem, and determined thermodynamic conditions

that guarantee its unique solution. In this section, we will
build on these results by describing the equations that
determine fluid flow, and outlining an approach for solving the
system of equations. Our goals are to determine the
mathematical structure of the flow equations and to separate
them according to type, so that we can apply numerical methods
that are appropriate to each.

Our first flow equation is volume balance, which requires that
the sum of the phase volumes must equal the pore volume:

$$e^{\mathsf{T}} w(n,p,v) = 1 \quad \text{or} \quad e^{\mathsf{T}} \bar{w}(n,p,v) = 1 \tag{17}$$

for saturated or undersaturated fluids, respectively. These
equations can be viewed as determining the pressure as a
function of the fluid component masses and internal energy. By
linearizing these equations we will develop a parabolic
equation for the time evolution of the pressure, as part of our
sequential approach to the flow equations. We also note that if
volume balance is satisfied then the entries of w are the phase
saturations. However, we will allow volume discrepancy errors
to occur in the sequential approach; this requires that we
distinguish w, for which the entries may not sum exactly to
one, from the entries of the saturation vector s, defined by

$$s = \frac{w}{e^{\mathsf{T}} w} \quad \text{or} \quad \bar{s} = \frac{\bar{w}}{e^{\mathsf{T}} \bar{w}}$$

for saturated or undersaturated fluids, respectively.

The next step in our discussion of the flow equations is to
describe Darcy's law. For simplicity, we will limit the
description to saturated flow. We assume that we are given
relative permeability functions $\kappa_l(s)$, $\kappa_v(s)$ and $\kappa_a(s)$. These
functions must be nonnegative, and must be zero when the
corresponding phase saturation is zero. Note that these very
reasonable physical requirements are not necessarily satisfied
by Stone's second model [22], or by Baker's model [9]. (Our
choice for these functions is described in the Appendix.) We
also assume that we are given positive viscosity functions
$\mu_l(p,T)$, $\mu_v(p,T)$ and $\mu_a(p,T)$. From these functions, we determine
the diagonal matrix of phase mobilities

$$\Lambda(s,p,T) = \begin{bmatrix} \frac{\kappa_l(s)}{\mu_l(p,T)} & 0 & 0 \\ 0 & \frac{\kappa_v(s)}{\mu_v(p,T)} & 0 \\ 0 & 0 & \frac{\kappa_a(s)}{\mu_a(p,T)} \end{bmatrix}.$$

Then Darcy's law states that the vector of phase velocities is
given by

$$v = -\Lambda \left[e \frac{\partial p}{\partial x} - Re\delta \right] \kappa. \tag{18}$$

Here, κ is the total permeability, which is taken to be a function of position x, and δ is the gravitational acceleration times the depth gradient, also a function of position.

The second flow equation is component/energy conservation, which says that the mass of each component and the total energy must be conserved:

$$\frac{\partial}{\partial t}\begin{bmatrix} n\phi \\ v \end{bmatrix} + \frac{\partial}{\partial x}\begin{bmatrix} X \\ h^\mathsf{T} \end{bmatrix} R\mathsf{v} = 0. \tag{19}$$

This form of component/energy conservation is correct only in the absence of diffusion terms and source terms (such as wells and heat losses to the surrounding rock). It also assumes that the kinetic energy of the fluid can be ignored, and that no chemical reactions occur.

Our system of equations is now complete. It consists of phase equilibrium, Darcy's law, volume balance and component/energy conservation. We shall use phase equilibrium as a system of constraints to determine N, T and w (or $\bar{w}$) as functions of n, p and v. Next, we shall linearize volume balance to derive an evolution equation for pressure in a saturated fluid:

$$\frac{e^\mathsf{T} w - 1}{\Delta t}\phi \approx -\phi e^\mathsf{T}\frac{\partial w}{\partial p}\frac{\partial p}{\partial t} - \phi e^\mathsf{T}\left[\frac{\partial w}{\partial n\phi}, \frac{\partial w}{\partial v}\right]\frac{\partial}{\partial t}\begin{bmatrix} n\phi \\ v \end{bmatrix}$$

$$= -\phi e^\mathsf{T}\frac{\partial w}{\partial p}\frac{\partial p}{\partial t} + \phi e^\mathsf{T}\left[\frac{\partial w}{\partial n\phi}, \frac{\partial w}{\partial v}\right]\frac{\partial}{\partial x}\begin{bmatrix} XR\mathsf{v} \\ h^\mathsf{T} R\mathsf{v} \end{bmatrix}$$

$$= -\phi e^\mathsf{T}\frac{\partial w}{\partial p}\frac{\partial p}{\partial t} \tag{20}$$

$$+\phi e^\mathsf{T}\left[\frac{\partial w}{\partial n\phi}, \frac{\partial w}{\partial v}\right]\frac{\partial}{\partial x}\left\{\begin{bmatrix} X \\ h^\mathsf{T} \end{bmatrix}R\Lambda\left(e\frac{\mathsf{v}_T}{e^\mathsf{T}\Lambda e} + \left[I - \frac{ee^\mathsf{T}\Lambda}{e^\mathsf{T}\Lambda e}\right]Re\delta\kappa\right)\right\},$$

where the total fluid velocity is defined by

$$\mathsf{v}_T = e^\mathsf{T}\mathsf{v} = \left[-\frac{\partial p}{\partial x} + \frac{e^\mathsf{T}\Lambda Re}{e^\mathsf{T}\Lambda e}\delta\right]e^\mathsf{T}\Lambda e\kappa. \tag{21}$$

In our sequential approach, we use the system (20) and (21) in those regions of the reservoir with three fluid phases, in order to determine pressure and total fluid velocity. Here, we consider n and v to be independent of p.

In order to emphasize mechanical stability, we shall employ a slightly different linearization in the undersaturated case. Here, we consider the phase volumes / pore volume $\bar{w}$ to be functions of n, p and T, and recall that phase equilibrium determined T as a function of p, n and v. This allows us to derive

$$\frac{e^{\mathsf{T}}\bar{w}-1}{\Delta t}\phi \approx -\phi e^{\mathsf{T}}\frac{\partial \bar{w}}{\partial p}\frac{\partial p}{\partial t} - \phi e^{\mathsf{T}}\left[\frac{\partial \bar{w}}{\partial n\phi}, \frac{\partial \bar{w}}{\partial T}\right]\frac{\partial}{\partial t}\left[\begin{array}{c}n\phi \\ T\end{array}\right]$$

$$= -\phi e^{\mathsf{T}}\frac{\partial \bar{w}}{\partial p}\frac{\partial p}{\partial t} - \phi e^{\mathsf{T}}\left[\frac{\partial \bar{w}}{\partial n\phi}, \frac{\partial \bar{w}}{\partial T}\right]\left[\begin{array}{c}\frac{\partial n\phi}{\partial t} \\ \frac{\partial T}{\partial p}\frac{\partial p}{\partial t} + \frac{\partial T}{\partial n\phi}\frac{\partial n\phi}{\partial t} + \frac{\partial T}{\partial v}\frac{\partial v}{\partial t}\end{array}\right]$$

$$= -\phi\left[e^{\mathsf{T}}\frac{\partial \bar{w}}{\partial p} + e^{\mathsf{T}}\frac{\partial \bar{w}}{\partial T}\frac{\partial T}{\partial p}\right]\frac{\partial p}{\partial t} - \phi e^{\mathsf{T}}\left[\frac{\partial \bar{w}}{\partial n\phi} + \frac{\partial \bar{w}}{\partial T}\frac{\partial T}{\partial n\phi}, \frac{\partial \bar{w}}{\partial T}\frac{\partial T}{\partial v}\right]\frac{\partial}{\partial t}\left[\begin{array}{c}n\phi \\ v\end{array}\right]$$

$$= -\phi\left[e^{\mathsf{T}}\frac{\partial \bar{w}}{\partial p} + e^{\mathsf{T}}\frac{\partial \bar{w}}{\partial T}\frac{\partial T}{\partial p}\right]\frac{\partial p}{\partial t} + \phi e^{\mathsf{T}}\left[\frac{\partial \bar{w}}{\partial n\phi} + \frac{\partial \bar{w}}{\partial T}\frac{\partial T}{\partial n\phi}, \frac{\partial \bar{w}}{\partial T}\frac{\partial T}{\partial v}\right]\frac{\partial}{\partial x}\left[\begin{array}{c}\bar{R}\bar{v} \\ \bar{h}^{\mathsf{T}}\bar{R}\bar{v}\end{array}\right]$$

$$= -\phi e^{\mathsf{T}}\left[\frac{\partial \bar{w}}{\partial p} + \frac{\partial \bar{w}}{\partial T}\frac{\partial T}{\partial p}\right]\frac{\partial p}{\partial t} \quad (22)$$

$$+\phi e^{\mathsf{T}}\left[\frac{\partial \bar{w}}{\partial n\phi} + \frac{\partial \bar{w}}{\partial T}\frac{\partial T}{\partial n\phi}, \frac{\partial \bar{w}}{\partial T}\frac{\partial T}{\partial v}\right]\frac{\partial}{\partial x}\left\{\left[\frac{\bar{R}}{\bar{h}^{\mathsf{T}}\bar{R}}\right]\bar{\Lambda}\left(e\frac{v_T}{e^{\mathsf{T}}\Lambda e} + \left[I - \frac{ee^{\mathsf{T}}\bar{\Lambda}}{e^{\mathsf{T}}\Lambda e}\right]\bar{R}e\delta\kappa\right)\right\}.$$

Here, v_T is defined by (21) with the obvious modifications. The principal difference between the saturated and undersaturated cases involves the variables being held fixed while the phase volumes / pore volume are varied with pressure. However, the end result is the same: the coefficient of $\frac{\partial p}{\partial t}$ is again the pressure derivative of the total fluid volume, for constant n and v.

After the pressure equation has been solved, we use the results to rewrite the phase velocity vector in terms of the total fluid velocity:

$$\mathbf{v} = \Lambda\left\{e\frac{v_T}{e^{\mathsf{T}}\Lambda e} + \left[I - \frac{ee^{\mathsf{T}}\Lambda}{e^{\mathsf{T}}\Lambda e}\right]Re\delta\kappa\right\}, \quad (23)$$

and use this form in the component/energy flux. We then solve (19) for n and v, treating p and v_T as independent of n and v.

5. PARABOLICITY OF THE PRESSURE EQUATION

In the previous section, we described the sequential approach to the flow equations. We found an evolution equation for pressure and total fluid velocity, and a conservation law for component mass and energy. In this section, we will analyze the pressure equation to show that it is parabolic. In the next section, we will perform a characteristic analysis to show that the component/energy conservation law is hyperbolic.

First, we consider the saturated case, in which we consider w to be a function of n, p and v. By inserting the definition

(21) of the total fluid velocity into the pressure equation (20), we obtain the following equation for p:

$$\frac{e^T w - 1}{\Delta t}\phi \approx -\phi e^T \frac{\partial w}{\partial p}\frac{\partial p}{\partial t} + \phi e^T \left[\frac{\partial w}{\partial n\phi}, \frac{\partial w}{\partial v}\right]\frac{\partial}{\partial x}\left\{\begin{bmatrix} X \\ h^T \end{bmatrix} R\Lambda \left(-e\frac{\partial p}{\partial x} + Re\delta\right)\kappa\right\}.$$

Note that this equation involves a first-order derivative of pressure with respect to time, and a second-order derivative with respect to space. The coefficient of the time derivative is minus the pressure derivative of the total fluid volume while n and v are fixed, and is positive because of inequalities (13) and (14) imposed on the model functions in §3 above. The coefficient of the second-order spatial derivative is seen from (15) to be

$$-\phi e^T \left[\frac{\partial w}{\partial n\phi}, \frac{\partial w}{\partial v}\right]\begin{bmatrix} X \\ h^T \end{bmatrix} R\Lambda e\kappa = -\phi e^T \left[\frac{\partial w}{\partial n\phi}, \frac{\partial w}{\partial v}\right]\left(\begin{bmatrix} XR \\ u^T R \end{bmatrix} + \begin{bmatrix} 0 \\ pe^T \end{bmatrix}\right)\Lambda e\kappa$$

$$= -e^T \Lambda e\kappa (1 + e^T \frac{\partial w}{\partial v}\phi p).$$

In order to show that the pressure equation is parabolic and properly posed for increasing time, we need only show that $1 + e^T \frac{\partial w}{\partial v}\phi p$ is positive. Since

$$w = R^{-1}N^T e = R^{-1}(D_n Y - Z n_{wa})^T e,$$

and

$$\frac{\partial n_{wa}}{\partial v} = \frac{1}{e^T Z u \phi},$$

we have

$$\frac{\partial e^T w}{\partial v} = \frac{\partial e^T w}{\partial n_{wa}}\frac{\partial n_{wa}}{\partial v} = \frac{e^T R^{-1} Z^T e}{e^T Z u \phi} = \frac{\frac{1}{\rho_v} - \frac{1}{\rho_a}}{u_v - u_a} \geq 0,$$

by virtue of inequalities (4) and (5). This shows that the pressure equation is properly parabolic in the saturated case.

In the undersaturated cases, we take $\bar{w}$ to be a function of n, p and T, where T is that function of n, p and v defined by phase equilibrium. Then

$$\frac{e^T \bar{w} - 1}{\Delta t}\phi \approx -\phi e^T \left[\frac{\partial \bar{w}}{\partial p} + \frac{\partial \bar{w}}{\partial T}\frac{\partial T}{\partial p}\right]\frac{\partial p}{\partial t}$$

$$+\phi e^T \left[\frac{\partial \bar{w}}{\partial n\phi} + \frac{\partial \bar{w}}{\partial T}\frac{\partial T}{\partial n\phi}, \frac{\partial \bar{w}}{\partial T}\frac{\partial T}{\partial v}\right]\frac{\partial}{\partial x}\left\{\begin{bmatrix} \bar{R} \\ h^T \bar{R} \end{bmatrix} \bar{\Lambda}\left(-e\frac{\partial p}{\partial x} + \bar{R}e\delta\right)\kappa\right\}.$$

Note here that (3) shows that

$$\frac{\partial T}{\partial v} = \frac{1}{n^T \frac{\partial \bar{u}}{\partial T}\phi + \rho_s \frac{\partial u_s}{\partial T}(1 - \phi)},$$

and then that

$$\frac{\partial T}{\partial n\phi} = -\frac{\partial T}{\partial v}\bar{u}^T,$$

and finally that

$$\frac{\partial T}{\partial p} = -\frac{\partial T}{\partial v}\left[n^\top \frac{\partial \bar{u}}{\partial p}\phi + \rho_s(1-\phi)\frac{\partial u_s}{\partial p}\right].$$

These inequalities show that the coefficient of the second-order spatial derivative is

$$-\phi e^\top \left[\frac{\partial \bar{w}}{\partial n\phi} + \frac{\partial \bar{w}}{\partial T}\frac{\partial T}{\partial n\phi}, \frac{\partial \bar{w}}{\partial T}\frac{\partial T}{\partial v}\right]\left[\frac{\bar{R}}{\bar{h}^\top \bar{R}}\right]\bar{\Lambda}e\kappa$$

$$= -\phi e^\top \left[I\frac{1}{\phi} + \frac{\partial \bar{R}}{\partial T}\bar{R}^{-2}n\frac{\partial T}{\partial v}(\bar{u}-\bar{h})^\top \bar{R}\right]\bar{\Lambda}e\kappa$$

$$= -e^\top \bar{\Lambda}e\kappa(1 - e^\top \frac{\partial \bar{R}}{\partial T}\bar{R}^{-2}n\frac{\partial T}{\partial v}\phi p)$$

$$= -e^\top \bar{\Lambda}e\kappa \frac{\phi n^\top \frac{\partial \bar{u}}{\partial T} + \rho_s \frac{\partial u_s}{\partial T}(1-\phi) - \phi n^\top \bar{R}^{-2}\frac{\partial \bar{R}}{\partial T}ep}{n^\top \frac{\partial \bar{u}}{\partial T}\phi + \rho_s \frac{\partial u_s}{\partial T}(1-\phi)}$$

$$= -e^\top \bar{\Lambda}e\kappa \frac{\phi n^\top \frac{\partial \bar{h}}{\partial T} + \rho_s \frac{\partial u_s}{\partial T}(1-\phi)}{n^\top \frac{\partial \bar{u}}{\partial T}\phi + \rho_s \frac{\partial u_s}{\partial T}(1-\phi)}.$$

This expression is negative by virtue of inequalities (6) and (7).

The coefficient of the time derivative of p consists of two pieces. The first, $-e^\top \frac{\partial \bar{w}}{\partial p}$, is positive because of mechanical stability. However, parabolicity requires that the sum of the two terms be positive:

$$0 \leq -e^\top \frac{\partial \bar{w}}{\partial p} - e^\top \frac{\partial \bar{w}}{\partial T}\frac{\partial T}{\partial p}$$

$$= \frac{n^\top \bar{R}^{-2}\left[\frac{\partial \bar{R}}{\partial p}e\phi \frac{\partial \bar{u}^\top}{\partial T} - \frac{\partial \bar{R}}{\partial T}e\phi \frac{\partial \bar{u}^\top}{\partial p}\right]n + n^\top \bar{R}^{-2}\left[\frac{\partial \bar{R}}{\partial p}e\phi \frac{\partial u_s}{\partial T} - \frac{\partial \bar{R}}{\partial T}e\phi \frac{\partial u_s}{\partial p}\right]\rho_s(1-\phi)}{n^\top \frac{\partial \bar{u}}{\partial T}\phi + \rho_s \frac{\partial u_s}{\partial T}(1-\phi)}.$$

The denominator in this expression is positive, because of (7). By varying parameters, we find that proper parabolicity requires that the symmetric part of

$$\bar{R}^{-2}\left[\frac{\partial \bar{R}}{\partial p}e\frac{\partial \bar{u}^\top}{\partial T} - \frac{\partial \bar{R}}{\partial T}e\frac{\partial \bar{u}^\top}{\partial p}\right]$$

have nonnegative entries, and that

$$\bar{R}^{-2}\left[\frac{\partial \bar{R}}{\partial p}e\frac{\partial u_s}{\partial T} - \frac{\partial \bar{R}}{\partial T}e\frac{\partial u_s}{\partial p}\right] \geq 0.$$

These inequalities are equivalent to

$$\frac{\partial \rho_l}{\partial p}\frac{\partial h_l}{\partial T} - \frac{\partial \rho_l}{\partial T}\left[\frac{\partial h_l}{\partial p} - \frac{1}{\rho_l}\right] \geq 0, \text{ for all } T;$$

$$\frac{\partial \rho_v}{\partial p}\frac{\partial h_v}{\partial T} - \frac{\partial \rho_v}{\partial T}\left[\frac{\partial h_v}{\partial p} - \frac{1}{\rho_v}\right] \geq 0, \text{ for all } T > T_b(p);$$

$$\frac{\partial \rho_a}{\partial p}\frac{\partial h_a}{\partial T} - \frac{\partial \rho_a}{\partial T}\left[\frac{\partial h_a}{\partial p} - \frac{1}{\rho_a}\right] \geq 0, \text{ for all } T < T_b(p); \quad (24)$$

$$\frac{1}{\rho_v^2}\left[\frac{\partial \rho_v}{\partial p}\frac{\partial h_l}{\partial T} - \frac{\partial \rho_v}{\partial T}\left(\frac{\partial h_l}{\partial p} - \frac{1}{\rho_l}\right)\right]$$
$$+\frac{1}{\rho_l^2}\left[\frac{\partial \rho_l}{\partial p}\frac{\partial h_v}{\partial T} - \frac{\partial \rho_l}{\partial T}\left(\frac{\partial h_v}{\partial p} - \frac{1}{\rho_v}\right)\right] \geq 0, \text{ for all } T > T_b(p);$$

$$\frac{1}{\rho_a^2}\left[\frac{\partial \rho_a}{\partial p}\frac{\partial h_l}{\partial T} - \frac{\partial \rho_a}{\partial T}\left(\frac{\partial h_l}{\partial p} - \frac{1}{\rho_l}\right)\right]$$
$$+\frac{1}{\rho_l^2}\left[\frac{\partial \rho_l}{\partial p}\frac{\partial h_a}{\partial T} - \frac{\partial \rho_l}{\partial T}\left(\frac{\partial h_a}{\partial p} - \frac{1}{\rho_a}\right)\right] \geq 0, \text{ for all } T < T_b(p);$$

and
$$\frac{\partial \rho_l}{\partial p}\frac{\partial h_s}{\partial T} - \frac{\partial \rho_l}{\partial T}\left[\frac{\partial h_s}{\partial p} - \frac{1}{\rho_s}\right] \geq 0, \text{ for all } T;$$

$$\frac{\partial \rho_v}{\partial p}\frac{\partial h_s}{\partial T} - \frac{\partial \rho_v}{\partial T}\left[\frac{\partial h_s}{\partial p} - \frac{1}{\rho_s}\right] \geq 0, \text{ for all } T > T_b(p); \quad (25)$$

$$\frac{\partial \rho_a}{\partial p}\frac{\partial h_s}{\partial T} - \frac{\partial \rho_a}{\partial T}\left[\frac{\partial h_s}{\partial p} - \frac{1}{\rho_s}\right] \geq 0, \text{ for all } T > T_b(p).$$

Thus, if inequalities (24) and (25) hold, then the pressure equation is properly parabolic. Note that a sufficient set of conditions for (24) and (25) to hold is that for each phase the density should increase with pressure and decrease with temperature, and $\rho\frac{\partial h}{\partial p}$ should be greater than 1.

6. HYPERBOLICITY OF THE COMPONENT/ENERGY CONSERVATION LAWS

Our next task is to show that the component/energy conservation equations are hyperbolic. We shall do so by constructing the characteristic speeds and directions. As in the analysis of the pressure equation, it is necessary to consider the saturated and undersaturated cases separately.

Let us begin with the saturated case. Recall that in the sequential approach we consider p and v_T to be functions only of x and t during the evolution of the conservation law. Also recall from phase equilibrium that $T = T_b(p)$, so the densities R and specific internal energies u are functions of x only. Only the phase velocities v, defined by (21), depend on n and v, this through the saturation dependence of the mobilities. Note that v also depends on p, v_T and δ. We can use the fact that the total fluid velocity is independent of s to derive

$$h^T R \frac{\partial v}{\partial s} = (u + R^{-1}ep)^T R \frac{\partial v}{\partial s} = u^T R \frac{\partial v}{\partial s}.$$

Now, we consider the conserved quantities and the flux to be functions of w and p, in order to derive

$$\frac{\partial}{\partial t}\begin{bmatrix} n\phi \\ v \end{bmatrix} = \begin{bmatrix} XR \\ u^T R \end{bmatrix}\frac{\partial w}{\partial t}\phi + \frac{\partial}{\partial p}\begin{bmatrix} n\phi \\ v \end{bmatrix}\frac{\partial p}{\partial t},$$

and

$$\frac{\partial}{\partial x}\begin{bmatrix} XRv \\ h^T Rv \end{bmatrix} = \begin{bmatrix} XR \\ u^T R \end{bmatrix}\frac{\partial v}{\partial s}\frac{\partial s}{\partial w}\frac{\partial w}{\partial x}$$
$$+\frac{\partial}{\partial p}\begin{bmatrix} XRv \\ h^T Rv \end{bmatrix}\frac{\partial p}{\partial x} + \frac{\partial}{\partial v_T}\begin{bmatrix} XRv \\ h^T Rv \end{bmatrix}\frac{\partial v_T}{\partial x} + \frac{\partial}{\partial \delta}\begin{bmatrix} XRv \\ h^T Rv \end{bmatrix}\frac{\partial \delta}{\partial x}.$$

At this point, we need only use first-degree homogeneity (15) to discover that

$$0 = \begin{bmatrix} \frac{\partial w}{\partial n\phi}, & \frac{\partial w}{\partial v} \end{bmatrix}\left\{\frac{\partial}{\partial t}\begin{bmatrix} n\phi \\ v \end{bmatrix} + \frac{\partial}{\partial x}\begin{bmatrix} XRv \\ h^T Rv \end{bmatrix}\right\}$$

$$= \frac{\partial w}{\partial t} + \frac{\partial v}{\partial s}\frac{\partial s}{\partial w}\frac{\partial w}{\partial x}\frac{1}{\phi} + \begin{bmatrix} \frac{\partial w}{\partial n\phi}, & \frac{\partial w}{\partial v} \end{bmatrix}\frac{\partial}{\partial p}\begin{bmatrix} n\phi \\ v \end{bmatrix}\frac{\partial p}{\partial t} \qquad (26)$$

$$+\begin{bmatrix} \frac{\partial w}{\partial n\phi}, & \frac{\partial w}{\partial v} \end{bmatrix}\left(\frac{\partial}{\partial p}\begin{bmatrix} XRv \\ h^T Rv \end{bmatrix}\frac{\partial p}{\partial x} + \frac{\partial}{\partial v_T}\begin{bmatrix} XRv \\ h^T Rv \end{bmatrix}\frac{\partial v_T}{\partial x} + \frac{\partial}{\partial \delta}\begin{bmatrix} XRv \\ h^T Rv \end{bmatrix}\frac{\partial \delta}{\partial x}\right).$$

This has the form of a first-order system of evolution equations for w, with source terms due to the spatial and temporal variations of p, v_T and δ. The hyperbolicity of the system is determined by the linearized coefficient matrix

$$\frac{\partial v}{\partial s}\frac{\partial s}{\partial w}\frac{1}{\phi} = \frac{\partial v}{\partial s}(I - se^T)\frac{1}{\phi e^T w}.$$

This matrix is also the linearized coefficient matrix for three-phase Buckley-Leverett flow. In practice, it has real eigenvalues and eigenvectors for *almost all* saturations in three-phase flow; however, it has real eigenvalues and eigenvectors for *all* saturations only under special circumstances [6,13,21,23]. We have chosen Corey-type relative permeabilities for this work (see the Appendix) so that we are guaranteed to have real eigenvalues, even under the influence of gravity [18].

Next, we turn to the undersaturated characteristic analysis. Recall that we can consider n and v to be functions of $\bar{w}$, T and p. The phase velocities $\bar{v}$ depend on these variables, as well as

v_T and δ. The variables p, v_T and δ are in turn functions of x and t. This allows us to derive

$$\frac{\partial}{\partial t}\begin{bmatrix} n\phi \\ v \end{bmatrix} = \begin{bmatrix} \bar{R}\phi & \frac{\partial \bar{R}}{\partial T}\bar{w}\phi \\ \phi \bar{u}^T \bar{R} & \phi \bar{w}^T \frac{\partial \bar{R}\bar{u}}{\partial T} + \rho_s \frac{\partial u_s}{\partial T}(1-\phi) \end{bmatrix} \frac{\partial}{\partial t}\begin{bmatrix} \bar{w} \\ T \end{bmatrix} + \frac{\partial}{\partial p}\begin{bmatrix} n\phi \\ v \end{bmatrix}\frac{\partial p}{\partial t},$$

and

$$\frac{\partial}{\partial x}\begin{bmatrix} \bar{R}\bar{v} \\ \bar{h}^T \bar{R}\bar{v} \end{bmatrix} = \begin{bmatrix} \bar{R}\frac{\partial \bar{v}}{\partial \bar{s}}\frac{\partial \bar{s}}{\partial \bar{w}} & \frac{\partial \bar{R}\bar{v}}{\partial T} \\ \bar{u}^T \bar{R}\frac{\partial \bar{v}}{\partial \bar{s}}\frac{\partial \bar{s}}{\partial \bar{w}} & \frac{\partial \bar{h}^T \bar{R}\bar{v}}{\partial T} \end{bmatrix} \frac{\partial}{\partial x}\begin{bmatrix} \bar{w} \\ T \end{bmatrix}$$

$$+\frac{\partial}{\partial p}\begin{bmatrix} \bar{R}\bar{v} \\ \bar{h}^T \bar{R}\bar{v} \end{bmatrix}\frac{\partial p}{\partial x} + \frac{\partial}{\partial v_T}\begin{bmatrix} \bar{R}\bar{v} \\ \bar{h}^T \bar{R}\bar{v} \end{bmatrix}\frac{\partial v_T}{\partial x} + \frac{\partial}{\partial \delta}\begin{bmatrix} \bar{R}\bar{v} \\ \bar{h}^T \bar{R}\bar{v} \end{bmatrix}\frac{\partial \delta}{\partial x}.$$

Using first-degree homogeneity, the conservation law can be written

$$0 = \frac{\partial}{\partial t}\begin{bmatrix} \bar{w} \\ T \end{bmatrix} + \begin{bmatrix} \frac{\partial \bar{v}}{\partial \bar{s}}\frac{\partial \bar{s}}{\partial \bar{w}}\frac{1}{\phi} & \bar{R}^{-1}(\frac{\partial \bar{R}\bar{v}}{\partial T} - \frac{\partial \bar{R}}{\partial T}\bar{w}\alpha\phi)\frac{1}{\phi} \\ 0 & \alpha \end{bmatrix}\frac{\partial}{\partial x}\begin{bmatrix} \bar{w} \\ T \end{bmatrix}$$

$$+ \begin{bmatrix} \frac{\partial \bar{w}}{\partial n\phi} & \frac{\partial \bar{w}}{\partial v} \\ \frac{\partial T}{\partial n\phi} & \frac{\partial T}{\partial v} \end{bmatrix}\frac{\partial}{\partial p}\begin{bmatrix} n\phi \\ v \end{bmatrix}\frac{\partial p}{\partial t} \qquad (27)$$

$$+ \begin{bmatrix} \frac{\partial \bar{w}}{\partial n\phi} & \frac{\partial \bar{w}}{\partial v} \\ \frac{\partial T}{\partial n\phi} & \frac{\partial T}{\partial v} \end{bmatrix}\left(\frac{\partial}{\partial p}\begin{bmatrix} \bar{R}\bar{v} \\ \bar{h}^T \bar{R}\bar{v} \end{bmatrix}\frac{\partial p}{\partial x} + \frac{\partial}{\partial v_T}\begin{bmatrix} \bar{R}\bar{v} \\ \bar{h}^T \bar{R}\bar{v} \end{bmatrix}\frac{\partial v_T}{\partial x} + \frac{\partial}{\partial \delta}\begin{bmatrix} \bar{R}\bar{v} \\ \bar{h}^T \bar{R}\bar{v} \end{bmatrix}\frac{\partial \delta}{\partial x}\right),$$

where

$$\alpha = \frac{\frac{\partial \bar{u}^T}{\partial T}\bar{R}\bar{v}}{\frac{\partial \bar{u}^T}{\partial T}\bar{R}\bar{w}\phi + \frac{\partial u_s}{\partial T}\rho_s(1-\phi)}.$$

These results show that the characteristic speeds are either the weighted particle velocity α, or are eigenvalues derived from two-phase Buckley-Leverett flow. Like the polymer model [1,15,20], undersaturated flow with this model has an eigenvector deficiency when α coincides with one of the Buckley-Leverett wavespeeds.

7. COMPUTATIONAL APPROACH

In the preceding sections of the paper, we described the solution of the phase equilibrium and guaranteed its unique solution by means of thermodynamic constraints on the model functions. We also outlined the sequential splitting of the flow equations into a pressure equation and component/energy conservation laws, and then showed that the pressure equation is properly parabolic, while the component/energy conservation laws are hyperbolic, subject to careful selection of the three-phase relative permeability model. This gives us the firm foundation needed to develop a numerical method for the solution of the equations of thermal fluid flow. Unfortunately, limitations of space prohibit us from describing our proposed numerical method in detail.

We would begin each timestep by computing the phase compositions and temperature satisfying phase equilibrium. In addition, we would compute the phase volumes / pore volume, volume discrepancy error, saturations, first-degree homogeneity arrays, partial derivatives of the flux and perform the characteristic analysis. All of this is straightforward, given the analysis above.

The next step would be to solve the pressure equation for the pressure and total fluid velocity. Note that the pressure equation (20) and the total fluid velocity equation (21) are in the form described in [1] for the general structure of pressure equations in compressible fluid flow; thus the description of block-centred finite differences (or, equivalently, the lowest-order mixed finite element method) is applicable to the problem in this paper. This method is second-order in space and first-order in time, the latter because the coefficients of the pressure equation are evaluated only at the previous time level. The assumption is that tertiary recovery is generally engineered in order to maintain pressure levels in the reservoir; thus the pressure is slowly varying in time, when compared to the component masses and fluid energy. For more details, see the cited lecture notes.

We must remark that the greatest difficulties in this numerical approach seem to be due to the linearization of volume balance to form the pressure equation. It is likely that steam flooding would produce much larger volume discrepancy errors than in black-oil simulations. This suggests that an iteration to resolve the volume discrepancy errors may be needed. One possibility is to use the first pass through the pressure equation and conservation laws to get first-order accuracy in time for p, and then to use these results as a predictor for a second-order accurate correction. It may even be possible that further iteration on the corrector is needed.

The remaining step is to solve the component/energy conservation laws. Here, we would use the second-order Godunov techniques in [1,4].

9. SUMMARY

One of the most frustrating aspects of dealing with reservoir
fluid flow is that there are so few models that are completely
and properly described in the literature. When we worked
previously on the black-oil model, we had the benefit of access
to practical but proprietary models, from which we could
extract sanitized but thermodynamically-consistent examples.
Here, we had no such information to draw upon. The correlations
published in the literature, such as those in [10], are not
designed to be used directly in reservoir simulators; they
merely reproduce certain limited aspects of the steam tables.
Perhaps one of the most useful accomplishments of this work has
been to present a consistent model, together with the
thermodynamic principles constraining it. We believe that the
modifications we have made to the published correlations
subtract little from their accuracy, while contributing greatly
to their thermodynamic consistency.

The other principal accomplishment in this paper has been to
examine the mathematical structure of the flow equations. The
thermal model provides extra evidence for the postulate that
three-phase flow produces characteristic speeds related to the
three-phase Buckley-Leverett model, while two-phase flow
produces one Buckley-Leverett speed and one speed that is a
weighted sum of particle velocities. While this observation has
been supported by all of the models we have examined so far
[1,3,5,4,25,24], the thermodynamic and mathematical principles
used to discover the characteristic speeds have been different
in each case. It would be interesting for a student, presumably
better-versed in thermodynamics, to determine the unifying
thermodynamic principles that decide the characteristic
structure of the conservation laws of multi-phase flow in
porous media.

APPENDIX

Throughout this work we have adopted a consistent set of units:
distance in **metres**, time in seconds, mass in kilograms, energy
in megajoules, viscosity in millipascal-seconds, and
temperature in degrees Kelvin. These imply that velocities are
in metres/second, density is in kilograms/cubic metre, pressure
is in megapascals, enthalpy is in megajoules/kilogram,
permeability is in 10^{-9} square metres, and the gravitational
force is in 10^6 metres/second squared. We fix $\phi = 0.1$,
$\rho_s = 2650 kg/m^3$ and $\kappa = .5$ Darcies $= 4.9349 10^{-4} m^2$.

The first of our functions is the bubble-point curve, taken from [10]:
$$T_b(p) = 180.89 p^{0.2350} + 273.15.$$
Note that $T_b(p)$ is strictly increasing, so we may invert it to find the bubble-point pressure given temperature.

We also need to describe the mass densities of the phases. The density of vapour is given by the ideal gas law, with a correction factor to match the correlation in [10] at the bubble point:
$$\frac{p}{T_b^{-1}(T)} \frac{1}{\rho_v(p,T)} = \begin{cases} .211075/p - 0.00294, & p \le 10.2 \\ .237483/p - 0.005537, & p > 10.2 \end{cases}$$
From the steam tables, we found that the compressibility of water is approximately $4.15 10^{-4}\text{MPa}^{-1}$, and assumed this to be constant in modifying the correlations in [17]:
$$\rho_a(p,T) = \frac{A_0 + A_1 T_C + A_2 T_C^2 + A_3 T_C^3 + A_4 T_C^4 + A_5 T_C^5}{1 + A_6 T_C} \exp(4.15\,10^{-4}(p-10.2)),$$
where
$$A_0 = 999.83952,\ A_1 = 16.945176,\ A_2 = -7.987\,10^{-3},\ A_3 = -46.170461\,10^{-6},$$
$$A_4 = 105.56302\,10^{-9},\ A_5 = -280.54253\,10^{-12},\ A_6 = 16.87985\,10^{-3},$$
and $T_C = T - 273.15$ is the temperature in centigrade. Similarly, the density of oil was computed by performing a polynomial fit to predictions by a Soave-Redlich-Kwong equation of state correlation for nC_{20}, and assuming a constant compressibility of approximately $4.35 10^{-3}\text{MPa}^{-1}$:
$$\rho_l(p,T) = \left[1132.59 + T\left(-.750946 + 3.37391\,10^{-4} T\right)\right] \exp(4.35\,10^{-3}(p - 10.2)).$$

We took the enthalpy of solid to be
$$h_s(T) = 4.7\,10^{-6}(t - 485.57),$$
as suggested by [26]. The enthalpy of saturated vapour was taken from [10] and an enthalpy departure function based on a constant heat capacity of 2.5 btu / lbmole °F was included:
$$h_v(p,T) = 2.8024 - 2.6265\,10^{-3}(p - 3.1249)^{1.73808} + 1.046\,10^{-2}(T - T_b(p)).$$

For aqua, we used the correlations in [10] for both the enthalpy of saturated condensate and for the bubble-point temperature to rewrite the function in terms of the temperature:
$$h_a(p,T) = \begin{cases} 0.04061(T_b^{-1}(T) - 10.2) + 0.732898(10.2)^{0.28302} + 0.002p, & T \ge T_b(10.2) \\ 0.732878(T_b^{-1}(T))^{0.28302} + 0.002p, & T < T_b(10.2) \end{cases}$$

Here, it was necessary to add a small dependence on pressure in order to satisfy inequalities (24). Finally, the enthalpy of liquid was found by correlation with a Soave-Redlich-Kwong equation of state predictions to be

$$h_l(p,T) = \left[(4.4343\ 10^{-10}T - 6.3563\ 10^{-7})T + 7.7331\ 10^{-4}\right]T - 0.18104 + p*.0014\ .$$

Again, a small pressure dependence was necessary in order to make the pressure equation properly parabolic. With these choices of the densities and enthalpies, the inequalities of this paper were satisfied for $3.4 < p < 17.2$, and $300 < T < 600$.

The liquid viscosity was taken from [26]:

$$\mu_l(p,T) = 0.132683\exp(821.29/T)\ .$$

The vapour viscosity is taken from [10]:

$$\mu_v(p,t) = 0.0049402 + 5.0956\ 10^{-5}T_C + 2.92233\ 10^{-6}p^{2.5077}$$

The viscosity of water is due to Miller [19]:

$$\mu_a(p,T) = \exp(\exp((-0.057139\ln(T_F) + 0.30841)\ln(T_F) + 1.3926))/208.9\ ,$$

where $T_F = 1.8(T - 273.15) + 32$ is the temperature in Fahrenheit. Finally, we chose the relative permeabilities to be of Corey-type:

$$\kappa_l(s) = s_l^2\ ,\ \kappa_v(s) = s_v^2\ ,\ \kappa_a(s) = s_a^2\ .$$

ACKNOWLEDGMENTS

We would like to thank Ananda Wiejesinghe, Gary Pope and Ken Brantferger for help in getting this project off the ground. A special note of thanks goes to Ken Brantferger for using his equation of state algorithms to compute realistic correlations when none could be found. Ken was also very useful in supplying thermodynamic insight into the initial modelling crisis.

References

[1] Allen, M.B. III, Behie, A. and Trangenstein, J.A. [1988], *Multi-Phase Flow in Porous Media: Mechanics, Mathematics and Numerics*, volume 34 of *Lecture Notes in Engineering*, Springer-Verlag.

[2] Aziz, K., Ramesh, B. and Woo, P.T. [1985], Fourth SPE comparative solution project: A comparison of steam injection simulators. In *Proceedings of the Eighth SPE Symposium on Reservoir Simulation*, 441-454.

[3] Bell, J.B. and Shubin, G.R. [1985], Higher order Godunov methods for reducing numerical dispersion in reservoir simulation, In *Eighth SPE Symposium on Reservoir Simulation*, Dallas.

[4] Bell, J.B., Colella, P. and Trangenstein, J.A. [1989], Higher-order Godunov methods for general systems of hyperbolic conservation laws, *J. Comp. Phys.*, 82:362-397.

[5] Bell, J.B., Shubin, G.R. and Trangenstein, J.A. [1986], A method for reducing numerical dispersion in two-phase black-oil reservoir simulation, *J. Comp. Phys.*, 65:71-106.

[6] Bell, J.B., Trangenstein, J.A. and Shubin, G.R. [1986], Conservation laws of mixed type describing three-phase flow in a porous media. *SIAM J. Appl Math.*, 46:1000-1017.

[7] Coats, K.H. [1978], A highly implicit steamflood model, *Soc. Pet. Eng. J.*, 18:369-383.

[8] Crawford, F.H. and van Vorst, W.D. [1968], *Thermodynamics for Engineers*, Harcourt, Brace and World.

[9] Delshad, M. and Pope, G.A. [submitted 1988], Comparison of the three-phase relative permeability models. *Transport in Porous Media*.

[10] Ejiogu, G.C. and Fiori, M. [1987], High-pressure saturated steam correlations, *J. Pet. Tech.*, 39:1585-1590.

[11] Forsyth, P.A. [1989], Adaptive implicit criteria for two-phase flow with gravity and capillary pressure, *SIAM J. Sci. Stat. Comput.*, 10:227-252.

[12] Fryxel, B.A., Woodward, P.R., Colella, P. and Winkler, K.-H. [1986], An implicit-explicit hybrid method for lagrangian hydrodynamics, *J. Comp. Phys.*, 63:283-310.

[13] Holden, L. [1988], On the strict hyperbolicity of the Buckley-Leverett equations for three-phase flow in a porous medium, Norwegian Computing Center, P.O. Box 114 Blindern, 0314 Oslo 3, Norway.

[14] Holing, K. [1990], *A Conservative Front-Tracking Method for Two-Dimension Polymer Flooding*, PhD thesis, Norges Tekniske Hgskole Trondheim.

[15] Isaacson, E., Marchesin, D. and Plohr, B. [unpublished], The structure of the Riemann solution for non-strictly hyperbolic conservation laws, preprint.

[16] Ishimoto, K. [1985], One-dimensional fully implicit compositional model for steam flooding, Master's thesis, Department of Petroleum Engineering, University of Texas at Austin.

[17] Kell, G.S. [1975], Density, thermal expansivity, and compressibility of liquid water from 0.deg. to 150.deg.. correlations and tables for atmospheric pressure and saturation reviewed and expressed on 1968 temperature scale, *Journal of Chemical and Engineering Data*, 20:97-105.

[18] Marchesin, D. and Medeiros, H.B. [Maio, 1988], A note on gravitational effects in multiphase flow, Relatorio de Pesquisa e Desenvolvimento, Departamento de Matematica, Pontificia Universidade Catolica do Rio de Janeiro, Rua Marquês de Sao Vicente, 225-CEP 22453, Rio de Janeiro – Brasil.

[19] Marx, J.W. and Langenheim, R.H. [1959], Reservoir heating by hot fluid injection, *Trans. AIME*, 216:312.

[20] Pope, G.A. [1980], The application of fractional flow theory to enhanced oil recovery, *S.P.E.J.*, 20:191-205.

[21] Shearer, M. and Trangenstein, J.A. [1989], Loss of real characteristics for models of three-phase flow in a porous medium. *Transport in Porous Media*, 4:499-525.

[22] Stone, H.L. [1973], Estimation of three-phase relative permeability and residual oil data, *J. Cdn. Pet.*, 12:53-61.

[23] Trangenstein, J.A. [1989], Three-phase flow with gravity, *Contemporary Mathematics* 100:147-159.

[24] Trangenstein, J.A. and Bell, J.B. [1989], Mathematical structure of compositional reservoir simulation, *SIAM J. Sci. Stat. Comput.*, 10:817-845.

[25] Trangenstein, J.A. and Bell, J.B. [1989], Mathematical structure of the black-oil model for petroleum reservoir simulation, *SIAM J. Appl. Math.*, 49:749-783.

[26] Wingard, J.S. [1988], *Multicomponent, Multiphase Flow in Porous Media with Temperature Variation*, PhD thesis, Department of Petroleum Engineering, Stanford University.

Operator Splitting and Domain Decomposition Techniques for Reservoir Flow Problems

Helge K. Dahle   Magne S. Espedal
Øystein Pettersen   Ove Sævareid

Dept. of Applied Mathematics,
University of Bergen, Norway

Abstract

An algorithm for solving the (near) hyperbolic saturation equation of reservoir flow dynamics numerically is described. The procedure is split into several sub-problems, namely operator splitting, tracking of discontinuous modes, and diffusion correction. To resolve rapid variations and reduce processor time, grid refinement, domain decomposition, and parallel computations have been used.

1 Introduction

Miscible or immiscible reservoir flow problems are often characterized by a discontinuous or near-discontinuous concentration or saturation profile. Most commercial simulators perform badly when calculating the position and shape of this profile, with unsatisfactory prediction of breakthrough and production of invading fluid as a result. This problem is particularly noticed in heterogeneous layered rocks, which frequently occur in North Sea reservoirs.

In order to reduce or eliminate undesired numerical errors, such as smearing of rapidly changing solutions, false oscillations of Gibbs type, and grid orientation effects, the problem of numerical treatment of fluid front movement has attracted considerable interest during the last decade.

For pure hyperbolic problems the most significant contributions have been the Uniform Sampling Method for 1-dimensional problems including systems (Concus, Proskurowski 1979); the Front Tracking Algorithm developed at the Courant Institute (Glimm et. al. 1983), and the Method of

Modified Characteristics introduced in this context by Douglas and Russell (1982).

Being based on characteristics of hyperbolic problems, these methods cannot always be extended to diffusive flow in a straightforward manner. On the other hand, classical methods for diffusion equations lack the ability to capture shocks which is inherent in the tailor-made methods. In order to retain the features of hyperbolic methods and still allow for diffusion we have chosen to handle the diffusion term as a correction to pure convective flow by an operator splitting procedure. When the nonlinear convection is split into a transport and a balance term, build-up of false discontinuities is prevented, allowing for large time steps.

When diffusion effects are reproduced on a finite grid one has to take consistency into consideration. On mesh sizes typically used in reservoir simulation the diffusion will in general be exaggerated. On the other hand a global sufficiently small grid size that can reproduce both small-scale front movement and a diffused front is prohibited by the processor time that would result. Frontal movement in porous rocks is a typical example of a localizable problem; rapid changes occur in the vicinity of the front, whereas the main part of the computational area is characterized by slow variation of solution data. The problem is therefore well suited for local grid refinement, concentrating effort to the domains where needed. Further, the algorithm lends itself to domain decomposition and parallel computations on multi-processor computers.

1.1 Model Equations

Saturation Equation:

$$\frac{\partial S}{\partial t} + \nabla \cdot \mathbf{u} F(S) = \varepsilon \nabla \cdot \{D(S)\nabla S\}, \tag{1}$$

Pressure/velocity Equation:

$$\nabla \cdot \mathbf{u} = -\nabla \cdot \{a(S) \cdot \nabla p\} = q(\mathbf{x}, t), \tag{2}$$

where S is saturation (or concentration) of invading fluid, $\mathbf{u}$ is the total filtration velocity, the nonlinear convective term $F(S)$ is the common fractional flow function (ratio of invading fluid to total fluid mobilities, possibly with addition of a gravity term), ε is the inverse Peclet number, and $D(S)$ is the diffusion function. p is total fluid pressure, as defined by Chavent (1981). For a more detailed description of the model system, see e.g. Ewing (1983).

When $\varepsilon \ll 1$, eqn (1) is almost hyperbolic, degenerating to a pure convective flow problem for $\varepsilon = 0$.

Concentrating on the saturation equation, our objective is to resolve the (near) discontinuous fluid/fluid interface characterized by a rapid change in S over a short interval for vanishing to moderate diffusion. Moreover, we want to advance the solution with time steps greatly exceeding the Courant-Friedrichs-Lewy-condition (as measured on the smallest scale involved). To achieve these goals we will utilize the following tools:

- Operator Splitting
- Specialized Hyperbolic Solver
- Adaptive Local Grid Refinement
- Domain Decomposition
- Parallel Computations

Each of these steps will be described separately in the following.

2 Operator Splitting

Following Espedal and Ewing (1987), we split the convective term in eqn (1) in two parts,

$$F(S) = F^*(S) + S\, b(S), \qquad (3)$$

where F^* is the concave or convex hull to F so that solutions to

$$\frac{\partial S}{\partial t} + \nabla \cdot \mathbf{u} F(S) = 0, \qquad (4)$$

with the entropy condition satisfied, and

$$\frac{\partial S}{\partial t} + \nabla \cdot \mathbf{u} F^*(S) = 0 \qquad (5)$$

are identical throughout the following time step.

For a decreasing profile:

$$F^* = \begin{cases} \dfrac{F_m - F_R}{S_m - S_R}(S - S_R) + F_R, & \text{if } S_R \le S \le S_m \\ F(S), & \text{if } S > S_m. \end{cases} \qquad (6)$$

Defining the splitting function $b(S)$ by

$$S\,b(S) = F(S) - F^*(S), \qquad (7)$$

and

$$\mathbf{b}(S,\mathbf{x}) = S\,b(S)\mathbf{u}, \qquad (8)$$

we may write the split saturation equation as

$$\underbrace{\frac{\partial S}{\partial t} + \nabla \cdot \mathbf{u} F^*(S)}_{A} = \underbrace{\varepsilon \nabla \cdot \{D(S)\nabla S\}}_{B} - \underbrace{\nabla \cdot \mathbf{b}(S,\mathbf{x})}_{C} \qquad (9)$$

with

A: Convective part (hyperbolic)

B: Capillary diffusion (elliptic behaviour)

C: Balance term.

The construction of the splitting function is depicted in fig.1 for a fully developed front and a typical shape of the fractional front function.

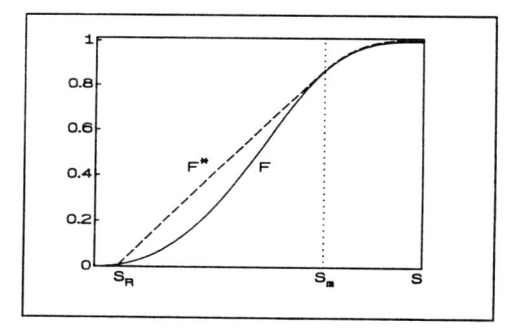

Figure 1: Splitting of fractional flow function

In practice the right state, identified by saturation S_R will be the minimal saturation S_{wc} of the invading fluid. For a fully developed front the left state S_L will exceed the Buckley-Leverett saturation, implying that the point S_m will be found where the tangent line to $F(S)$ through the point $(S_R, F(S_R))$ touches $F(S)$. Hence F^* will not change with time. During a period of frontal build-up we encounter the case $S_L < S_m$. The definition of F^* is unchanged, but since the front height will increase during the next time step, the splitting has to be re-evaluated at the beginning of each step until the front is fully developed. At out-flow boundaries or by front collisions the profile need not be decreasing in the flow direction. This complicates the splitting procedure, as another type of discontinuities is introduced, and the splitting defined by eqn (6) is no longer valid.

2.1 Splitting procedure

Assume S, $\mathbf{u}$, p have been computed at time t^n. To advance the solution one time step (see (Dahle, Espedal, Ewing 1988) for details),

1. Solve for intermediate solution $\bar{S}$:

$$\frac{\partial \bar{S}}{\partial \tau} = \frac{\partial \bar{S}}{\partial t} + \nabla \cdot \mathbf{u} F^* = 0 \qquad (10)$$

2. With initial data $\bar{S}$, solve

$$\frac{\partial S}{\partial \tau} - \varepsilon \nabla \cdot \{D(S)\nabla S\} + \nabla \cdot \mathbf{b}(S, \mathbf{x}) = 0, \qquad (11)$$

where

$$\frac{\partial S}{\partial \tau} = \frac{S^{n+1} - \bar{S}}{\Delta t} \qquad (12)$$

3 Solution of the hyperbolic part

We seek a solution procedure for the convective part of the problem that takes care of the highly irregular front shapes that can occur, especially for heterogeneous problems. At the same time the procedure should easily be adaptable to the diffusion correction part that is to follow. Whatever solution method is used, the front will be smeared or dispersed if it is interpolated to a fixed grid. This statement is valid even for locally exact solutions, which on a fixed grid will experience a smearing which is a function of time alone, independent of mesh or time step size. Something "special" is therefore called for to resolve the front.

Several methods for tracking of discontinuous fronts have been suggested during the last decade. We list some which we have investigated, together with advantages/disadvantages concerning our implementation of these:

Front Tracking (Glimm et. al. 1983).

This algorithm still gives the best front resolution for hyperbolic problems. An unfortunate feature for our purpose is the frontal grid, which captures irregular front shapes excellently, but cannot easily be expanded to a local mesh for the diffusive correction. It would be necessary to increase the number of nodal front points to an unmanagable level or work with unacceptable grid aspect ratios. Generalization of the method to three dimensions and/or systems presently lacks understanding.

Uniform Sampling (Random Choice) (Concus, Proskurowski 1979). The procedure has had great success for one dimensional problems, and is easily adapted to the diffusive part of the splitting. The mass balance problem (see below) is particularly serious for this method — the main reason for us to abandon it. Some attempts to generalize the algorithm to multi-dimensional problems by dimension-splitting have been proposed (Colella, Concus, Sethian 1983), but seem to lack the ability to resolve irregular fronts. For one-dimensional *systems* the Random Choice Method is, nevertheless, still a favourite.

Method of Modified Characteristics (Douglas, Russell 1982). Our present procedure is based on finite difference MOMC, which resolves the front reasonably well, is suitable for the splitting and grid refinement, and relatively simple to code in two or three dimensions. The method is, however, not easily extendable to systems.

Higher Order Godunov Methods (2-D) (Bell, Dawson, Shubin 1988) In order to increase robustness we are currently investigating Godunov type methods. For the convective part this procedure is less accurate than the other mentioned methods, but combined with diffusion splitting this inconvenience is to a large extent removed. The algorithm is extremely suitable for grid refinement and domain decomposition, and seems simple to extend.

3.1 Mass balance

Unfortunately, a characteristic feature of many specialized hyperbolic solvers, is lack of material balance. Since the balance is never used as a convergence criterion by these methods, as opposed to standard techniques, we have to accept that the increased front resolution is obtained at the expense of mass balance.

This fact will often imply rejection of the methods by reservoir managers, since conservation of mass will be looked upon as a necessity within any simulator. On the other hand we have often experienced that the forced balance criterion has deteriorated a solution, either by increasing smearing or by delivering unphysical data values.

Our attitude is that a minor deviation from exact mass balance can be accepted as a trade-off to obtain what we feel is a "better" solution.

4 Diffusion Correction

Equation (11) is solved along approximate characteristics, implying that transport is (almost) absent in the equation, hence the resulting problem is elliptic of nature.

By splitting the fractional flow function we have ensured that the established part of the front will be moved with constant velocity, and no "false" discontinuities will originate.

The b—term balances the established shock, but unfortunately also introduces a non-symmetric element in the equation. In order to proceed with a Petrov-Galerkin solution procedure, it is necessary to symmetrize the equation. This is done by the method developed by Barrett and Morton (1984).

5 Local Grid Refinement

With the tailormade hyperbolic solvers we can compute the front shape to an accuracy not resolvable on the coarse grid. As minor irregularities may grow with time it is vital to represent the front on a mesh that can handle small scale variations. However, a global fine grid is prohibited by the resulting processor time. Moreover, since the solution varies slowly on the major part of the computational grid, a uniform grid would mean a waste of processing power, as accuracy in smooth areas would be unnecessary high.

Ideally, the mesh should be constructed such that a globally uniform discretization error is obtained. Since this error is fairly proportional to $|\nabla^2 S| \cdot h^2$, where h is a typical grid block measure, the mesh size should be chosen inversely proportional to $|\nabla^2 S|$.

In our present algorithm, the decision to refine a coarse grid block is taken on basis of an explicit estimate of the front position at the following time step. A chosen block will be refined in a uniform and fixed manner, e.g. such that each refined coarse block will consist of a local 10 by 10 grid.

An example of a saturation front, depicted as level curves, for displacement in a heterogeneous reservoir, and the computational grid used at the present time step is shown in fig.2.A.

5.1 Pressure/velocity equation

Equation (2) has singular behaviour near wells, but is elsewhere smooth. The demand for consistent accuracy when computing the velocity field

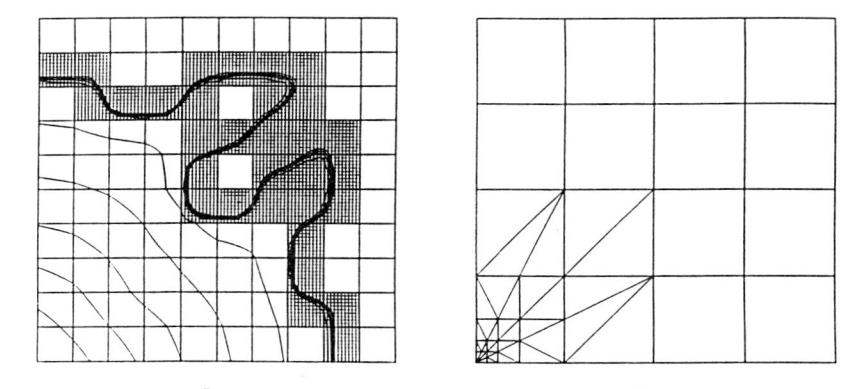

Figure 2: Grid Refinement: A. Grid with level curves for saturation solution. B. Near-well refinement for pressure solution.

has motivated the use of the method of mixed finite elements (Russell, Wheeler 1983) to solve eqn (2).

Alternatively, we have investigated the possibility to solve this equation by elementary methods, utilizing dynamic grid generation to resolve the velocity near wells. (See (Bramble, Ewing, Pasciak, Schatz 1988) for details.) An example of the type of grid used is shown in fig.2.B, and although this algorithm has not yet been fully tested, the results are promising.

6 Domain Decomposition

To advance the solution on the composite grid, we split the computational area into three subdomains; an outer area Ω_o, characterized by slow variation, an inner area Ω_i, coinciding with the refined blocks, and the common boundary $\partial \Omega_b = \Omega_o \cap \Omega_i$. The decomposition is schematically shown in fig.3.

Knowledge of the saturation profile from the previous time step allow us to advance the solution in an efficient manner, see Espedal and Ewing (1987). The iterative procedure, which is an accelerated linear solver consists of the following steps:

1. The coarse grid solution gives values at the fine grid corners

2. The fine grid edges are computed, giving start data for

3. Computation of solution on interior of fine grid

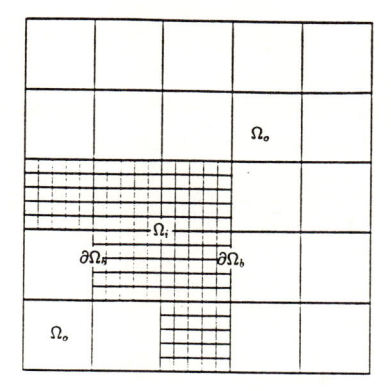

Figure 3: Substructuring of computational domain

4. Common points are adjusted

5. Procedure is iterated until convergence.

Symbolically the linear problem can be written:

$$\left\{ \begin{array}{ccc} A_{oo} & A_{ob} & 0 \\ A_{bo} & A_{bb} & A_{bi} \\ 0 & A_{ib} & A_{ii} \end{array} \right\} \left\{ \begin{array}{c} S_h^o \\ S_h^b \\ S_h^i \end{array} \right\} = \mathbf{d}^T, \qquad (13)$$

where

S_h^o = coarse grid solution on Ω_o,
S_h^b = solution on interface $\partial \Omega_b$,
S_h^i = inner solution defined on Ω_i.

The linear equations are solved by preconditioned conjugate gradients.

Since the most elaborate part of solving eqn (13) is the inversion of the submatrix A_{ii}, this problem is treated separately. The fine grid domain Ω_i is substructured further, such that each coarse grid block defines one subdomain. The inversion is then performed by the domain decomposition preconditioner introduced by Bramble, Pasciak, Schatz (1986).

7 Parallelization

The domain decomposition procedure *is* parallel by nature:
Each fine grid problem is solved on a separate processor, the results are

pooled and adjusted, after which they are handed back to the subprocessors for the next iteration.

The code has been run on an Alliant FX\8 with 8 processors — typical achievements we have experienced are shown below:

No. of refined blocks	Speed-up	Efficiency
7	5.64	0.71
15	6.16	0.77
19	5.57	0.70
23	6.44	0.81
23	6.27	0.78

8 Conclusion

By combining some existent and some novel techniques we have constructed a robust code where we have been able to reduce the mass balance error to a lower lever than other tailormade hyperbolic solvers, and still retain a high degree of front resolution, even for flow in heterogeneous rocks.

The widely referred grid orientation effect is absent, or at least less than plotter pen thickness on comparison figures. In spite of locally small mesh sizes the code permits very large time steps, a typical time step being of the magnitude dictated by the coarse blocks.

9 Acknowledgements

This work has been supported by VISTA, a research cooperation between the Norwegian Academy of Science and Letters and Den norske stats oljeselskap a.s. (Statoil), and NAVF, the Norwegian Research Council for Science and Humanities.

References

Barrett, J.W. and Morton, K.W. (1984). Approximate Symmetrization and Petrov-Galerkin Methods for Diffusion-Convection Problems. *Computer Methods in Applied Mechanics and Engineering* **45**, 97–122.

Bell, J.B., Dawson, C.N., and Shubin, G.R. (1988). An Unsplit, Higher Order Godunov Method for Scalar Conservation Laws in Multiple Dimensions. *Journal of Computational Physics* **74**, 1–24.

Bramble, J.H., Pasciak, J.E., and Schatz, A.H. (1986). The Construction of Preconditioners for Elliptic Problems by Substructuring. *Journal of Mathematics of Computation* **47**, 103–134.

Bramble, J.H., Ewing, R.E., Pasciak, J.E., and Schatz, A.H. (1988). A Preconditioning Technique for the Efficient Solution of Problems with Local Grid Refinement. *Computer Methods in Applied Mechanics and Engineering* **67**, 149–159.

Chavent, G. (1976). A new Formulation of Diphasic Incompressible Flows in Porous Media. Lecture Notes in Mathematics **503**, Springer-Verlag, Berlin, 258–270.

Colella, P., Concus, P., and Sethian, J. (1983). Some Numerical Methods for Discontinuous Flows in Porous Media. The Mathematics of Reservoir Simulation, (R. E. Ewing, ed.) SIAM, Philadelphia, 161–186.

Concus, P. and Proskurowski, W. (1979). Numerical Solution of a Nonlinear Hyperbolic Equation by the Random Choice Method. *Journal of Computational Physics* **30**, 153–166.

Dahle, H.K., Espedal, M.S., and Ewing, R.E. (1988). Characteristic Petrov-Galerkin Subdomain Methods for Convection-Diffusion Problems. The IMA Volumes in Mathematics and its Applications **11**, Springer-Verlag, 77–88.

Douglas, J. and Russell, T.F. (1982). Numerical Methods for Convection-dominated Diffusion Problems based on Combining the Method of Characteristics with Finite Element or Finite Difference Procedures. *SIAM Journal of Numerical Analysis* **19**, 871–885.

Espedal, M.S. and Ewing, R.E. (1987). Characteristic Petrov-Galerkin Subdomain Methods for Two-Phase Immiscible Flow. *Computer Methods in Applied Mechanics and Engineering* **65**, 113–135.

Ewing, R.E. (1983). Problems Arising in the Modeling of Processes for Hydrocarbon Recovery. The Mathematics of Reservoir Simulation, (R. E. Ewing, ed.) SIAM, Philadelphia, 3–34.

Glimm, J., Lindquist, B., McBryan, O., and Padmanabhan, L. (1983) A Front Tracking Reservoir Simulator, Five-Spot Validation Studies and the Water Coning Problem. The Mathematics of Reservoir Simulation, (R. E. Ewing, ed.) SIAM, Philadelphia, 107–136.

Russell, T.F. and Wheeler, M.F. (1983). Finite Element and Finite Difference Methods for Continuous Flows in Porous Media. The Mathematics of Reservoir Simulation, (R. E. Ewing, ed.) SIAM, Philadelphia, 107–136.

SIMULATION OF COMPOSITIONAL RESERVOIR PHENOMENA ON A HYPERCUBE

John E. Killough and Rao Bhogeswara
(University of Houston, Department of Chemical Engineering)

ABSTRACT

   Techniques to improve the efficiency of numerical simulation for the prediction of compositional reservoir processes are described. In particular, parallel computational approaches to the numerical modelling of the cycling of a gas condensate reservoir have been investigated on a hypercube parallel computer (INTEL IPSC/2). A commercial reservoir simulation model was used as the initial basis for the seven component simulation. Parallelization of the coefficient and saturation function routines was accomplished using a technique to minimize the overhead of messages in this distributed memory environment. For the linear equation solution a domain decomposition approach was implemented along with multigrid techniques to obtain more robust parallel solutions. The initial goal of the research to develop efficient parallel numerical models for the prediction of compositional processes was achieved. Greater than ninety-eight per cent of the CPU time of the compositional model was parallelized. Future work to achieve further parallelization will involve both the linear equation solution and the coefficient routines.

1. INTRODUCTION

   Computer performance over the past decade has advanced significantly. These advances have been such that current single processor performances are limited by speed-of-light considerations. These limitations indicate that to achieve speedups greater than an order of magnitude above current technology, significant architectural changes must be employed. The emerging parallel computer architectures appear to be the most likely avenue for this advancement in computing.

Several publications in the recent literature have dealt with the application of parallel computing to petroleum reservoir simulation in shared memory parallel environments. Scott, et al.[1] investigated the parallelization of the coefficient routines and linear equation solvers for a black-oil model on a Denelcor HEP. Chien, et al.,[2] investigated compositional modeling in parallel on a CRAY X-MP 4/16. Barua, et al.[3] applied parallel computing using a non-linear equation solver for the black-oil case on the Encore Multimax. Killough, et al.[4] looked at parallel linear equation solvers on both the CRAY X-MP and IBM 3090. Each of these applications involved the use of a shared-memory parallel computer. The question still remained as to whether a distributed memory architecture could be efficiently utilized for simulation of petroleum reservoirs.

This work deals with the application of compositional reservoir modelling to the distributed memory, message passing, INTEL IPSC/2 Hypercube. In particular a commercial compositional reservoir simulator (VIP-COMP) has been adapted to run in parallel on the IPSC/2. The key issues in this parallelization involve the data structure of the program, message passing of data among processors, and development of parallel linear equation solvers for the model.

2. ARCHITECTURE OF THE INTEL IPSC/2

The architecture of the INTEL IPSC/2 Hypercube is illustrated on the bottom of Figure 1 for the case of sixteen (2^4) processors.

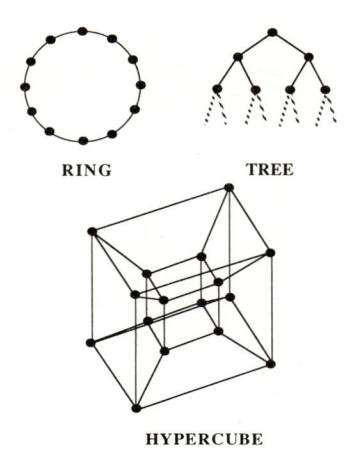

Fig. 1 Configurations for Distributed Memory Computers

As shown in the figure for an n-dimensional hypercube with 2^n processors, each processor has connections to n other processors. If these processors are numbered in a binary fashion, the nearest neighbours sharing connections differ by only one binary digit. Currently, the IPSC/2 can have up to 128 processors with up to sixteen megabytes each. As shown in Figure 2, each processor is cabable of up to 0.2 double precision MFLOPS with the 16 MHz 80387 chip, 0.6 MFLOPS with the Weitek 1167, and 6.0 MFLOPS when using the vector option.

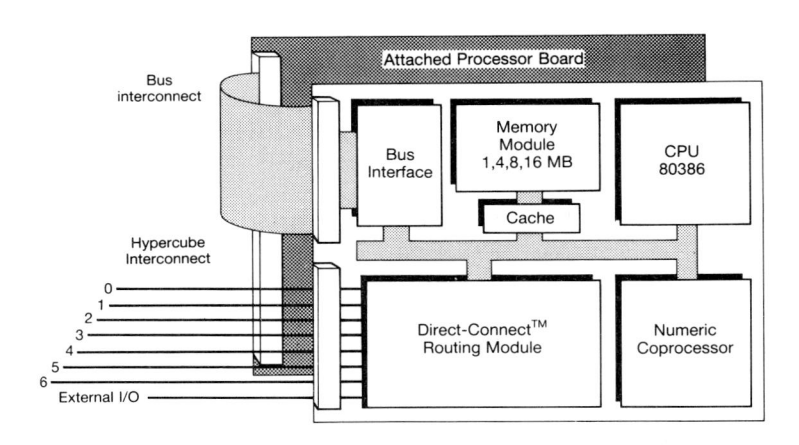

Fig. 2   Diagram of IPSC/2 Hypercube Node

The INTEL IPSC/2 has two unique features which make it particularly useful for reservoir modelling. First, as shown in Figure 1, communication among processors does not require a knowledge of the hypercube topology to achieve efficiency. Through what is called a "direct-connect" scheme, messages can pass from one processor to any other processor in the system with only a few percent overhead for each connection which must be crossed. Data transfer rate is at 2.8 megabytes per second for message passing. Latency for each message setup is approximately 300 microseconds.

Second, the local memory size available for programs may be up to sixteen megabytes. This allows an entire reservoir simulator of reasonable size to fit on a single node. As will be discussed later, this memory size allows a straightforward method for parallelization of existing models.

3. THE COMPOSITIONAL RESERVOIR SIMULATOR

The commercial n-component, 3-phase, equation-of-state model VIP-COMP was chosen for porting to the IPSC/2. Although the model contains provision for fully-implicit capabilities, only the IMPES option was investigated.

The model is based on the Young-Stephenson[5] formulation. Basically, this formulation involves the use of a Newton-Raphson iteration for the solution of the overall component material balances using an equation-of-state. The equation-of-state is used for calculation of both fugacities (K-values) and densities.

At each time-step a Jacobian is formed for each of the cells in the model. The unknowns for the Jacobian are as follows:

$$x_{ij}, y_{ij}, z_{ij}, v_j, F_j, W_j, p_j \quad \begin{array}{l} i=1, \ldots, n_c \\ j=1, \ldots, n_{cells} \end{array}$$

where,

x_{ij}, y_{ij}, z_{ij} = mole fractions of component i in block j for liquid, vapour, and overall compositions, resp.

v_j = vapour fraction for block j

F_j = overall hydrocarbon mass

W_j = mass of water = $\rho_w S_w$

p_j = pressure of oil phase in block j

A pressure equation is derived for the IMPES case through a forward elimination of the Jacobian. Back substitution then yields the values for the $3n_c+4$ unknowns at each grid block in the model.

A typical flow chart for the compositional reservoir model is shown in Figure 3.

```
          ┌──────────▶│ INPUT DATA │
          │           ▼
          │     │ ROCK PROPERTIES │
          │           ▼
          │     │ PVT PROPERTIES │
          │           ▼
          │     │ WELL RATES │
          │           ▼
          │  ┌──▶│ COEFFICIENT CALCULATIONS │
          │  │        ▼
          │  └──│ LINEAR EQUATION SOLUTION │
          │           ▼
          └─────│ VARIABLE UPDATE │
                      ▼
                   │ END │
```

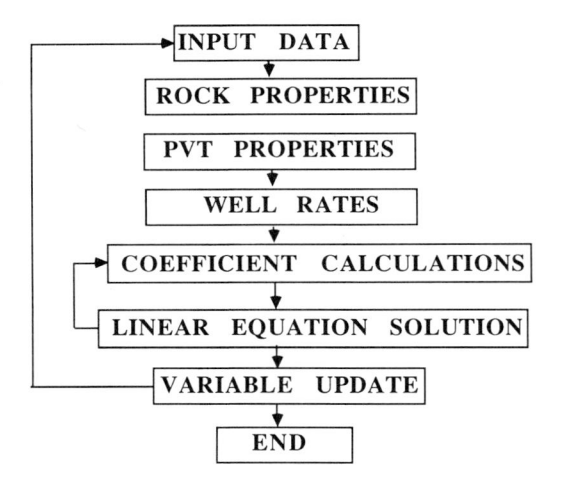

Fig. 3   Flowchart for Typical Compositional Model

After input data such as reservoir description and well rates are read, rock properties (relative permeabilities) and PVT properties are evaluted (viscosities). Well rates are then distributed among the cells in the model. The Jacobian is then formed and the forward elimination is performed to yield a pressure equation. An iterative linear equation solver is applied to this equation. After solution for the pressures, back substitution yields the updated values for all variables.

4. PARALLELIZATION OF THE COMPOSITIONAL MODEL

The first step in parallelization of the model was the profiling of the segments of the model to determine the portions which consumed the most CPU time. As shown on the flowchart in Figure 4, the majority of the CPU time was spent in the formation of the Jacobian and the forward elimination process for the formation of the pressure equation (EQUATION SETUP). The equation solution and the variable update were next in order of computational work requirements with 33 and 7 percent of the overall CPU time.

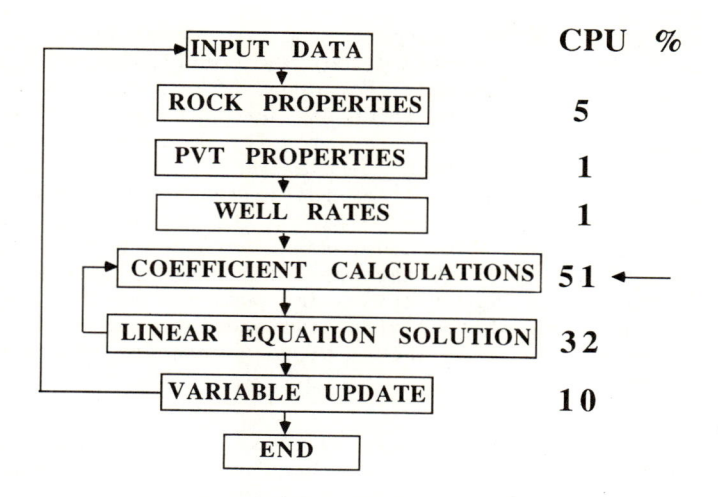

```
                    ┌─►│ INPUT DATA   │           CPU  %
                    │  │ ROCK PROPERTIES │          5
                    │  │ PVT PROPERTIES  │          1
                    │  │ WELL RATES      │          1
                    │┌►│ COEFFICIENT CALCULATIONS │ 51 ◄──
                    ││ │ LINEAR EQUATION SOLUTION │ 32
                    │└─│ VARIABLE UPDATE │        10
                    │  │ END │
```

Fig. 4 Flowchart for Compositional Model Showing Distribution of Serial Computational Work

Amdahl's law describes how much efficiency a parallel program can achieve:

$$\text{Speedup} = \frac{1.0}{s + p/n}$$

where,

s = percent serial portion of program
p = percent parallel portion of program
n = number of CPU's available

For the case of medium grained parallelism of 32 processors, the maximum speedup for 5% serial overhead is 12.54. This indicates that almost all portions of the model must be parallelized to achieve large parallel computing efficiencies.

The initial goal of this work has been to reduce the serial component to below 5% for the compositional model. To achieve this, all portions of the program were parallelized with the exception of the well routines. Input and output processing were assumed to be insignificant for most large-scale simulation problems.

5. INITIAL STAGES OF PARALLELIZATION

The parallelization of the coefficient routines was

accomplished in three steps. From the computational work analysis above, it appeared that the coefficient routines should be first attacked. For proof of concept, the coefficient routines in which the Jacobian setup and forward elimination occurred were isolated for parallelization. A separate program for this calculation was set up on each node in the hypercube. After initialization, the main program executing on node #0 passed the required data to each of the nodes. The nodes then performed the elimination process and passed back the resulting residual terms, flow coefficients, and upper triangular portions of the Jacobian. The question was whether efficient speedups could be obtained for the large amount of data transfer required to and from the nodes.

The results shown for SPE #3 Comparative Solution[6] in Figure 5 indicate that the overhead due to message passing was equivalent to three percent of the original CPU time. This remained constant when the number of parallel processes was increased from 4 up to 16.

NCPUS	SPEEDUP	OVERHEAD (%)
4	3.65	3.04
8	6.60	3.11
16	11.03	3.20

Fig. 5 Results for Parallelization of the Coefficient Calculations

With these parallel efficiencies on a portion of the model, it appeared that overall parallelization could result in significant parallel efficiencies. The second stage of the process was then begun. In this stage all portions of the coefficient routines, the relative permeability and PVT calculations, and the equation update were parallelized. Simultaneously, the development on the parallel linear equation solver was begun. The final stage of parallelization involved the combining of all portions of the coefficient routines and solver into a single "node" program with limited data communication to the main reservoir simulator. At the end of this stage the main program would only be used for model initialization, material balance reporting, and array output.

6. THE PARALLEL LINEAR EQUATION SOLVER

The parallel linear equation solver is a combined domain decomposition/multigrid solver. This algorithm consists of three parts. The first part is to obtain initial guesses for the interfacial values of subdomains using z-line corrections. (Watts[7]) The convergence of this algorithm basically depends on how accurately these boundary values of subdomains are predicted. A multigrid[8] technique was used to solve the 2-dimensional line corrections in parallel. The second step involves domain decomposition. For this initial parallelization a simplistic approach was taken. Subdomains were obtained by block-Jacobi type domain decomposition with homogeneous boundary conditions. The subdomain problems are solved using a reduced system incomplete factorization approach (RS/ILU(0))[9]. Finally the overall convergence is accelerated with ORTHOMIN(k).[10]

Figure 6 depicts the z-line correction steps in an iteration of the parallel solver. The grid is collapsed to a two-dimensional areal problem by summing the coefficients areally. The equations are solved using multigrid by a V-cycle in which successively coarser grid problems are solved followed by successively finer grid problems. Each of these subproblems is solved using a simple, but highly parallel, solver such as red-black point Gauss-Seidel. Only two V-cycles with two levels of grid were used in the iteration. Finally, the z-line correction results are applied to the overall three-dimensional grid to begin the domain decomposition step.

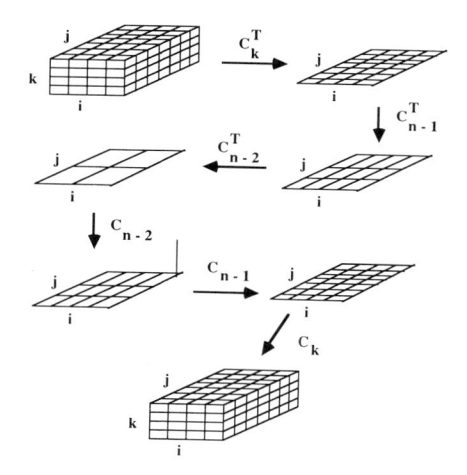

Fig. 6 Schematic of Parallel Linear Equation Solution

The implementation of the algorithm on the hypercube was designed to minimize the message-passing overhead. To accomplish this it is important to isolate those parts of the algorithm which require message passing. Each subdomain, obtained by areal splitting, is assigned to each node of the hypercube. These subdomains possess the same identity before and after line corrections in the z-direction. This makes message passing across the nodes much easier. Parts of the multigrid algorithm that require message passing are the Gauss-Seidel iteration and calculation of residuals after each multigrid V-cycle. Each processor must communicate with its four neighbouring processors while the Gauss-Seidel iteration progresses. Similarly, each node must communicate with the neighbouring nodes to calculate residuals. Multicolour, asynchronous message passing schemes were attempted for this portion of the algorithm, but little advantage over a simplistic nearest neighbour scheme was found. In the asynchronous message passing, a message is passed but execution of the program does not wait for the message to be received. A global communication library was used to find the residual sums for the L^2 norm. The only message passing during the domain decomposition step was for the communication of boundary values of neighbouring domains. Again, global communication routines were used to pass values required by the inner product calculations of the ORTHOMIN(k) acceleration.

7. GLOBAL PARALLELIZATION OF THE MODEL

The steps toward global parallelization of the model proceeded as shown in Figure 7. All major groups of tasks were initially made into separate processes on each of the nodes of the hypercube. All common areas and subroutine arguments were passed as messages between the main reservoir simulation program and the node programs at each stage of the simulation. Figure 7 is a schematic of this procedure. Although inefficient, this technique allowed a step-by-step parallelization to be performed and substantially reduced debugging time. At each stage of the parallelization process the results could be compared to the original program to determine if data integrity had been maintained.

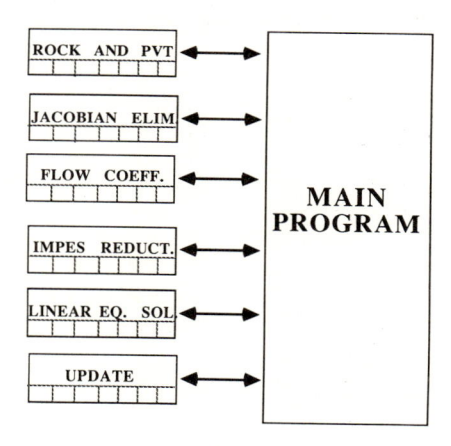

Fig. 7 Schematic of Initial Model Parallelization Procedure

In the next step all routines with the exception of the linear equation solver were combined to form a single parallel process on each of the nodes for calculation of relative permeabilities, PVT properties, coefficient setup, and variable update. All communication with the main program was then eliminated except for the passing of material balance data for each of the phases, well rates, and input/output.

Communication among the node programs was required for the calculation of the finite difference flow coefficients. Rather than adapt a specific data structure, global communication or concatenation of several key variables was used for this calculation. Currently, this global communication represents the major area in which communication overhead is non-optimal; however, as shown later in the results section, even this overhead is still small relative to the overall computing time.

The final parallelization step involved incorporating the parallelized multigrid/domain decomposition solver into the node processes. A global concatenation was also required for this process since the coefficients of the pressure equation had a different data structure from that of the solver. Figure 8 shows a schematic of the final parallelized model. Small amounts of data are passed at the begining timestep and for well rates at each step during the simulation. The major overhead remains the global data concatention before the solver and for the flow coefficients.

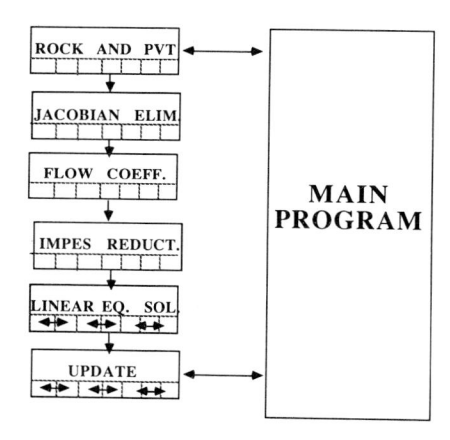

Fig. 8 Schematic of Final Parallelization Process

8. RESULTS

Results for the parallelized 16x16x7 SPE #3 comparative solution[6] model are shown in Figures 9-13. Figure 9 compares the speedups for the four processor case at the initial parallelization stage with all processes communicating all required data with the main program.

	ELAPSED TIMES		
	SERIAL	**PARALLEL**	**SPEEDUP**
ROCK, PVT	11.86	7.13	1.66
COEF. SETUP	96.25	57.72	1.67
LINEAR EQ.	36.71	48.75	0.75
UPDATE	16.25	16.36	0.99
OVERALL	165.96	134.86	1.23

Fig. 9 Results for the Initial Parallelization

As anticipated, this naive approach led to only small gains in efficiency. Figure 10 represents the 4 processor case after all parallel procedures had been combined and most message passing eliminated. As shown in the figure speedups in excess of 3.5 were obtained for all segments of the model. The coefficient calculation was the most efficient of the segments due to its large granularity. Overall speedup was 3.77 implying that 98% of the original CPU was parallelized. The two percent overhead is due to message passing and movement of data into buffers before message passing.

	ELAPSED TIMES		
	SERIAL	PARALLEL	SPEEDUP
ROCK, PVT	442.68	117.20	3.77
COEF. SETUP	3374.67	862.96	3.91
LINEAR EQ.	2212.65	626.81	3.53
UPDATE	587.08	150.92	3.89
OVERALL	6617.08	1757.89	3.77 (98 %)

Fig. 10 Final Parallelization - Four Processor Results

Figure 11 shows results for the 32 processor case. Again the coefficient routine had extremely high parallel efficiency. The overall speedup of 20.51 corresponds to less than 2% overhead and indicates that the overhead does not increase with number of processors. Figure 12 summarizes results for 4-32 processors. Again overhead remained constant at about 2% of orignal CPU time.

	ELAPSED TIMES		
	SERIAL	PARALLEL	SPEEDUP
ROCK, PVT	442.68	22.85	19.37
COEF. SETUP	3374.67	126.20	26.74
LINEAR EQ.	3358.61	191.41	17.54
UPDATE	587.08	38.11	15.40
OVERALL	7763.04	378.57	20.51

PARALLEL COMPONENT = 98.2 % OF CPU TIME

Fig. 11 Results for Parallelization on 32 Processors

	SPEEDUP NUMBER OF PROCESSORS			
	4	8	16	32
ROCK, PVT	3.81	7.02	12.31	19.37
COEF. SETUP	3.91	7.64	14.51	26.74
LINEAR EQ.	3.52	6.52	11.08	17.54
UPDATE	3.89	7.12	11.21	15.40
OVERALL	3.77	7.40	12.87	20.51

Fig. 12 Summary of Results for 4-32 Processors

The results for Figures 9-12 compare the parallel code to the serial version of the same code excluding well calculations. Obviously, the goal is to achieve speedups on the entire model over the original unchanged program. Figure 13 compares the parallel models with the original program timings. Also shown are the speedups with the well routine timings included.

	SPEEDUP NUMBER OF PROCESSORS			
	4	8	16	32
PARALLEL CODE	3.77	7.40	12.87	20.51
W/ WELLS	3.73	7.16	11.83	18.62
VS ORIGINAL	3.04	5.57	9.40	13.35

Fig. 13 Comparison of Speedups With Original Model

The anticipated degradation of the block-Jacobi iteration with numbers of domains is seen in the reduction of the speedup of the parallel model over the original to a factor of 13.35. Inclusion of the well timings had a small effect on reducing speedups. For models with large numbers of wells efficiency gains in parallel would be possible for the well routines, but this was not attempted in the current version of the code.

9. CONCLUSIONS AND FUTURE DIRECTIONS OF THE RESEARCH

This work has demonstrated that a highly efficient parallel model can be generated for a commercial n-component, three-phase, equation-of-state reservoir simulator in a distributed memory parallel computer. Greater than 98% of the CPU time for the SPE #3 comparative solution problem was parallelized. Overhead was shown to be constant near 2% of the original CPU time for a wide range of numbers of processors. The linear equation solver using multigrid, domain decomposition, and z-line corrections can be efficiently parallelized.

Future work for this research involves the implementation of more efficient, but equally parallelizable, domain decomposition and multigrid algorithms. Data structure can be improved such that interprocessor communication is significantly reduced. Finally, investigations will be made on more powerful node processors - first with the Weitek 1167 and, at some point in the future, with the next generation of more powerful processors which may offer a twenty fold increase over current node capabailities.

10. ACKNOWLEDGEMENT

This work was supported by Intel Scientific Computers and the Texas Advanced Technology Program. We are grateful to Western Atlas Integrated Technologies for the use of their simulator in this work.

11. REFERENCES

1. Scott, S. L., Wainwright, R. L., Raghavan, R., and Demuth, H., "Application of Parallel (MIMD) Computers to Reservoir Simulation", SPE 16020 presented at the 9th SPE Symposium on Reservoir Simulation, San Antonio, Texas, Feb 1-4, 1987.

2. Chien, M. C. H., Wasserman, M. L., Yardumian, H. E., and Chung, E. Y., "The Use of Vectorization and Parallel Processing for Reservoir Simulation", SPE 16025 presented at the 9th SPE Symposium on Reservoir Simulation, San Antonio, Texas, Feb. 1-4, 1987.

3. Barua, J., and Horne, R. N., "Improving the Performance Parallel (and Serial) Reservoir Simulators", SPE 18408 presented at the 10th SPE Symposium on Reservoir Simulation, Houston, Texas, Feb. 6-8, 1989.

4. Killough, J. E., and Wheeler, M. F., "Parallel Iterative Linear Equation Solvers: An Investigation of Domain Decomposition Algorithms for Reservoir Simulation", SPE 16021 presented at the 9th SPE Symposium on Reservoir Simulation, San Antonio, Texas, Feb. 1-4, 1987.

5. Young, L. C. and Stephenson, R. E., "A Generalized Compositional Approach for Reservoir Simulation," SPEJ (Oct. 1983), 727-742.

6. Kenyon, D. E., and Behie, A., "Third SPE Comparative Solution PRoject: Gas Cycling of Retrograde COndensate Reservoirs", SPE 12278 in Proceedings of the Seventh SPE Symposium on Reservoir Simulation", San Francisco, November, 1983.

7. Watts, J. W., "An Iterative Matrix Solution Method Suitable for Anisotropic Problems", SPEJ (March, 1971), 47-51.

8. Brandt, A., "Multi-Level Adaptive Solutions to Boundary-Value Problems", Math. Comp., Vol. 31 (1977), 333-390.

9. Wallis, J. R., "Vectorization of Preconditioned Generalized Conjugate Residual Methods" in *Mathematical and Computational Methods in Seismic Exploration and Reservoir Modeling*, SIAM, 1987, 250-251.

10. Vinsome, P.K.W., "Orthomin: An Iterative Method for Solving Sparse Banded Sets of Simultaneous Equations", SPE 5729 presented at the SPE 4th Symposium on Reservoir Simulation, Los Angeles, Ca., Feb. 1976.

The Parallelisation of Bosim, Shell's Black/Volatile Oil Reservoir Simulator [*]

D.T. van Daalen, P.J. Hoogerbrugge, J.A. Meijerink, R.J.A. Zeestraten

Koninklijke/Shell Exploration and Production Laboratory
Rijswijk, The Netherlands

Abstract

It is reported how the IMPEC version of Bosim, the black/volatile oil variant of Shell's Multisim family of reservoir simulators has been adapted to run on a medium-scale (10–100 processors) Meiko Computing Surface, a local-memory parallel MIMD computer based on Inmos transputers.

Bosim is widely used and currently takes up considerable amounts of CPU time on vector supercomputers and large mainframes. Since most of its computations are inherently parallel, reservoir simulation makes a good candidate for parallelisation, now that parallel processing has become a practical reality.

The strategy chosen is coarse grain parallelisation using 2-D domain decomposition: the reservoir is divided among a 2-D grid of processors. First we have parallelised the local (e.g. PVT) and semi-local (e.g. matrix set up) computations. Afterwards we have handled the linear solver and the well computations (both requiring global communication).

Parallel Bosim is in Fortran 77. The communication between the Fortran programs running on different processors is provided by a library of communication subroutines in cooperation with a small Occam "harness". A source code analyser has helped us generate the calls to the communication routines that had to be inserted.

Currently over 98% (for a small model) of the time-step computations run in parallel, and on 60 processors the time steps run at about half the Cray speed — at a fraction of the cost.

---

[*] Project funded by Shell UK Exploration and Production

1 Introduction

Numerical oil reservoir simulation is an extremely demanding computational task. Size and complexity of the problems that could reasonably be solved have always been restricted by the capacity of the available hardware. Powerful as the current vector supercomputers are, one still wants more: to run larger jobs than currently possible and to run the present ones faster and cheaper.

Now that sequential computers appear to be approaching their limit speed and parallel computers of various types are commercially available it becomes attractive to start exploiting the long-claimed potential of parallelism. Parallelism indeed looks promising for a reservoir simulator because most of its computations are inherently parallel.

Most efforts up to now have concentrated on simple simulators or on parts of a simulator [2] [4], whereas we have taken Bosim, a fully fledged production simulator, to parallelise it as a whole.

Our first aim was performance, but a secondary aim was to find out what performance can be achieved by how much effort. Reprogramming large parts of the simulator was out of the question, the more so because, for reasons of maintainability, its parallel and sequential versions should differ as little as possible. Parallelisation of existing programs (particularly large ones) has rarely been described in the literature. As far as code size is concerned, Parallel Bosim is probably the largest application ever implemented on a parallel machine.

Earlier attempts to build parallel reservoir simulators either considered SIMD parallelism [7] or small-scale shared memory machines [4],[8]. Instead, we have chosen the local-memory MIMD approach. MIMD is more flexible and, hence, more generally applicable than SIMD. Local memory enables scaling of the machine size to massive parallelism (as demonstrated by, for example, [3]), though at first, we are rather aiming at medium-scale machines, say with 10–100 processors.

Parallelising compilers for local-memory machines are only in the development stage [11]. Hence, the parallelism and the required communication between processors have to be specified explicitly. In a large program, it is not easy to find out what has to be communicated and where, so we have developed some tools that help us with this work.

Our implementation runs on a Meiko Computing Surface, but porting the code to different machines of the same type of architecture (e.g. Ncube or Intel IPSC) is intended to be fairly easy. To achieve this we have hidden machine dependent features inside a library of communication routines.

The organisation of the paper is as follows. In section 2 we give some characteristics of Bosim that are relevant to the present work. Section 3 describes the target architecture for Parallel Bosim. Then (section 4) we describe the parallelisation strategy (domain decomposition: each processor is responsible for a part of the reservoir) and the communication infrastructure it requires. In section 5 we show how the source has been modified. In most cases only calls to communication subroutines had to be inserted, a large part of which could be done in a semi-automated fashion. The linear solver clearly required some more radical adaptation and this is documented in section 6.

Finally, in section 7 and 8 we present the speed-up achieved and evaluate the work.

2   Bosim

Bosim is the black/volatile oil version of Multisim, a family of simulators in use within the Shell group. More specifically, it is an isothermal, three component, three-phase simulator that can be run in IMPEC [1] and in Fully Implicit mode. (We have parallelised the IMPEC version). It uses a cell-centred 7-point finite difference discretisation. Non-linearities are handled by a Newton scheme, and a variety of linear solvers for the resulting asymmetric linear equations can be called. Most often used are CGS [9] or Orthomin [10] preconditioned by incomplete factorisations such as line Jacobi or IBLU [5]. Available features are adaptive time stepping, extra (irregular) gridblock connections, segregated flow, non-Darcy flow, well management and the handling of platform and field constraints.

Bosim is written in Fortran 77, and contains some 100,000 lines of source code. It is in use with many operating companies all over the world. It runs on various machines but is most often used on Crays and large VAXes. Considerable parts of the code vectorise well on the Cray. The size of models studied by Bosim normally ranges from 3,000 to 30,000. Numerous long-term simulation studies are continually being carried out, and this takes up several hours of Cray CPU time a day.

Outside the solver, Bosim packs the non-void gridblock data into one long one-dimensional array. The position of each block in this array does not necessarily reflect its position in reservoir geometry. Hence, with each grid block there are associated pointers to its neighbours along the grid axes.

3   The Meiko parallel computer

Meiko is one of several manufacturers building computer systems ("Computing Surfaces") from Inmos transputers. The T800 transputer is a 32-bit RISC-like microprocessor with a floating-point unit and four bidirectional communication links on a single chip. For practical purposes, the T800 can be rated at 0.5 Mflop/s; the link start-up time is about equal to the time needed to send one double precision real, viz. around 5 microseconds. It is specifically designed for use in parallel computers of the local-memory (message-passing), MIMD type.

The Computing Surfaces come with up to hundreds of transputers, each optionally equipped with a sizable amount (up to 48 MByte) of memory. In the current project we normally use one processor with such a large memory (for the master process, see below) and many slave processors with 4 MBytes of memory. Programming can be done in *Occam*, the native language of the transputer [6], but also in ordinary sequential Fortran or C.

In a Meiko machine, the network connecting the processors is electronically reconfigurable. The

Meiko runs under a Unix-like operating system. The machine can be equipped with internal disks, or it can access the filing system of an external host. In a multi-user environment, the system can be configured to have several domains, to be divided among different users.

The favourable communication performance figures of the transputer (compared with other systems) have contributed considerably to the success of the current project.

4 Domain decomposition and communication

In Parallel Bosim the reservoir grid is divided over a two-dimensional rectangular grid of processors, see Figure 1. Each *grid processor* is responsible for the gridblock and well data belonging to one *subdomain*. In addition to these grid processors there is a *master* processor, which handles initialisation and input/output.

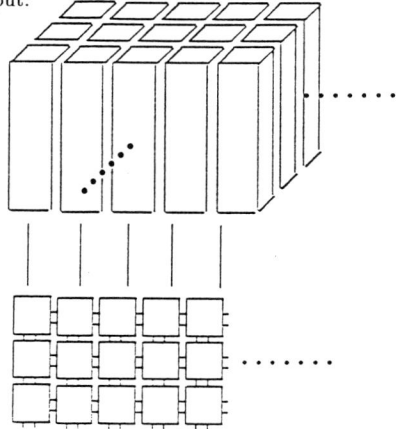

Figure 1: Two dimensional domain decomposition

In order to start the distributed computation, master-to-grid communication has to take place. Data that have to be known by all processors are *broadcast*, data that have to be partitioned among them are *distributed*. Later, results obtained on the grid have to be sent back to the master.

Part of the computation (e.g. PVT) is of a local character and does not induce any communication. However, to compute flow into a subdomain, for instance, data from outside that subdomain are required. To implement such *semi-local* computations the subdomains proper are extended with *exterior* data: data needed by one processor and belonging to another. The associated communication functions are *move* (to refresh exterior data) and *merge* (to combine contributions from different processors). The data needed are not always held at neighbouring processors: computing the inflow into wells penetrating several subdomains requires non-local communication.

Global communication is needed to implement *global operations*. For instance, to decide about convergence, each processor at first determines the maximum residual in its domain, after which the global maximum of these local maxima has to be determined. Or, within an inner product computation (inside the linear solver), local processors' sums have to be combined into one global sum.

Local indexing is used for the distributed data sets: to a grid processor it seems as if it has the whole reservoir to itself. Therefore the major part of the original Bosim source was left unmodified and all grid processors run the same code. In addition, the communication of gridblock data and well data can be built upon one single set of primitive communication functions.

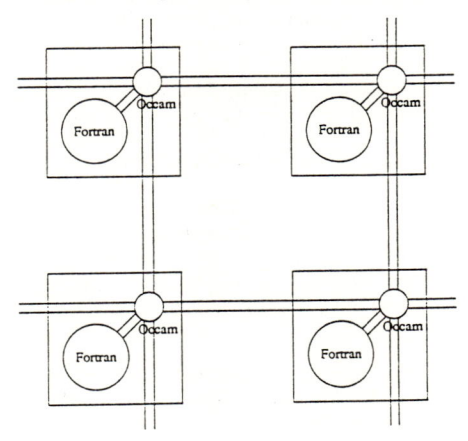

Figure 2: An Occam harness takes care of the communication between the Fortran programs running on different processors

In Parallel Bosim the bulk of the programming is done in standard Fortran 77. On each processor, the Fortran program runs in parallel with an Occam process, see Figure 2. The Occam processses on different processors cooperate to form an Occam *harness* that takes care of the communication. When Fortran on a given processor decides that a communication has to take place, it activates the corresponding Occam function and hands over the data to be communicated. Occam sees to it that the messages arrive at their destinations and are returned to Fortran on the receiving processors. After a message has been copied from Fortran to Occam, the Fortran program can continue with its work until its next communication. In this way communication and computation can to some extent be overlapped.

5 Semi-automated generation of communication calls

It can be a tedious job to find out which data have to be communicated, once the program is divided into a program running on the master and one running on the grid. Furthermore, writing all calls to the communication library routines with the correct message types and lengths is error-prone.

These two problems have been tackled by the construction of a tool for *data flow analysis* and *code generation* based on a LISP data base and some functions operating on it. The data base describes, for all Bosim routines, which other routines it calls, which variables and arrays it uses, and which variables and arrays it sets.

Let it be specified where each program part is going to be executed. At each point where program execution switches from master to grid or vice versa, a number of variables and arrays have to be communicated. Our tool automatically determines which ones. For example, a variable has to be sent from the master to the grid when it is first set on the master and afterwards used on the grid.

The data base also contains a description of all Bosim global variables, where global means: stored in a COMMON block. For each variable and array, the type, dimension (1 for variables) and the name of the COMMON block in which it resides is specified. This information is used to generate automatically the subroutine calls for communicating certain variables or arrays.

To indicate the importance of such a tool, the following approximate figures are given. The communication requirements have to be determined by keeping track of some 2000 global variables in a 100,000 line Fortran source code. We found that at one stage in the program execution, after the reading of the initialisation data and before the calculation starts, some 400 variables had to be communicated from the master to the grid, resulting in 12 pages of subroutine calls to the communication library.

Our tool has only been used for determining the master-grid communication. Move and merge operations for gridblock data have been identified by tracing the use of the neighbour pointers (see section 2). Use of possibly non-local gridblock information in the well calculations is found in a similar manner. The global operation calls have been inserted manually, but we also have a program that searches the grid processor program and finds the places where such calls are needed.

6 Solution of the linearised pressure equation

The pressure equation is solved by means of preconditioned conjugate gradient type methods, ordinary CG for symmetric equations and CGS or Orthomin for asymmetric ones. The calculations in such methods can be divided into three types, each of which will be discussed separately.

Vector operations: These comprise vector additions, scalar times vector multiplications, inner products and maximum calculations. The inner products and maximum calculations are global operations available in the communication library.

Matrix vector multiplication: The equations consist of the linearised volume balances for each grid block and the linearised well inflow and constraint equations. Matrix multiplication therefore requires the same kind of data transfer (move and merge) between processors as the calculation of the residuals for the non-linear equations. By keeping the data structure in the linear solver the same as the overall data structure, the communication library could be used without modification. Most communication can be overlapped with calculations.

Preconditioning: The standard preconditioning of sequential Bosim, a block incomplete factorisation, is not well suited for parallel computers because it is strongly recurrent. We have replaced it with a block Jacobi preconditioning in which one matrix block corresponds with the gridblock and well unknowns belonging to one subdomain. This preconditioning is fully local. The non-zero structure of the matrix is illustrated by Figure 3. The small local system is decomposed (at most) once per Newton step, and solved once per iteration using forward/backward substitution. By neglecting some of the elements in (i.e. by incompletely factorising) the local matrices, the number of calculations per iteration can be reduced at a cost of more iterations. In some cases line Jacobi preconditioning with the lines perpendicular to the processor axes is sufficient.

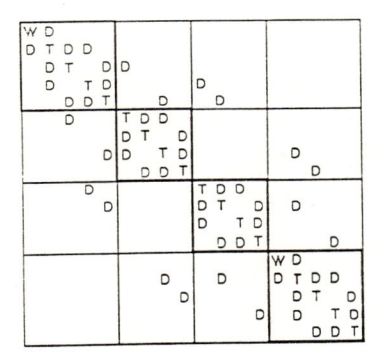

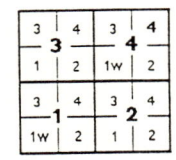

Figure 3: Non-zero structure of linear equations (left) for an example configuration (right) with a 4 x 4 x * reservoir grid decomposed over a 2 x 2 processor grid. Subdomains 1 and 4 have a well in their bottom left corner. W = well equations, D = diagonal matrix, T = tridiagonal matrix (corresponding to vertical lines). The part of the matrix inside the boldface diagonal blocks is accounted for in the preconditioner.

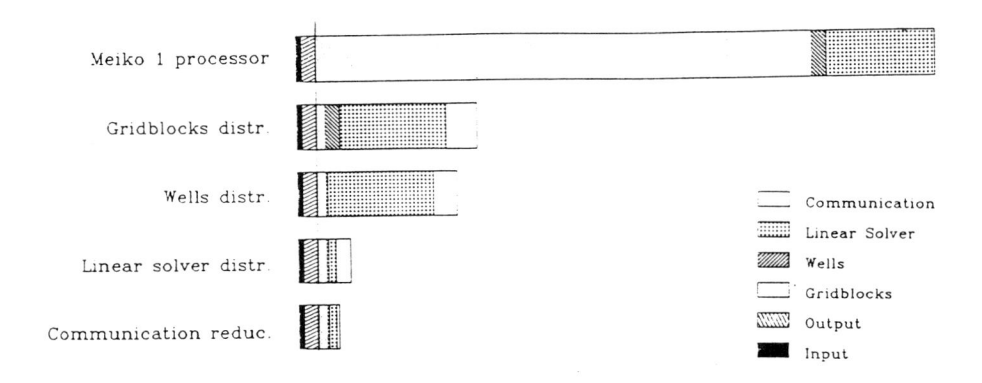

Figure 4: Bosim timings after the various phases of the parallelisation. Data deck described in the text.

7 Results

At the beginning of the parallelisation project, we considered that Bosim was too large and too complex to be fully parallelised at one go. We therefore started off with a single processor version running on the master processor. Subsequently we parallelised the code step by step, each time moving calculations from the master program to the grid program.

It took some time to get the single processor version running, because of problems with the Fortran compiler. After these problems had been resolved, timings showed that the computation speed (for this program) of a T800 transputer was approximately 1/50-th that of a Cray XMP-1. Removing some optimisations specially introduced for the vector computer improved this: rewriting one particular routine saved some 20% of the total run time on a single transputer.

The various stages in the parallelisation, with their timings, are shown in Figure 4. The figures are based on 20 year simulations with a real data set containing 2988 grid blocks in an irregularly shaped black oil reservoir model. The model further comprises 7 production wells and 7 injection wells, each with additional platform constraints. The parallel timings were carried out on a processor grid of 60 transputers. Each bar in Figure 4 corresponds to a stage in the parallelisation process.

In the first phase the gridblock calculations were distributed as indicated in section 4. Much attention had to be paid to master-grid communication, for which the software tool described earlier proved very useful.

In the second phase we parallelised the well computations. Each well is handled by one grid processor using data from other processors when needed (see section 4). Note that the domain can be decomposed in such a way that *vertical* wells are local to one processor. Calculations involving more than one well, e.g. determination of platform constraints, still take place on the master.

The work on a parallel linear solver (see section 6) was carried out more or less concurrently with the distribution of well calculations. A non-optimised general purpose sparse solver was employed for the local preconditioning. The timings shown in bar 4 of Figure 4 were obtained with line Jacobi preconditioning in the vertical direction.

At first, all results calculated on the grid were sent back to the master each time step. This took considerable time which was reduced (Figure 4, bar 5) in the next stage by collecting results only when needed for output.

The final 60-processor version is, for this problem, a factor 29 faster than the original single processor version (hence, the efficiency is roughly 50%.) Note that this includes a 20% reduction in overall calculation time as mentioned previously, as well as the effect of using a different linear solver than in the single processor version.

A comparison of the, now fully parallel, version of Bosim on 60 transputers with timings on the Cray-XMP is shown in Figure 5. *Apart from input/output*, the performance achieved is about half that of the Cray.

The effect of using more processors can be analysed as follows. The time spent in gridblock and solver calculations will simply reduce. However, there is a limit to the parallelisation of well calculations, which is reached when every processor handles at most one well. Generally, communication overhead increases with the size of the processor grid. But in our case, the communication costs are largely dominated by master-grid communication, which is restricted by the capacity of the master-grid connections[1]. Hence, the time spent in communication is relatively insensitive to the number of processors. We conclude that there still is scope for further speed-up by employing more processors.

Input and output processing is much slower on the Meiko machine than on the Cray. As a consequence, the *total* ratio (i.e. including I/O) Cray speed to 60-transputer speed is four rather than two. Some improvement in input/output processing is expected from planned modifications of the Meiko Fortran compiler. Furthermore, we are currently investigating the possibility of overlapping output by the master with calculations on the grid.

---

[1] For that matter, detailed timings indicate that the speed of master-grid communication can be improved considerably.

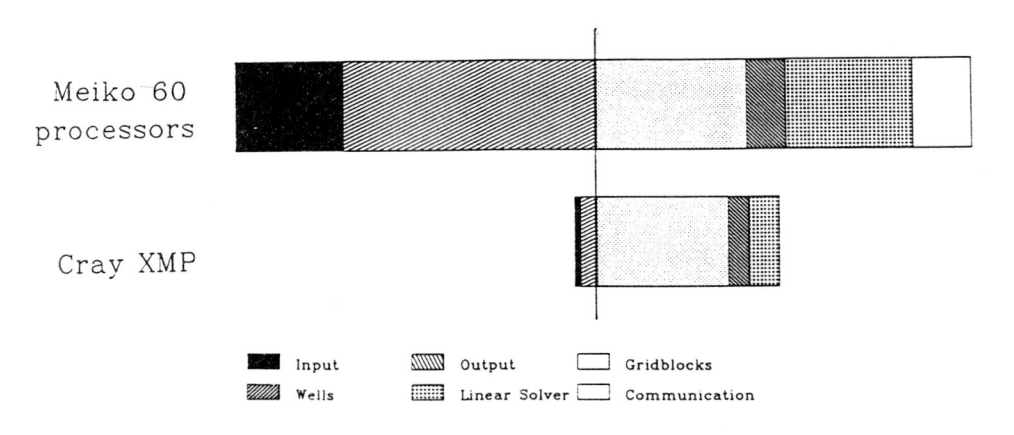

Figure 5: Cray-Meiko comparison. Same deck as previous figure.

8 Conclusions

The IMPEC version of Bosim, Shell's black/volatile-oil reservoir simulator, has been successfully ported to a Meiko Computing Surface. The speed obtained on its timestep, for a medium-size problem, is on a 60-transputer Meiko about half that of a single processor Cray XMP. For larger problems the performance can be improved by employing more transputers.

A source code analyser has been used to assist in the parallelisation. The majority (95%) of the original Fortran routines could be used without modification. Communication is dealt with by means of a communication library, which makes the code fairly easy to port to other MIMD computers.

The efforts going with MIMD parallelisation are large, but will decrease when better analysis, monitoring and debugging tools for parallel computers become available. This will make parallelisation also attractive for applications which are less compute-intensive or less frequently used than Bosim.

References

[1] Acs, G., Doleschall, S. and Farkas, E., *A General Purpose Compositional Model*, SPE Paper 10515, Proc. 6th Symp. Reservoir Simulation (New Orleans, 1982), 385–404.

[2] Barua, J. and Horne, R.N., *Improving the Performance of Parallel (and Serial) Reservoir Simulators*, Proc. 10th SPE Symp. Reservoir Simulation (Houston, 1989), 7–18.

[3] Gustafson, J.L., Montry, G.R. and Benner, R.E., *Development of Parallel Methods for a 1024-Processor Hypercube*, SIAM J. Sci. Stat. Comp., 9, 1988.

[4] Killough, J.E. and Wheeler, M.F., *Parallel Iterative Linear Equation Solvers: An Investigation of Domain Decomposition Algorithms for Reservoir Simulation*, SPE Paper 16021, Proc. 9th SPE Symp. Reservoir Simulation (San Antonio, Texas, 1987), 293–312.

[5] Meijerink, J.A., *Iterative Methods for the Solution of Linear Equations Based on Incomplete Block Factorisation of the Matrix*, SPE Paper 12262, Proc. 7th SPE Symp. Reservoir Simulation (San Francisco, 1983).

[6] Inmos Ltd., *Occam 2 Reference Manual*, Prentice Hall Int. Series Comp. Sc., 1988.

[7] Scott, A.J., Sutton, B.R., Dunn, J., Minto, P.W., Thomas, C.L., Habib, S., Oakley, C.A. and Krzeczkokowski, A.J., *An Implementation of a Fully Implicit Reservoir Simulation on an ICL Distributed Array Processor*, SPE Paper 10525, Proc. 6th SPE Symp. Reservoir Simulation (New Orleans, 1982), 523–533.

[8] Scott, S.L., Wainwright, R.L., Raghavan, R. and Demuth, H., *Application of Parallel (MIMD) Computers to Reservoir Simulation*, SPE Paper 16020, Proc. 9th SPE Symp. Reservoir Simulation (San Antonio, Texas, 1987), 281–292.

[9] Sonneveld, P., *CGS, A Fast Lanczos-Type Solver for Non-Symmetric Linear Systems*, SIAM J. Sci. Stat. Comp., 10, 1989, 36–52.

[10] Vinsome, P.K.W., *Orthomin, an Iterative Method for Solving Sparse Sets of Simultaneous Linear Equations*, SPE Paper 5729, Proc. 4th SPE Symp. Reservoir Simulation (Los Angeles, 1976).

[11] Zima, H.P., Bast, H.-J., Gerndt, H.M. and Hoppen, P.J., *SUPERB: A Tool for Semi-Automatic MIMD/SIMD Parallelization*, Parallel Comp., 6, 1988, 1–18.

MONTE CARLO SIMULATION OF LITHOLOGY FROM SEISMIC DATA IN A CHANNEL-SAND RESERVOIR

P. Doyen, T. Guidish and M. de Buyl

Western Geophysical, a Division of Western Atlas International
3600 Briarpark Drive, Houston, Texas 77042

ABSTRACT

In areas of rapid lithologic variations, the areal extent of sand and shale units usually cannot be inferred from sparse well data alone. Seismically derived interval velocity data can be used to help predict lithologic variations away from wells. However, in general, the overlap of the velocity ranges for sands and shales is such that seismic discrimination of lithology is ambiguous. We present a Monte Carlo technique for numerically simulating the spatial arrangement of sand/shale units. This technique accounts for the ambiguous nature of the seismic velocity information. Rather than calculating a unique sand/shale model, the Monte Carlo method provides a family of alternative lithologic images, all of which are consistent with the data. The range of models reflects the uncertainty of the lithologic classification and is used to assess risk in reservoir development. The sand/shale simulation technique is illustrated using a data set from an oil-producing channel-sand reservoir. Sand/shale cross-sectional simulations are generated along a seismic traverse that intersects three wells. The simulated models reproduce the log-derived lithologic sequences at the wells; they are conditioned by interval velocities inverted from the seismic amplitude data and are consistent with the spatial autocorrelation and crosscorrelation structures of the seismic and well data. The seismically derived lithologic models of the reservoir are better spatially constrained than models solely conditioned by well data. However, in keeping with the inherent ambiguity of the seismic information, the exact location of the lateral truncation of the channel sand is not precisely defined in the Monte Carlo lithologic simulations.

1. INTRODUCTION

In reservoirs characterized by alternating shale and sand bodies, prediction of the flow behaviour is conditioned by our ability to delineate lithologic variations away from the wells. For instance, shale lenses may act as permeability barriers and inhibit the vertical drainage of oil; they also can control gas coning at the wells. Unfortunately, in areas of rapid lithologic variations, the lateral continuity of sand and shale units can rarely be inferred from sparse well-log measurements alone. Stratigraphic interpretation of seismically derived impedances or velocities can be used to help differentiate sands from shales and determine their lateral continuity.

We present a Monte Carlo technique that simulates vertical sand/shale cross sections from seismic interval velocity profiles. This technique is based on the recent work of Alabert (1987) and Journel (1988) in the area of conditional simulation. The simulated models have the following properties:

- They reproduce the sand/shale vertical sequences interpreted at the wells.
- The areal extent and thickness of the simulated sand/shale units are constrained by the spatial autocorrelation structure of the lithologic data.
- The simulations are consistent with the spatial crosscorrelation existing between lithology and seismic velocity.

Compared with previous methods for simulating sand/shale sequences [1, 3, 4, 6], our approach provides better spatially constrained models by systematically incorporating densely sampled seismic data indirectly related to lithology.

In the following paragraphs, the Monte Carlo simulation method is briefly described. The technique is then used to predict the lateral variations of lithology in an oil-bearing channel-sand reservoir in the Taber-Turin area of Alberta, Canada.

2. SIMULATION OF SAND/SHALE SEQUENCES

In the geostatistical framework adopted here, the unknown vertical sand/shale cross section is interpreted as a particular realization of a two-dimensional, statistically homogeneous binary random field $B(\tilde{x})$. At each point $\tilde{x} = (x, z)$ of the cross section, the random variable B is defined by

$$B(\tilde{x}) \begin{cases} = 1 \text{ if } \tilde{x} \text{ is in shale} \\ = 0 \text{ if } \tilde{x} \text{ is in sand} \end{cases} \quad (2.1)$$

with univariate probability distribution given by

$$\begin{aligned} \text{Prob } \{B(\tilde{x}) &= 1\} &= \text{ shale volume fraction} \\ \text{Prob } \{B(\tilde{x}) &= 0\} &= \text{ sandstone volume fraction}. \end{aligned} \quad (2.2)$$

At each point $\tilde{x}$ of the cross section, the unknown lithology must be inferred from observations of the variable B in nearby wells and from the knowledge of the seismic interval velocity, $V(\tilde{x})$, at that point. In practice, interval velocities are derived from reflection seismic data by converting the recorded amplitudes to velocity values. Figure 1 shows an example of velocity cross section derived from 90 seismic amplitude traces, spaced 20m apart in the horizontal direction. This velocity section may be seen as a collection of closely spaced, pseudo-sonic logs. The advantage of using seismic velocities in inferring lithology is that the seismic data provide spatially dense information between boreholes. However, in general, seismic interval velocities are ambiguously related to lithology. That is, the velocity ranges of sands and shales often overlap. For instance, depending on the presence of gas, pore pressure, age, mineralogy and porosity, reservoir sands can exhibit higher or lower velocities than those of adjacent shales. Here, we do not assume that there is a unique correspondence between B and V. Instead, we model their dependence statistically using a spatial crosscorrelation function.

An important step in the Monte Carlo simulation is to obtain a linear-mean-square (LMS) estimate of the following conditional probability at each point $\tilde{x}$ of the cross section:

$$\text{Prob }\{B(\tilde{x}), = 1 | B(\tilde{x}_1), ..., B(\tilde{x}_n), V(\tilde{x})\}$$

$$=$$

$$\text{Prob }\{B(\tilde{x}) | \text{data}\} . \quad (2.3)$$

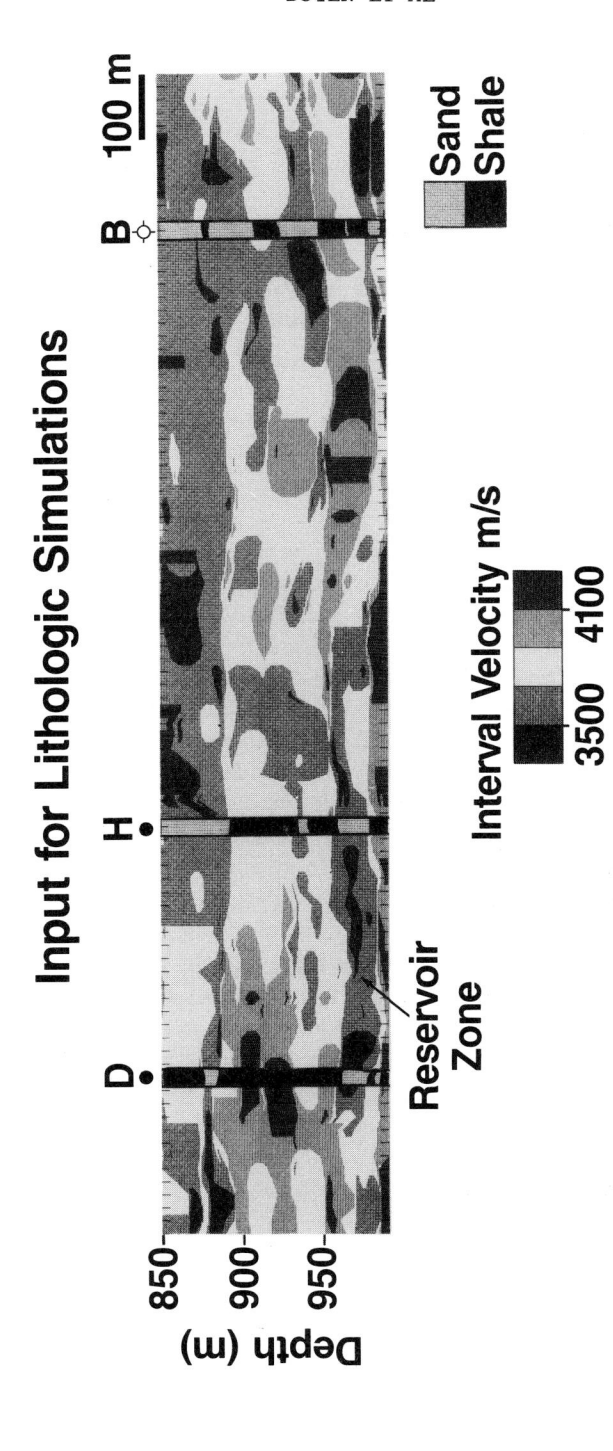

Fig. 1 Seismic internal velocity profile and interpreted lithology at three wells.

In the right hand side of equation (2.3), the term 'data' represents the data $B(\tilde{x}_1), ..., B(\tilde{x}_n)$, which are n binary-valued observations of the lithology at spatial locations $\tilde{x}_1, ..., \tilde{x}_n$ in the wells, and $V(\tilde{x})$, which is the seismic interval velocity at $\tilde{x}$. By definition of the binary variable B, this probability also is equal to the conditional expectation of $B(\tilde{x})$, given the data. The LMS estimate, $B^*(\tilde{x})$, is given by

$$B^*(\tilde{x}) = \text{Prob}^* \{B(\tilde{x}) = 1 \mid \text{data}\}$$

$$= \sum_{i=1}^{n} \omega_i(\tilde{x}) B(\tilde{x}_i) + \alpha(\tilde{x}) V(\tilde{x}) + c(\tilde{x}), \qquad (2.4)$$

where weights $\omega_1, ..., \omega_n$ and α assigned to the data are determined by minimizing the mean square error $E\{[B(\tilde{x}) - B^*(\tilde{x})]^2\}$. This minimization only requires knowledge of the spatial autocorrelation and crosscorrelation structures of the variables B and V. It is performed by solving a system of normal equations analogous to a cokriging system. In (2.4), the constant, $c(\tilde{x})$, is determined from the condition that the estimate is globally unbiased; i.e., $E[B^*] = E[B]$ = shale volume fraction. Note that the relative magnitude of weight α in (2.4) depends on the degree of local crosscorrelation existing between seismic velocity and lithology; i.e., the stronger the correlation, the larger the weight value. In practice this crosscorrelation is determined by comparing the lithology interpreted in wells with velocity data derived from seismic traces which are in the direct vicinity of the wells.

The probability estimate given in equation (2.4) mixes binary-valued observations with velocity data, which are continuous and can extend beyond the interval [0,1]. In practice, this scaling problem is solved by performing the LMS estimation with the variable V transformed so that its values lie between 0 and 1. At this stage, one limitation of the method is that it assumes error-free interval velocity observations. In reality, the velocities extracted from band-limited and noise-contaminated seismic data are inherently non- unique. However, if the error in the seismic velocity inversion can be quantified, then the Monte Carlo simulation technique can be constrained by bounds on the velocities rather than by single velocity estimates [5].

The sand/shale simulation method involves the sequential estimation of the conditional probability [equation (2.4)] at all sample points of the seismic velocity profile. The algorithm can be summarized as follows.

For all points $\tilde{x}$ in the cross section:

1. Obtain the LMS estimate
   Prob* $\{B(\tilde{x}) = 1 \,|\, \text{data}\} = B^*(\tilde{x})$,

2. At $\tilde{x}$, draw a simulated value B_s that is equal to 1 or 0 with probability $B^*(\tilde{x})$ and $1 - B^*(\tilde{x})$, respectively, and

3. Add the simulated value $B_s(\tilde{x})$ to the conditioning data set of the probability in Step 1 above.

Note that, when the first location $\tilde{x}$ is considered in the simulation, the conditioning data set contains only the observations of B at the wells. However, when the ith point is selected, the data set also includes the $(i-1)$ previously simulated values.

Using the above algorithm, it can be shown that, in theory, the simulated binary-valued field $\{B_s(\tilde{x})\}$ has the same autocorrelation and crosscorrelation structures as the real field $\{B(\tilde{x})\}$ itself. This property directly follows from the fact that, in the LMS estimation, the error $B - B^*$ is orthogonal to the data. A derivation of the equality of the autocorrelation functions is given in [1]. Also, it is easy to show that the simulated lithology honours the well information; i.e., $B_s(\tilde{x}) = B(\tilde{x})$ for $\tilde{x}$ in the wells.

3. CASE STUDY OF A CHANNEL-SAND RESERVOIR

The simulation technique is applied to predict the spatial arrangement of sands and shales in a section of the Upper Mannville Formation in the Taber-Turin area of Alberta, Canada. Figure 1 shows a seismically derived interval velocity cross section for the depth window of interest. This velocity profile was extracted from a seismic line using the Seismic Lithologic Modeling (SLIM^R)[1] process. SLIM is a forward modelling method in which the parameters (interval velocity, density and

---

[1]SLIM is a registered trademark of Western Atlas International, Inc.

thickness) of a 2-D layered model are perturbed iteratively. The convergence criterion for the SLIM iterative process is the improvement of the match between the model-derived seismic amplitude traces and the stacked/migrated seismic traces. The parameter perturbation process is performed over several iteration cycles through the full set of model parameters. It terminates when a satisfactory match is obtained between the seismic amplitude response of the model and the recorded data traces. More details on the derivation of the SLIM velocity model shown in Figure 1 can be found in reference [2].

At the base of the displayed velocity section is the 15- to 30-m-thick Glauconitic reservoir sand, which produces oil at wells D and H in the northern portion of the field. The velocity profile shows that the reservoir interval exhibits lower velocities than the overlying siltstones. Laterally, to the south, where dry well B was drilled, the channel sand is truncated by high velocity shales and tight siltstones.

Figure 1 also shows the major sand/shale intervals interpreted from log measurements at the three well locations along the seismic profile. Since only two lithologic classes are modelled in the simulation, the "shale class" includes not only actual shales, but also the tight siltstones and shaly sands in the clastic sequence. Comparison of the binary lithology with the velocity cross section demonstrates that the rocks belonging to the shale class on the average tend to exhibit higher velocities than the sands. Mathematically, this implies that the variables B and V are positively crosscorrelated. In this case, the coefficient of correlation between B and V, which was calculated at the wells, is equal to 0.7.

Note that the sign of this correlation may be reversed, for example, when dealing with low porosity sands or overpressured shales. Also, the magnitude of the correlation coefficient would decrease if the velocity ranges of the two lithologies overlapped more strongly. In this latter case, conditioning the simulations with seismic data would not necessarily enhance the sand/shale discrimination.

A geometrically anisotropic correlation model was selected for the lithologic variable B, with major and minor axes oriented in the horizontal and vertical directions, respectively. The autocorrelation function for B is defined by

$$C_{BB}(h_x, h_z) \propto \exp\left\{-[(\frac{h_x}{r_x})^2 + (\frac{h_z}{r_z})^2]^{\frac{1}{2}}\right\}, \qquad (3.1)$$

which is a function of the horizontal and vertical distances, h_x and h_z, in the plane of the cross section. In equation (3.1), $r_x = 100\,\text{m}$ and $r_z = 10\,\text{m}$ can be considered one third of the practical correlation ranges in the horizontal and vertical directions, respectively. In the vertical direction, this exponential correlation model was inferred from observations of B at the wells. In the horizontal direction, the wells are too sparse to allow direct estimation of the correlation structure. An approximate correlation length, $3r_x$, was therefore determined indirectly from the interval velocity data. By thresholding the velocity section at $V \approx 3500$ m/s, a coarse sand/shale cross-sectional model was obtained. This cross section was then used to estimate the horizontal correlation length.

Figure 2 shows two sand/shale models simulated from the SLIM velocity section and from the lithologic data at wells D and B (Figure 1). Well H is used as a test well at which the accuracy of the lithology prediction can be evaluated. The simulations were performed at the sample points of the velocity profile, which consists of 90 traces, each containing 71 samples. The spacing between samples in the horizontal and vertical directions is, respectively, 20 and 2 m. The simulated images represent alternative sand/shale distributions that are consistent with the geophysical and log information at hand:

- The simulated models exactly reproduce the sand/shale units observed in the two wells D and B where the lithology is assumed to be known.

- The simulations approximately reproduce the autocorrelation model [equation (3.1)] inferred from the reservoir data. The solid lines in Figures 3a and 3b represent the exponential autocorrelation model C_{BB} as a function of the horizontal and vertical distances, respectively.

The triangle and circle symbols correspond to experimental correlations calculated from the two simulations displayed in Figure 2. By

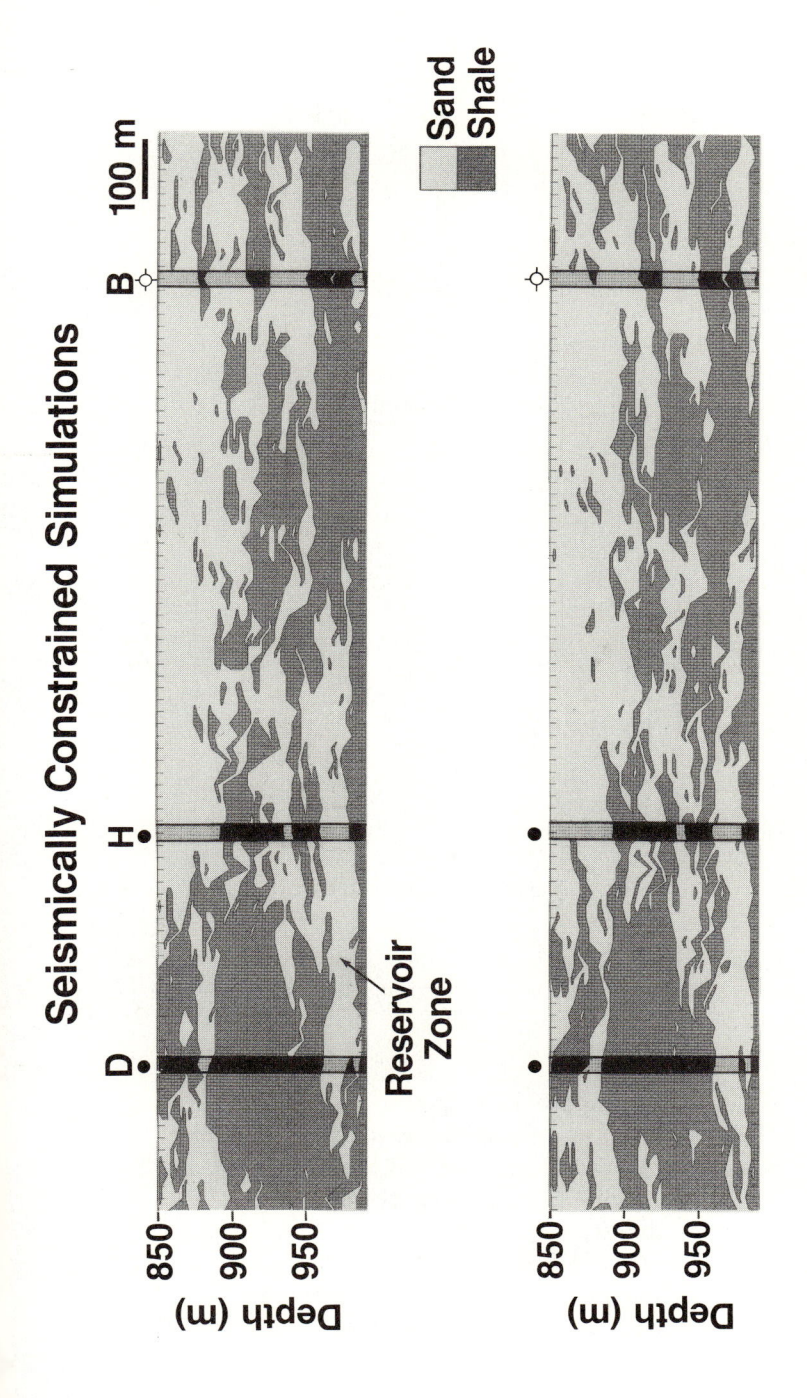

Fig. 2 Monte Carlo sand/shale models simulated from the SLIM velocity profile and the lithologic data at wells D and B.

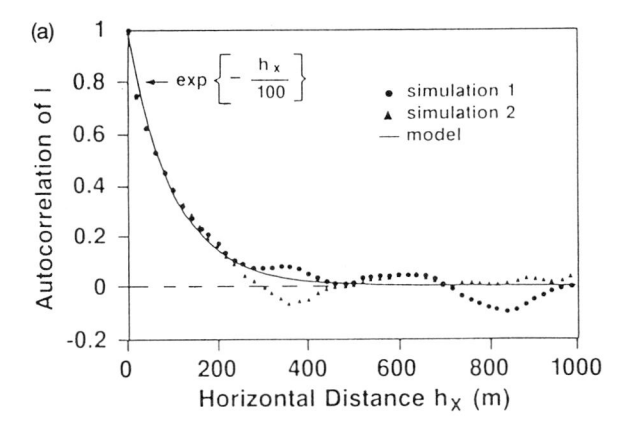

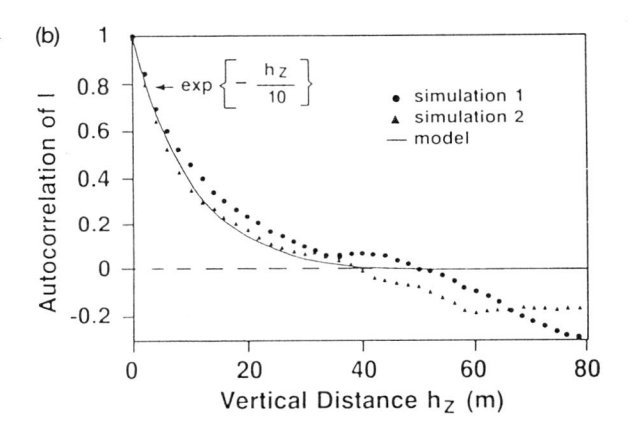

Fig. 3 Experimental and model autocorrelations of lithology as a function of the horizontal distance (a) and the vertical distance (b).

constraining the lithologic images to reproduce the correlation model, we are actually constraining the areal extent and thickness of the simulated sand/shale units. For instance, the distance at which the exponential curve drops to a level close to zero (i.e., $h_x = 3r_x = 300$ m) in Figure 3a is on the order of the maximum lateral dimension of the shale bodies.

- The simulations are also consistent with the spatial crosscorrelation existing between B and V. In particular, the coefficient of correlation between the simulated fields and the velocity profile is approximately equal to that inferred from the data.

Examination of Figure 2 shows that the morphology of the principal sand/shale units is similar in the two Monte Carlo models. In particular, both models depict the lateral truncation of the channel sand in the southern portion of the profile. Also, both models provide accurate lithology prediction at test well H, which was not used in conditioning the simulations. The stability of the simulated lithologic features directly results from constraining the simulations with the seismic data. To demonstrate this point, Figure 4 shows two lithology models simulated from the same well data set, but without using the seismic data to guide the interpolation away from wells D and B. In contrast to the seismically guided simulations, the two well-derived models are widely different and do not provide good estimates of lithology at well H. Also, as expected from the large distance separating the two control wells, the shape of the reservoir sand is poorly defined in the well-derived simulations.

Although the gross layering of the seismically derived lithology simulations is clearly defined, the models depicted in Figure 2 still differ in the detailed connectivity of the sand compartments. The variability from one simulation to the other is greatest in the areas of the seismic profile where the velocity value lies in the overlap region between the velocity ranges of the sand and shale classes. For instance, the exact position of the southern lateral truncation of the channel sand is not precisely defined in the simulations of Figure 2. This reflects the ambiguity of the seismic information in the lateral transition zone from sand to shale.

4. CONCLUSION

We have introduced a Monte Carlo technique for simulating sand/shale sequences from seismic interval velocity profiles and lithologic observations in wells. This technique yields alternative lithologic models that are conditioned by both the seismic and well data and by their autocorrelation and crosscorrelation structures. Traditionally, lithologic simulations are conditioned by sparse well observations alone. Successive simulations

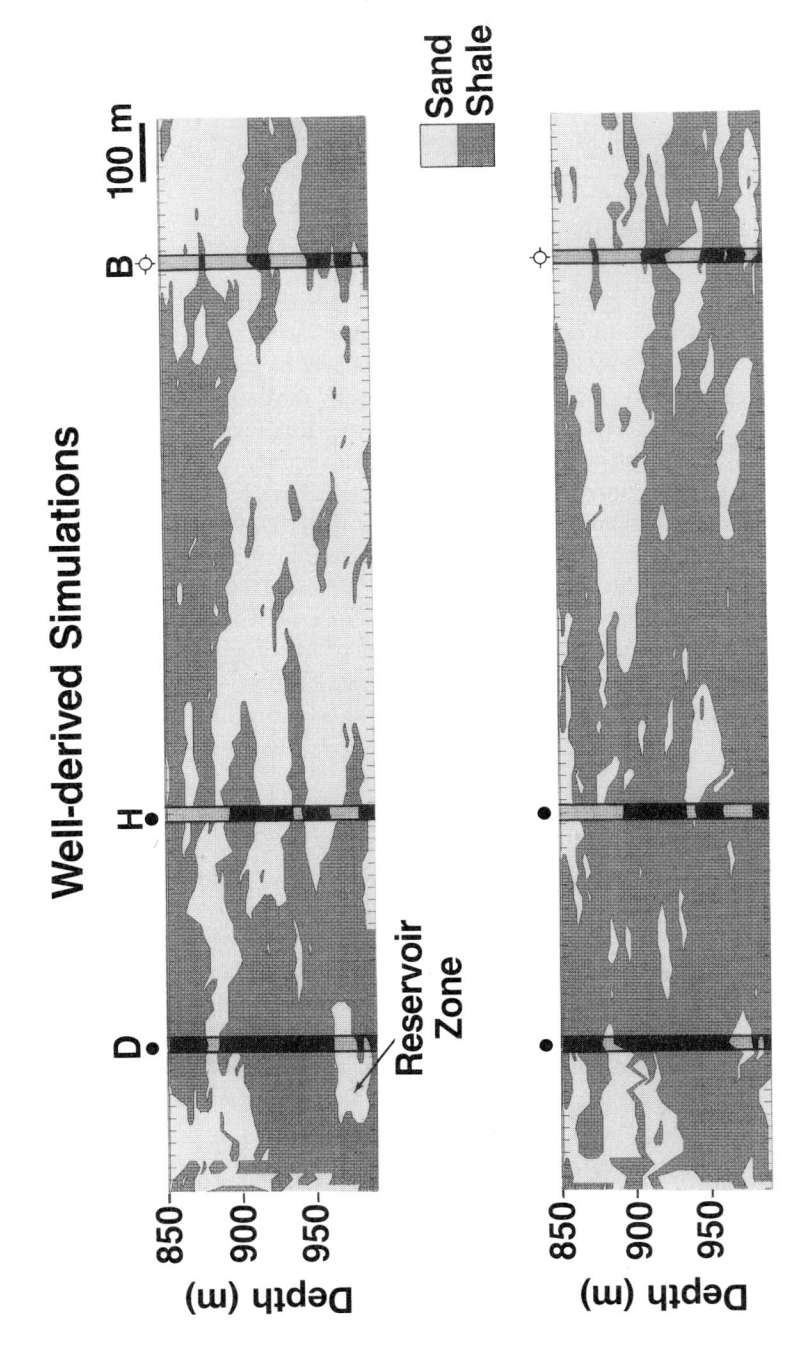

Fig. 4 Monte Carlo sand/shale models simulated from the lithologic data at wells D and B.

then are widely different in uncontrolled areas, revealing the underconstrained nature of the models (Figure 4). By contrast, the seismically consistent lithologic models displayed in Figure 2 are better constrained spatially and therefore exhibit more stable gross lithologic features.

In particular, these two models show the truncation of the reservoir sand by shales in the southern region. However, in keeping with the ambiguity of the velocity data, the seismically consistent simulations still differ in the connectivity and areal extent of the sand units. Analyzing the variability among successive simulations is useful for assessing the uncertainty associated with the subsurface models. For example, different simulated models, selected by the geologist, can be used in a flow simulator to forecast reservoir behaviour during production. By comparing the range of forecasts based on the selected models, we can determine the uncertainty in predicting the flow behavior that stems from an imperfect description of lithologic heterogeneities. Moreover, we can assess the need for additional geophysical or geological data to further constrain the reservoir models.

REFERENCES

[1] Alabert, F., 1987, Stochastic imaging of spatial distributions using hard and soft information: M.S. thesis, Stanford Univ.

[2] de Buyl, M., Guidish, T., and Bell, F., 1988, Reservoir description from seismic lithologic parameter estimation: J. Petrol. Technol., **40**, 4, 475-482.

[3] Desbarats, A. J., 1987, Numerical estimation of effective permeability in sand-shale formations: Water Resources Research, **23**, 273-286.

[4] Haldorsen, H. H., and Lake, L. W., 1982, A new approach to shale management in field scale simulation models: SPE Paper 10976, presented at the 57th Annual Fall Technical Conference and Exhibition of the SPE, New Orleans.

[5] Journel, A. G., and Alabert, F., 1988, Focusing on spatial connectivity of extreme- valued attributes: Stochastic indicator models of reservoir heterogeneities, SPE Paper 18324, presented at the 63rd Annual Technical Conference and Exhibition of the SPE, Houston.

[6] Matheron, G., Beucher, H., de Fouquet, C., Galli, A., Guerillot, D., and Ravenne, C., 1987, Conditional simulation of the geometry of fluvio-deltaic reservoirs: SPE Paper 16753, presented at the 62nd Annual Technical Conference and Exhibition of the SPE, Dallas.

Numerical Rocks

C.L. Farmer
AEA Petroleum Services

Abstract A new reservoir characterisation technique is developed which uses the frequency of rock types, the frequency of selected rock type pairs and the correlations, in several directions. A robust simulated annealing algorithm is described for generating 'numerical rocks' with a given description.

1 Introduction

An outstanding problem in mathematical geology has been to provide an algorithm for the synthesis of patterns with a prescribed histogram and correlation function [5]. In the following we describe a flexible and robust technique based on an application of simulated annealing [9,1] which solves this synthesis problem.

There has been some discussion in the literature [7, 8] concerning the adequacy of correlation functions as a means of capturing local structure. Standard correlation function methods can only be made to converge (in the sense that any pattern can be matched as closely as desired) by the trivial technique of increasing the number of points at which the pattern is conditioned. Techniques such as constrained Gibbs-Markov random fields [3] or Boolean methods [13] are able to improve on some aspects of the correlation function technique but they require the solution of difficult inverse problems to identify their control parameters.

By introducing a generalisation of the correlation function, the 'two-point histogram', we provide a technique which (i) can be proved to converge (ii) involves no inverse problems and (iii) has control parameters with an intuitive interpretation.

2 The Characterisation of Patterns

Let Ω be an array of cells, labelled $\underline{i} = (1, \ldots, i_d, \ldots, i_D)$, in D-dimensional space such that:

$$\Omega = \{\underline{i} : 1 \leq i_d \leq N_d, 1 \leq d \leq D\} \qquad (2.1)$$

where N_d is an integer specifying the number of cells in the $d-th$ direction. This restriction to rectangular systems is made only for convenience; all the techniques generalise to an arbitrary arrangement of cells.

Associate with each cell, $\underline{i}$, an integer, $r_{\underline{i}}$, with:

$$0 \leq r_{\underline{i}} \leq R - 1 \tag{2.2}$$

where R is an integer called the 'number of rock types'. Denote by Z, the set:

$$Z = \{0, 1, 2, \ldots, R-1\} \tag{2.3}$$

and by Z_p, the set of ordered pairs:

$$Z_p = \{(a, b,) : a \epsilon Z, b \epsilon Z\} \tag{2.4}$$

The field:

$$\underline{r} = \{r_{\underline{i}} : \underline{i}\epsilon\Omega\} \tag{2.5}$$

will be called a 'pattern' or a 'numerical rock'.

If we introduce a real valued function, k, with an integer argument, then the field:

$$\underline{k} = \{k_{\underline{i}} = k(r_{\underline{i}}) : \underline{i}\epsilon\Omega\} \tag{2.6}$$

can approximate any piecewise-constant function such as the permeability, porosity or other properties in a system of reservoir grid blocks.

2.1 One-point Histograms

Partition Ω into J disjoint subsets Ω_j such that:

$$\Omega = \bigcup_{j=1}^{J} \Omega_j \tag{2.7}$$

The one-point histogram $h^j(a;\underline{r})$ of the pattern $\underline{r}$ on the subset Ω_j is defined by:

$$h^j(a;\underline{r}) = \sum_{\underline{i}\epsilon\Omega_j} \delta_{a,r_{\underline{i}}} \quad \text{for } a\epsilon Z \tag{2.8}$$

where $\delta_{\alpha,\beta}$ is the Kronecker delta-function defined by:

$$\begin{aligned}\delta_{\alpha,\beta} &= 1 & \alpha = \beta \\ &= 0 & \alpha \neq \beta\end{aligned} \tag{2.9}$$

Thus $h^j(a;\underline{r})$ is the number of cells with the value a in Ω_j. The J subsets are introduced so that one-point histograms can be assigned to J sub-regions of a reservoir. This still admits the case $J = 1$.

The global one-point histogram $h(a;\underline{r})$ is defined by:

$$h(a;\underline{r}) = \sum_{j=1,J} h^j(a;\underline{r}) \qquad (2.10)$$

Define the average, $\bar{k}$, and the variance, σ^2, of $\underline{k}$ by

$$\bar{k} = \sum_{a \epsilon Z} k(a) h(a;\underline{r})/N \qquad (2.11)$$

$$\sigma^2 = \sum_{a \epsilon Z} (k(a) - \bar{k})^2 h(a;\underline{r})/N \qquad (2.12)$$

where N is the total number of cells in Ω.

2.2 Two-Point Histograms

The two-point histogram in the direction $\underline{m}$, $h_{\underline{m}}(a,b;\underline{r})$, of the pattern $\underline{r}$ is defined by:

$$h_{\underline{m}}(a,b;\underline{r}) = \sum_{\underline{i},\underline{i}+\underline{m} \epsilon \Omega} \delta_{a,r_{\underline{i}}} \delta_{b,r_{\underline{i}+\underline{m}}} \qquad (2.13)$$

$$\text{for } (a,b) \epsilon Z_p$$

where $\underline{m}$ is the non-zero D-dimensional vector $(m_1, \ldots, m_d, \ldots, m_D)$ with $0 \leq m_d \leq N_d$.

The summation in Equation (13) is over all pairs of cells $(\underline{i}, \underline{i} + \underline{m})$ such that both cells are in Ω. $h_{\underline{m}}(a,b;\underline{r})$ is the number of pairs (a,b) in the direction $\underline{m}$. The two-point histogram is assigned to the whole reservoir in contrast to the possibility of assigning individual one-point histograms to each sub-region.

The two-point histogram generalises the notion of 'grey-tone spatial-dependence matrices' introduced by Haralick et al [6] for classification, but not simulation, purposes. Journel and Alabert [7, 8] introduce a similar definition for the continuous case. The formalism of Serra [12] includes two-point histograms as a special case.

2.3 Correlations

We define the correlation, $c_{\underline{m}}$, for the field $\underline{k}$, in the direction $\underline{m}$, by the equation:

$$c_{\underline{m}}(\underline{r}) = \sum_{(a,b)\epsilon Z_p} k'(a)k'(b)h_{\underline{m}}(a,b;\underline{r})/\sigma^2 N_{\underline{m}} \qquad (2.14)$$

where $N_{\underline{m}}$ is the number of distinct pairs of cells, $(\underline{i}, \underline{i}+\underline{m})$, separated by the vector $\underline{m}$, such that both members of the pair are in Ω, and

$$k'(a) = k(a) - \bar{k} \qquad (2.15)$$

3 A Simulated Annealing Approach to the Generation of Numerical Rocks

3.1 A Distance Function Based on Two-Point Histograms and Correlations

Define a distance function Δ by:

$$\Delta^2(\underline{r}) = \sum_h \left[h_{\underline{m}}(a,b;\underline{r}) - h_{\underline{m}}^o(a,b) \right]^2 / \sum_h N_{\underline{m}}^2$$
$$+ \sum_c \left[c_{\underline{m}}(\underline{r}) - c_{\underline{m}}^o \right]^2 / N_c \qquad (3.1)$$

where $\sum_h$ is a summation over a (possibly empty) subset of values of $(a,b)\epsilon Z_p$ and a set of directions, L_h. $\sum_c$ is a summation over L_c, another (possibly empty) set of N_c directions (which may be different from L_h). The $h_{\underline{m}}^o(a,b)$ and $c_{\underline{m}}^o$ are either (i) control parameters or (ii) the two-point histograms and correlations of some control pattern $\underline{r}^o$.

In the following we describe a technique which solves the problem: Find a pattern (or patterns) $\underline{r}$ such that:

$$h^j(a;\underline{r}) = h_o^j(a) \text{ for all } a\epsilon Z, j=1,J \qquad (3.2)$$

$$\Delta(\underline{r}) \text{ is minimised for given control parameters} \qquad (3.3)$$

or control pattern, subject to the constraint

$$r_{\underline{i}} = r_{\underline{i}}^c \text{ for all } \underline{i}\epsilon\Omega_c \qquad (3.4)$$

where the h_o^j are prescribed one-point histograms. The conditioning set, Ω_c, is a subset of Ω and the $r_{\underline{i}}^c$ are specified conditioning values.

3.2 Simulated Annealing

Our approach to the problem stated above is to (i) generate a 'starting pattern' which exactly satisfies Equation (3.2) by assigning, for each j and

a, $h_o^j(a)$ values of a in Ω_J, in any convenient order, (ii) rearrange the starting pattern by successively interchanging or 'swapping' the values in pairs of cells until Δ is minimised. According to the theory of permutations we can obtain any pattern from any other by an appropriate sequence of such transpositions.

The method of simulated annealing is used to determine an appropriate sequence of transpositions. To control the annealing algorithm we introduce (i) the initial temperature, a positive real number T_o, (ii) the reduction factor, a real number $\lambda, 0 < \lambda < 1$, (iii) the maximum attempted number of swaps, a positive integer K_A (of the order of 100 times the number of cells), (iv) the acceptance target, a positive integer K (of the order of 10 times the number of cells) such that $K \leq K_A$, (v) the stopping number, a positive integer E (typically set to 2 or 3), and (vi) the stopping level $\Delta_E \geq 0$.

We call the parameters, T_o, λ, K_A, K, E and Δ_E an 'annealing schedule'.

During the simulated annealing algorithm we attempt to swap the values in randomly chosen pairs of cells, where neither cell is conditioned and such that the values are different. The criterion of acceptance is that

$$\Delta^2(\underline{r}^t) \leq \Delta^2(\underline{r}) - T \ln u \qquad (3.5)$$

where $\underline{r}$ and $\underline{r}^t$ are the numerical rocks before and after swapping respectively. T is the current temperature and u a random number from the uniform distribution on the interval $(0, 1)$.

The algorithm proceeds by starting at a high temperature T_o, chosen empirically so that all swaps are accepted. At the $n - th$ stage the system is 'cooled' by reducing the temperature T_n to T_{n+1} by

$$T_{n+1} = \lambda T_n \qquad (3.6)$$

At each temperature sufficient swaps are attempted so that either K swaps are accepted or K_A swaps have been attempted. If the required number of acceptances, K, is not achieved at a total of E different temperatures, or Δ drops to, or below, the threshold Δ_E, then the annealing is stopped. A good set of parameters needs to be chosen empirically. We have found that such a choice is not difficult.

An extensive discussion of simulated annealing can be found in Aarts and Korst [1]. Other useful discussions may be found in the original paper of Kirkpatrick et al [9], Kirkpatrick and Toulouse [10], Bounds [2] and Dolan et al [4].

The computationally intensive stage of the algorithm is in the evaluation of $\Delta(\underline{r}^t)$. However, if this is coded with care it is possible to find $\Delta(\underline{r}^t)$ from $\Delta(\underline{r})$ using relatively few arithmetic operations.

Implementation of the numerical rocks algorithm requires the generation of reliable uniform random numbers. This is done using the ACORN generator of Wikramaratna [14], which has the useful feature that systematic improvements in the level of randomness can be obtained by increasing an integer control parameter.

4 Discussion

4.1 Convergence of Simulated Annealing

We have found the simulated annealing algorithm to be surprisingly robust.

For problems whose control two-point histograms and/or correlations are obtained from a control pattern, we have always been able to devise an annealing schedule which results in patterns whose two-point histograms differ from the controls rarely by more than 3 pairs. Often several of the two-point histogram values are exact. The correlations are generally within two significant figures or better. Note that two-point histograms and correlations are generally in just a few directions.

If the two-point histogram control parameters are set without reference to a control pattern then we do not, generally, expect convergence to small values of Δ. Failure to find annealing schedules which give a small Δ is interpreted as implying that no pattern with the prescribed parameters exists. For example, it is impossible to produce a two-dimensional pattern with continuous stripes in both co-ordinate directions.

If only correlations are prescribed, with or without reference to a control pattern, then convergence is generally achieved to within two or more significant figures.

4.2 Convergence of the Characterisation Technique

It can be proved that if the two-point histograms, obtained from a control pattern, are specified in all possible directions then this information is sufficient to reconstruct the control pattern exactly. One can also show that, provided the pattern possesses particular reflection symmetries, the correlations (alone) suffice to reconstruct the control pattern exactly. Through extending a pattern in all co-ordinate directions by joining reflections together one can thus completely specify a pattern using just correlations.

We do not have room for these proofs in this paper. However, they use mathematical induction on the directions. The proofs begin with one of the corner pairs for which the values of the rock types are uniquely specified by the two-point histogram in one case and the correlation function in the other.

The utility of these results is indirect, since the amount of information in all the two-point histograms or all the correlations greatly exceeds the number of cells. The results do show that if one includes enough directions then one can specify a given pattern as accurately as one wishes. Numerical experiments indicate that this convergence is rapid, even when many length scales are present. However, practical success is dependent upon specifying a set of directions and some trial and error is needed for this. Although we regard convergence as an important property, our aim is not to provide a data compression technique, even if this is sometimes achievable. The primary aim is to provide a means for modifying or extending existing patterns.

4.3 Some Possible Applications of the Numerical Rocks Algorithm

We envisage that the numerical rocks technique may be used to:

(i) build models of existing patterns with controllable degrees of similarity;

(ii) patch a sub-region into an existing model such that a controllable degree of continuity between the surroundings and the patch can be attained;

(iii) extend/truncate existing patterns by extrapolating the two-point histograms;

(iv) extend patterns in dimensionality by postulating correlations and/or two-point histograms in the added dimension;

(v) modify existing patterns (in addition to adjustment to the single-point histograms) by changing correlations and two-point histograms during history matching.

4.4 An Example

To provide an illustration of the above remarks we generated two numerical rocks with the same global one-point histograms and control correlation functions. Each rock is composed of equal fractions of eight rock types. Table 1 displays the control parameters and results. In this example it is coincidental that the number of rock types equals the number of directions. In general any number of rock types and directions can be specified, limited only by the memory capacity and speed of the computer upon which the algorithm is implemented. Figure 1 is a density plot of the resulting pattern when only correlations in the chosen directions are controlled. Figure 2 shows the results when the two-point histograms $h_{\underline{m}}(7,7;\underline{r})$ are controlled, in addition to the correlations, for each of the eight values of $\underline{m}$. For comparison purposes Table 1 gives the $h_{\underline{m}}(7,7;\underline{r})$ for Figure 1 in which these parameters are not controlled.

In Figures 1 and 2 the lightest grey corresponds to a value of 1.0 with a linear variation up to the darkest grey whose value corresponds to 8.0. This defines the function, k, of Equation (2.6).

	$c_{\underline{m}}^o$ Figs 1 & 2	$c_{\underline{m}}$ Fig 1	$c_{\underline{m}}$ Fig 2	$h_{\underline{m}}^o(7,7)$ Fig 2	$h_{\underline{m}}(7,7;\underline{r})$ Fig 1	$h_{\underline{m}}(7,7;\underline{r})$ Fig 2
1, 0	0.8307	0.8298	0.8300	169	103	168
2, 0	0.5015	0.5016	0.5014	140	48	139
3, 0	0.1481	0.1484	0.1484	111	28	110
4, 0	-0.1511	-0.1512	-0.1514	82	15	82
5, 0	-0.4684	0.4680	-0.4682	53	0	54
0, 1	0.5000	0.4999	0.4997	59	64	59
0, 2	0.1000	0.0999	0.0997	18	32	20
0, 3	-0.1000	-0.1002	-0.1000	12	16	15

Table 1
Control Parameters and Results for Figures 1 and 2

From Figures 1 and 2 we see that control of the $h_{\underline{m}}(7,7;\underline{r})$ two-point histograms causes longer streaks of the high (i.e. black) values and yet the correlations are essentially identical.

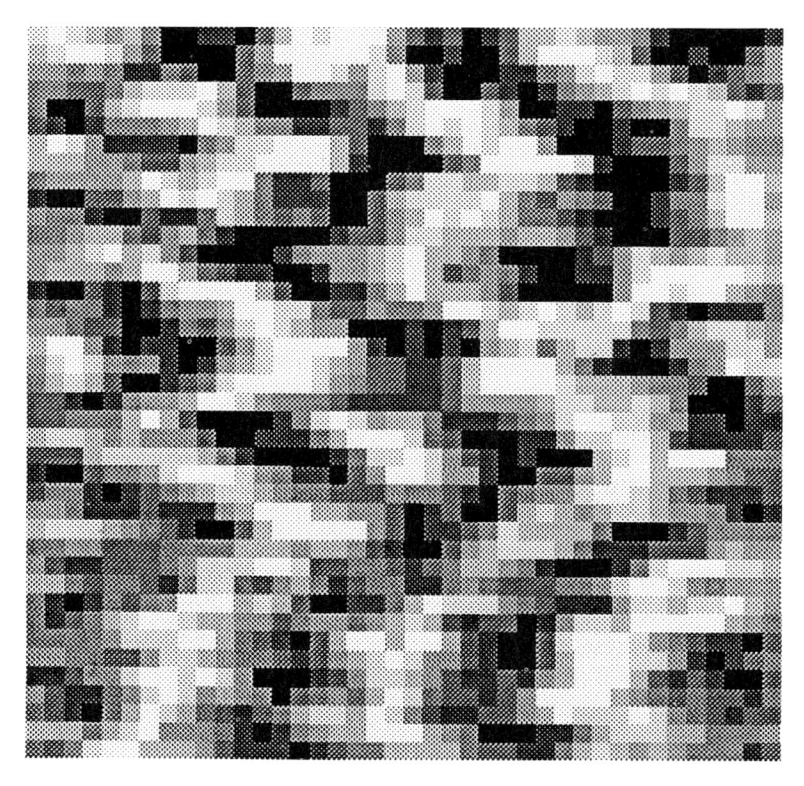

Figure 1. Correlations Controlled in 8 Directions

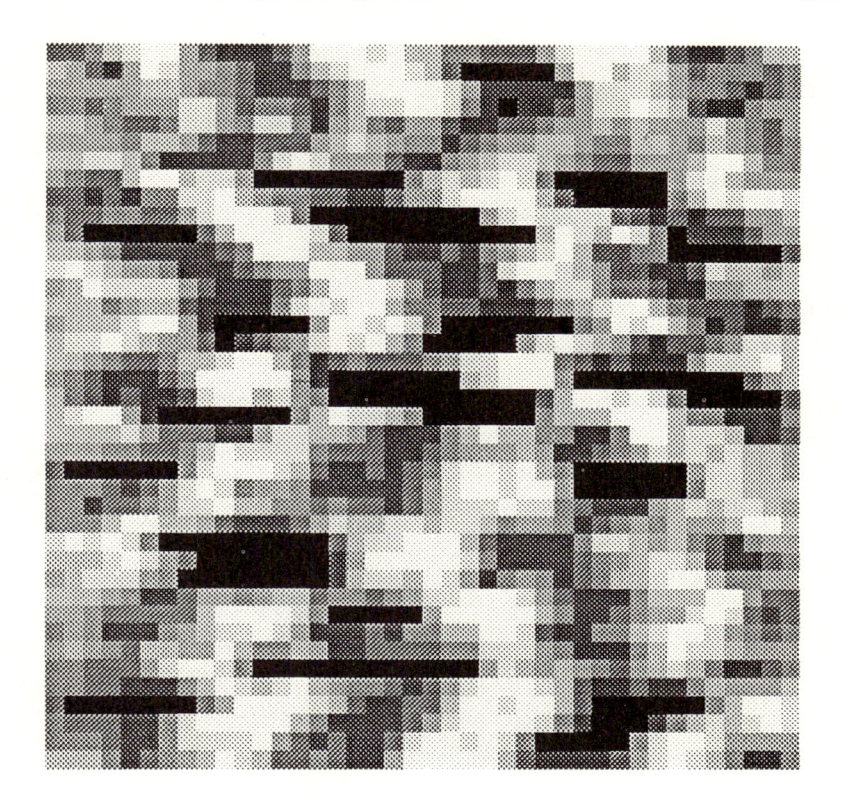

Figure 2. Correlations and (7, 7) Two-Point Histograms Controlled in 8 Directions

In both cases we used the annealing schedule

$$(T_o, \lambda, K_A, K, E, \Delta_E) = (1.0, 0.1, 16.10^4, 10^4, 2, 10^{-18}). \qquad (4.1)$$

5 Conclusions

A systematically controllable and convergent reservoir characterisation technique has been introduced. This method uses two-point histograms and correlations for description and a robust simulated annealing algorithm for generating patterns or 'numerical rocks' with the required properties.

6 Acknowledgements

I would like to thank S.G. Goodyear, J.M.D. Thomas, and R.S. Wikramaratna for their helpful comments. I would also like to thank the Editors for suggesting improvements in the notation and exposition. This research was supported by the AEA Technology Underlying Research Programme.

References

1. Aarts, E., and Korst, J., (1989), "Simulated Annealing and Boltzmann Machines", J. Wiley & Sons, Chichester.

2. Bounds, D.G., (1987), New Optimisation Methods from Physics and Biology, Nature 329, 215-219.

3. Cross, G.R., and Jain, A.K., (1983), Markov Random Field Texture Models, IEEE Transactions on Pattern Analysis and Machine Intelligence, Vol. PAMI-5, No. 1, 25-39.

4. Dolan, W.B., Cummings, P.T., and LeVan, M.D., (1989), Process Optimisation via Simulated Annealing: Application to Network Design, AIChE Journal, 35(5), 725-736.

5. Farmer, C.L., (1988), The Generation of Stochastic Fields of Reservoir Parameters with Specified Geostatistical Distributions. In "Mathematics in Oil Production", Ed. Sir Sam Edwards and P.R. King, Clarendon Press, Oxford, 235-252.

6. Haralick, R.M., Shanmugam, K., and Dinstein, I., (1973), Textural Features for Image Classification, IEEE Transactions on Systems, Man and Cybernetics, SMC-3, 610-621.

7. Journel, A.G., and Alabert, F.G., (1988), Focussing on Spatial Connectivity of Extreme-Valued Attributes: Stochastic Indicator Models of Reservoir Heterogeneities, SPE Paper No. 18324.

8. Journel, A.G., and Alabert, F.G., (1989), Non-Gaussian Data Expansion in the Earth Sciences, Terra Nova, 1 (2), 123-134.

9. Kirkpatrick, S., Gelatt, Jr., C.D., and Vecchi, M.P., (1983), Optimisation by Simulated Annealing, Science, 220 (4598), 671-680.

10. Kirkpatrick, S., and Toulouse, G., (1985), Configuration Space Analysis of Travelling Salesman Problems, J. Physique, 46, 1277-1292.

11. Ripley, B.D., (1988), "Statistical Inference for Spatial Processes", Cambridge University Press.

12. Serra, J., (1982), "Image Analysis and Mathematical Morphology", Academic Press, London.

13. Stoyan, D., Kendall, W.S., and Mecke, J., (1987), "Stochastic Geometry and its Applications", J. Wiley & Sons, Chichester.

14. Wikramaratna, R.S., (1989), ACORN - A New Method for Generating Sequences of Uniformly Distributed Pseudo-random Numbers, J. Computational Physics, 83, 16-31.

Modelling Flow through Heterogeneous Porous Media using Effective Relative Permeabilities Generated from Detailed Simulations

A.H.Muggeridge

BP Research, Sunbury Research Centre, Sunbury-on-Thames, Middx TW16 7LN ENGLAND

ABSTRACT

The amount of oil recovered from waterflooding an oil reservoir can be significantly altered by variations in the reservoir rock permeability. However the typical grid block dimensions used in conventional reservoir simulation cannot generally resolve the details of the permeability variations and hence accurately model the displacement. The most widely used method of representing the effects of heterogeneities is to replace the measured rock relative permeabilities in a reservoir model with pseudo relative permeabilities which have been chosen so that they reproduce the altered fluid flow in a coarse grid model.

Pseudo functions can be derived analytically for flow through layered reservoirs, but for more complex permeability distributions they have to be generated by matching coarse grid and fine grid simulations. The problem with most numerical methods for generating pseudo functions is that they seek to represent a number of different effects which makes it difficult to determine how the functions depend upon any one variable. In this paper we use *effective* relative permeabilities, rather than pseudo functions, to represent the effect of heterogeneities without compensating for different levels of numerical diffusion.

The aim of the paper is to use detailed simulation methods to study the properties and variability of effective relative permeabilities generated for three different types of permeability distribution. The effects of spatial correlations are investigated by comparing the effective curves obtained for the different permeability distributions, while the effects of anisotropy are investigated by comparing simulations of displacements conducted in different directions through each distribution.

From this work we conclude that

1) A fine, uncorrelated log-normal permeability distribution does not substantially alter the fluid flow.

2) The shape of effective relative permeabilities differs significantly from the rock curves for more correlated permeability distributions.

3) The shape of the effective relative permeability curves is not necessarily independent of flow direction.

4) The flow through a given permeability distribution is not necessarily linearly scalable. This means that effective relative permeabilities can not be generated for flow through sections of such a distribution and then used to reproduce the flow through the whole system.

1. INTRODUCTION

The amount of oil recovered from waterflooding an oil reservoir can be significantly altered by variations in the reservoir rock permeability. Spatial variations in both the relative permeability and the absolute permeability can mean that the water finds preferred flow channels resulting in an early breakthrough and a reduction in the amount of oil contacted. It is therefore important to be able to model permeability variations and understand how they alter the flow.

The typical grid block dimensions used in conventional reservoir simulation cannot generally resolve the details of the permeability variations and hence accurately model the displacement. One way of addressing this problem is to alter the input data used by the reservoir model so that simulations reproduce the average fluid flow observed in the heterogeneous reservoir.

The most widely used method of representing the effects of heterogeneities is to replace the measured rock relative permeabilities in a reservoir model with pseudo relative permeabilities. These pseudo functions have been chosen so that they reproduce the altered fluid flow in a coarse grid model.

Pseudo functions can be derived analytically for flow through layered reservoirs [1,2,3], but for more complex permeability distributions they have to be generated by matching coarse grid and fine grid simulations [4]. The problem with generating pseudos by this method is that most simulation codes cannot use the very fine grids required to represent all scales of heterogeneity. Instead a successive scaling up procedure has to be applied. This was the approach adopted by Lasseter *et al* [5] and Kossack *et al* [6] to enable them to investigate how different types of permeability distribution altered the fluid flow. However in the absence of suitable fine grid simulations techniques they could not check how accurately the pseudos generated by this procedure reproduced the average properties of the real displacement.

The problem with most numerical methods for generating pseudo functions is that they seek to represent a number of different effects which makes it difficult to determine how the functions depend upon any one variable. Clearly the number of variables represented can be reduced by eliminating physical effects, such as gravity and capillary pressure, from the fine grid simulations, but the pseudoization process also compensates for the increased levels of numerical diffusion present in the coarse grid model. Hence the pseudo functions will depend upon the averaging volume over which they are generated. However the effect of permeability variations on the pseudo functions may also change with the averaging volume as the larger the coarse grid block the greater the number of permeability fluctuations that need to be represented.

In this paper we use *effective* relative permeabilities, rather than pseudo functions. These effective relative permeabilities are derived from simulations in the same way as rock curves are derived from displacement experiments in the laboratary. They only represent the average effects of heterogeneities on a displacement and do not attempt to compensate for increased levels of numerical diffusion present in a coarser grid model. Using these effective functions we can represent a displacement through a two-dimensional, heterogeneous system on a one-dimensional, homogeneous, fine-grid model. This is equivalent to generating pseudo functions for a one grid-block model without having to compensate for numerical diffusion. The method has the advantage that the effective relative permeabilities can then be compared directly with the original rock curves.

The aim of the paper is to use detailed simulation methods to study the properties and variability of effective relative permeabilities generated for three different types of permeability distribution. The effects of spatial correlations are investigated by comparing the effective curves obtained for the different permeability distributions, while the effects of anisotropy are investigated by comparing simulations of displacements conducted in different directions through each distribution. The way in which the effective relative permeabilities are altered by the averaging volume used in their generation has already been described in a companion paper [7].

2. METHOD

High resolution reservoir simulation was used to understand the effects of the permeability variations on the detailed flow through the different permeability models and to generate effective relative permeabilities which represented the average properties of the flow. The numerical methods used in the simulation program have been fully described by Christie [8] and will not be repeated here. The program was originally developed to model viscous fingering in miscible displacements, but the combined requirements of both speed and accuracy necessary to model this type of flow apply equally well to the simulation of immiscible flow in heterogeneous media. Conventional black oil simulators would not have been sufficiently fast or sufficiently accurate to model the flow through permeability distributions containing small-scale variations and extreme permeability contrasts.

We modelled a constant rate line drive through the chosen permeability distribution, maintaining a constant pressure along the outlet face. The volume of water injected, the volume of oil produced, the injection rate and the mean pressure drop across the model were recorded as a function of time.

We used the method of Jones and Roszelle [9] to compute effective relative permeabilities from these data. This method is generally used to derive rock relative permeability curves from displacement experiments through cores. It is based on the analytical method described by Johnson, Bossler and Naumann [10]. The way in which we derive these effective functions is described in the Appendix. We call the curves generated at the end of this step *effective* relative permeabilities as they are not the same as the *pseudo* functions used by Lasseter et al [5] and Kossack et al [6]. They only represent the effects of the heterogeneities on the average fluid flow. They cannot be used directly in a coarse grid model as they do not compensate for the increased levels of numerical diffusion present in such a model.

The shape of the effective curves was not determined exactly for all water saturations, as to do so we would have had to continue the simulation of each displacement until all the oil originally in place had been produced. This would have been computationally expensive so we chose to halt the simulations at 2 pore volumes injected when determining the effective curves for the whole of a permeability distribution and at 5 pore volumes injected when generating effective functions for sections of the distribution near the inlet face. The shape of the curves between the saturation corresponding to this cutoff and the endpoint was then obtained by linear interpolation. The loss in accuracy resulting from this approximation was minimal as the amount of oil displaced towards the end of a waterflood is small.

The effect of correlation length on the flow was examined by comparing the results of simulations of displacements through three different permeability distributions with different correlation lengths. This gave us an understanding of how the detailed fluid flow was altered by the presence of heterogeneities. We could then see how these changes were reflected in the effective relative permeabilities by comparing these curves with the original rock curves used in the fine grid simulations.

We also investigated the effect of anisotropy in the permeability distribution by comparing the effective relative permeabilities generated when modelling line drives in different directions through the permeability distributions.

3. PERMEABILITY DISTRIBUTIONS

We modelled the flow through three different permeability distributions, each with a different correlation length. Each distribution was modelled on a 100×100 grid with an overall system length $1\frac{1}{2}$ times the width.

The first model reservoir consists of an uncorrelated, log-normal distribution (see figure 1). The distribution was isotropic with a geometric mean permeability of 100mD and a variance in $\log_e K$ of 20%.

The second distribution (illustrated in figure 2) was generated to represent a sandbody distribution containing five different types of sand, each with a different permeability and correlation length. The highest permeability sand has a correlation length in the x-direction that approaches the length of the model, while the lower permeability sands have correlation lengths of the order of one-fifth the reservoir length. The y-direction correlation lengths of all the rock types are of the order of one-sixth the width of the model. The permeability distribution within each sand-type was isotropic and the total range of permeabilities in the whole model covers some four orders of magnitude. This compares with a standard deviation of 52mD in the first model.

The third distribution (figure 3) represents a hypothetical channel sand made up of two rock types, each with a different mean permeability and

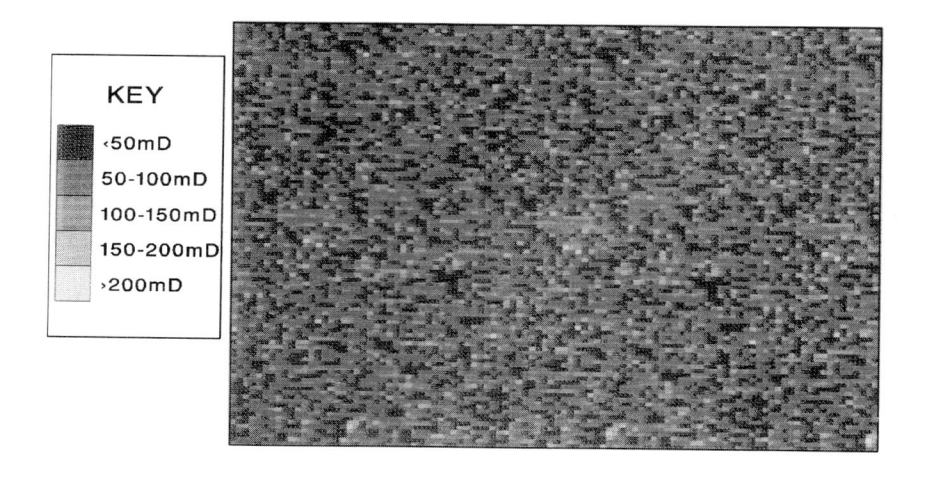

Figure 1: Log-normal, uncorrelated permeability distribution

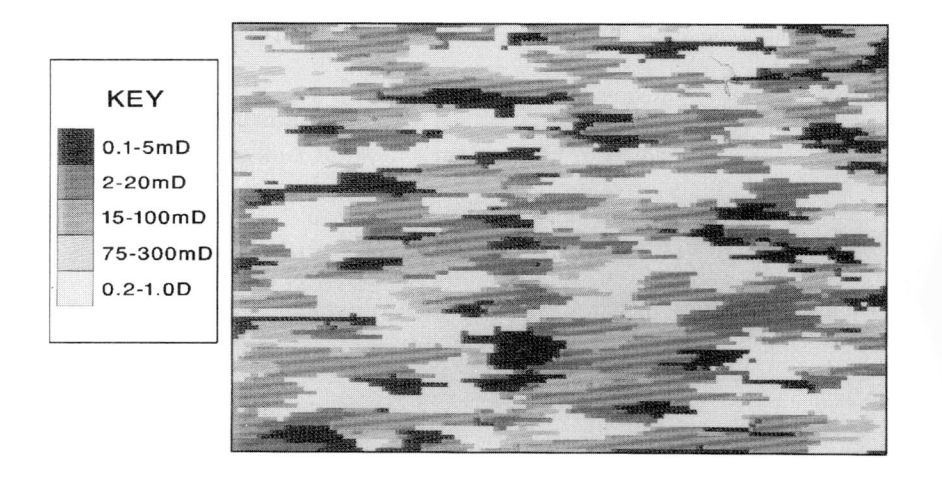

Figure 2: Sandbody permeability distribution

Figure 3: Channel sand permeability distribution

different permeabilities in the x-direction and y-direction. It can be seen that the distribution is highly correlated with obvious high permeability channels through the reservoir, although the difference in permeabilities (500mD in the high permeability sand and 50mD in the low permeability sand) between the two sands is much less than that seen in the sandbody distribution. In addition the amounts of high and low permeability sands is not constant across the model - there is an increase in the amount of low permeability rock towards the right of the distribution. This contrasts with the sandbody model in which there is no discernible trend in rock type along the distribution.

In the first two models the rock relative permeability curves were chosen to be independent of the absolute permeability. They were based on those observed in high permeability sands in a North Sea reservoir. Using a power law parameterization and normalized water saturations they can be written as,

$$k_{rw} = 0.45 S_w^2 \tag{1}$$
$$k_{ro} = (1 - S_w)^5 \tag{2}$$

In the third model there were two rock types with different permeabilities and these were assigned different relative permeabilities. The high

Table I: Effective permeabilities for different models.

	Log-normal distribution	Sandbody model	Channel sand model
x-direction	99.4	134.	249.5
y-direction	99.5	40.4	55.3

permeability rock curves were the same as those used in the first two permeability distributions (eqs. (1,2)). The low permeability rock curves were based on those observed in a low permeability sand from the same North Sea reservoir from which the high permeability rock curves were derived. They are given by,

$$k_{rw} = 0.3 S_w^3 \qquad (3)$$
$$k_{ro} = (1 - S_w)^3 \qquad (4)$$

We assumed that the critical water saturations and residual oil saturations were constant throughout each reservoir in order to simplify the problem. In practice different rock types would also have different critical water saturations and residual oil saturations. Similarly there should be different capillary pressure curves in rocks with different permeabilities, but again as we wished to concentrate on how the effective curves changed with permeability distribution we ignored this variation and assumed that capillary pressure was negligible. We used a constant oil/water viscosity ratio of 6 and a constant porosity throughout each permeability distribution.

The effective absolute permeability of each of the models for single-phase, one-dimensional flow in the $\pm x$-direction and $\pm y$-direction is shown in Table I. These were evaluated using the pressure solver method of Begg, Carter and Dranfield [11]. This technique uses the finite difference solution of the pressure equation to calculate the effective permeability.

It can be seen the effective permeability of the uncorrelated, log-normal distribution is essentially the same for flow in the x and y directions. The sandbody model has a ratio of k_y/k_x of 0.3 and the channel sand model a ratio of 0.2. This suggests that the effective flow along each of these models will be different from the effective flow across the models.

4. RESULTS

4.1 Random Permeability Distribution

Figure 4 shows the water saturation distribution predicted by high resolution simulation after 0.4 pore volumes of water have been injected into the random permeability distribution. Some small scale fingering of the water into the oil is evident.

The effective relative permeability curves that we generated for this displacement are compared with the rock curves in figure 5. They are virtually identical, as might be expected from the low level of perturbations in the oil-water front. Differences arise for saturations lower than the shock front saturation as there is insufficient information in a single displacement to determine the correct shape for the curves below this point. There is also no information concerning the shape of the curves for saturations greater than 0.5 as the displacement was only simulated until 2 pore volumes had been injected.

4.2 Sandbody Distribution

The fingering of water through the higher permeability flow channels in the sandbody distribution is evident in figure 6, which shows the saturation distribution obtained at 0.4 pore volumes injected. These preferred flow channels are a result of the larger correlation length of this distribution and the more extreme permeability contrasts.

The average saturation profiles observed in the sandbody model at different times are compared in figure 7. The distance moved by each saturation has been normalized so that the area under each average saturation profile is the same. This enables us to examine how the shape of the profile changes with time. It can be seen that the profile is very spread out when compared with that predicted for a homogeneous reservoir as the water channels through the high permeability paths in the rock. The profile shape is essentially constant with time – there are some small variations but each saturation moves with approximately constant speed. This means that the average motion can be satisfactorily matched with a fractional flow model *ie.* a hyperbolic first order equation.

Figure 8 shows the effective water and oil relative permeability curves calculated for the whole sandbody distribution for line drives in the x-direction and the y-direction. The oil curve for flow in the x-direction is remarkably similar to the rock curve considering the large difference in the mean saturation profiles for homogeneous and heterogeneous media seen in

Figure 4: Saturation distribution in log-normal permeability distribution at 0.4 (PVI) pore volumes injected.

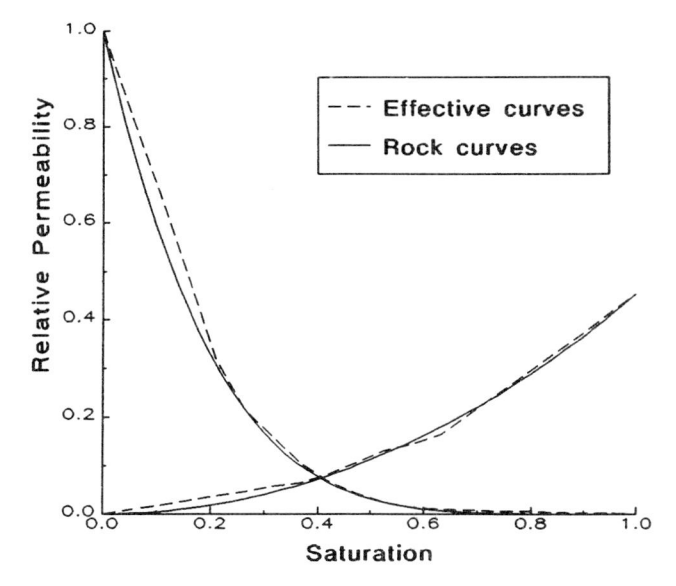

Figure 5: Effective relative permeability curves for log-normal permeability distribution

EFFECTIVE RELATIVE PERMEABILITIES 459

Figure 6: Saturation distribution in sandbody model at 0.4 PVI.

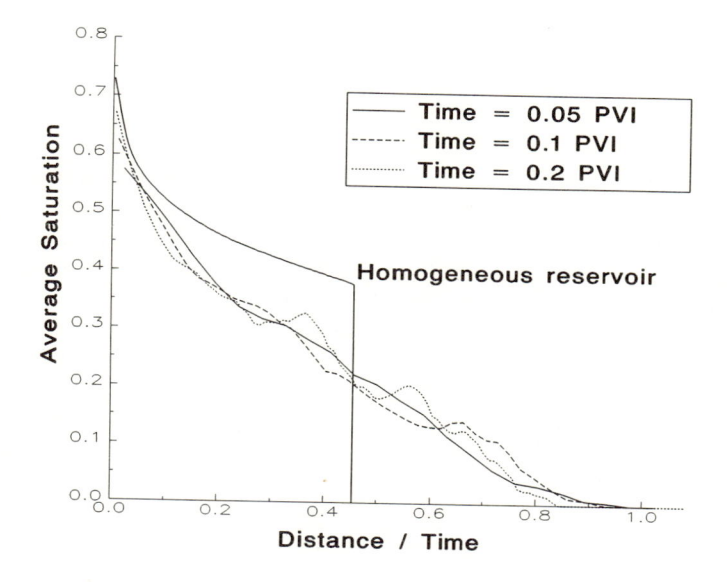

Figure 7: Saturation profiles along the sandbody model.

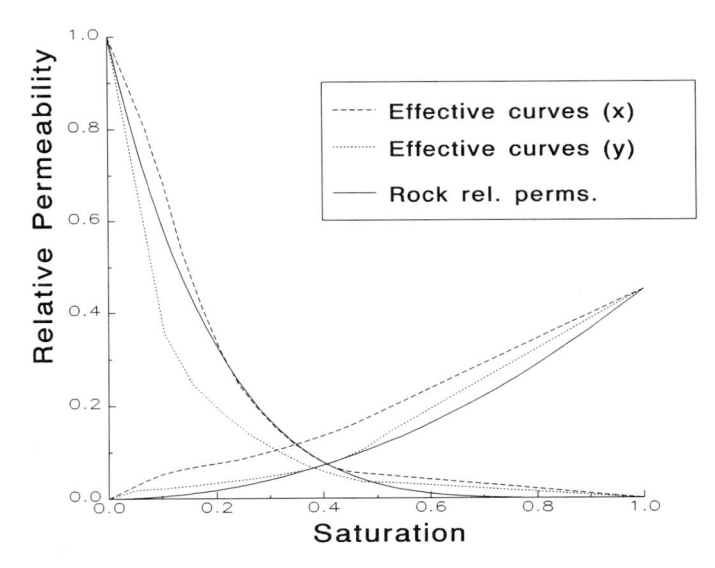

Figure 8: Effective relative permeability curves for whole sand-body model.

figure 7. The largest difference occurs between the rock and effective curves for water for flow in this direction. The effective water relative permeability is greater at all saturations for which the curves were calculated.

There is a considerable difference between the effective oil and water relative permeabilities generated for flow in the x and y directions. The y-direction effective water curve is closer to the rock curve in shape while at low saturations the effective oil relative permeability is rather less than the rock value. These directional variations in the effective relative permeabilities correspond to the less diffuse mean saturation profile observed in the y-direction displacement. There was no change in the shape of the effective curves when the flow direction was reversed in either the x or y-direction line drives.

4.3 Channel Sand Distribution

The saturation distribution predicted by fine grid simulation of a waterflood through the channel sand is shown in figure 9. The effect of the high permeability channels on the displacement is even more obvious than in the simulation through the sandbody distribution. The patches of higher water saturation in the picture show the buildup of water in the lower permeability sands. This buildup occurs because of the different rock relative curves that we used in the high and low permeability sands. There is a higher shock front saturation moving with a lower velocity in the lower permeability rock.

Figure 10 compares the shapes of the average saturation profiles along the model at different times. The distance scale has been normalized in the same way as in figure 7. Although the shapes of the curves are broadly similar at different times it can be seen that different mean saturations do not move with a constant speed. Saturations between 0.3 and 0.6 slow down as the front progresses along the model, while lower saturations (below 0.3) seem to speed up.

The effective relative permeability curves generated for a $+x$ direction line drive through the whole channel sand distribution are compared with the rock curves used for the two different rock types in figure 11. The effective curves are very similar to those used in the higher permeability sand and bear little resemblance to those for the lower permeability sand. The same higher water relative permeability at lower saturations is observed as was noted in the sandbody effective relative permeabilities. The endpoint of the effective water relative permeability curve was calculated so that the effective permeability of the system to water at residual oil saturation was correct.

In figure 12 the effective curves generated for line drives in different directions through the channel sand model are compared. There is a considerable difference between the effective curves for oil generated for flow in the $+x$ and $+y$ directions. There is also a significant difference between the curves generated for flow in the $+x$ and $-x$ directions.

5. DISCUSSION
5.1 Directional Dependence of Effective Curves

The fine grid simulations showed that the different permeability distributions modified the flow in a line drive in different ways and this was reflected in the different effective relative permeabilities generated.

The uncorrelated random permeability distribution did not really alter the flow at all despite the relatively high variance because the correlation length of the distribution was negligible. As a result the effective relative permeability curves for the displacement were virtually indistinguishable from the rock curves.

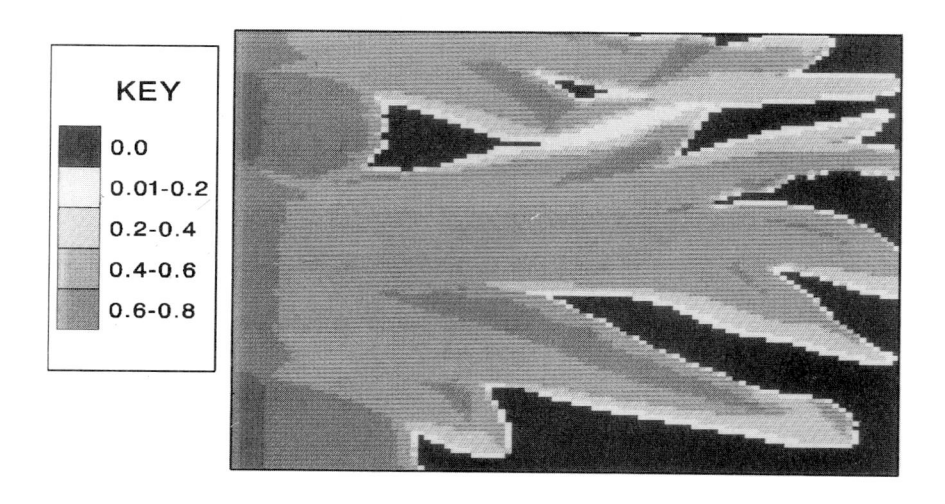

Figure 9: Saturation distribution in channel sand at 0.4PVI.

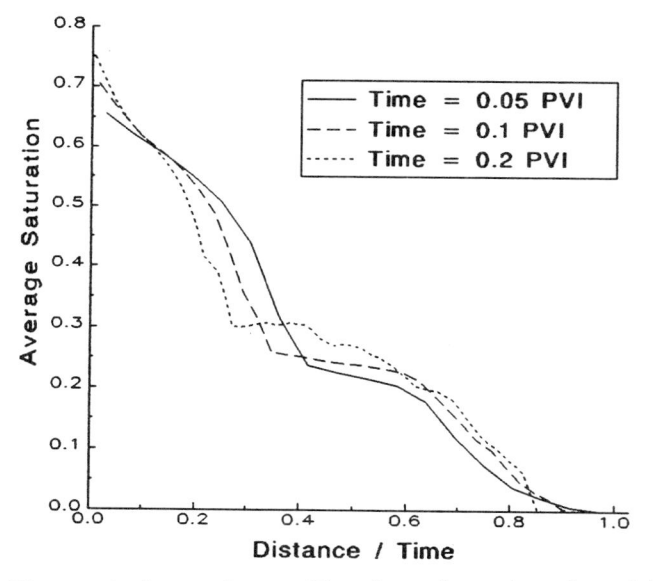

Figure 10: Saturation profiles along channel sand model.

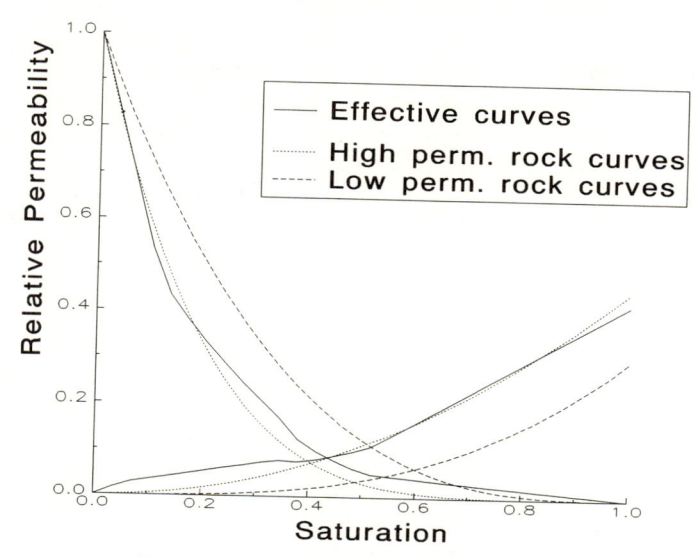

Figure 11: Effective relative permeabilities for whole channel sand compared with rock curves.

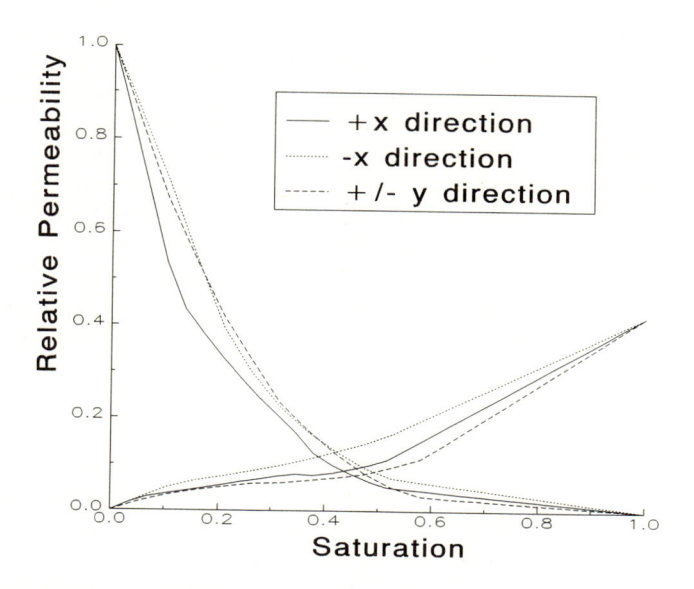

Figure 12: Effective relative permeabilities for whole channel sand for flow in different directions.

The x-direction waterflood through the sandbody model was significantly less efficient because of the greater correlation length and the more extreme range of permeabilities. The high permeability channels through the distribution meant that the mean saturation profile along the model was more diffuse. The effective relative permeability curves for the distribution are consistent with this – the oil curve is very similar to the rock curve while the effective water relative permeability is increased with respect to the rock curve, particularly at low saturations. This means that the water is more mobile, especially at low saturations, and hence moves more quickly through the system. The total mobility of the two fluids is also greater at low saturations in order to model the flow through the high permeability channels.

The difference in the effective curves generated for flow in the x and y directions can be attributed to the anisotropy in the sandbody distribution. All the rock types have a longer correlation length in the x-direction than in the y-direction. In addition the correlation length in the y-direction is the same for all rock types. The highest permeability sand has a much greater correlation length in the x-direction than the other rock types.

The channel sand also altered the displacement efficiency, again because of the greater correlation length of the distribution and the large difference in permeabilities between the two rock types. The effective relative permeability curves are very similar to the rock curves for the higher permeability sand because the flow occurs principally through this rock type. The higher effective relative permeability for water at low water saturations is again needed to represent the more diffuse mean water saturation profile.

The difference in the effective curves generated for flow along and across the channel sand model can again be attributed to the anisotropy in the permeability distribution. The difference in the curves generated for the $+x$ and $-x$ directions results from the variation in the distribution of low and high permeability sands along the model. This trend in the amount of low permeability rock is less significant for flow in the y-direction, explaining why the y-direction effective curves do not change when the flow is reversed.

5.2 Linear Scaling

When we generate effective relative permeabilities for a representative section of the reservoir we want to ensure that when we use these functions in a full field model they will accurately reproduce the flow through that model. We want to ensure that the effective curves are not unique to the particular displacement from which they were generated. One way to investigate how dependent these functions are on the way in which they were generated is to use the concept of linear scaling.

A displacement is linearly scalable if the saturation at any point along the mean flood front is a function only of the number of pore volumes injected with respect to that point (Rapoport [12]). This means that the shape of the mean saturation profile in any given displacement does not change with time. The effective relative permeabilities that we generate from such a displacement should be accurate when applied to problems other than the particular displacement from which they were generated.

In a linearly scalable displacement the effective curves will not depend upon the inlet and outlet boundary conditions used in their generation. Hence we will be able to generate effective curves from sections of a permeability distribution in which the flow has this property and still be able to use this function to reproduce the average flow through the whole model. If flow through the distribution did not have this property then we would have to generate effective functions from a fine grid simulation of a displacement through the entire reservoir.

From an examination of the mean saturation profiles shown in figures 7 and 10 we can infer that the displacement through the sandbody distribution is approximately linearly scalable whilst that through the channel sand is not. The shape of the mean saturation profile in the channel sand changes as the displacement progresses whilst the shape of the profile in the sandbody distribution does not.

We should be able to generate effective relative permeabilities from displacements either through the whole of the sandbody distribution or through separate sections of it and still reproduce the results of the fine-grid simulation. The reason for this is that the flow through large subsections of the model is very similar to the flow through the whole reservoir.

In contrast the effective relative permeabilities generated for subsections of the channel sand will not reproduce the flow observed in the fine grid model as the flow changes along the model. The mean saturation at any given position is not solely dependent upon the number of pore volumes injected with respect to that point, but also on the position of that point with respect to the inlet. Hence the flow in each subsection of the channel sand will be different depending upon whether that section is isolated (as is the case when the effective relative permeabilities are generated) or in its correct position within the whole distribution. Thus the effective relative permeabilities generated for each subsection by modelling a line drive through each individual section would not reproduce the flow observed in a displacement through the whole model.

6. CONCLUSIONS

We have used high resolution simulation to simulate a waterflood through three permeability distributions with different correlation lengths, thus giving us an understanding as to how heterogeneities altered the detailed flow. By examining how the mean saturation profiles in these simulations change with time we have been able to decide whether these displacements are linearly scalable and hence predict how the corresponding effective relative permeabilities will vary with the scale length of the permeability distribution.

The data from these simulations was used to generate effective relative permeability curves for the whole of each model for flow in the $\pm x$ and $\pm y$ directions. This reduced each heterogeneous two dimensional reservoir model to a homogeneous one dimensional problem. By comparing these effective relative permeabilities with the rock curves we were able to deduce how the shape of these curves depended on the principal direction of flow through the permeability distributions.

From this work we can draw the following conclusions:-

1) A fine, uncorrelated log-normal permeability distribution with a log variance of 20% does not substantially alter the fluid flow. The effective relative permeability curves generated for such a displacement are very similar to the rock curves, so the displacement can be adequately modelled in one dimension using the rock curves and a single effective permeability.

2) The shape of effective relative permeabilities differs significantly from the rock curves for more correlated permeability distributions. In particular the water relative permeability is increased at low saturations because of the channelling of the displacing fluid through the high permeability paths.

3) The shape of the effective relative permeability curves is not necessarily independent of flow direction. For the sandbody model the curves used to represent flow in the x-direction were different to the curves needed to represent flow in the y-direction. For the channel sand model the curves also changed when the flow was reversed.

4) The flow through a given permeability distribution is not necessarily linearly scalable. This means that effective relative permeabilities can not be generated for flow through sections of such a distribution and then used to reproduce the flow through the whole system.

In summary, different permeability distributions alter the flow in a line drive in different ways and this is reflected in the properties of the effective relative permeabilities. In addition the shape of the effective relative permeabilities generated to represent the flow can be sensitive to the direction of flow. This work did not investigate how different realizations of the models affected the effective functions. Further work is also needed to investigate the implications for two-dimensional flow and the successive rescaling of permeability distributions.

ACKNOWLEDGEMENTS

Permission to publish this work has been given by the British Petroleum Company plc. I would like to thank Dr. Steve Begg for providing the sandbody permeability distribution and the channel sand model used in this paper. I would also like to thank Dr. Mike Christie and Dr. John Fayers for helpful discussions during this work.

REFERENCES

1) Pande K.K, Ramey Jr., H.J., Brigham W.E. and Orr Jr., F.M., 'Frontal Advance Theory for Flow in Heterogeneous Porous Media', SPE 16344, 1987.

2) Reznik A.A., Enick R.M. and Panvelkar S.B., 'An Analytical Extension of the Dykstra-Parsons Vertical Stratification Discrete Solution to a Continuous Real-Time Basis', *Society of Petroleum Engineers Journal*, pp 643-655, December 1984.

3) Tompang R. and Kelkar B.G., 'Prediction of Waterflood Performance in Stratified Reservoirs', SPE 17289 presented at the SPE Permian Basin Oil and Gas Recovery Conference, March 10-11th 1988.

4) Kyte J.R. and Berry D.W., 'New Pseudo Functions to Control Numerical Dispersion', *Journal of the Society of Petroleum Engineers*, pp 269-276, August 1975.

5) Lasseter T.J., Waggoner J.R. and Lake L.W., 'Reservoir Heterogeneities and their Influence on Ultimate Recovery', NIPER Reservoir Characterization Conference 1986.

6) Kossack C.A., Aasen J.O. and Opdal S.T., 'Scaling up Laboratory Relative Permeabilities and Rock Heterogeneities with Pseudo Functions for Field Simulations', SPE 18436, presented at the SPE Reservoir Simulation Symposium, Houston, Texas, 6-8th February 1989.

7) Muggeridge A.H., 'Generation of Pseudo Relative Permeabilities from Detailed Simulation of Flow in Heterogeneous Porous Media', presented at the Second International Reservoir Characterization Technical Conference held in Dallas, Texas from June 25^{t}h-28^{t}h 1989.

8) Christie M.A., 'High-Resolution Simulation of Unstable Flows in Porous

Media', SPE *Reservoir Engineering*, August 1989.
9) Jones S.C. and Roszelle W.O., 'Graphical Techniques for Determining Relative Permeability from Displacement Experiments', *Journal of Petroleum Technology*, pp 807-817, May 1978.
10) Johnson E.F., Bossler D.P. and Naumann V.O., 'Calculation of Relative Permeability from Displacement Experiments', *Trans* AIME, *216*, pp370-372, 1959.
11) Begg S.H., Carter R.R. and Dranfield P., 'Assigning Effective Values to Simulator Grid Block Parameters in Heterogeneous Reservoirs', SPE 16754, presented at the 62nd Annual Technical Conference and Exhibition of the Society of Petroleum Engineers, September 27-30th 1987.
12) Rapoport L.A., 'Scaling Laws for Use in Design and Operation of Water-Oil Flow Models', *Trans.* AIME, *204*, pp 143-150, 1955.

APPENDIX
The Derivation of Effective Relative Permeabilities

Ignoring gravity and capillary pressure, effective relative permeabilities for linear flow through a chosen model are given by,

$$k_{rw} = \mu_w f_w \lambda_t \tag{1}$$

$$k_{ro} = \mu_o f_o \lambda_t \tag{2}$$

where μ_w and μ_o are the water and oil viscosities, λ_t is the effective total mobility and f_w and f_o are the effective fractional flows of oil and water. We need to determine the effective fractional flows and effective total mobilities in order to be able to calculate the effective relative permeabilities. We evaluate these quantities at the outlet of the reservoir model.

a) The calculation of effective fractional flow

The mean oil saturation in the model is given by,

$$\overline{S}_o = 1 - V_{op} \tag{3}$$

where V_{op} is the pore volumes of oil produced.

The saturation at a particular fractional distance $x(= \frac{l}{L})$ along the model is given by,

$$S_o(V_{wi}, x) = \lim_{\Delta x \to 0} \frac{\int_x^{x+\Delta x} S_o dx}{\Delta x} \tag{4}$$

$$= \overline{S}_{ox}(V_{wi}, x) + x \frac{d\overline{S}_{ox}}{dx} \tag{5}$$

where V_{wi} is the number of pore-volumes of water injected into the whole model and $\overline{S}_{ox}$ is the mean saturation in the model between the inlet at $x = 0$ and some point a distance x from the inlet.

Writing $\varsigma = \frac{V_{wi}}{x}$,

$$S_o(V_{wi}, x) = \overline{S}_o - \frac{d\overline{S}_o}{d\varsigma}\varsigma \tag{6}$$

At the outflow face of our model $x = 1$ (as we are using normalized coordinates) so the saturation at this point is given by,

$$S_o = \overline{S}_o - V_{wi}\frac{d\overline{S}_o}{dV_{wi}} \tag{7}$$

As the fractional flow of oil at the outlet is given by the gradient of the oil recovery curve,

$$f_o = \frac{d\overline{S}_o}{dV_{wi}} \tag{8}$$

we can obtain the effective fractional flow from,

$$f_o = \frac{\overline{S}_o - S_o}{V_{wi}} \tag{9}$$

b) Calculation of effective total mobility

From Darcy's Law, the effective total mobility at the any point along our model is given by,

$$1/\lambda_t = \frac{kA}{q}\frac{dp}{dl} \tag{10}$$

The effective absolute permeability for single phase flow is given by,

$$kA = \frac{q_o\mu_o L}{-\Delta p_o} \tag{11}$$

so we can write,

$$1/\lambda_t = \left(\frac{q_o\mu_o}{\Delta p_o}\right)\frac{1}{q}\frac{dp}{dl} \tag{12}$$

Hence the mean effective total mobility from the inlet to a position x in the model is given by,

$$1/\bar{\lambda}_t = \frac{q_o \mu_o}{\Delta p_o} \frac{\Delta p}{qx} \tag{13}$$

The point effective total mobility can then be derived in the same way as the point saturation from,

$$1/\lambda_t = 1/\bar{\lambda}_t - V_{wi} \frac{d(1/\bar{\lambda}_t)}{dV_{wi}} \tag{14}$$

VISCOUS/GRAVITY SCALING OF PSEUDO RELATIVE PERMEABILITIES FOR THE SIMULATION OF MODERATELY HETEROGENEOUS RESERVOIRS

John Killough and You Pei Fang
(University of Houston, Department of Chemical Engineering)

ABSTRACT

Pseudo relative permeabilities are derived based on a dimensionless parameter G, a measure of the ratio of viscous to gravity forces acting in the reservoir. For given reservoir fluids, G essentially represents the variation of viscous to gravity forces as a function of the average displacement velocity, horizontal permeabilities, and vertical permeabilities.

Pseudo relative permeabilities which are scaled by the parameter G are shown to have a limited sensitivity to reservoir descriptions for a reasonable variation in reservoir permeabilities. The derivations of the pseudos were based on simulations of several reservoir descriptions simulated at a number of different velocities.

The scaling of gas-oil pseudo relative permeabilities by G can be thought of as an approximation to partial gravity segregation in gas-oil displacement processes. The concept of partial gravity segregation can in turn be used to derive a dynamic pseudo gas-oil capillary pressure with viscous-gravity scaling. The pseudo capillary pressure is calculated from the average apparent residual oil saturation left behind a rapidly advancing gas-oil contact.

Finally, the concept of partial gravity segregation and viscous gravity scaling is validated through matches between one-dimensional and two-dimensional simulations.

1. INTRODUCTION

Pseudo saturation functions have been used in reservoir simulation for over twenty years to reduce grid sizes and improve model accuracy.[1-6] Despite efforts to increase the accuracy of numerical models through increased numbers of finite difference cells, the need for pseudo functions persists. One major difficulty in the use of pseudos is the need for a priori simulation of representative cross-sectional models. Mattax et al.[4] showed that pseudo relative permeabilities are sensitive to the velocity of displacement in a peripheral waterflood. They showed that through the use of velocity-dependent pseudos, only a limited number of cross-sectional simulations at different flood rates were required. Even with this velocity-dependence it was still necessary to simulate several cross-sections if reservoir description varied areally.

The following paper describes a technique for incorporating scaling of pseudo relative permeabilities by a dimensionless viscous-gravity ratio. The work emphasizes the difficult situation in which gravity override is not dominant and the vertical equilibrium assumption is no longer valid. Through the use of viscous-gravity scaling, the number of representative cross-sectional simulations and resultant pseudo functions can be reduced. In turn, viscous-gravity scaling allows modifications to be made to permeabilities during the history matching phase of simulation without destroying the underlying pseudo development.

3. MATHEMATICAL BASIS FOR VISCOUS/GRAVITY SCALING OF PSEUDOS

The effect of gravity segregation on recovery from secondary recovery processes has been recognized for many years. The correspondence between rate and gravity segregation has also been recognized for both gas and waterflooding of oil reservoirs. Work by Spivak[7] gives perhaps the best theoretical derivation of the effect of gravity. In this article Spivak shows that the degree of gravity segregation may be scaled by a dimensionless parameter G which relates viscous to gravity forces in the displacement process. The following derivations are minor modifications of the orignal Spivak work.

The three-dimensional, two phase equations in Buckley-Leverett form for flow in porous media, ignoring capillary pressure and solution gas effects, can be written as follows (See nomenclature for description of variables used.):

$$-\nabla \cdot \left(\frac{f_o}{B_o} v_t + \frac{K k_{ro}}{\mu_o B_o} f_g \Delta \rho \nabla Z \right) = \frac{\partial}{\partial t} \left(\frac{\phi S_o}{B_o} \right) - q_o \quad (1)$$

$$-\nabla \cdot \left(\frac{f_g}{B_g} v_t + \frac{K k_{rg}}{\mu_g B_g} f_o \Delta \rho \nabla Z \right) = \frac{\partial}{\partial t} \left(\frac{\phi S_g}{B_g} \right) - q_g \quad (2)$$

where,

$$\Delta \rho = \rho_o - \rho_g$$

For the two-dimensional case in a non-dipping reservoir, equation (1) can be written as follows:

$$\frac{\partial}{\partial x} \left(\frac{f_o v_t}{B_o} \right)_x + \frac{\partial}{\partial z} \left(\frac{f_o v_t}{B_o} \right)_z + \frac{\partial}{\partial z} \left(\frac{K_z}{\mu_o B_o} \Delta \rho k_{ro} f_g \right)$$
$$= \frac{\partial}{\partial t} \left(\frac{\phi S_o}{B_o} \right) - q_o \quad (4)$$

The derivation of the dimensionless gravity term G requires the definition of several other dimensionless quantities as follows:

$$x_D = \frac{x}{L_x}$$

$$y_D = \frac{y}{L_x} \sqrt{K_x/K_y}$$

$$z_D = \frac{z}{L_x} \sqrt{K_x/K_z}$$

$$t_D = \frac{v^* t}{\phi L_x}$$

where v^* is a characteristic superficial velocity, for example, q/A in a linear flood (ft /day). L_x is a characteristic length in the x-direction, (i. e., the total length of the reservoir), and K_i represents permeability in the i-direction. Assuming that flow due to the superficial velocity is only in the horizontal direction (only buoyant forces affect vertical flow) and substituting the dimensionless quantities into equation (4) (oil continuity equation) yields the following:

$$\frac{1}{L_x} \frac{\partial}{\partial x_D} \left(\frac{f_o v_t}{B_o} \right)_x + \frac{1}{L_x} \frac{\partial}{\partial z_D} \left(\frac{\sqrt{K_x K_z}}{\mu_o B_o} \Delta\rho \; k_{ro} f_g \right) + q_o$$

$$= \frac{v^*}{L_x} \frac{\partial}{\partial t_D} \left(\frac{\phi S_o}{B_o} \right) \tag{5}$$

Assuming that the total velocity is constant and equal to v^* in the x-direction and zero otherwise and rearranging yields:

$$\frac{\partial}{\partial x_D} \frac{f_o}{B_o} + \frac{\Delta\rho \sqrt{K_x K_z}}{\mu_o v^* B_o} \frac{\partial}{\partial z_D} (k_{ro} f_g)$$

$$= \frac{\partial}{\partial t_D} \left(\frac{\phi S_o}{B_o} \right) - q_o \left(\frac{L_x}{v^*} \right) \tag{6}$$

The scaling factor for the gravity group in equation (6) represents the ratio of gravity forces to viscous forces for these assumptions. With the additional assumption of small oil compressibility effects and addition of the appropriate conversion factor to make the group dimensionless for permeability in millidarcies, $\Delta\rho$ in psi/ft, μ_o in centipoise, and v^* in RB/(D·ft^2) the final expression for the viscous to gravity scaling parameter is given by the following equation:

$$G = 887.31 \frac{\mu_o q_t}{\Delta\rho A \sqrt{K_h K_v}} \quad (7)$$

In this equation the total flow rate q_t in RB/D and cross-sectional area perpendicular to flow, A, have been substituted into the expression for the superficial velocity v^*. The generic terms K_v (vertical permeability) and K_h (horizontal permeability) have also replaced the terms for K_z and K_x.

Although other derivations are possible involving terms such as gas viscosity in the parameter G, experimentation by Spivak showed that the term given in equation (7) was the best at scaling viscous to gravity forces. Intuitively this seems correct since the rate of gravity drainage for oil should only depend on the oil mobility and not the infinite-acting gas phase mobility.

An analysis of the parameter G gives further insight into the viscous to gravity force ratio. As permeability in both the horizontal and vertical directions increases, the gravity forces increase with respect to the viscous forces. The case of the vertical permeability seems obvious; for the horizontal permeability the case is not as clear. At a given flow rate q, as the horizontal permeability increases, the pressure drop in the horizontal direction of flow diminishes. The lowering of pressure drop results in a lessening of the viscous forces compared to the gravity forces. Because oil viscosity acts in both horizontal and vertical directions in a manner inversely proportional to the permeabilities, the appearance of viscosity in the numerator as a first power quantity appears reasonable. Finally, the superficial velocity q/A describes the superficial linear flow velocity in the horizontal direction. Intuitively, the greater the linear velocity in the horizontal direction, the less should be the relative effect of gravity.

4. DESCRIPTION OF THE CROSS-SECTIONAL MODELS FOR PSEUDO RELATIVE PERMEABILITIES

Four cross-sectional models were built to study the gas-oil pseudo relative permeabilities. These models are described in more detail in Tables 1-2. Basically, from ten to twelve layers were used for each of the models. Rock properties varied from near homogeneous to mildly heterogeneous layered systems. Relative permeability curves are shown in Figure 1.

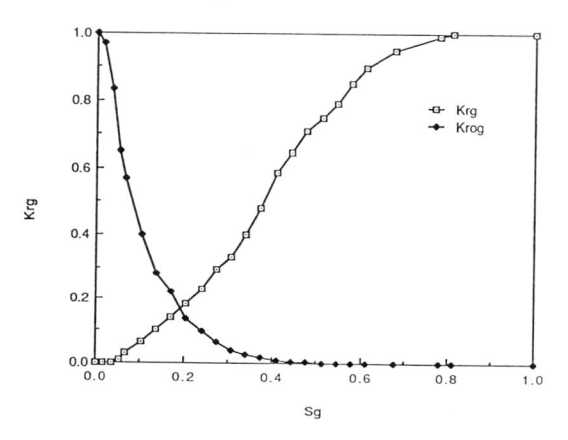

Fig. 1 Relative Permeabilities for the Cross-Sections

Figure 2 is a schematic of the grid used in the cross-sectional models.

TABLE 1
Reservoir Properties for the Cross-Sectional Models

Cross-Section 1 ($K_v/K_h = 0.24$, $\sqrt{K_v K_h} = 177.59$)

Layer	Permeability (md)	Porosity	Thickness (ft)
1	400.	0.151	15.0
2	400.	0.151	15.0
3	400.	0.176	15.0
4	228.	0.187	15.0
5	293.	0.235	15.0
6	293.	0.235	15.0
7	310.	0.236	15.0
8	310.	0.236	15.0
9	303.	0.241	15.0
10	361.	0.245	15.0
11	427.	0.224	15.0
12	625.	0.222	15.0

TABLE 1 (Continued)

Cross Section 2 ($K_v/K_h = 0.35$, $\sqrt{K_v K_h} = 406.70$)

Layer	Permeability (md)	Porosity	Thickness (ft)
1	615.	0.167	15.0
2	615.	0.167	15.0
3	664.	0.188	15.0
4	664.	0.188	15.0
5	757.	0.236	15.0
6	757.	0.236	15.0
7	725.	0.239	15.0
8	725.	0.239	15.0
9	676.	0.231	15.0
10	676	0.231	15.0

Reservoir Properties for the Cross-Sectional Models

Cross Section 3 ($K_v/K_h = 0.20$, $\sqrt{K_v K_h} = 215.11$)

Layer	Permeability (md)	Porosity	Thickness (ft)
1	498.	0.212	8.0
2	498.	0.212	8.0
3	498.	0.212	8.0
4	498.	0.212	8.0
5	472.	0.219	8.0
6	472.	0.219	8.0
7	472.	0.219	8.0
8	472	0.219	8.0
9	473.	0.221	8.0
10	473.	0.221	8.0
11	473.	0.221	8.0
12	473.	0.221	8.0

Cross Section 4 ($K_v/K_h = 0.14$, $\sqrt{K_v K_h} = 194.38$)

Layer	Permeability (md)	Porosity	Thickness (ft)
1	570.	0.236	5.0
2	570.	0.236	5.0
3	570.	0.236	5.0
4	570.	0.236	5.0
5	570.	0.236	5.0
6	570.	0.236	5.0
7	469.	0.235	5.0
8	469.	0.235	5.0
9	469.	0.235	5.0
10	469.	0.235	5.0
11	469.	0.235	5.0
12	469.	0.235	5.0

Table 2
Reservoir Fluid Properties for the Example Simulations

B_o = 1.378 RB/STB
B_g = 0.740 RB/MCF
μ_o = 0.900 cp
μ_g = 0.025 cp
R_s = 735.0 CF/STB

TABLE 3
Rates for Various Viscous To Gravity Parameters

Cross Section 1
G	Rate (RB/D)
0.1	1216.
0.5	6079.
1.0	12157.
2.0	24314.
5.0	60787.
10.0	121573.

Cross Section 2
G	Rate (RB/D)
0.1	2320.
0.5	11601.
1.0	23203.
2.0	46406.
5.0	116015.
10.0	232030.

Cross Section 3
G	Rate (RB/D)
0.1	785.
0.5	3927.
1.0	7853.
2.0	15706.
5.0	39265.
10.0	157060.

Cross Section 4
G	Rate (RB/D)
0.1	443.
0.5	2217.
1.0	4434.
2.0	8868.
5.0	22170.
10.0	44340.

CROSS-SECTIONAL MODELS FOR PSEUDOS

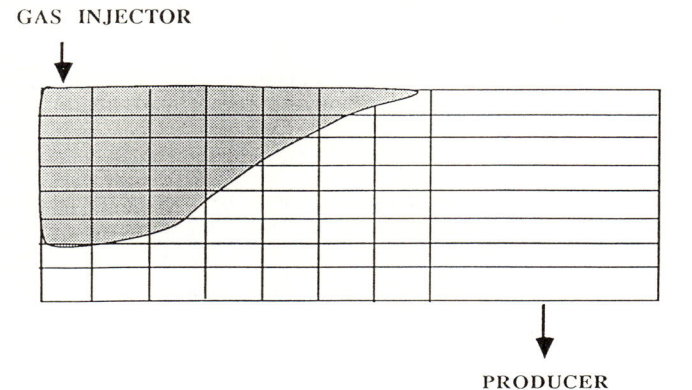

GAS INJECTOR

Fig. 2 Schematic of Grid for Cross-Section Simulations

The grid was 15 cells in the x-direction with 10-12 layers. As shown in Figure 2, simulation of the movement of gas across the reservoir involved injection of gas at the top of one end of the system and production of fluids at the bottom of the far end of the system. Since a gravity drainage process was being simulated, a large pore volume multiplier was applied to the far end of the system in an attempt to avoid the influence of boundary conditions on the fluid flow.

Scaling of the pseudo relative permeabilities was accomplished by simulating the displacement process for three different rates for each of the cross-sectional models. These rates corresponded to G values of 1.0, 5.0, and 10.0. Table 4 lists the specific rates which were used for the simulations. To derive these rates certain values in the expression for G were assumed to be as follows:

$$A = 1320 \text{ feet times total thickness}$$
$$\rho_o = 0.320 \text{ psi/ft}$$
$$\rho_g = 0.090 \text{ psi/ft}$$
$$\mu_o = 0.90 \text{ cp}$$

For example, using these values in cross-section 1 resulted in a reservoir barrel flow rate of gas of 12157.0 or about 16429.0 mcf/D at 4000 psi. The cross-sections were simulated with these rates for time periods long enough so that recoveries were forty to fifty percent of oil-in-place.

Table 4
Reservoir Properties for the One-dimensional Models

Cross Section 1

Layer	Permeability (md)	Porosity	Thickness (ft)
1	362.5	0.2116	180.0

Pseudo relative permeabilities were derived from these simulations using the standard techniques originally developed by Martin[1], Coats[2], and Hearn[3]. For each column and each time step of the simulations, the pseudo k_{rog} (oil relative permeability in the gas-oil two phase system) was derived as follows:

$$pk_{rog}(\tilde{S}_l) = \frac{\sum_{layers} K_{hi} \cdot H_i \cdot k_{rog}(S_{li})}{\sum_{layers} K_{hi} \cdot H_i} \quad (13)$$

The average liquid saturation $\tilde{S}_l$ is derived as the porosity thickness weighted saturation as follows:,

$$\tilde{S}_l = \frac{\sum_{layers} \phi \cdot H_i \cdot S_{li}}{\sum_{layers} \phi_i \cdot H_i} \quad (14)$$

Several observations can be made concerning the resultant pseudo curves. First, experiments showed that the choice of the column for averaging does not change the result as long as the column is at least two to three columns from the ends of the model (Figure 3).

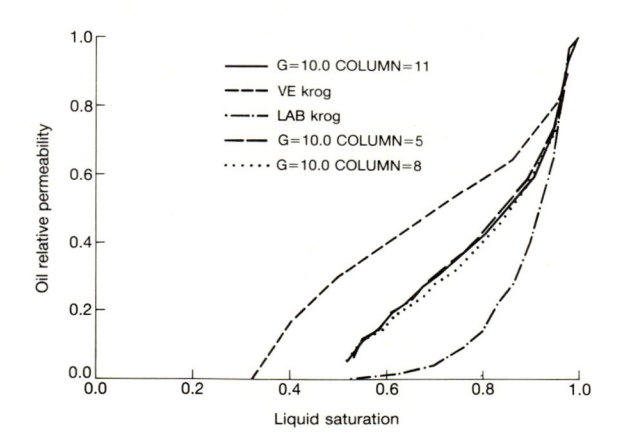

Fig. 3 Comparison of Pseudo k_{rog} for Different Columns

The choice of columns near the ends of the models showed differences due to boundary effects. Increasing the grid refinement in the vertical direction had little effect on the resultant pseudos; a coarser grid led to substantial deviations (Figure 4).

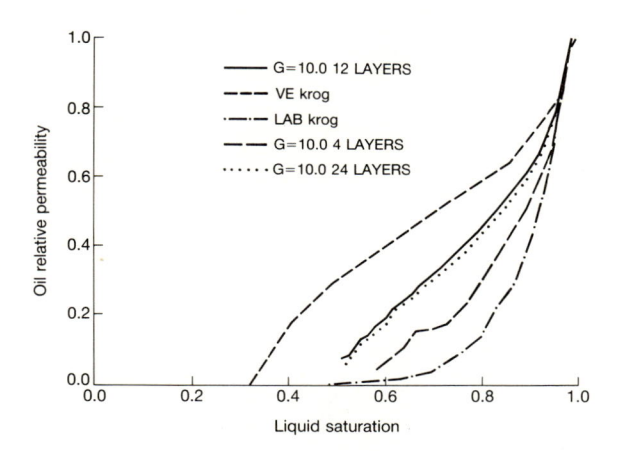

Fig. 4 Effect of Vertical Grid Refinement on Pseudo k_{rog}

The next section discusses the resultant pseudo k_{rog}

curves and the use of the parameter G for scaling viscous/gravity forces.

5. DISCUSSION OF THE PSEUDO GAS-OIL RELATIVE PERMEABILITIES

Figures 5 through 8 present the pseudo k_{rog} curves which were derived from averaging the simulation results of the four cross-sectional models.

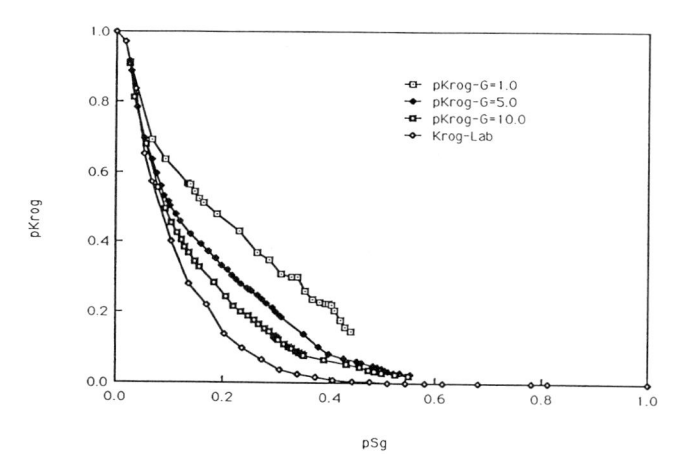

Fig. 5 Pseudo k_{rog} s for Cross-Section 1

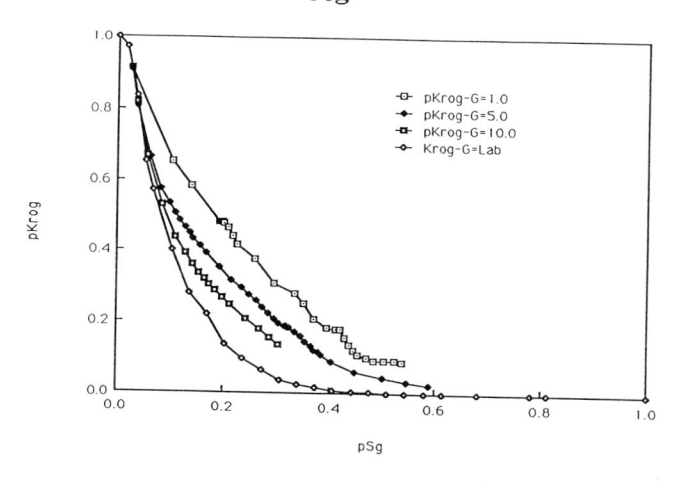

Fig. 6 Pseudo k_{rog} s for Cross-Section 2

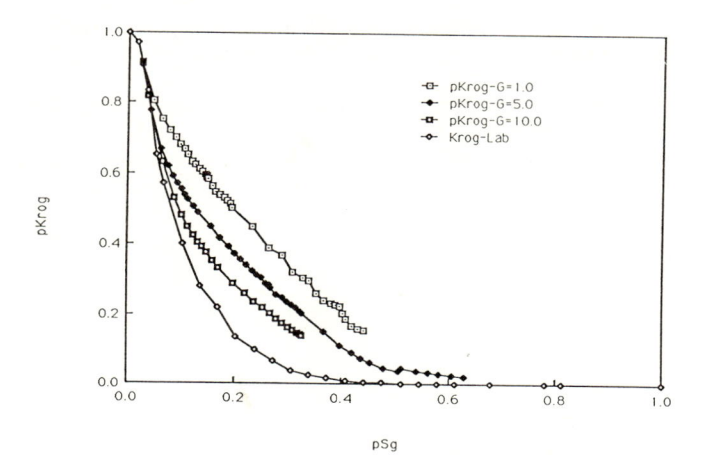

Fig. 7 Pseudo k_{rog}s for Cross-Section 3

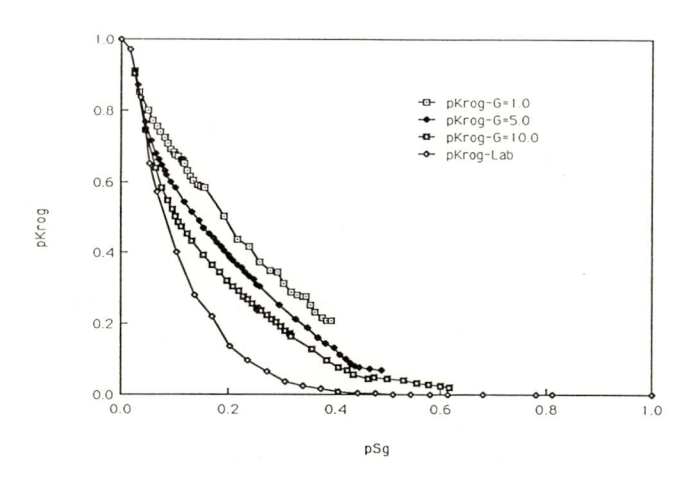

Fig. 8 Pseudo k_{rog}s for Cross-Section 4

For comparison the laboratory and vertical equilibrium relative permeabilities are also shown. As can be seen in these figures, a definite effect of viscous to gravity forces is evident; the higher the factor G, the more the pseudo approaches the laboratory curve. Conversely, for smaller values of G the pseudos approach the completely gravity segregated or vertical equilibrium curves.

To compare the validity of the use of the parameter G for scaling the pseudos over a range of reservoir characteristics the pseudos for G=1.0 were plotted together in Figure 9.

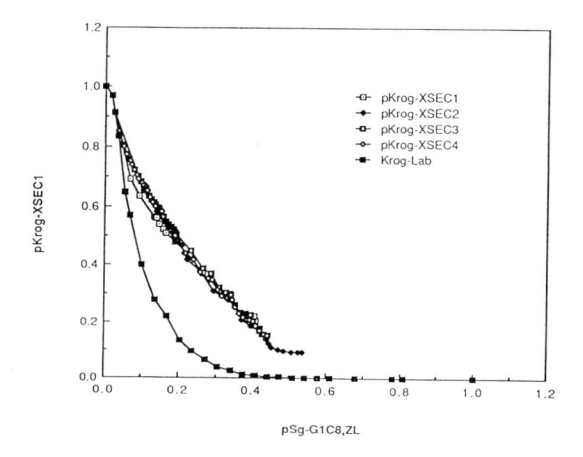

Fig. 9 Comparison of k_{rog} for G=1.0

As shown in this figure the deviation between the pseudos for different descriptions is small.

From this comparison it can be reasonably concluded that a single set of pseudo k_{rog} curves can be used for this variation in reservoir description given that a single laboratory relative permeability applies everywhere.

Figure 10 shows a comparison of the pseudos for three cross-sections with different properties but the same value of G.

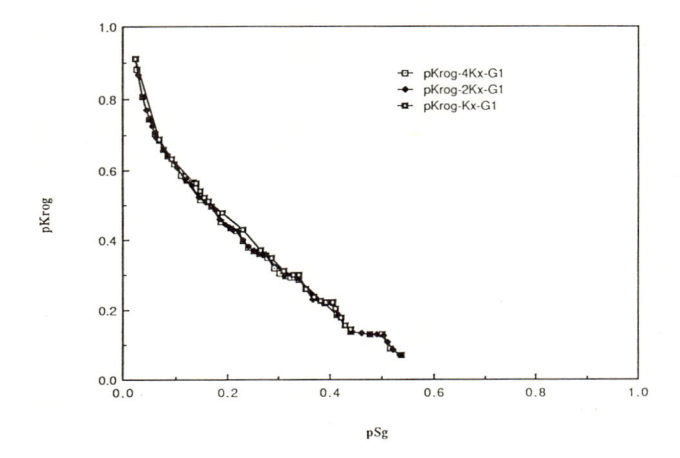

Fig. 10 Comparison of X-sec 1 with 2 and 4 Times K_h

The standard cross-section 1 is compared with the same cross-section having twice and four times the horizontal and vertical permeabilities and flow rate q increased by factors of two and four to maintain the same value of G. The resulting pseudos in the figure are essentially identical.

Further validation of the scaling of pseudos was performed on cross-section 1 by randomly permuting the permeabilities in each column of the cross-section. As shown in Figure 11, the scaling with G=1.0 for three different permeability multiplying factors appears identical.

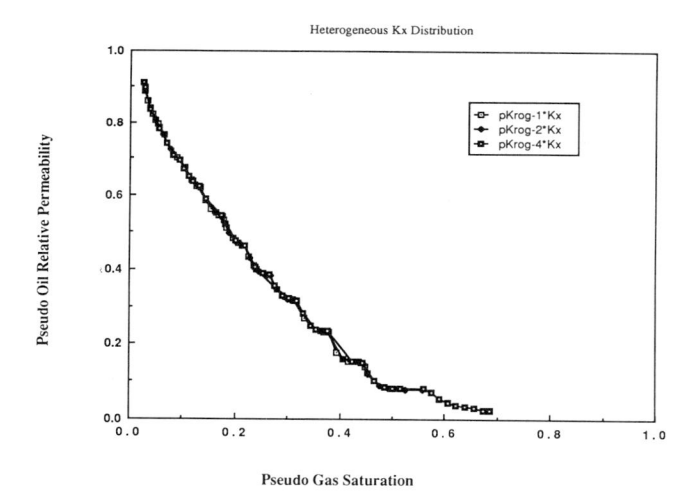

Fig 11. Comparison of Scaling for Random Permeabilities

6. VELOCITY-DEPENDENT S_{org} (S_{orv})

The cross-sectional results indicate the concept of a velocity-dependence can be extended to S_{org}, the residual oil to gas flooding for gravity drainage mechanisms. The variation with time of residual oil saturation left behind an advancing gas tongue or gas cap depends on the velocity of displacement.

To verify this concept, velocity-dependent S_{org} ($S_{orv} = S_{org}(v)$) values were calculated for each of the displacement velocities used in the development of the cross-section 1 pseudo k_{rog} s. S_{orv} was taken to be the pore-volume-weighted average of the oil saturations above the gas-oil contact. The gas-oil contact is defined as the point at which the gas saturation in a column is above 5%. As shown in Figure 12, not only do the values of S_{orv} vary with velocity (or G) but also with time; however, a medium term asymptotic limit is approached for each value of G.

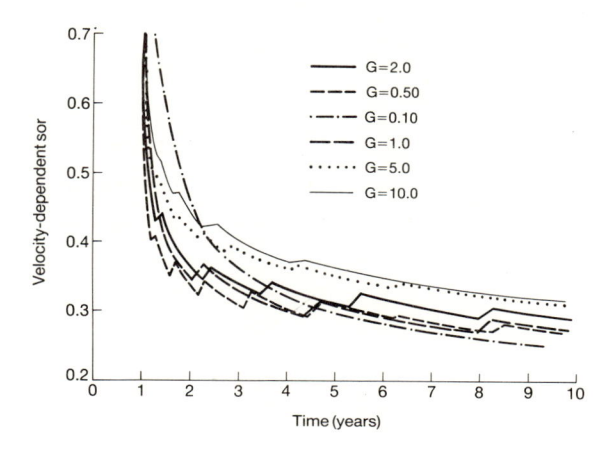

Fig. 12 Comparison of Apparent Residual Oil Saturations

Of course, the ultimate residual oil for all cases will be the value of zero oil relative permeability for the laboratory curves (approximately 13%).

These results indicate that the use of a pseudo gas-oil capillary pressure equal to the product of the oil-gas density difference times the height of the gas cap or gas tongue invasion results in improved potentials for both vertical and horizontal flow. The height of the gas cap or tongue can be determined from the value of S_{orv}.

Pseudo gas-oil capillary pressures have been shown to improve recovery predictions for drainage of oil from shales and gas tongue movement in large gridblock models[5]. To extend the concept of partial gravity segregation to pseudo capillary pressures all that is required is the combination of the ideas of pseudo capillary pressure (as defined above) and the concept of velocity-dependent S_{org} (S_{orv}). Inclusion of S_{orv} allows the calculation of the thickness of invasion of the gas-oil contact. The thickness of invasion is simply given by the following formula:

$$H_{inv} = \frac{(S_g - S_{gc} - S_w)}{(1.0 - S_{orv} - S_{gc} - S_w)} \cdot H$$

where S_g is the overall grid cell gas saturation, H is the cell thickness and S_{gc} is the critical gas saturation. Pseudo capillary pressure pP_{cgo} is then given by the following:

$$pP_{cgo} = \Delta\rho \, H_{inv}$$

The use of S_{orv} in the formula for H_{inv} brings velocity-dependence into pP_{cgo} since S_{orv} is dependent on velocity (or G). S_{orv} is only used in the calculation of the pseudo capillary pressures and is not used to modify pseudo relative permeabilities.

7. VALIDATION OF THE PSEUDO GAS-OIL SATURATION FUNCTIONS

To validate the gas-oil saturation functions (both relative permeabilities and capillary pressures) a classical approach was taken. In this approach the cross-sectional model used to generate the pseudo curves is collapsed into a one-dimensional model. The one-dimensional model is simulated with the pseudo curves and the results of the one and two-dimensional models are compared. If the pseudos are correct, the two results should be nearly identical.

The specific case for the comparison was the G=2.0 cross-sectional 1 simulation. Properties of the model are given in Table 4. In this case a constant value of S_{orv} equal to 0.334 was chosen for the pseudo capillary pressures. The pseudo k_{rog} was taken from the cross-sectional simulation for G=2.0. Vertical equilibrium data were used as further comparisons with the cross-sectional simulations.

Three cases with different gas-oil saturation functions were simulated for ten years to compare with the original cross-sectional model. These cases used laboratory saturation functions, cross-section 1 vertical equilibrium pseudos, and the G=2.0 cross-section 1 pseudo curves.

Figures 13 and 14 compare the gas saturation profiles for the four different cases - the original cross-section and the three one-dimensional models at times four and ten years.

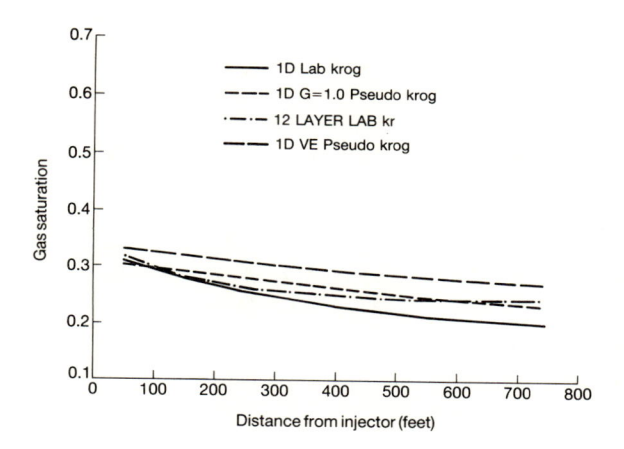

Fig. 13 Comparison of 1-D and X-sec Simulations at 4 Years

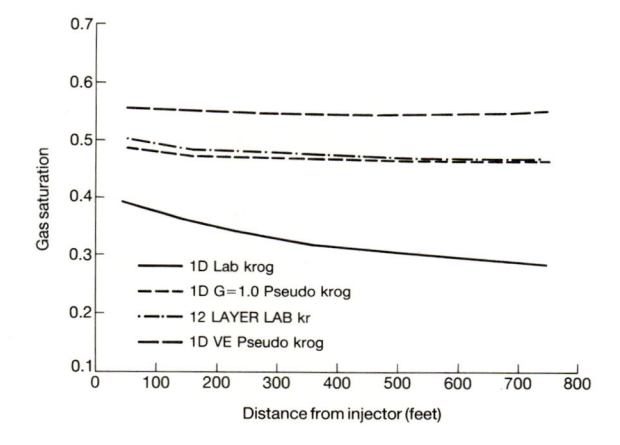

Fig. 14 Comparison of 1-D and X-sec Simulations at 10 Years

Only the half of the model near the injector (columns 1-8)

was analyzed to avoid potential difficulties from boundary effects of the large pore volume cell on the opposite end of the system.

The gas saturation profiles shown in Figure 13 at time 4.0 years indicate the performance of the one-dimensional model with the cross-sectional 1 G=2.0 pseudos is superior to either the laboratory or vertical equilibrium pseudo curves. The difference between the two-dimensional results and the pseudo results is probably due to the use of a single value of S_{orv} for the entire simulation rather than a time varying S_{orv} as was observed in the cross-sectional models.

The comparison at ten years in Figure 14 again shows that the pseudo treatment with viscous to gravity ratio scaling gives excellent results. The differences between the models due to the single value of S_{orv} is again evident but is not large enough to warrant the further complication of an S_{orv} that varies with both velocity and time.

8. CONCLUSIONS

The concept of partial gas-oil gravity segregation was validated for the derivation of pseudo saturation functions Viscous to gravity scaling of a single set of pseudo gas-oil relative permeabilities was shown to be consistent for varying levels of heterogeneities.

Velocity-dependent apparent residual oil to gas flood (S_{orv}) and pseudo gas-oil capillary pressures were derived. Validation of these gas-oil pseudo saturation functions was accomplished by comparing a one-dimensional model with pseudos and a finely-gridded two-dimensional cross-sectional model with laboratory saturation functions. The comparisons showed that the scaled velocity-dependent pseudos gave superior results to models using either vertical equilibrium pseudos or laboratory saturation functions in the one-dimensional models.

9. ACKNOWLEDGEMENT

The author wishes to express his appreciation to ARCO Oil and Gas Company and ARCO Alaska for permission to publish this work.

10. NOMENCLATURE

A = cross-section perpendicular to flow, ft^2
B_g = gas formation volume factor, MCF/RB
B_o = oil formation volume factor, STB/RB
f_g = gas fractional flow
f_o = oil fractional flow
G = dimensionless viscous to gravity ratio
H = grid block thickness, ft
H_i = grid block thickness of block i, ft
H_{inv} = thickness of gas invaded zone, ft
K = permeability, md
K_h = horizontal permeability, md
K_{hi} = horizontal permeability of layer i, md
k_{rg} = gas relative permeability
k_{rg}^{gc} = gas cap gas relative permeability
k_{ro} = oil relative permeability
k_{rog} = oil relative permeability in the gas-oil system
k_{rog}^{gc} = oil relative permeability in the gas cap
k_{rog}^{oz} = oil relative permeability in the oil zone
k_{rw} = water relative permeability
k_{row} = oil relative permeability in the oil-water system
K_v = vertical permeability, md
K_x = permeability in the x-direction, md
K_y = permeability in the y-direction, md
K_z = permeability in the z-direction, md
L_x = characteristic length in the x-direction, ft
P = pressure, psia
P_c = capillary pressure, psia
P_{cgo} = gas-oil capillary pressure, psia
P_{cow} = oil-water capillary pressure, psia
P_g = gas phase pressure, psia
P_o = oil phase pressure, psia

pk_{rog} = pseudo oil relative permeability
pP_{cgo} = pseudo gas-oil capillary pressure, psia
q = flow rate, RB/D
q_o = oil flow rate, RB/(ft^3 D)
q_w = water flow rate, RB/(ft^3 D)
S_g = gas phase saturation
S_{gc} = critical gas saturation
S_o = oil phase saturation
S_{org} = residual oil to gas flooding
S_{orv} = velocity-dependent apparent residual oil to gas flood
S_w = water phase saturation
S_{wi} = initial water saturation
v = superficial velocity, ft/D
v^* = characteristic velocity, ft/D
v_g = gas phase superficial velocity, ft/D
v_o = oil phase superficial velocity, ft/D
v_t = total superficial velocity, ft/D
t = time, D
t_D = dimensionless time
x_D = dimensionless x-direction length
y_D = dimensionless y-direction length
Z = depth, ft
z_D = dimensionless z-direction length
$\Delta \rho$ = oil-gas density difference, psi/ft
μ_g = gas phase viscosity, cp
μ_o = oil phase viscosity, cp
ϕ = porosity
ρ_g = gas phase density, psi/ft
ρ_o = oil phase density, psi/ft

11. REFERENCES

1. Martin, J. C., "Partial Integration of Equations of Multiphase Flow", SPEJ (December, 1968) 370-376.

2. Coats, K. H., Demsey, J. R., Henderson, J. H.,
   "The Use of Vertical Equilibrium in Two-Dimensional
   Simulation of Three-Dimensional Reservoir Performance",
   SPEJ (March, 1971), 63-71.

3. Hearn, C. L., "Simulation of Stratified Waterflooding
   By Pseudo Relative Permeability Curves, J. Pet. Tech.
   (July, 1971) 805-813.

4. Jacks, H. H., Smith, O. J. E., and Mattax, C. C.,
   "The Modeling of a Three-Dimensional Reservoir with a
   Two-Diemsnional Reservoir Simulator - The Use of Dynamic
   Pseudo Functions", SPEJ (June, 1973) 175-185.

5. Killough, J. E., Pavlas, E. J., Martin C., and
   Doughty, R. K., "The Prudhoe Bay Field: Simulation
   of a Complex Reservoir", SPE 10023 presented at the
   International Petroleum Exhibition and Technical
   Symposium, Beijing, China, March 18-26, 1982.

6. Starley, G. P., "A Material Balance Method for Deriving
   Interblock Water/Oil Pseudofunctions for Coarse Grid
   Reservoir Simulation", SPE 15621 prsented at the 61st SPE
   Fall Conference and Exhibition, New Orleans, October,
   1986.

7. Spivak, A., "Gravity Segregation in Two-Phase
   Displacement Processes", SPEJ (December, 1974),
   619-632.

Microscopic flow and generalized Darcy's equations

F. Kalaydjian and C-M. Marle

Institut Français du Pétrole, France

Abstract In view of a better understanding of the macroscopic behaviour of multiphase flows in porous media, we first discuss equations which govern these flows at the pore level. Difficulties arising from boundary conditions on moving fluid-fluid interfaces and fluid-fluid-solid contact lines are stressed. Molecular dynamics and lattice gas models offer new approaches to the discussion of the validity of these boundary conditions; some recent works along these lines are briefly reviewed. Another approach uses capillary tubes with a square cross section as simple geometric models of porous media. Approximate solutions of the flow equations obtained for these models suggest, for real porous media, some modifications of the widely used generalized Darcy's equations. Results drawn from the analysis of more complicated models (networks of interconnected capillary tubes), about the effects of the capillary number and the viscosity ratio of the displacing and displaced fluids, are reviewed. Finally, methods for the derivation of macroscopic equations governing multiphase flow in porous media, starting from equations valid at the pore level, are discussed.

1 Introduction

Multiphase flows in porous media are mathematically described, at the macroscopic level, by the well known generalized Darcy's equations, which are of empirical origin. In these relatively simple equations, properties of the porous medium appear only through a few macroscopic quantities, mainly the relative permeabilities and the capillary pressure, which are assumed to be functions of the fluid saturations. At the pore level, these flows involve several phenomena which interact in a very complicated way: viscous flow, interfacial phenomena, interactions between two different fluid phases and between each fluid phase and the solid, and even more complex interactions between two different fluid phases and the solid along a moving contact line.

In view of the opposition between the simplicity of the macroscopic description and the complexity of the underlying phenomena at the pore level, one may ask several questions. Do the macroscopic equations give a faithful description of all the macroscopic effects of phenomena which occur at the pore level? How are macroscopic concepts, such as the residual saturations in each phase, the relations between relative permeabilities, capillary pressure, and the fluid saturations, related with the pore structure and with other concepts valid at the pore level? Can a careful examination of phenomena which occur at the pore level lead to a better understanding of the macroscopic behaviour of fluid flows in porous media? In particular, starting from laws valid at the pore level, is it possible to derive laws valid at the macroscopic level? This would either confirm the validity of the empirical generalized Darcy's equations, or suggest to replace them by other, better suited equations. This paper describes some of the work done to answer these questions.

In the first part, we discuss phenomena which occur at the pore level, when two immiscible fluid phases are simultaneously flowing in a porous medium. If the relevant equations at that level in each fluid phase are well known, it seems that boundary conditions along moving fluid-fluid-solid contact lines are still incompletely understood. We briefly describe attempts to circumvent this difficulty in which the fluid phases are treated not as continuous media, but as collections of particles. We also describe numerical approximate solutions of two-phase flow equations in capillary tubes with a square cross-section.

The second part is devoted to the studies of fluid flows in simplified models of porous media: bundles of capillary tubes, networks of interconnected capillaries. Such models do explain many properties of fluid flows in actual porous media, and can be used for obtaining qualitative indications about the effects of various parameters, such as the capillary number and the viscosity ratio of the two fluid phases.

The third part deals with the derivation of flow equations valid at the macroscopic level, starting from flow equations valid at the pore level. Three types of methods are briefly described: statistical methods, volume averaging methods and multi-scale methods. We discuss the closure problem, which is encountered in the application of all these methods, and we indicate results obtained about macroscopic equations for two-phase flows.

The conclusion gathers some significant results and open questions.

2 Fluid flows at the pore level

In this section we focus on the pore level aspects of fluid flows in a porous medium.

2.1 Flow equations in the framework of continuum mechanics

We consider the case when the pore space contains two isothermal, totally immiscible, homogeneous fluid phases, and when no adsorption phenomena occur at the interfaces. More complicated cases can be treated along similar lines. The case of totally miscible fluids (when a single, non-homogeneous fluid phase fills the pore space) is easier, because in that case there is no moving fluid-fluid interface.

The classical equations of fluid mechanics, which are generally assumed to be valid at the pore level, are the mass balance equations (2.1), the momentum balance equations (2.2) and the expressions of the stress tensors (2.3), for each fluid phase. They are recalled below. We denote the two fluid phases by the subscripts a and b, the solid by the subscript s, and we use the subscript i which may take the values a or b.

$$\frac{\partial}{\partial t}\rho_i + div(\rho_i \vec{v_i}) = 0. \qquad (2.1)$$

$$\frac{\partial}{\partial t}(\rho \vec{v_i}) + \overrightarrow{\text{Div}}(\rho_i \vec{v_i} \otimes \vec{v_i} - \overline{\overline{\tau_i}}) - \rho_i \vec{g} = 0 \qquad (2.2)$$

$$\overline{\overline{\tau_i}} = (-p_i + \lambda_i \text{div}\,\vec{v_i})\,\overline{\overline{I}} + \mu_i \,\overline{\text{Grad}_s\,\vec{v_i}}. \qquad (2.3)$$

We have denoted by ρ_i the mass per unit volume, $\vec{v_i}$ the velocity, $\overline{\overline{\tau_i}}$ the stress tensor, p_i the pressure, λ_i the volume viscosity and μ_i the shear viscosity of phase i; $\overline{\overline{I}}$ is the Kronecker (unit) tensor, and $\vec{g}$ the gravity acceleration.

Boundary conditions must be added, along the fluid-solid interfaces Σ_{as} and Σ_{bs}, as well as along the fluid-fluid interface Σ_{ab}. They are written below (see [28], appendix 2, for details and generalizations).

Boundary conditions along Σ_{is} ($i = a$ or b):

$$\vec{v_i} = 0, \qquad \text{(no slip condition)}. \qquad (2.4)$$

Boundary conditions along Σ_{ab}:

$$\vec{v_a} = \vec{v_b} = \vec{v_{ab}}, \qquad \text{(no slip condition)} \qquad (2.5)$$

$$(\overline{\overline{\tau_b}} - \overline{\overline{\tau_b}}) \cdot \vec{n_a^b} + \sigma_{ab} r_a^b \, \vec{n_a^b} + \overline{\text{grad}}_{\Sigma_{ab}} \, \sigma_{ab} = 0. \qquad (2.6)$$

We have denoted by $\vec{v_{ab}}$ the velocity of interface Σ_{ab} (equal, by the no slip condition (2.4), to the velocities of both fluid phases), $\vec{n_a^b}$ the unit vector normal to Σ_{ab}, oriented towards b, σ_{ab} the interfacial tension, r_a^b the mean curvature of Σ_{ab} (each principal radius of curvature being taken as positive if the corresponding centre of curvature is on the side b, and negative in the opposite case). The symbol $\overrightarrow{\text{grad}}_{\Sigma_{ab}}$ denotes the surface gradient along Σ_{ab}.

Moreover, boundary conditions must be written along the contact line L, i.e., the line along which the fluid-fluid interface Σ_{ab} meets the solid surface Σ_s. It is generally assumed that the interfacial tensions σ_{ab}, σ_{as} and σ_{bs} of the fluid-fluid and fluid-solid interfaces are related with the contact angle θ (measured in phase a) by the formula

$$\sigma_{bs} - \sigma_{as} - \sigma_{ab}\cos\theta = 0. \tag{2.7}$$

When $|\sigma_{bs} - \sigma_{as}| > \sigma_{ab}$, it is assumed that one fluid phase only (a if $\sigma_{bs} > \sigma_{as}$, b in the opposite case) is in contact with the solid, so there is no contact line.

2.2 Discussion

For isothermal flows, in addition to the equations written above, one may assume that properties of the fluid phases, such as their mass per unit volume and viscosity coefficients, are known functions of the pressure. One may assume too that the fluid-fluid and fluid-solid interfacial tensions are known functions of the pressures in the adjacent fluid phases. With these assumptions, one may ask whether the resulting system of equations is mathematically well posed. As far as we know, no clear answer has been given to that question, but several observations lead to the belief that this answer is probably negative.

The main difficulties come from the existence of a moving interface between the two fluid phases, which intersects the solid surface along a moving contact line. Dussan and Davis [10] have analyzed the velocity fields in the fluid phases in a neighbourhood of that line, and have pointed out that the no slip assumption leads to unbounded forces.

Several attempts have been made to solve this difficulty ([7], [15], [36]). In the gas-liquid case, de Gennes and Joanny have shown that a thin liquid film develops along the solid surface, and goes beyond what seems to be the "macroscopic" contact line. The existence of this film leads to modified boundary conditions along that line. However, the mathematical expression of these modified boundary conditions remains to be explicitly written and the well-posedness of the resulting system of equations is still to be discussed.

2.3 The discrete modelization of fluid flows

As seen in the previous section, the validity of continuum mechanics equations and classical boundary conditions may be questioned for fluid flows in porous media, particularly near the fluid-fluid-solid contact line. In order to investigate their validity, one should start from more "fundamental" equations, valid at a still smaller level than the pore level, *i.e.*, at the molecular level, reflecting the discrete nature of the fluid and solid phases. Such an approach is well known for bulk fluid phases: starting from kinetic theory and using the method of Chapman and Enskog [5] or that of Grad [13], one can derive the Navier-Stokes equations. It would be nice to be able to derive, in a similar way, the boundary conditions along interfaces and contact lines.

Rather than such a theoretical derivation, several attempts have been made to investigate interfacial phenomena by using numerical simulations of the behaviour of collections of particles. This approach, which is called *Molecular dynamics*, appeared more than thirty years ago, but its practical application for investigating interfaces, made possible by the power growth of available computers, is relatively recent. The method has been used for the study of gas-liquid interfaces [4], solid-fluid interfaces [43], [26], [38], and the displacement of a meniscus along solid walls with a molecular structure [20]).

Several significant results have been obtained. In particular, it has been shown [20] that in the displacement of a meniscus along solid walls, the no slip condition does not hold near the contact line, and the apparent contact angle varies with the velocity.

Another approach in which the discrete structure of the fluid and solid phases are taken into account is based on the lattice gas formalism. This approach, which leads to less heavy calculations than the molecular dynamics approach, was initiated by Hardy, de Pazzis and Pomeau [14]. It may be used for two-dimensional [12] and three-dimensional flows [8]. The particles are compelled to be placed at the vertices of a fixed lattice. Their velocities are such that at each time step, they move from their initial position to one of the adjacent vertices. Collisions occur only at vertices. Knowing the velocities of the colliding particles before a collision, given rules are applied to determine the velocities after the collision, hence the vertices where the particles will go at the next time step. These rules must satisfy the conservation of total momentum and energy, and are chosen in such a way that the macroscopic behaviour of the model is similar to that of a real fluid. Such a model may be considered as a cellular automaton, since at each value of the discretized time, the status of each vertex depends on the status, at the preceding value of the time, of its immediate neighbours.

The method has been used for modelling the flow of a single fluid phase or of two immiscible fluid phases in a porous medium [33] [34].

2.4 The displacement of one fluid phase by another in a capillary tube

A fluid particle flowing in a porous medium encounters, along its way, a succession of cavities and throats of irregular shapes. This geometry may be schematically represented by a capillary tube of varying cross section. The use of such simplified models offers another mean for investigating fluid flow phenomena at the pore level. This approach is of course less fundamental than those described in the previous sections, but it may be more suitable for obtaining qualitative trends and for comparing calculations with experimental results.

For these studies, capillary tubes with a polygonal (triangular or square) cross-section seem to be more suitable than capillary tubes with a circular cross section, because stable two-phase flows can exist in a long part of the tube: the wetting fluid flows along the walls of the tube, in the corners of the cross-section, while the non-wetting fluid flows along the axis of the tube, in the central part of the cross-section.

Capillary tubes with a triangular cross-section were considered by Singhal and Somerton [39]. Kalaydjian and Legait [17] [18] [19] [22] used capillary tubes with a square cross-section for the study of several phenomena at the pore level: the entrapment of a droplet of non-wetting fluid displaced by a wetting fluid, against a reduction of the cross-section; the mobilization of a droplet of non-wetting fluid, under dominant capillary forces, by the displacing wetting fluid. Let us briefly describe their work.

The following assumptions are done: Transverse components of fluid phase velocities are much smaller than the corresponding longitudinal component. The flow is slow enough to make inertial and time-dependent terms negligible. The fluid-fluid interface curvature depends only on the coordinate z, along the axis of the tube. In each cross section, the line which separates the two fluids is made of four arcs of a circle, which meet the walls of the tube under a contact angle θ, taken as a parameter (figure 1).

Under these assumptions, the flow equations have been numerically solved by a finite element method. For each value of the coordinate z along the axis of the tube, the flow rate q_i of fluid i ($i = a$ or b) is calculated by integrating its velocity over the part of the cross section occupied by that fluid. Results show that these flow rates can be expressed as

$$\begin{cases} q_a = -k \left(\frac{k_{r_a}^a}{\mu_a} \frac{\partial p_a}{\partial z} + \frac{k_{r_a}^b}{\mu_b} \frac{\partial p_b}{\partial z} \right), \\ q_b = -k \left(\frac{k_{r_b}^a}{\mu_a} \frac{\partial p_a}{\partial z} + \frac{k_{r_b}^b}{\mu_b} \frac{\partial p_b}{\partial z} \right). \end{cases} \quad (2.8)$$

The coefficient k, which plays the role of the permeability, depends only on the local transverse dimensions of the capillary, while the coefficients $k_{r_a}^a$, $k_{r_a}^b$, $k_{r_b}^a$ and $k_{r_b}^b$ depend on the viscosity ratio of the two fluids and on the contact angle θ, and are functions of the fraction of the cross

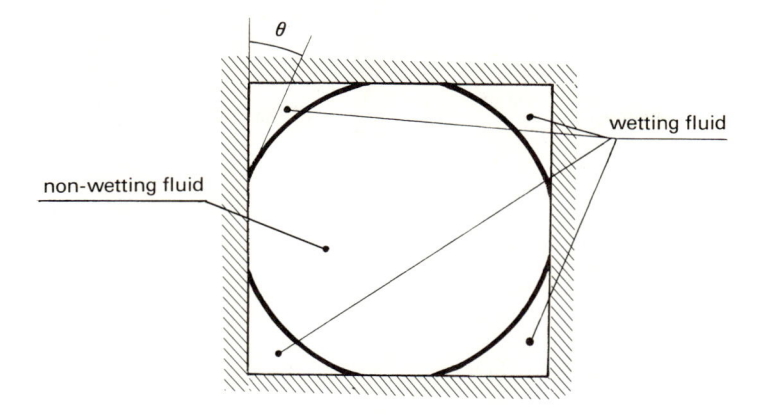

Figure 1. Traverse section of the capillary tube

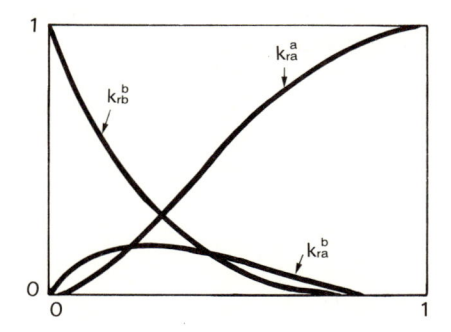

Figure 2. The coefficients k_{ri}^j for a viscosity ratio of 1

section occupied by each fluid, which have been explicitly calculated. The diagonal terms k_{ra}^a and k_{rb}^b are similar to the classical relative permeability coefficients, while the non-diagonal terms k_{ra}^b and k_{rb}^a represent the viscous coupling effect exerted between the two fluid phases. The numerical results show that

$$\frac{k_{ra}^b}{\mu_b} = \frac{k_{rb}^a}{\mu_a}. \tag{2.9}$$

Figures 1 and 2 present the values of the coefficients k_{ri}^j, i and $j = a$ or b, for a zero contact angle and for two values of the viscosity ratio: 1 and 10, respectively. We observe that in the latter case, the coefficient k_{ra}^a can take values larger than 1. This is explained by a lubricating effect of the less viscous, wetting fluid b on the flow of the more viscous, non wetting fluid a. This effect was observed experimentally [6].

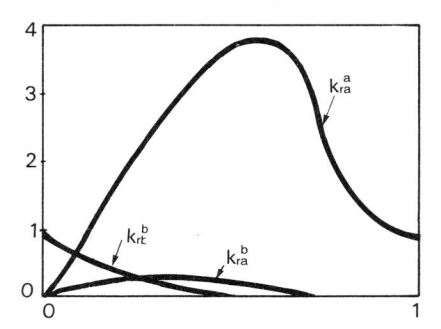

Figure 3. The coefficients $k_{r_i}^j$ for a viscosity ratio of 10

The same numerical model was used also for investigating spontaneous and for determining the effect of the viscosity ratio on the untrapping of a droplet of non-wetting fluid trapped by a shrinkage of the cross section of the capillary. For this last problem, it was shown that the critical capillary number (for the flow of the wetting fluid, around the trapped droplet of non-wetting fluid) above which that droplet is untrapped, is lower when the viscosity ratio is higher; it means that droplets of non-wetting fluid are untrapped more easily when their viscosity is higher.

All these results were confirmed by experiments on flow in capillary tubes with a square cross section.

3 Fluid flow in simplified models of porous media

Studies of fluid flows in single capillaries, described in the previous section, cannot account for the effects of the existence, in a real porous medium, of pores of many different dimensions and orientations, interconnected in a complicated way. In order to account for these effects, many authors have considered models of porous media more complicated than single capillaries, yet simple enough to make calculations tractable. Most of these models use, instead of the true flow equations valid at the pore scale, simplified equations such as Poiseuille equation for the flow of a single fluid phase, and Washburn equation for the displacement of a fluid phase by another one in a cylindrical tube.

3.1 The simplest models

Apart from using a single capillary tube, discussed in the previous section, the simplest geometric model for representing a porous medium is a bundle of cylindrical capillary tubes of various radii and orientations. With such models, several macroscopic properties, such as permeability, capillary pressure as a function of saturation, can be related to the tube radii frequency distribution. See the references to the works of Childs and Collis-George (1948), Purcell (1949), Burdine et al. (1950), Fatt and Dykstra (1951) in [11]. Such models allow the derivation of explicit formulae for these macroscopic quantities. They have been recently used [1] for discussing the effects of nonstationarity and interfacial viscosities on relative permeabilities. However, these models do not consider interconnections between the various tubes, and cannot account for residual saturations. For multiphase flows, the conclusions which may be drawn from their use seem to be, at best, of qualitative value only.

3.2 Network models

Recognizing the limitations of bundles of cylindrical capillaries, Fatt [11] introduced network models of porous media in 1956. Many important properties, such as the coordination number, are due to Fatt, who created the main techniques: construction of the network by choosing randomly the sizes of the capillaries, step by step calculations of the displacement of one fluid by another in the network, using linear laws relating pressure drops with flow rates in each capillary. Generalizations and extensions have been made by many authors. Let us review some of the recent ones

Singhal and Somerton [40] considered a network of interconnected channels, put along some of the sides and diagonals of a regular square lattice. The cells where several channels are interconnected are assumed to have a negligible volume, but a finite radius, which defines their capillary properties. The channels have a triangular cross section, which allows the simultaneous flow of two immiscible fluid phases, under several different flow regimes (single phase, displacement, slug or pendular), depending on the local saturation. The model was used to compute capillary pressure and relative permeabilities as functions of the saturation.

Koplik and Lasseter [21] have used a model made of circular "pores" of variable radii, whose centres form a regular square lattice. These pores are randomly connected with each other by straight channels of variable width and length, set along the sides or the diagonals of the square lattice. The number of channels connected with a given pore may vary, with a fixed mean value. Two immiscible fluids are present in the network, one displacing the other. Displacement in each channel and in each pore is described using the Washburn approximation. Precise rules are introduced for describing what happens when a meniscus travels through the junction between a pore and some adjacent channel. These rules are in agreement with the elementary laws of capillarity and with experimental evidence. For instance, a meniscus may remain constrained at a junction when the pressure difference between the two fluids is such that the meniscus cannot travel into the channel nor into the pore. In order to simplify the calculations, the authors have assumed that the viscosities of the two fluids are equal. However, difficulties arising from constrained menisci make the calculations rather heavy and limit the practical use of the model to relatively small networks.

Dias and Payatakes [9] have used randomly sized elementary cells, whose shape is that of a constricted tube, put along the sides of a regular square lattice. When a meniscus separating two fluid phases is present in a cell, the calculation of the flow through this cell takes into account the pressure difference between the two sides of the meniscus, which varies with the position of the meniscus in the cell. At each time step, the fluid distribution being known, the pressure field is obtained by resolving a system of linear equations. Then the flow rates through the various cells are obtained and used to calculate the fluid distribution at the next time step. The model has been used for investigating the effects of the viscosity ratio and of the capillary number on various aspects of the flow, in particular the size of the droplets of residual oil and the remobilization of these droplets.

Lenormand et al. [23] [24] [25] have used both numerical and physical network models of porous media for investigating the effects of capillary and viscous forces on the displacement of a non-wetting phase by a wetting phase. They succeeded in distinguishing three qualitatively different regimes of flow, depending on the values of the capillary number and the viscosity ratio: *capillary fingering* which occurs when capillary forces are dominant, *viscous fingering* occurring at unfavourable viscosity ratio, and *stable displacement*.

4 The derivation of macroscopic flow equations

The works described in the previous sections of this paper can give useful indications about the macroscopic properties of two-phase flows in porous media. Studies of phenomena at the pore level described in section 2 may help for identifying the most important phenomena and for predicting, at least qualitatively, their macroscopic effects. However, they cannot yield much more than qualitative trends, and do not account for the effects of the complicated geometry of pores. Simplified models of porous media, as described in section 3, do account for such effects and for complex interactions between several phenomena. However, they rest on very simple geometries and on simplified flow equations whose general validity may be questioned. For these reasons, results drawn from such studies may have a limited range of validity.

In this section we describe some attempts for the derivation of flow equations valid at the macroscopic level, starting from equations valid at the pore level. In contrast with attempts described in the previous sections, these studies do not rest on simple geometric models of porous media, nor on simplified equations for the flow at the pore level. They keep the pore geometry as general as possible (although in some cases, they assume a spatial periodicity) and use, for the description of flows at the pore level, mass, momentum and energy balance equations whose validity is very general. Such methods can indicate what is the precise meaning of the various quantities which are used at the macroscopic level, and how these quantities are related with the corresponding quantities valid at the pore level. They can also show what types of equations may be reasonably expected as valid at the macroscopic level. Some of these methods even allow, in principle, the effective calculation of coefficients which occur in macroscopic flow equations, such as the permeability or the relative permeabilities as functions of the fluid saturations; however, these results are obtained up to now only with simplifying assumptions about the pore geometry.

4.1 Statistical methods

These methods rest on the idea [37] that flow phenomena in a porous medium can be mathematically modelled by stochastic processes. The true flow which occurs really in a given porous medium is looked at as a particular realization of a stochastic process, *i.e.*, as a particular point in a probability space. The various quantities which describe the flow, such as the velocity vector of a given fluid phase at a particular point in space and a particular time, are regarded as random variables. By using the flow equations valid at the pore level and the mathematical tools of probability theory, these methods aim at obtaining results about the probability laws of these variables; in particular, their mean values and standard deviations. Macroscopic laws appear as relations between the probability distributions of the various quantities used for the description of the flow. For example, in such a probabilistic framework, Darcy's law appears as a linear relationship between the mean value of the velocity of a fluid phase and the mean value of the pressure gradient of that phase.

The physical interpretation of results drawn from statistical theories raises difficult questions. For example, one may ask how the flow rate of a fluid phase measured in a flow experiment is related with the mean value of the random variable used for the mathematical representation of the velocity of that phase. The measured value appears as a mean value over the cross-section of the sample of porous medium used for the experiment; the mean value of the random variable which represents the velocity is taken over a probability space, not over the physical space. Relations between these two different concepts come from ergodicity assumptions: according to such assumptions, for processes which are stationary in space, mean values over a large volume in space converge, in a precise mathematical sense, towards the mean value over the probability space.

Statistical methods have been used for modelling fluid flows in porous media mainly for single phase flows, *i.e.*, for the derivation of Darcy's law, and for miscible displacement in the tracer approximation, in which the hydrodynamic properties of the fluid phase do not depend on the tracer concentration.

Single phase flows

Matheron [30] has laid down the mathematical foundations of the statistical theory of fluid flow in porous media with great care and mathematical rigour. For single phase flows, he obtained a proof of Darcy's law and a proof of the symmetry of the permeability tensor of a non-isotropic medium.

Other results due to Matheron are about the large scale permeability of an heterogeneous porous medium. He proved that for a single phase flow which is linear in the large, the heterogeneous medium is, in a certain sense, equivalent to a homogeneous medium whose permeability is comprised between the arithmetic mean and the geometric mean of the local permeabilities. But for two-dimensional radial flows, there is no homogeneous medium equivalent to the heterogeneous one, the flow behaviour being too strongly dependent on the local values of the permeability in an immediate neighbourhood of the well.

Miscible displacement in the tracer approximation

In miscible displacement the pore space of the medium is filled up with a single fluid phase, whose composition is not uniform in space. In the tracer approximation, this fluid phase is a binary mixture of a carrier fluid and a tracer, whose hydrodynamic properties are assumed to be independent of the tracer concentration. The system of equations which governs the phenomena at the pore level splits into two separate subsystems:
  - the hydrodynamic subsystem, which governs the flow of the fluid phase; this subsystem does not involve the tracer concentration, and is identical to the system governing the flow of a single fluid phase;
  - the concentration subsystem, which governs the evolution of the tracer concentration; in this subsystem, the velocity of the fluid phase should be considered as a coefficient, which is a known function of the space and time coordinates, once the hydrodynamic subsystem has been solved.

In most statistical methods which have been used for the derivation of macroscopic equations, the velocity vector field, solution of the hydrodynamic subsystem, is assumed to be a random vector field. In fact, it is even assumed that the motion of the fluid particles, which should be obtained by integrating this vector field, can be represented by a random walk process. By looking at the time evolution of the spatial distribution of the tracer concentration, it has been shown that under suitable assumptions, this concentration obeys a diffusion-convection equation.

Saffman [35] succeeded in deriving the dispersion coefficient as a function of the Péclet number; his derivation rests on the use, for representing the pore geometry, of a network model with simplified statistical properties.

4.2 Volume averaging methods

These methods rest on a very natural idea: at each point of the porous medium, the local value of any macroscopic quantity should be defined as the average value, over some elementary volume around that point, of the corresponding pore level quantity. By averaging the pore level flow equations over an elementary volume around each point, one should obtain macroscopic flow equations involving macroscopic quantities only.

A simple example

Let us consider a porous medium containing two fluid phases. Subscripts s, a and b represent the solid and the fluid phases a and b, respectively. We note χ_i the indicating function of phase i (with $i = s$, a or b), *i.e.*, the function equal to 1 at points which belong to i, and equal to 0 elsewhere. By convention, all quantities pertaining to a phase are assumed to be equal to zero outside that phase. For example, the mass per unit volume of fluid phase a, ρ_a, is everywhere defined and equal to 0 at points which belong to b or to s.

We choose, once and for all, an elementary volume V around the origin, for example a ball of radius R centred at the origin. Its volume is denoted by $\text{vol}(V)$. For every point x in space, we take as elementary volume around that point the volume V_x deduced from V by the translation which brings the origin onto x, *i.e.*, $V_x = \{x + y \mid y \in V\}$. At point x we define the porosity, $\Phi(x)$, the saturation in fluid a, $S_a(x)$, the macroscopic mass per unit volume of fluid a, $\overline{\rho_a}(x)$, by

$$\Phi(x) = \frac{1}{\text{vol}(V)} \int_{V_x} (1 - \chi_s(z))\, dz ;$$

$$\Phi(x) S_a(x) = \frac{1}{\text{vol}(V)} \int_{V_x} \chi_a(z)\, dz ;$$

$$\Phi(x) S_a(x) \overline{\rho_a}(x) = \frac{1}{\text{vol}(V)} \int_{V_x} \rho_a(z)\, dz \qquad (4.1)$$

We observe that the definition of the macroscopic mass per unit volume $\overline{\rho_a}$ is chosen in such a way that $\overline{\rho_a} = \rho_a$ at point x when the pore level mass per unit volume ρ_a is locally constant around x.

Similarly, if $\vec{v_a}$ is the pore level velocity of fluid phase a, we define the macroscopic level velocity of fluid phase a at point x, $\vec{V_a}(x)$, by

$$\Phi(x)S_a(x)\overline{\rho_a}(x)\overrightarrow{V_a}(x) = \frac{1}{\text{vol}(V)} \int_{V_x} \rho_a(z)\overrightarrow{v_a}(z)\, dz \,. \tag{4.2}$$

Let us now show how the macroscopic mass balance equation for fluid phase a is derived, starting from the pore level mass balance equation (2.1). By averaging both sides of this equation over the elementary volume V_x around x, we obtain

$$\frac{1}{\text{vol}(V)} \int_{V_x} \left(\frac{\partial}{\partial t}\rho_a\right)(z)\, dz + \frac{1}{\text{vol}(V)} \int_{V_x} div(\rho_a \overrightarrow{v_a})(z)\, dz = 0\,. \tag{4.3}$$

We recall that ρ_a is zero outside fluid phase a, thus has a jump discontinuity across the surfaces Σ_{ab} and Σ_{as} which separate fluid phase a from fluid phase b and from the solid s, respectively. Similarly, $\overrightarrow{v_a}$ has a jump discontinuity across Σ_{ab}, but is continuous across Σ_{as}, taking into account the no-slip condition (2.4). Since Σ_{as} does not move, while Σ_{ab} moves at the velocity $\overrightarrow{v_{ab}}$, we may write

$$\frac{\partial}{\partial t}\left(\frac{1}{\text{vol}(V)}\int_{V_x}\rho_a(z)\,dz\right) = \frac{1}{\text{vol}(V)} \left(\int_{V_x}\left(\frac{\partial}{\partial t}\rho_a\right)(z)\,dz + \int_{V_x\cap\Sigma_{ab}}\rho_a\overrightarrow{v_{ab}}.\overrightarrow{n_a^b}\,d\Sigma\right), \tag{4.4}$$

where the second integral in the right hand side is a surface integral over the part of Σ_{ab} contained in V_x, and where $\overrightarrow{n_a^b}$ is the unit vector normal to Σ_{ab}, oriented towards b. We observe that the left hand side is equal to $\frac{\partial}{\partial t}(\Phi S_a\overline{\rho_a})$. By using Stokes' formula, we obtain

$$\int_{V_x} div(\rho_a\overrightarrow{v_a})\,dz = div_x\left(\int_{V_x}\rho_a\overrightarrow{v_a}\,dz\right) + \int_{V_x\cap\Sigma_{ab}}\rho_a\overrightarrow{v_a}.\overrightarrow{n_a^b}\,d\Sigma\,. \tag{4.5}$$

This result is a special case of the so-called spatial averaging theorem; in fact, it is a consequence of Stokes' formula, which was pointed out independently by several people in 1967 [2] [27] [41] [44].

Taking into account the no slip condition (2.5) on Σ_{ab}, we see that when expressions (4.4) and (4.5) are put in (4.3), the surface integrals cancel. We obtain the macroscopic mass balance equation for fluid phase a:

$$\frac{\partial}{\partial t}(\Phi S_a\overline{\rho_a}) + \div(\Phi S_a\overline{\rho_a}\overrightarrow{V_a}) = 0\,. \tag{4.6}$$

Averaging with a weight function

Instead of averaging over an elementary volume, some authors use for averaging a technique introduced by G. Matheron [29]. Let m be a bounded function of the space variable x, with compact support, such that

$$\int m(x)\,dx = 1. \tag{4.7}$$

This function will be called a *weight function*. By definition, the average value of any function f of the space variable, at a given point x in space, is the convolution product

$$f * m\,(x) = \int f(z) m(x-z)\,dz. \tag{4.8}$$

Such a definition is in fact a generalization of that used in the previous section. Indeed, let V be an elementary volume around the origin. We set

$$m(x) = \begin{cases} 1/\text{vol}(V) & \text{if } -x \in V, \\ 0 & \text{if } -x \notin V. \end{cases} \tag{4.9}$$

For this particular function, we have

$$f * m\,(x) = \frac{1}{\text{vol}(V)} \int_{V_x} f(z)\,dz,. \tag{4.10}$$

Averaging with a weight function is not fundamentally different from averaging over an elementary volume. But combined with the use of distributions (generalized functions) for expressing the balance equations, it makes calculations much easier. For example, it may be seen that the so-called spatial averaging theorem expresses simply the well known fact that the convolution product with the weight function m commutes with derivation operators in the sense of distributions.

4.3 Multi-scale methods

These methods are closely related with the volume averaging methods. They rest on the idea that any quantity a, defined at the pore level, may be split into a sum

$$a = \langle a \rangle + \tilde{a}, \tag{4.11}$$

where $\langle a \rangle$ is the local mean value of a, and $\tilde{a}$ its deviation from that local mean value. This splitting should be made in such a way that the space scales over which the two terms $\langle a \rangle$ and $\tilde{a}$ vary, are very different: $\langle a \rangle$ varies smoothly across macroscopic distances, while $\tilde{a}$ varies rapidly at the pore level.

Multi-scale methods may, in some cases, offer more flexibility than volume averaging methods. Combined with asymptotic expansions for evaluating limits when the ratio of the pore scale to the macroscopic scale vanishes, they have given rise to homogenization theory. They have been used [3] for studying two-phase flows in periodic porous media, in the case when the interface between the two fluid phases remain steady.

4.4 The closure problem

There is a loss of information when, for specifying the state of a system at a given time, macroscopic quantities alone are used instead of the corresponding pore level quantities. In most cases, the time evolution of the macroscopic quantities depends on that lost information. Therefore, it is in general impossible to find a well-posed, complete system of equations, involving only these macroscopic quantities, which rigorously and fully governs their time evolution. For obtaining such a system of equations, some kind of approximation must be made. This difficulty, called the *closure problem*, is encountered under various forms in all the previously described methods for the derivation of macroscopic equations.

For example, the assumptions of stationarity and ergodicity which are made in some applications of statistical methods [30] may be regarded as a way of solving the closure problem. In multiscale methods, the use of asymptotic expansions in powers of a small parameter ϵ (which is the ratio of the pore dimension to the macroscopic dimension of the system), and the search for limit equations valid when ϵ vanishes, is also a way of solving the closure problem. Whitaker [45] has recently introduced a general closure method, in which equations governing the deviations of the various quantities involved from their mean value are solved on a domain with periodic boundaries, regarded as representative of the pore structure. See also the research comments by Torres [42]. Quintard and Whitaker [32] have used this method for the large-scale averaging of two-phase flow in heterogeneous porous media.

Another way of solving the closure problem, which uses the methods of thermodynamics of irreversible processes, proposed in [28], was used in [16] and [31] for the derivation of the capillary pressure and the generalized Darcy's equations.

4.5 Some results

Most attempts at the derivation of macroscopic equations for two-phase flow in porous media lead to equations of the same form as traditional generalized Darcy's equations. These equations are similar to equations (2.8) but with no extra-diagonal terms, *i.e.*, with $k_{r_a}^{\ b} = k_{r_b}^{\ a} = 0$. However, the methods of thermodynamics of irreversible phenomena lead to equations similar to the full (2.8) equations, with extra-diagonal terms [16] [31]. A similar result was found in [3]. Equations obtained in [45] contain additional terms which, just like the extra-diagonal terms in (2.8), represent the viscous drag of each fluid phase on the other; but these terms have a different form, and Whitaker concludes that they may probably be neglected except when the viscosity ratio is close to unity.

In most of these attempts, the contact line between the fluid-fluid interface and the solid was assumed to be fixed. In [16] and [31], where that assumption was not done, additional terms involving the time derivative of the saturation were obtained in the capillary pressure equation.

5 Conclusion

The foundations of macroscopic laws for multiphase flow in porous media have been investigated for many years, and from several different approaches: study of phenomena at the pore level, use of simplified models of porous media, attempts at the derivation of macroscopic equations from equations valid at the pore level. The main phenomena which are involved are known. More and more realistic models are built, which can explain, in a certain sense, many properties of fluid flows in porous media. Studies of two-phase flow in capillaries with a square cross-section, and some attempts at the general derivation of macroscopic equations, suggest that extra-diagonal terms should be added in the traditional generalized Darcy's equations. But, in our opinion, a precise knowledge of what happens at a moving contact line between a fluid-fluid interface and a solid, and how such phenomena can be expressed in the equations or the boundary conditions governing the flow, are still lacking

References

1. Alemán-Gomez, M., Ramamohan, T. R., Slattery, J. C., A statistical structural model for unsteady-state displacement in porous media, Preprint SPE 13265, 1984.

2. Anderson, T. B., Jackson, R., A fluid mechanical description of fluidized beds, Ind. Eng. Chem. Fundam., 6, 527–538, 1967.

3. Auriault, J. L., Sanchez-Palencia, E., Remarquas sur la loi de Darcy pour les écoulements diphasiques en milieu poreux, J. Theor. Appl. Mech., special issue, 141–156, 1986.

4. Chapela, G. A., Saville, G., Thompson, S. M., Rowlinson, J. S., Computer simulation of a gas-liquid surface, part 1, J.C.S. Faraday II, **73**, 1133–1144, 1977.

5. Chapman, S., Cowling, T. G., "The mathematical theory of non-uniform gases", Cambridge University Press, Cambridge, 1939.

6. Danis, M., Jacquin, C., Influence du contraste de viscosités sur les perméabilités relatives lors du drainage; expérimentation et modélisation, Revue de l'Institut Français du Pétrole, **38**, 6, 723–733, 1983.

7. de Gennes, P.-G., Wetting: statics and dynamics, Reviews of Modern Physics, **57**, 3, part I, 827–863, 1985.

8. D'Humières, D., Lallemand, P., Lattice gas models for 3D hydrodynamics, Europhysics Letters, **2**, 4, 291–297, 1986.

9. Dias, M. M., Payatakes, A. C., Network models for two-phase flow in porous media, part 1: Immiscible microdisplacement of non-wetting fluids, part 2: motion of oil ganglia, J. Fluid Mech., **164**, 305–336 and 337–358, 1986.

10. Dussan, E. B., Davis, S. H., On the motion of a fluid-fluid interface along a solid surface, Journal of Fluid Mechanics, **65**, 71–95, 1974.

11. Fatt, I., The network model of porous media, I: Capillary pressure characteristics; II: Dynamic properties of a single size tube network; Trans. AIME, **207**, 111–159 and 160–181, 1956.

12. Frisch, U., Hasslacher, B., Pomeau, Y., Lattice-gas automata for the Navier-Stokes equation, Phys. Rev. Lett., **56**, 1505–1507, 1986.

13. Grad, H., On the kinetic theory of rarefied gases, Comm. Pure and Appl. Math., **2**, 4, 331–407, 1949.

14. Hardy, J., de Pazzis, O., Pomeau, Y., Molecular dynamics of a classical lattice gas: transport properties and time correlation functions, Phys. Rev., Ser. A, **13**, 1949–1961, 1976.

15. Joanny, J.-F., Le mouillage: quelques problèmes statiques et dynamiques, Thèse de doctorat d'état, Université Pierre et Marie Curie, Paris, 1985.

16. Kalaydjian, F., A macroscopic description of Multiphase flow in porous media involving spacetime evolution of fluid-fluid interface, Transport in Porous Media, **2**, 537–552, 1987.

17. Kalaydjian, F., Legait, B., Écoulement lent à contre-courant en imbibition spontanée de deux fluides non miscibles dans un capillaire présentant un rétrécissement, C. R. Acad. Sc. Paris, Sér. II, **304**, 15, 869–872, 1987.

18. Kalaydjian, F., Legait, B., Perméabilités relatives couplées dans des écoulements en capillaire et en milieu poreux, C. R. Acad. Sc. Paris, Sér. II, **304**, 17, 1035–1038, 1987.

19. Kalaydjian, F., Legait, B., Effets de la géométrie des pores et de la mouillabilité sur le déplacement diphasique à contre-courant en capillaire et en milieu poreux, Rev. Phys. Appl., **23**, 1071–1081, 1988.

20. Koplik, J., Banavar, J. R., Willemsen, J. F., Molecular dynamics of Poiseuille flow and moving contact lines, Phys. Rev. Lett., **60**, 13, 1282–1285, 1988.

21. Koplik, J., Lasseter, T. J., Two-phase flow in random network models of porous media, Preprint SPE 11014, 1982.

22. Legait, B., Laminar flow of two phases through a capillary tube with variable square cross-sectrion, J. Colloid Interface Sci., **96**, 1, 28–38, 1983.

23. Lenormand, R., Touboul, E., Zarcone, C., Numerical models and experiments on immiscible displacements in porous media, J. Fluid Mech., **189**, 165–187, 1988.

24. Lenormand, R., Zarcone, C., Sarr, A., Mechanisms of the displacement of one fluid by another in a network of capillary ducts, J. Fluid Mech., **135**, 337–353, 1983.

25. Lenormand, R., Zarcone, C., Two-phase flow experiments in a two-dimensional permeable medium, P.C.H., **6**, 5/6, 497–506, 1985.

26. Magda, J. J., Tirrell, M., Davis, H. T., Molecular dynamics of narrow, liquid-filled pores, J. Chem. Phys., **83**, 4, 1888–1901, 1985.

27. Marle, C.-M., Écoulements monophasiqus en milieu poreux, Rev. Inst. Français du Pétrole, **22**, 10, 1471–1509, 1967.

28. Marle, C.-M., On macroscopic equations governing multiphase flow with diffusion and chemical reactions in porous media, Int. J. Engng Sci., **20**, 5, 643–662, 1982.

29. Matheron, G., Les variables régionalisées et leur estimation, Thèse de doctorat d'état, Faculté des sciences de Paris, 1965.

30. Matheron, G., "Éléments pour une théorie des milieux poreux", Masson, Paris, 1967.

31. Pavone, D., Lois de Darcy et de pression capillaire d'un ecoulement biphasique en milieu poreux deduites par prise de moyenne spatiale, C.R. Acad. Sc. Paris, Ser. II, 308, 508, 1989.

32. Quintard, M., Whitaker, S., Two-phase flow in heterogeneous porous media: the method of large-scale averaging, Transport in Porous Media, **3**, 357–413, 1988.

33. Rothman, D. H., Lattice-gas automata for immiscible two-phase flow, Geophysics, **53**, 4, 509–518, 1988.

34. Rothman, D. H., Keller, J. M., Immiscible cellular-automaton fluids, J. Stat. Phys., **52**, 3/4, 1119–1127, 1988.

35. Saffman, P. G., Dispersion due to molecular diffusion and macroscopic mixing in flow through a network of capillaries, J. Fluid Mech., **7**, 2, 194–208, 1960.

36. Seppecher, P., Étude d'une modélisation des zones capillaires fluides: interfaces et lignes de contact, Thèse de doctorat de l'Université Pierre et Marie Curie, Paris, 1987.

37. Scheidegger, A. E., Statistical hydrodynamics in porous media, J. Appl. Phys., **25**, 994, 1954.

38. Sikkenk, J. H., Indekeu, J. O., Van Leeuwen, J. M. J., Vossnack, E. O., Bakker, A. F., Simulation of wetting and drying at solid-fluid interfaces on the Delft molecular dynamics processor, J. Stat. Phys., **52**, 1/2, 23–44, 1988.

39. Singhal, A. K., Somerton, W. H., Two-phase flow through a non-circular capillary at low Reynolds numbers, J. Can. Pet., 197–205., 1970.

40. Singhal, A. K., Somerton, W. H., Quantitative modelling of immiscible displacement in porous media; a network approach, Rev. Inst. Français du Pétrole, **32**, 6, 897–920, 1977.

41. Slattery, J. C., Flow of viscoelastic fluids through porous media, AIChE Journal, **15**, 6, 1066–1071, 1967.

42. Torres, F. E., Closure of the governing equations for immiscible, two-phase flow: a research comment, Transport in Porous Media, **2**, 383–393, 1987.

43. 43. Toxvaerd, S., The structure and thermodynamics of a solid-fluid interface, J. Chem. Phys., **74**, 3, 1998–2005, 1981.

44. Whitaker, S., Diffusion and dispersion in porous media, AIChE Journal, **13**, 3, 420–427, 1967.

45. Whitaker, S., Flow in porous media, I: A theoretical derivation of Darcy's law, II: The governing equations for immiscible, two-phase flow, Transport in Porous Media, **1**, 3–25 and 105–125, 1986.

PREDICTION OF PERMEABILITY FROM A COMBINATION OF MERCURY INJECTION AND PORE IMAGE ANALYSIS DATA

David MacGowan

*BP Research, Sunbury Research Centre,
Chertsey Road, Sunbury-on-Thames,
Middlesex TW16 7LN, U.K.*

ABSTRACT

A series-parallel (simplified network) model has been developed for which invasion percolation calculations modelling mercury injection can be carried out analytically. A considerable strength of the series-parallel model is its inclusion of a variable pore connectedness since this has as strong an influence on permeability as does the pore size distribution. Pore sizes are assigned to the model according to a distribution obtained from digital image analysis and mercury injection calculations are carried out for various values of connectivity. Agreement between experimental and modelled mercury injection curves is optimised by choice of a connectivity parameter with which the model permeability is then calculated. The result has been found to be a good prediction of directly measured permeability for a wide range of rocks.

1. INTRODUCTION

The accurate characterisation of fluid flow in porous media is of fundamental importance to successful oil exploration and production. Among the measurements routinely made which are relevant to this problem are porosities, permeabilities and mercury injection curves. Lack of direct permeability measurements often leads to the use of empirical porosity-permeability correlations which are not, in general, very satisfactory. There is therefore enduring interest in reliable indirect permeability estimates which can supplement or complement existing direct methods.

Much useful qualitative insight into the relation between permeability and microscopic void geometry of porous rocks has come from network modelling. Calculation of quantitative permeability results from microscopic data for real porous rocks has been rarer, partly due to lack of detailed information about pore structure. The present attempt at permeability prediction has been inspired by the increasing amount of data becoming available from pore image analysis.

The structure of this paper is as follows: In §2, the interpretation of mercury injection is discussed and basic definitions and terminology of pore size distributions are introduced. In §3, after reviewing existing models of porous media, mercury injection and permeability are calculated for the series-parallel model introduced in this report.

Qualitative results and permeability predictions for various rock samples are given in §4. After discussion of possible improvements in §5, conclusions are summarised in §6.

2. INTERPRETATION OF MERCURY INJECTION

In the following discussion it will be assumed that the pore space consists entirely of uniform cylindrical pores but that these may be connected arbitrarily. Then a probability density function p(D) for the capillary diameters D (where p(D)dD represents the fraction of the total number of capillaries with diameters between D and D+dD) completely defines the pore size distribution. It is convenient to define the cumulative probability

$$y(D) = \int_D^\infty p(x)dx, \qquad (2.1)$$

the number fraction of pores with diameter $\geq D$ (normalisation: y(0)=1) and

$$z(D) = \int_D^\infty p(x)x^2 dx. \qquad (2.2)$$

An alternative definition of pore size distribution is $\alpha(D)$, where $\alpha(D)dD$ gives the fraction of total void space occupied by capillaries of diameters between D and D+dD. Clearly, for cylindrical pores of equal length,

$$\alpha(D) = D^2 p(D)/z(0). \qquad (2.3)$$

$\alpha(D)$ will be referred to here as the pore volume fraction and the cumulative volume fraction is $z(D)/z(0)$.

Mercury porosimetry is a widely used technique for getting an idea of pore size distributions in reservoir rocks. A cleaned and evacuated core sample is placed in a bath of mercury and the volume of mercury injected into the pore space is recorded as the externally applied pressure P is gradually increased. Assuming that the injection process is purely determined by capillary forces (which is quite reasonable for a well controlled injection), the raw pressure/saturation curve for mercury injection can be transformed to a pore-entry-diameter/saturation curve, using the equation

$$D_e = 4\gamma |\cos\theta|/P, \qquad (2.4)$$

where γ is the mercury surface tension and θ the contact angle. The incremental volume at pressure P represents the void volume just enterable by a meniscus of mean curvature $2|\cos\theta|/D_e$. The pore volume fraction with entry diameter D_e is now given by

$$\alpha_e(D_e) = - dS/dD_e, \qquad (2.5)$$

where S is the mercury saturation (as a fraction of total void volume). For any given pore, D and D_e need not be related in any way. The filling of a wide pore can be totally controlled by a set of narrow pores surrounding it, possibly at quite some distance.

In order to permit quantitative estimation of the pore size distribution from mercury injection it is necessary to make some simplifying assumption about the connectivity of the pore space. The conventional model is a set of parallel uniform nonintersecting cylindrical capillaries of length equal to the sample thickness. This allows a trivial determination of what is loosely referred to as the 'pore size distribution' but which will here be called an apparent pore volume fraction $\alpha_{app}(D)$. Since pore blocking effects are completely absent from the parallel capillary model,

$$\alpha_{app}(D) = - dS/dD. \qquad (2.6)$$

Clearly $\alpha_e(D_e)$ and $\alpha_{app}(D)$ are numerically equivalent, but their physical interpretations are quite different.

The relation of $\alpha_{app}(D)$ to the true pore volume fraction $\alpha(D)$ associated with pores of diameter D is most easily seen from Figure 1. Figure 1a shows $\alpha(D)$, the real pore volume fraction for some imagined porous medium. Also shown is $\alpha(D;E)$, where $\alpha(D;E)dD$ is the fraction of total void volume represented by pores of diameters between D and D+dD which have entry diameters $D_e \geq E$. This is the type of bivariate distribution used by Dullien and Dhawan [1]. Clearly $\alpha(D;E)$ is zero for D<E and $\alpha(D;E) \leq \alpha(D)$.

In Figure 1b, $\alpha(D;E-dE)$ has been added to Figure 1a and the shaded area represents the fraction of total void volume having entry diameter between E-dE and E, that is $\alpha_e(D_e=E)dE$. Thus the point $\alpha_{app}(D=E)$ [or $\alpha_e(D_e=E)$] of the mercury injection pore volume fraction may be obtained by distorting the shaded area of Figure 1b into the vertical strip shown in Figure 1c. The full mercury injection pore volume fraction can be constructed similarly.

Whenever both $\alpha(D)$, from image analysis and equation (2.3), and $\alpha_{app}(D)$, from mercury injection and equation (2.6), have been obtained for the same rock sample, the latter distribution has been considerably narrower and peaked at considerably lower D than the former [2] (as in Figure 1). Experimentally, this shift is partly due to image analysis accessing pore bodies while mercury injection is controlled by pore throats. In the above discussion, since uniform capillaries were assumed from the start, the only effect contributing to the shift is pore blocking. This is, however, believed to be the predominant cause of the experimental shift.

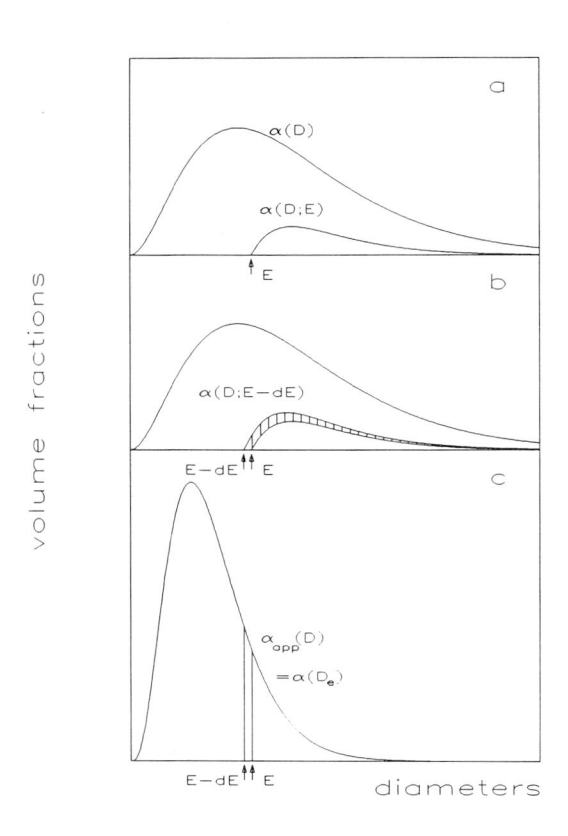

Figure 1. Schematic relation of mercury injection to image analysis.

3. MODELLING

Earlier models

The simple parallel nonintersecting capillary model remains in common use for mercury porosimetry, despite its shortcomings. These include failure to account for either hysteresis or trapping on withdrawal, drawbacks shared by the 'random cut and rejoin' model [3]. Series models [4,5] exhibit hysteresis but no trapping. The best current models are capillary networks which, though still idealised, improve upon the parallel and series models by accounting (at least qualitatively) for both hysteresis and trapping.

Capillary dominated displacements in network models are closely related to the mathematical concept of invasion percolation [6]. There is a variant of invasion percolation including trapping which is appropriate to displacement of an incompressible fluid (e.g. waterflood of an oil reservoir). There is no trapping in displacement of a highly compressible fluid (as in mercury injection) and so trapping need not be considered in this work. The early attempts at stochastic network simulations specifically aimed at the study of immiscible fluid displacements are well summarised by Dullien [7]. In the work of Chatzis and Dullien [8], the injection was from one side (in three dimensions, face) of a network. Mann and co-workers [9,10] have treated injection from all sides (closer to mercury injection) and have also made the most serious attempt to reproduce an experimental mercury injection curve from network modelling. This is, however, of limited value since they arbitrarily fix connectivity by assuming a two-dimensional square lattice.

Mercury Injection for the Series-parallel Model

A model intermediate in complexity between the early capillary models and network models is illustrated in Figure 2. The capillaries are represented simply by bonds connecting nodes, though they should be thought of as having a distribution p(D) of diameters. There are ν bonds in parallel within each of N segments, all of which reconnect at the next node. If mercury reaches any node, it is assumed that all bonds at the next level are accessible to it (given sufficiently high injection pressure). This series-parallel model includes the simpler parallel nonintersecting capillary model as the special cases N=1 and $\nu\to\infty$, and the series model as the case ν=1. The complete series-parallel model consists of an ensemble of channels, each having N segments of ν uniform cylindrical capillaries of equal length l. Thus it is reasonable to regard L=Nl as the sample length.

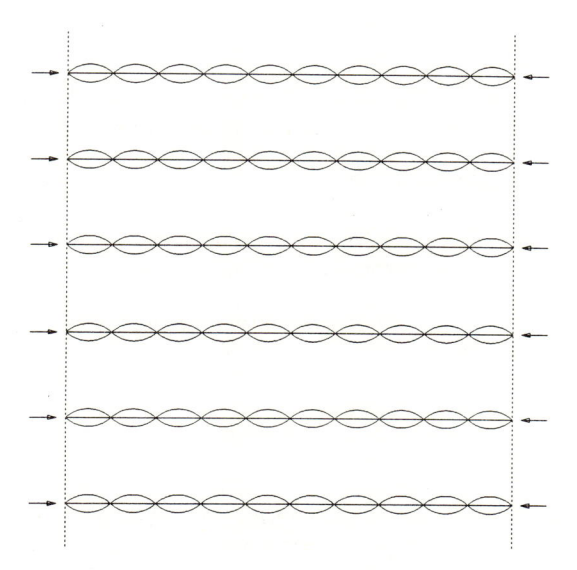

Figure 2. Illustration of a set of series-parallel channels (N=10, ν=3).

Consider a statistical ensemble of series-parallel channels. Suppose the pressure is increased from zero to P, with corresponding minimum diameter of capillary entered by mercury, D, given by equation (2.4). Then the probability of entering m segments from one end of a channel and n segments from the other end is

$$\Xi(m,n) = b^2 e^{m+n}, \quad m+n<N-1, \tag{3.1}$$

$$\Xi(m,n) = be^{N-1}, \quad m+n=N-1, \tag{3.2}$$

where

$$b = [1 - y(D)]^\nu, \tag{3.3}$$

$$e = 1 - b. \tag{3.4}$$

e and b are the entry and blocking probabilities for a single segment at injection pressure P. Two blocking segments are necessary for most configurations but there is only a single unentered segment when n+m=N-1. The probability of entering all N segments is e^N.

The expectation value for the injected volume of mercury at pressure P is

$$V = \left[\sum_{m=0}^{N-1} \sum_{n=0}^{N-m-1} (m+n)\Xi(m,n) + Ne^N \right] v_s, \tag{3.5}$$

where v_s is the expected value of the volume of a single <u>entered</u> segment during mercury injection to pressure P. Now, because diameters are distributed randomly,

$$v_s = n_c v_c = \pi l z(D)\nu/4e, \tag{3.6}$$

where $n_c = \nu y(D)/e$ is the expected number of <u>filled</u> capillaries in an entered segment and $v_c = \pi l z(D)/4y(D)$ is the expected volume of a filled capillary.

Carrying out the summations in (3.5), yields the following result for the saturation, $S = 4V/[\pi \nu N l z(0)]$, during injection:

$$S = \left[\frac{2(1 - e^N)}{N(1 - e)} - e^{N-1} \right] \frac{z(D)}{z(0)} \tag{3.7}$$

As expected, the parallel capillary result S=z(D)/z(0) holds when N=1; it is also valid for N=2, with the obvious physical interpretation that blocking the channels at their centres would have no effect on filling so that the N=2 model is equivalent for mercury injection to N=1 channels open only at one end. The parallel capillary result is also approached as $\nu \to \infty$, since then $e \to 1$ for any nonzero D.

Permeability Calculation for the Series-parallel Model

Permeability is calculated by a Kirchhoff's law type calculation where the pore network is treated like an electrical resistor network with hydraulic conductance replacing electrical conductance. Details of fluid motion at the nodes are completely ignored. In single phase Darcy flow of an incompressible fluid, the hydraulic conductance of a regular cylindrical pore of diameter D and length l is

$$Q/\Delta P = \pi D^4/128\eta l, \qquad (3.8)$$

where Q is the volume rate of flow, ΔP is the pressure difference across the pore and η is the fluid viscosity. For an ensemble of n series-parallel channels,

$$Q = \frac{n\pi \Delta P}{128 \eta L} \left[\int_0^\infty dG \, \frac{r(G)}{G} \right]^{-1}, \qquad (3.9)$$

where L=Nl is the sample length. The random variable G is a sum of ν independent random variables each identically distributed as $g=D^4$. Thus r(G), the probability density function for G, is obtained by ν-fold convolution of the probability density function q(g) for g, where q(g)dg=p(D)dD.

For the series-parallel model, the void cross section is

$$A_v = \frac{n\nu\pi}{4} \int_0^\infty dD \, D^2 \, p(D) = \frac{\phi A}{3}, \qquad (3.10)$$

where ϕ is the porosity. In obtaining the last equality it is assumed that equal numbers of channels run in each of three spatial dimensions; A is the total cross section (void and solid). Combining equations (3.9) and (3.10), it is seen by comparison with Darcy's law

$$Q = kA\Delta P/\eta L, \qquad (3.11)$$

that the permeability k for the series-parallel model is given by

$$k = \frac{\phi}{96\nu} \left[\int_0^\infty dD \, D^2 \, p(D) \int_0^\infty dG \, \frac{r(G)}{G} \right]^{-1} \qquad (3.12)$$

For $\nu=1$, equation (3.12) reduces to

$$k = \frac{\phi}{96} \left[\int_0^\infty dD \, D^2 \, p(D) \int_0^\infty dD \, \frac{p(D)}{D^4} \right]^{-1}, \qquad (3.13)$$

in agreement with the expression given by Dullien [11] for the permeability of series capillaries.

4. RESULTS

Qualitative Effects

For an exploratory study of the series-parallel model, lognormal distributions of pore diameter were used. The unconventional form

$$p(D) = \left[\frac{\mu}{\pi}\right]^{1/2} \frac{1}{cD^{3/2}} \exp\left[-\frac{(\ln D - \ln \mu)^2}{c^2} - \frac{c^2}{16}\right], \quad D \geq 0, \quad (4.1)$$

of lognormal distribution function is chosen so that μ is the mean and the standard deviation is $\mu\sqrt{[\exp(c^2/2) - 1]}$. In order to present results of mercury injection calculations for wide pore size distributions, it is best to plot

$$\beta_{app}(D) = -dS/d\log D = \ln 10\, D\, \alpha_{app}(D). \quad (4.2)$$

Correspondingly, the real pore volume fraction is plotted as

$$\beta(D) = \ln 10\, D\, \alpha(D). \quad (4.3)$$

Examples of $\beta_{app}(D)$ are shown in Figure 3. The shift of $\beta_{app}(D)$ relative to $\beta(D)$ becomes more pronounced as ν decreases. Other factors leading to more pronounced shifts are broadening of p(D) (increasing c) and increasing N.

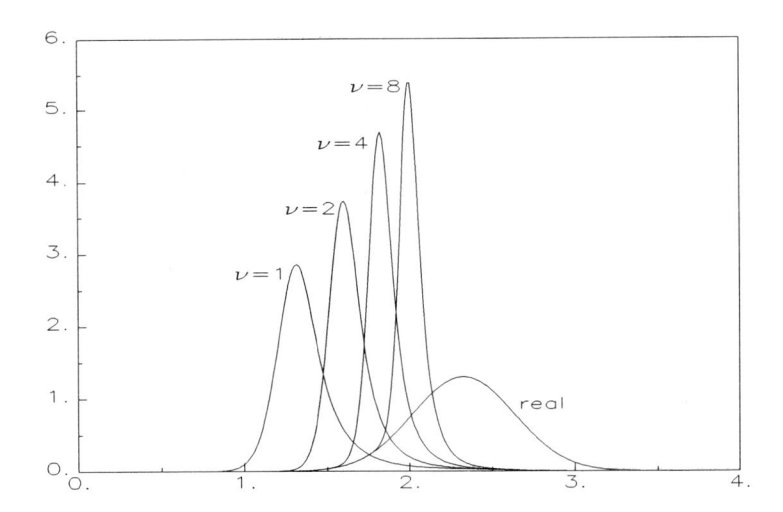

Figure 3. Effect of varying ν on series-parallel mercury injection for μ=100, c=1.0.

The explicit $1/\nu$ dependence shown in equation (3.12) may be misleading, since it is reversed by the implicit dependence. In fact k increases with increasing ν due to the greater probability of fluid flow avoiding narrow pores. This is illustrated by the $k/\phi\mu^2$

results shown in Figure 4 for lognormal p(D) with c=1.0 and c=1.5. The most rapid relative increase occurs between $\nu=1$ and $\nu=2$ and this is greater for the broader p(D). The series-parallel model predicts permeabilities intermediate between those of the series and parallel models in a manner which reflects increasing connectivity of pores. The potential range of values of k with changing connectivity easily exceeds the effect of a doubling or more of all pore sizes.

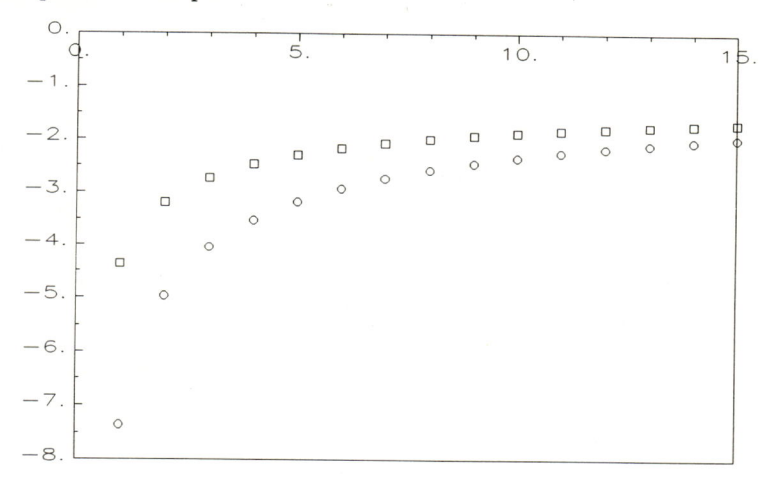

Figure 4. Series-parallel model permeability variation with ν for c=1.0 (squares) and c=1.5 (circles).

Permeability Predictions

Image analysis results are obtained from thin sections cut from the ends of plugs used for mercury injection and are therefore relevant to horizontal flow paths. The image analysis data used are the lengths and breadths of pores, defined as the longest and shortest chords through the centroid of the pore. The approximation is made of identifying these results from two-dimensional sections as lengths and diameters of cylindrical pores. (The latter identification is exact for a truly cylindrical pore.) In the model, all capillaries are of equal length and so the distribution of lengths from image analysis is not used; just the mean length determines the number N of pores necessary to span a 25.4mm mercury injection plug. The capillary diameter distribution is approximated by a lognormal having the mean and standard deviation of the image analysis breadth distribution.

Model mercury injection curves are calculated for a series of values of ν. A value of ν to represent the plug under consideration is determined by finding the best match between one of the model curves and the experimental mercury injection curve (see Figure 5). Results obtained in this way for a suite of core samples — all sandstones apart from C which is a carbonate — are shown in column 2 of Table 1. Where a range of ν is given a small change in connectivity caused a large change in the model mercury injection curve and the experimental injection curve lay between two of the model curves.

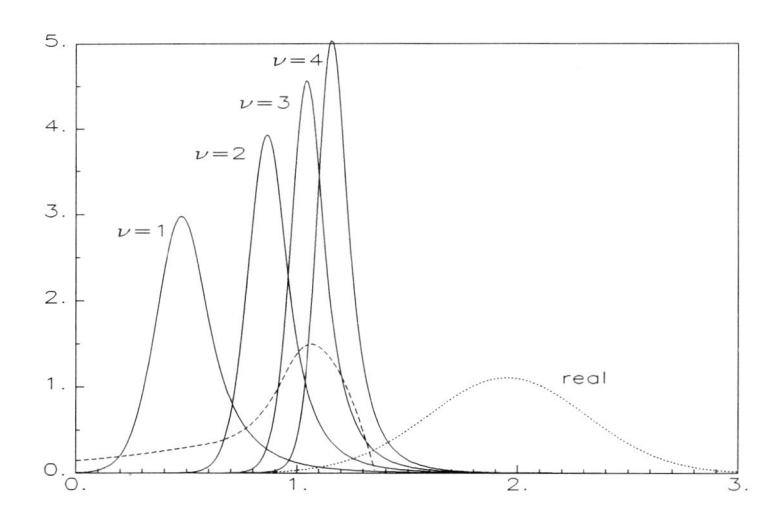

Figure 5. Selection of connectivity $v=3$ for sample F4 (experimental results, broken curve).

The permeability results obtained from equation (3.12) are shown in column 4 of Table 1. For plugs with a range of v, logk is linearly interpolated between the values for the nearest integers. It is emphasised that these results are fully 'a priori' predictions with no adjustable parameters. However, since some of the assumptions made in the calculation (e.g. ignoring details of flow at pore junctions and identifying p(D) with a two-dimensional breadth distribution) are expected to cause errors roughly of a constant factor, it is worth including such a factor. Agreement between experiment and theory is optimised for samples A1-A8 (where the image analysis magnification was ×100) by dividing all model results by a factor 1.3. The same correction factor has been applied to subsequent results where ×100 image analysis data was available. The results are shown in column 5 of Table 1 and Figure 6. The blanks in the table indicate that ×50 magnification was used.

5. DISCUSSION

The series-parallel model is a simple example of invasion percolation without trapping. The qualitative differences between $\alpha_{app}(D)$ and $\alpha(D)$ for the model are very similar to those found in experiments when pore size distributions from mercury porosimetry are compared to those from quantitative micrography for the same rock [5]. The use of the model for permeability prediction has generally been successful. The only discrepancy significantly exceeding a factor of 2 is for sample F12, where the plug showed obvious longitudinal heterogeneity.

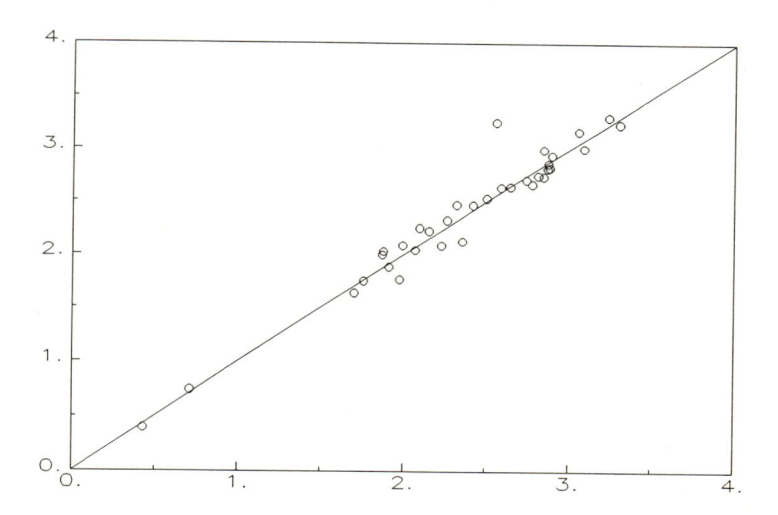

Figure 6. Comparison of experimental and predicted permeabilitie

However, at least two aspects of the series-parallel model are unrealistic: the partial lack of cross connections and the complete lack of correlations between capillary diameters. The former fault is believed to be the more serious. It leads to a trivial one-dimensional percolation problem related to the filling probabilities which is quite unlike its two- and three-dimensional counterparts [12]. Consequently, there is no limiting form of the apparent distribution as $N \to \infty$.

A number of improvements may be envisaged which retain the essential character of the model: A continuous connectivity parameter should be introduced into the model in place of the discrete parameter ν. This can be done by fixing only an average value of ν without insisting that it be the same for each segment and will be particularly important at low connectivity. A proper transformation from two-dimensional pore image analysis data to distributions of three-dimensional features should be attempted. The pore image analysis distribution function should be represented accurately rather than crudely approximated by a lognormal of the correct mean and variance.

Table 1: Permeability Predictions

Comparison of predictions of permeabilities based on the series-parallel (sp) model and rescaling of it by 1.3 (where ×100 image analysis data was used) with values measured in core flow experiments. Column 2 gives the connectivity parameter ν best able to reconcile the image analysis and mercury injection data within the sp analysis for each sample.

sample	best ν	permeabilities (milliDarcy)		
		experiment	sp model	rescaled
A1	8	183	276	212
A2	2-3	2.7	3.2	2.5
A3	4	5.1	7.2	5.5
A4	6	57	73	56
A5	5	50	56	43
A6	8	74	129	99
A7	5	94	76	58
A8	6	117	143	110
B	2	9.0	4.0	-
C	4	72	88	-
D	4	891	556	-
E1	3	109	101	-
E2	14	1220	1320	1013
E3	9	388	555	427
E4	11	782	1130	868
E5	8	318	439	338
E6	9	442	566	435
E7	8	209	382	294
E8	7	263	380	292
E9	8	549	655	504
E10	5	81	99	76
F1	3-4	75	139	107
F2	16	1720	2650	2040
F3	3-4	124	233	179
F4	3	142	218	168
F5	7-8	760	857	659
F6	15	2010	2270	1750
F7	7-8	697	706	543
F8	7	745	942	725
F9	3-4	227	174	134
F10	8-9	738	841	647
F11	10	696	1280	985
F12	12	360	2370	1820
F13	12	1130	1940	1490
F14	3	98	159	122
F15	3-4	169	159	122
F16	6-7	647	720	554
F17	5-6	597	597	459

It is of interest to compare the series-parallel method to Dullien's work [5]. As indicated in §2, Dullien's model is inferior in respect of its inability to account even qualitatively for trapping. From the point of view of experimental input, Dullien's method requires bivariate pore size distributions [1] which are obtained by injecting Wood's metal to various different pressures and, at each pressure, image analysing just the invaded pores. This is a difficult and time-consuming procedure compared to the single mercury injection and single image analysis of the whole pore space which are required for the series-parallel method. The accuracy of permeability prediction achieved in this work is good, considering the preliminary nature of the calculations and the crude level of numerical analysis, but is not as good as that obtained by Dullien.

6.   CONCLUSIONS

A study has been carried out of mercury injection for the series-parallel model. It is believed that this new exactly calculable model gives helpful insights into the interpretation of porosimetry experiments since it shows all the qualitative features of mercury intrusion data. Pore connectivity is inversely related to the shift of the mercury injection pore size distribution compared to the corresponding one obtained from image analysis.

A new method of estimating permeabilities has been presented. An approximate prediction of sandstone (plus one carbonate) permeabilities has been carried out on the basis of the series-parallel model. Results are encouraging and make clear how this method can distinguish between permeabilities of samples with similar pore sizes and porosities but very different connectedness of pores.

ACKNOWLEDGEMENTS

G.R. Duckett, O.P. Whelehan, A.C. Brayshaw, M.K. Dosanjh and C.H. Harris are thanked for providing image analysis data; S.J. Morrison and M.Z. Kalam for permeability data; and G.S. ter Kuile for mercury porosimetry data. The British Petroleum Co. plc is thanked for permission to publish this work.

REFERENCES

[4] Androutsopoulos, G.P. and Mann, R., Modelling of the skin effect in catalyst pellets, Chem. Eng. Sci., 31, 1131-1138, 1976.

[9] Androutsopoulos, G.P. and Mann, R., Evaluation of mercury porosimeter experiments using a network pore structure model, Chem. Eng. Sci., 34, 1203-1212, 1979.

[8] Chatzis, I. and Dullien, F.A.L., Modelling pore structure by 2-D and 3-D networks with application to sandstones, J. Can. Petr. Tech. 16, 97-108, 1977.

[5] Dullien, F.A.L., New network permeability model of porous media, AIChE J., 21, 299-307, 1975.

[7] Dullien, F.A.L., "Porous Media: Fluid Transport and Pore Structure", pp 44-48, Academic, New York, 1979.

[11] Dullien, F.A.L., "Porous Media: Fluid Transport and Pore Structure", p 175, Academic, New York, 1979.

[2] Dullien, F.A.L. and Dhawan, G.K., Characterisation of pore structure by a combination of quantitative photomicrography and mercury porosimetry, J. Coll. Interf. Sci., 47, 337-349, 1974.

[1] Dullien, F.A.L. and Dhawan, G.K., Bivariate pore-size distributions of some sandstones, J. Coll. Interf. Sci., 52, 129-135, 1975.

[10] Mann, R., Androutsopoulos, G.P. and Golshan, H., Application of a stochastic network pore model to oil-bearing rock with observations relevant to oil recovery, Chem. Eng. Sci., 36, 337-346, 1981.

[3] Millington, R.J. and Quirk J.P., Permeability of porous solids, Trans. Faraday Soc., 57, 1200-1207, 1961.

[12] Stauffer, D., "Introduction to percolation theory", Taylor and Francis, London, 1985.

[6] Wilkinson, D. and Willemsen, J.F., Invasion percolation: a new form of percolation theory, J. Phys. A, 16, 3365-3376, 1983.

THE INCLUSION OF MOLECULAR DIFFUSION EFFECTS IN THE NETWORK MODELLING OF HYDRODYNAMIC DISPERSION IN POROUS MEDIA

K. S. Sorbie
(Heriot-Watt University, Edinburgh)

P. J. Clifford
(BP Research Centre, Middlesex)

ABSTRACT

Hydrodynamic dispersion of tracers in flow through porous media has attracted a considerable amount of experimental and theoretical study over the past thirty years.

A number of mathematical approaches have been used to analyse dispersion, which take account of its stochastic origins. These include methods based on ensemble averaging of particle motions in conjectural networks of tubes or idealised particle beds or alternatively, Monte Carlo simulations of the motions of large samples of particles through a particular network structure. All of these methods attempt to evaluate appropriate statistics from distance travelled or arrival time distributions of the marker particles, from which the dispersion coefficient may be calculated.

In this paper, we derive a network model of hydrodynamic dispersion in porous media which takes into account the effects of molecular diffusion of the particles propagated through the network. Our approach is based on a description of the particle jump statistics in the network elements (capillaries) which takes proper account of the molecular diffusion of the particles over all Peclet number, Pe, and aspect ratio ranges of the tube. We demonstrate that the convection-diffusion equation in the "Taylor limit" is inappropriate to use over very wide flow regimes for calculating these single element jump statistics; other (Monte Carlo) methods are used to obtain these quantities. When the statistical parameters derived in this way are incorporated into a full network model, the calculated dispersion behaviour predicted over a very wide range of network Peclet numbers, Pe_N, agrees very well with experimental results presented in the literature. By plotting the calculated quantities in an appropriate way, the diffusion-dominated, mixed and convection dominated flow regimes are clearly reproduced by our model. In particular, we note that in the flow regime which is mainly convection dominated, but where molecular diffusion still plays a role (the "mixed" regime), our model predicts that D increases faster than linearly with fluid flow velocity, U; D varies as U^α where α is in range 1.19-1.25.

Most of the data on dispersion presented in the literature have been for low-molecular weight species in both consolidated and unconsolidated packs. Recently, some data have been published on both polymer and tracer dispersion in the same flow experiment. It appears that polymer dispersion is larger than that of tracer by a factor of between 2 and 4 in certain flow regimes. In this paper we analyse this effect in terms of the widely different diffusion constants of the polymer and tracer molecules.

1. INTRODUCTION

The transport of molecules through porous media is important in oil industry studies of tracer, salinity and chemical EOR. In a porous medium, molecules of a trace component are transported by a complex interaction between convection in the solvent fluid and molecular diffusion through Brownian motion. If two molecular species possess different diffusion coefficients, their dispersive behaviour in a porous medium will be clearly distinguishable at both high and low flow rates, even in the absence of further factors such as rheological effects or adsorption.

There are ample experimental data on the dispersion of low molecular weight tracers in porous media (Gunn and Pryce, 1969; Fried and Combarnous, 1971; Perkins and Johnston, 1963), which demonstrate tracer behaviour in both low velocity (diffusion-dominated) and high velocity (convection-dominated regimes. Some of these studies have demonstrated that dispersivity, defined as (dispersion coefficient/flow velocity), is not constant as a function of flow velocity but in fact, shows different flow regimes. At very low flow rates dispersivity decreases with velocity and in the convection-dominated regime it increases slowly with velocity. A recent experimental study (Sorbie et al, 1987) examined dispersion in a sandstone core of both a chloride tracer and xanthan biopolymer over a range of flow rates. The results, as shown in Figure 1, demonstrate clearly a greater dispersivity for the xanthan than chloride, and a dispersivity for both species which increases slowly with velocity. The polymer concentration in the dispersion experiments presented in Figure 1 is sufficiently low (50 ppm) that the fluid is Newtonian; this avoids any additional viscosity and rheological effects which are known to be present in higher concentration polymer solutions (Sorbie et al, 1987; Sorbie et al, 1989).

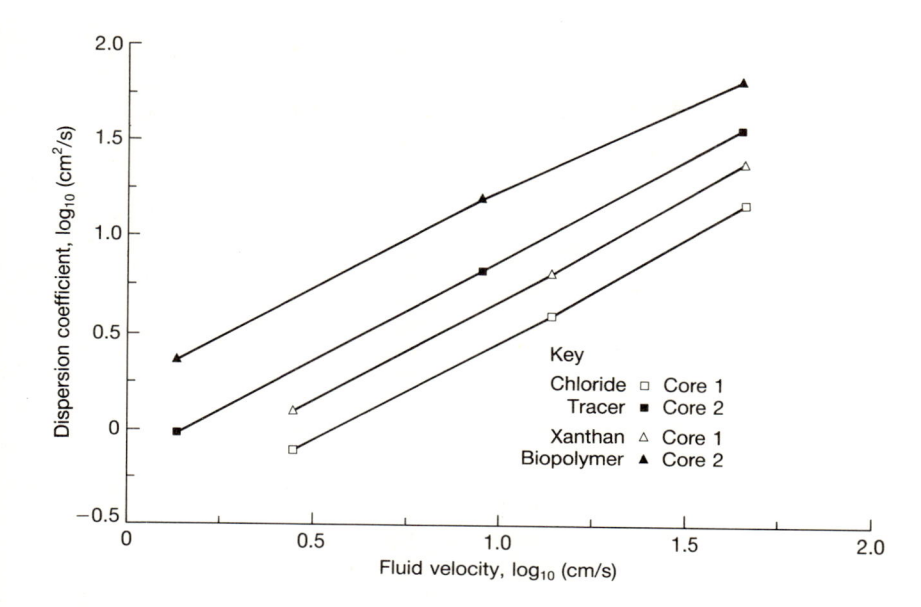

Figure 1. Measured dispersion coefficients for chloride Tracer (^{36}CL) and ranthan biopolymer in flow through clashach sandstone cores for a low concentration (50ppm) polymer solution; data replotted from Sorbie et al (1987).

The objective of this paper is to present a network model which allows us to calculate the hydrodynamic dispersion of a system over all flow regimes. This must incorporate an appropriate model of the porous medium, through network heterogeneity, and the molecular diffusion behaviour of the tracer particle itself. This general approach to the calculation of hydrodynamic dispersion will also allow us to explain the specific finding on tracer and polymer dispersion noted above. We will demonstrate in this paper that this is associated with the relative magnitudes of the molecular diffusion coefficients of the two species. There is a factor of nearly 1000 between the molecular diffusion coefficients of xanthan and chloride, which would tend to place the xanthan in a more "convection-dominated" regime in terms of Peclet number than has normally been investigated for tracers of low molecular weight.

Calculations of dispersion in porous media are generally based either on ensemble averages of particle motion through a conjectural network of tubes (Saffman, 1959; de Jong, 1958) or a bed of particles (Koch and Brady, 1985) or on Monte Carlo simulations which follow the motions of many particles through a particular network structure (Dullien 1979; de Arcangelis et al, 1986; Sahimi et al, 1983, 1986; Torelli, 1972). The work of Saffman (1959) was based on highly simplified representations of the transit times of particles through individual tubes, all of which were assumed to be of identical radius, but random in direction, but nevertheless showed mechanisms whereby the dispersivity could increase with velocity at high Peclet number, due to the limited radial motions of particles within tubes. Koch and Brady (1985) examine certain cases of "non-mechanical" contributions to dispersion, including terms from boundary layers and regions of closed streamlines, but in the context of particle beds.

In this paper, we will mainly be concerned with calculations of dispersion in porous media using network modelling. This approach is based on representing the porous medium by an idealised 2D or 3D random network, usually of capillaries but sometimes of more complex elements (Dullien, 1979), in which the flows are calculated and dispersion is evaluated by tracking many particles through the system and calculating certain statistics. A number of workers have used this type of method to calculate dispersion in both single and two phase flow (Torelli, 1972; Dullien, 1979; Mohanty and Salter, 1983; Sahimi et al, 1983, 1986; de Arcangelis et al, 1986; Koplik et al, 1986). The particles perform a weighted random walk through the lattice where the jump probability at a node is usually determined by the convective flows at that node. In this form, the lattice walk is a type of Markov process since particles have no memory of their previous history when they jump from one node to another.

However, in its simplest form, the network calculation of dispersion with jumps based on convected flow only, can only predict a _linear_ relationship between dispersion and velocity at all pressure gradients (flow rates) in the network (Sahimi et al, 1983). In order to reproduce dispersion/velocity behaviour which is observed experimentally, it is necessary to incorporate the effects of molecular diffusion of the transported marker particle into the network calculation. This has been attempted although no previous work has provided a model which is both mechanistically correct and is applicable over a wide range of Peclet numbers. Both Torelli (1972) and Sahimi et al (1986) explain in the slow increase of dispersivity with velocity by using a model based on "streamline splitting" at network junctions. Sahimi et al (1986) also describe a model for particle hopping between streamlines. However, although this approach does incorporate some of the features of tracer transport in convection dominated flow, it does not relate the jump statistics across a bond (capillary) to the molecular diffusion constant or the specific flow regime in the tube (ie the tube Peclet number, Pe and aspect ration, (R/L) - see below). Neither of these models is applicable at very low (diffusion dominated) rates or over a very wide range of system Peclet numbers. An alternative method of introducing the effects of molecular diffusion into network modelling is proposed by Koplik and co-workers (de Arcangelis et al, 1986; Koplik et al, 1987) which uses a Laplace transform method to calculate transit time distributions of particles. This method avoids possible problems with inadequate sampling of flow-paths which can arise from "slow" bonds in the network. Interesting results are derived for cases of low and medium Peclet number. However, their method uses the one-dimensional convection-diffusion equation to represent particle transit times across individual tubes, which is particularly inappropriate at high velocity, as we demonstrate later in this paper.

In this paper, we calculate dispersion by tracking particles through networks of tubes of random radius. In this process, it is essential to have a good model to calculate the transit time distribution within network elements in all flow regimes. This is done here in a considerably moe detailed manner than in previous studies using Monte Carlo simulations of particle motions within individual tubes. The results of the single tube Monte Carlo are incorporated into a full calculation in which the network used is sufficiently random to demonstrate the very wide range of Peclet number experienced by different tubes for any given overall pressure gradient. The importance that these configurational effects have on the dispersion and its dependence on the mean flow velocity are also discussed in detail. This is the first network study to model and interpret dispersion behaviour over a very wide range of flow regimes (below turbulent conditions) using such detailed microscopic jump statistics.

2. CALCULATION OF TRANSIT THROUGH INDIVIDUAL NETWORK ELEMENTS

2.1 Monte Carlo Method

The distribution of transit times for each element is calculated from a knowledge of the mean fluid velocity $\bar{v}$, (with direction of flow from the node in question), the tube aspect ratio (ratio of radius, R, to length, L) and the molecular diffusion coefficient, D_0. Fluid flow in each tube is assumed to obey Poiseuille's law. The method of calculation is very similar across orders of magnitude in flow rate, expressed through a Peclet number as follows:

$$Pe = \frac{\bar{v} L}{4 D_0}$$

The method follows individual test particle motions through the fluid, subject to the local convective velocity, and to randomised three-dimensional jumps in space associated with molecular diffusion. *It is not appropriate in any limit to derive transit times by straight forward solution of the convection-diffusion equation in one dimension.* For high velocity, dispersion is controlled by the radial motion of particles within tubes; for low velocity, a pulse of particles will be homogenised over radius, but particles will then (for typical aspect ratios considered) be significantly affected by longitudinal molecular diffusion, which will mean that substantial particle motions occur against the direction of flow. The convection-diffusion equation can provide only a net particle flux, which is insufficient information to predict particle passage times in cases where particles are moving in both directions simultaneously.

The calculation considers a cylindrical tube of radius, R, and which is infinite in the positive and negative directions. A particle is assigned a starting position in the plane = 0, at a randomly calculated starting radius r_0, where the distribution function for r_0 is taken to be:-

$$f(r_0) = \frac{4 r_0}{R^2} \left(1 - \left(\frac{r_0}{R}\right)^2\right)$$

corresponding to the Poiseuille flux into the tube. This is not exact for cases of low Pe, but in those cases particles will move rapidly in the radial direction due to molecular diffusion and the value of r_0 is, therefore, less important.

The particle is then allowed to jump in three dimensions within the capillary according to the distribution function (Chandrasekhar, 1943):

$$\phi(\Delta\underline{r}) = \left(4\pi\, D_0\, \Delta t\right)^{-\frac{3}{2}} \exp\left\{-\frac{\Delta r^2}{4D_0\, \Delta t}\right\}$$

expressing the displacement in 3D, $\Delta\underline{r}$, occurring due to Brownian motion over a time interval Δt. To this displacement is added a second term due to convective fluid flow in the direction which is given by:

$$\Delta z_{000} = 2\bar{v} \cdot \left(1-\left(\frac{r}{R}\right)^2\right) \cdot \Delta t$$

Boundary **conditions** are imposed to prevent particles from moving beyond the radius R.

The calculation is continued until a sufficient number of timesteps has been taken for the particle to travel to a distance $\pm$ L from the origin for the first time. The transit time is recorded and this process is repeated for a large number of particles (at least 1000). The fraction of particles emerging in the positive (f_p) and negative (f_n) direction with respect to flow velocity is recorded. For each of these two groups, the mean ($<>$) and standard deviation (σ_t) of transit times is calculated. In many cases the distribution function of transit times is also examined which allows us to develop simplified distribution functions for the tube transit time statistics as discussed below.

2.2 Results for Single Tube Statistics

The behaviour of the statistical average of particle motion depends on two dimensionless quantities only. The normalised transit time

$$\langle t \rangle^* \equiv \frac{\langle t \rangle \bar{v}}{L}$$

and standard deviation

$$\sigma_t^* \equiv \frac{\sigma_t \bar{v}}{L}$$

and the fractions f_p and f_n of particles emerging in the positive and negative directions depend simply on the Peclet number defined in equation (2.1) and the aspect ratio (R/L). The Peclet number, Pe, represents the ratio of transit time through the element from longitudinal molecular diffusion to the transit time from convection. The dimensionless average and standard deviations of transit times along with the quantities f_p and f_n calculated using the single tube Monte Carlo method are shown in Figures 2 and 3.

Before discussing the results in Figures 2 and 3 in detail, it is useful to define a third dimensionless number, Pe_T, which

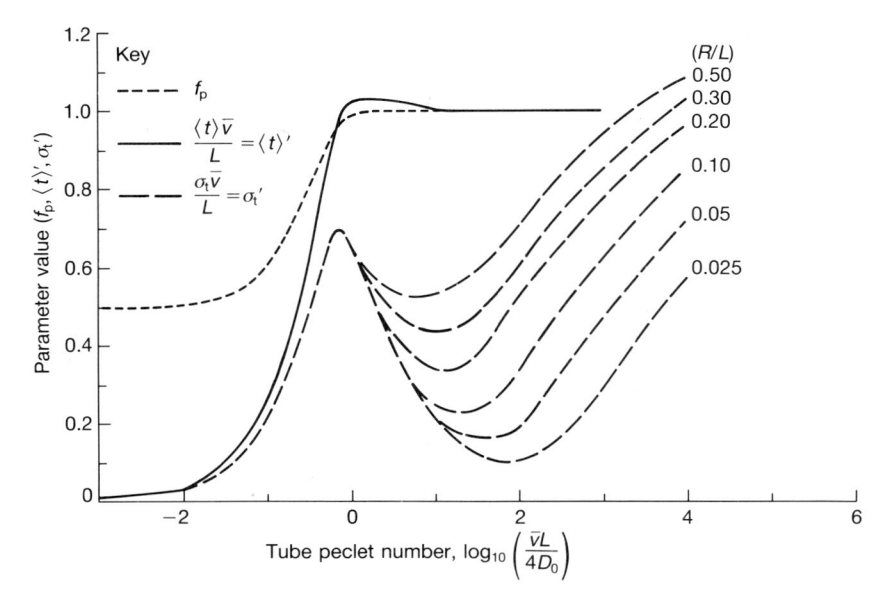

Figure 2. Parameters for positive convection in tube (f_p, $\langle t \rangle'$, σ_t') vs the tube Peclet number and aspect ratio, (R/L).

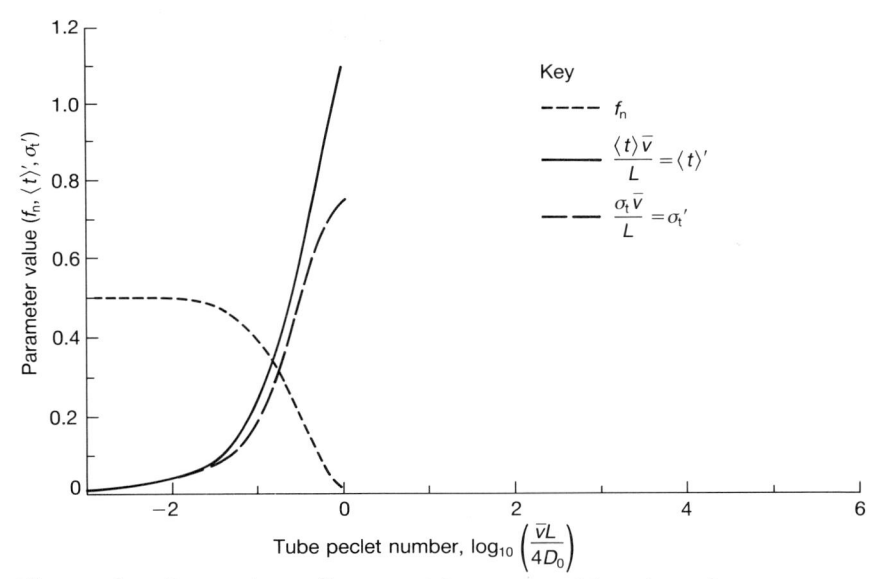

Figure 3. Parameters for negative convection in tube (f_n, $\langle t \rangle'$, σ_t') vs the tube Peclet number.

we call the Taylor Peclet number, as follows:

$$Pe_T = \frac{\bar{v} L}{4 D_T} \quad \text{(a)}$$

where

$$D_T = \frac{\bar{v}^2 R^2}{48 D} \quad \text{(b)}$$

such that

$$(Pe)(Pe_T) = 3\left(\frac{R}{L}\right)^{-2} \quad \text{(c)} \quad (2.5)$$

and

$$Pe_T = \frac{12 D \cdot L}{\bar{v} R^2} \quad \text{(d)}$$

The quantity D_T in equation 2.5(b) above is the dispersion formula derived by Taylor (1953) for tubes of certain dimensions. The Taylor result would apply with reasonable accuracy for the case of tubes with $Pe_T \gg 1$, and simultaneously $Pe \gg 1$, which requires therefore that $(R/L) \ll 1$; the condition is in fact quite stringent, and aspect ratios of order 0.01 are needed for the Taylor formula to be applied directly. Since porous media are characterised by significantly larger aspect ratios ($0.025 \leq (R/L) \leq 0.5$ is used in this study), it is not valid to claim as many authors have done, that the high velocity limit can be treated as a case of "Taylor dispersion". The Taylor formula does not apply for the individual elements at any velocity, and is quite inappropriate for the very high velocity case, where $Pe_T \ll 1$. For the "Taylor dispersion" limit (defined above), it is expected that;

$$\sigma_i^{\cdot \cdot} = \frac{2 D_T^{\frac{1}{2}}}{\bar{v} L} = (2 Pe_T)^{-\frac{1}{2}}$$

From figure 2 it can be shown that this behaviour is correctly reproduced in our modelling for cases with $Pe_T \geq 10$ and (R/L) sufficiently small (certainly $R/L < 0.05$) that longitudinal diffusion has not yet become significant at the value of Pe_T in question.

We now consider the detailed results for the jump statistics from the Monte Carlo model presented in Figures 2 and 3. For the diffusion-dominated limit ($Pe \leq 0.01$), particles move radially at such a rate that they sample the full range of flow velocities in the tube, and move convectively at the mean velocity $\bar{v}$.

The transit time in this limit is related to the longitudinal diffusion timescale (L^2/D_ℓ), and we find:

$$\langle t \rangle = 0.63(L^2/D_\ell), \text{ and } \sigma_t = 0.51(L^2/D_\ell) \quad \text{(a)} \quad (2.6)$$

or in dimensionless groups:

$$\langle t \rangle = 2.52 \text{ Pe and } \sigma_t = 2.04 \text{ Pe} \quad \text{(b)} \quad (2.7)$$

where the effective distribution of transit times goes from zero to a value several times greater than $\langle l \rangle$. In this limit, 50% of the particles emerge in either direction ($f_p = f_n = 0.5$) and the tube transit statistics are, as we expect, very similar for both positive and negative convection (cf Figures 2 and 3 at Pe $\leq$ 0.01).

As Pe increases from 10^{-2} to 10, the assumption of efficient radial diffusion still applies, but longitudinal convection and diffusion are similar in magnitude. This leads to the fraction of particles moving in the positive flow direction, f_p, increasing from about 0.52 to 0.98 over this range of Pe, which $\langle l \rangle'$ approaches 1.0 from below as Pe approaches unity. The normalised standard deviation σ_t for positive convection reaches a peak value of about 0.7 at a Peclet number somewhat less than 10 (Figure 2). For the negative convection case, shown in Figure 3, the values of $\langle l \rangle$ and σ_t are similar to those for positive convection over the regime in which f_n is significant. Figure 3 also shows that virtually no particles emerge in the negative direction when the tube Peclet number is >1.

For Pe in range 10^0 to 10^1, almost 100% of particles emerge in the direction of flow, but with $\langle l \rangle'$ slightly greater than 1.0 due to diffusive "tail" of emerging particles.

Above Pe=1, we need only consider positive convection. For Pe $\leq$ 3, particles do not diffuse radially at a sufficient rate to sample the entire tube cross-section, during their residence within it. The parameter (R/L) becomes significant and causes substantial deviations between σ_t' values calculated for tubes of different shape with identical Pe as shown in Figure 2. At sufficiently high velocity (Pe>10^2), longitudinal molecular diffusion becomes insignificant compared to convection and σ_t again comes to depend on a single parameter; the Taylor Peclet number, Pe_T. In this regime, $\langle l \rangle'$ is close to unity, as would apply to pure convection, but as $\bar{v} \to 0$ σ_t' diverges. The divergence is slow, with $(\sigma_t')^2$ behaving roughly as the logarithm of $(Pe_T)^{-1}$, as would be expected for purely convective flow, but with a cut-off in transit time due to the escape of particles from very slow velocity streamlines near the outer tube radius, due to radial molecular diffusion. For $Pe_T \leq 4.0$ (that is, Pe > 0.75 $(L/R)^2$), when particles are tied to streamlines, the minimum possible transit time is <t>/2, corresponding to motion along the r = 0 streamline, and a tail of longer transit times is formed.

The cases with low Pe_T or high Pe_T are therefore dominated in their transit time behaviour by the small number of particles which move along low velocity streamlines close to pore walls. As the velocity increases, radial diffusion becomes increasingly less effective in removing particles from these streamlines, and the ration $(\sigma_t/<t>)^2$, which is approximately proportional to (dispersion/$\bar{v}$) increases. Therefore for individual tubes, dispersion will tend to increase with velocity at a rate higher than the first power, and a similar result might be expected for an entire network.

For a real porous medium, particle behaviour is of course somewhat more complex than calculated in this section. Pores are not cylindrical tubes, and in addition to stagnant regions at pore walls there may be stagnant "dead-end" pores. However, these effects would constitute modifications to the behaviour outlined here, and would not require a fundamental reappraisal of the model used.

3. CALCULATION OF DISPERSION IN A NETWORK

3.1 Basic Network Model

A two-dimensional square network of total length, L_N, was used, consisting of cylindrical tubes parallel and perpendicular to the principal direction of flow. All tubes were of equal length, L, and meet at volume-less nodes with a coordination number of 4. This choice was based on simplicity, without attempting to represent accurately the pore structure of particular media. However, given the complexity of porous media, it would be unwise to claim without good evidence that any particular alternative model is better a priori at representing the dispersion phenomenon.

The tubes in the network were assigned radii, R, randomly from a distribution function with well-defined upper and lower limits: here (R/L) lies between 0.025 and 0.500. It should be sufficiently versatile to allow wide or narrow distributions of tube radius, and peaks at different radius values. For this purpose a Haring-Greenkorn distribution was chosen, with distribution function.

$$f(r) = K (R_{max}-r)(r-R_{min}) \exp(-(r-\bar{R})^2/2\sigma^2) \qquad (3.1)$$

where R_{max} and R_{min} are the maximum and minimum radii, $\bar{R}$ and σ are parameters related to the mean and distribution width respectively of the function and K is a normalisation constant (Haring and Greenkorn 1970). A grid of (40 x 10) tubes was used in all calculations presented in this paper. It was made cyclic both parallel and transverse to the flow direction, and allowed for particle motion in the negative direction, as required in the diffusion dominated limit.

The next stage is the calculation of a flow field in the network. Pressures are defined at each end of the (40 x 10) grid, and are then calculated at each intermediate node using network analysis. At node (i,j) the pressure equation for incompressible flow takes the form:

$$\sum_{k=1}^{4} Q_{ij.k} = 0 \qquad (3.2)$$

where the values of k correspond to flow from the node in the four directions. The flows, Q, are given by the Poiseuille formula as follows:

$$Q_{ij.k} = \frac{\pi R^4 \Delta p}{8 \eta L} \qquad (3.3)$$

where R and Δp are the radius of and pressure across the individual tube in direction k from node (i,j) and h is the fluid viscosity. The full set of equations requires a matrix inversion to obtain node pressures, which was performed numerically. From the calculated nodal pressures, the flow and mean velocity in each tube may be calculated.

The dispersion in the network is calculated by following the paths of individual particles through the grid. The transit time of a particle through a tube is obtained from the distribution as calculated in section 2, accessed randomly for each passage. However, it is also necessary to contruct an algorithm for choice of subsequent jump direction for a particle arriving at any given node. For convection-dominated flow (high Peclet number), the choice of jump direction is simply in proportion to the flows Qijk away from the node, with zero probability of crossing a tube with flow towards the node. At very low flow rates when the movement of the particle from node to node is diffusion dominated (very low Peclet number), the convected flows are unimportant, but it is not clear without simulation in a detailed geometry how precisely the jump probability should be calculated. One possibility is to let it depend on tube cross-sectional area; however, the radii, R, correspond more closely to an average radius over the length of tube rather than to the actual radius at the point of junction. Therefore, cases have been considered with jump probability in this limit both independent of and proportional to cross-sectional area, represent extreme cases. The algorithm adopted for all flow rates for the jump probability from node ij along bond k, P_{ijk} is as follows:

$$P_{ijk} \propto f_p (q+\beta D.) \qquad \text{(a)}$$

for a jump in the convected flow direction

(3.4)

$$P_{ijk} \propto f_n \beta D. \qquad \text{(b)}$$

where f_p, f_n are the fractions of particles entering the tube under the given flow condition (tube Peclet numbers) as calculated in Section 2; b is a quantity which can be set to L or (pR^2/L) for the cases independent of and proportional to cross-sectional area respectively. The formula (3.4) is, of course, an approximation.

3.2 Dispersion Calculation from Spatial and Temporal Statistics

In order to calculate the dispersion coefficient for a particle in a network, particles are placed randomly at nodes on the line X = 0, and their motion through the network is followed from node to node, according to the jump probabilities and transit times as outlined above. The resulting dispersion may be calculated in two ways:

Method (i)   by following a large number of particles for a fixed time, T_N, and measuring their displacements from the origin at this time;

Method (ii)   by measuring the times taken for a large number of particles to cross the network from X = 0 to a fixed distance x = L.

The mathematical conditions for random processes, such as those simulated by tracking particles through the network, to be diffusive are discussed by Chandrasekhar (1943) and Sahimi et al (1983; 1986) and will not be repeated here.

However, in a diffusive process, where the fluid velocity is non-zero, then the dispersion constant calculated from methods (i) and (ii) above will in the long distance or time limit be identical. The condition of non zero fluid velocity is necessary for method (ii) so that the moments of the arrival time distribution, $<t>$ and $<t^2>$, do not become infinite. In the purely diffusive limit, where the mean fluid velocity is zero, method (ii) cannot be used. In addition to these rigorous demands from statistical theory of random processes, there are also some practical computational limitations where either (i) or method (ii) should be used. It will be shown that only method (i) is appropriate at very low flow rates, as we expect from the above remarks, and that method (ii) is more appropriate in practice at very high flow rates. In the intermediate flow regime, we find that both methods give very similar results.

For method (i), the dispersion, D, is calculated as:

$$D = \frac{\sigma_x^2}{2T_N} \qquad (3.5)$$

where σ_x is the standard deviation of distance travelled by particles in fixed time T_N2. In the diffusion-dominated cases (low tube Pe), the mean and standard deviation of transit time are similar for all tubes in the network. This approach should then yield an accurate measure of dispersion coefficient. In the convection-dominated (high Pe) limit, where the dispersion coefficient is heavily influenced by the retardation of a small number of particles in a few tubes low fluid velocity, the approach of method (i) becomes inappropriate. If the transit times for low velocity tubes are greater than the time T_N, this method cannot distinguish the effect of any further increase in those long transit times, even though such an increase could significantly influence the dispersion coefficient.

When the dispersion is calculated from the distribution of arrival times at the network outlet (method (ii) above) the following formula is used:

$$\tfrac{1}{2} \frac{\sigma_t^2}{\langle t \rangle} = \frac{D}{UL_N} \left[1 + 2/3 \frac{D}{UL_N} \left[1 + \frac{D^2}{UL_N} \right] \right] \qquad (3.6)$$

Where $\langle t \rangle$ and σ_t are the mean and standard deviation of time taken for particles to travel distance, L_N, across the entire network, and U is the mean fluid velocity in the network; $\langle t \rangle$ is given by the formula:

$$\langle t \rangle = \frac{L_N}{U} \left[1 + \frac{D}{UL_N} \right] \qquad (3.7)$$

These results are derived from the macroscopic one-dimensional convection-diffusion equation, with mass-conserving boundary conditions and correspond to the first passage times at $X = L_N$ of an appropriate number of particles.

For most cases investigated, $(D/UL_N) \ll 1$, so that a small percentage error:

$$D \approx \cdot \sigma_t^2 \cdot \frac{U^3}{L_N} = \frac{L_N U}{2} \frac{\sigma_t^2}{\langle t \rangle} \qquad (3.8)$$

If an experimental core is assumed to consist of M successive networks of the type used in these calculations, then the dispersion across the core may be derived from equation (3.8) as equal to the mean of the dispersions calculated for the networks individually.

3.3 Results from the Network Calculations

As the pressure gradient across the network is increased, the motion of tracer molecules transfers from a regime dominated by molecular diffusion to a regime in which convective movement is much more rapid than diffusion. There is a corresponding monotonic increase in the dispersion coefficient. However, different tubes in the network may have Peclet numbers differing by orders of magnitude, and the uneven transition of flow regime through the network is responsible for important and observable features in the curve of calculated dispersion coefficient, D, versus flow velocity U. In addition, even when the pressure gradient is sufficiently high for all individual tubes to have high Peclet numbers molecular diffusion is still an important factor in regulating the retention time of tracer particles in slower moving bonds, and can thus continue to influence the observed dispersion coefficient. For the calculations presented below, the data on the networks and the diffusion coefficients of the particles, corresponding to chloride tracer and xanthan biopolymer, are summarised in Table 1.

TABLE 1

PROPERTIES OF THE NETWORKS AND PARTICLES USED IN THE DISPERSION CALCULATIONS IN THIS WORK

Networks:
- 2D networks 40 (in flow direction) x 10
- All bond lengths, $L = 40\mu m$
- Bond radius distributions have:

    $R_{min} = 1\mu m$ and $R_{max} = 20\mu m$

- Haring-Greenkorn distribution (equation 3.1) has:

    $R = 6.69\ \mu m$ and $s = 5.00\mu m$

Particles:
- Diffusion constants are:

    Tracer (chloride) = $2.0 \times 10^{-5}\ cm^2/s$
    Polymer (xanthan) = $2.5 \times 10^{-8}\ cm^2/s$

The results of the network dispersion calculations may be plotted in three ways as follows:

(a) D versus fluid velocity, U

(b) D/D_0 versus network Peclet number, $Pe_N = (UL/D_0)$

(c) D/UL versus network Peclet number, $Pe_N = (UL/D_0)$

where D is the calculated hydrodynamic dispersion coefficient, L in the above groups is the length of one bond in the network, corresponding to the porous medium "grain size" which is frequently used in defining these quantities (Fried and Combarnous, 1971). In plots of type (a), the average fluid velocity, U, is the same for all transported species, and separate curves are therefore obtained for tracer and polymer using this method of representing results. This corresponds to the way that the tracer and polymer experimental data are plotted in Figure 1 (Sorbie et al, 1987). This method also displays the actual magnitude of calculated dispersions and flow velocities in a given set of units. Plots of type (b) and (c) above are in dimensionless form and are frequently used in the literature (Fried and Combarnous, 1971). Using these, all calculations for polymer and tracer are collapsed onto single curves in each case. The polymer dispersion at a lower flow rate corresponds to the tracer dispersion but at a much higher flow rate. Plots of type (c) emphasise more clearly the changing slope of the D/U relationship and hence the changing flow regimes. The most useful types of plot, for our purposes, are types (a) and (c) and dispersion calculations in the base case network (Table 1) are presented in these ways in Figures 4 and 5 respectively.

First, consider the calculated dispersion/velocity curves for tracer and polymer in Figure 4. The calculations indicate that there is a considerable range of intermediate to high velocities over which the polymer dispersion is larger than that of the tracer by a factor of 2-3. This region corresponds approximately to that for which experimental data are available (Figure 1; Sorbie et al, 1987) and thus our model satisfactorily reproduces the observations. The region of higher polymer dispersion corresponds to a convection dominated flow regime for the polymer and "mixed" (convective-diffusive) flow regime for the tracer.

DISPERSION MODELLING 545

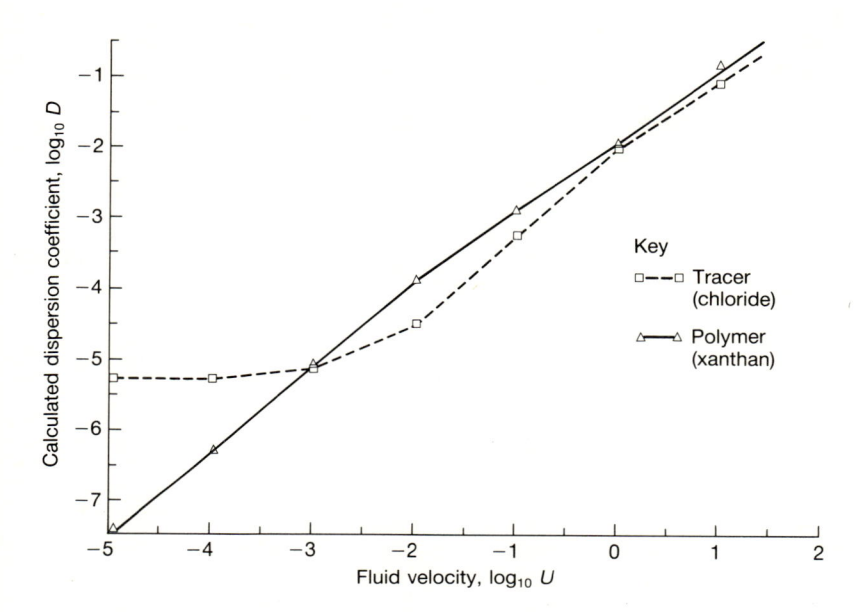

Figure 4. Calculated dispersion coefficient vs fluid velocity for tracer & polymer.

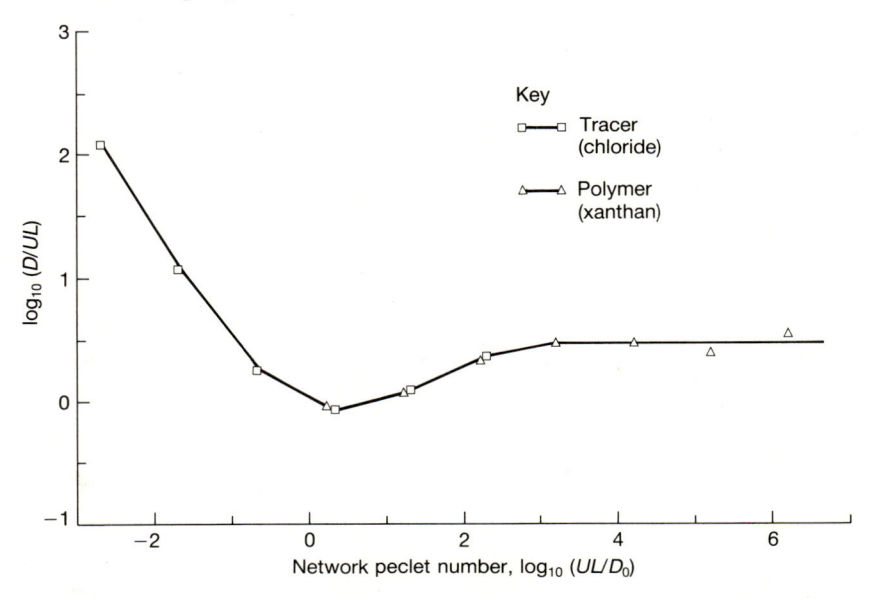

Figure 5. Calculated dispersion/velocity group vs Network Peclet number for both tracer & polymer; Haring--Greenkorn distribution.

Another important feature of our model is that, for both tracer and polymer, it predicts regions where the velocity exponent of dispersion is greater than unity. Previous work on calculating dispersion in networks has either omitted molecular diffusion hence has obtained a linear D/U relationship (Sahimi et al, 1983) or incorporated the effect through an incorrect mechanism (Torelli, 1972). In the "mixed" flow regime, where convection is dominant but molecular diffusion effects are important, we find that $D \sim U^\alpha$ where α is in the range 1.19 - 1.25 for both polymer and tracer. This calculated value of α is consistent with experimental observations.

An alternative way to discuss the flow regimes in porous medium or network flow is in terms of the slopes of the (D/UL) versus Pe_N plots (type (c)) as shown in Figure 5 for the base case network. The predicted curves in these figures distinguish between the network calculations which were performed on tracer molecules and those which had the diffusion constant of the polymer. Note that, in the overlap flow region ($2 < Pe_N < 200$), the calculations agree very well. The overall form of the curve in Figure 5 compares well with the experimental results compiled by Fried and Combarnous (1971). At values of $Pe_N < 1$, the curve has a negative slope since the network dispersion is constant in the diffusion dominated regime. A minimum in the value of D/UL is predicted in the range $1 < Pe_N < 10$ which is found experimentally (Fried & Combarnous, 1971). Above Pe_N values of about 2, the slope of (D/UL) is positive corresponding to the regime where mechanical or convection dominated dispersion is prevalent but where molecular diffusion cannot be ignored. It is, of course, the contribution of D_q that causes the slope of this curve to be positive. At Pe_N values greater than ~ 200 the slope of the D/UL approaches zero; this is where dispersion is purely mechanical and D is linear in U. In summary, all of the flow regimes and the minimum in the D/UL versus Pe_N curve are correctly reproduced in our model. We note that from Figure 2, the value of σ'_t is increasing at very high tube Peclet number. This causes a second high flow regime where (D/UL) increases with flow rate (Pe_N) in the network. Some evidence of this can be seen in Figure 5 at high Pe_N. However, this is at flow velocities in the porous medium where other non-Darcy effects may be very important (Fried and Combarnous, 1971) and is not thought to be an important predic

4. DISCUSSION AND CONCLUSIONS

In this paper, we have described a general model for calculati dispersion in porous media over a very wide range of Peclet numbe The main departure in our approach from network models proposed previously (Torelli, 1972; Dullien, 1979; Sahimi et al 1983, 198 de Arcangelis et al, 1986; Koplik et al, 1987), is that we corre incorporate the molecular diffusion into the jump statistics in the individual network elements (capillaries). This is done using Monte Carlo methods to characterise the statistics of the first arrival time distribution in the bonds (ie $\langle t \rangle$ and σ_t') this is shown to depend on both the tube Peclet number Pe and the tube aspect ratio, (R/L). Only at very low (R/L), much smaller than is typical in porous media, is it valid to take the Taylor limit (Taylor, 1953) and to then calculate the jump statistics based on the corresponding ID convection - diffusion equation. Our Monte Carlo approach can also be used to determine the fraction of particles travelling in the positive and negative directions relative to the convected flow.

The algorithms for the microscopic statistics have been incorporated into a network calculation of dispersion based on distribution of distance travelled or on arrival time distributions. This overall model correctly reproduces the form of the D/UL vs Pe_N curve over the entire range from diffusion dominated flow to pure mechanical dispersion. The model also predicts a minimum in D/UL in the range of network Peclet number $1 < Pe_N < 10$ as is found experimentally (Fried and Combarnous, 1971; Gunn and Pryce, 1969).

In particular flow regimes, it is computationally appropriate to use different types of calcluation of the network dispersion, D, based on either the distribution of distance travelled (method (i)) or on the distribution of first arrival times (method (ii)). We present a full discussion of how these should be applied, with examples, in our calculations.

Neglecting certain properties of polymers associated with polymer-pore wall interactions, such as adsorption and excluded volume effects, the network model treats both polymer and tracer dispersion in a very similar way. When plotted as calculated D vs U it is clear that, in the intermediate to high flow rate regime, the polymer dispersion is predicted to be larger than that of tracer by a factor of 2 to 3. At very low flow rates, the tracer curve (corresponding to chloride in our calculations) tends to a constant D which is a fraction of between 0.25 and 0.5 of the D_q depending on assumptions about the jump algorithm. Because the polymer D (corresponding to xanthan biopolymer) is about three orders of magnitude lower than that of the tracer, the calculated D for the polymer goes below the tracer dispersion at very flow rates. This is as we might expect although no experimental data has been presented in this flow regime for a polymer/tracer system. We note that the polydispersity (spread in molecular weight distribution) of the polymer may also play a role in contributing to the larger dispersion coefficient of the polymer relative to the tracer (Lecourtier and Chauveteau, 1984; Brown and Sorbie, 1989).

This is thought to arise because of the larger surface exclusion effect on higher molecular weight species compared to smaller species, and hence the larger molecules travel at a higher average velocity through the porous media (Lecourtier and Chauveteau, 1984). The interaction of the surface exclusion and dispersive spreading mechanisms in the flow of polydisperse polymers through porous media has been examined by Brown and Sorbie (1987). It appears that the dispersion coefficient measured for low viscosity solutions of polymers propagating through porous media may be affected by both the low D_0 of the polymer molecule and by polydispersity. The precise contribution of each mechanism has yet to be established experimentally.

We note that there are some differences between experiment and theory, principally in the actual magnitude of the dispersion coefficient, which is greater in the experiments by a factor of ∿10. This also effects the characteristic velocity for the onset of diffusive behaviour for the tracer. This may be due to the fact that the "pore size" (or tuve size) distribution which we used is not sufficiently wide to represent a realistic porous medium. It may also indicate that dispersion in a laboratory core of porous material may depend in part on heterogeneities on a larger scale than the pore size. However, the experimental velocity dependence of dispersion is sufficiently similar to the network model to indicate that the casual mechanism is the same. In a core, the existence of dead-end pore space could also extend the velocity regime over which longitudinal molecular diffusion is significant. In this latter respect, it may be quite appropriate that we have chosen to work with rectangular networks with the y-bonds perpendicular to the principal flow direction. Such a net work gives the very wide range of bond Peclet numbers which captures some of the features which are expected when dead-end pores are present.

REFERENCES

[1]   de Arcangelis L., Koplik J., Redner S. and Wilkinson D. (1986). Hydrodynamic dispersion in network models of porous media. Phys. Rev. Lett. 57, 966-999.

[2]   van Brakel J. (1975). Pore space models for transport in porous media: Review and evaluation with special enphasis on capillary liquid transport. Powder Techonology 11, 205-236.

[3]   Brown W.D. and Sorbie K.S., (1989). Dispersion and polydispersity effects in the transport of xanthan biopolymer in porous media, Macromol, 22, 2835-2845.

[4]   Chandrasekhar S., (1943). Stochastic problems in physics and astronomy. Rev. Mod. Phys. 15, 1-89.

[5]   Fried J.J. and Combarnous M.A., (1971). Dispersion in porous media. Adv. Hydrosci. 7, 169-282.

[6]   Gunn D.J. and Pryce C. (1969). Dispersion in packed beds. Trans. Inst. Chem. Engrs. 47, T341-T359.

[7]   Haring R.E. and Greenkorn R.A., (1970). A.I.Ch.E. Journal 16, 477 -xxx.

[8]   de Josselin de Jong G., (1985). Longitudinal and transverse diffusion in granular deposits. Trans.Am.Geophys. Union 39, 67-74.

[9]   Koch D.L. and Brady J.F., (1985). Dispersion in fixed beds. J. Fluid Mech. 154, 399-427.

[10]   Koplik J., Redner S. and Wilkinson D., (1987). Transport and dispersion in random networks with percolation disorder. Submitted to Phys.Rev.A.

[11]   Lecourtier J. and Chauveteau G. (1984). Macromolecules, 17, 134.

[12]   Mohanty K.K. and Salter S.J., (1983). Advances in pore level modelling of flow through porous media. Presented at 1983 Al Ch E Annual Fall Meeting, Washington DC, Oct 30 Nov 4.

[13]   Perkins T.K. and Johnston O.C., (1963). A review of diffusion and dispersion in porous media, Soc. Petrol Engrs. J. 3, 70-84.

[14]   Saffman P.G., (1959). A theory of dispersion is a porous media. J. Fluid Mech. 6, 321-349.

[15]   Sahimi M., Davis H.T. and Scriven L.E., (1983). Dispersion in disordered porous media. Chem. Eng. Commun. 23, 329-341.

[16]   Sahimi M., Hughes B.D., Scriven L.E. and Davis H.T., (1986). Dispersion in flow through porous media I. one-phase flow. Chem. Eng. Sci. 41, 2103-2122.

[17]   Sorbie K.S., Parker A. and Clifford P.J., (1987). Experimental and theoretical study of polymer flow in porous media. SPE Journal (Reservoir Engineering) 2, 281-304.

[18] Sorbie K.S., Clifford P.J. and Jones E.R.W., (1989). The rheology of pseudoplastic fluids in porous media using network models. J. Coll. Int. Sci., 130, 508-534.

[19] Taylor G.I. (1953), Dispersion of soluble matter in solvent flowing slowly through a tube. Proc.Roy.Soc.A219, 186-203.

[20] Torelli L. (1972). Computer simulation of the dispersion phenomena occurring during flow through porous media using a random maze model. Pure Appl. Geophys. 96, 75-88.

The Network Simulation of Displacement Processes in Fractured Reservoir Matrix Blocks

M.D. Santiago and R.A. Dawe

Imperial College, London

Abstract A novel approach to the study of fluid displacement from a single matrix block in a fractured porous medium is presented - the use of network modelling techniques.

The model consists of a two or three dimensional grid composed of tubes of varying radius which simulate pore throats. The ends of the tubes form volumeless nodes, which are used as reference points at which the pressure equations are solved. Capillary and hydrostatic pressure effects can also be incorporated and the trapping of fluid blobs during the displacement process is accounted for. The network is compared with existing dynamic and conceptual models.

The ultimate recovery results for imbibition displacement processes in a single block obtained from the network model are compared with those of laboratory experiments using micro-visual models. This allows a quantitative evaluation of the accuracy of the network simulator to be made. Additionally, transfer (source) functions derived from the simulator are presented and the effect of block dimensions, fluid viscosity ratios and wettability are examined.

1 Introduction

Fractured reservoirs account for over 50% of the world's petroleum reserves [5]. To achieve an efficient exploitation of such oil fields it is necessary to develop a thorough understanding of the production mechanisms involved.

Fractured systems are an extreme example of heterogeneity, with the flow between the rock matrix and the fracture being frequently the dominant influence on reservoir production characteristics and ultimate recovery. The matrix blocks may be regarded as hydrodynamically isolated volumes of rock surrounded by fissures and are only in point contact with each other. In a matrix-fracture system these blocks account for the majority of the system porosity. The fractures generally account for the majority of the reservoirs' permeability. Thus the overall production mechanism relies on oil being transferred from the matrix through the fracture system to the wellbore.

The lack of continuity of flow between blocks allows the problem of fluid displacement to be considered as the interaction between those fluids saturating the matrix and those which saturate the fracture. For this reason valuable information may be gained from the study of a single isolated block. The displacement mechanism in a single block results from the interplay of gravity and capillary forces. For water-wet blocks imbibition displacement occurs when blocks are partially or totally immersed in water. This paper describes simple imbibition models and links physical models in the form of 2-dimensional micromodels with numerical network simulation.

The equations describing oil displacement from the matrix are those of simultaneous fluid flow, relative permeability and capillary pressure. For the general case there is no known analytic solution to these equations, three approaches to the problem have been followed: the development of dynamic models, simple conceptual models and the use of numerical simulation techniques. The first two of these approaches require significant simplification and are described briefly in sections 2 and 3. Numerical simulation of single matrix blocks assumes Darcy flow both through the matrix and the fracture and is in effect identical to a conventional simulation where the boundary conditions imposed on the matrix block reflect the pressures and saturations in the fracture surrounding it [7].

The continuum assumptions made in the simple dynamic model and in conventional simulation depend on a knowledge of relative permeability, a concept with little scientific basis, particularly in the case of counter-current flows where wetting and non-wetting fluids flow in opposite directions simultaneously [8, 11]. Thus these methods do not provide the ideal means of studying the underlying physics of imbibition phenomena, especially those that occur at the pore scale (1-1000μm). It is however, at this scale that flow actually takes place and must be accurately modelled.

At the scale of a few hundred pores (1-100mm), mathematical network modelling combined with micromodel experiments provides a novel technique for the study of displacements in fractured media. The micromodels allow fluid behaviour to be visualised and can be used to validate the network results. The network model can then be used to investigate the effect on production of single system parameters such as fluid viscosity ratios, block dimensions and geometry independently of each other.

The use of such networks may allow the development of improved transfer or source functions so that the flow of oil from the rock matrix into the fracture system can be modelled more closely to the real physical processes that occur and hence allow more accurate simulations to be made.

2 The Simplified Dynamic Model of Imbibition in a Single Reservoir Block

2.1 Model Assumptions

The simplified dynamic model [7] assumes a uni-dimensional flow through the matrix block, which implies that the lateral faces of the block are impermeable. A piston-like displacement is postulated, with irreducible wetting and non-wetting fluid saturations on either side of the interface separating the two immiscible phases. This implies that only constant end-point relative permeabilities need be considered in the subsequent calculations and (assuming the matrix block is homogeneous) that the capillary pressure at the interface has a constant value. A consequence of the former is that the mobility ratio, M, is a constant for the displacement.

The pressure in the fracture system which envelops the matrix block is hydrostatic. This is reasonable since the fracture system generally has a high permeability and externally applied driving pressures are correspondingly low. Consequently the fluid potential in the fracture system is defined here as:

$$\Phi = P + \rho g z \qquad (2.1)$$

The flow rate per unit area for each phase is given by Darcy's law:

$$U_w = -\frac{K K_{rw}}{\mu_w} \frac{\partial \Phi_w}{\partial z} \qquad (2.2)$$

$$U_{nw} = -\frac{K K_{rnw}}{\mu_{nw}} \frac{\partial \Phi_{nw}}{\partial z} \qquad (2.3)$$

The flux U across each boundary is constant, therefore:

$$U_w = U_{nw} = U \qquad (2.4)$$

The matrix block and the two fluids are considered to be incompressible so that:

$$\frac{\partial U_w}{\partial z} = \frac{\partial U_{nw}}{\partial z} = 0 \qquad (2.5)$$

There are two situations in which a block can undergo imbibition; a) when the block is partially immersed, that is to say its top face is in contact with a non-wetting fluid filled fracture, and b) when the entire fracture system around the block is saturated with wetting fluid and both lower and upper faces of the block are in contact with wetting fluid (Figure 1).

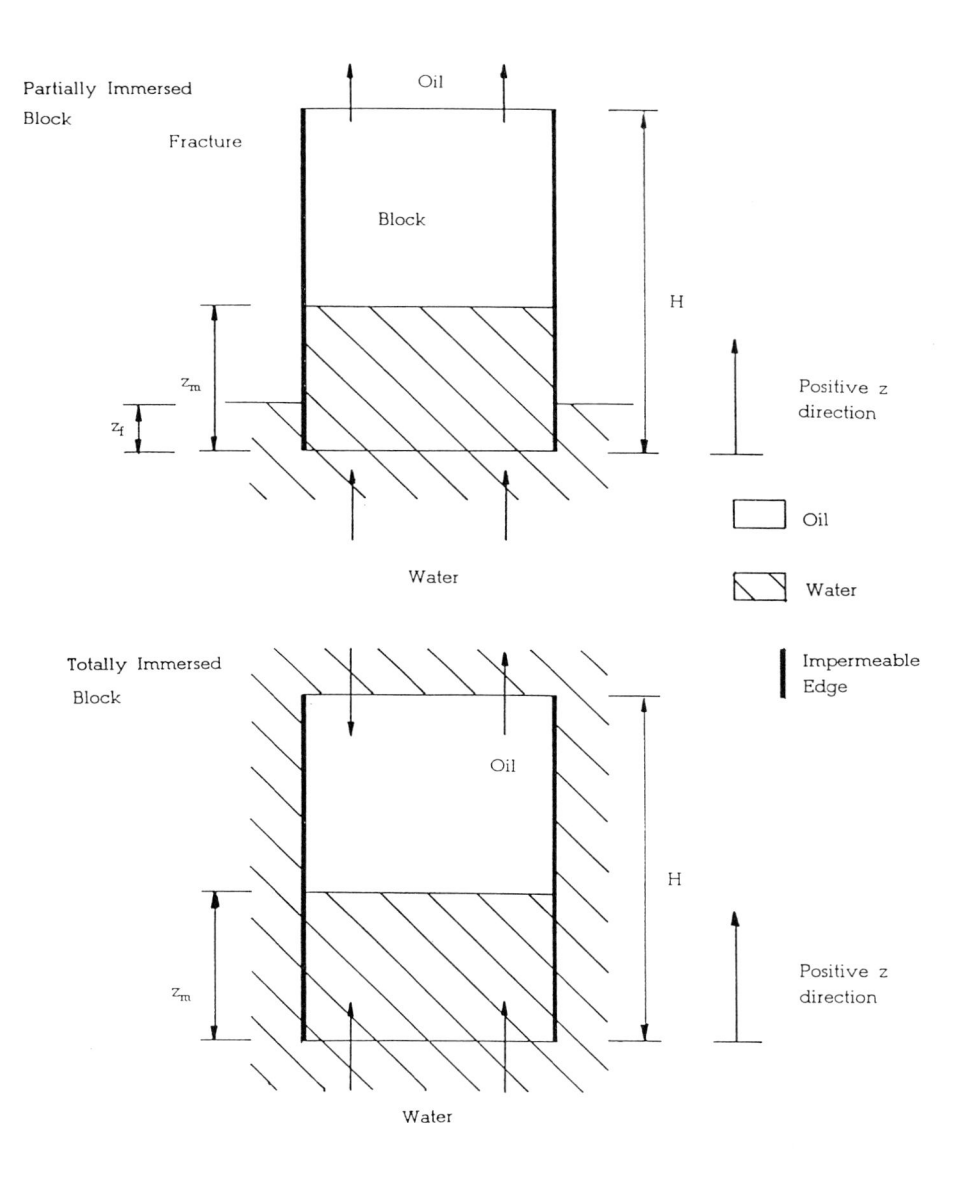

Figure 1. Dynamic Model of a Single Reservoir Matrix Block

2.2 General Production Equation from a Matrix Block

The general production equation for a single block under the conditions stated above is given by:

$$U = \frac{P_{cm} - P_{ce}\aleph - P_{cf}(1-\aleph) - \Delta\rho g(z_m - z_f)(1-\aleph) + \Delta\rho g(H - z_f)\aleph}{\frac{\mu_w}{KK_{rw}}\{MH + (1-M)z_m\}} \quad (2.6)$$

where $\aleph$ is a unit step function:

$$\aleph = \aleph(z_f - H) = \begin{cases} 0 \text{ if } z < H \\ 1 \text{ if } z > H \end{cases} \quad (2.7)$$

and P_c the capillary pressure is:

$$P_c = P_{nw} - P_w \quad (2.8)$$

P_{ce} is the capillary pressure at the exit face of the matrix block, P_{cm} is the capillary pressure in the matrix and P_{cf} is the capillary pressure in the fractures. Experiments [13] show that non-wetting fluid production from a block face is an intermittent process, with non-wetting fluid droplets forming on the surface and floating off as buoyancy forces overcome adhesion. P_{ce} varies with the inverse of droplet radius with a maximum value of P_{cm}. The time period over which P_{ce} is of significant magnitude compared to P_{cm} is small (Figure 2), consequently P_{ce} can be neglected. The capillary pressure in the fracture system, P_{cf}, is normally an order of magnitude smaller than P_{cm} and therefore may also be neglected.

The case where a matrix block is totally immersed in water is of particular interest, as counter-current imbibition will take place, principally on the top face of the block. For this case a simplified equation for the production rate per unit area can then be written:

$$U = \frac{P_{cm} - \Delta\rho g(H - z_m)}{\frac{\mu_w}{KK_{rw}}\{MH + (1-M)z_m\}} \quad (2.9)$$

From equation (2.9) it can be seen that the rate of flow of oil from the matrix block depends on the interplay between capillary and hydrostatic forces and on the mobility ratio of the fluids. The rate of advance of the fluid interface within the matrix can be derived from the porosity of the homogeneous matrix (φ):

$$U = \varphi \cdot (1 - S_{nwr} - S_{wr})\frac{dz_m}{dt} \quad (2.10)$$

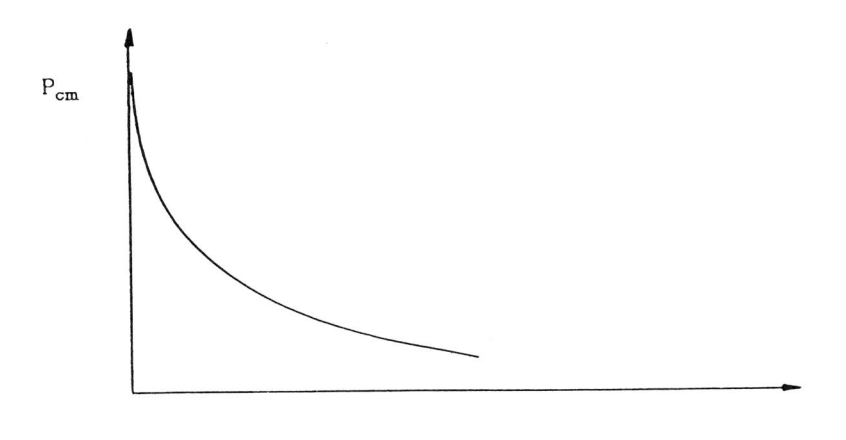

Figure 2. Variation of Exit Face Capillary Pressure with Time

2.3 Displacement Dominated by Capillary Pressure

When the displaced and displacing fluid are of similar density, the matrix blocks are small or the interface has advanced in the matrix such that $(H - z_m)$ is small, then the hydrostatic pressure term in (2.9) becomes negligible and the displacement is dominated by the capillary pressure. Then, the dimensionless expression for the time taken to achieve a fraction z_D of the ultimate recovery is:

$$t_{D,P_c} = M z_D + \frac{(1-M)z_D^2}{2} \tag{2.11}$$

where:

$$t_{D,P_c} = \frac{P_c K K_{rw}}{\mu_w \Phi H^2} t \tag{2.12}$$

and:

$$z_D = \frac{z_m}{H} \tag{2.13}$$

Equations (2.13) and (2.14) are the transfer (source) functions for single reservoir blocks, these represent the quantity of oil produced from a matrix block as a function of time. In these equations the dimensionless distance z_D may be regarded as the fraction of ultimate recovery achieved in a time t_{D,P_c}.

3 A Conceptual Model of Imbibition into a Single Reservoir Block

3.1 The Use of Conceptual Models

The analytical attempts which have been made to describe displacement from the matrix blocks of fissured reservoirs require a source function term. In some of these models this term must be assumed [1], in others it must be measured experimentally or derived from experimental data coupled with mathematical modelling [4]. The conceptual model of the block displacement process is a capillaric model developed by Bear [3] which allows a source function to be generated for use in certain analytical models.

3.2 The Derivation of a Transfer Function from a Conceptual Model

The matrix block is modelled as a bundle of randomly orientated straight, cylindrical capillary tubes of uniform radius crossing the block from face to face. During imbibition, the non-wetting fluid within each tube is displaced by wetting fluid. A sharp interface perpendicular to the tube axis separates the two fluid phases.

The mean velocity in a single tube neglecting the gravitational terms is given by:

$$\bar{V} = \frac{a^2}{8\{\mu_{nw}(L - L_w) + \mu_w L_w\}}(P_{i,w} - P_{j,nw} + P_{cm}) \qquad (3.1)$$

where i and j refer to the inlet and outlet ends of the tube respectively. The wetting fluid pressure at the inlet end of the tube is related to the non-wetting fluid pressure at the inlet by the capillary pressure in the fracture:

$$P_{w,i} = P_{nw,i} - P_{cf} \qquad (3.2)$$

Expression (3.1) can be simplified further by the introduction of six terms:

$$\Delta P_{nw} = P_{nw,i} - P_{nw,j} \qquad (3.3)$$

$$\Delta P_c = P_{cm} - P_{cf} \qquad (3.4)$$

$$S_w = L_w/L \qquad (3.5)$$

$$S_{nw} = 1 - S_w = (L - L_w)/L \qquad (3.6)$$

$$\Delta\mu = \mu_{nw} - \mu_w \tag{3.7}$$

$$K_{nw} = a^2/8\mu_{nw} \tag{3.8}$$

$$f(S_w) = \frac{1}{1 - \dfrac{S_w \Delta\mu}{\mu_{nw}} \dfrac{1}{\mu_{nw}}} \tag{3.9}$$

This gives:

$$\bar{V} = K_{nw} f(S_w) \left\{ \frac{\Delta P_{nw}}{L} + \frac{\Delta P_c}{L} \right\} \tag{3.10}$$

The non-wetting fluid production rate from a single block is then given by:

$$Q_{nw} = \int_{A_{o,nw}} \bar{V} dA \tag{3.11}$$

where $A_{o,nw}$ is the total area of capillary tubes through which non-wetting fluid is expelled from the blocks contained within a representative elementary volume Ω of the combined fracture-matrix system. $A_{o,nw}$ is related to the total surface area, A, of the blocks by a function α_{nw} which is uniform over A:

$$A_{o,nw} = \alpha_{nw} A \tag{3.12}$$

The source function can then be expressed as:

$$q_{nw}^* = \frac{Q_{nw}}{\Omega} = \int_A \frac{\alpha_{nw} \bar{V} dA}{\Omega} \tag{3.13}$$

or:

$$q_{nw}^* = \bar{A} K_{nw} \{F_a(S_w) + F_b(S_w)\} \tag{3.14}$$

where:

$$\bar{A} = A/\Omega \tag{3.15}$$

$$F_a(S_w) = \alpha_{nw} f(S_w) \frac{\Delta P_{nw}}{L} \tag{3.16}$$

$$F_b(S_w) = \alpha_{nw} f(S_w) \frac{\Delta P_c}{L} \tag{3.17}$$

The functions α_{nw}, F_a and F_b are functions of fluid saturations (in this particular case they are expressed as functions of the wetting fluid saturation) and $f(S_w)$ is constant over A. Both F_a and F_b must be determined experimentally but in the case of displacement due to imbibition only, where the pressure in the fracture system is hydrostatic, F_a becomes zero and the transfer function becomes:

$$q_{nw}^* = AK_{nw} F_b(S_w) \qquad (3.18)$$

Having determined the source function (3.18) experimentally as a function of wetting phase saturation the source term may be found as a function of time from:

$$\frac{\partial S_{nw,m}}{\partial t} = q_{nw}^*(S_{w,m}) \qquad (3.19)$$

It also follows for incompressible fluids that:

$$q_w^* = -q_{nw}^* \qquad (3.20)$$

A source function of this type can be used in analytical models of flow and production from fractured porous media [4].

3.3 The Limitations of the Conceptual Model

The capillaric conceptual model has some advantages over the dynamic approach in that it can be extended to 3-dimensions and the lateral sides of the matrix block (which in a cube accounts for two thirds of its surface area) are no longer considered to be impermeable. The conceptual model also has the advantage of not being restricted to piston-type displacements. Although a piston-type displacement occurs within each individual capillary tube the overall shape of the displacement front is not predetermined.

The capillary tube bundle which comprises the conceptual model cannot account for fluid trapping except in a rudimentary form such as when both ends of a capillary are in contact with a water filled fissure. There is no real indication therefore of ultimate recovery. Since there are no interconnections between tubes no fluid by-pass mechanisms or non-wetting fluid blob trapping can be reproduced by the model. The random orientation of the bundle of tubes restricts the model from accounting for any degree of anistropy. An additional limitation of this model is its inability to account for countercurrent flows.

The conceptual model is a useful tool for generating source functions. However since it makes no attempt to mimic the actual processes of imbibition and fluid flow within the labyrinth of pores and pore throats which comprise porous media, it does not add greatly to the understanding of the physical processes underlying the source function of a matrix block.

4 Description of the Network Simulator

4.1 Introduction

Various types of network model have been developed [6]. The model presented here overcomes some of the inadequacies of the dynamic and conceptual models. The model can simulate externally applied, hydrostatic and capillary pressures in both the matrix and the fractures. The network grid can be made to account for antisotropy. The model also incorporates a mechanism to account for the trapping of oil blobs within the matrix. By allowing the properties of a single 'pore' to be controlled and observed the network simulator can provide an insight into flow at the microscopic level.

4.2 Geometrical Structure of the Network

The network simulator models porous media as a two dimensional square grid of tubes connected at their ends to form nodes. The nodes are used as points of reference at which the pressure field is solved and fluid saturations evaluated.

Linear flooding geometries are implemented in the network model by designating the first row of the network as inlet nodes. The last row of the grid consists only of outlet or production nodes (Figure 3). Lateral fractures can be superimposed on the model network. The fracture and matrix properties can be set independently of each other.

○   Inlet Nodes

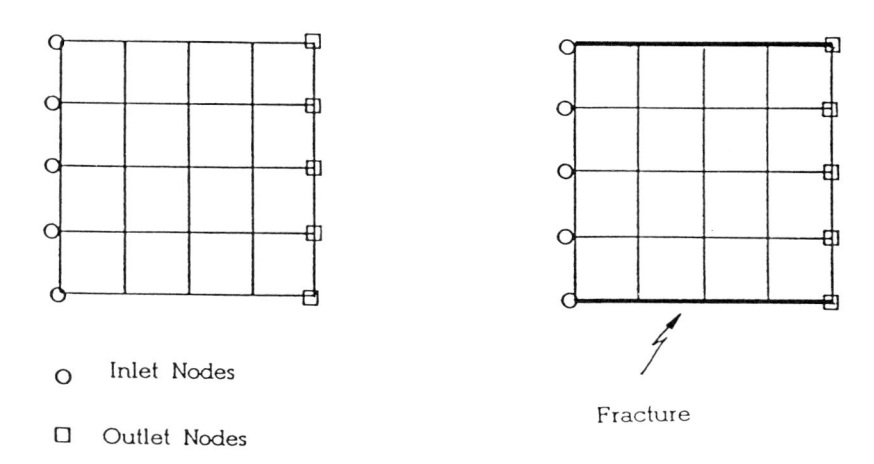

□   Outlet Nodes

Fracture

Figure 3. Schematic of Network Model showing Unfractured and Fractured Geometries

Cylindrical tube geometries have been chosen in this work because of the relative simplicity of the resulting flow equations. This simplification has the disadvantage of being unable to model the effects of channel roughness and fluid flow in grooves which may result in complex bypass and trapping mechanisms observed in micromodels and real porous media.

Uniform square networks are prone to persistent grid effects [2] which may be reduced by using an appropriate disorder parameter (λr) for the radii of the tubes comprising the network. The approximate equation used to calculate the normal radii distribution used is stated below [10]. In these equations c is a random number chosen from a uniform distribution in the range 0 to 1. In the distribution λ_r is taken as the standard deviation (in this study taken as 0.5) and $\bar{a}$ is the mean radius ($98\mu m$).

$$f = (-\pi\lambda_r \ln(1 - (2c-1)^2))^{1/2} \tag{4.1}$$

$$a = \bar{a}.(1-f) \quad c < 0.5 \tag{4.2}$$

$$a = \bar{a}.(1-f) \quad c \geq 0.5 \tag{4.3}$$

The length of all the network tubes was $2500\mu m$.

4.3 The Mathematical Description of Fluid Flow

In order to simulate any real physical system it must be made amenable to mathematical representation. This invariably requires a degree of simplification. In the network model a number of simplifying assumptions are made:
i) The model system (consisting of tubes and fluid) is considered to be incompressible.
ii) The tube lengths are considered to be long compared to their radii, thus the pressure drops associated with nodes can be neglected and the flow is assumed to be fully developed.
iii) It is assumed that fluids are Newtonian and that an immiscible piston-like displacement mechanism occurs in each tube of the network.
iv) Nodes can contain only one type of fluid at any instant.
v) Inertial effects can be neglected.
vi) Any oil displaced from the matrix block into the fracture no longer interacts with the matrix.

4.4 The Flow and Pressure Equations

The network model simulates fluid flow between inlet and outlet nodes held at fixed pressures. The equations of flow are based on:
i) The conservation of mass at each node of the network.

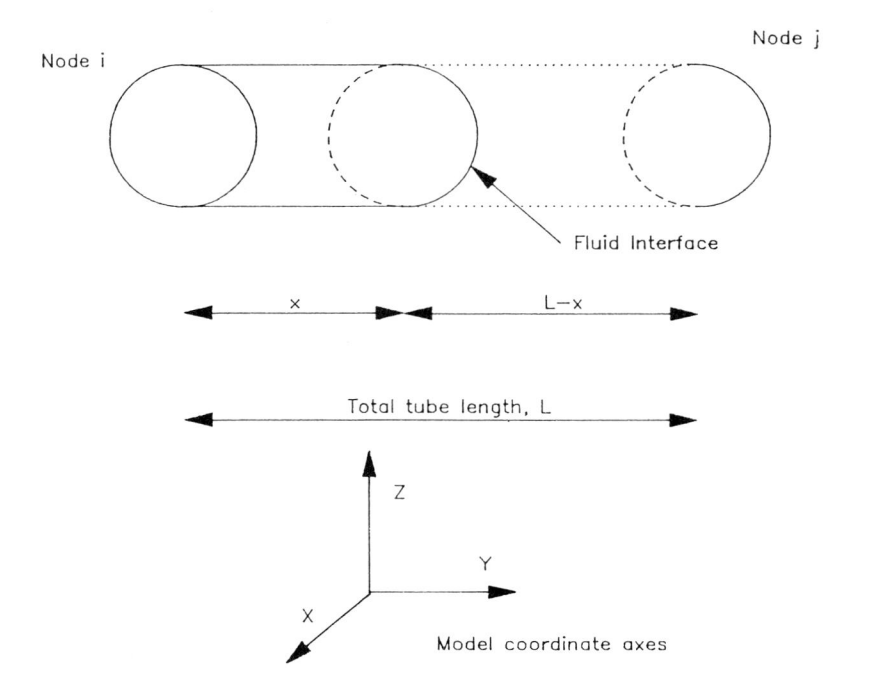

Figure 4. Schematic of a Network Tube

ii) The Hagen-Poiseuille law for fluid flow in circular tubes.

For a tube containing displaced and injected fluid (Figure 4) the flow rate is given by the Hagen-Poiseuille equation:

$$Q = \frac{\pi a^4 (P_i - P_j)}{8L\bar{\mu}} + \frac{\pi a^4 \bar{\rho} g_y}{8L\bar{\mu}} \qquad (4.4)$$

where:

$$\bar{\mu} = \frac{\mu_1(L-x) + x\mu_2}{L} \qquad (4.5)$$

and:

$$\bar{\rho} = \frac{\rho_1(L-x) + x\rho_2}{L} \qquad (4.6)$$

where a is the tube radius, x represents the length of the tube invaded by the injected fluid and subscripts 1 and 2 refer to displaced and injected fluids respectively, g_y is the component of gravitational acceleration parallel to the axis of the tube, L is the tube length, and P_i and P_j are the pressures at the nodes at the entrance and exit of the tube.

The effects of capillary pressure can be incorporated in the form of a step function by the inclusion of a term P_c in the tubes containing a fluid interface:

$$P_c = \frac{2\gamma\cos\vartheta}{a_c} \qquad (4.7)$$

where ϑ is the (real) contact angle and γ the interfacial tension. The capillary radius a_c varies identically with radius a, but can have a different mean value. In this way the network can model correctly the capillary pressure in the non-circular etched channels of the micromodels. Combining expressions (4.4) and (4.7) gives:

$$Q = \frac{\pi a^4 \{P_i - (P_j + P_c)\}}{8L\bar{\mu}} + \frac{\pi a^4 \bar{\rho} g_y}{8L\bar{\mu}} \qquad (4.8)$$

The sign of the P_c term is positive or negative depending on the fluid saturation of the nodes at either end of the tube. If the wetting fluid is being injected from node i towards node j, then the capillary pressure term is negative. If non-wetting fluid is being injected then the term is positive.

The flow equation assumes that fluid flows at a constant velocity within each network tube during each time step. No account is taken of the kinetic energy assimilated in the acceleration of fluid as streamlines converge when they pass from a node into a tube.

4.5 Derivation of the Pressure Equations

The pressure equation is derived by applying the principle of conservation of mass at each node:

$$\sum_{i=1}^{n} Q_{ki} = 0 \qquad (4.9)$$

where Q_{ki} is the fluid flux entering a node k through tube i and n is the coordination number of the node (in a two dimensional square network n has a maximum value of 4, in a three dimensional model n has a maximum value of 6). Combining expressions (4.8) and (4.9) leads to the following expression for mass conservation of each node.

$$\sum_{i=1}^{n} \frac{\pi a_i^4 . P_k}{8\bar{\mu}_i L_i} = \sum_{i=1}^{n} \frac{\pi a_i^4 \{(P_i + P_{ci})\bar{\rho}_i g_y\}}{8\bar{\mu}_i L_i} \qquad (4.10)$$

This equation is then solved for pressure (P_k) using the Incomplete Cholesky Conjugate Gradient Method [19,14]. The accuracy to which the pressures are evaluated at each stage of the simulation varies to achieve the material balance accuracy required in the least possible time.

4.6 Calculation of the Time Step Size and Ending the Simulation

The maximum size of the time step taken at each iteration is calculated so that a single fluid interface reaches a node at a time (either by moving forward to a new node or by retreating to the node saturated with injected fluid from which it originated). At each iteration all the fluid interface positions are updated

The viscosity of the fluids within a tube (as defined in (4.5)) varies linearly with the length of injected fluid penetration into it, thus an error is caused by using the mean viscosity of the fluids in the tube at the beginning of each time step. This error may be significant at the start of a simulation of a very small network if the viscosity ratio of the fluids is much greater than or much smaller than unity. However, after a number of time steps many interfaces will have formed and it will take numerous iterations to flood any particular tube. Thus the linear function (A in Figure 5) is no longer represented by a single step (B) but by a series of step functions (C). When simulating fluids of very different viscosities, the maximum timestep size can be limited so that a greater number of increments are used to model the 'average' fluid viscosity. Generally the effect of not limiting the maximum timestep size does not lead to significant errors and is preferable because it requires less computing effort.

The simulation ends when all the interfaces within the network become stationary. This occurs when all the outlet nodes of the network become saturated with displacing fluid or when capillary pressures prevent further movement of the fluids.

4.7 Calculation of Injection and Production Rates

The rate of fluid injection into the network is calculated by summing together all the fluid fluxes of tubes radiating from the inlet nodes. Production rates are calculated by the summation of fluxes entering the outlet nodes. The saturation of the node indicates which type is being produced. Until a node is saturated with the injected fluid it produces only displaced fluid. It then starts to produce injected fluid at the time iteration following that at which its saturation changed.

4.8 Obtaining Source Functions from the Network Model

The source or transfer functions for a network are generated automatically by the program at the end of the simulation accompanied by the ultimate recovery of the displacement process. An example of a transfer function is presented in the results section.

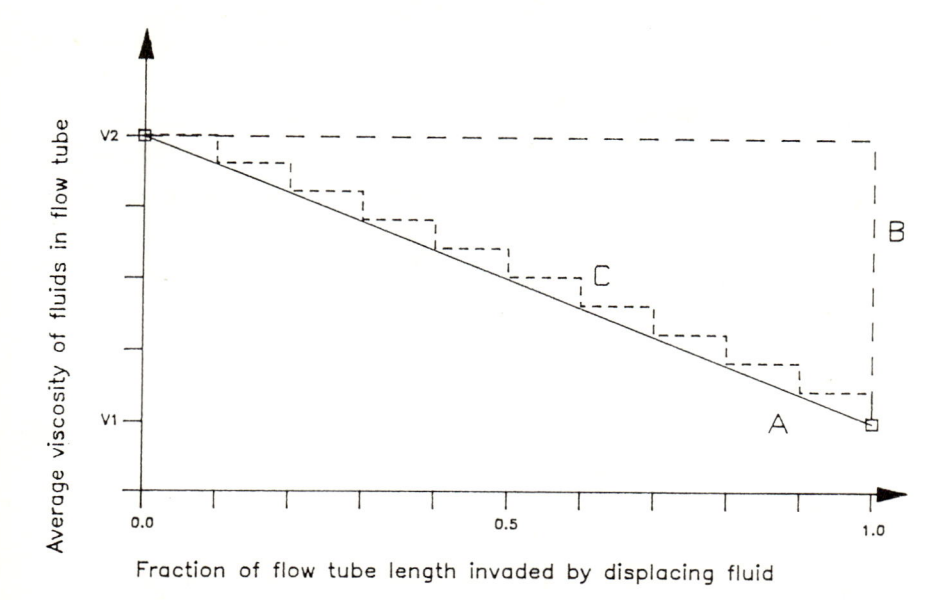

Figure 5. Network Tube Viscosity Function. V1 - Displaced Fluid Viscosity. V2 - Displacing Fluid Viscosity

5 Experimental Micromodel Work

Micromodels are artificial porous media made from transparent materials, in this particular study etched glass was used. The experimental methodolgy and fabrication technique are explained elsewhere [12,15].

The visual observations of fluid in micromodels aid the interpretation and understanding of flow phenomena. Fluid saturations within the model can be estimated by inspection of still photographs taken at intervals. This data can then be used to calibrate and evaluate the accuracy of the network simulator.

5.1 Micromodel Designs and Fluid Properties

The micromodel patterns used in the experiments had a Gaussian channel distribution characterised by the following parameters:

$$\text{Mean Radius} = 98 \mu m$$
$$\text{Standard Deviation}(\lambda_r) = 0.5 \qquad (5.1)$$

This distribution was approximated in the etched model by using five different channel sizes (with equivalent radii of 44, 76, 98, 116, and $131 \mu m$)

and in the case of fracture geometries the fracture radius was 200 μm. The etching process does not reproduce the channel sizes exactly, as a result the micromodel has a more variable channel size distribution than that actually designed, giving a closer approximation to the normal distribution. The length of all the channels was $2500\mu m$.

The fluids used were distilled water, n-decane and Nujol (mineral oil), their relevant properties are given in Tables 1 and 2. The glass micromodels were highly water-wet with the contact angle being, effectively, zero.

Fluid	Viscosity ($\times 10^{-3}$ Pa s)
Dyed Distilled Water	0.98
n - Decane	1.26
Mineral Oil (Nujol)	181.15

Table 1
Fluid Viscosities (measured at laboratory temparature 21-22°C)

Fluid System	Interfacial Tension ($\times 10^{-3}$ N/m)
Dyed Water / n - Decane	33.2
Dyed Water / Mineral Oil	32.0

Table 2
Fluid Interfacial Tensions (measured at laboratory temperature 21-22°C)

6 Results

6.1 Transfer Functions

An example of a typical transfer function as generated by the network simulator is given in Figure 6. This corresponds to a 5 node by 5 node fractured network, where water is displacing mineral oil solely under capillary forces. The curve shows the increasing production rate as the less viscous water imbibes into the system. The first kink on the curve indicates the breakthrough point.

6.2 The Effect of Flow Rate on Production

As may be seen from the results in Table 3 for the displacement of mineral oil by water, the flow rate across the block-fracture system has a significant effect on production behaviour. For the small networks examined (only 5 by 5 nodes with fractures on either side), higher flow rates, indicated by a higher applied pressure difference across the model, result in improved recovery inspite of an adverse viscosity ratio. The reason for this is that when the pressure in the fracture is hydrostatic water will tend to imbibe

Network Simulation Results
Transfer Curves

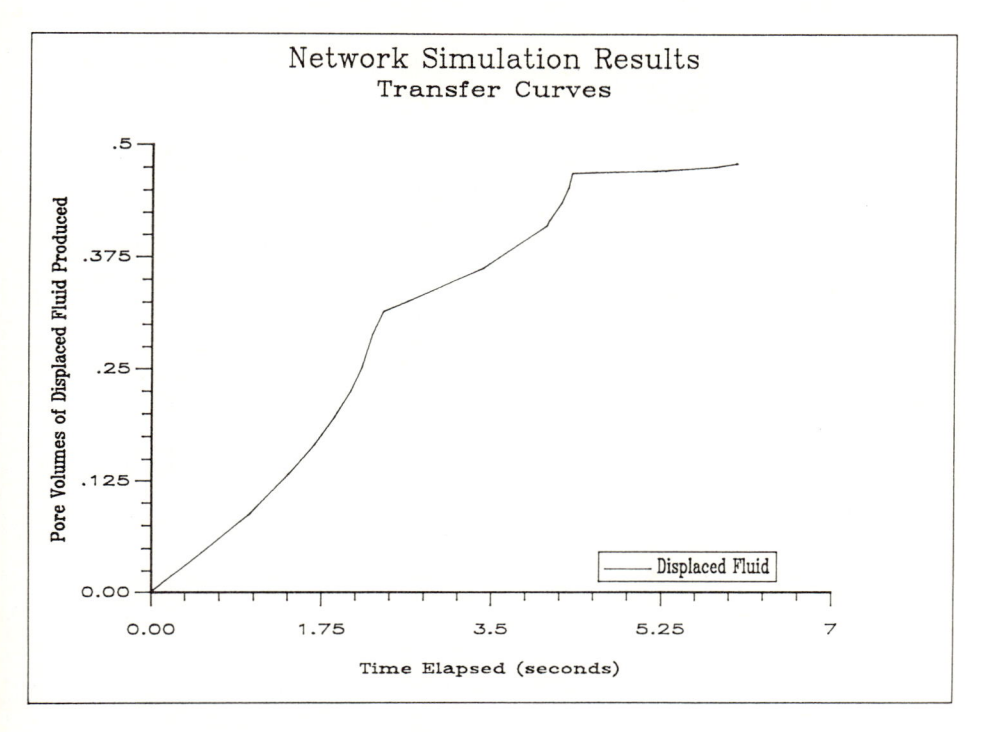

Figure 6. Example of a Typical Transfer Function

Pressure Difference (Pa)	Breakthrough Recovery (%)	Breakthrough Time (Seconds)	Ultimate Recovery (%)	Ultimate Recovery Time (Secs)
0	28.91	2.27	47.96	6.03
100	29.53	2.05	49.00	5.93
1000	31.66	1.05	53.49	4.17
10000	38.40	0.20	63.63	0.34

Table 3
Production Data at Varying Flow Rates (Applied Pressure Differences)
(5 x 5 node models with lateral fractures, contact angle 0°, water displacing mineral oil)

into the matrix whilst little forward movement of the water-oil interface in the fracture will take place. Having swept most of the matrix water will reach the outlet nodes trapping a significant volume of oil in the fractures. Whether this effect occurs in real reservoirs is open to question, although in narrow fractures or microfractures it may occur.

6.3 The Effect of Wettability on Production

Simulations were carried out varying the wettability of the network (Table 4). It was observed that as the wettability of the matrix was reduced the proportion of oil recovered at breakthrough and at ultimate recovery increased, particularly when the contact angle (measured through the wetting phase) increased from $0°$ to $30°$. In the small size of networks considered, the fracture system accounts for a significant part of the pore volume (approximately 58%), and as wettability is reduced the trapping process described in section 6.2 no longer occurs. The fracture is then swept more efficiently resulting in a higher recovery. In networks where the fracture does not account for a large fraction of the pore volume, the opposite effect occurs, with higher wettabilities generally leading to improved ultimate recovery.

Contact Angle (Degrees)	Breakthrough Recovery (%)	Breakthrough Time (Seconds)	Ultimate Recovery (%)	Ultimate Recovery Time (Secs)
90°	39.83	22.04	67.70	41.83
60°	29.99	3.67	49.04	11.26
30°	28.79	2.31	48.81	6.59
0°	29.53	2.05	49.00	5.93

Table 4
Network Simulation Production Results for Varying Wettability
(5 x 5 node models with lateral fractures, 100 Pa driving pressure difference, water displacing mineral oil)

6.4 The Effect on Production of Matrix Dimensions and an Evaluation of Network Simulator Accuracy

The results shown in Table 5 indicate that viscosity ratios have little effect on ultimate recovery in the case of imbibition with no external driving pressure. The network dimensions do have an effect on recovery. Maximum recovery is achieved when the length to width ratio is 1.0. The data show that lateral fractures tend to improve ultimate recovery.

The experimental micromodel data appears to follow the same general trends as the simulation results. Although both sets of results compare reasonably well, it is very difficult to measure saturation accurately in micromodels (the error may be up to 10% of the pore volume), consequently a strict comparison is invalid.

Other discrepancies between the experimental and network results arise from the assumption that micromodel channels can be approximated by cylindrical network tubes, whereas in reality the micromodel channels have a semi-circular or rectangular cross-section [15]. Overall the network simulation results agree well qualitatively with experimental data as reported in the literature.

Network Dimensions Nodes W-L	Length/Width Ratio	Geometry	Viscosity Ratio	Simulated Ultimate Recovery (%)	Experimental Ultimate Recovery (%)
10 x 19	0.5	Unfractured	1.286	32.4	50
10 x 10	1.0	Unfractured	1.286	43.5	70
19 x 10	2.0	Unfractured	1.286	27.4	30
10 x 19	0.5	Fractured	1.286	43.5	50
10 x 10	1.0	Fractured	1.286	49.0	70
19 x 10	2.0	Fractured	1.286	27.4	30
10 x 19	0.5	Unfractured	184.845	41.4	50
10 x 10	1.0	Unfractured	184.845	47.8	65
19 x 10	2.0	Unfractured	184.845	18.3	30
10 x 19	0.5	Fractured	184.845	44.3	50
10 x 10	1.0	Fractured	184.845	48.0	70
19 x 10	2.0	Fractured	184.845	29.5	35

Table 5
Production Results for Varying Fluid Viscosity Ratios, Varying Network Dimensions and Geometry and Check on Network Simulator Accuracy
(No applied external pressure gradient, water displacing n-decane and mineral oil)

7 Conclusions

Network models overcome some of the limitations of the dynamic and conceptual capillaric models previously described. The network model needs no assumption of a boundary condition at the matrix-fracture interface, other than the requirement that the produced oil is carried away from the matrix and does not interact with it further. The network simulator is capable of modelling features such as viscous fingering within matrix blocks and allows pore scale effects to be modelled. In this way it is possible to investigate the effect of pore scale behaviour on macroscopic production from fractured reservoir blocks. The network model allows source functions to be generated from pore size data and fluid properties without excessive calibration with experimental results.

As greater computer power becomes available (particularly with the advent of parallel processing) it will become feasible to simulate networks with hundreds of thousands of nodes and complex channel geometries. Ultimately whole reservoir fracture blocks may be modelled. The transfer functions so derived may then be used in conventional simulators to give a more realistic portrayal of the reservoir production process, leading to improved field development plans.

Nomenclature

Section 2

P Pressure
g Gravitational acceleration
z Vertical distance from the pressure datum
U Flow rate per unit area
K Absolute permeability
K_r Relative permeability
H Total height of matrix block
$\aleph$ Unit step function
φ Matrix porosity
Φ Fluid potential
ρ Density
M Mobility ratio (equivalent to $K_d \mu_{dd}/K_{dd}\mu_d$ subscript dd refers to displaced fluid and subscript d refers to displacing fluid)
$S_{..r}$ Residual saturation
$P_{c..}$ Capillary pressure
z_m Height of oil-water interface in the matrix
z_f Height of oil-water interface in the fracture
z_D Dimensionless distance of interface advance
$t_{D,G,Pc}$ Dimensionless time for a general displacement
$t_{D,Pc}$ Dimensionless time for a displacement dominated by capillary forces

Subscripts:

D Dimensionless

nw Non-wetting
w Wetting
e At exit face of block
f In the fracture
m In the matrix

Section 3

$\bar{V}$ Mean fluid interface velocity
μ Fluid viscosity
L Total capillary tube length
P_i Tube inlet pressure
P_j Tube outlet pressure
a Capillary tube radius
L_w Length of capillary tube filled with wetting fluid
Q Fluid production rate
A Total surface area of blocks
Ω Representative elementary volume of the combined matrix-fracture system

q^* Fluid production rate per unit elementary volume
S Saturation
P_{cf} Capillary pressure in the fracture
P_{cm} Capillary pressure in the capillary tube (i.e. matrix)

Subscripts:

nw Non-wetting
w Wetting
i Tube Inlet
j Tube Outlet

Section 4

c Random number from a uniform distribution between 0 and 1
λ_r Standard deviation of radii distribution
Q Flow rate in a network tube
$\bar{a}$ Mean tube radius
a_c Tube capillary radius
L Total tube length
P Pressure
μ Fluid viscosity
ρ Fluid density
g_y Component of gravity parallel to the tube axis
x Length of network tube invaded by displacing fluid
P_c Capillary pressure
ϑ Contact angle
γ Interfacial tension

Acknowledgments
We are indebted to Jane Gray, Dale Wong and Julian Phillips for their assistance with this paper.
We are also grateful to BP for providing financial support.

References

1. Aronofsky, J.S., Masse, L. and Natanson, S.G., (1958), A Model for the Mechanism of Oil Recovery from the Porous Matrix due to Water Invasion in Fractured Reservoirs, Trans, AIME, Vol. 213, 17.

2. Ball, .R.C, (1986), Fractals in Physics and Chemistry, Society of Chemistry Conference.

3. Bear, J. and Braester, C., (1972), On the Flow of Two Immiscible Fluids in Fractured Porous Media, "Fundamentals of Transport Phenomena in Porous Media" Elsevier Pub. Co., Amsterdam, 177-202.

4. Bokserman, A.A., Zheltov, I.P. and Kocheskov, A.A., (1964), Motion of Immiscible Liquids in a Cracked Porous Medium, Soviet Physics Doklady, 9,4, 285-287.

5. BP Statistical Review of World Energy, (June 1987), British Petroleum, London.

6. Brakel, J. van, (1975), Pore Space Models for Transport Phenomena in Porous Media Review and Evaluation with Special Emphasis on Capillary Liquid Transport, Powder Technology, 11, 205-236.

7. Golf-Racht, T.D. van, (1982), "Fundamentals of Fractured Reservoir Engineering", 2nd Ed., Elsevier Pub. Co., Amsterdam.

8. Hinch, E.J., (1988), The Recovery of Oil from Underground Reservoirs, "Mathematics in Oil Production", Clarendon Press, Oxford, 313-341.

9. Kershaw, D.S., (1978), The Incomplete Cholesky Conjugate Gradient Method for the Iterative Solution of Linear Equations, Journal of Computational Physics, 26, 43-65.

10. King, P., BP Research Centre, personal communication to the principal author.

11. Lefebvre du Prey, E., (June 1978), Gravity and Capillarity Effects on Imbition in Porous Media, SPE 6192, Society of Petroleum Engineering Journal, 195-206.

12. McKeller, M. and Wardlaw, N.C., (July-August 1982), A Method of Making Two-Dimensional Glass Micromodels of Pore Systems, Journal of Canadian Petroleum Technology, 21, No. 4, 39-41.

13. Parsons, R.W. and Chaney, P.R., (March 1966), Imbibition Model Studies on Water-Wet Carbonate Rocks, SPE 1091, Society of Petroleum Engineering Journal, 6, 26-34.

14. Reid, J.K., (1971), On the Method of Conjugate Gradients for the Solution of Large Sparse Sets of Linear Equations, "Proceedings of the Conference on Large Sparse Systems of Linear Equations", Academic Press, London and New York, 21-254.

15. Santiago, M.D. and Dawe, R.A., Network Modelling of Immiscible Flow in Porous Media - Simulation Compared with Experimental Micromodel Data, Submitted to Chemical Engineering Science, 1989.

AN ANALYTICAL SOLUTION TO A MULTIPLE-REGION MOVING BOUNDARY PROBLEM NONISOTHERMAL WATER INJECTION INTO OIL RESERVOIRS

Reidar Brumer Bratvold
(IBM European Petroleum Application Center, Norway)

and

Roland N. Horne
(Dept. of Petroleum Eng., Stanford University, California)

ABSTRACT

This work is concerned with the mathematical modelling of nonisothermal water injection problems using analytical techniques. The solutions derived incorporate the effects of temperature and saturation gradients and give the pressure, saturations and temperature as functions of distance and time in a one dimensional radial reservoir.

A two-step procedure is used to derive the solutions to the moving boundary injection problem. Assuming the fluids to be incompressible, we obtain a non-strictly hyperbolic system for saturation and temperature (Riemann problem). This problem is solved by using the method of characteristics together with the appropriate shock-admissibility criteria. With the knowledge of the saturation and temperature distributions, and hence the mobilities and diffusivities, from step one, the second order pressure diffusivity equation is solved assuming slightly compressible fluids. Both a similarity and a quasi-stationary method is used to solve the resulting moving boundary problem.

The falloff solution is derived both by solving the complete initial-boundary value problem with the injection solution as the initial condition and by superposing solutions to the stationary variable coefficients problem.

It is demonstrated that the saturation and temperature gradients have significant effects on the pressure data for both the injection and falloff periods.

1 INTRODUCTION

Numerous full field waterflooding projects are currently underway throughout the world to improve oil recovery. In many large fields, particularly in the North Sea, water injection is initiated during the early stages of reservoir development.

Pressure transient testing can provide valuable information about the parameters of an injection-falloff scheme. These tests are usually run to give information about the progress of the flood (i.e. frontal advance), residual oil saturation, the flow characteristics of the virgin formation and near wellbore damage.

In a water injection well test, the injected fluid usually has a temperature different from the initial reservoir temperature, and during injection, both a saturation and a temperature front propagates into the reservoir. Furthermore, due to differences in the oil and water properties, a saturation gradient is established in the reservoir. The water saturation will be highest close to the well and will continuously decrease with distance from the well. Ahead of this invaded region is the unflooded oil bank at initial water saturation. The fluid mobilities will change continuously in the invaded region and this will have to be accounted for in the reservoir modelling and data interpretation.

Many different models have been introduced for the analysis of water injection and falloff tests. Typically, these models neglect the temperature effects, the saturation gradient or both[1].

In this paper we present analytical solutions which includes the most important effects in a nonisothermal water injection-falloff scheme. Specifically, we consider the pressure behaviour at the well due to the simultaneous flow of oil and water in a cylindrical reservoir with a temperature gradient. Solutions for linear reservoirs are presented by *Bratvold* and *Larsen* [3].

In a typical injection test the fracture emanating from a water-injection well is not stationary but propagates. Solutions accounting for this propagation, and the subsequent fracture closure during the falloff period, are presented by *Bratvold* and *Larsen* [4].

2 MATHEMATICAL MODEL

The reservoir considered is assumed to be cylindrical with the well at the centre of the cylinder. The well penetrates the entire thickness of the formation, and fluid is injected at a constant rate. The reservoir is assumed to be a uniform, homogeneous porous medium completely saturated with oil and water. Neglecting effects of gravity, as well as heat transfer to the surrounding formation permits the use of a one-dimensional radial model.

2.1 Discussion of the Solution Method

The transient, nonisothermal two-phase flow of oil and water requires that the coupled energy and mass conservation equations be solved simultaneously. Furthermore, since the injection of cold water into a hot oil reservoir is a moving boundary problem, it cannot be solved using standard linear techniques such as eigenfunction expansion, integral transforms or Green's function methods.

To circumvent these complications we will derive alternative approximate solutions to the injection problem using a two-step procedure:

---

[1] A review of previous work is given in Refs. [1] and [2].

1. Assume incompressible fluids. Solve the resulting first order coupled energy and mass conservation equations using fractional flow theory [5,6,7,8]. This essentially amounts to decoupling the equations for saturation and temperature from the equation for pressure. The saturation profile obtained is a *Buckley-Leverett* [9] type profile including temperature effects.

2. With the saturation and temperature profiles, and hence the mobilities and diffusivities, known as functions of time and distance from step 1, solve the diffusion equation for pressure assuming the compressibilities of the fluids to be small and constant. Hence, the pressure distribution in the system is obtained by superimposing pressure transient effects on an a-priori known saturation profile.

The key assumption in this procedure is that for a typical water injection-falloff test, the fluid compressibilities will not significantly affect the saturation and temperature distributions. The pressure response, however, is strongly affected by the fluid compressibilities.

3 SATURATION AND TEMPERATURE PROFILES

Step 1 of the solution procedure consists of obtaining the nonisothermal saturation profile assuming incompressible fluids. The solution to this nonstrictly hyperbolic system was presented by *Fayers* [5]. Following the development of the required mathematical theory by *Keyfitz* and *Kranzer* [6] several authors have presented articles on Buckley-Leverett type problems that include temperature effects [7,8]. Here we will only present the model and its solution. For a more complete description of the solution procedure the reader may consult e.g. *Hovdan* [8].

Under the assumption of incompressible fluids the mass conservation equation for water is given by[2]

$$\phi \frac{\partial S}{\partial t} + q \frac{\partial f}{\partial A} = 0 \qquad (3.1)$$

where S is the water saturation and where $S + S_o = 1$. $A = \pi r^2$ is the area and $q = 2\pi r u$ is the flow rate. The fractional flow of water, at any point in the reservoir, is defined as

$$f = f(S,T) = \frac{u_w}{u} = \left\{ 1 + \frac{k_{ro}\mu_w}{k_{rw}\mu_o} \right\}^{-1} \qquad (3.2)$$

where $u = u_w + u_o$ is the total fluid velocity. Since the fractional flow of water is a function of both saturation and temperature, we can expand Eq. 3.1 to

---
[2]Nomenclature in Appendix A.

get

$$\phi\frac{\partial S}{\partial t} + q\left(\frac{\partial f}{\partial S}\frac{\partial S}{\partial A} + \frac{\partial f}{\partial T}\frac{\partial T}{\partial A}\right) = 0 \qquad (3.3)$$

In deriving the energy conservation equation we will assume that conduction and radiation can be neglected. The fluids will be assumed to be incompressible with constant densities and heat capacities. For a radial system we obtain

$$[\phi(\rho_w C_{Vw} S + \rho_o C_{Vo} S_o) + (1-\phi)\rho_s C_{Vs}]\frac{\partial T}{\partial t}$$
$$+q\left[f(\rho_w C_{Vw} - \rho_o C_{Vo}) + \rho_o C_{Vo}\right]\frac{\partial T}{\partial A} = 0 \qquad (3.4)$$

Introducing the dimensionless parameters

$$\alpha = \frac{\rho_o C_{Vo}}{\rho_w C_{Vw} - \rho_o C_{Vo}} \qquad (3.5)$$

$$\beta = \frac{\rho_o C_{Vo} + \frac{1-\phi}{\phi}\rho_w C_{Vw}}{\rho_w C_{Vw} - \rho_o C_{Vo}} \qquad (3.6)$$

$$T_D = \frac{T - T_i}{T_{ir} - T_i}, \qquad t^* = \frac{qt}{\phi r_w}$$

and

$$A_D = \frac{A}{r_w}, \qquad g = \frac{f + \alpha}{S + \beta}$$

we can rewrite the conservation equations as

$$\frac{\partial S}{\partial t^*} + \frac{\partial f}{\partial S}\frac{\partial S}{\partial A_D} + \frac{\partial f}{\partial T_D}\frac{\partial T_D}{\partial A_D} = 0 \qquad (3.7)$$

and

$$\frac{\partial T_D}{\partial t^*} + g\frac{\partial T_D}{\partial A_D} = 0 \qquad (3.8)$$

As discussed by *Hovdan* [8] these conservation equations with appropriate auxiliary conditions constitute a non-strictly hyperbolic Riemann problem and can be solved by the method of characteristics (or Riemann invariants). A physically reasonable solution is obtained by tracing the characteristic curves and introducing appropriate discontinuities to obtain a unique solution.

As an example of how to generate the solution we will consider the fractional flow curves shown in Fig. 1. The left curve corresponds to the cold fluids at $T_D = 0$ while the right curve is generated by using the hot fluids at $T_D = 1$. On the same figure we have added the characteristic curves with slopes g and f_S that are tangents to the hot fluids curve. The solution is generated by tracing the slope of the fractional flow curve. The appropriate discontinuities represented by the tangent points between the characteristic

MOVING BOUNDARY PROBLEM

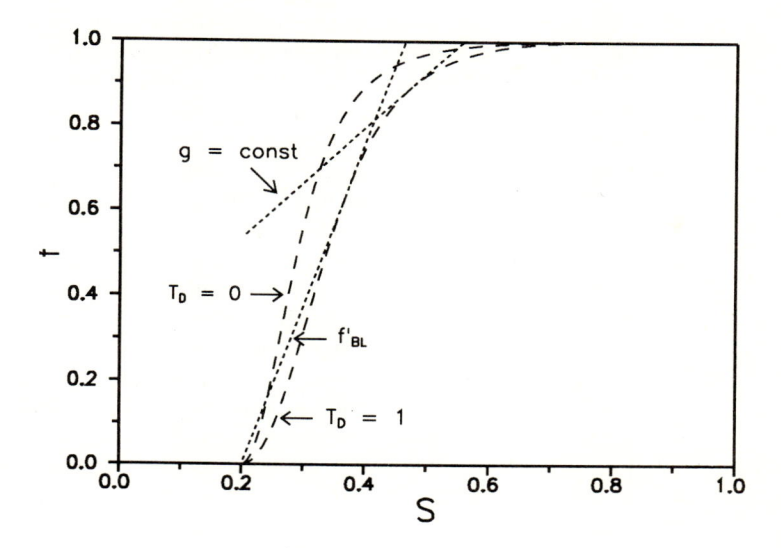

Figure 1: Fractional flow curves.

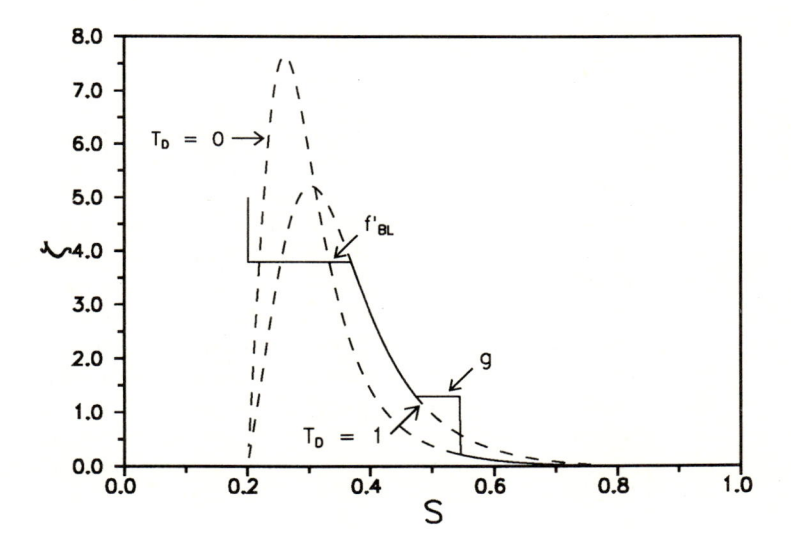

Figure 2: Derivatives of the fractional flow curves.

curves and the $f(T_D = 1)$ vs. S curve have to be included to ensure uniqueness. Fig. 2 shows the slopes of the two fractional flow curves plus g and f'_{BL}. Starting at $S = 1 - S_{or}$ we follow the cold fractional flow curve until we reach the point where g intersects. This represents the transition point from the cold to the hot curve; i.e., the step from $T_D = 0$ to $T_D = 1$ and is a discontinuity in the slope shown in Fig. 2. Now follow the $g = constant$ curve until the hot fractional flow curve is reached and then continue along this curve until reaching the tangent $f_S = f'_{BL} = constant$. This point represents the discontinuity between the invaded and uninvaded regions; i.e., the standard *Buckley-Leverett* [9] shock. The inverse of the velocity profile in Fig. 2 is shown as the saturation profile in Fig. 3 and Fig. 4 shows the corresponding temperature profile.

To summarize, the solution to the conservation equations for incompressible fluids **gives** the saturation and temperature profiles in the reservoir. The saturation profile will typically contain two discontinuities; one corresponding to the single discontinuity in the temperature profile. The temperature profile will consist of a simple Heaviside step function while the saturation profile will have continuous changes in addition to the two discontinuities. Right behind the temperature front there will be a region where the saturation is constant. The size of this region will vary with the fluid properties; it can be infinitely small or cover most of the region between the injector and the temperaturefront.

4 PRESSURE SOLUTION

In the second step of the outlined solution procedure we need to solve the diffusion equation for pressure. In this work we limit our discussion to cold water injection into a hot oil reservoir. Note, however, that the solutions derived apply to any system in which the saturation profile can be described a-priori.

With the saturation distribution known as a function of time and space from step 1, the total fluid mobility and total system compressibility become functions of time and space. This is a moving boundary problem since the locations of the saturation and temperature fronts are time dependent.

The basic multiphase flow equations form the basis for the mathematical modelling of injection and falloff behaviour. A description of the general case with three separate phases and the appropriate simplifications for the injection-falloff situation may be found in e.g. *Matthews* and *Russell* [10].

Assuming that the reservoir consists of two different regions separated by a moving discontinuity in fluid saturation as illustrated in Fig. 3 and Fig. 4, we obtain the following dimensionless mathematical model for an infinite system with a line source well:

Governing equations:

$$\frac{1}{r_D}\frac{\partial}{\partial r_D}(\lambda r_D \frac{\partial p_{D_1}}{\partial r_D}) = \frac{1}{\eta}\frac{\partial p_{D_1}}{\partial t_D}, \qquad 0 < r_D < r_{D_{BL}}(t_D) \qquad (4.1)$$

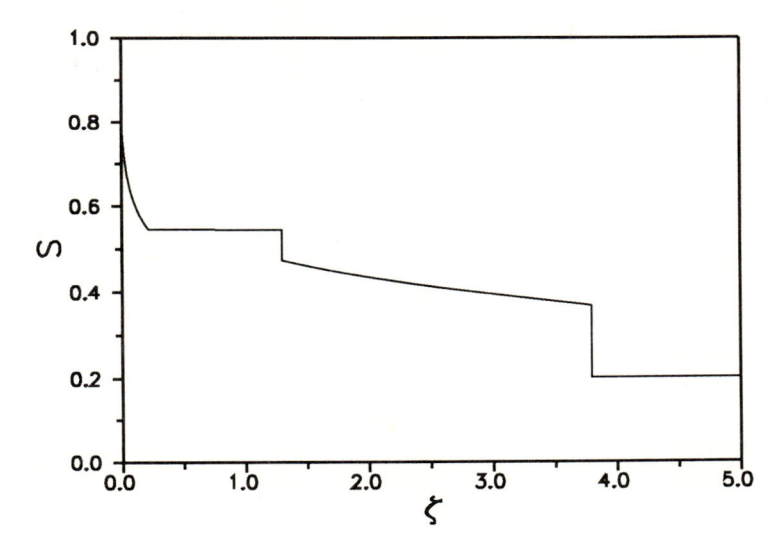

Figure 3: Saturation profile.

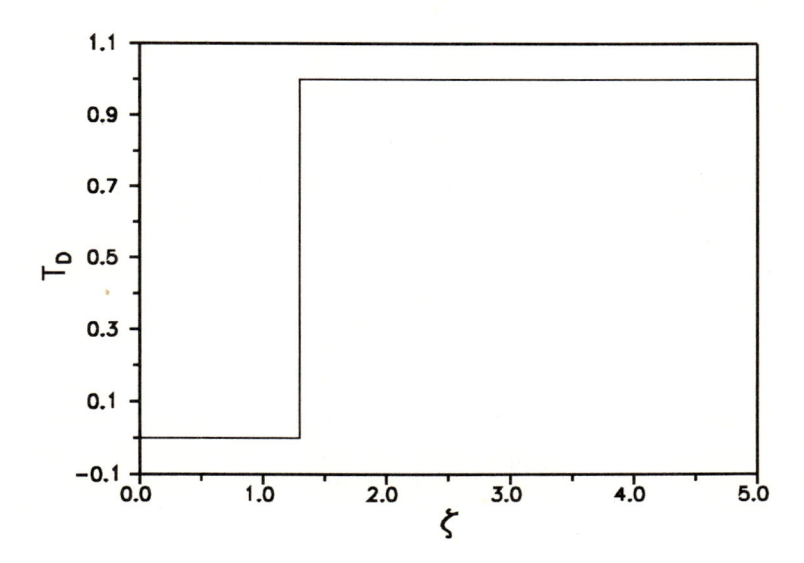

Figure 4: **Temperature profile.**

$$\frac{1}{r_D}\frac{\partial}{\partial r_D}(r_D\frac{\partial p_{D_2}}{\partial r_D}) = \frac{\partial p_{D_2}}{\partial t_D}, \quad r_{D_{BL}}(t_D) < r_D < \infty \quad (4.2)$$

where

$$\eta[S(r_D, t_D)] = M\frac{c_{to}}{c_t} = \frac{M c_{to}}{Sc_w + (1-S)c_o + c_s} = \eta(r_D, t_D)$$

and

$$\lambda[S(r_D, t_D)] = \frac{\lambda_o + \lambda_w}{\hat{\lambda}_w} = \lambda(r_D, t_D)$$

Initial conditions:

$$p_{D_1} = p_{D_2} = 0, \quad t_D = 0 \; \forall r_D \quad (4.3)$$

$$r_{D_{BL}} = 0, \quad t_D = 0 \quad (4.4)$$

Boundary conditions:

$$\lim_{r_D \to 0} r_D \frac{\partial p_{D_1}}{\partial r_D} = -1 \quad (4.5)$$

$$\lim_{r_D \to \infty} p_{D_2} = 0 \quad (4.6)$$

Moving boundary conditions:

$$p_{D_1} = p_{D_2}, \quad r_D = r_{D_{BL}}(t_D) \quad (4.7)$$

$$\lambda \frac{\partial p_{D_1}}{\partial r_D} = \frac{1}{M}\frac{\partial p_{D_2}}{\partial r_D}, \quad r_D = r_{D_{BL}}(t_D) \quad (4.8)$$

All variables and parameters are dimensionless as defined in the nomenclature. p_{D_1} and p_{D_2} are the pressures in the invaded and uninvaded regions respectively. $r_{D_{BL}}(t_D)$ is the position of the moving interface between the two regions. Note that there is a second moving discontinuity in the reservoir; the temperature discontinuity. In the above model this is accounted for through the time and space dependent total mobility.

The mathematical model can be solved by two different methods: (i) Similarity transform approach and (ii) Quasi-stationary approach. The similarity transform solution is exact as opposed to the quasi-stationary solution which can only be approximate for compressible fluids. However, the advantage of the quasi-stationary solution procedure is the possibility of including finite boundaries. The similarity transform solution is limited to an infinite reservoir with a line source well. For the problem at hand the two solution approaches give identical quantitative results for all practical purposes [1].

5 SIMILARITY TRANSFORM METHOD

The moving boundaries problem may be transformed into a two-region (fixed interfaces) variable coefficient problem in one variable by introducing the Boltzmann variable

$$y = \frac{r_D^2}{4t_D} \tag{5.1}$$

From the nonisothermal Buckley-Leverett theory

$$\left(r_D \frac{dr_D}{dt_D}\right)_S = \epsilon f'|_S = constant \tag{5.2}$$

where

$$\epsilon = \frac{qc_{to}}{2\pi\hat{\lambda}_o h}$$

The saturation distribution is a unique function of $y \propto \zeta$. By transforming our problem from the independent variables t_D and r_D to the variable y, we obtain a two-region problem where the saturation distribution is a unique function of y. The solution is obtained by integration as

$$\begin{aligned} p_{D_1}(y) &= \frac{1}{2}\int_y^{y_{BL}} \frac{dy'}{\lambda y'} exp(-\int_0^{y'} \frac{dy''}{\lambda \eta}) \\ &+ \frac{M}{2}exp(y_{BL} - \int_0^{y_{BL}} \frac{dy}{\lambda \eta})E_1(y_{BL}), \quad 0 \le y \le y_{BL} \end{aligned} \tag{5.3}$$

and

$$p_{D_2}(y) = \frac{M}{2}exp(y_{BL} - \int_0^{y_{BL}} \frac{dy}{\lambda \eta})E_1(y), \quad y_{BL} \le y < \infty \tag{5.4}$$

where $E_1(x)$ is the exponential integral.

For a typical water injection test the reservoir and fluid parameters are such that ϵ is small and $y_{BL} = \epsilon f'/2$ is of order $10^{-2} - 10^{-3}$ while $\lambda\eta$ is > 1. This makes the exponential terms in the solution ≈ 1 after the first few seconds and the fluids in the invaded region essentially behave as if they were incompressible. Furthermore, since

$$y = \frac{r_D^2}{4t_D} = \frac{\phi c_o r^2}{4\hat{\lambda}_o t} = \frac{1}{2}\epsilon f' \tag{5.5}$$

we can transform back to the original variables (note that $y_{BL} = const.$)

$$p_{D_2}(r_D, t_D) = \frac{M}{2}E_1(\frac{r_D^2}{4t_D}), \quad t_D \le \frac{r_D^2}{4y_{BL}} \tag{5.6}$$

and

$$\begin{aligned} p_{D_1}(r_D, t_D) &= \frac{1}{2}E_1(\frac{r_D^2}{4\hat{\eta} t_D}) - \frac{1}{2}E_1(\frac{y_{BL}}{\hat{\eta}}) + \frac{M}{2}E_1(y_{BL}) \\ &+ \frac{1}{2}\int_{S(r_D,t_D)}^{S_{BL}} \frac{f''}{f'}(\frac{1}{\lambda} - 1)dS, \quad t_D \ge \frac{r_D^2}{4y_{BL}} \end{aligned} \tag{5.7}$$

Evaluating the pressure at the wellbore we note that for long times $S(1,t_D) \to 1 - S_{or}$, and the last integral in Eq. 5.7 becomes negligible. At these times we can also use the approximation $-E_1(x) \approx \ln x + \gamma$ to obtain the wellbore pressure as

$$p_{wD} = \frac{1}{2}(\ln t_D + 0.80907) + s_a + s \tag{5.8}$$

where the apparent skin factor s_a is given by

$$s_a = \frac{1}{2}\int_{1-S_{or}}^{S_{BL}} \frac{f''}{f'}(\frac{1}{\lambda} - 1)dS + \frac{1}{2}(\ln y_{BL} + \gamma)(1 - M) \tag{5.9}$$

That is, the wellbore pressure is given by the familiar *Theis* solution [11] plus an apparent skin factor caused by the saturation gradient and the propagating temperature and phase discontinuities. In this expression for the wellbore pressure we have added a mechanical skin factor s to account for any formation damage or stimulation near the wellbore.

If the relative permeabilities and viscosities of the fluids are such that the injection results in piston-like displacement, $\lambda = 1$, $\eta = \hat{\eta}$ and from Eqs. 5.3 and 5.4 we obtain the *Verigin* [12,13] solution

$$\begin{aligned} p_{D_1} &= \frac{1}{2}E_1(\frac{r_D^2}{4\hat{\eta}t_D}) - \frac{1}{2}E_1(\frac{y_{BL}}{\hat{\eta}}) \\ &+ \frac{M}{2}e^{y_{BL}(1-\frac{1}{\hat{\eta}})}E_1(y_{BL}), \quad 0 \leq r_D \leq r_{D_{BL}}(t_D) \end{aligned} \tag{5.10}$$

$$p_{D_2} = \frac{M}{2}e^{y_{BL}(1-\frac{1}{\hat{\eta}})}E_1(y), \quad r_{D_{BL}}(t_D) \leq r_D < \infty \tag{5.11}$$

5.1 Injection into a Finite Cylindrical Reservoir

When injecting into an infinite reservoir, only the compressibility in the uninvaded zone is significant [1]. Consequently, it is likely that the outer boundary of a finite cylindrical reservoir will start influencing the wellbore pressure at the time given by the radius of investigation concept [10]

$$t_{eD} = \frac{1}{4}r_{eD}^2 \tag{5.12}$$

The Buckley-Leverett phase front position at time t_{eD} is given by

$$r_{D_{BL}}^2 = 2\epsilon f'_{BL} t_{eD} = \frac{\epsilon f'_{BL}}{2}r_{eD}^2 \tag{5.13}$$

According to this equation, the invaded region is still only occupying a small part of the total reservoir at the end of the infinite-acting period. Hence, it is also to be expected that the compressibility in the large uninvaded region is dominating also after the outer boundary is felt.

Verigin [12] and *Ramey* [14] discussed approximate solutions to moving boundary problems in finite domains. Applying their ideas we can add an outer region to obtain an additional moving boundary in an infinite reservoir. The two outer regions have constant (but different) mobilities while the inner region has a saturation gradient and a temperature discontinuity as before. The interfaces move according to mass and energy balances and hence the saturation and temperature distributions remain constant at any constant value of the Boltzmann variable. An approximation to a closed outer boundary can be obtained by taking the limit as the mobility in the outermost region approaches zero while its interface is kept at r_{eD}. The approximate closed outer boundary solution is

$$p_{D_2}(r_D, t_D) = \frac{M}{2}\left[E_1(\frac{r_D^2}{4t_D}) - E_1(\frac{r_{eD}^2}{4t_D})\right]$$
$$+ 2M\frac{t_D}{r_{eD}^2}e^{-r_{eD}^2/4t_D}, \quad t_D \leq \frac{r_D^2}{4y_{BL}} \qquad (5.14)$$

and

$$p_{D_1}(r_D, t_D) = \frac{1}{2}\left[E_1(\frac{r_D^2}{4\hat{\eta}t_D}) - E_1(\frac{y_{BL}}{\hat{\eta}})\right] + \frac{1}{2}\int_{S(r_D,t_D)}^{S_{BL}} \frac{f''}{f'}(\frac{1}{\lambda} - 1)dS$$
$$+ \frac{M}{2}\left[E_1(y_{BL}) - E_1(\frac{r_{eD}^2}{4t_D})\right] + 2M\frac{t_D}{r_{eD}^2}e^{-r_{eD}^2/4t_D},$$
$$t_D \geq \frac{r_D^2}{4y_{BL}} \qquad (5.15)$$

The constant pressure outer boundary approximation is obtained by taking the limit as the mobility in the outermost region approaches infinity. The solution to this problem is given by Eqs. 5.14 and 5.15 without the exponential terms.

6 QUASI-STATIONARY METHOD

A quasi-stationary method for solution of moving boundary problems was introduced by *Leibenzon* [15]. The method is reviewed by *Rubinsthein* [16] who suggests the algorithm:

1. Solve the associate problem with a *stationary* boundary between the zones. Let the solution of this problem be $\hat{p}_{D_i} = \hat{p}_{D_i}(r_D, t_D; R_D)$ where $r_D = R_D$ is the position of the stationary boundary.

2. Use the solution $\hat{p}_{D_i}$ in the Stefan condition to construct an explicit equation for $\varrho(t_D)$, which is an approximation to the position of the *moving* boundary:

$$\varrho' = -\frac{\epsilon}{\Delta S}\frac{\partial \hat{p}_{D_1}}{\partial r_D}(\varrho, t_D; \varrho) \qquad (6.1)$$

3. Substitute $R_D = \varrho(t_D)$ into $\hat{p}_{D_i}$ and use this as an approximation for the solution to the moving boundary problem:

$$p_{D_1} \approx \hat{p}_{D_1}(r_D, t_D; \varrho(t_D)) \tag{6.2}$$

$$p_{D_2} \approx \hat{p}_{D_2}(r_D, t_D; \varrho(t_D)) \tag{6.3}$$

Rubinsthein [16] states that the method "gives a qualitatively correct result, although quantitatively it contains errors". Obviously, the difference between the similarity transform and the quasi-stationary solution will be a function of the compressibilities of the fluids in the invaded region.

The first step in the quasi-stationary algorithm requires the stationary solution to the variable coefficient problem. This solution can be obtained by discretizing the saturation profile and use a multizone model such that a sufficiently accurate representation of the mobility and diffusivity as a function of radius is obtained. Following this procedure we obtain the governing equations and auxiliary conditions given in Appendix B. The solution in Laplace space is

$$\mathcal{L}\{\hat{p}_{D_{1i}}\} = x_{i1} K_0(\sqrt{\kappa_i z}r_D) + x_{i2} I_0(\sqrt{\kappa_i z}r_D), \; i = 1, \ldots, n \tag{6.4}$$

and

$$\mathcal{L}\{\hat{p}_{D_2}\} = x_{(n+1)1} K_0(\sqrt{z}r_D) + x_{(n+1)2} I_0(\sqrt{z}r_D). \tag{6.5}$$

The associated problem for the coefficients x_{i1} and x_{i2} is the same as for the falloff problem (to be discussed later) and is detailed by *Bratvold* [2,1]. The solution is inverted numerically by using the Stehfest inversion algorithm [17].

6.1 Long Time Approximation

As illustrated in the previous section, the stationary multibank solution to the variable mobility and diffusivity problem is algebraically complex. Although it is computationally tractable it does not lend itself easily to an analytical analysis of the properties of the solution.

An approximate long time solution to the same problem can be derived by ignoring the compressibility of the fluids in the invaded region. The appropriate problem is then given by Eqs. 4.1 through 4.8 with $r_{D_{BL}} = R_D$ and by setting

$$\frac{1}{\eta}\frac{\partial p_{D_1}}{\partial t_D} = 0$$

in Eq. 4.1. In Laplace space the solution with wellbore storage and skin is given by

$$\mathcal{L}\{\hat{p}_{D_1}\} = \frac{z^2 C_D \mathcal{L}\{\hat{p}_{D_2}(R_D; z)\} - M}{z^2 C_D [I_\lambda(1) - s] - zM} I_\lambda(r_D) + \mathcal{L}\{\hat{p}_{D_2}(R_D; z)\} \tag{6.6}$$

where the integral I_λ is given by

$$I_\lambda(x) = \int_x^{R_D} \frac{dr}{\lambda r}. \tag{6.7}$$

$\mathcal{L}\{\hat{p}_{D_2}(R_D; z)\}$ will vary according to the outer boundary condition. If t_D/R_D^2 is large enough, we get an approximate real space solution for an infinite reservoir without storage and skin

$$\hat{p}_{D_1} \approx \frac{M}{2} E_1(\frac{R_D^2}{4t_D}) + \int_{r_D}^{R_D} \frac{dr}{\lambda r} \quad 1 \leq r_D \leq R_D \tag{6.8}$$

6.2 Injection Problem

With basis in the solution derived in the previous section, we can now proceed with the second and third step in the quasi-stationary method.

Since we are using the nonisothermal Buckley-Leverett theory to describe the motion of the saturation profile, we approximate the velocity of any saturation by Eq. 5.2 as

$$\varrho \varrho'|_S = \epsilon f'|_S \tag{6.9}$$

where $\varrho^2 = 2\epsilon f'_{BL} t_D$. An approximate solution for the pressure in the invaded region is now constructed from Eq. 6.8 with $p_{D_1} \approx \hat{p}_{D_1}(r_D, t_D; \sqrt{2\epsilon f'_{BL} t_D})$

$$p_{D_1} \approx \frac{M}{2} E_1(\frac{\epsilon f'_{BL}}{2}) + \int_{r_D}^{\sqrt{2\epsilon f'_{BL} t_D}} \frac{dr}{\lambda r}. \tag{6.10}$$

A closer inspection of Eq. 5.7 reveals that for times when the log-approximation can be used for the exponential integral, the simplified similarity solution, Eq. 5.7 is identical to the quasi-stationary solution Eq. 6.10. Note, however, that since the stationary variable mobility and diffusivity problem is linear, we can include wellbore storage and skin in the quasi-stationary solution as opposed to the similarity solution which was derived with the assumption of a line-source well.

7 FALLOFF SOLUTION

In deriving the saturation distribution for the injection problem we neglected the fluid compressibilities and used the nonisothermal Buckley-Leverett theory to obtain the solution. If the fluids are incompressible, the saturation and temperature distributions will stop their propagation into the reservoir immediately upon shut-in and remain stationary during the falloff period. Using the injection solution as the initial condition for the falloff problem the mobilities and diffusivities are given by their value at the end of the injection period ($t_D = t_{D_i}$, $\Delta t_D = t_D - t_{D_i} = 0$). The falloff problem is then a variable coefficient two-region problem. This problem can be solved by dividing the invaded region into n-regions where the coefficients are evaluated at the average saturation of each region. Since we do not rely upon the use of a similarity variable in solving the falloff problem, we can easily implement a finite wellbore radius with storage and skin as well as a finite outer boundary. The inclusion of wellbore storage will result in continued water flow across the wellbore sandface into the reservoir after wellhead shut-in. We will assume, however, that the propagation of the temperature and phase fronts after shut-in will be negligible compared with the distance the fronts travelled during the injection period. The governing equations and auxiliary conditions in dimensionless form for the (n+1)-region composite reservoir are given in Appendix A.

The problem is solved by using the Laplace transformation. The nonzero initial condition results in a nonhomogeneous problem in Laplace space. In real space the solution is given by

$$p_{D_{1_i}}(r_D, \Delta t_D) = \frac{\kappa_i}{2\Delta t_D} \int_{a_{i-1}}^{a_i} \xi \psi_1(\xi) e^{\left\{-\frac{\kappa_i(r_D^2+\xi^2)}{4\Delta t_D}\right\}} I_0\left(\frac{\kappa_i r_D \xi}{2\Delta t_D}\right) d\xi$$
$$+\mathcal{L}^{-1}\left\{x_{i1} K_0(\sqrt{\kappa_i z} r_D)\right\} + \mathcal{L}^{-1}\left\{x_{i2} I_0(\sqrt{\kappa_i z} r_D)\right\},$$
$$a_{i-1} < r_D < a_i, \quad i = 1, \ldots, n \quad (7.11)$$

and

$$p_{D_2}(r_D, \Delta t_D) = \frac{1}{2\Delta t_D} \int_{r_{D_{BL}}}^{r_{eD}} \xi \psi_2(\xi) e^{\left\{-\frac{(r_D^2+\xi^2)}{4\Delta t_D}\right\}} I_0\left(\frac{r_D \xi}{2\Delta t_D}\right) d\xi$$
$$+\mathcal{L}^{-1}\left\{x_{(n+1)1} K_0(\sqrt{z} r_D)\right\} + \mathcal{L}^{-1}\left\{x_{(n+1)2} I_0(\sqrt{z} r_D)\right\},$$
$$r_{D_{BL}} < r_D < r_{eD} \quad (7.12)$$

The constants are determined by applying the boundary and interface conditions. In doing this we need to solve a $(n+1) \times (n+1)$ block tridiagonal system of equations for the constants x_{i1} and x_{i2}, $i = 1, \ldots, n+1$:

$$\mathbf{\Gamma} \vec{x} = \vec{d}. \quad (7.13)$$

The coefficients of the matrix $\mathbf{\Gamma}$ and the right hand side $\vec{d}$ depends on the auxiliary conditions and are described in detail by *Bratvold* and *Horne* [2].

8 SUPERPOSITION

The principle of superposition is commonly used in the modelling of the falloff period or to account for the effect of a general change in rate. The superposition theorem applies only if the problem is linear. In our model the movement of the phase and temperature fronts are governed by mass and temperature balances using incompressible fluids. Hence, if the well is shut-in, the fronts will stop immediately. The problem is then linear and provided appropriate expressions to construct the total solution can be found, superposition can be used. Due to the close relationship between the similarity solution, Eq. 5.7, and the solution for a reservoir with a stationary mobility and diffusivity distribution, Eq. 6.10, such expressions are now readily found.

8.1 Falloff Problem

Let us consider the falloff problem in an infinite reservoir. Let the well be closed in at time $t_D = t_{D_i}$ and let the phase front position at this time be $R_D = \sqrt{2\epsilon f'_{BL} t_{D_i}}$. Since we assume that the saturation and temperature distributions are stationary during the entire shut-in period we need ensure that the solution obtained by superposition does not change the distributions during the falloff period. Because of the close relationship between the moving boundary problem and the stationary variable mobility and diffusivity problem, the pressure during falloff can be described as a superposition of the solutions to the stationary problems. The principle of superposition gives

$$p_{wD}(1, \triangle t_D) \approx \hat{p}_{D_1}(1, t_D; \sqrt{2\epsilon f'_{BL} t_{D_i}}) - \hat{p}_{D_1}(1, \triangle t_D; \sqrt{2\epsilon f'_{BL} t_{D_i}}) \qquad (8.1)$$

8.2 Changes in Rate

The principle of superposition can also be used to model any change in rate as long as we use the stationary variable mobility and diffusivity solution. Consider the case of a well producing at a series of constant rates q_i beginning at time t_{i-1}. Define $q_0 = 0$ and $t_0 = 0$. The principle of superposition gives

$$\frac{2\pi \hat{\lambda}_w h(p - p_{ir})}{q_n} = \sum_{i=1}^{n} \frac{\triangle q}{q_n} \hat{p}_D(t_{D_n} - t_{D_{i-1}}), \quad t_{D_{n-1}} < t_D < t_{D_n} \qquad (8.2)$$

where $\hat{p}_D$ is the stationary variable coefficient solution.

9 VERIFICATION AND DISCUSSION OF SOLUTIONS

To verify the analytical results developed in this paper a large number of injection tests, covering a wide range of reservoir parameters, have been numerically simulated. In this paper we will present only a few of the results. A more complete discussion is given by *Bratvold* [1].

The numerical simulations were performed with a two phase, two dimensional, black-oil simulator developed by *Nyhus* [18] The simulator is a single well thermal model which numerically solves a more detailed model of the physical situation. The appropriate partial differential equations are solved by using a finite difference technique and the solution procedure is fully implicit with respect to pressure, saturation and temperature.

The numerical model has been tested and found to accurately simulate the heat and mass transfer effects [18]. Any discrepancies between analytical and numerical results may be caused both by the simplifications inherent in the analytical models and by the approximations used when constructing the analytical solutions. The fact that the numerical solutions contain numerical

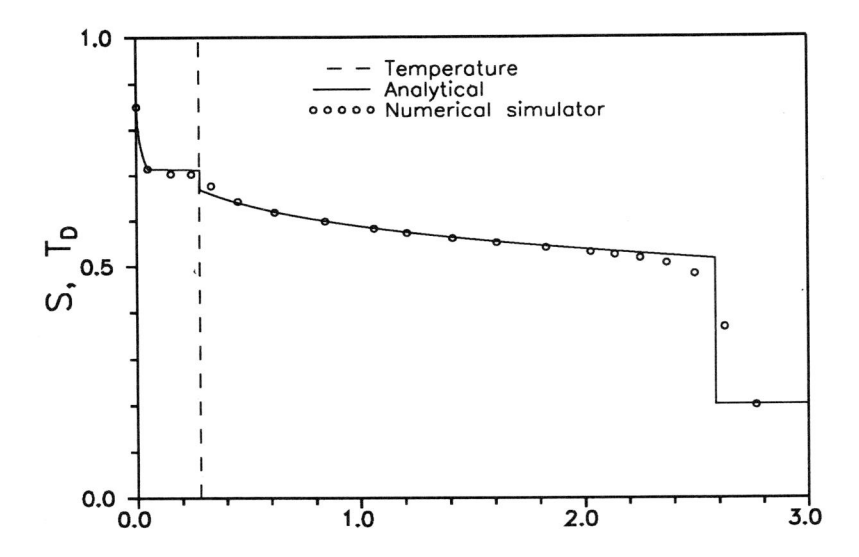

Figure 5: Comparison of analytical and numerical solutions for the saturation profile.

dispersion and other inaccuracies caused by the finite grid may also lead to discrepancies between the solutions.

The most critical assumption used in deriving the solutions presented earlier, is that of a nonisothermal Buckley-Leverett type saturation profile. In using the Buckley-Leverett theory we assume that the saturation distribution is independent of the pressure distribution by ignoring the compressibilities of both the injected and displaced fluids.

In Fig. 5 we show the water saturation distribution calculated from the nonisothermal Buckley-Leverett model. Saturation and temperature is plotted versus the similarity variable $\zeta = \pi \phi h r^2/qt$. Superimposed on the saturation profile is the saturation distribution obtained from the numerical simulator which includes the effects of the fluid compressibilities. Note that by plotting the water saturation versus ζ, the profile will be the same for any time and radius. As seen from the figure, the saturation distribution obtained numerically is very close to the analytical solution for the chosen fluid properties which are considered to be typical for water injection tests.

Once the saturation profile is given, assumed independent of pressure, the similarity solution Eqs. 5.3 and 5.4 is exact for a line source reservoir.

Fig. 6 presents injection data for two different data sets in an infinite system. The uppermost curve corresponds to the saturation profile presented in Fig. 5. The two cases shown differ in the values of the viscosities at injection

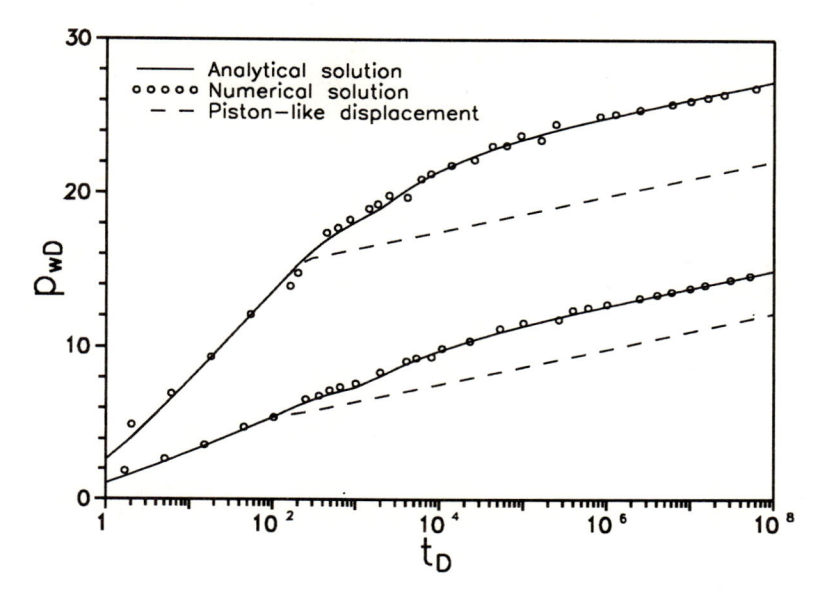

Figure 6: Comparison of analytical and numerical solutions for the injection period.

and reservoir temperature and hence in the mobility ratios. As was the case for the saturation profile, the numerical results compare very well with the analytical solution for both cases.

Also presented in Fig. 6 is the wellbore pressure solution for a moving boundary problem where the saturation gradient is ignored. It should be clear from the figure that neglecting the variable saturation profile has a major effect on the wellbore pressure behaviour.

Fig. 7 **presents** falloff results for the two injection cases shown in Fig. 6, with injection time $t_i = 100$ days. Again, the analytical solutions derived here compare well with the numerically obtained results as opposed to the piston-like displacement solution.

In addition to the examples presented here, numerous other cases were investigated. In all cases where typical water injection data were used, the agreement between the numerical and analytical solutions was excellent. Hence the validity of the mathematical model is established and the use of the analytical solutions for well test interpretation is justified.

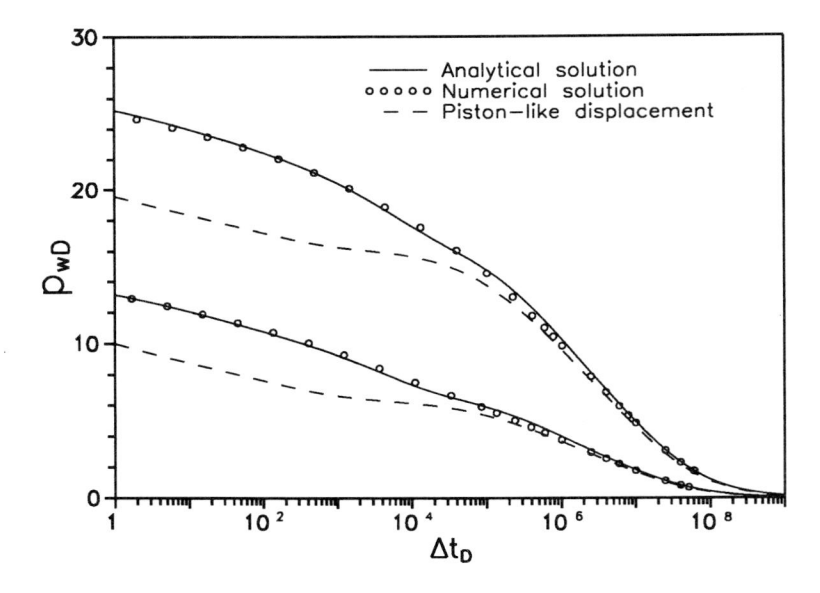

Figure 7: Comparison of analytical and numerical solutions for the falloff period.

10 CONCLUDING REMARKS

We have attempted in this work to present coherent mathematical models and solutions for the testing of nonisothermal injection wells. Though the results for falloff tests are probably the more significant in terms of practical applicability, the foundation of our approach, for both injection and falloff, is seen to be the injection response. This statement is true not only for the theory presented here but for the general theory of testing for liquid systems.

Based on this study, the following summary is warranted:

1. Analytical solutions describing the pressure, saturation and temperature distribution during nonisothermal water injection in an infinite reservoir have been developed. Approximate solutions for finite reservoirs have been derived. By comparing with the numerical simulations, these solutions have been verified to give good results for typical water injection data.

2. The quasi-stationary method is an approximate method which was found to produce accurate results for typical water injection problems in infinite reservoirs. The method gives reasonable results for finite reservoirs during the period following the end of the infinite-acting period [1].

3. Analytical falloff solutions have been derived both by solving the complete initial-boundary value problem and by superposing the solutions to the linear stationary variable mobility and diffusivity problem. Adding the nonlinear similarity solutions with different arguments does not reflect the mathematical definition of superposition and will not give the correct solution to the falloff problem.

4. The solution technique presented is applicable to a wide range of injection-falloff problems. In this paper we have considered only cylindrical geometries. *Bratvold* and *Larsen* present injection-falloff solutions for linear geometries [3] and propagating fractures [4].

References

[1] R. B. Bratvold. *An Analytical Study of Reservoir Pressure Response Following Cold Water Injection.* PhD thesis, Stanford U., Stanford, CA, March 1989.

[2] R. B. Bratvold and R. N. Horne. Analysis of Pressure Falloff Tests Following Cold Water Injection. SPE 18111, presented at the 63rd Annual Meeting of the Society of Petroleum Engineers held in, Houston. Oct., 1988.

[3] R. B. Bratvold and L. Larsen. Effects of Linear Boundaries on Pressure-Transient Injection and Falloff Data. SPE 19830, presented at the 64rd Annual Meeting of the Society of Petroleum Engineers held in, San Antonio. Oct., 1989.

[4] R. B. Bratvold and L. Larsen. Effects of Propagating Fractures on Pressure-Transient Injection and Falloff Data. SPE 20580, presented at the 65th Annual Meeting of the Society of Petroleum Engineers held in, New Orleans. Sept., 1990.

[5] F. J. Fayers. Some Theoretical Results Concerning the Displacement of a Viscous Oil by a Hot Fluid in a Porous Medium. *J. Fluid Mech.*, 13:65–76, 1962. Part 1.

[6] B. L. Keyfitz and H. C. Kranzer. *A System of Non-Strictly Hyperbolic Conservation Laws Arising in Elasticity Theory.* Volume 72, Springer Verlag, New York, 1980. Archive for Rational Mechanics and Analysis.

[7] M. Karakas, S. Saneie, and Y. Yortsos. Displacement of a Viscous Oil by the Combined Injection of Hot Water and Chemical Additive. *SPE Reservoir Engineering*, 391–402, July 1986.

[8] M. Hovdan. *Water Injection - Incompressible Analytical Solution with Temperature Effects.* Technical Report MH - 1/86, Statoil, RESTEK, PETEK, Statoil, Postboks 300, 4001 Stavanger, Norway, Sept. 1986. (In Norwegian).

[9] S. E. Buckley and M. C. Leverett. Mechanism of Fluid Displacement in Sands. *Transactions of the AIME*, 146–149, 1942.

[10] C. S. Matthews and D. G. Russell. *Pressure Buildup and Flow Tests in Wells*. Volume 1, SPE, 1967. Monograph Series.

[11] C. V. Theis. The Relationship Between the Lowering of Piezometric Surface and Rate and Duration of Discharge of Wells Using Ground-Water Storage. *Transactions of the AGU*, II:519, 1935.

[12] N. N. Verigin. On the Pressurized Forcing of Binder Solutions into Rocks in Order to Increase the Strength and Imperviousness to Water of the Foundations of Hydrotechnical Installations. *Akademija Nauk SSR Izvestija Odt. Tehn. Nauk*, 5:674–687, 1952. (In Russian).

[13] T. Barkve. *A Study of the Verigin Problem with Application to Analysis of Water Injection Wells*. The University of Bergen, Bergen, Norway, March 1985. Dr.scient. thesis.

[14] H. J. Ramey Jr. Approximate Solutions for Unsteady State Liquid Flow in Composite Reservoir. *Journal of Canadian Petroleum Technology*, 32–37, Jan.-March 1970.

[15] L. S. Leibenzon, editor. *Handbook on Petroleum Mechanics*. GNTI, Moscow, 1931.

[16] L. I. Rubinshtein. *The Stefan Problem*. Amer. Math. Soc., 1971. Translation of Mathematical Monograph Vol. 27.

[17] H. Stehfest. Numerical Inversion of Laplace Transforms. *Communications of the ACM*, 13, Jan. 1970. No. 1, Algorithm 368.

[18] E. Nyhus. *Modelling of Thermal Injection and Falloff Tests*. Technical Report, Rogaland Research Institute, Postboks 2503 Ullandhaug, 4004 Stavanger, Norway, Oct. 1987. (In Norwegian).

A Nomenclature

A	area, πr^2
C	specific heat capacity
C_D	dimensionless wellbore storage coefficient based on the injected water properties
c	compressibility
c_t	total compressibility, $c_t = c_t(S)$
c_{to}	total compressibility of oil region
$E_1(x)$	exponential integral, $\int_x^\infty e^{-u}/u \, du$
f	fractional flow of water
f'	df/dS
f''	d^2f/dS^2
f_S	df/dS
g	characteristic velocity (Riemann invariant), $(f+\alpha)/(S+\beta)$
h	formation thickness
$I_0(x)$	modified Bessel function of first kind of order zero
$I_1(x)$	modified Bessel function of first kind of order one
K_0	modified Bessel function of second kind of order zero
K_1	modified Bessel function of second kind of order one
k	permeability
M	mobility ratio, $\hat{\lambda}_w/\hat{\lambda}_o$
m_i	mobility ratio, $\bar{\lambda}_i/\bar{\lambda}_{i+1}$
p_D	dimensionless pressure, $2\pi\hat{\lambda}_w(p-p_{ir})/q$
p_{ir}	initial reservoir pressure
$\hat{p}$	pressure solution to stationary variable mobility and diffusivity problem
q	injection rate
R	stationary phase boundary
r_e	exterior reservoir boundary radius
r_w	wellbore radius
S	water saturation
S_{or}	residual oil saturation
S_{wi}	initial water saturation
s	skin factor
T_D	dimensionless temperature, $(T-T_i)/(T_{ir}-T_i)$
T_i	injection temperature
T_{ir}	initial reservoir temperature
t_D	dimensionless time, $\hat{\lambda}_o t/\phi c_{to} r_w^2$
t_{eD}	time at which outer boundary is "felt" in wellbore pressure
t_i	total injection time
u	Darcy (superficial) velocity
y_{BL}	Boltzmann variable evaluated at r_{BL}, $r_{D_{BL}}^2/4t_D$

γ Euler number, $0.57722\ldots$
ζ similarity transform, $\pi r^2 h\phi/qt$
$\hat{\eta}$ endpoint diffusivity ratio, Mc_{to}/c_{tw}
$\bar{\eta}$ average diffusivity ratio, $\eta(\bar{S})$
$\hat{\lambda}_o$ endpoint oil mobility, $kk_{ro}(S_{wi})/\mu_{oh}$
$\hat{\lambda}_w$ endpoint water mobility, $kk_{rw}(1-S_{or})/\mu_{wc}$
$\bar{\lambda}$ average mobility, $\lambda(\bar{S})$
μ viscosity
ρ density
ϱ approximate phase front position
ϕ porosity
ψ pressure distribution at shutin time

Subscripts

a apparent
BL Buckley-Leverett
D dimensionless
e external
o oil
r relative
s reservoir rock
t total
V volumetric
w water
1 invaded region
2 uninvaded region

B Multizone Problem

In this appendix we present the governing equations and auxiliary conditions for the multizone problem. The reservoir consists of $n+1$ regions. The invaded region is discretized into n regions, all with a constant saturation $\bar{S}_i$ which is the volume average saturation within region i. Note that for the falloff problem $t_D = \Delta t_D$ where $\Delta t_D = t_D - t_{D_i}$ and t_{D_i} is the injection time. The governing equations and boundary conditions in dimensionless form for the (n+1)-region composite reservoir are

Governing equations:

$$\frac{1}{r_D}\frac{\partial}{\partial r_D}(\lambda_i r_D \frac{\partial p_{D_{1i}}}{\partial r_D}) = \frac{1}{\eta_i}\frac{\partial p_{D_{1i}}}{\partial t_D}, \quad a_{i-1} < r_D < a_i, \quad i=1,\ldots,n \quad (B.3)$$

MOVING BOUNDARY PROBLEM

$$\frac{1}{r_D}\frac{\partial}{\partial r_D}\left(r_D\frac{\partial p_{D_2}}{\partial r_D}\right) = \frac{\partial p_{D_2}}{\partial t_D}, \quad a_n < r_D < r_{eD} \ (or \ \infty) \tag{B.4}$$

Initial conditions:

$$p_{D_{1i}} = \begin{cases} 0, & Injection \\ \psi_1(r_D), & Falloff \end{cases} \quad i = 1,\ldots,n, \quad t_D = 0 \tag{B.5}$$

$$p_{D_2} = \begin{cases} 0, & Injection \\ \psi_2(r_D), & Falloff \end{cases} \quad t_D = 0 \tag{B.6}$$

where $\psi_i(r_D)$ is the injection solution with the appropriate outer boundary condition evaluated at $t_D = t_{D_i}$.

Inner boundary conditions:

$$\frac{C_D}{M}\frac{dp_{wD}}{dt_D} - \left[r_D\frac{\partial p_{D1}}{\partial r_D}\right]_{r_D=1} = \begin{cases} 1, & Injection \\ 0, & Falloff \end{cases} \tag{B.7}$$

$$p_{wD} = \left[p_{D1} - sr_D\frac{\partial p_{D1}}{\partial r_D}\right]_{r_D=1} \tag{B.8}$$

Outer boundary conditions:

$$Infinite \quad \lim_{r_D \to \infty} p_{D_2} = 0 \tag{B.9}$$

$$Closed \quad \left.\frac{\partial p_{D2}}{\partial r_D}\right|_{r_{eD}} = 0 \tag{B.10}$$

$$Constant \ pressure \quad p_{D2}|_{r_{eD}} = 0 \tag{B.11}$$

Interface conditions:

$$p_{D_{1i}} = p_{D_{1(i+1)}}, \quad r_D = a_i, \ i = 1,\ldots,n-1 \tag{B.12}$$

$$m_i\frac{\partial p_{D_{1i}}}{\partial r_D} = \frac{\partial p_{D_{1(i+1)}}}{\partial r_D}, \quad r_D = a_i, \ i = 1,\ldots,n-1 \tag{B.13}$$

$$p_{D_{1n}} = p_{D_2}, \quad r_D = a_n \tag{B.14}$$

$$\bar{\lambda}_n M\frac{\partial p_{D_{1n}}}{\partial r_D} = \frac{\partial p_{D_2}}{\partial r_D}, \quad r_D = a_n \tag{B.15}$$

where

$$m_i = \frac{\bar{\lambda}_i}{\bar{\lambda}_{i+1}} \tag{B.16}$$

STATISTICAL AND EXPERIMENTAL CALCULATION
OF EFFECTIVE PERMEABILITIES
IN PRESENCE OF OBLIQUE SHALES

O.B. Abu-elbashar, T.S. Daltaban,
C.G. Wall and J.S. Archer
(Department of M.R.E., Royal School of Mines, U.K.)

ABSTRACT

In this paper, a mathematical model for calculation of effective vertical and horizontal permeabilities of formations containing discontinuous shales is presented. It accounts for the phenomena of shales convergence and sand compartmentalisation, therefore it addresses these aspects of geology more closely than previous work. The distribution of flow barriers is estimated probabilistically incorporating analogue outcrop data for fluvial deposits.

Numerical validation with a three-dimensional model was performed using a super computer. A method of experimental validation was devised using a three-dimensional electrical resistivity model. Both numerical and experimental results agreed well with predictions from the mathematical model.

A 3-D experimental rig is also used in testing the statistical streamtube method and since it has drawbacks these are also discussed in our paper.

1. INTRODUCTION

Characterisation of absolute permeability is one of the important and challenging tasks of reservoir description prior to full scale reservoir simulation.

Absolute permeability 'averaged' over a Representative Elementary Volume [REV] results in an 'effective' permeability. The appropriate REV for field scale simulators is a grid block, and an objective is to determine this effective permeability over such a volume. This may include discontinuous sand bodies, shales and even more than one depositional unit. Simple models which appeared in the recent oil literature [9] [14] [7] [8] [1] [2], can be misleading, and result in unrealistic permeability predictions since they often lack detailed geological considerations and neglect sand compartmentalisation and shale convergence.

The main objective of this paper is to introduce methods of predicting the effective vertical and horizontal components of absolute permeability in the presence of geological heterogeneities over a REV representing the grid block sizes. Although oil reservoirs are intrinsically deterministic, since it is not possible at present to describe all the geological heterogeneities with known data acquisition techniques, it is highly tempting to resort to statistical means. This also forms one of the aspects of the methods presented in this paper. However, in addition to the identification of the most likely distributions, any such technique should also be conditioned with the realistic description of the geological features involved, and to this end, vertical, horizontal and oblique orientation of flow barriers (shales), their convergence, and the sand body compartmentalisation are included in the REV for predicting the effective permeabilities.

The theoretical techniques of predicting the effective permeabilities presented in this paper cover detailed single phase simulation procedures and analytical/statistical methods. Also included in the paper is an experimental technique by which the flow characteristics in the presence of 'U' shaped sheets, flat sheets, and closed parallelograms consistent with the analogue outcrop data for fluvial depositional environments are investigated. The performance of the numerical and analytical procedures is compared. The statistical streamtube method is also modelled by an experiment using only flat sheets and the results show that this technique is highly conservative.

Finally, the method presented by Haldorsen [5] is used to compare horizontal permeability predictions with those provided by the method of this paper. The results show that the former technique appears to provide very optimistic effective permeabilities especially in the presence of high shale frequencies.

2. SHALE GEOMETRY AND CONTINUITY IN FLUVIAL DEPOSITS

A conventional practice adopted by many researchers [2] [5] [13] [3] is to generate a random horizontal shale distribution in a given sand body. The objective is to honour given statistical data such as observed shale frequency and to represent this synthetic heterogeneous porous medium as the *in situ* rock. In reality, shales can be oblique and the adjacent shales may intercept resulting in 'Y' shaped geometries. They may also intercept at both sides trapping sand between them to form a discontinuous sand unit (DCSU) as shown in Figure 1. This phenomenon is called shale convergence and sand compartmentalisation, and it is well documented in the literature [3] [10] [11] [12]. Zeito [15], for example, noted the importance of shale convergence especially in fluvial environments. He reported that in fluvial environments about 62-70% of the shale breaks converged within less than 250 ft. The areal fraction of the cross-sections occupied by the DCSUs to the observed shale frequency in wells is shown in Figure 2.

Weber [13] combined the limited available statistical data on shale continuity into a single relationship relating the probability of shale breaks extending over a certain length to the type of depositional environment. Later, Martin and Cooper [10] extended Weber's work by including Continuity Distribution Functions (CDF) for two new environments, namely, Distal Glaciodeltaic and Proximal Glaciodeltaic (see Figure 3). These two latter reports form the present state of public domain data for shale quantification in relevant depositional environments.

3. METHODS FOR PREDICTION OF EFFECTIVE PERMEABILITIES

In this section both analytical/statistical and numerical methods for predicting effective vertical and horizontal permeability will be presented. An experimental technique is also included as a means to validate the theoretical results.

3.1 Analytical/Statistical Methods

[A] Effective Vertical Permeability: For the purpose of this study, Figure 1 is used as a basis where the lengths of DCSUs are considered to be the same as shale lengths but their thicknesses vary between 3.6 ft and 10 ft [15]. The following assumptions are also introduced:-

(i) Assumptions:

* shale data from wells and outcrops are statistically representative of the reservoir under consideration,
* shale permeability is zero,
* steadystate incompressible single phase flow,
* no-flow boundaries at the walls of the model, and
* gravity and capillary effects are negligible.

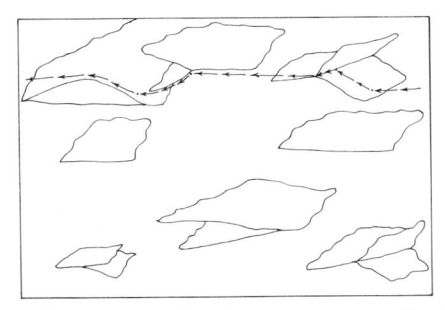

Fig.1 A horizontal streamtube in presence of oblique shales

(ii) Brief Description of Various Steps of Calculation: In earlier published work (1,2) the effective vertical permeability is given by,

$$K_{ve} = \frac{(1-F_s)H^2 K_e}{N_s} \sum_{i=1}^{N_s} \frac{1}{S_i^2} \qquad [1]$$

where,
- F_s = Shale fraction met by the well.
- H = Thickness of the block.
- K_e = The harmonic mean of the streamtube permeabilities
- S_i = The length of the steamtube.
- N_s = An arbitrary number of streamtubes.

The above equation ignores the shale convergence and sand compartmentalisation phenomena. Therefore, to account for such phenomena, the following modifications are introduced:

1- Consideration of sand compartmentalisation: Using the observed shale frequency, enter Figure 2 to find the fraction of the cross-section occupied by DCSUs (F_{dc}). Since this fraction is not available for fluid flow, it is deducted from the term between brackets in equation 1. Furthermore if we use an averaged streamtube length, S_e, the equation will become,

$$K_{ve} = \frac{(1-F_s-F_{dc})H^2 K_e}{S_e^2} \qquad [2]$$

2-Calculation of the average length of the streamtubes: To account for the shale convergence phenomenon – which affects the number of barriers met by a vertical streamtube – the calculation of streamtube lengths is performed as shown below.

EFFECTIVE PERMEABILITIES

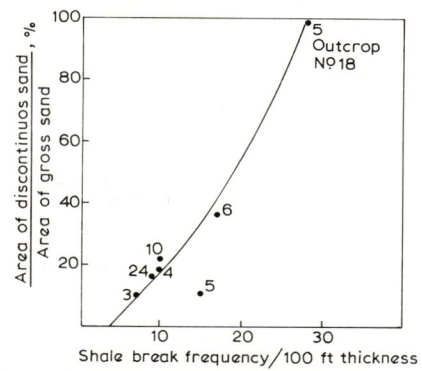

Fig.2 Relationship between shale frequency and magnitude of discontinuities in fluvial deposits (after Ref.15)

If N_{dcsu} is the number of discontinuous sand units and 'f' is the shale frequency per foot, the number of barriers encountered by the streamtube (N_B) is given by:

$N_B = N_{dcsu}$ + No of 'Y' shaped barriers + No of isolated shales

$$N_B = N_{dcsu} + (f*H*0.66 - 2*N_{dcsu})/2 + f*H*0.34 \quad [3]$$

$$N_B = 0.67*f*H$$

where the number 0.67 [15] represents the fraction of shales that converge. If the sand's horizontal permeability is constant, then the length of the streamtube is given by

$$S_e = H + \sum_{i=1}^{N_B} \frac{1}{2} \min(R_{1i}W_{shi}, R_{2i}L_{shi}) \quad [4]$$

where W_{shi} and L_{shi} are the width and length of shale in ft chosen at random from a given CDFs, and R_{1i} and R_{2i} are random numbers between 0 and 1. In equation 4 the streamtube travels horizontally a random fraction of half the barrier's length or width (not the whole length or width as suggested by Begg and King (1,2)) since it follows the direction of the least resistance. K_e is given by,

$$\frac{S_e}{k_e} = \frac{S_v}{k_v} + \frac{S_x}{k_x} + \frac{S_y}{k_y} \quad [5]$$

where S_v is the summation of vertical distances, S_x and S_y are the summation of horizontal distances travelled in x and y directions, and k_x, k_y and k_z are uniform directional permeabilities.

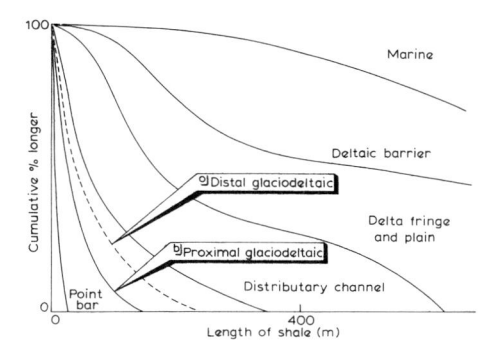

Fig.3 Continuity of shale breaks as a function of depositional environment (after Ref. 10, 13)

In order to apply streamtube theory to anisotropic porous media, the following coordinate transformation is necessary as it reduces the flow equation into Laplace form:-

$$\bar{x} = x, \quad \bar{y} = y\left(\frac{k_x}{k_y}\right)^{0.5} \text{ and } \bar{z} = z\left(\frac{k_x}{k_z}\right)^{0.5}$$

where x, y and z are the transformation variables. Therefore, Equation [4] will take the following form:-

$$S_e = \bar{H} + \sum_{i=1}^{N_B} \frac{1}{2} \min(R_{1i}\bar{W}_{shi}, R_{2i}L_{shi}) \qquad [6]$$

where $\bar{H} = H\dfrac{\bar{z}}{z}$ and $\bar{W} = W\dfrac{\bar{y}}{y}$

[B] Effective Horizontal Permeability: The effective horizontal permeability k_{eh} is calculated by a modified form of Equation [2] as follows:-

$$K_{ve} = \frac{(1-F_s-F_{dc})L^2 K_{eh}}{S_{eh}^2} \qquad [7]$$

Where, S_{eh} is the length of the horizontal streamtube, K_{eh} is the effective permeability of the streamtube. The travel distance of the horizontal streamtube may be increased due to transverse movements caused by 'Y' shaped barriers and DCSUs. Therefore, the length of horizontal streamtube is given by

S_h = L+(increase of distance caused by the barriers Y_c and Y_o) [8]

where Y_c is the number of 'Y' shaped barriers having the 'Y' tail against the flow direction, plus the number of DCSUs, N_{dm}, met by the streamtube, and Y_o is the number of 'Y' shaped barriers - having the 'Y' tail to the direction of flow - met the by the streamtube.

To find the number of different types of flow barriers, if the total number of shales generated in a cross-section is N_{sh}, and the number of converging shales is N_{shc}, then the number of 'Y' shaped barriers is given by

$$N_y = \frac{N_{shc} - 2N_{dcsu}}{2} \qquad [9]$$

At this stage, it is reasonable to assume that half of N_y may have the tail in the direction of flow and the rest may have the tail against it. This then implies a calculation of arithmetic mean thickness of DCSU, that is S_{th}, followed by the division of cross-section into horizontal streaks according to this mean thickness. The number of barriers encountered by a horizontal streamtube is regarded as the number of barriers contained in one horizontal streak, therefore

$$Y_o = N_y * \frac{S_{th}}{H} \text{ and } Y_c = Y_o + N_{dm}$$

The streamtube travels a random distance of the length 'C' as shown in Figure 4 for all Y_o and Y_c. Therefore, the total number of distances is $Y_{ct} = Y_o + Y_c$. From Figure 4, the increase in travel distance caused by each barrier of 'Y_{ct}' is given by,

$$\text{Increase} = R_i\left(\frac{S_{thi}}{2\text{Tan}\theta} - \frac{S_{thi}}{2\text{Sin}\theta}\right) \quad [10]$$

where R_i is a random number. Therefore, the length of the streamtube is calculated by

$$S_e = L + \sum_{i=1}^{Y_o} \frac{1}{2} R_i W_{shi} + \sum_{i=1}^{Y_{ct}} R_i\left(\frac{S_{thi}}{2\text{Tan}\theta} - \frac{S_{thi}}{2\text{Sin}\theta}\right) \quad [11]$$

Usually the third term of Equation [11] is relatively small, therefore we can use an average value for S_{thi} and multiply by 0.5 instead of a random number.

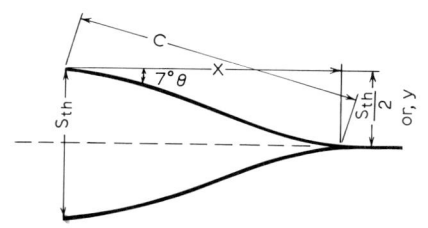

Fig.4 Increase of the horizontal travel distance caused by the DCSUs

If the cross-section is normal to the centre line of the channel, K_e is calculated by,

$$\frac{S_e}{K_e} = \frac{L}{K_x} + \frac{(S_e - L)}{K_y} \quad [12]$$

where K_y is the sand permeability in the direction of the centre line of the channel. Swap the positions of Kx and Ky in Equation [12] if the cross-section is parallel to the centre line.

3.2 Numerical Method

The detailed discretisation of heterogeneous reservoirs may require an excessive number of grid blocks, and hence computer capacity and time. Analytical approximations are then a valuable means of obtaining solutions, or of reducing the grid block requirements in a realistic manner. The analytical techniques, and the degree of approximation involved can be validated by relatively simple numerical studies. This is the case in this project. The simulation method used in this paper is similar to that reported by Begg and King [1] with the exception that the number of dimensions has been increased to 3 and the application of the method is extended to cover domains including shale convergence and sand compartmentalisation. A block composed of 6400 (20* 16*20) fine grid blocks is used for the calculations. Each fine grid is 50 ft long, 50 ft wide, and 2.5 ft thick. Figure 5 demonstrates a 3-D realisation of a model (20*20*20) fine grids.

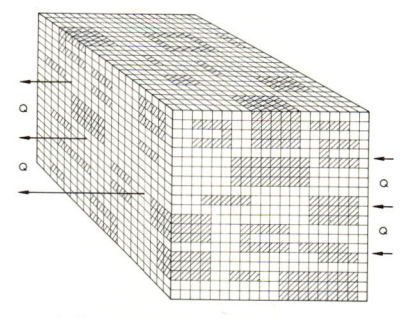

Fig 5 Simulation model considering shale convergence (set-up for calculation of horizontal permeability)

Constant pressure boundaries are assumed at the inlet and outlet faces of the model and no flow boundaries are assumed for other exterior faces of the domain. Constant upstream and downstream pressures (P_1, P_2 = 0) are maintained by adding top and bottom layers with constant pressures.

Figure 5 demonstrates the various types of flow barriers used in the simulation model. For shale convergence representation, a 'U' shaped discretisation is appropriate, and for DCSUs, parallelograms are used. Subsequent to pressure calculations, an expression for the flow rate throughout the model is obtained. The flow rate through an equivalent homogeneous medium can be calculated by Darcy's law. Equating the two expressions can give the effective vertical permeability, Kve.

$$k_{ve} = \frac{H+\Delta Z}{LW(P_2 - P_1)} \sum_{i=1}^{N_x} \sum_{j=1}^{N_y} k_{i'j'nz} (P_2 - P_{i'j'nz}) \quad [13]$$

Where

$k_{i'j'nz}$ = grid block permeability,
N_x, N_y, N_z = number of blocks in x,y,z direction,
ΔZ = grid block thickness, and
W = Width of the model.
$P_{i'j'nz}$ = grid block pressure in layer NZ.

The effective horizontal permeability is calculated similarly by effectively turning the model 90 degrees in the vertical direction.

3.3 Experimental Technique

In the literature numerical techniques dominate the prediction of the effects of flow barriers on permeability. The only public domain experimental report is published by Dupuy and Le Febure [4]. This does not include shale convergence and sand compartmentalisations.

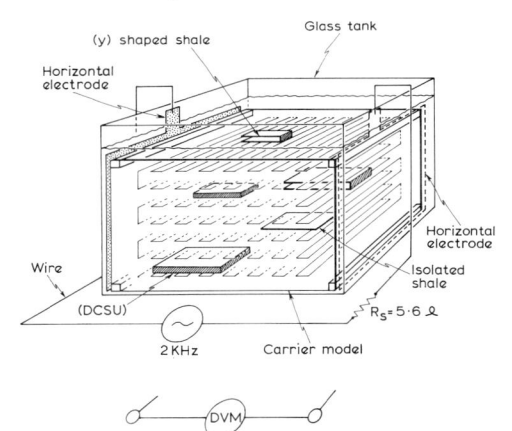

Fig.6 Diagram of the experimental set-up (Display for measuring horizontal permeability)

In order to gain insight into the flow processes in the presence of complex flow barriers, an electrolytic experimental simulation study has been undertaken as follows:-

* Sand is simulated by brine with fixed conductivity and shale is simulated by sheets of perspex having different sizes analogous to the (CDF) of shales (see Figure 3).

* The perspex sheets are suspended in brine by means of a carrier model which is composed of two parallel vertical

sheets of perspex held apart by means of perspex bars. The sheets are screened and a thin plastic wire is netted between them through their rows of holes to serve for suspension of the sheets representing the barriers. Figure 6 shows the schematic representation. An alternating current (AC) is generated in the circuit and the vertical and horizontal conductivities are measured by a digital voltmeter.

* Comparison is carried out between readings with and without perspex sheets to find the reduction in conductivity caused by the sheets. The results are scaled up to determine the analogous reduction in sand permeability due to stochastic shales.

4. RESULTS AND DISCUSSIONS

4.1 Performance of Analytical Methods

[A] Effective Vertical Permeability Predictions: Table 1 shows the estimated vertical permeabilities for different shale frequencies using Equation 2. Also shown in the same table are the predictions of the statistical streamtube method of Refs [1] [2]. There is a clear discrepancy between the predictions of the analytical method presented in this paper and the statistical streamtube method. As can be seen from the table, although both methods show a decreasing permeability trend with increasing shale frequency, for a shale frequency of 12/100 ft the analytical method yields vertical permeability of 21.8 md whereas the streamtube method gives 6. There are two reasons for this discrepancy; firstly, this latter method ignores the phenomenon of shale convergence; secondly, it assumes that the streamtube travels horizontally a random fraction of the barrier's length or width, although in reality it travels a fraction of half the length or width.

[B] Effective Horizontal Permeability Predictions: Table 1 also shows the results of effective horizontal permeability using different shale frequencies (column 4). At low shale frequencies (2, 4 shales/100 ft), the rate of effective horizontal permeability decrease is very low because the only reduction factor is the transverse movement of streamtubes due to the 'Y' shaped barriers. At higher shale frequencies, the DCSUs begin to appear resulting in further reduction of horizontal permeability.

[C] Effects of Random Number Seeds: The numbers generated by the computers are pseudo random numbers and depend on the seed used to generate them. Table 2 shows the effects of various random number seeds on the permeability predictions.

There is an appreciable seed dependency of the predictions (minimum being 27.30 and maximum being 67.16 with an average value of 47.15). To overcome this apparent non-uniqueness in the permeability values, an arbitrarily large number of seeds is used in the computer code developed for the calculation methods to generate large numbers of streamtube lengths (150) and then their arithmetic average is taken. By this way, the variation of the results have become very small.

TABLE 1

$K_{hx} = K_{hy} = 300$ md

Shales 100 ft	Sand Fr. Occupied by DCSU	Kve (Analy Meth.)	K with Conv.	Kve Begg & King	K Ref [7]
2	0	93.99	297	45.48	299.1
4	0	59.71	294	24.87	298.5
6	0.075	51.73	286.33	15.44	291
8	0.115	36.18	253.22	10.61	286.8
10	0.18	24.012	230.66	7.62	.
12	0.23	21.78	212.61	5.93	.

Calculated Values of Effective Vertical Permeability Considering Shale Convergence and Corresponding Values Using statistical Streamtube Method [1] [2].

4.2 Results of Numerical Simulation Method

Two models for different shale frequencies were used; one for 8 shales/100 ft and the other for 12 shales/100 ft, a representative grid is shown in Figure 5. The results of this study for vertical and horizontal permeabilities are shown in Table 3.

[A] Effective Vertical Permeability Predictions: The two results shown in Table 3 conform with reasonable accuracy to the analytical and experimental results. Figure 8 is a plot of the analytical, experimental, and numerical vertical permeability predictions, which show similar results. This similarity validates our analytical technique.

[B] Effective Horizontal Permeability: Column 5 of Table 3 shows the results of effective horizontal permeability predictions and Figure 8 shows that the experimental, analytical, and numerical results give reasonable agreement confirming that the analytical technique is adequate.

4.3 Discussion of Experimental Results

Two different brines (different in salt concentrations) were used in this study. Also considered in the experimentation are the variations in shale concentrations. The results are shown in Table 4 and Figure 7. The observations from the results displayed are as follows:

[A] General Observations

. At low shale frequencies (0-6 shales/100 ft) a small amount of shale results in considerable vertical permeability reduction. Further increase in shale frequency leads to creations of DCSUs which reduce the conductivity to flow. This phenomenon (compartmentalisation) may be offset - to some degree - by decrease in tortuosity of the streamtubes. They appear to bypass more than one barrier as they travel horizontally. The net result is then the stabilisation of the rate of decrease of vertical permeability with increasing shale frequency (> 6 shales/100 ft).

TABLE 2

Random Number Seed	Effective Vertical K
1234	52.84
2234	29.46
3234	53.09
4234	44.02
5234	56.53
6234	27.3
7234	67.26
Arithmetic Average	47.15

Different Results for the Same Shale Frequency (4 shales/100 ft) Using Different Random Number Seeds

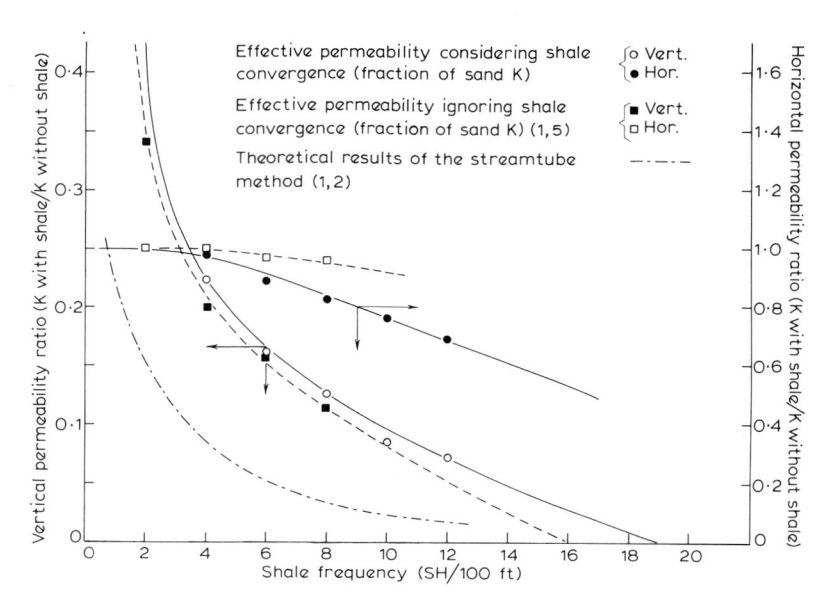

Fig.7 Permeability ratios versus shale frequency

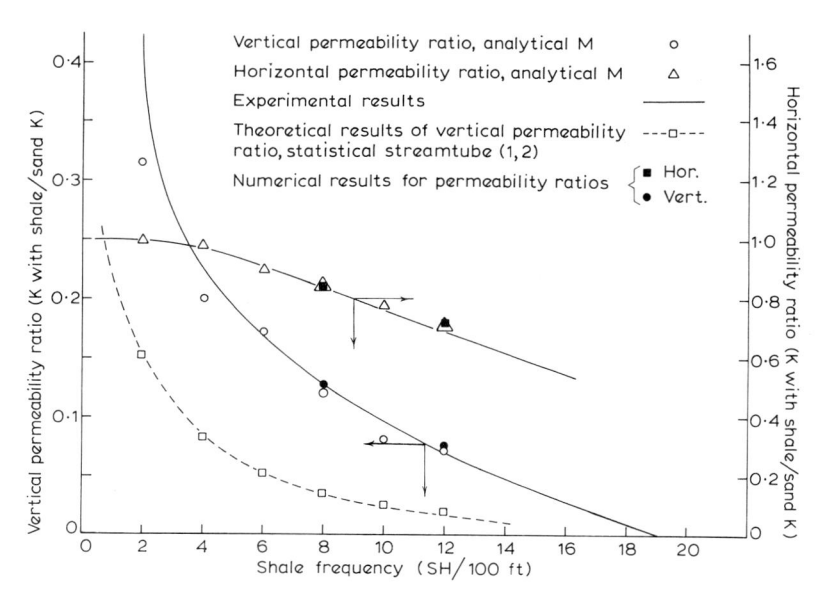

Fig.8 Permeability ratios for Analytical, Numerical, and Experimental approaches; and the streamtube method [1,2]

TABLE 3

Model No.	Shale Frequency/ft.	Fraction of DCSUs	Kve	Khe
A	0.08	0.115	38.16	251.4
B	0.12	0.23	22.27	218.03

Results of Kve for all Models Using the Numerical Simulation Method (with Shale Convergence).

TABLE 4

Shales/ 100 ft.	Av. Kve/ K(sand)	Av. Khe/ K(sand)	Av. Anisotropy
12	0.00714	0.69	0.1022
10	0.884	0.76	0.1113
8	0.1252	0.823	0.152
6	0.1595	0.887	0.176
4	0.223	0.977	0.228
2	0.462	0.989	0.468

Average Vertical and Horizontal Permeability Anisotropy from the Results of the two Concentrations

TABLE 5

Shale Freq/ 100 ft	Vertical Resis.	Pure R/ V.R.	Horiz. Resis.	Pure R/ H.R.	Anisotropy
[1]	[2]	[3]	[4]	[5]	[3]/[5]
8	1192.3	0.1156	134.03	0.956	0.121
6	876.7	0.1576	132.05	0.97	0.162
4	690.58	0.1996	128.63	0.995	0.201
2	404.15	0.341	128.25	0.997	0.342

Calculation of the Anisotropy Factor and the Permeability Ratios Ignoring the Shale Convergence Phenomenon

. Negligible reduction is observed on effective horizontal permeability for shale frequencies ranging between 0-6/100 ft. Beyond this limit, the horizontal permeability decreases steadily (the curve shown in Figure 7 exhibits an almost straight line).

[B] Sensitivity Studies

(i) Sensitivity of the Measurements to the Flow Barrier Distribution: The experiments for each shale frequency were repeated three times using different perspex sheet configurations. The difference between the measurements was found to be negligibly small (less than 1%) which fulfils the random shale distribution assumption. Then, the arithmetic average of three measurements was retained as the reading for that frequency.

(ii) Sensitivity of the Measurements to Different Brine Resistivities: The experiments were run twice using brines with different concentrations (5 and 10 gm/cc) to investigate the effect of brine concentration on the measurements. As expected, the differences were very small.

[C] Comparative Study

In this section the results obtained from experiments considering shale convergence are compared with the analytical methods of Refs [1] [2] [5]. The predictions of the analytical techniques are first validated by comparing them with the experimental findings based on SSTM (Statistical Streamtube Method), i.e. no DCSUs.

(i) Experimental versus Analytical Permeability Predictions of Refs [1] [2] [5]: To examine the validity of SSTM for the effective vertical permeability calculation and Halderson's method [5] for effective horizontal permeability calculation over a three dimensional domain, a set of four experiments were run using different shale frequencies. The shales were represented by plain sheets of perspex (no DCSUs or U shaped barriers). The results of this study are presented in Table 5 and Figure 7 (represented by dashed lines).

Examination of Figure 7 immediately reveals that the vertical permeability is not a linear function of shale frequency which is contrary to what has been reported in the literature [2] [5]. The curve flattens out to give permeability values higher than expected from linear trends. The great discrepancy observed between the analytical and experimental vertical permeability results is due to the following reasons:-

. At high shale frequencies, streamtubes can bypass several flow barriers at one step due to the existence of a long barrier which may shadow several smaller ones.

. SSTM has been validated only for two dimensional cases [2] [5] [6] which may lead to erroneous results, because in reality the flow barriers can also show variation in three dimensions, and the third dimension can provide the shortest path, resulting in higher vertical permeabilities.

. The SSTM assumes that the horizontal distance travelled along each barrier to be random fraction of the whole length or width, whereas a more realistic value is a random fraction of half the length or width, as the streamtube follows the direction of the least resistance.

(ii) Experimental Results (with and without DCSUs): The experimental permeability curves of Figure 7 reveal two distinct parts based on the shale frequency ranges:-

(1) At Low Shale Frequencies (< 6 shales/100 ft): The SSTM considers all the observed shale frequency, whereas the actual effective frequency is much lower (66% [15]). Therefore the vertical permeabilities predicted by this method come out to be lower than the actual. As for horizontal permeabilities, the prediction technique proposed by Haldorsen [5] yields similar results to the one considering shale convergence (see column 6 of table 1) This is due to the small fraction of the horizontal streamtube's length caused by transverse movements owing to the low shale frequency.

(2) At High Shale Frequencies (> 6 shale/100 ft): In this case there is an increasing discrepancy between SSTM and the case with DCSUs. In addition to the reasons stated above, the net effect of shale convergence is also responsible from this discrepancy as it slows down the rate of decrease of vertical permeability resulting in further divergence between the two curves.

For horizontal permeabilities, the increase of shale frequency and the appearance of DCSUs result in an increase in the length of streamtube due to transverse movements, and decrease in the sand area available for fluid flow. Therefore, the two curves diverge and the model ignoring shale convergence gives very optimistic results.

5. CONCLUSIONS

1. Small amounts of shale can considerably reduce the effective vertical permeability of the formations, whereas

their effect is negligible on horizontal permeability.

2. Shale convergence and sand compartmentalisation phenomena must be considered in calculation of effective vertical and horizontal permeabilities.

3. The permeability decrease with increase in shale frequency is linear only at high shale frequencies when the DCSUs begin to appear (> 6 shales/100 ft).

4. The statistical streamtube method is very conservative and can estimate erroneous low permeabilities.

5. Haldorsen's method [5] for calculation of effective horizontal permeability is very optimistic and can estimate erroneous high values due to neglect of the shale convergence.

6. Further efforts are needed to investigate the relation between shale frequency and the resulting sand compartmentalisation. For the time being the main source of data is Ref [15].

ACKNOWLEDGEMENT

This work is sponsored by the Government of Sudan.

REFERENCES

[1] Begg, S.H. et al (1985): 'Modelling the Effects of Shales on Reservoir Performance:', SPE Paper No. 13529, presented at the SPE Reservoir Simulation Symposium, Dallas, Texas, Feb. 10-13.

[2] Begg, S.H. et al (1985): 'A Simple Statistical Method for Calculating the Effective Vertical Permeability of a Reservoir Containing Discontinuous Shales', SPE Paper No. 14271, presented at the 60th Technical Conference and Exhibition of the SPE, Las Vegas, Nevada, Sept. 22-25.

[3] Delhomme, Abd. El. K. and Giannesini, J.F. (1979): 'New Reservoir Description Technics Improve Simulation Results in Hassi-Massaoud Field - Algeria', SPE Paper No. 8435, presented at 54th Annual Fall Technical Conference and Exhibition of the SPE, Las Vegas, Nevada, Sept. 23-26.

[4] Dupuy, M. and Lefebure Du Prey, E. (1968): 'L'anisotropic d'ecolument en Milieu Poreux Presentant des Intercalation Horizontale Discontinues', Communication No. 34 Troisieme Colloque De L'association De Recherch sur les Techniques De Farage et De Production, June 10-14, PAU,

France.

[5] Haldorsen, H.H. (1983): 'Reservoir Characterisation Procedures for Numerical Simulation', University of Texas, PhD Dissertation.

[6] Haldorsen, H.H., Chang, and Begg, S.H. (1987): 'Discontinuous Vertical Permeability Barriers: A challenge to Engineers and Geologists', North Sea Oil and Gas Habitat, Edited by J. Kleppe et al, Published by Graham & Trotman Limited.

[7] Haldorsen, H.H. et al (1987): 'Review of the Stochastic Nature of Reservoirs', Presented at a Seminar entitled The Mathematics of Oil Production, Robinson College, Cambridge, July 6-7.

[8] Haldorsen, H.H. and Lake, L.W. (1984): 'A New Approach to Shale Management in Field-Scale Models', SPEJ August, Volume 14, pp 447-452.

[9] Haldorsen, H.H. and Chang, D.M. (1986): 'Notes on Stochastic Shales; From Outcrop to Simulation Model', appeared in Reservoir Characterisation, Edited by Lake, L.W. and Carroll, H.B. Jr., Academic Press. Inc., Orlando Fl., USA.

[10] Martin, J.H. and Cooper, J.A. (1984): 'An Integral Approach to the Modelling of Permeability Barrier Distribution in a Sedimentologically Complex Reservoir', SPE Paper No. 13051, presented at the 29th Annual Technical Conference and Exhibition of the SPE, Houston, Texas, Sept. 16-19.

[11] Pryor,W.A. (1972): 'Reservoir Inhomogeneities of Some Recent Sand Bodies', SPEJ, June, Vol. 12, pp. 229-245.

[12] Verrien et al (1967): 'Application of Production Geology Methods to Reservoir Characteristics Analysis From Outcrops Observations', Seventh World Petroleum Congress, Volume 2, pp. 425-446.

[13] Weber, K.K. (1982): 'Influence of Common Sedimentary Structures on Fluid Flow in Reservoir Models', J. Pet. Tech. March, pp. 665-672.

[14] Willis, S. (1986):'Effective Vertical Permeability in the Presence of Stochastic Shales: Simulation and Estimation', MSc Thesis, Imperial College, London.

[15] Zeito, G.A. (1965): 'Interbedding of Shale Breaks and Reservoir Heterogeneities', J. Pet. Tech., Oct., 1223-1228.

RESERVOIR SIMULATION USING MIXED METHODS, A MODIFIED METHOD CHARACTERISTICS, AND LOCAL GRID REFINEMENT

M.S. Espedal and O. Sævareid
Department of Mathematics
University of Bergen
Allegt. 55, N5007 Bergen, Norway

R.E. Ewing
University of Wyoming
Enhanced Oil Recovery Institute
P.O. Box 3036
Laramie, WY 82071

Thomas F. Russell
University of Colorado at Denver
P.O. Box 170
Denver, CO 80204

Abstract

The simulation of multiphase and multicomponent fluid flows often requires large coupled systems of nonlinear partial differential equations. These equations are convection-dominated with important local diffusive effects. An operator-splitting technique is used to address these different phenomena. Convection is treated by time stepping along the characteristics of the associated pure convection problem, and diffusion is modelled via Galerkin method for miscible displacement and a Petrov-Galerkin method for immiscible displacement. Accurate approximations of the fluid velocities needed in the modified method of characteristic time-stepping procedure are obtained by mixed finite element methods.

Adaptive local grid refinement methods are then presented to resolve the moving internal boundary layers which arise and often govern the mass transfer between the phases. Local grid refinement techniques are also described for mixed finite element methods around wells and other fixed singularities. Due to the large size of many applications, efficiency in computation is critical. Concepts of domain decomposition for dynamic adaptive grid refinement are presented. Numerical results for both miscible and immiscible displacements will be presented.

1. INTRODUCTION

Mathematical models for reservoir flow are governed by partial differential equations whose solution may exhibit fine structure within small regions of the computational domains. Large-scale simulation of models for enhanced-oil-recovery processes normally involves domains of such great size that uniform gridding on the length scale of these local phenomena would lead to discrete problems that are too large to be solved on even the most powerful computers. A promising solution to this problem lies in local refinement techniques, where points may be added to the global grid in regions of rapid change.

In this paper, we will describe some operator splitting and preconditioned iterative techniques which allow grid refinement in areas where we need to resolve local phenomena (see [3,4,5,8,11]). These techniques are also highly amenable to vectorization and parallelization of the algorithms. The parallel properties and domain-decomposition concepts allow them to be easily incorporated in existing large-scale simulators without disrupting the basic solution process.

For definiteness, we shall consider two-dimensional models for immiscible and miscible flow. Neglecting gravity and compressibility, two-phase immiscible flow in a porous medium can be described by the following set of partial differential equations:

$$\nabla \cdot \mathbf{v} = q_1(\mathbf{x}, t), \tag{1}$$

$$\mathbf{v} = -\mathbf{A}(S, \mathbf{x})\nabla p, \tag{2}$$

$$\phi \frac{\partial}{\partial t} S + \nabla \cdot (f(S)\mathbf{v}) - \epsilon \nabla \cdot (\mathbf{D}(S, \mathbf{x})\nabla S) = q_2(\mathbf{x}, t), \tag{3}$$

where $\mathbf{v}$ is the total Darcy velocity, p the total fluid pressure [6], S denotes the saturation of water, $\mathbf{A}$ is a transmissibility tensor, and ϵ is a parameter scaling the diffusion term $\mathbf{D}$. We restrict ourselves to two space dimensions and assume the absolute permeability tensor to have the form

$$\mathbf{K}(\mathbf{x}) = k_x(\mathbf{x})\mathbf{ii} + k_y(\mathbf{x})\mathbf{jj}. \tag{4}$$

Let λ_i, $i = w, o$, denote the mobility (relative permeability divided by viscosity) of water and oil, respectively. We then have

$$\mathbf{A}(S, \mathbf{x}) = \mathbf{K}(\mathbf{x})(\lambda_w + \lambda_o), \tag{5}$$

$$f(S) = \frac{\lambda_w}{\lambda_w + \lambda_o}, \tag{6}$$

and

$$\mathbf{D}(S, \mathbf{x}) = \mathbf{K}(\mathbf{x}) \frac{\lambda_w \lambda_o}{\lambda_w + \lambda_o} \frac{dp_c}{dS}, \tag{7}$$

where p_c is the capillary pressure between the phases. The mobilities and the capillary pressure are assumed to be known functions of the water saturation.

In order to specify a concrete computational problem, we need boundary and initial conditions. The repeated five-spot well pattern, see for instance [1] or [12], is frequently used as a test case for these equations. In this context, it is natural to impose the boundary conditions

$$\mathbf{v} \cdot \mathbf{n} = 0, \quad \mathbf{x} \in \partial\Omega, \tag{8}$$

$$\nabla S \cdot \mathbf{n} = 0, \quad \mathbf{x} \in \partial\Omega, \tag{9}$$

where Ω is our computational domain with boundary $\partial\Omega$ and $\mathbf{n}$ denotes the unit outward normal vector to $\partial\Omega$. The equations for miscible flow are basically of the same form [1,12], with different properties for f and $\mathbf{D}$.

The solution process for these models may introduce the need for both fixed and dynamic local grid refinement techniques. Near wells, faults, and fractures certain fixed local refinement may be needed. It is well known that equation (3) can produce near shock-like solutions where patches of refinement, moving with the fluid interface, will prove useful. We illustrate both of these classes of local grid refinements by the numerical examples given in the last section.

2. APPROXIMATION STRATEGY

We adopt a sequential strategy to handle the system (1)–(3) by solving equations (1)–(2) implicitly in the first step and equation (3) by an operator-splitting technique in the second step. This procedure defines a global time-step restricted by the accuracy wanted in the treatment of the interaction between pressure-velocity and saturation.

2.1 Pressure – Velocity

The weak formulation of equations (1)–(2) and (8), is given by:

$$A(p, \phi) = (q_1, \phi), \tag{10}$$

where $A(u,v) = \iint_\Omega \mathbf{A}\nabla u \cdot \nabla v \, dx \, dy$ and $(u,v) = \iint_\Omega uv \, dx \, dy$. We model the wells using point sources and sinks represented by delta functions

$$q_1(\mathbf{x},t) = \sum_{\text{wells}} q_n(t)\delta(x - x_n, y - y_n). \tag{11}$$

This approach introduces logarithmic singularities in the pressure solution at the wells. For the diagonal grid of the five-spot pattern, we have used a radial refinement ([5,8]) at the injection well. In the interior, where we may have large variations in the permeability, we are using a regular grid refinement (see Figure 1).

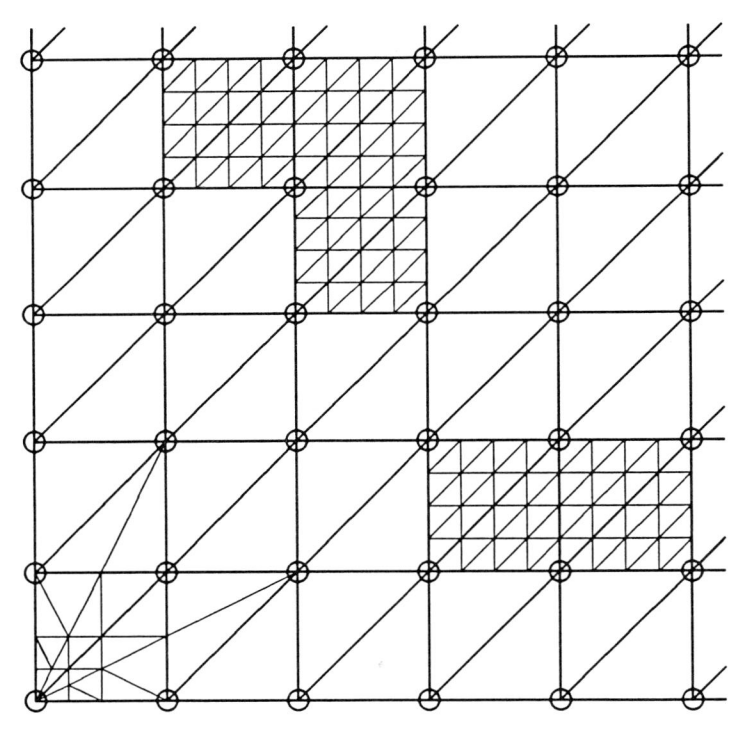

Figure 1: Composite mesh for velocity computations.

We solve the pressure equation using a preconditioned iterative procedure, based on domain decomposition [5,8]. The procedure starts with an original coarse grid and allows the possibility of refining each of the coarse-grid blocks independently. Most of the calculations on each of the coarse-grid blocks can be done independently of the others, which means that we get an algorithm with good parallel properties. The local grid refinement and the BEPS- type preconditioner [5] are implemented both for diagonal and parallel grids on the five-spot pattern. The iteration is very fast and gives no noticeable grid effects [8].

This solution gives a good resolution of the pressure distribution. The velocity is determined from equation (2), imposing mass conservation on each grid cell [8]. The loss of accuracy in the last step is compensated by the refinement in the areas where we have large pressure gradients.

A more accurate approach may be considered by using a mixed finite element method to approximate the fluid pressure and the Darcy velocity simultaneously. Standard approximations for the pressure have difficulty in approximating accurate fluid velocities in the presence of discontinuous rock permeabilities in heterogeneous media. Mixed finite element methods have proven to be very effective in treating random permeabilities [9,14]. They have been described in some detail in the context of miscible displacement by Douglas et al. [9]. The numerical results presented in Section 3 for miscible displacement utilize mixed finite elements for the pressure and velocity and modified method of characteristics Galerkin methods for the saturation [9,14].

2.2 Saturation

To solve the system (1)–(3) accurately in regions where steep saturation gradients are present, we must be able to handle the dominating convective part of equation (3) appropriately. This may be done by an operator-splitting technique [10,11], where a modified hyperbolic part of equation (3) is solved as a first step in an iterative procedure, i.e

$$\phi \frac{\partial}{\partial \tau} \bar{S}^{m+1} \equiv \phi \frac{\partial}{\partial t} \bar{S}^{m+1} + \mathbf{v} \cdot \nabla \bar{f}(\bar{S}^{m+1}) = 0, \quad t \in [t^m, t^{m+1}], \quad (12)$$

subjected to the initial condition

$$\bar{S}^{m+1}(\mathbf{x}, t^m) = S_h^m(\mathbf{x}), \tag{13}$$

where $S_h^m(\mathbf{x})$ denotes the discrete saturation approximation at time t^m. Here

$$f(S) = \bar{f}(S) + b(S)S, \tag{14}$$

and

$$\bar{f}(S) = \begin{cases} \dfrac{f(S^\star)}{S^\star} S, & 0 \le S \le S^\star, \\ f(S), & S^\star < S \le 1, \end{cases} \qquad (15)$$

where $S^\star$ is maximum saturation in the saturation shock.

We want to solve the equations in such a manner that the conservation properties are maintained and such that grid-orientation effects are negligible. In order to achieve this, we have to solve the modified hyperbolic equation (12) accurately. The procedure developed allows very long time-steps in areas where the velocity $\mathbf{v}(\mathbf{x}, t)$ is slowly varying [7]. We want to maintain this long global time-step, partly because we get a very fast code and partly because we want to keep the numerical dispersion effects small compared to the physical diffusion effects. Therefore, if the velocity field is varying too much on the global time step Δt, we have to use micro time-steps when we are tracing the saturation along the characteristics. We define:

$$\Delta t = t^{m+1} - t^m = \Delta t_1 + \Delta t_2 + \ldots + \Delta t_k. \qquad (16)$$

Using the time-stepping (16), we integrate the saturation backwards along the modified characteristics. The procedure given by Ewing et al. [13] for the miscible model has been extended to the immiscible case [17]. We may note that in these critical regions, the velocity field is solved on a refined grid, so that an accurate solution should be obtained.

Using the split version of the fractional flow function and the operator defined in (12), equation (3) can be written on the form

$$\phi \frac{\partial}{\partial \tau} S + \nabla \cdot (\mathbf{v} b(S) S) - \epsilon \nabla \cdot (\mathbf{D}(\mathbf{x}, S) \nabla S) = q_2(\mathbf{x}, t). \qquad (17)$$

The modified characteristic solution provides an approximation of the time-derivative of equation (17), and also approximates the nonlinear saturation dependence of the coefficients, see [11]. Further, the solution $\bar{S}^{m+1}(\mathbf{x}, t^{m+1})$ localizes regions where the saturation has large gradients, such as at saturation fronts. Thus we can apply a strategy of dynamic grid refinement where, on each time-level, we construct the grid from information given by the modified characteristic solution. In this way, the solution of the modified characteristic problem leads to both a simplification and discretization of (17), giving the following Petrov-Galerkin formulation for the discrete saturation approximation at time t^{m+1}.

$$A(S_h^{m+1}, \psi) \equiv (S_h^{m+1}, \psi) - \left(\frac{\Delta t}{\phi} vb S_h^{m+1}, \nabla \psi\right)$$
$$+ \left(\epsilon \frac{\Delta t}{\phi} \mathbf{D} \cdot \nabla S_h^{m+1}, \nabla \psi\right) \qquad (18)$$
$$= (\bar{S}_h^{m+1}, \psi), \qquad \forall \psi \in \mathbf{T}_h^{m+1}(\Omega). \qquad (19)$$

Here $\mathbf{T}_h^{m+1}$ is an appropriate finite-dimensional test space.

We are using a solution procedure developed by Espedal and Ewing [11] to solve the equation. Appropriate test functions are determined by a symmetrization technique developed by Barrett and Morton [2,15]. Again, a preconditioned iterative procedure, based on domain decomposition [4,11], is developed.

Here too, the algorithms have good parallel properties [8]. Figure 2 shows a typical grid, with local grid refinement, used in the computation. Most of the computations on each of the coarse-grid blocks are done in parallel.

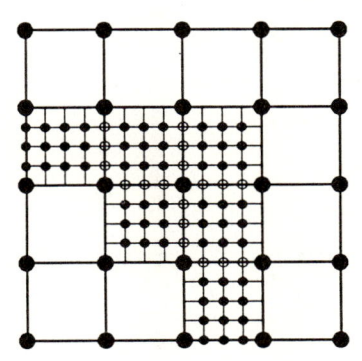

Figure 2: Typical composite grid used in the saturation computations.

3. NUMERICAL RESULTS

In this section we will provide numerical results for the miscible and two-phase immiscible displacement. In the last case we will present results from a model with heterogeneous permeability.

Example 1: Model specifications

We have used the following computational forms of the coefficients: $\phi = .2$, $\lambda_w = S^2$, $\lambda_o = (1-S)^2$ and $\mathbf{D} = S(1-S)(\mathbf{ii}+\mathbf{jj})$. The well rates needed in equation (11) are set to $q(t) = \pm .25$ for injectors and producers, respectively. We choose our domain Ω to have a rectangular mesh with an injector in the lower left corner and a producer in the upper right corner; i.e., we consider the diagonal grid of the five-spot pattern. The initial condition for the saturation equation is given as an approximately established shock somewhat away from the injector in order to focus on the transport mechanism instead of well-related effects.

Figure 3 gives the permeability distribution used in the calculations. The base permeability is $k = 1.0$.

Figure 4, Figure 5 and Figure 6 give saturation contours for 0.27, 0.37 and 0.47 (pore-volume injected) respectively. The saturation front is moved along almost half a coarse-grid block in each global time-step (0.05 pore-volume).

The calculations show that the code is able to resolve the dynamics caused by the change in permeability. The adaptive refinements used at the fluid interfaces in the calculation, are shown in the figures. As reported earlier [8], the preconditioned iteration based on this domain decomposition gives a very fast code.

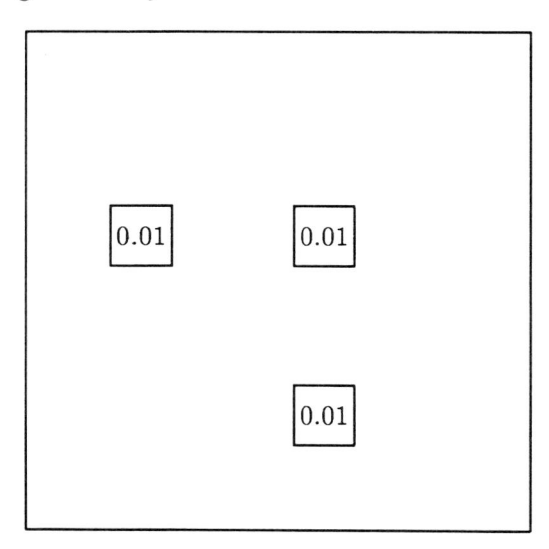

Figure 3: Permeability distribution.

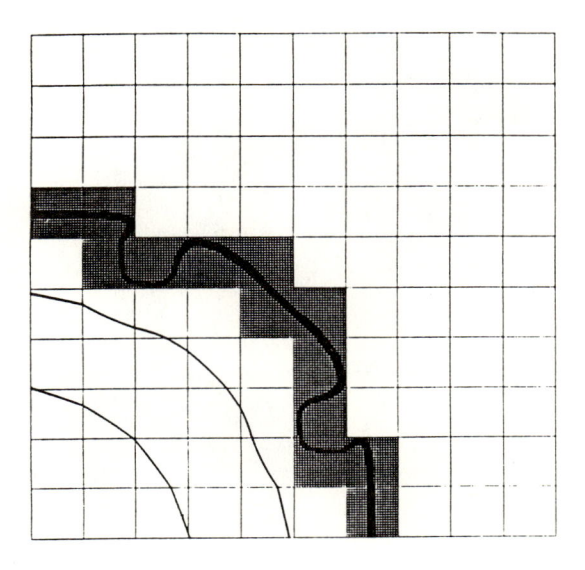

Figure 4. Saturation contours at 0.27 pore-volume injected.

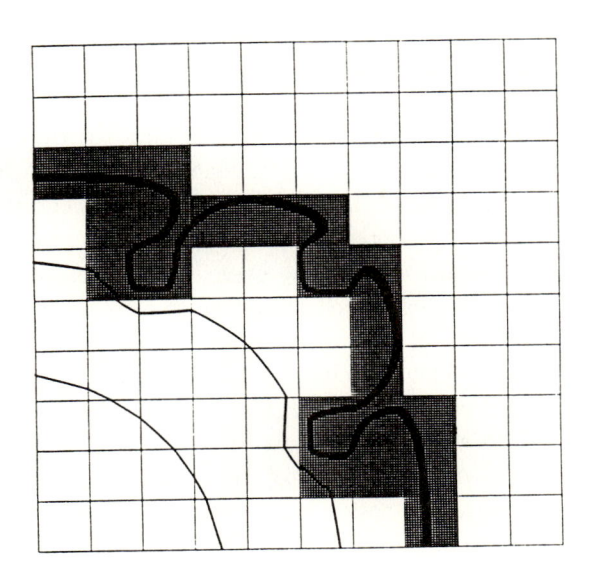

Figure 5. Saturation contours at 0.37 pore-volume injected.

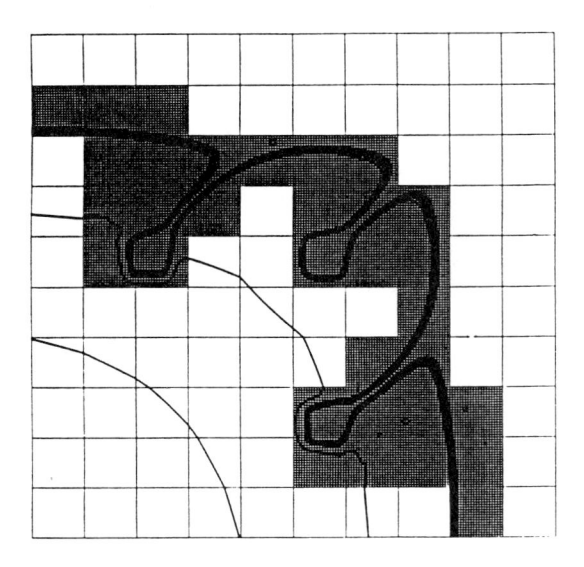

Figure 6. Saturation contours at 0.47 pore-volume injected.

Example 2: Miscible Displacement

This is an example of miscible displacement in porous media with essentially random permeabilities, with a standard deviation of 50%, generated on a 100 × 100 grid. The model has a dispersion tensor where longitudinal dispersion is approximately ten times transverse dispersion; therefore, when the permeability heterogeneity causes an unstable viscous finger to form, this finger will grow preferentially in the direction of the local fluid velocity. For this reason, it is extremely important to calculate the local fluid velocities accurately, even in the presence of discontinuous permeabilities. We have used mixed finite element methods following Raviart and Thomas [16] to approximate the Darcy velocity directly as a separate variable. The velocities are also extremely important in the modified method of characteristics. The code used was described in [13]. It uses mixed finite element methods for velocities and pressures, and a Galerkin procedure with modified method of characteristics time stepping. The calculations with this code display very little grid orientation effects or numerical dispersion; hence the code is valid for the viscous fingering study presented in [14].

The model problem had a mobility ratio of 41. The time step was 0.0⁚ pore volumes through the computation. The choices of the dispersion velocity, and grid sizes for this example yielded longitudinal and transverse Peclet numbers of 100 and 1000, respectively. This difficult problem with high Peclet numbers and discontinuous permeabilities could not have been solved with similar time-steps and accuracy without the power of the modified method of characteristics and mixed finite element method combinations. See Figure 7 for computational results for this problem.

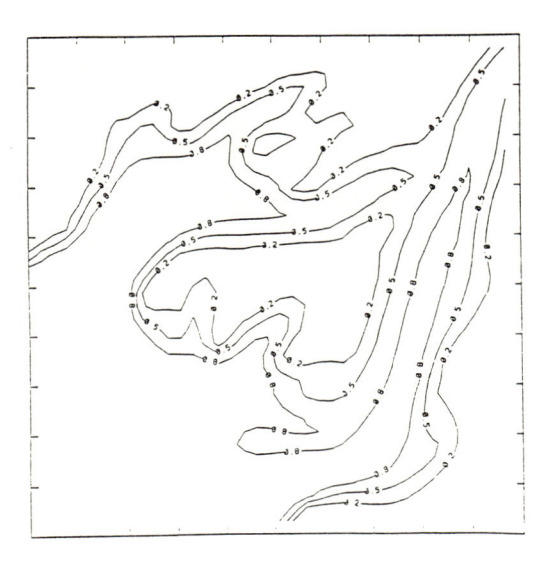

Figure 7.

Conclusions:

The algorithms for both the characteristic and the elliptic parts of the equations are very efficient on a computer with parallel architecture. This has been demonstrated on an Alliant FX/8 [8]. We may summarize the properties of the the general results of the method in the following way:

- Almost no grid orientation effects are present.

- Numerical diffusion is small compared to the physical diffusion in the model.

- We may use long global time-steps with acceptable accuracy.

- Fixed and adaptive local grid refinement capabilities are included.

- The algorithms have good parallel properties.

Acknowledgments

This research was supported in part by National Science Foundation Grant No. RII–8610680, by Office of Naval Research Contract No. 0014-88-K-0370, by the Pittsburgh and Minnesota Supercomputer Centers, by the Enhanced Oil Recovery Institute, by the Norwegian Research Council of Science and Humanities (NAVF), and by VISTA, a research cooperation between the Norwegian Academy of Science and Letters and Den norske stats oljeselskap a.s (Statoil).

REFERENCES

1. K. Aziz and A. Settari, *Petroleum Reservoir Simulation*, Elsevier Applied Science Publishers, 1979.

2. J.W. Barrett and K.W. Morton, Approximate symmetrization and Petrov-Galerkin methods for diffusion-convection problems, *Comp. Meth. in Appl. Mech. and Eng.*, *45* (1984), 97–122.

3. A. Behie and P.K.W. Vinsome, Block iterative methods for fully implicit reservoir simulation, *SPEJ, 22* (1982), 658–668.

4. J.H. Bramble, J.E. Pasciak, and A.H. Schatz, The construction of preconditioners for elliptic problems by substructuring, *J. Math. Comp.*, *47* (1986), 103–134.

5. J.H. Bramble, R.E. Ewing, J.E. Pasciak, and A.H. Schatz, A preconditioning technique for the efficient solution of problems with local grid refinement, *Comp. Meth. in Appl. Mech. and Eng.*, *67* (1988), 149–159.

6. G. Chavent, G. Cohen, and J. Jaffré, Discontinuous upwinding and mixed finite elements for two-phase flows in reservoir simulation, *Comp. Meth. in Appl. Mech. and Eng.*, *47* (1984), 93–118.

7. H.K. Dahle, M.S. Espedal, R.E. Ewing, and O. Sævareid, Characteristic adaptive sub-domain methods for reservoir flow problems, to be published in *Numerical Methods for Partial Differential Equations*

8. H.K. Dahle, M.S. Espedal, and O. Sævareid, Domain decomposition for a convection diffusion problem, presented at the SIAM Conference on Domain Decomposition, Houston, Texas, April, 1989 (to be published).

9. J. Douglas, Jr., R.E. Ewing, and M.F. Wheeler, A time-discretization procedure for a mixed finite element approximation of miscible displacement in porous media, *R.A.I.R.O. Analyse Numerique, 17* (1983), 249–265.

10. J. Douglas, Jr. and T.F. Russell, Numerical methods for convection-dominated diffusion problems based on combining the method of characteristics with finite element or finite difference procedures, *SIAM J. Numer. Anal., 18* (1982), 871–885.

11. M.S. Espedal and R.E. Ewing, Characteristic Petrov-Galerkin subdomain methods for two-phase immiscible flow, *Comp. Meth. in Appl. Mech. and Eng., 64* (1987), 113–135.

12. R.E. Ewing, ed., The mathematics of reservoir simulation, *Frontiers in Applied Mathematics 1*, SIAM, Philadelphia, Pennsylvania, 1983.

13. R.E. Ewing, T.F. Russell, and M.F. Wheeler, Simulation of miscible displacement using mixed methods and a modified method of characteristics, *SPE 12441, Proceedings 7th SPE Symposium on Reservoir Simulation*, San Francisco, California, November, 1983.

14. R.E. Ewing, T.F. Russell, and L.C. Young, An anisotropic coarse-grid dispersion model of heterogeneity and viscous fingering in five-spot miscible displacement that matches experiments and fine-grid simulations, *SPE 18441, Proceedings 10th SPE Symposium on Reservoir Simulation*, Houston, Texas, February 1989, 447–466.

15. J.C. Heinrich, P.S. Huyakorn, A.R. Mitchell and O.C. Zienkiewicz, An upwind finite element scheme for two-dimensional convective transport equations, *Inter. J. Numer. Engrg., 11* (1977), 131–143.

16. P.A. Raviart and J.M. Thomas, A mixed finite element method for 2nd order elliptic problems, in *Mathematical Aspects of the Finite Element Method, Lecture Notes in Math., Vol. 606*, Springer-Verlag, Berlin and New York, 1977.

17. O. Stava, Masters Thesis, Department of Mathematics, University of Bergen.

TRACKING AN IN-SITU COMBUSTION FRONT USING THE THIN FLAME TECHNIQUE

Richard Davies
Stanford University

Petroleum Engineering Department
Stanford University
CA 94305
USA

ABSTRACT

One of the major problems when modelling a field scale in-situ combustion process is that of grid block size effects. Since the grid block thickness is far greater than the flame thickness the length scale in our finite difference model will be too great to observe the sharp gradients occurring over a very short region in the field.

The method introduced here overcomes this problem by decoupling the flame from the finite difference model and replacing it with an interface which moves through the region acting as a moving heat source and displacing pump. This representation of the flame in a finite difference model is known as the thin flame technique. In order to apply the thin flame technique we need to determine the velocity and reaction rate of the flame by a separate solution for the decoupled region. A steady state analysis is developed to solve for these parameters. This solves a reduced set of conservation equations over the burning region using a Lagrangian grid and the shooting method to determine the steady state solution.

1 INTRODUCTION

In-situ combustion is an oil recovery technique in which air, or oxygen enriched air is injected into a reservoir in order to displace the oil. Under suitable conditions the oxygen will burn with part of the oil, raising the temperature of the reservoir and reducing the viscosity of the oil, hence allowing it to flow more easily.

A serious problem with mathematical modelling of in-situ combustion is that of flame extinction due to grid block size effects. When modelling a field scale process using finite difference techniques the grid block size will be far larger than the flame length. Since parameters such as temperature and saturations are averaged over a grid block they will be misrepresented in the Arrhenius reaction rate equation,

$$r = Ae^{(-E/RT)} f(S, P, Y, X) \tag{1.1}$$

since this term is an exponential function of temperature, any error in the temperature used will produce serious errors in the computed reaction rate. In certain circumstances this can lead to the flame in the simulation dying out.

This grid block size effect is a well known problem and has been addressed by several authors. Hwang *et al* [5] attacked the problem using a model in which the flame is viewed as an interface which generates heat and displaces the reservoir fluids. The reaction is governed by the presence of reactants rather than temperature. Fuel consumption and residual oil saturation have to be specified (using results of a laboratory combustion tube test). The flame front is positioned by allocating to each cell a number proportional to the amount of fuel remaining unburnt (between 0 and 1). It is then advanced by considering the amount of fuel consumed during the current time step.

Coats [3] has reported an extended version of his previous thermal simulator [2] in which he introduces the 'activation temperature' concept. This involves stipulating a minimum temperature for use in the Arrhenius term of the oil oxidation equation, ensuring that whenever air and oil are in contact a certain amount of combustion will occur. When oxygen inflow begins in a cell this will ensure combustion continues until the block temperature reaches the activation temperature, then the actual block temperature takes over, allowing the burning to become more vigorous as the temperature continues to increase. The main disadvantage of this method is that it cannot predict fuel consumption accurately, while retaining the need for the complex input data associated with the Arrhenius term. Also, the choice of activation temperature can have a significant effect on the results.

The approach taken here to overcome the problem is to decouple the flame from a conventional finite difference simulator and solve separately for the reaction rate and flame velocity. This is achieved using a steady state analysis that applies a moving frame to a reduced set of the conservation equations over the flame region, and solves the resulting eigenvalue problem using the shooting method. The reaction rate and flame velocity determined by the steady state analysis are then used to apply the 'thin flame' technique [1] to the conventional simulator. This treats the flame as a moving heat source and displacing pump, travelling through the domain with the velocity obtained by the steady state analysis. This approach has been developed for a one dimensional simulator.

2 THE CONVENTIONAL SIMULATOR

The model is a one-dimensional finite difference code capable of handling five components (water, heavy oil, light oil, inert gas and oxygen) in three phases (water, oil and gas). Upstream weighting is used spatially and the time discretization is implicit as the improved stability that it offers has, for thermal simulation, usually resulted in more economic models (in terms of computer time) than explicit methods [2],[4]. The use of implicit methods requires the iterative solution of the system of algebraic equations and this is achieved using Newton's method.

The equations have been written in a form allowing treatment of capillary pressure, conduction and convection of heat and one chemical reaction describing the burning of heavy oil. Routines have also been included to model heat loss and the injection of heat from a band heater. This allows the simulator to be used to model combustion tube experiments as well as field projects. Since the simulator is a standard design, similar to that described in [4] it is not described in detail here.

3 STEADY-STATE ANALYSIS FOR FLAME PARAMETERS

The method used to determine the flame profile is to apply a set of mass and energy balances over the burning zone, and, by imposing a moving grid on the region, reduce the problem to a boundary value problem which can be solved using the shooting method [8]. This works by taking guesses for the unknown parameters and solving the equations as an initial value problem. The estimates of the unknowns are then updated using Newton's method until the results of the initial value problem match the known boundary values.

The main assumption made by the analysis is that there is only one liquid phase, namely a heavy oil. Any water present, and any other oil components are assumed to be in the gas phase. The results of fine grid runs justify this simplification and show that it is accurate. This assumption leads to a set of three mass and energy balances being used to model the burning zone, namely an energy balance and mass balances for oxygen and liquid oil. The chemical reaction scheme consists of a single heavy oil burning reaction which occurs for oil in the liquid phase only,

$$\text{Heavy Oil} + \text{Oxygen} \longrightarrow \text{Light Oil} + \text{Inert Gas} + \text{Water}$$

The PDE's to be solved are as follows:-

OXYGEN MASS BALANCE

$$\phi \frac{\partial}{\partial t}(S_g \rho_g Y_5) + \frac{\partial}{\partial x}(V_g \rho_g Y_5) = -s_5 r \tag{3.1}$$

OIL MASS BALANCE

$$\phi \frac{\partial}{\partial t}(S_o \rho_o X_2) + \frac{\partial}{\partial x}(V_o \rho_o X_2 + V_g \rho_g Y_2) = -s_2 r \tag{3.2}$$

ENERGY BALANCE

$$\frac{\partial}{\partial t}(\rho C_p T) + (V_g \rho_g C_{pg} + V_o \rho_o C_{po}) \frac{\partial T}{\partial x} = \lambda \frac{\partial^2 T}{\partial x^2} + r \Delta H_r \tag{3.3}$$

The next step is to apply a moving grid $\xi = x - vt$ to the region, where v is the flame velocity. The flame velocity is not yet known and will be determined by the shooting method from the set of steady state equations that result from the transformation. Using the chain rule with $\xi = x - vt$; $\tau = t$ and writing for steady state where $\frac{\partial}{\partial \tau} = 0$, we can rewrite our equations as

$$-\phi v \frac{\partial}{\partial \xi}(S_g \rho_g Y_5) + \frac{\partial}{\partial \xi}(V_g \rho_g Y_5) = -s_5 r \qquad (3.4)$$

$$-\phi v \frac{\partial}{\partial \xi}(S_o \rho_o) + \frac{\partial}{\partial \xi}(V_o \rho_o + V_g \rho_g Y_2) = -s_2 r \qquad (3.5)$$

$$-v(\rho C_p)\frac{\partial T}{\partial \xi} + (V_g \rho_g C_{pg} + V_o \rho_o C_{po})\frac{\partial T}{\partial \xi} = \lambda \frac{\partial^2 T}{\partial \xi^2} + r\Delta H_r \qquad (3.6)$$

This version of the energy balance is a second order equation and is converted to first order by letting,

$$\alpha = \frac{\partial T}{\partial \xi} \qquad (3.7)$$

which allows us to write (3.6) as,

$$-v(\rho C_p)\alpha + (V_g \rho_g C_{pg} + V_o \rho_o C_{po})\alpha = \lambda \frac{\partial \alpha}{\partial \xi} + r\Delta H_r \qquad (3.8)$$

The four equations (3.4,3.5,3.7,3.8) are a set of ODE s which constitute a boundary value problem,

$$\frac{dy}{dx} = f(x, y)$$

which is solved using the Shooting method [8].

The system of equations (3.4,3.5,3.7,3.8) has a total of ten parameters. They are, temperature, α, oil saturation and oxygen mole fraction at each boundary, the length of the region L, and the flame velocity v. The unknowns are L, v and temperature at the upstream (inlet) end of the region T^{lhs}. The other parameters are defined by the user or obtained from the conventional simulation when applying the thin flame technique.

The solution of the initial value problem is computed using Euler's method, from the upstream to downstream end of the region, using initial guesses for the three unknowns. When the solution at the unburnt end has been found, the three boundary values T^{rhs}, S_o^{rhs} and Y_5^{rhs} are compared to the correct boundary values which are known. The values of T^{lhs}, v and flame length (the shooting method unknowns) are then updated using Newton's method, in which the Jacobian matrix (of order three) is calculated numerically. The three unknowns T^{lhs}, velocity and flame length are aligned to the energy, oil and oxygen equations respectively. Calculations were also made with the velocity aligned with the oxygen equation and length with the oil. This produced identical results.

This question of existence and uniqueness of solutions to these equations has been addressed by Norbury and Stuart [6,7] who consider a similar set of equations and make some simplifying assumptions. As we have done here, they reduce the system to a two-point boundary value problem and, by applying local bifurcation theory, prove the existence of non-trivial solutions. The existence of solutions is clearly what we would expect. However, what is not so clear is whether the solution is stable and unique. They showed that flames that are governed by the presence of reactants (as is an in-situ combustion flame) are stable within certain parameter regions. Their findings indicate that high inlet gas velocities and high solid specific heat have destabilising effects on the flame. This agrees with observations made on the performance of the shooting method presented here, although no conditions have been found that cause the solution to be so unstable as to prevent the shooting method converging. No proof of uniqueness can be given, however the parameters are constrained to be within the physically realistic and 'expected' range and the analysis has proved extremely consistent.

3.2 Example Computation

An example computation is presented to illustrate the analysis. The test parameters and results are given in Table 1. All boundary conditions except T^{lhs} are specified. The analysis calculates T^{lhs}, L, v and the profiles of all parameters at interior points. The total reaction rate is computed from the rates at interior points.

The temperature and saturation profiles of the flame region are shown in Figures 1 and 2. It can be seen that the majority of the reaction occurs over a relatively small part of the region, the remaining part of the region is largely dominated by conduction and convection. Figure 3, which shows the rate of energy release ($r \cdot \Delta H_r$) as a function of distance, emphasizes this.

Sensitivity studies show that the temperature boundary condition is not important, as is the effect of increasing the number of grid blocks used by the initial value solver. They also show that the oxygen mole fraction at the inlet end of the region (Y_5^{lhs}) is the most important value for an accurate estimation of the flame speed. The oxygen mole fraction bypassing the flame and unburnt oil saturation (Y_5^{rhs} and S_o^{lhs} respectively) are less important, but still need to be determined with care. The gas injection rate also has a very important effect on the flame. Higher rates result in higher temperatures, lower fuel consumption and higher flame velocity.

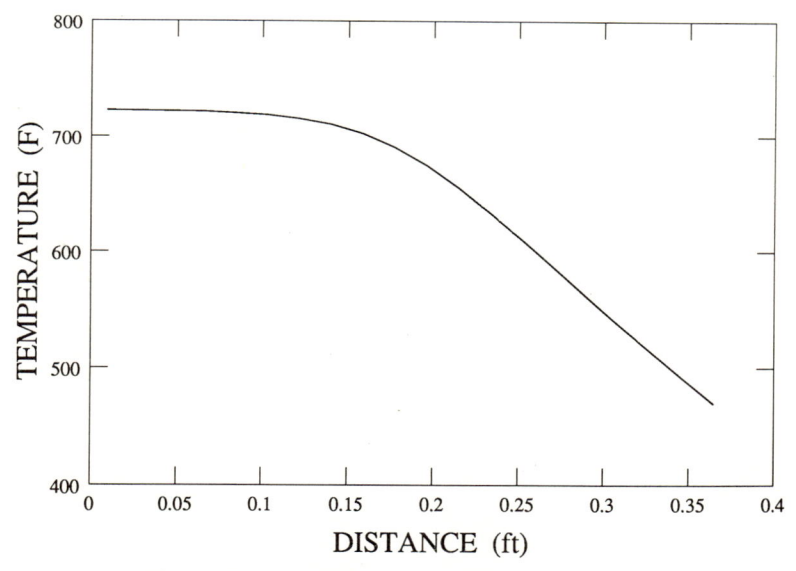

Fig. 1 Temperature profile of steady state flame.

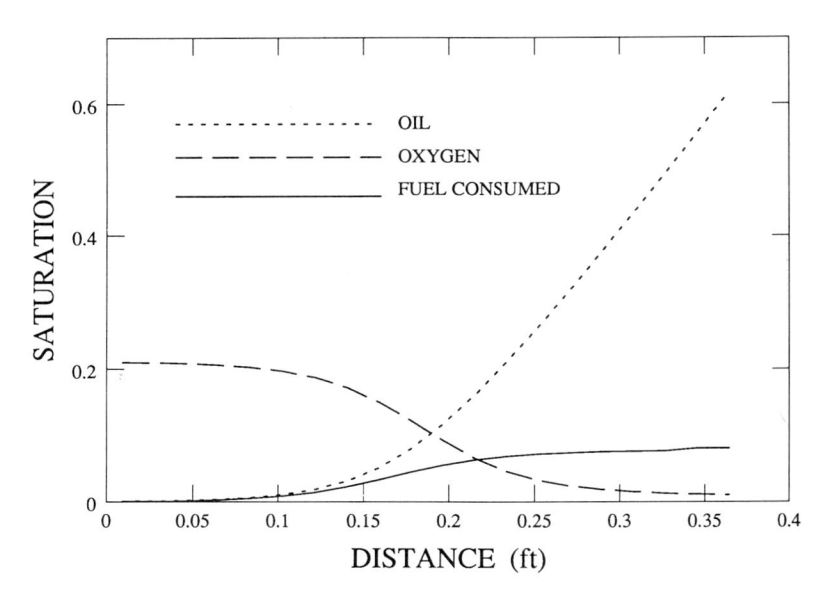

Fig. 2 Saturation profiles of steady state flame

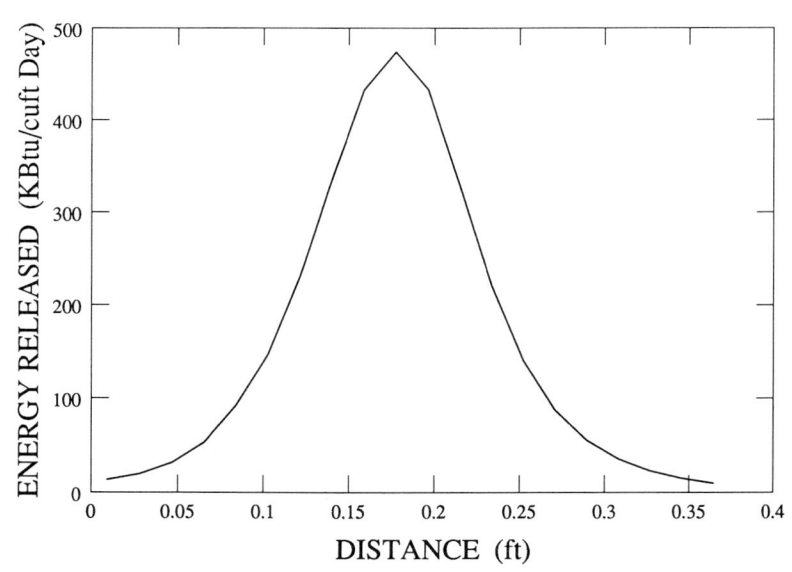

Fig. 3 Profile of energy release rate at steady state.

Table 1. Dataset for Example of Steady State Analysis

Pressure (psi) = 300
Porosity = 0.4
Air Flux (scf/ft^2 day) = 384

Boundary Values	Upstream	Downstream
Oil Saturation	0.001	0.62
Oxygen Mole Fraction	0.21	0.001
Temperature	Unknown	470

Results

Peak Temp °F	725	Flame Speed ft/day	4.4
Fuel Burned lb/cuft	1.29	Flame Width in.	1.5

4 THE THIN FLAME TECHNIQUE

The movement of an interface travelling through a fluid medium is a difficult phenomenon to simulate numerically. Traditional finite difference and element techniques suffer from numerical diffusion and dispersion which lead to numerical smearing of the fluid interface and requires the use of large numbers of grid blocks to prevent it losing its definition. Although methods exist for reducing numerical diffusion and dispersion they only go part of the way to solving the problem, and the difficulty of representing the fluid boundary within a grid block still remains.

The thin flame technique is a method of tracking a front through a finite difference grid. It treats the flame as an interface between two distinct fluids (in this case injection gas and oil and exhaust gases). The interface has no length and moves through the region with an unknown velocity, which is determined here using the steady state analysis. Use of the thin flame algorithm to model a propagating reaction was first reported by Chorin [1] who modelled the motion of a flame in a two dimensional region under the influence of turbulence.

The application of the thin flame algorithm in a one dimensional scheme is quite straightforward. We know that the flame will be moving steadily along the grid in a direction corresponding to the fluid flow. Therefore we can simply advance the flame position by the relevant distance on each time step. The basis of the method is to replace the reaction rate expression in the conservation equations for the burning block with the value obtained by the steady state analysis, the reaction rate in all other grid blocks being equal to the normal Arrhenius term. The flame position is advanced through the burning block by the appropriate distance on each time step, and has two uses. The first is as a pointer to the grid block occupied by the flame. The second is as a 'displacing pump' which separates the injection gas from the oil and exhaust gases. All oil in the block, apart from that which remains unburnt, is placed in the section ahead of the flame and all oxygen, apart from Y_5^{rhs}, is kept in the section behind the flame (Fig 4). It should be emphasised that these adjusted saturations are used only in the flow terms and not the accumulation terms. This redistribution of components ensures that the flow terms realistically represent the physical processes taking place. It limits the amount of oxygen bypassing the flame to Y_5^{rhs} (which we specify), and produces better estimates of the rates at which components pass through to the next block. The oxygen and oil concentrations on either side of the flame resulting from this modification are supplied to the steady state method as boundary values. The method is therefore self adjusting in that if oxygen builds up behind the flame then a larger oxygen supply is specified in the steady state analysis and the reaction rate will increase. Conversely, if the oxygen concentration behind the flame is reduced then the reaction rate will fall.

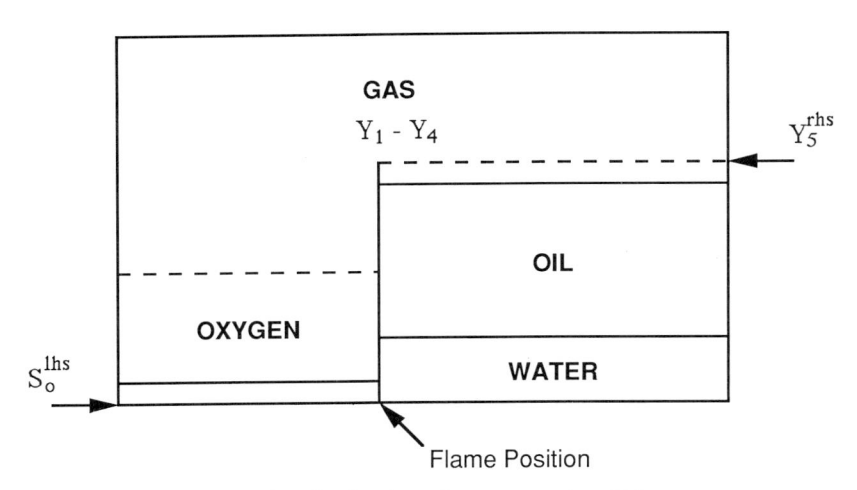

Fig. 4 Redistribution of components within burning grid block.

When implementing the technique it was noted that performing the steady state analysis on each iteration of each time step and for every derivative calculation of the Jacobian construction for the burning grid block would be very expensive. It was found sufficient to perform the analysis once per time step (i.e. use an explicit reaction rate) since this has only a slight effect on the time step size that is taken.

One other restriction that has been necessary is the adjustment of time step sizes so that the flame stops on the grid block boundary rather than passing over it during a time step. This avoids complicating the calculation of oil flowing out of the grid block ahead of the flame and is included as a convenience rather than a necessity.

5 RESULTS

In order to examine the performance of the thin flame code it was run against the conventional model on a combustion tube problem. The problem chosen was a six foot tube which was preheated for five hours prior to air injection. The length of the tube allowed the conventional simulation to reach a steady state and the preheating encouraged early ignition. The full dataset is given in Table 2.

The model was run four times using the conventional flame representation (with 5, 10, 20 and 40 grid blocks), and once using the thin flame method with 10 blocks. Fig 5 shows the advancement of the flame front for each of the five cases. This indicates that the predicted motion of the flame converges to a particular path as the number of grid blocks is increased. The first 1.2 ft of the tube is burnt quickly since this region has been preheated. As the flame passes into the unheated region it requires some time for the conventional model to reach a steady state. This process of reaching steady state takes more time as the number of grid blocks is decreased, and below 10 blocks appears not to reach a steady state at all for this tube length. In contrast, when the thin flame simulation moves from the preheated to unheated region it immediately allows for the different temperature, and the flame proceeds smoothly.

Table 2. Dataset for Comparison of Models

Initial Conditions
Pressure = 400 psi
Temperature = $150°F$
Oil Saturation = 0.25
Water Saturation = 0.2
Gas Saturation = 0.55
$X_2 = 1.0$

Core Dimensions
Length = 6 ft
Cross Section = 0.04 ft^2
Porosity = 0.26
Permeability = 1000mD
Thermal Cond. = 38.4 Btu/ft.day

Pre-heating of tube
First 20% of the tube is heated for 5 hours at a rate equal to,
$H_{inj} = 410(450 - T)$ Btu/(cuft. Day °F)

Injection and Production
Injection rate = 14 cu ft air/day
Injection gas $Y_5 = 0.214$
Injection Temperature = $100°F$
Production Pressure = 400psi
Productivity Index = 10.0

Reaction Parameters
$$s_2(\text{heavy oil}) + s_5(\text{oxygen}) \Longrightarrow s_1(\text{water}) + s_3(\text{light oil}) + s_4(\text{inert gas})$$

$s_1 = 12.0$
$s_2 = -1.0$
$s_3 = 3.0$
$s_4 = 15.0$
$s_5 = -18.0$

$A = 1.0 * 10^6 \text{day}^{-1}\text{psi}^{-1}$
$E = 33300.0$ Btu/lbmol
$\Delta H_r = 3.49 * 10^6$ Btu/lbmol

The convergence of the steady state flame speed as the number of grid blocks used in the conventional method increases is shown in Fig 6. A quadratic polynomial was fitted to the points corresponding to 10, 20 and 40 grid blocks. It suggests that a velocity of 2.0 ft/Day would be reached if the number of grid blocks were increased indefinitely. This compares well with the average thin flame velocity of 2.07 ft/Day.

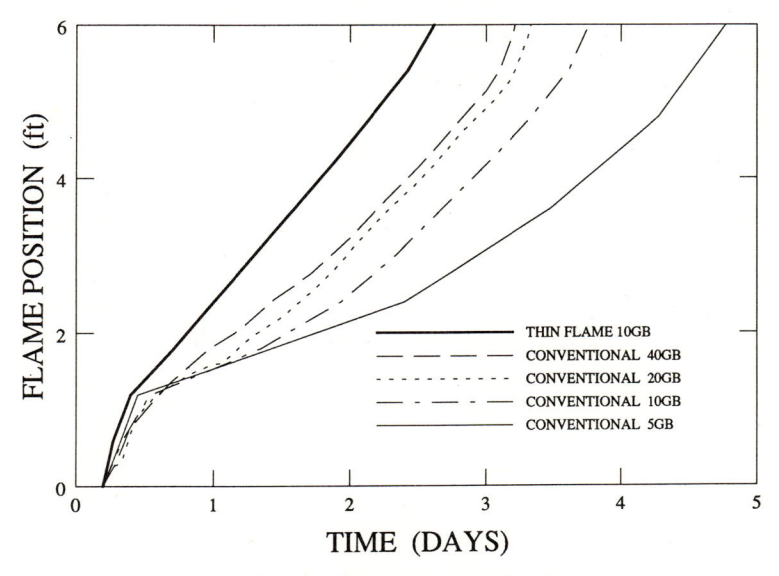

Fig. 5 Flame position in time.

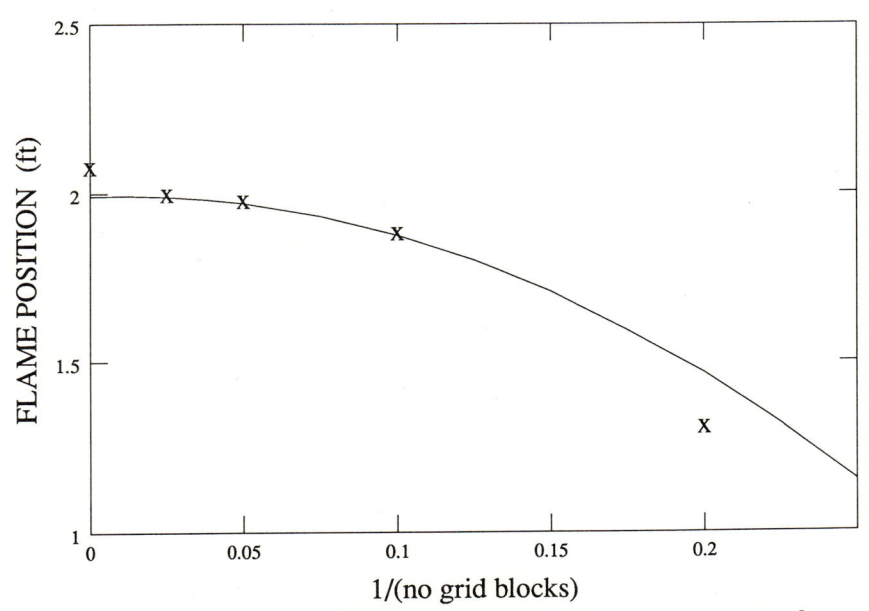

Fig. 6 Asymptotic convergence of flame speed as no. of grid blocks increase.

When running the thin flame method on this problem it was found necessary to set the mole fraction of oxygen passing through the flame (Y_5^{rhs}) to 0.1 in order to match the amount of oxygen passing through the flame region in the conventional simulations. Figures 7 and 8 show oxygen and temperature profiles of the different simulations and indicate that the displacement process taking place in the simulations are the same, but that the burning front in the thin flame simulation moves approximately half a day ahead of the conventional model due to the time taken for the conventional simulation to reach a steady state.

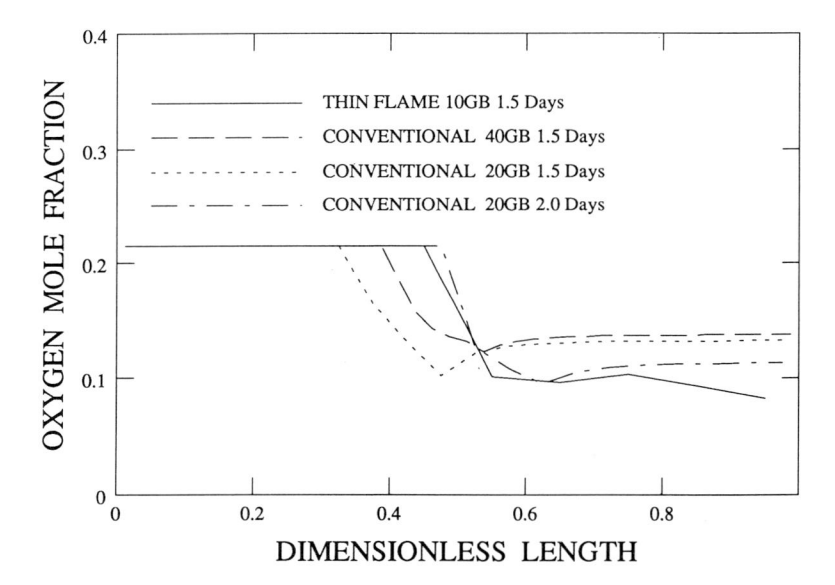

Fig. 7 Oxygen mole fraction profiles.

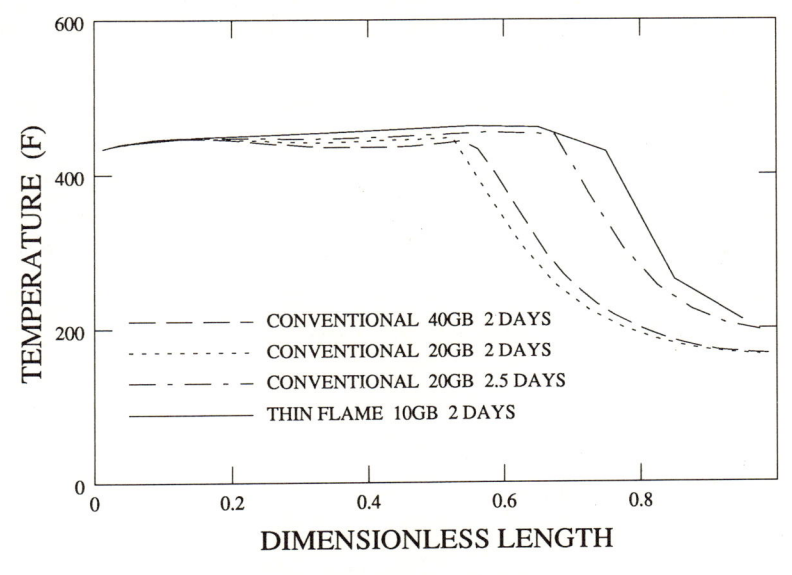

Fig. 8 Temperature profiles.

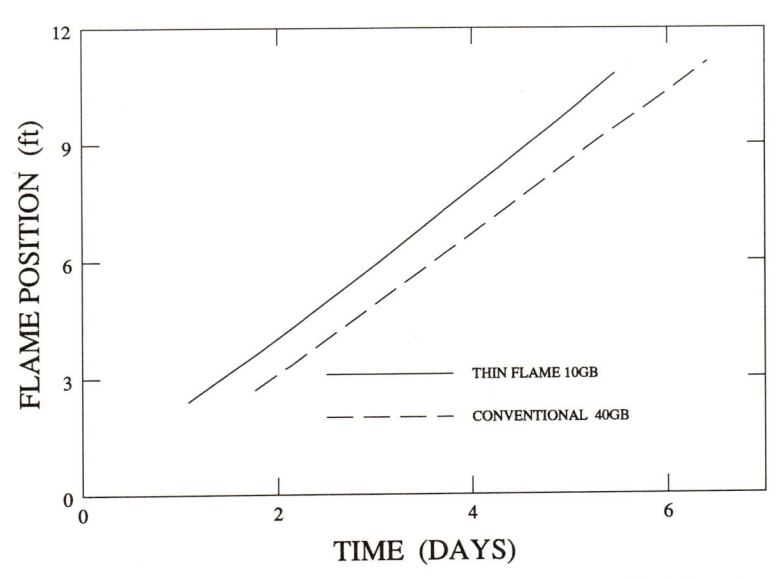

Fig. 9 Flame position in time for steady period in 12ft tube.

In order to examine more closely what happens when the simulations have reached a steady state two more computations were performed on the above example, but this time with a tube length of 12 ft. The runs performed were a 40 grid block conventional and 10 grid block thin flame simulation. The flame positions in time for the steady portion of the runs are shown in Fig 9 and from this it can be seen that the two slopes are almost parallel, implying that the flames are moving with the same velocity, and their horizontal separation implies that the flame in the conventional simulation lags half a day behind the flame in the thin flame simulation (due to the initial time required for ignition).

6 CONCLUSIONS

1 The steady state analysis models the burning zone of an in-situ combustion flame providing detailed information on the velocity, fuel consumption and saturation profiles of the flame.

2 The thin flame technique accurately tracks a flame through a finite difference grid. It also results in the use of more accurate flow rates from the burning block as a result of the component redistribution.

3 The thin flame technique and steady state analysis complement each other excellently since the analysis provides the flame speed and reaction rate needed by the thin flame technique, and the thin flame approach supplies the boundary values needed by the analysis.

4 The application of the thin flame technique enables in-situ combustion simulations to be performed accurately with far fewer grid blocks than with a conventional simulator. This results in great savings in computational costs.

5 In order for the technique to be useful for detailed field studies two improvements to the work presented here are necessary. Firstly the steady state analysis needs to be developed for four phase, multiple chemical reaction systems. Secondly, a front tracking technique is needed to track the flame in two and three dimensions.

ACKNOWLEDGEMENTS

This work was performed while the author was a postgraduate student at the University of Bath. The author would like to express his sincere thanks to Dr. Mike Christie of B.P. Research International and Dr. M. Greaves of Bath University for their help and encouragement while this work was in progress.

NOMENCLATURE

SYMBOL	DESCRIPTION	UNITS
A	Frequency Coefficient in Reaction Rate Expression	day^{-1} psi^{-1}
C_p	Heat Capacity	Btu/lbmol °F
E	Activation Energy	Btu/lbmol
P	Pressure	psi
R	Ideal Gas Constant	Btu/°R lbmol
r	Reaction Rate	lbmol/cuft sec
s	Stoichiometric Coefficient	-
S	Saturation	-
T	Temperature	°R
t	Time	secs
V	Velocity	ft/sec
v	Flame Velocity in Steady State Analysis	ft/sec
X	Oil Phase Mole Fraction	-
Y	Gas Phase Mole Fraction	-
ΔH_r	Heat of Reaction	Btu/lbmol
λ	Thermal Conductivity	Btu/ft day °R
ρ	Density	lbmol/cuft
ϕ	Porosity	-

SUBSCRIPTS

- o Oil Phase
- g Gas Phase
- 2 Heavy Oil Component
- 5 Oxygen Component

SUPERSCRIPTS

- rhs value at right hand (downstream) boundary
- lhs value at left hand (upstream) boundary

REFERENCES

1 Chorin A J "Flame Advection and Propagation Algorithms" *J. Comp. Phys.* **35** (1980)1-11

2 Coats K H "In-Situ Combustion Model" *Soc. Pet. Eng. J.* (Dec 1980)533-554

3 Coats K H "Some Observations on Field-Scale Simulation of the In-Situ Combustion Process" paper SPE 12247 presented at the Reservoir Simulation Symposium, San Francisco Nov 1983

4 Grabowski J W, Vinsome P K, Lin R C, Behie A and Rubin B "A Fully Implicit General Purpose Finite Difference Thermal Model for In-Situ Combustion and Steam" paper SPE 8396 presented at the SPE-AIME 54th Annual Fall Technical Conference and Exhibition, Las Vegas, Sept 1979

5 Hwang M K, Jines W R, and Odeh A S "An In-Situ Combustion Process Simulator with a Moving Front Representation" *Soc. Pet. Eng. J.* (April 1982)271-279

6 Norbury J and Stuart A M "Travelling Combustion Waves in a Porous Medium Part 1 - Existence" *SIAM J. Appl. Math.* **48** No. 1 (Feb 1988)155-169

7 Norbury J and Stuart A M "Travelling Combustion Waves in a Porous Medium Part 2 - Stability" *SIAM J. Appl. Math.* **48** No. 2 (April 1988)374-392

8 Stoer J and Bulirsch R "Introduction to Numerical Analysis" Springer-Verlag 1980

AVERAGING OF RELATIVE PERMEABILITY IN COMPOSITE CORES

Magnar Dale

Rogaland Research Institute /Rogaland Regional College, Norway

Abstract

We compute and compare the relative permeability functions obtained by applying the steady state and the unsteady state measuring techniques to a composite core. Special emphasis is given to the problem in what sense, if any, the measured relative permeability curves represent an average of the relative permeability curves for the different rock types constituting the core.

1. INTRODUCTION

Both the steady state (SS) and unsteady state (USS) methods for measuring relative permeability assume that the core sample measured is <u>homogeneous</u>. In this paper we study mathematically the properties of relative permeability curves obtained by applying these standard measuring techniques to a <u>composite</u> <u>core</u>. A composite core is considered a deterministically heterogeneous medium, consisting of a small number of distinct homogeneous parts, each part characterized by its length, porosity, absolute permeability and relative permeability curves. Thus the correlation length of the parameter variation is large compared to the core length, hence traditional volume averaging techniques do not apply. Instead we use the Buckley-Leverett theory, which makes an exact study of these relative permeability concepts possible.

It turns out that the SS-measured relative permeability depends only on the rock characteristics and length of each part of the core. On the other hand, the USS-measured relative permeability will in addition depend on the ordering of the parts inside the core, as well as on the fluid viscosities. However, the fractional flow curves built on the two distinct pairs of measured relative permeabilities will be identical for saturations where they are both defined. Explicit formulas are presented, and an example is calculated.

Special emphasis is given to the problem of in what sense, if any, the measured relative permeability curves represent an average of the relative permeabilities for the rock types constituting the core. We show that the SS-measured relative permeability is in a natural and precise way an average, homogenizing the composite core with respect to two-phase SS-flow. On the other hand, the USS-measured relative permeability will in general not homogenize the composite core with respect to two-phase USS-flow, not even with respect to the particular displacement from which it is constructed.

In general, the USS-relative permeability for a heterogeneous medium results from performing a displacement experiment on the medium, and interpreting the displacement characteristics via the standard Johnson-Bossler-Naumann formalism, see [6]. This USS-relative permeability concept has been studied by several authors, both numerically ([3], [5],[8], [10]) and formally ([1], [4], [9]). However, apart from some results contained in the paper [9], it seems that analytical studies of its properties are lacking.

2. TWO-PHASE FLOW IN A COMPOSITE CORE

2.1 N-part composite core

An N-part composite core C is a 1D porous medium of constant cross section A, consisting of $N \geq 1$ homogeneous parts in series. Neglecting the saturation dependent capillary curves, the composite core is completely characterized by specifying the length l_j, porosity φ_j, absolute permeability k_j and relative permeability functions k_{rij} (i = oil, water) for each of its j = 1,..,N homogeneous parts. Flow in the core takes place in the direction of increasing j. Let L_j denote the total length of the j first parts of the core. Put L_N = L, the total core length. For j < N the phrase " x lies in the j-th homogeneous part" means $x \in [L_{j-1}, L_j)$, the half-open interval. The N-th homogeneous part also includes the end point x = L .

Implicit in the specification of relative permeability functions for the j-th part are the irreducible water- and oil saturations S_{wcj} and S_{orj}, as well as their end point relative permeability values $k'_{roj} = k_{roj}(S_{wcj})$, $k'_{rwj} = k_{rwj}(1-S_{orj})$. Also, $k_{rwj}(S_{wcj}) = 0.$, and $k_{roj}(1-S_{orj}) = 0.$.

2.2 General equations

Consider incompressible two-phase flow of oil and water through the composite core. We neglect capillarity and gravity. Within the j-th homogeneous part the process is governed by the following equations ($t \geq 0$, $x \in [L_{j-1}, L_j]$, i = oil, water) :

$$q_i(x,t) = - \frac{k_j k_{rij}(S_w(x,t))}{\mu_i} \frac{\partial p}{\partial x}(x,t) \qquad (2.1)$$

$$\frac{\partial q_i}{\partial x}(x,t) + \varphi_j \frac{\partial S_i}{\partial t}(x,t) = 0 \qquad (2.2)$$

$$S_w(x,t) + S_o(x,t) = 1 \qquad (2.3)$$

Here q_i is the Darcy-velocity of phase i (= flow rate of phase i divided by cross sectional area A), which will be assumed continuous at the boundary between two homogeneous parts of the core. Further, S_i is the saturation of phase i, μ_i is the viscosity of phase i and p is the pressure. The functions $S(x,t)$, $p(x,t)$ are the water saturation and the pressure along the core at time t. By equation (2.3) we can select water saturation $S = S_w$ to be the sole variable on which k_{rwj} and k_{roj} depend. For any x in the composite core let

$$q_T(x,t) = q_w(x,t) + q_o(x,t) \qquad (2.4)$$

denote the <u>total velocity</u>, and let

$$f_i(x,t) = q_i(x,t)/q_T(x,t) \qquad (2.5)$$

be the <u>fractional flow of phase i</u>. If x lies in the j-th homogeneous part we have

$$f_w(x,t) = f_{wj}(S) = \frac{k_{rwj}(S)}{k_{rwj}(S) + M k_{roj}(S)} \qquad (2.6)$$

where $M = \mu_w/\mu_o$ is the viscosity ratio and $S = S_w(x,t)$.

2.3 The discontinuity functions $\Delta_{kl}(S)$

We assume that the fractional flow functions $f_{wj}(S)$ are strictly monotonically increasing on the closed interval $I_j = [S_{wcj}, 1-S_{orj}]$, $j = 1,\ldots,N$. For any pair of integers $k,l \leq N$ we define the discontinuity function $\Delta_{kl}(S)$, $S \in I_l$, by

$$\Delta_{kl}(S) = f_{wk}^{-1}(f_{wl}(S)) \qquad (2.7)$$

The discontinuity function measures the saturation discontinuity gap necessarily occurring at the boundary between two rock types whose fractional flow functions differ.

3. THE STEADY STATE CASE

The purpose of this section is to determine the steady state measured relative permeability of a composite core as a function of its rock parameters. We also discuss in what sense it is an effective relative permeability with respect to steady state flow in the core.

3.1 The saturation distribution

Consider two-phase <u>steady state</u> flow through the composite core. Thus, at any fixed location $x \in [0,L]$ along the core, the fluid pressure and saturations are assumed to be independent of time : $S(x,t) = S(x)$, $p(x,t) = p(x)$. Then equation (2.2) implies that q_w and q_o are both constant within each homogeneous part, hence constant throughout the core by the assumed continuity of Darcy velocities. Hence also

$$f_w(x) = f_w = \text{constant} \qquad (3.1)$$

By (2.6) and (3.1) it follows that each number $f_w \in [0,1]$ uniquely determines a saturation distribution along the core; namely, the water saturation S_j within the j-th homogeneous part is given by

$$S_j = f_{wj}^{-1}(f_w) \qquad (3.2)$$

Thus the saturation distribution $S(x)$ along the core is piecewise constant, with a discontinuity jump at each boundary $x = L_j$ determined by

$$S_j = \Delta_{j,j-1}(S_{j-1}) \qquad (3.3)$$

3.2 The differential pressure

From the equations (2.1), (3.1) and (3.2) we obtain the following expression for the differential pressure Δp over the composite core in the steady state case (i = oil, water):

$$\Delta p = p(0) - p(L) = - \int_0^L \frac{\partial p}{\partial x} dx = \mu_i q_T f_i \sum_{j=1}^N \frac{l_j}{k_j k_{rij}(S_j)} \qquad (3.4)$$

3.3 The steady state measured relative permeability

Denote by $\overline{k}_{ri}$, i = oil, water the relative permeability functions obtained by applying the <u>steady state measuring technique</u> to the composite core: During steady state flow we measure the average water saturation $\overline{S}$ in the core, and the differential pressure Δp across the core. Then the function $\overline{k}_{ri}$ evaluated in $S = \overline{S}$ is computed from Darcy's law by

$$\bar{k}_{ri}(\bar{S}) = -\frac{\mu_i q_T f_i L}{\bar{k} \Delta p} \qquad (3.5)$$

Here $\bar{k}$ is the one-phase effective permeability for the composite core, which in the 1D case is given by the harmonic mean:

$$\frac{L}{\bar{k}} = \sum_{j=1}^{N} \frac{l_j}{k_j} \qquad (3.6)$$

The saturation distribution is given by (3.2), hence the average saturation $\bar{S}$ in the core is

$$\bar{S} = \frac{1}{L\bar{\varphi}} \sum_{j=1}^{N} l_j \varphi_j S_j \qquad (3.7)$$

where

$$L\bar{\varphi} = \sum_{j=1}^{N} l_j \varphi_j \qquad (3.8)$$

In particular, for $f_w = 0$ and $f_w = 1$ equation (3.7) defines the average irreducible saturations $\bar{S}_{wc}$ and $\bar{S}_{or}$ in the composite core.- Inserting (3.4) into (3.5) we obtain the following expressions for the functions $\bar{k}_{rw}(S)$ and $\bar{k}_{ro}(S)$ evaluated in $S = \bar{S}$:

$$\frac{L}{\bar{k}\bar{k}_{rw}(\bar{S})} = \sum_{j=1}^{N} \frac{l_j}{k_j k_{rwj}(S_j)} \qquad (3.9a)$$

$$\frac{L}{\bar{k}\bar{k}_{ro}(\bar{S})} = \sum_{j=1}^{N} \frac{l_j}{k_j k_{roj}(S_j)} \qquad (3.9b)$$

As f_w varies continuously in $[0,1]$, the average saturation $\bar{S}$ varies continuously in the interval $[\bar{S}_{wc}, 1-\bar{S}_{or}]$. Thus the functions $\bar{k}_{rw}$ and $\bar{k}_{ro}$ are uniquely determined on this interval by (3.9).

3.4 Discussion; Remarks

(3.4.1) Apparently, the functions $\bar{k}_{rw}$ and $\bar{k}_{ro}$ will depend on the viscosities μ_i, through the dependence of S_j on the fractional flow function f_{wj}. However, note that by (3.2), (S_j), $j=1,..,N$ is an allowable N-tuple of saturation values, corresponding to some f_w, if and only if

$$f_{w1}(S_1) = \ldots = f_{wN}(S_N) \; ; \qquad (3.10)$$

or, equivalently, by definition of the functions f_{wj}

$$\frac{k_{ro1}}{k_{rw1}}(S_1) = \ldots\ldots = \frac{k_{roN}}{k_{rwN}}(S_N) \qquad (3.11)$$

This shows that the saturation distribution is dependent on the relative permeability ratios only, thus the <u>steady state measured relative permeability functions are dependent on properties of the composite core only</u>.

(3.4.2) Denote by $\bar{f}_w(S)$ the water fractional flow function based on the steady state measured relative permeability functions :

$$\bar{f}_w(S) = \frac{\bar{k}_{rw}(S)}{\bar{k}_{rw}(S) + M\,\bar{k}_{ro}(S)} \qquad (3.12)$$

Fix any value $f_w \in [0,1]$ and let $\bar{S}$ be determined by (3.2) and (3.7). Then by (3.5) we get

$$\frac{\bar{k}_{ro}(\bar{S})}{\bar{k}_{rw}(\bar{S})} = \frac{1}{M} \left(\frac{1}{f_w} - 1 \right) \qquad (3.13)$$

which gives

$$\bar{f}_w(\bar{S}) = f_w \qquad (3.14)$$

or equivalently

$$L\bar{\varphi}\,\bar{f}_w^{-1}(f_w) = L\bar{\varphi}\,\bar{S} = \sum_{j=1}^{N} l_j \varphi_j f_{wj}^{-1}(f_w) \qquad (3.15)$$

Since (3.15) holds for any f_w, we obtain the following equality of functions defined on [0,1] :

$$\bar{f}_w^{-1} = \frac{1}{L\bar{\varphi}} \sum_{j=1}^{N} l_j \varphi_j f_{wj}^{-1} \qquad (3.16)$$

Equation (3.16) shows that the "average" functions $\bar{k}_{ro}$, $\bar{k}_{rw}$ of the given j pairs of relative permeability functions obtained by steady state measurement on the composite core, equation (3.9), determine a fractional flow function whose <u>inverse</u> is equal to the length-times-porosity weighted <u>pointwise</u> arithmetic average of the <u>inverses</u> of the j given fractional flow functions.

(3.4.3) Note that the relative permeability functions $\bar{k}_{ri}$ depend on all the parameters characterizing the composite core. For instance, $\bar{k}_{rw}$ will depend on the k_{roj}, k_j, φ_j, for all j = 1,..,N.

(3.4.4) The relative permeability functions $\bar{k}_{ri}$ are independent of the <u>order</u> of the homogeneous parts inside the composite core. Thus, if several of the homogeneous parts of the core have the same parameter set, $\bar{k}_{ri}$ will depend on the total length of these parts only.

(3.4.5) The formulas (3.9) show that in general it is not possible to express the function $\bar{k}_{ri}$ as a pointwise average of the functions k_{rij}, i.e., express $\bar{k}_{ri}(S)$ as an average of k_{rij} evaluated in the same saturation S. To make this quite explicit, choose $S_1 = S$, say, as the independent variable. Then formula (15a) reads

$$\frac{L}{\bar{k}\bar{k}_{rw}(S)} = \frac{l_1}{k_1 k_{rw}(S)} + \sum_{j=2}^{N} \frac{l_j}{k_j k_{rwj}(\Delta_{j1}(S))} \qquad (3.17)$$

This shows that $\bar{k}_{rw}$ (resp. $\bar{k}_{ro}$) is a pointwise average of the j relative permeability functions k_{rwj} (resp. k_{roj}) if and only if the discontinuity functions Δ_{j1} are all equal to the identity function, or equivalently, the fractional flow functions f_{wj} are all identical, or equivalently, the relative permeability ratios k_{roj}/k_{rwj} are all identical.

'(3.4.6) Steady state determined relative permeabilities are also studied in [11]. The relevance of the steady state relative permeabilities for a displacement process through the composite core is studied in the paper [2].

RELATIVE PERMEABILITY IN COMPOSITE CORES 659

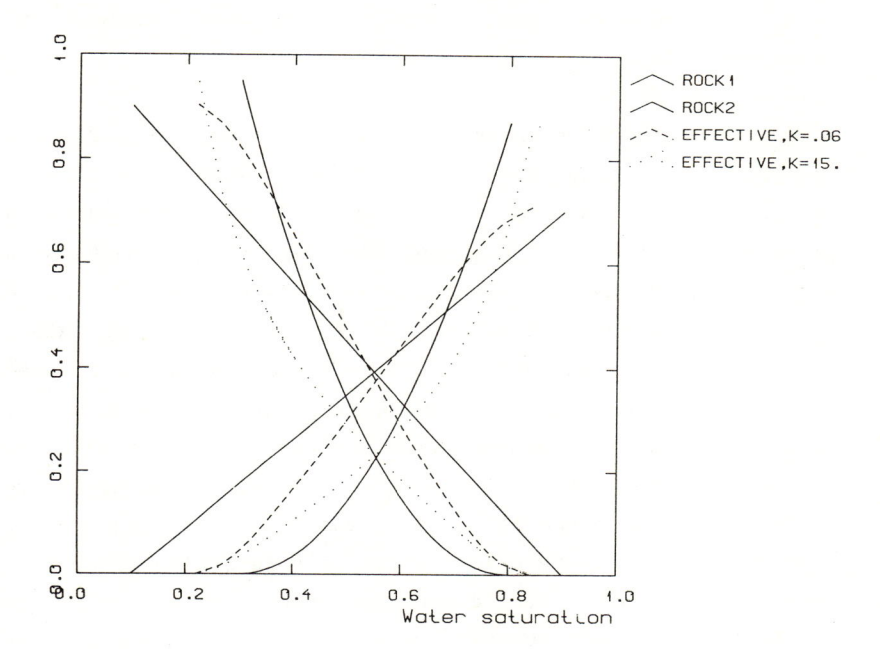

FIG. 1 STEADY STATE MEASURED (EFFECTIVE) RELATIVE PERMEABILITY

Fig.1 shows an example of steady state measured relative permeability (termed effective) for a composite core with N = 2 homogeneous parts of equal length. The result for two different absolute permeability ratios K = k_1/k_2 are plotted. The porosity ratio φ_1/φ_2 is equal to 0.6 for both cases.

3.5 Homogenization of a composite core with respect to steady state flow

Given a composite core C with data as in 2.1, the <u>homogenization problem</u> with respect to steady state flow in the core may be formulated as follows : Does there exist a homogeneous core $\overline{C}$ <u>with properties depending on those of C only</u>, such that steady state two-phase flow through C and $\overline{C}$ with the same constant value f_w of fractional flow of water, determines the same value of average saturation and differential pressure in C and $\overline{C}$, for any f_w? The porosity, absolute permeability and relative permeability functions for the homogeneous core $\overline{C}$ will then be called the <u>effective</u> properties of the composite core C, and $\overline{C}$ is the <u>equivalent homogeneous core</u> of the given composite core (with respect to steady state flow). Our discussion above shows that the homogenization problem with respect to steady state flow is indeed solvable. Explicitly, the effective porosity $\overline{\varphi}$, effective absolute permeability $\overline{k}$ and effective relative permeability functions $\overline{k}_{ri}$ solving the homogenization problem are given by (3.8), (3.6) and (3.9).

4. THE UNSTEADY STATE CASE

4.1 The unsteady state measured relative permeability

Recall the equations from which one determines the relative permeability functions for (a composite) core C by performing a displacement experiment on the core (see [6], or [7], Chapter 18): Initially the core is saturated with oil and water, with water everywhere at its irreducible saturation. At time t = 0 one starts injecting water into the core at the inlet end x = 0 with a Darcy velocity q_T remaining constant in time. During the displacement of oil from the core, one measures the cumulative production of water at the outlet end up till time t, $Q_w(t)$, as well as the differential pressure $\Delta p(t)$ across the core. Clearly

$$\frac{dQ_w}{dt}(t) = A q_T f_w(t) \qquad (4.1)$$

Here $f_w(t)$ is the value of fractional flow of water at the outlet end $x = L$ at time t. Thus we have $f_w(t) = f_{wN}(S(L,t))$ where $S(L,t)$ is the true outlet water saturation at time t. The <u>apparent value</u> $\underline{S}(L,t)$ of outlet water saturation at time t is then defined by and computed from the equation

$$\underline{S}(L,t) = \bar{S}(t) - \frac{q_T t}{L\bar{\varphi}}\left[1 - \frac{1}{A q_T}\frac{dQ_w}{dt}(t)\right] \qquad (4.2)$$

where $\bar{S}(t)$ is the average water saturation in the core at time t, obviously given by

$$\bar{S}(t) = \bar{S}_{wc} + \frac{A q_T t - Q_w(t)}{AL\bar{\varphi}} \qquad (4.3)$$

The relative permeability functions $\underline{k}_{rw}$ and $\underline{k}_{ro}$ for the composite core, evaluated in the apparent saturation $\underline{S}(L,t)$, are now determined from the following two linear, algebraic equations:

$$\frac{\underline{f}_w}{\underline{k}_{rw}} = \frac{1}{\underline{k}_{rw} + \frac{\mu_w}{\mu_o}\underline{k}_{ro}} = \frac{\bar{k}}{L q_T \mu_w}\left[\Delta p(t) - t\frac{d\Delta p}{dt}(t)\right] \qquad (4.4)$$

$$\underline{f}_w = \cfrac{1}{1 + \cfrac{\mu_w \underline{k}_{ro}}{\mu_o \underline{k}_{rw}}} = \frac{1}{A\,q_T}\frac{dQ_w}{dt}(t) \qquad (4.5)$$

Here $\underline{f}_w$ is the fractional flow function based on the unsteady state measured relative permeability functions, see (2.5). If the core C is homogeneous, then (4.2) gives the <u>true</u> value of water saturation at $x = L$ at time t, see [7], p. 69. Furthermore, the relative permeability curves calculated will be identical to the rock curves. To see this, one computes from Buckley-Leverett theory

$$\Delta p(t) - t\frac{d\Delta p}{dt}(t) = \frac{q_T \mu_w L}{k}\frac{f_w}{k_{rw}}(S(L,t)) \qquad (4.6)$$

Inserting (4.6) into (4.4) results in $\underline{k}_r = k_r$. In general, we refer to the relative permeabilities $\underline{k}_{rw}$, $\underline{k}_{ro}$ determined by this procedure as the <u>apparent relative permeability curves</u> for the composite core. Thus, we obtain these curves by interpreting the measured quatities $Q_w(t)$ and $\Delta p(t)$ as if they were produced from displacement on a <u>homogeneous</u> core.

4.2 The apparent relative permeability for a composite core

Our first basic problem is to understand the nature of the dependence of the apparent relative permeabilities on the rock parameters and lengths of the homogeneous parts constituting the core. Unfortunately, for the general case we have obtained simple analytical formulas with respect to saturation behaviour only, see (4.7). Therefore, with respect to differential pressure behaviour we state only a result that is necessary for a discussion of the homogeneization problem, see (4.8).

To study the apparent relative permeabilities, there are two basic equations valid for the homogeneous core case which need to be generalized to the composite core case, namely (4.2) and (4.6). First, a generalization of (4.2) is given by the following result, relating the "true" outlet saturation in the composite core at time t, S(L,t), and the apparent outlet saturation $\underline{S}(L,t)$ computed by the equation (4.2) after measuring $Q_w(t)$:

$$\underline{S}(L,t) = \frac{1}{L\bar{\varphi}} \sum_{j=1}^{N} l_j \varphi_j \Delta_{jN}(S(L,t)) \qquad (4.7)$$

The equation (4.7) holds for t ≥ 0. - Secondly, the expression generalizing (4.6) we give only for the simple case N = 2 and $\Delta_{12}(S)$ = identity (or equivalently, $f_{w1} = f_{w2} = f_w$):

$$\Delta p(t) - t \frac{d\Delta p}{dt}(t) = q_T \mu_w \left[\frac{\bar{\varphi} L}{\varphi_2 k_2} \frac{f_w}{k_{rw2}}(S(L,t)) \right. \qquad (4.8)$$

$$\left. - L_1 \left(\frac{\varphi_1}{\varphi_2 k_2} \frac{f_w}{k_{rw2}}(S(L_1,t)) - \frac{1}{k_1} \frac{f_w}{k_{rw1}}(S(L_1,t)) \right) \right)$$

which holds for t ≥ t_{BT} = breakthrough time in the composite core. The derivation of these expressions is given in the Appendix.

4.3 Example We apply the generalized equations (4.7) and (4.8) to explicitly compute the apparent relative permeability curves for a very simple example. Namely, we take N = 2, and we let the two homogeneous parts of the composite core have the same rock relative permeability curves k_{rw}, k_{ro} which we assume to be straight lines. Moreover, for simplicity we assume the irreducible saturations to be zero. Finally, we set M = μ_w/μ_o < 1. With these assumptions no shock front will occur, and we have

$$f_w(S) = \frac{S}{(1-M)S + M} \qquad (4.9)$$

By (4.6), note that for this case the apparent outlet saturation $\underline{S}(L,t)$ will be identical to the true outlet saturation $S(L,t)$, since Δ_{12} is equal to the identity. Thus it follows from (4.5) that $\underline{f}_w(S) = f_w(S)$ for all S, hence

$$\underline{k}_{ro} = \frac{k_{ro}}{k_{rw}} \underline{k}_{rw} \qquad (4.10)$$

which shows that it is sufficient to compute the apparent relative permeability to water.

To compute $\underline{k}_{rw}$ we start from (4.4) with $S = S(L,t)$:

$$[\underline{k}_{rw}(S)]^{-1} = \frac{\bar{k}}{f_w(S)q_T L \mu_w} [\Delta p(t) - t \frac{d\Delta p}{dt}(t)] \qquad (4.11)$$

Next, an expression for the factor involving $\Delta p(t)$ is given by (4.8), which in its turn involves the saturation $S_1 = S(L_1,t)$. Now, as is easy to see, an analytical expression for S_1 will depend on whether $t_{BT} \geq t_1$ or $t_{BT} \leq t_1$, where t_1 is the time when steady state is reached throughout the first homogeneous part of the composite core Thus, by definition of t_1 we have $t \geq t_1$ implies $S(x,t) = 1$ for $x \leq L_1$.

First assume $t_{BT} \geq t_1$. Then $S_1 = 1$ for $t \geq t_{BT}$, and a combination of (4.8) and (4.11) gives

$$[\underline{k}_{rw}(S)]^{-1} = B[k_{rw}(S)]^{-1} + C[f_w(S)]^{-1}, \quad S \in (0,1] \qquad (4.12)$$

where the constants B, C are given by

$$B = \frac{\bar{k}\,\bar{\varphi}}{k_2 \varphi_2}, \quad C = \frac{L_1 \bar{k}}{L}\left[\frac{1}{k_1} - \frac{\varphi_1}{\varphi_2 k_2}\right] \qquad (4.13)$$

Next assume $t_{BT} \leq t_1$. Then we obtain by Buckley-Leverett theory the following formula for S_1 as a function of $S(L,t)$:

$$(1-M)S_1 + M = \left(\frac{L\,\bar{\varphi}}{L_1 \varphi_1}\right)^{\frac{1}{2}}[(1-M)S + M] \qquad (4.14)$$

Combination of (4.14), (4.8) and (4.11) finally produces

$$\underline{k}_{rw}(S(L,t)) = \frac{L}{\bar{k}\,D}\,k_{rw}(S(L,t)), \quad t \in [t_{BT}, t_1] \qquad (4.15)$$

where the constant D is given by

$$D = \frac{\bar{\varphi}\,L}{\varphi_2 k_2} - L_1\left(\frac{\varphi_1}{\varphi_2 k_2} - \frac{1}{k_1}\right)\left(\frac{L_1 \varphi_1}{L\,\bar{\varphi}}\right)^{\frac{1}{2}} \qquad (4.16)$$

It is easy to compute

$$S(L,t_1) = \frac{\left(\frac{L_1 \varphi_1}{L\,\bar{\varphi}}\right)^{\frac{1}{2}} - M}{1 - M} \stackrel{\text{Def.}}{=} S_0 \qquad (4.17)$$

hence we have the following result: If $S_0 > 0$, then the apparent relative permeability is given by (4.15) for $0 \leq S \leq S_0$ and by (4.12) for $S \geq S_0$. If $S_0 < 0$, then the apparent relative permeability is given by (4.12) for all S.

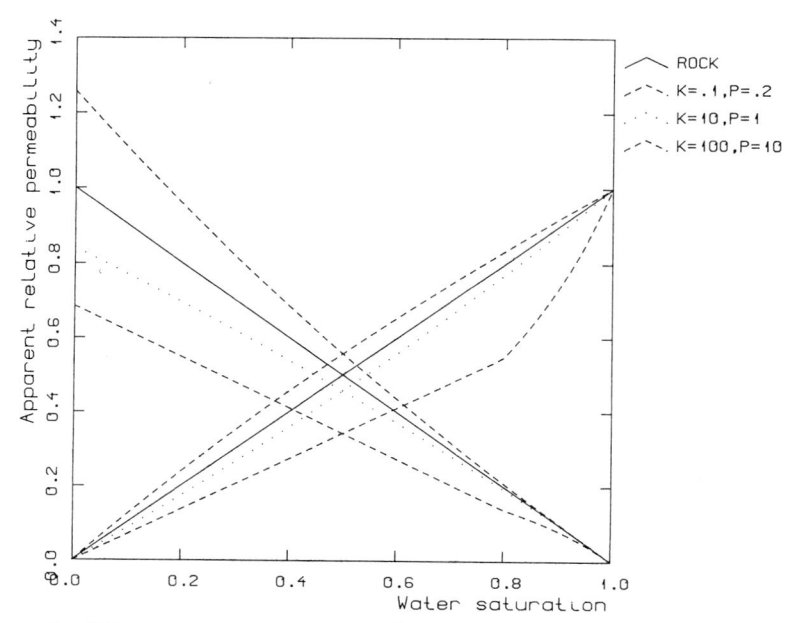

FIG.2 UNSTEADY STATE MEASURED (APPARENT) RELATIVE PERMEABILITY.
DEPENDENCE ON ABSOLUTE PERMEABILITY- AND POROSITY RATIO.

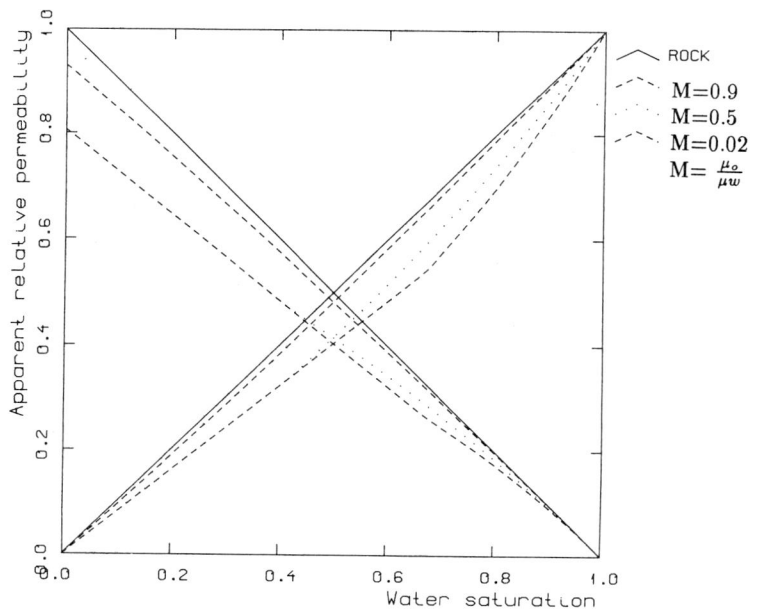

FIG.3 UNSTEADY STATE MEASURED (APPARENT) RELATIVE PERMEABILITY.
DEPENDENCE ON VISCOSITY RATIO.

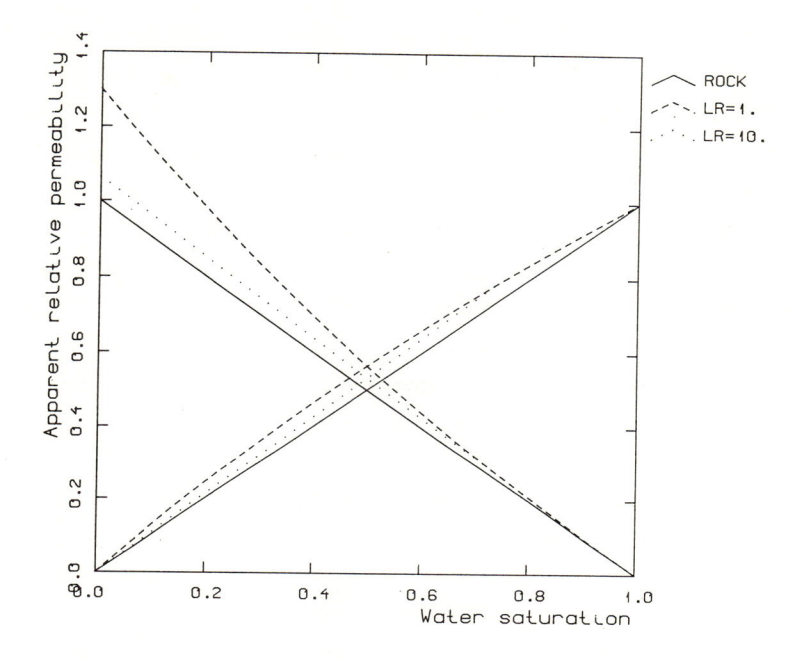

FIG.4 UNSTEADY STATE MEASURED (APPARENT) RELATIVE PERMEABILITY.
DEPENDENCE ON LENGTH RATIO.

Fig.2 - Fig.4 show the apparent relative permeability curves for this example for different values of the parameters. Here $K = k_1/k_2$ denotes the absolute permeability ratio, $P = \varphi_1/\varphi_2$ the porosity ratio and $LR = l_1/l_2$ the length ratio. Note that the apparent relative permeability for the displacing phase (water) always has the value 1. for the maximum saturation S = 1., while $\underline{k}_{ro}(S=0) \neq 1$ in general. The reason is that $\Delta p(t) - t(d\Delta p/dt)$ at $t = t_{BT}$ in general will be different from the value of Δp for steady state flow of oil through the composite core, and equality of these quantities is exactly the condition for $\underline{k}_{ro}(S=0) = 1$ to hold. - Finally we remark that the apparent relative permeabilty curves shown here have characteristics very similar to those computed numerically for some 2D cases in [3].

4.4 Homogenization of a composite core with respect to unsteady state flow; Statement of results

Consider a standard displacement experiment on the composite core C: Initially the core is saturated with oil and water, with water everywhere at its irreducible saturation. At time t = 0 one starts injecting water into the core at the inlet end x = 0 with a Darcy velocity q_T remaining constant in time. As the "global behaviour" of the displacement one measures the two functions

$f_w(t)$ = fractional flow of water at the outlet end x = L at time t ;

$\Delta p(t)$ = $p(0,t) - p(L,t)$ = differential pressure across the core at time t ;

The homogenization problem for C with respect to unsteady state flow may now be formulated as follows : Does there exist a homogeneous core $\underline{C}'$ with properties depending on those of C only, such that displacement experiments in C and $\underline{C}'$, with the same water injection velocity q_T and same fluid viscosity ratio M result in identical global behaviours of C and $\underline{C}'$, for any values of q_T and M ?

We will define the absolute permeability and porosity of
$\underline{C}'$ to be determined by (3.6) and (3.8). Thus, the real problem
consists in determining the relative permeability functions
for $\underline{C}'$, which will be called the effective relative
permeability functions for C with respect to unsteady state
flow. Promising candidates are of course the apparent
relative permeability functions for the composite core C.
However, we have seen that these relative permeability
functions depend in an essential way on the viscosity ratio
for the <u>fluids</u> used in generating them, hence they cannot
strictly solve the homogenization problem as formulated
above. In the following we state our main results concerning
the unsteady state homogenization problem. In particular we
discuss to what extent the problem is solved by the apparent
relative permeabilities.

Thus, for the rest of the paper we now let $\underline{C}$ denote the
homogeneous core of length L and cross section A, whose
porosity $\overline{\varphi}$ and absolute permeability $\overline{k}$ are determined by (3.6)
and (3.8). Further, the relative permeability curves of $\underline{C}$ are
the apparent relative permeabilities of the composite core C,
determined during a displacement experiment on C with total
rate q_T and viscosity ratio M. Let this displacement
experiment on C result in a breakthrough time t_{BT}, pressure
drop function $\Delta p(t)$ and fractional flow function $f_w(t)$ at the
outlet end $x = L$. Also consider a displacement experiment on
the homogeneous core $\underline{C}$, with total rate $\underline{q}_T$ and viscosity
ratio $\underline{M}$, giving a breakthrough time $\underline{t}_{BT}$, pressure drop funtion
$\underline{\Delta p}(t)$ and water fractional flow function $\underline{f}_w(t)$ at the outlet.
We now compare the output data from these two experiments.

4.4.1 Homogenization with respect to fluid production
(a) Assume that $q_T = \underline{q}_T$. Then in general we have $t_{BT} \neq \underline{t}_{BT}$,
even if $M = \underline{M}$.

Remark To show that the homogeneous core $\underline{C}$ will not in general reproduce the breakthrough time in the composite core C, it is sufficient to give a counterexample. For this, pick a composite core C with the property that its breakthrough time depends on the ordering of the homogeneous parts inside the core. For instance, a two-part composite core whose two pairs of relative permeability curves are straight lines and parabolas, respectively, satisfies this requirement, as one may check by Buckley-Leverett theory. (The paper [1] also studies such cores.) Now, by (4.20) it follows that the fractional flow function $\underline{f}_w$ based on the apparent relative permeabilities for C is independent of the ordering of the homogeneous parts inside C. Therefore, there is some ordering of C whose t_{BT} the fractional flow curve $\underline{f}_w(S)$ cannot reproduce.

(b) On the other hand, if $M = \underline{M}$ and we let $r = \underline{q}_T/q_T$, then we do have

$$\underline{f}_w(t) = f_w(rt) \quad \text{for } t \geq \max(t_{BT}, \underline{t}_{BT}) \qquad (4.18)$$

Thus, in particular, under identical fluid/injection conditions, C and $\underline{C}$ will have identical fractional flow behaviours from the moment both cores produce water.

Remark This result shows that apart from reproduction of breakthrough time, $\underline{C}$ will in fact exactly reproduce the outlet fractional flow behaviour of C, at least for a viscosity ratio equal to that used in the generation of $\underline{f}_w$. The proof of this result is given in the Appendix, Section 9.2.

(c) The homogeneous core $\underline{C}$ will not reproduce the outlet saturation value $S(L,t)$ in C, not even for $t \geq \max(t_{BT}, \underline{t}_{BT})$.

However, this will be the case if the fractional flow functions f_{wj} are all identical. In this case we have in fact $t_{BT} = \underline{t}_{BT}$. In general, however, $\underline{C}$ will reproduce the apparent outlet saturation $\underline{S}(L,t)$ in C for $t \geq \max(t_{BT}, \underline{t}_{BT})$ when $M = \underline{M}$, $q_T = \underline{q}_T$.

Remark That $\underline{C}$ will reproduce the apparent outlet saturation of C is proved in the Appendix, Section 9.2.

(d) There will always exist relative permeability curves to be put into $\underline{C}$ such that (b) holds for all $t \geq 0$.

Remark More precisely, one may show that there exist an infinite number of different pairs of relative permeability curves exactly reproducing any given outlet fractional flow vs. time behaviour. We omit the proof of this purely formal result.

4.4.2 Homogenization with respect to differential pressure

(e) In general, $\underline{C}$ will not reproduce the differential pressure function in C. In fact, no homogeneous core whatsoever can reproduce the differential pressure function for a displacement in a composite core.

Remark This result is based on a characteristic property of the differential pressure function $\Delta p(t)$ in homogeneous cores. Namely, $\Delta p(t)$ is a linear function of time for $t \leq t_{BT}$ (see [12]). On the other hand, it is not difficult to specify a composite core for which $\Delta p(t)$ does not have this property. For instance, one may use the core given in the Example above.

(f) On the other hand, at least for two-part composite cores which are homogeneous with respect to the fractional flow functions, we do have

$$\underline{\Delta}p(t) = \Delta p(t) \text{ for } t \geq t_1 \quad (4.19)$$

where t_1 denotes the time when the first part of C have reached steady state flow of water only.

Remark To prove this claim we apply the generalized formula (4.8). We omit the proof.

4.4.3 Relationship between the steady state measured and unsteady state measured relative permeability

(g) Let C be a composite core with fractional flow functions f_{wj}. As usual, denote by $\bar{k}_{rw}$, $\bar{k}_{ro}$ and $\underline{k}_{rw}$, $\underline{k}_{ro}$ the steady state and unsteady state measured relative permeability functions for C, respectively. Denote by $\bar{f}_w$ and $\underline{f}_w$ the saturation dependent fractional flow functions based on $\bar{k}_r$ and $\underline{k}_r$, respectively. Then we have

$$\bar{f}_w^{-1} = \underline{f}_w^{-1} = \frac{1}{L\bar{\varphi}} \sum_{j=1}^{N} l_j \varphi_j f_{wj}^{-1} \quad (4.20)$$

or equivalently

$$\frac{\bar{k}_{rw}}{\bar{k}_{ro}} = \frac{\underline{k}_{rw}}{\underline{k}_{ro}} \quad (4.21)$$

in their common domain of definition.

Remark To prove (g), recall that by definition of $\underline{f}_w$ we have $\underline{f}_w(\underline{S}(L,t)) = f_{wN}(S(L,t))$, see (4.5) and the remark following (4.1). Thus, since Δ_{NN} = identity, the relation (4.7) immediately implies that

$$f_w^{-1} = \frac{1}{L\bar{\varphi}} \sum_{j=1}^{N} l_j \varphi_j f_{wj}^{-1} \tag{4.22}$$

Now compare (4.22) with (3.16), and the claims follow.

5. ACKNOWLEDGEMENTS

This research has been supported by a British-Norwegian research program financed by Den Norske Stats Oljeselskap a.s. (Statoil).

6. CONCLUSIONS

1. We have studied properties of the two pairs of relative permeability curves obtained by applying the steady state and unsteady state measuring techniques to a composite core, assuming that its homogeneous parts have different pairs of rock relative permeability curves.

2. We show that the relative permeability curves obtained from a steady state measurement only depend on the rock characteristics of the homogeneous parts constituting the core. An explicit formula is given. Moreover, we show that these relative permeabilities solve the homogenization problem for steady state flow in the core, with respect to both saturation and pressure behaviour.

3. On the other hand, we show that the relative permeabilities obtained by an unsteady state experiment on the composite core will in general not solve the homogenization problem for displacement in the core. We also discuss in some detail to what extent they fail to solve this problem.

4. Our main technical results are:

(a) The formula (4.7), giving the relation between the apparent and the true outlet saturation during displacement in the composite core.

(b) The formulas (3.16) and (4.22), showing that the steady state - and the unsteady state determined relative permeabilities have identical ratios, (4.22).

(c) The formulas (3.9), giving the dependence of the steady state determined relative permeabilities on the rock characteristics and the discontinuity functions of the core.

7. REFERENCES

[1] Codreanu,D. et al.: Le Deplacement de l´Huile par l´Eau dans les Roches Heterogenes et l´Evaluation de l´Heterogeneite dans les Calculs de "Reservoir Engineering"", Revue de IFP, (Jan. 1966).

[2] Dale,M.: "Effective Relative Permeability for a 1-Dimensional Heterogeneous Reservoir," Paper presented at the NIPER/DOE Second International Reservoir Characterization Technical Conference, held in Dallas, July 1989.

[3] Haldorsen, H.H. :" Reservoir Characterization Procedures for Numerical Simulation," Dissertation, The University of Texas at Austin, May 1983.

[4] Huppler,J.D.: "Waterflood Relative Permeabilities in Composite Cores," JPT (May 1969) 539-40.

[5] Huppler,J.D.: "Numerical Investigation of the Effects of Core Heterogeneities on Waterflood Relative Permeabilities," SPEJ (Dec. 1970) 381-92.

[6] Johnson,E.F.,Bossler, D.P. and Naumann,V.O.: "Calculation of Relative Permeability from Displacement experiments," Trans. AIME, 1959, Vol.216, p.370.

[7] Marle, M.G.: Multiphase Flow in Porous Media.

[8] Muggeridge,A.H. : "Generation of Pseudo Relative Permeabilities from Detailed Simulation of Flow in Heterogeneous Porous Media," Paper presented at the NIPER/DOE Second International Reservoir Characterization Technical Conference, held in Dallas, July 1989.

[9] Pande,K.K. et al. : " Frontal Advance Theory for Flow in Heterogeneous Porous Media," SPE Paper No. 16344, April 1987.

[10] Sigmund, P.M. and McCaffery, F.G.: " An Improved Unsteady-State Procedure for Determining the Relative Permeability Characteristics of Heterogeneous Porous Media," SPEJ (Feb. 1979) 15-28.

[11] Smith,E.H. : "The influence of correlation between capillary pressure and permeability on the average relative permeability of reservoirs containing small scale heterogeneity," Paper presented at the NIPER/DOE Second International Reservoir Characterization Technical Conference, held in Dallas, July 1989.

[12] Watson,A.T. et al.: "A Test for Detecting Rock Property Non-Uniformities in Core Samples," SPEJ, (Dec. 1985)

8. NOMENCLATURE
Roman letters:

A	core cross section area
B	constant, defined in (4.13)
C	constant, defined in (4.13)
C	composite core
C	homogeneous core equivalent to C with respect to steady state flow, see section 3.5
$\underline{C}'$	homogeneous core equivalent to C with respect to unsteady state flow
$\underline{C}$	homogeneous core with properties derived from those of C: porosity = $\bar{\varphi}$, absolute permeability = $\bar{k}$, relative permeability curves $\underline{k}_{rw}$, $\underline{k}_{ro}$ = apparent (unsteady state measured) relative permeabilities of C.
f_i	fractional flow of phase i, i = o(il),w(ater)
$f_w(t)$	outlet water fractional flow of C
$\underline{f}_w(t)$	outlet water fractional flow of $\underline{C}$
f_{wj}	saturation dependent water fractional flow function based on the rock relative permeabilities for C's j-th homogeneous part
$\underline{f}_w(S)$	saturation dependent water fractional flow function based on the apparent relative permeabilities $\underline{k}_{rw}$, $\underline{k}_{ro}$
$\bar{f}_w(S)$	saturation dependent water fractional flow function based on the steady state determined relative permeabilities $\bar{k}_{rw}$, $\bar{k}_{ro}$ of C
i	phase index, i = o,w
j	index for homogeneous part of composite core, j=1,...,N
k_j	absolute permeability for j-th homogeneous part of C
k_{rwj}	rock relative permeability to water for C's j-th homogeneous part
k_{roj}	rock rel. perm. to oil for C's j-th homogeneous part
k'_{rw}	end point rel. perm. to water, see section 2.1
k'_{ro}	end point rel. perm. to oil, see section 2.1

$\bar{k}$	effective absolute permeability for C, see (3.6)
$\bar{k}_{rw}$	steady state determined water rel. perm. for C
$\bar{k}_{ro}$	steady state determined oil rel. perm. for C
$\underline{k}_{rw}$	unsteady state determined water rel. perm. for C
$\underline{k}_{ro}$	unsteady state determined oil rel. perm. for C
L_j	total length of j first homogeneous parts of C
l_j	length of j-th homogeneous part of C
M, $\underline{M}$	viscosity ratio μ_w/μ_o
N	number of homogeneous parts in C
p	pressure (water pressure = oil pressure)
$\underline{q}_T, q_T$	total Darcy velocity (= water injection velocity)
q_i	Darcy velocity for phase i
Q_w	cumulative production of water at outlet end of C
S_i	saturation of phase i
S	water saturation
S_{wcj}	connate water saturation for C's j-th homogeneous part
S_{orj}	residual oil saturation for C's j-th homogeneous part
S_j	steady state water saturation distribution in C
$\bar{S}$	average water saturation in C
S(L,t)	"true" outlet water saturation in C
$\underline{S}$(L,t)	"apparent" outlet water saturation in C, see (4.2)
S_1	$= S(L_1,t)$
S_0	$= S(L,t_1)$
t	time coordinate
t_1	time when steady state is reached in first homogeneous part of C
t_{BT}	breakthrough time in C
$\underline{t}_{BT}$	breakthrough time in $\underline{C}$

Greek letters:

Δ_{kl}	discontinuity function, see (2.7)
Δp	pressure drop across composite core C
$\underline{\Delta p}$	pressure drop across homogeneous core $\underline{C}$

μ_i viscosity for phase i
φ_j porosity for C's j-th homogeneous part
$\bar{\varphi}$ average porosity for C (= porosity for $\underline{C}, \bar{C}$)

9. APPENDIX

9.1 Derivation of (4.7) and (4.8)

Consider a standard displacement experiment on the composite core C: Initially the core is saturated with oil and water, with water everywhere at its irreducible saturation. At time t = 0 one starts injecting water into the core at the inlet end x = 0 with a Darcy velocity q_T remaining constant in time. We will derive the generalized formulas (4.7) and (4.8) under these assumptions. The basic idea is the same as that used in [7] for a homogeneous core, with the modifications necessary to treat the composite core case.

First consider (4.7). For simplicity we will consider the case N = 2 only. The general case is only notationally more complicated. Denote by S(t) the average water saturation in the core at time t. Then clearly

$$\bar{S}(t) = \frac{\varphi_1}{\bar{L}\varphi} \int_0^{L_1^-} S(x,t)dx + \frac{\varphi_2}{\bar{L}\varphi_1} \int_L^L S(x,t)dx \quad (9.1)$$

Here S(x,t) is the saturation distribution along the core at time t. Note that by the required continuity of Darcy velocities, we have

$$\lim_{x \to L_1^-} S(x,t) = \Delta_{12}(S(L_1,t)) \quad (9.2)$$

RELATIVE PERMEABILITY IN COMPOSITE CORES 679

We will determine expressions for the two integrals in (9.1) separately. The first (improper) integral is treated exactly as in the homogeneous case, see [7], by partial integration and application of the basic Buckley-Leverett relation

$$xdS = \frac{q_T t}{\varphi_1} \frac{df_{w1}}{dS}(S) dS \quad (9.3)$$

This gives

$$\int_0^{L_1^-} S(x,t) = L_1 \Delta_{12}(S(L_1,t)) - \frac{q_T t}{\varphi_1} [\lim_{x \to L_1} f_{w1}(S(x,t)) - 1] \quad (9.4)$$

Next consider the second integral in (9.1). Partial integration leads to an integral of the form

$$\int_{S(L_1,t)}^{S(L,t)} xdS \quad (9.5)$$

To compute (9.5), define the function $t(S)$, $S \in (S_{wc2}, 1-S_{or2})$, by $S(L_1, t(S)) = S$. Thus $t(S)$ = "time when the saturation S arrives at $x = L_1$. For x lying in medium 2, the saturation $S = S(x,t)$ then satisfies

$$xdS = [L_1 + \frac{q_T(t - t(S))}{\varphi_2} \frac{df_{w2}}{dS}(S)] dS \quad (9.6)$$

Inserting (9.6) into (9.5), we need finally to determine the integral of the function $t(S)(df_{w2}/dS)$. For this, we remark that $t(S)$ may also be defined as " $t(S)$ = time needed for the saturation $\Delta_{12}(S)$ to travel the distance L_1 in medium 1." Hence, by Buckley-Leverett theory there is the relation

$$L_1 = \frac{q_T t(S)}{\varphi_1} \frac{df_{w1}}{dS}(\Delta_{12}(S)) \qquad (9.7)$$

On the other hand we have $f_{w1}(\Delta_{12}(S)) = f_{w2}(S)$, see (2.7), hence by the chain rule we obtain

$$t(S) \frac{df_{w2}}{dS}(S) = \frac{L_1 \varphi_1}{q_T} \frac{d\Delta_{12}}{dS}(S) \qquad (9.8)$$

Application of (9.8) now produces

$$\bar{S}(t) = S(L,t) + \frac{q_T t}{L \bar{\varphi}}(1 - f_{w2}(S(L,t))) + \frac{L_1 \varphi_1}{L \bar{\varphi}}(\Delta(S(L,t)) - S(L,t)) \qquad (9.9)$$

Now note that the difference between the left hand side and the second term of the right hand side of (9.9) is by definition equal to the apparent effluent saturation in the composite core, see (4.2). Thus rearrangement of (9.9) gives exactly (4.7).

Now consider (4.8). A closed analytical expression for $\Delta p(t) - t(d\Delta p/dt)$ in the composite core case will be derived only when $f_{w1} = f_{w2} = f_w$, or equivalently, $\Delta_{12}(S) =$ identity. We start from the equation (3.4):

$$\Delta p(t) = \mu_w q_T [\frac{1}{k_1} \int_0^{L_1} \frac{f_w}{k_{rw1}}(S(x,t))dx + \frac{1}{k_2} \int_{L_1}^{L} \frac{f_w}{k_{rw2}}(S(x,t))dx] \qquad (9.10)$$

Denote the first integral by $A(t)$, the second one by $B(t)$. We first treat the integral $A(t)$ exactly as in [7] : Change independent variable from x to df_w/dS, using (9.3). Differentiate the resulting integral with respect to t, to obtain

$$A(t) - t\frac{dA}{dt}(t) = L_1 \frac{f_w}{k_{rw1}}(S(L_1,t)) \qquad (9.11)$$

Similarly, to treat B(t), change variable from x to df_w/dS, now using (9.6) and (9.8) with $\Delta_{12}(S)$ = identity. Computation of $B(t)-t(dB/dt)$ is now straightforward. Combined with (9.11) it results in the expression (4.48) for $A(t) - t(d\Delta p/dt)$.

9.2 Further proofs

Here we give the proofs of part (b) and part (c) of section 4.4.1 . Part (b) is an immediate consequence of part (c), hence we first prove that

(c) For $t \geq \max(t_{BT}, \underline{t}_{BT})$, the homogeneous core $\underline{C}$ reproduces the apparent outlet saturation in C (for M = $\underline{M}$, $q_T = \underline{q}_T$).

Proof: To show (c), let $S'(L,t)$ denote the true outlet saturation in $\underline{C}$. Recall that by Buckley-Leverett theory in the homogeneous core $\underline{C}$, $S'(L,t)$ is uniquely determined for $t \geq \underline{t}_{BT}$, by

$$\frac{\underline{q}_T t}{\overline{\varphi}} \frac{d\underline{f}_w}{dS}(\underline{S}'(L,t)) = L \qquad (9.12)$$

Thus (c) will follow if we can show that the apparent saturation function $\underline{S}(L,t)$ also satisfies (9.12) for $t \geq t_{BT}$. To check this, first note that from (4.5) we have for $t \geq t_{BT}$

$$\frac{d\underline{f}_w}{dS}(\underline{S}(L,t))\frac{d\underline{S}}{dt}(t) = \frac{1}{A\,\underline{q}_T}\frac{d^2\underline{Q}_w}{dt^2}(t) \qquad (9.13)$$

and by (4.2) and (4.3)

$$\frac{d\underline{S}}{dt}(t) = \frac{t}{AL\bar{\varphi}} \frac{d Q_w^2}{dt^2}(t) \qquad (9.14)$$

This gives

$$\frac{d\underline{f}_w}{d\underline{S}}(\underline{S}(L,t)) = \frac{L\bar{\varphi}}{q_T t} \quad \text{for } t \geq t_{BT} \qquad (9.15)$$

and the claim is proved. Part (b) now follows immediately, since

$$\begin{aligned}\underline{f}_w(t) &= \underline{f}_w(\underline{S}(L,t)) &&\text{by part (c)} \\ &= \underline{f}_w(t) &&\text{by definition of } \underline{f}_w\end{aligned}$$

A Reservoir Simulator Based on Front-Tracking

F. Bratvedt[1], K. Bratvedt[1], C. Buchholz[1], D. T. Rian[2] and N.H. Risebro[1].

1) Dep. of Mathematics, University of Oslo, P.O. Box 1053, 0316 OSLO 3, Norway.
2) Scandpower A/S, P. O. Box 3 N-2007 Kjeller, Norway.

0. ABSTRACT

A method of solving the saturation equations (first order hyperbolic conservation laws) by approximating the fractional flow function by a piecewise linear function has been developed. The method makes it possible to solve the saturation equations without stability problems. In this new front-tracking method, the pressure equation is solved by a finite element method. The grid for the pressure equation can therefore be fitted to the reservoir geometry with great flexibility.

A reservoir simulator based on the new methods for the solution of saturation equations is under development. The simulator is able to simulate two-phase immiscible flow in two dimensions at present. Both gravity and compressibility are included, as well as heterogeneities in the geological data. The simulator has been used for field simulation problems. It has proved to be insensitive to grid orientation and numerical dispersion. In addition, for larger grid systems, the simulator is computationally more efficient than finite difference simulators.

A simpler version of the front-tracking simulator can also be used as a streamline simulator. The streamline simulator has the same CPU-efficiency and front-tracking capabilities as traditional streamtube simulators without the need to transform the reservoir into a set of one-dimensional regions.

1. INTRODUCTION

The reservoir simulators most commonly used in industry today are based on finite difference methods to approximate the partial differential equations that describe fluid flow in porous media. Usually a five point difference scheme is used. Since the pressure and the saturations are solved implicitly in most industrial simulators and stable solutions require restricted time step sizes, such models can be CPU-consuming when simulating larger grid systems. Unwanted grid effects can also occur [28]. Hence, such simulators have their limitations when modeling irregular reservoir geometries. In addition, numerical dispersion can be dominating when using large grid block dimensions. This creates a need for larger grid systems to obtain reliable solutions. Some of these problems can be solved by alternative methods described in [5] through [9], [11], [12] and [27].

The front-tracking simulator described in this paper is based on a different technology. The theory behind the numerical method is explained in [3], [4], [10], [16] through [22] and [26]. The equations used to describe the flow of fluids in porous media are the traditional equations of mass conservation and Darcy´s law, which by combination yield a set of coupled, nonlinear partial differential equations. These equations are split into an equation for pressure and equations for saturation. A solution is found to the pressure equation first, and then with basis in the established velocity field, the saturation equations are solved by a front-tracking method.

The pressure solution is obtained with a finite element method, which is accurate with respect to a general grid architecture. The grid system is based on triangular grid cells, which can adapt to almost any geometry (node-based grid).

The method for solving the saturation equations is based on an approximation of the fractional flow function by a piecewise linear function. This method makes it possible to solve the saturation equations without stability problems and unbound by the timestep length limiting CFL condition.

A reservoir simulator based on this method is under development. The goal of this development work is a three-dimensional, three-phase simulator. Currently, the simulator is limited to two-dimensional, two-phase black oil problems.

As a special project, a streamline simulator has been developed. Its purpose is to supply the needs of industry which are normally satisfied by simulators based on a streamtube concept. The traditional streamtube simulators are based on methods that transfer the reservoir to a set of one-dimensional problems (streamtubes) [13], [14], [15], [23] and [25], and the convective part of the saturation equation is solved for these one-dimensional streamtubes. The result is mapped back to the two-dimensional grid.

2. EQUATIONS AND SOLUTIONS

The equations that govern the flow in porous media, Darcy's law and the mass conservation equations are found in Appendix A. By combining Darcy's law and the mass conservation equations we get a set of coupled, non-linear partial differential equations. These can be separated into a pressure equation and saturation equations.

Solution of the Pressure Equation

The well terms q_n and q_w are set to zero. The production / injection terms are modeled by using the divergence theorem for the element formulation of the equation. We also set the capillary pressure P_c to zero, so that the phase pressures are equal to the average pressure. We assume a black oil formulation [24]. The pressure equation can be written:

$$-\nabla [\alpha\, v_t] = \alpha \Phi C_t\, \partial P/\partial t + \alpha \{c_n v_n + c_w v_w\} \nabla P \quad\quad\quad (2.1)$$

The pressure Equation (2.1) is a parabolic (almost elliptic for small compressibility) second order partial differential equation. The saturations are represented by a set of constant values divided by fronts. The fronts, which can have very complex shapes (fingering), are not directly connected to the grid. To get a pressure field solution which is as correct as possible, it is important that the saturations are correctly mapped to the pressure grid. The mapping of the saturation fronts to the pressure grid is called triangulation of the fronts. This procedure is more thoroughly explained in F. Bratvedt et al. [3]. We solve the pressure equation by using a finite element method. The finite element method used is based on first order triangular elements. Triangular elements are chosen due to the need to fit the grid to the saturation fronts and to allow high grid flexibility when modeling irregular and heterogeneous geological models.

The linear equation system set up by the finite element method is solved with a preconditioned conjugate gradient method.

Solution of the Saturation Equation.

Since the capillary effects are ignored, the right hand term of the saturation equation (diffusion term) is set to zero and the saturation equation of the black oil model is written (see Appendix for definitions):

$$\alpha[\Phi \varrho_w S_w]_t + \nabla[\alpha\varrho_w f_w \{v_t + (\varrho_n - \varrho_w)\lambda_n G\nabla D\}] = 0 \quad \ldots\ldots\ldots\ldots\ldots\ldots\ldots (2.2)$$

Make the definition of a flow function F_w:

$$F_w = f_w [v_t + \lambda_n (\varrho_n - \varrho_w) G\nabla D] \quad \ldots\ldots\ldots\ldots\ldots\ldots\ldots\ldots\ldots\ldots\ldots\ldots (2.3)$$

Using this definition in Equation (2.2), the saturation equation is written:

$$\alpha[\Phi \varrho_w S_w]_t + [\alpha\varrho_w F_w]_x = 0 \quad \ldots\ldots\ldots\ldots\ldots\ldots\ldots\ldots\ldots\ldots\ldots\ldots\ldots\ldots (2.4)$$

By making a set of simplifications that only is valid in each grid cell [3], the saturation equation in each grid cell is written:

$$\Phi[S_w]_t + \nabla F_w = 0 \quad \ldots\ldots\ldots\ldots\ldots\ldots\ldots\ldots\ldots\ldots\ldots\ldots\ldots\ldots\ldots\ldots\ldots (2.5)$$

By using the approximation that the flow function F_w is piecewise linear, the solution of this equation consists of the solution of a set of Riemann problems. The numerical procedure requires solution of the Riemann problem in one and two dimensions in addition to solving the equation when moving across a grid cell boundary. The solution procedures are described in [3], [18], [20] and [26].

There are several advantages to this solution procedure. The non-linearities in the saturation equations are solved analytically so that the most time-consuming part per timestep is to solve the linear equation system for the pressure equation. This equation system consists of one unknown in contrast to the system set up by an implicit formulation, where a system of at least two unknowns must be solved more than once per timestep. The solution of the saturation equation is independent of the grid, so there is no limitation on the timestep length (the fronts can move more than one grid cell without producing non-physical solutions).

3. THE STREAMLINE CONCEPT

Streamline simulators were introduced by Higgins and Leighton [14],[15] in 1962 as an aid to forecast the performance of water floods in oil/gas reservoirs. The streamtube concept made this possible by simplifying the problem to a set of one dimensional problems (streamtubes, defined in [1]). Complex simulations could then be performed consuming less CPU time compared to traditional simulators.

Today streamline concepts of various complexity exist, but most streamline models are based on standard finite difference techniques to approximate the differential equations describing the flow in the reservoir [25]. However, the concept used in the model described herein is different, since the two-dimensional front-tracking method described in Section 2 forms the basis.

The streamline concept involves certain restrictions in the mathematical foundation. By simplifying the physics of the system, the complexity of a general reservoir simulator is reduced, and an extremely CPU-efficient tool for approximate studies of fluid flow processes is established.

The standard simplifications that are made in a streamtube concept are the following:

$\varrho = \varrho_n = \varrho_w$ (no density difference between the phases) .... (3.1A)

$\partial \Phi / \partial t = 0, \; \partial \varrho / \partial t = 0$ (incompressible) .... (3.1B)

Condition (3.1A) sets $(\varrho_n - \varrho_w) = 0$ in the definition of a flow function F_w (2.3), which can then be written:

$$F_w = f_w \, v_t \qquad (3.2)$$

Condition (3.1B) simplifies the pressure equation (2.1) to:

$$-\nabla [\alpha \, v_t] = 0 \qquad (3.3)$$

Using Equation (3.3) and condition (3.1B) in Equation (2.5), the saturation equation is written:

$$\Phi [S_w]_t + v_t [f_w]_x = 0 \qquad (3.4)$$

Given the physical assumptions above, streamlines are defined as flow paths orthogonal to the potential lines [1]. A streamtube is defined as the cross sectional area between two neighboring streamlines [1]. The streamlines are computed from a pressure field and the saturation fronts are moved along the streamlines to create a drainage picture. The pressure field is computed every time the production/injection rates are changed.

The total velocity field resulting from solution of the pressure will remain nearly constant throughout the whole period of simulation, given constant production conditions. The pressure equation is in principle solved only once during a simulation run, and this is the main reason for the extreme CPU-efficiency of the streamline simulator. The saturation equation is solved with basis in the established total velocity field.

The main advantage of this concept is that there is no transformation of the reservoir geometry into one-dimensional streamtubes. In traditional streamtube simulators this transformation results in a loss of information that could be important for the flow dynamics. Yet, the typical advantages of the traditional concepts, CPU-efficiency and accurate front description, are kept.

4. SIMULATOR APPLICATIONS

For verification purposes, the front tracking reservoir simulator (FTS) described above has been applied to a set of test cases [2], [3]. Field studies illustrating the FTS capabilities have also been performed. Compared to a standard five-point finite difference simulator (FDS), several advantages exist.

The pressure solution, which is obtained by means of a finite element method, gives high accuracy and freedom with respect to grid architecture. The grid is based on triangular cells, which can adapt to almost any geometry (node based grid). Geological details and complex reservoir geometries can thereby be reproduced in great detail. A grid preprocessor is developed for automatic generation of irregular grid systems.

The solution of the saturation equation is to a large extent independent of grid size and geometry, since the grid is not directly used for the solution. The fronts are tracked in the total velocity field established from the pressure solution. A typical Buckley-Leverett saturation profile is obtained, so no numerical dispersion takes place.

The solution of the saturation equation is fully stable for all time step sizes and the solution method is fast (direct solution). This, in combination with the grid geometry advantages, makes the concept very CPU efficient compared to traditional finite difference methods, particularly for larger grid systems.

4.1 Test Cases

To verify the FTS simulator and to demonstrate typical advantages, a set of comparative runs were performed on the FTS and on a commercial FDS simulator. Three catagories of typical simulation problems were highlighted:

i)   Grid Effects   ii)   Numerical Dispersion   iii)   CPU-consumption

Grid Effects

Standard FDS simulators based on the five-point scheme tend to give distorted saturation solutions when operating on irregular grid systems [28]. This imposes a restriction on the user when modeling complex reservoir geometries. The FTS and the streamline simulator, however, give the user freedom with respect to definition of grid cell geometries.

To demonstrate how the saturation solution of the FDS simulator and the FTS simulators (general and streamline) react to grid cell geometry, two areal grid systems, one orthogonal and one skewed, were set up. A symmetric well pattern with a water injector and two producers was defined in both grid systems, as shown in Figure 4.1. The model input data are given in Table 4.1.

Figures 4.2a, 4.3a and 4.4a show the saturation picture in the orthogonal grid system for the three different simulators at 1200 days. In the orthogonal grid system a symmetric saturation picture is observed in all three simulators, and water breakthrough occurs simultaneously in the two producers. Due to numerical dispersion, water breakthrough occurs earlier in the FDS simulator than in the FTS simulator, which reflects a shock front behavior. The FTS and the streamline simulators behave similarly. The water cut behavior resulting from the FDS and the streamline simulator for the orthogonal grid system is given in Figure 4.5a.

In Figure 4.2b, 4.3b and 4.4b the corresponding saturation behavior in the skewed grid systems is shown. A severe distortion of the fronts, with a preferred flow direction orthogonal to the grid faces, is observed in the FDS simulator. This is also reflected in Figure 4.5b, where the water breakthrough time differs significantly in the two production wells.

In the FTS and in the streamline simulator grid effects scarcely occur, and as shown in Figure 4.5b, breakthrough in the two wells is almost simultaneous. The breakthrough time in the skewed grid corresponds well to the time observed in the orthogonal grid. The grid-independency of the saturation solution in the FTS simulator is clearly demonstrated.

CPU Consumption and Numerical Dispersion

The CPU-efficiency of the FTS simulator is a result of four factors in combination. First, the solution of the saturation equation is explicit and direct, which makes it fast. Second, it is fully stable for all timestep sizes, which allows the user to specify larger timesteps. Since no numerical dispersion occurs, grid cell sizes needn't be considered. Overall, this implies that reliable solutions can be obtained with coarser grid systems. Finally, the flexible grid option can be utilized to reduce the number of active cells when modeling complex reservoir geometries.

To demonstrate these features, an areal rectangular model was set up with two grid systems, one with dimensions 21 x 40 and one with dimensions 63 x 120. A water injector was located in each of the lower corners of the grid and a producer was located in the upper middle cell (Figure 4.6). During the first simulation period the injectors were injecting at equal rates (corresponding to field voidage). Then the right injector (II) was shut in, while injector I was doubling its rate. Model input data are given in Table 4.2.

Figure 4.6 shows the saturation contours from the FDS simulator for both grid systems immediately following shut in of injector II (1560 days). Figure 4.8 shows the results at a later time step (2740 days). In the coarse grid, the effect of numerical dispersion is clearly observed. Mobile saturations are observed ahead of the theoretical shock at both time steps. Since the finite difference technique converges towards the theoretically correct solution with higher grid density, the fine grid system reflects more of a shock behavior. It can be seen from the saturation pictures that the breakthrough times in the two cases are significantly different. This demonstrates that reliable solutions require finer grid systems in a FDS simulator.

Figures 4.7 and 4.9 show the corresponding results from the FTS simulator. No numerical dispersion occurs, and the grid-independency is confirmed by fronts that behave almost identically in both the coarse and the fine grid system.

By comparing the results from the FDS and the FTS simulators, one can see that the fronts in the fine FDS run correspond quite well with the fronts in both the FTS runs. Consequently, the results obtained with the coarse FTS grid are of the same quality as the results obtained with the fine FDS grid.

The above examples were run in the compressible simulation mode. Since the pressure was kept constant by voidage water injection and the densities of the fluids were equal, the streamline option (ref. Section 3) was applied as a sensitivity.

The CPU-consumption in the referred cases are given in the table below:

Grid	Simulator	Time steps	Sim. time (Days)	CPU-time (s)
21 x 40	FDS	84	3000	436
63 x 120	"	201	3000	19932
21 x 40	FTS	24	3000	210
63 x 120	"	24	3000	1437
21 x 40	STREAM	24	3000	35
63 x 120	"	24	3000	328

In compressible mode, comparing the coarse FTS run with the fine FDS run, the FTS simulator proved 95 times faster CPU-wise. Although if we disregard the accuracy of the results, the FDS simulator is faster than the FTS simulator on smaller grid systems.

4.2 Flexible Grid Systems Applied to Field Cases

The grid flexibility of the simulator is demonstrated through two field scale cases. The objectives of both studies were to simulate water injection and sweep efficiency in geologically and geometrically complex reservoir sands. In the first study, an areal system of interbedded channel sands (fluvial system) with moderate permeability contrasts was modeled. In the second study, a cross-sectional model was constructed with basis in a channel sand consisting of eight sand types with contrasting permeabilities.

In the simulator the individual grid block dimensions can be specified manually in the input file, but when a geometrically complex model is to be constructed, this can be cumbersome. Therefore, to fully utilize the triangular grid block geometries, a grid preprocessor has been developed. The grid preprocessor enables automatic adaption of the grid nodes to contour lines of any kind, for example a map of sand bodies. Starting with a regular grid system and a set of contour lines, the grid nodes in the vicinity of a contour line are automatically moved to fit the contour, and a distorted grid is formed. This means that both the external reservoir boundary and the internal reservoir sand bodies can be reproduced in the grid with high accuracy and with a moderate number of grid cells.

Areal Fluvial Channel Sand System

A geological model which reflected a complex system of interacting channels and crevasse splays (fluvial system) was established by means of statistical data for sand body characteristics. Field data from wells and correlation sections were used to generate frequency curves for channel dimensions, channel density, crevasse volumes, sand body orientation etc. By means of random draws on the frequency curves and deterministic field data, a 3D geological model of the fluvial system was established in a rectangular grid system consisting of millions of cells.

From this 3D model, a 2D areal section was chosen for the simulation study of a water injection process. Due to the extremely high number of cells, the grid used for the geological modeling, could not be used directly for dynamical simulation purposes in practice. The above described grid preprocessor was therefore applied to generate a flexible grid for input to the FTS simulator.

The geology (permeability distribution) reproduced in the finite element grid by means of the grid preprocessor is shown in Figure 4.10. A total of 1000 grid cells (gross) were used. A close-up picture of the grid is given in Figure 4.11. It can be seen how the trianglar grid cells are used to optimise the fitting of the grid to the geological boundaries.

The channels were given a permeability of 1600 mD, while the crevasse splays were given 110 mD. A dip of seven degrees orthogonal to the main channel direction was applied. An initial oil-water contact was intersecting the reservoir. Three producers and three injectors were located as shown in Figure 4.12. The reservoir was produced under full pressure maintenance.

Figure 4.12 shows the areal sweep after 3400 days of water injection.

Cross-Sectional Channel Sand System

A geological cross section resulting from outcrop studies was established by the geologists. The model represents a section through a complex channel system. Eight sand types were identified, with permeabilities in the range of 0 to 2000 mD.

A set of contour lines (polygons) were generated for each sand type by interactive digitization of the cross section drawn by the geologists. The contours formed the basis for the irregular grid, which was automatically generated by means of the grid preprocessor (ref. above). Figure 4.13 shows the sand contours digitized from the geological cross section model. To separate the sand types vertically, a minimum of 40 layers had to be defined. Horizontally, 100 cells were used. Figure 4.14 shows the sand types as they were reproduced in the simulation grid.

Water flooding was imposed directly in the oil zone by an injector perforated partially on the left boundary of the model. A production well was perforated partially on the right boundary. The model was given a dip of ten degrees in the direction of flow. The pressure was maintained through voidage replacement. To avoid problems with gravity effects due to permeability discontinuities, the streamline simulator (ref. Section 3) was used. The water front behavior at two specified time steps is shown in Figure 4.15a and b. The figures demonstrate that the heterogeneous geology creates complex front geometries in time. In the high permeable sands, fingers of water are advancing rapidly towards the producer.

4.3 Streamline Options

The physical assumptions behind the streamline concept are described in Section 3. The streamline simulator can be accessed as an option in the general FTS simulator.

The streamline simulator serves several purposes. In this mode, the simulator serves as an extremely cost-effective tool for sensitivity studies. It allows simulation studies to be performed on economically lower priority projects. It can be a substitute for analytical methods and contributes to increased credibility of results usually obtained by hand calculations.

Typically, the simulator is used to establish an areal picture of well sweep efficiency. From the pressure solution, a streamline picture can be formed (Figure 4.16). Saddle points in the pressure solution are identified, and drainage areas for producers and sweep efficiency for injectors can be displayed (Figure 4.17). By combining the streamline pattern with the potential lines, a basis for curvelinear grids is formed (Figure 4.18). These can be used as input to standard FDS simulators.

From the water saturation front development, the time for water breakthrough and production profiles for wells is generated. The volumetric sweep/drainage efficiency is calculated for each well relative to both the field and the individual well's sweep/drainage area.

The pressure and thereby also the streamlines, are recalculated automatically when the production rate changes. In addition, the user can specify a recalculation explicitly. The material balance will be maintained.

The CPU-efficiency makes the simulator suitable in connection with automatic optimization of production parameters. An optimizer based on stochastic methods is linked to the simulator. By iteration on a set of simulation runs, an objective function (net present value) can be optimized. Currently the available optimization parameters are the location of wells, number of wells and start-time for wells.

5. CONCLUSIONS

a) A new numerical method has been developed based on a piecewise linear approximation of the flow function. This method enables a set of Riemann problems to be solved analytically. The numerical method is under implementation in a reservoir simulator.

b) The new reservoir simulator has been verified by comparing its results with the results of a standard finite difference simulator.

c) The FTS simulator shows no numerical dispersion. The simulator is to a high degree independent of grid cell geometries, and the flexible grid option is suitable for modeling of flow processes in geologically complex architectures.

d) The FTS simulator is fully stable for large timesteps. For larger grid systems described in this paper, the simulator demonstrated a reduction factor of 95 in CPU time compared to the commercial FDS simulator.

e) If the physical conditions are fulfilled, the streamline simulator, which exists as an option in the general FTS simulator, can be accessed, and additional CPU savings are obtained. An efficiency factor of 6 compared to the general FTS simulator was observed.

ACKNOWLEDGEMENT

We would like to thank Norsk Hydro a.s and Saga Petroleum a.s for their support. Thanks also to Knut Søvold, Scandpower a.s for his practical contributions. VISTA supports the project financially and with field reservoir data. All the authors except D.T. Rian have been supported financially by The Royal Norwegian Council for Technical and Industrial Reasearch (NTNF) and The Norwegian Council for Science and the Humanities (NAVF).

NOMENCLATURE

FDS Finite difference simulator
FTS Front tracking simulator
w   denotes the wetting phase
n   denotes the non-wetting phase
C_{nw} mass fraction of component n in phase w
B   volume factor
P_i phase pressure of phase i
S_i saturation, phase i
v_i Darcy fluid velocity, phase i
K   absolute permeability
k_i relative permeability of phase i
μ_i fluid viscosity of phase i
ϱ_i fluid density of phase i
$\alpha(x)$ reservoir thickness
λ_i phase mobilities, see definition below
Φ porosity
G   gravitation
q   production rate
D   depth
$[\ ]_x$ space derivative
$[\ ]_t$ time derivative

REFERENCES

1. Bear, J., "Dynamics of Porous Media." *American Elsevier Publishing Company*, New York 1967.

2. Bratvedt, K. and Rian, D.T. : "Modelling and Simulation of a Fluvial Geological System by the Application of Front Tracking Methods." The First International Forum on Reservoir Simulation, Alpbach Austria. 1988.

3. Bratvedt, F., Bratvedt, K., Buchholz, C., Holden, H., Holden, L. and Risebro, N.H. :"A New Front-Tracking Method for Reservoir Simulation," paper SPE 19805 presented at the 64th Annual Technical Conference and Exhibition of the Society of Petroleum Engineers, San Antonio, TX, October 8-11, 1989.

4. Dafermos, C.M.: "Polygonal Approximation of Solutions of the Initial Value Problem for a Conservation Law," *J. Math. Analysis Applic.* (38, 1972) 33-41.

5. Douglas, Jr.,J.:"Simulation of Miscible Displacement in Porous Media by a Modified Method of Characteristic Procedure." Numerical Analysis, Dundee 1981, *Lecture Notes in Mathematics*, 912, Springer-Verlag, Berlin 1982.

6. Douglas, Jr., J. and Russell, T.F.: "Numerical Methods for Convection-Dominated Diffusion Problems Based on Combining the Method of Characterestics with Finite Difference or Finite Element Procedures." *SIAM J. Numer. Anal.*, 19(5), (1982).

7. Ewing, R.E., Russell, T.F. and Wheeler, M.F.: "Simulation of Miscible Displacement Using Mixed Methods and a Modified Method of Characteristics," paper SPE 12241 presented at the 1983 Reservoir Simulation Symposium, San Fransisco, November 1983.

8. Ewing, R.E., Russell, T.F. and Wheeler, M.F.: "Convergence Analysis of an Approximation of Miscible Displacement in Porous Media by Mixed Finite Elements and a Modified Method of Characteristics," *Computer Methods in Applied Mechanics and Engineering* (47, 1984) 73-92.

9. Fayers, F.J. : "An Introduction to some of the Mathematical and Physical Problems of Modelling Oil Displacement in Porous Media." Mathematics in Oil Production. Clarendon Press. Oxford. 1988.

10. Gimse, T.:"A Numerical Method for a System of Modelling One-Dimensional Three-Phase Flow in a Porous Medium," Proceedings of the Second International Conference on Nonlinear Hyperbolic Problems. Aachen, March 1989.

11. Glimm, J., Lindquist, B., McBryan, O.M., Plohr, B., and Yaniv,S. : "Front Tracking for Petroleum Reservoir Simulation," paper SPE 12238 presented at the SPE 1983 Reservoir Simulation Symposium, SanFrancisco, November 15-18.

12. Hales, H.B. : "A Reservoir Simulator Based on the Method of Characteristics," SPE 13219. 1984.

13. Hewett, T.A. and Beherens, R. A.:"Scaling Laws in Reservoir Simulation and Their Use in a Hybrid Finite Difference/Streamtube Approach to Simulating the Effects of Permeability Heterogeneity," presented at the NIPER/DOE Second International Reservoir Characterization Technical Conference, Dallas TX, June 25-28, 1989.

14. Higgins, R.V. and Leighton, A.J., "A Computer Method to Calculate Two-Phase Flow in any Irregularly Bounded Porous Medium." Journal of Petroleum Technology, pp. 679-683, june 1961.

15. Higgins, R.V. and Leighton, A.J., :"Aids to Forecasting the Performance of Water Floods."

Journal of Petroleum Technology, pp. 1076-1082, sept 1964.

16. Holden, L.: "On the Strict Hyperbolicity of the Buckley-Leverett Equations for Three-Phase Flow in Porous Media," *Norwegian Computing Center Report.* 1988.

17. Holden, H. and Holden L.: "On the Riemann Problem for a Prototype of a Mixed Type Conservation Law. II," *Preprint Mathematics No.3/1988. Division of Mathematical Sciences, The University of Trondheim, Norway.* 1988.

18. Holden, H., Holden, L. and Høegh-Krohn, R.: "A Numerical Method For First Order Nonlinear Scalar Conservation Laws In One-dimension," *Comput. Math. Applic.* (Vol. 15, No. 6-8, 1988) 595-602.

19. Holden, H., Holden, L. and Høegh-Krohn, R.: "A Numerical Method For First Order Nonlinear Scalar Conservation Laws In One-dimension," *Preprint Series, Institute of Mathematics, University of Oslo.* 1988.

20. Holden, L. and Høegh-Krohn, R.: "A Class of N Nonlinear Hyperbolic Conservation Laws," *J. of Differential Equations.* To Appear.

21. Holden, H., Holden, L. and Risebro, N.H.: "Some Qualitative Properties of 2x2 Systems of Conservation Laws of Mixed Type," *Preprint Series, Institute of Mathematics, University of Oslo.* 1988.

22. Lucier, L.J.: "Error Bounds for the Methods of Glimm, Godunov and LeVeque,"*SIAM, Numerical Analysis* (22, 1985) 1074-1081.

23. Martin, J.C. and Wegner, R.E. : "Numerical Solution of Multiphase, Two-Dimensional Incompressible Flow Using Stream-Tube Relationships." SPE 7140. 1978.

24. Peaceman, D.W., *Fundamentals of Numerical Reservoir Simulation.*, Elsevier Scientific Publishing Company, (1977).

25. Renard, G. : "A 2D Reservoir Stream Tube EOR Model with Periodical Automatic Regeneration of Streamlines." Proceedings of 4th European Symposium on Enhanced Oil Recovery, pp. 711-722. Oct. 1987.

26. Risebro, N.H.: "The Solution of the Scalar Riemann Problem in Several Space Dimensions, A Numererical Method," Proceedings of the Fourth European Symposium on Enhanced Oil Recovery. Hamburg, October 1987.

27. Russell, T.F.: "Finite Elements with Characteristics for Two-Component Incompressible Miscible Displacement," paper SPE 10500 presented at the Sixth SPE Symposium on Reservoir Simulation, New Orleans, January-February 1982.

28. Yanosik, J.L. and McCraken, T.A.: "A Nine-Point, Finite Difference Reservoir Simulator for Realistic Prediction of Adverse Mobility Ratio Displacement," *SPEJ* (19, 1979) 253-262.

APPENDIX A

The Equations for Two-Phase Flow.

Darcy's law for the wetting and non-wetting phases and the mass conservation equations[24]:

$$v_n = -\lambda_n (\nabla P_n - \varrho_n G \nabla D) \qquad v_w = -\lambda_w (\nabla P_w - \varrho_w G \nabla D) \quad \dots\dots\dots\dots\dots \text{(A1)}$$

$$-\nabla[\alpha(C_{nn}\varrho_n v_n + C_{nw}\varrho_w v_w)] + \alpha q_n = \alpha[\Phi(C_{nn}\varrho_n S_n + C_{nw}\varrho_w S_w)]_t \quad \dots\dots\dots \text{(A2)}$$

$$-\nabla[\alpha(C_{wn}\varrho_n v_n + C_{ww}\varrho_w v_w)] + \alpha q_w = \alpha[\Phi(C_{wn}\varrho_n S_n + C_{ww}\varrho_w S_w)]_t \quad \text{.........................} \quad (A3)$$

In addition these relations are used:

$$S_n + S_w = 1, \quad C_{nn} + C_{wn} = 1, \quad C_{nw} + C_{ww} = 1 \quad \text{..................} \quad (A4)$$
$$\lambda_n = K k_n / \mu_n, \quad \lambda_w = K k_w / \mu_w, \quad f_w = \lambda_w / (\lambda_n + \lambda_w) \quad \text{..................} \quad (A5)$$
$$\varrho_n = B_n(\varrho_{ns} + R_s \varrho_{gs}), \quad \varrho_w = B_w \varrho_{ws} \quad \text{..................} \quad (A6)$$
$$v_t = v_n + v_w, \quad \lambda_t = \lambda_n + \lambda_w \quad \text{..................} \quad (A7)$$
$$c_n = 1/\varrho_n \partial\varrho_n/\partial P_n \quad c_w = 1/\varrho_w \partial\varrho_w/\partial P_w \quad c_r = (1/\Phi)\partial\Phi/\partial P \quad C_t = c_r + S_n c_n + S_w c_w \quad \text{..........} \quad (A8)$$

Table 4.1 - Reservoir Data Grid Effects Case

Grid Dimensions	51x12
$\Delta x = \Delta y$	15m
Thickness	10m
Permeability	50mD
Porosity	0.20

B_w (P_w =275.0bar)	1.0 Rm^3/Sm^3	B_o (P_o=250.0bar)	1.01 Rm^3/Sm^3
Viscosity water	0.50 cp	B_o (P_o=275.0bar)	1.00 Rm^3/Sm^3
Compress. water	4.0 bar^{-1}	Viscosity oil (P_o=250.0bar)	2.0 cp
Density water (surf)	1000.0 kg/m^3	Viscosity oil (P_o=275.0bar)	2.0 cp
Rock Compress.	0.0 bar^{-1}	Density oil (surf)	780.0 kg/m^3
R_s	100.0 m^3/m^3		

Table 4.2 - Reservoir Data Cpu-consumption and Numerical Dispersion Case

Grid	21x40	63x120
Δx	14.29m	4.76m
Δy	15.00m	5.00m
Thickness	10.0m	
Permeability	100.mD	
Porosity	0.25	

B_w (P_w =250.bar)	1.0 Rm^3/Sm^3	B_o (P_o=100.0bar)	1.015 Rm^3/Sm^3
Viscosity water	0.50 cp	B_o (P_o=250.0bar)	1.005 Rm^3/Sm^3
Compress. water	0.0 bar^{-1}	Viscosity oil (P_o=100.0bar)	0.49 cp
Density water	1000.0 kg/m^3	Viscosity oil (P_o=250.0bar)	0.50 cp
Rock Compress.	0.0 bar^{-1}	Density oil	1000.0 kg/m^3
R_s	100.0 m^3/m^3		

A RESERVOIR SIMULATOR

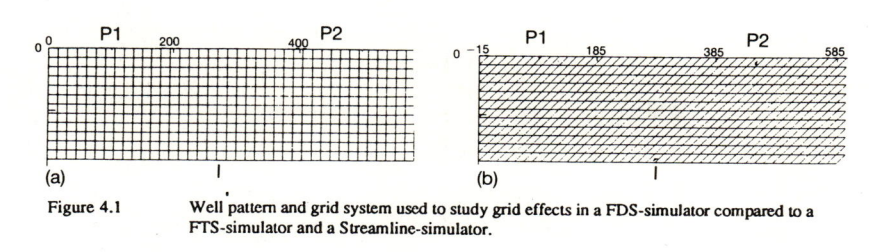

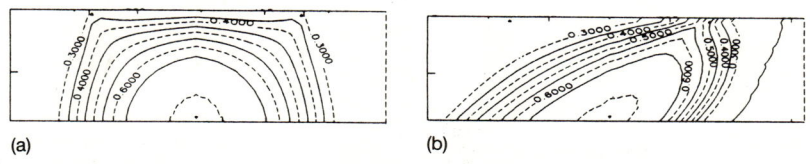

Figure 4.1 Well pattern and grid system used to study grid effects in a FDS-simulator compared to a FTS-simulator and a Streamline-simulator.

Figure 4.2 Saturation contours at 1200 days for the FDS-simulator, (a) rectangular grid and (b) skewed grid.

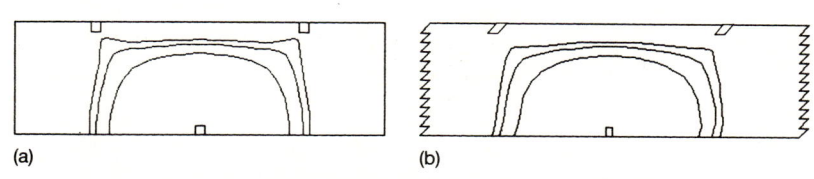

Figure 4.3 Saturation contours at 1200 days for the FTS-simulator, (a) rectangular grid and (b) skewed grid.

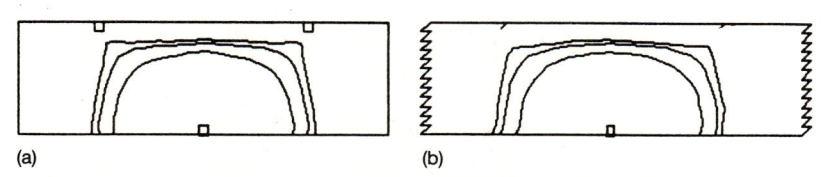

Figure 4.4 Saturation contours at 1200 days for the Streamline-simulator, (a) rectangular grid and (b) skewed grid.

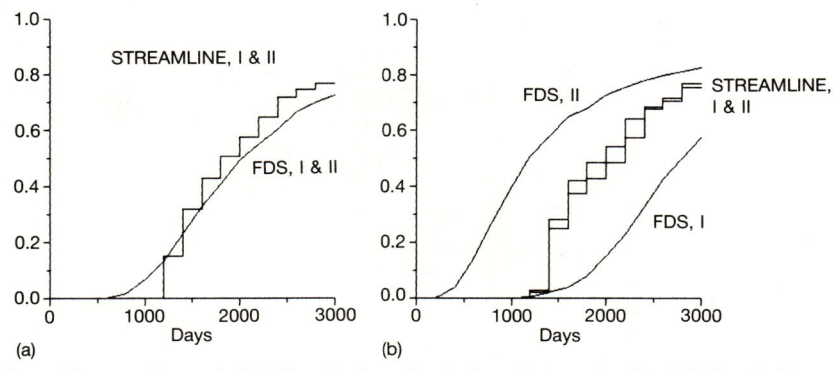

Figure 4.5 Water cut in the FTS- and the Streamline-simulator, (a) rectangular grid and (b) skewed grid.

Figure 4.6 Saturation contours from the FDS simulator at 1560 days for the CPU-consumption and Numerical Dispersion study.
Coarse grid left, fine grid right.

Figure 4.7 Saturation contours from the FTS simulator at 1560 days for the CPU-consumption and Numerical Dispersion study.
Coarse grid left, fine grid right.

Figure 4.8 Saturation contours from the FDS simulator at 2740 days for the CPU-consumption and Numerical Dispersion study.
Coarse grid left, fine grid right.

Figure 4.9 Saturation contours from the FTS simulator at 2740 days for the CPU-consumption and Numerical Dispersion study.
Coarse grid left, fine grid right.

Channel sand
Crevasse splay

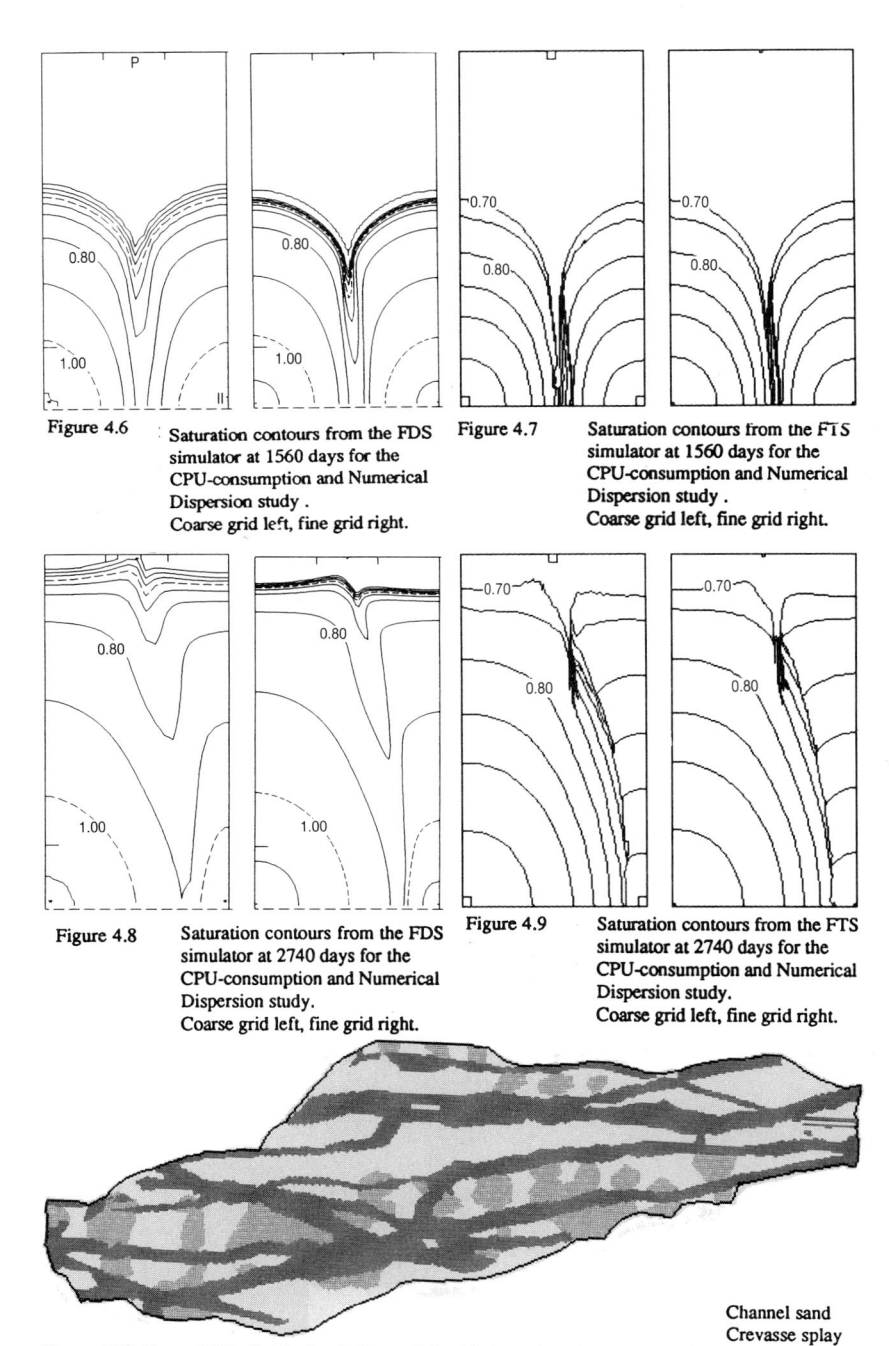

Figure 4.10 Permeability distribution in the areal fluvial channel sand system reproduced by the grid preprocessor.

A RESERVOIR SIMULATOR 697

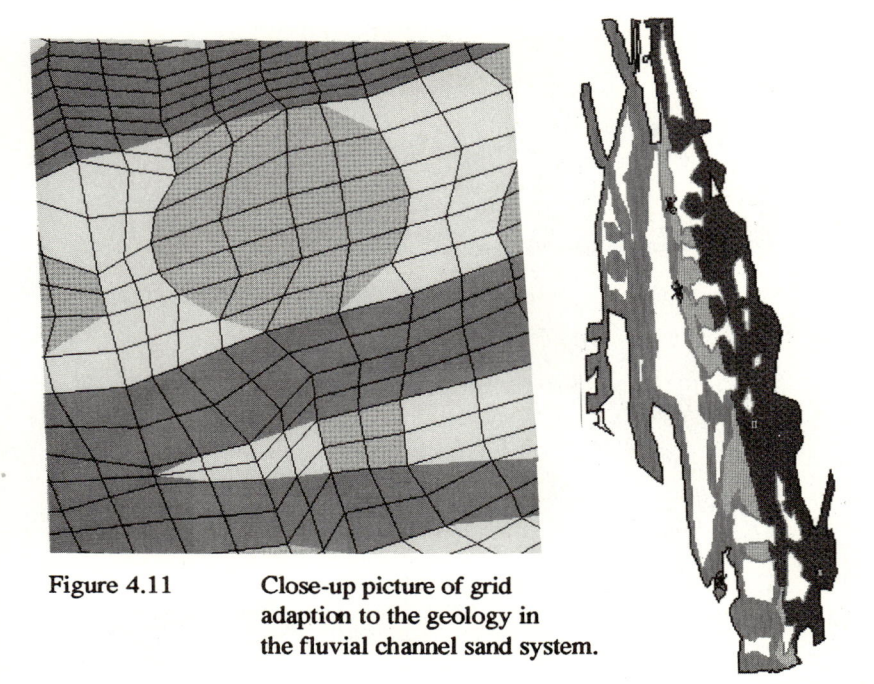

Figure 4.11 Close-up picture of grid adaption to the geology in the fluvial channel sand system.

Figure 4.12 Saturation fronts in the areal fluvial channel sand system after 3400 days.

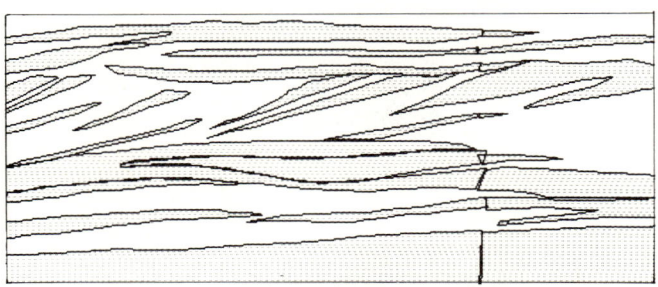

Figure 4.13 Digitized raw data (sand contours) of the cross-sectional channel sand system.

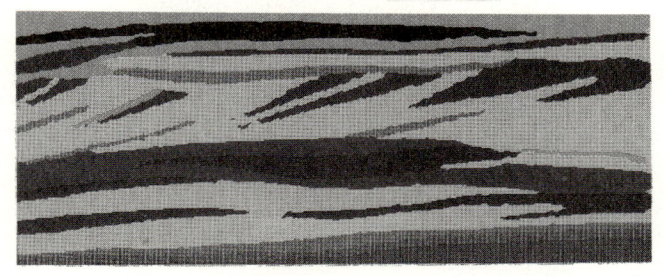

Figure 4.14 Permeability map for the cross sectional channel sand system.

Figure 4.15a   Saturation fronts timestep 1.

Figure 4.15b   Saturation fronts timestep 2.

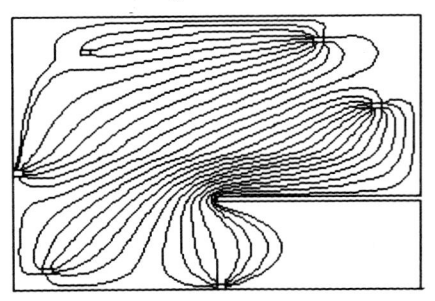

Figure 4.16   Streamline option -
Streamlines

Figure 4.17   Streamline option -
Drainage areas

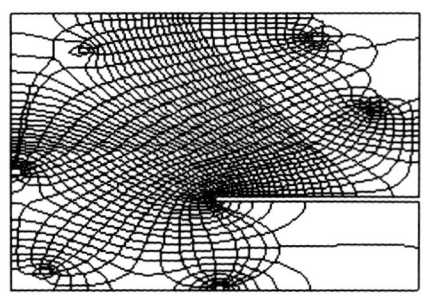

Figure 4.18 Streamline option -
Streamlines and potential lines
forming a curvilinear grid.

EFFECTIVE ABSOLUTE PERMEABILITY IN THE PRESENCE OF SUB-GRID HETEROGENEITIES: AN ANALYTICAL APPROACH

Hennie N.J. Poulisse
Koninklijke/Shell Exploratie en Produktie Laboratorium
P.O. Box 60
2280 AB Rijswijk
The Netherlands

ABSTRACT

An effective permeability tensor is derived, which can be assigned to a grid block under the assumption that the sub-grid permeability is a scalar, bounded random function in the spatial coordinates with given covariance function. The effective permeability tensor depends on both the nature of the spatial heterogeneities and the considered flow system operating within the grid-block flow domain. Examples are given substantiating the theoretical results. Attention is restricted to single-phase flow.

1. INTRODUCTION

Let us consider a portion of an oil reservoir. More specifically, consider the reservoir within the boundaries of a large grid block, such as being used in coarse-grid reservoir flow simulators. We are interested in the length scales of the rock heterogeneity. Computational feasibility and lack of a detailed geological description of the reservoir, particularly in terms of the permeability, prevent the grid blocks from being too small. At the same time, approximation of the solution of the flow equations and representation of the geology of the reservoir prohibit the grid blocks from being too large. This trade-off usually results in a situation in which the spatial variations in permeability range over orders of magnitude over smaller length scales than the length scales of the grid blocks. The sub-grid permeability variations influence the flow of fluids through the reservoir. These influences will usually not have been "averaged out" on the grid scale. Reservoir flow simulators should account for these sub-grid variations. Therefore, in numerical flow simulations an "effective permeability" has to be assigned to each grid block. Intuitively the effective permeability for each grid block can be viewed as a transformation of flow phenomena from the grid scale to the sub-grid scale. To put it bluntly, if the effective permeability is inserted every time the permeability appears in the equations used in a coarse-grid flow simulator, the simulated flow is representative of what happens inside the grid block as well as on a gridblock scale.

Many attempts have been made to tackle the problem of finding an effective permeability. A recent, elucidating survey of the research in this area is contained in [8]. Therefore we can confine ourselves to making only some observations based on King's paper [8].

Both numerical and analytical solutions have been proposed. The numerical solutions amount to either a simple estimate for the effective permeability, notably the geometric mean, or the determination of an effective value from a detailed numerical study of the single-phase flow equations. However, the simple estimates turn out to be too simple for many practical situations, while the extensive numerical simulation studies are hampered by in-core data storage limits or, if virtual memory techniques are used, increasing computer time.

The analytical methods are based on perturbation expansions and effective medium theory. In the perturbation approach estimates for the effective permeability are derived from perturbation expansions of the single-phase flow equations. In effective medium theory the heterogeneous medium is replaced by an effective homogeneous medium with constant effective permeability, such that the mean pressure fluctuations caused by the inclusion of a different permeability is zero. The analytical methods break down when the permeability fluctuations became large, which is almost invariably the case in practice.

In [8] King introduces a real-space renormalization technique for estimating the effective permeability. It is neither a numerical nor an analytical method, but rather a numerical method with a definite analytical flavour. The method can handle the situation of large fluctuations. It outperforms existing methods with respect to range of scales and computing time. The idea of the method is to calculate the effective permeability over a local region first. This leads to a reduction in the permeability fluctuations and gives a renormalized probability distribution of the scaled-up permeability. Successive application of the renormalization procedure gives a fixed point value for the effective permeability.

The method described in this paper is an analytical method with a computational flavour. It abandons the idea that the effective permeability in 2-D is always a scalar quantity, or that there are scalar-valued vertical and horizontal effective permeabilities in the anisotropic 3-D case. Instead, our effective permeability is a tensor, the entries of which can be arranged in a 2x2 matrix in 2-D and in a 3x3 matrix in 3-D. Our effective permeability may be a scalar only if the porous medium is isotropic. The intuitively appealing thought that goes with an effective permeability tensor is that the anisotropy of the porous medium is not lost when we scale up from the sub-grid to the grid scale; it may disappear when scaling up from the sub-grid scale proceeds well beyond the grid-scale.

The idea of an effective permeability tensor is not new. In [16] White and Horne give a numerical procedure for calculating an effective permeability tensor for each block in a coarse-grid simulator. Their method may be viewed as a sort of numerical counterpart of the present method. Because extensive numerical sub-grid calculations have to be performed, the method of White and Horne shares the disadvantages of the numerical techniques.

An important difference between the numerical treatment of White and Horne [16] and the present method is that they tacitly assume independence of the effective permeability of a grid block from the boundary value flow problem, defined for the grid block considered. However, it has been reported in earlier work in this area, notably by Smith and Freeze [14,15], who performed a conscientious analysis using Monte Carlo techniques, that, with respect to the effective permeability, both the nature of the spatial heterogeneities and the flow system operating within the flow domain must be considered. It follows from our analysis that the effective permeability is independent of the flow system within certain classes of boundary value problems only, but that different effective permeability tensors can be assigned to the same grid block when different classes of boundary value problems are considered. The effective permeability is no longer "just" a material property, as opposed to the permeability itself.

The method presented here is not computationally intensive. It can deal with large permeability variations, as will be shown in Section 4. Another possibility in this respect may be using our analytical approach complementary to King's renormalization procedure [8], in that a renormalized situation, which is still distant from the "renormalization fixed point", is used as input for the present method.

From the mathematical point of view our analytical treatment of the boundary value problem considered is an extension of earlier contributions, measured with respect to on the one hand the generality of the concrete function spaces in which we perform the analysis and the practical applicability of the results on the other.

This paper is organized as follows: in Section 2 we formulate the flow model; we restrict our attention to the flow of just one, incompressible fluid through an undeformable reservoir. In Section 3 we give the solution of the boundary value problem describing the flow through the porous medium within a grid block. In Section 4 we define our effective permeability tensor. The method is illustrated in Section 5 with some examples.

2. FORMULATION OF THE MODEL

We restrict ourselves to single-phase flow in 2-D of an incompressible fluid through an undeformable bounded porous medium. Since gravity effects are of little importance for single-phase flow, they have been ignored. Sources and sinks do not provide a principal contribution to the problem of finding an effective permeability in the single-phase flow situation, so we have left them out for the sake of simplicity.

Hence, we use Darcy's Law as the equation of motion – see, e.g. [2, Chapter 5] – in conjunction with the continuity equation:

$$\nabla \cdot V = 0 \qquad (2.1)$$

$$-k\nabla P = V \qquad (2.2)$$
in S

where V is the velocity, P the pressure and k the permeability. S is the bounded flow domain, in our context a coarse grid block, that may, without loss of generality, conveniently be chosen as the closed unit square in $\mathfrak{R}^2$, i.e.

$$S = [0,1] \times [0,1] \qquad (2.3)$$

Equations (2.1 and 2.2) are just a formal representation of our model; precise definitions of the velocity and the pressure will be given in Section 3.

Our main concern at this moment is the definition of the permeability k in (2.2). Following a widely used procedure in this area, we conceptualize the irregular spatial sub-grid variations in the permeability by considering the permeability as a random quantity. The realizations of the random permeability represent possible spatial distributions of permeability values over the grid block. These realizations will generally not be smooth functions in the spatial coordinates. Indeed, there will be an abrupt change in the value of the permeability when two different types of rock adjoin. This implies that our framework must be such that the realizations of the random permeability may have discontinuities on S. The following assumption with respect to the permeability meets this requirement.

ASSUMPTION 2.1 Let U(S) be the space of real-valued, bounded functions on S. Let $(\Omega, \mathcal{A}, \mu)$ be a probability measure space, Ω is the sample space, $\omega \varepsilon \Omega$ an elementary outcome, $\mathcal{A}$ is the σ-algebra of subsets of Ω and μ is a complete probability measure on $\mathcal{A}$, see e.g. Loève [9; pp. 151-152].

We assume that

$$k \varepsilon B_2(\Omega, U(S)) \qquad (2.4)$$

where – see Curtain and Pritchard [5; p. 89] – $B_2(\Omega, U(S))$ is the space of functions k: $\Omega \to U(S)$ such that $\int_\Omega \|k(\omega)\|_u^2 d\mu(\omega) < \infty$, $\|k(\omega)\|_u = \sup_{(x,y)\varepsilon S} k(\omega)(x,y)$, the sup-norm of $k(\omega)$ on U(S).

For some $\omega \varepsilon \Omega$, the function $k(\omega):(x,y)\varepsilon S \mapsto k(\omega)(x,y)\varepsilon \mathfrak{R}$ is called a sample function or a realization of the random permeability. ∎

Eliminating the velocity V in equations (2.1) and (2.2) gives

$$-\nabla \cdot k\nabla P = 0 \qquad (2.5)$$

We now have to specify the boundary conditions associated with equation (2.5). We shall treat the Neumann problem, for which there are given velocities across the boundaries of S. As evidenced in the sequel of this paper, the type of boundary conditions is not crucial.

We will formally define the flow of a fluid through S as a random boundary value problem. Similar definitions of random boundary value problems have been formulated in the past for a number of physical problems, one of the earlier contributions being the investigation of Beran and McCoy [3], who have also considered the elliptic equation (2.5).

DEFINITION 2.2. The sub-grid random boundary value problem, describing the flow of a fluid through S is

$$-\nabla \cdot k(\omega)(x,y) \nabla P(\omega)(x,y) = 0 \tag{2.6}$$

$$k(\omega)(x,0)\frac{\partial P(\omega)(x,0)}{\partial y} = -V_4(x,0)$$

$$k(\omega)(x,1)\frac{\partial P(\omega)(x,1)}{\partial y} = -V_3(x,1)$$

$$k(\omega)(0,y)\frac{\partial P(\omega)(0,y)}{\partial x} = -V_2(0,y)$$

$$k(\omega)(1,y)\frac{\partial P(\omega)(1,y)}{\partial x} = -V_1(1,y) \tag{2.7}$$

$\omega \varepsilon \Omega$; $(x,y) \varepsilon S$.

The boundary velocities V_i (i=1,..,4) are real-valued, bounded functions such that, in view of the incompressibility condition

$$\sum_{i=1}^{4} \int_{\partial S_i} V_i \cdot n_i \, d\ell = 0 \tag{2.8}$$

$$\partial S = \bigcup_{i=1}^{4} \partial S_i$$

where ∂S is the boundary of S and n_i the outward normal. Because of stochastic consistency – see e.g. Bharucha-Reid [4; p. 100] – the velocities V_i (i=1,..,4) in (2.7) assume their values with probability one, i.e.

$$\mu\{\omega : V_i(\omega) = V_i\} = 1 \quad (i=1,..,4) \tag{2.9}$$

The boundary conditions (2.7) are in fact deterministic, i.e. for every realization of the spatial distribution of permeabilities over S, the boundary conditions are the same. Strictly, the deterministic boundary condition assumption is a physical inconsistency. Mathematically random boundary conditions can be covered by our method, but the numerical evaluation of the results using random boundary conditions requires information that will usually not be available. This is why we have made the simplifying assumption on the boundary conditions from the outset.

3. SOLUTION OF THE RANDOM SUB-GRID BOUNDARY VALUE PROBLEM

Our strategy for arriving at an effective permeability is to derive the solution of the random sub-grid boundary value problem first. Using this solution, we subsequently construct a transformation between flow quantities on the grid scale and flow quantities on the sub-grid scale. This transformation is identified as an effective permeability in a natural way. The second step in the analysis is the subject of the next section.

The solution of the boundary value problem is established according to a familiar scenario: transformation of the original boundary value problem to a problem involving an integral equation, subsequent investigation of the corresponding integral expression as an operator acting on suitable function spaces, which are chosen such that the integral operator possesses sufficiently good properties, and finally the solution of the original problem by applying general methods of functional analysis to the integral equation, in particular in this case random fixed-point principles. Although the solution procedure sketched above may be familiar, the concrete form we give to it in our analytical treatment of the random boundary value problem of definition 2.2 is new.

We shall indicate the main steps.

The inverse of the Laplace operator $-\nabla^2$ under the boundary conditions (2.7) can be represented by an integral operator with kernel G, given by the mapping

$$G : S \times S \to \Re \tag{3.1}$$

The function G admits a Fourier-series representation.

Let the random permeability be decomposed in the following way

$$k(\omega)(x,y) = m_k + \hat{k}(\omega)(x,y) \tag{3.2}$$

where the real constant m_k is not necessarily the mean value of the permeability over S. We return to the choice of m_k later on in this section. We might as well add here that the splitting of the partial differential operator $-\nabla . k \nabla(.)$ that goes with equation (3.2) into a deterministic part - $m_k \nabla^2(.)$ with known inverse and a random part $-\nabla . \hat{k} \nabla(.)$ is the basis of the Adomian decomposition method [1].

The random boundary value problem of definition 2.2 can be transformed to the following random integral equation

$$\nabla_{xy} P(\omega)(x,y) = \mathcal{F}(\nabla_{xy} P)(\omega)(x,y)$$

$$= \nabla P_h(x,y) - \int\int_{[0,1][0,1]} N(\omega)(x,y;x',y') \nabla_{xy} P(\omega)(x',y') dx'dy' \qquad (3.3)$$

$$\int\int_{[0,1][0,1]} N(\omega)(x,y;x',y') \nabla_{xy} P(\omega)(x',y') dx'dy'$$

$$= m_k^{-1} \int\int_{[0,1][0,1]} \hat{k}(\omega)(x',y') \nabla_{xy} ((\nabla_{x'y'} G(x,y;x',y'))^T \nabla_{x'y'} P(\omega)(x',y')) dx'dy' \qquad (3.4)$$

where dx',dy' are Lebesgue measures and the superscript T denotes transposition. $P_h(x,y)$ represents the solution of what we will call the homogeneous medium boundary value problem, i.e. the solution of equation (2.6) with the random permeability $k(\omega)(x,y)$ replaced by the constant m_k under the boundary conditions (2.7). $P_h \varepsilon C^2(S) \times C^2(S)$, where $C^2(S)$ is the space of twice continuously differentiable functions on S, is given by

$$P_h(x,y)$$

$$= m_k^{-1} \left\{ \int_{[0,1]} G(x,y;1,y') V_1(1,y') dy' - \int_{[0,1]} G(x,y;0,y') V_2(0,y') dy' \right.$$

$$\left. + \int_{[0,1]} G(x,y;x',1) V_3(x',1) dx' - \int_{[0,1]} G(x,y;x',0) V_4(x',0) dx' \right\} \qquad (3.5)$$

The rigorous justification of the passage from the random boundary value problem of definition 2.2 to equation (3.3) is by no means trivial; it is, however, beyond the scope of the present paper.

With respect to the gradient of the random pressure field, we make the following assumption.

ASSUMPTION 3.1. Let $L_1(S)$ be the Lebesgue space of real-valued, absolutely integrable functions on S. We assume that

$$\nabla P \varepsilon B_1(\Omega, L_1(S)) \times B_1(\Omega, L_1(S)) \qquad (3.6)$$

where $B_1(\Omega, L_1(S))$ is the space of functions $\frac{\partial P}{\partial x}: \Omega \to L_1(S)$ such that

$$\int_\Omega \left\|\frac{\partial P}{\partial x}(\omega)\right\|_{L_1} d\mu(\omega) < \infty, \quad \text{where } \|\cdot\|_{L_1} \text{ is the } L_1 \text{ norm on } L_1(S).$$ ∎

The norm $\|\cdot\|_{B_1^2}$ on $B_1(\Omega, L_1(S)) \times B1(\Omega, L_1(S))$ is defined by

$$\|\nabla P\|_{B_1^2} = \|\partial P/\partial x\|_{B_1} + \|\partial P/\partial y\|_{B_1} \tag{3.7}$$

where

$$\|\partial P/\partial x\|_{B_1} = \int_\Omega \int_{[0,1]} \int_{[0,1]} |\partial P(\omega)(x,y)/\partial x| \, dx \, dy \, d\mu(\omega)$$

With respect to this norm, $B_1(\Omega, L_1(S)) \times B_1(\Omega, L_1(S))$ is a Banach space.

Assumption 3.1 is not restrictive, since the Lebesgue space L_1 contains a large class of functions.

Despite the fact that the kernel N in (3.3) is non-singular, the mapping $\mathcal{F}$ can be shown to be a well-behaved operator on the product space $B_1(\Omega, L_1(S)) \times B_1(\Omega, L_1(S))$ – see Kantorovich and Akilov [7; Chapter XI] for a treatment of integral operators on L_p spaces:

$$\mathcal{F} \in \mathcal{L}(B_1(\Omega, L_1(S)) \times B_1(\Omega, L_1(S))) \tag{3.8}$$

where $\mathcal{L}(B_1(\Omega, L_1(S)) \times B_1(\Omega, L_1(S)))$ is the space of linear continuous operators on $B_1(\Omega, L_1(S)) \times B_1(\Omega, L_1(S))$.

In particular, for $\omega \in \Omega$, $\mathcal{F}(\omega)$ qualifies as a random operator – see Bharucha-Reid [4; p. 72] – on $L_1(S) \times L_1(S)$.

Application of the random version of Banach's fixed point theorem – see Bharucha-Reid [4; theorems 3.4, 3.5] gives the solution of equation (3.3):
if the positive real-valued random variable $\xi(\omega)$ is such that

$$\xi(\omega) < m_k \text{ a.s.} \tag{3.9}$$

where

$$\xi(\omega) = \max\left\{\left(\left\|\hat{k}(\omega)\frac{\partial^2 G}{\partial x \partial x'}\right\|_\infty + \left\|\hat{k}(\omega)\frac{\partial^2 G}{\partial y \partial x'}\right\|_\infty\right), \left(\left\|\hat{k}(\omega)\frac{\partial^2 G}{\partial x \partial y'}\right\|_\infty + \left\|\hat{k}(\omega)\frac{\partial^2 G}{\partial y \partial y'}\right\|_\infty\right)\right\} \quad (3.10)$$

$\|\cdot\|_\infty$ being the essential sup. norm, then the solution of equation (3.3) can be represented in the following way.

$$\nabla_{xy} P(\omega)(x,y) = \nabla_{xy} P_h(x,y)$$

$$- m_k^{-1} \int\int_{[0,1][0,1]} \hat{k}(\omega)(x',y') \nabla_{xy}((\nabla_{x'y'} G(x,y;x',y'))^T \nabla_{x'y'} P_h(x',y')) dx' dy'$$

$$+ m_k^{-2} \int\int\int\int_{[0,1][0,1][0,1][0,1]} \hat{k}(\omega)(x',y') \hat{k}(\omega)(x'',y'')$$

$$\left(\nabla_{xy}(\nabla_{x''y''}((\nabla_{x'y'} G(x,y;x',y'))^T \nabla_{x'y'} G(x',y';x'',y'')))^T\right) \nabla_{x''y''} P_h(x'',y'')$$

dx'dy'dx"dy" .... (3.11)
$\omega \varepsilon \Omega$; $(x,y) \varepsilon S$.

The solution of the random boundary value problem of definition 2.2 is given by

$$P(\omega)(x,y) = \int^x \frac{\partial P(\omega)(x',y)}{\partial x'} dx' + g(\omega)(y)$$

$$= \int^y \frac{\partial P(\omega)(x,y')}{\partial y'} dy' + h(\omega)(x)$$
(3.12)

x,y $\varepsilon[0,1]$; $(x,y) \varepsilon S$; $\omega \varepsilon \Omega$.

Since $\partial P/\partial x$, $\partial P/\partial y \varepsilon B_1(\Omega, L_1(S))$, it follows that – see Rudin [12; theorem 8.17]

$$P \varepsilon B_1(\Omega, AC(S)) \quad (3.13)$$

where AC(S) is the space of absolutely continuous functions on S.
The random velocity follows from Darcy's law

$$V(\omega)(x,y) = -k(\omega)(x,y)\,\nabla P(\omega)(x,y)$$

$$V^T(\omega)(x,y) = (V_x(\omega)(x,y),\, V_y(\omega)(x,y)) \qquad (3.14)$$

$$(x,y)\,\varepsilon\,S\,;\,\omega\,\varepsilon\,\Omega$$

Invoking Hölder's inequality for integrals – see e.g. Rudin [12; theory 3.8] – gives

$$V\varepsilon\,B_1(\Omega,L_1(S)) \times B_1(\Omega,L_1(S)). \qquad (3.15)$$

The usefulness of equation (3.11) and hence equations (3.12) and (3.14) for practical applications depends on the speed of convergence of the series (3.11). This speed of convergence is controlled by the random variable $\xi(\omega)$, defined in (3.10): if $\xi(\omega) \ll m_k$ a.s., there will be rapid convergence of the series. From (3.10)

$$\xi(\omega) \le M(\omega)\left(\left\|\frac{\partial^2 G}{\partial x \partial x'}\right\|_\infty + \left\|\frac{\partial^2 G}{\partial y \partial x'}\right\|_\infty\right)$$

$$= M(\omega)\left(\left\|\frac{\partial^2 G}{\partial x \partial y'}\right\|_\infty + \left\|\frac{\partial^2 G}{\partial y \partial y'}\right\|_\infty\right) \text{ a.s.} \qquad (3.16)$$

where

$$M(\omega) = \sup_{(x,y)\varepsilon S}\left|\hat{k}(\omega)(x,y)\right| \qquad (3.17)$$

Numerical investigation of the second-order partial derivatives of G reveals that

$$\xi(\omega) < M(\omega).\alpha \quad \text{a.s.} \qquad (3.18)$$
$$\alpha < 1$$

From (3.9) and (3.16), it follows that

$$M(\omega) < m_k/\alpha \quad \text{a.s.} \qquad (3.19)$$

is a conservative upper bound for the admissible maximum variations in the permeability with respect to the chosen fixed value m_k – see (3.2). m_k need not be the mean value of $k(\omega)$ over S – assuming, for simplicity, for the moment that $k(\omega)$ has a constant mean value on S. In fact, if it is known that the involved probability distributions of the random permeability are skew, a value for m_k different from the mean value may speed up the convergence and may increase the covered range of scales. Alternatively, it may be useful in some cases to apply renormalization techniques first – see, e.g., Martin et al. [10], Frisch [6] – to decrease the permeability variations and, as follows from the above analysis, speed up the convergence. In particular, a combination of King's renormalization technique [8] and the present method may be useful.

It follows from equations (3.11) and (3.12) that the average pressure will be a function of the statistical moment functions of the random permeability. We can be sure that if any information about statistical moment functions of the permeability is available, it will be at best up to the second-order moment function – the two-point correlation function. Hence, for practical applications, only the approximate solution containing terms up to second order in the permeability is of interest. To appraise the usefulness of this second-order approximation, we have the following result

$$\frac{\left\|\nabla P(\omega) - \nabla P_2(\omega)\right\|_{L_1}}{\left\|\nabla P_h\right\|_{L_1}} < \frac{\left(\xi(\omega)/m_k\right)^3}{1 - \left(\xi(\omega)/m_k\right)} \quad \text{a.s.} \tag{3.20}$$

where $\nabla P_2(\omega)$ is the second order aproximate solution of equation (3.3). Equation (3.20) can be derived using the error estimate for contraction mappings – see Bharucha-Reid [4; theorem 3.1].

From equations (3.9) and (3.20) it follows that if

$$0.67\, m_k < \xi(\omega) < m_k \quad \text{a.s.} \tag{3.21}$$

the second-order approximation is useless. Hence, for the second-order approximation to be useful

$$M(\omega) < 0.67\, m_k / \alpha \quad \text{a.s.} \tag{3.22}$$

is a conservative upper bound for the admissible maximum variations in the permeability with respect to m_k.

Using the solution of the random boundary value problem of definition 2.2, we shall obtain an effective permeability tensor in the next section.

4. THE EFFECTIVE PERMEABILITY TENSOR

Because we assume that there is no sample function of the random permeability available, we have to resort to statistical averages to arrive at the effective permeability tensor. However, we wish to stipulate here that if a realization of the permeability were available, our results would also apply to this deterministic situation. If in this deterministic situation the sub-grid permeability function is available only in tabulated form, the integrals in our equations should be replaced by appropriate numerical approximations.

Invoking the Fubini theorem – see, e.g., Loève [9; pp. 136-137] – we have from (3.12) and (3.11)

$$E\{P(\omega)(x,y)\} = \int^x E\left\{\frac{\partial P(\omega)(x',y)}{\partial x'}\right\}dx' + E\{g(\omega)(y)\} \tag{4.1}$$

$$E\left\{\frac{\partial P(\omega)(x,y)}{\partial x}\right\} = \frac{\partial P_h(x,y)}{\partial x}$$

$$- m_k^{-1} \int\int_{[0,1][0,1]} \hat{a}(x',y') \frac{\partial}{\partial x} ((\nabla_{x'y'}G(x,y;x',y'))^T \nabla_{x'y'}P_h(x'y'))dx'dy'$$

$$+ m_k^{-2} \int\int\int\int_{[0,1][0,1][0,1][0,1]} \hat{R}(x',y';x'',y'') \frac{\partial}{\partial x} ((\nabla_{x''y''}((\nabla_{x'y'}G(x,y;x',y'))^T$$

$$\nabla_{x'y'}G(x',y';x'',y''))) T \nabla_{x''y''}P_h(x'',y'')dx'dy'dx''dy'' \tag{4.2}$$

where $E\{.\}$ denotes expectation with respect to the probability measure μ and

$$\hat{a}(x,y) = E\{\hat{k}(\omega)(x,y)\} \tag{4.3}$$

$$\hat{R}(x,y;x',y') = E\{\hat{k}(\omega)(x,y).\hat{k}(\omega)(x',y')\} \tag{4.4}$$

Note that $\hat{a} \in U(S)$ and $\hat{R} \in U(SxS)$ – see assumption 2.1.
Consider

$$\Delta P_x^a(y) = E\{P(\omega)(1,y) - P(\omega)(0,y)\} \tag{4.5}$$

$$\Delta P_y^a(x) = E\{P(\omega)(x,1) - P(\omega)(x,0)\} \tag{4.6}$$

Clearly, ΔP_x^a and ΔP_y^a can be associated with the grid scale. However, equations (4.1) and (4.2) show that the representations of ΔP_x^a and ΔP_y^a are in terms of the sub-grid flow system. Intuitively, the passage from the sub-grid scale to the grid scale can be related to a loss of detail. This intuitive idea can be formalized in the following way. It follows from equation (3.13) that $\Delta P_x^a \in AC([0,1])$ and $\Delta P_y^a \in AC([0,1])$. Hence and ΔP_x^a and ΔP_y^a attain their maximum and their minimum on [0,1] – see, e.g., Rudin [13; theorem 4.16]. If we define

$$\Delta P_x^a(y_1) = \min_{y \in [0,1]} \Delta P_x^a(y)$$

$$\Delta \overline{P}_x^a(y_2) = \max_{y \in [0,1]} \Delta P_x^a(y)$$

$$\Delta P_y^a(x_1) = \min_{y \in [0,1]} \Delta P_y^a(x)$$

$$\Delta P_y^a(x_2) = \max_{y \in [0,1]} \Delta P_y^a(x)$$

(4.7)

then we can write

$$<\Delta P_x^a> = \int_{[0,1]} \Delta P_x^a(y)\,dy$$

$$<\Delta P_y^a> = \int_{[0,1]} \Delta P_y^a(x)\,dx$$

(4.8)

The "extremes" $\Delta P_x^a(b)$ and $\Delta P_y^a(c)$ of ΔP_x^a and ΔP_x^a respectively are defined in the following way:

$$\text{extr}\{\Delta P_x^a\} = \Delta P_x^a(b)$$

$$b = y_1 \text{ if } |\Delta P_x^a(y_1)-<\Delta P_x^a>|>|\Delta P_x^a(y_2)-<\Delta P_x^a>|$$

$$= y_2 \text{ if } |\Delta P_x^a(y_2)-<\Delta P_x^a>|\geq|\Delta P_x^a(y_1)-<\Delta P_x^a>|$$

$$\text{extr}\{\Delta P_y^a\} = \Delta P_y^a(c)$$

$$c = x_1 \text{ if } |\Delta P_y^a(x_1)-<\Delta P_y^a>|>|\Delta P_y^a(x_2)-<\Delta P_y^a>|$$

$$= x_2 \text{ if } |\Delta P_y^a(x_2)-<\Delta P_y^a>|\geq|\Delta P_y^a(x_1)-<\Delta P_y^a>|.$$

(4.9)

If ΔP_x^a (ΔP_y^a) is constant, take b=0.5 (c=0.5). The following assumption gives concrete form to the loss of detail argument in going from the sub-grid scale to the grid scale.

ASSUMPTION 4.1. Only the extremes $\Delta P_x^a(b)$ and $\Delta P_y^a(c)$ of the functions ΔP_x^a and ΔP_y^a are of importance on the grid scale. ∎

At the end of this section we give another motivation for assumption 4.1 from a different point of view.

Now we must relate $\Delta P_x^a(b)$ and $\Delta P_y^a(c)$ to the sub-grid flow system. Equation (3.13) implies that the functions P_b^a and P_c^a, defined by

$$P_b^a(x) = P^a(x,b) = E\{P(\omega)(x,b)\}$$
$$P_c^a(y) = P^a(c,y) = E\{P(\omega)(c,y)\} \tag{4.10}$$

are continuous on [0,1]. Assume in the first instance that there exist points a, d $\varepsilon(0,1)$ such that

$$\Delta P_x^a(b) = \frac{\partial}{\partial x} P^a(a,b)$$
$$\Delta P_y^a(c) = \frac{\partial}{\partial y} P^a(c,d) \tag{4.11}$$

If the derivatives of the functions P_b^a and P_c^a exist at every point of (0,1), then according to the mean value theorem [13; theorem 5.10] there are sufficient conditions for the existence of these points a,dε(0,1). However, these conditions are not necessary, i.e. the points a,dε(0,1) may exist, despite the fact that the derivatives of P_b^a and P_c^a do not exist at every point of (0,1). In our general framework the derivatives of P_b^a and P_c^a exist almost everywhere and are elements of $L_1([0,1])$. Hence it is possible that the points a,dε(0,1) do not exist. Nevertheless, we maintain in this situation equation (4.11), albeit with a different interpretation: according to the classical Weierstrass theorem – see, e.g., [13; theorem 7.26] – the polynomials are dense in C([0,1]), the space of continuous functions on [0,1] with the supremum norm. If we accept a slight abuse in notation, the derivatives in (4.11) may be interpreted as the derivatives of polynomials that are arbitrarily close to the functions P_b^a and P_c^a with respect to the sup-norm. a,dε(0,1) are located respectively at points where P_b^a and P_c^a have a cusp. Note that it is not necessary to perform an approximation, only the numbers $\Delta P_x^a(b)$, $\Delta P_y^a(c)$ and their interpretation are of importance.

Following Darcy's law we decompose the numbers $\partial P^a(a,b)/\partial x$ and $\partial P^a(c,d)/\partial y$ in the following way:

$$-\frac{\partial P^a(a,b)}{\partial x} = L_{11}(a,b)V_x^a(a,b) + L_{12}(a,b)V_y^a(a,b) \tag{4.12}$$

$$-\frac{\partial P^a(c,d)}{\partial y} = L_{21}(c,d)V_x^a(c,d) + L_{22}(c,d)V_y^a(c,d) \tag{4.13}$$

where $L_{11}(a,b)$, $L_{12}(a,b)$, $L_{21}(c,d)$, $L_{22}(c,d)$ are real numbers and

$$(V_x^a, V_y^a) = V^a = E\{V(\omega)\} \tag{4.14}$$

With reference to the discussion concerning equation (4.11) we note that it is possible that the point values of $V_x^a(.,b)$, $V_y^a(.,b)$ and $V_x^a(c,.)$, $V_y^a(c,.)$ at $a\epsilon(0,1)$ and $d\epsilon(0,1)$ respectively, do not exist. In this case equation (4.12) should be interpreted in the following way. First of all we know that $V_x^a(.,b)$, $V_y^a(.,b)$, $V_x^a(c,.)$, $V_y^a(c,.) \in L_1([0,1])$. The space of continuous functions with compact support in [0,1], denoted by $C_c([0,1])$ – see e.g. Rudin [12; definition 2.9] – is dense in $L_1([0,1])$ in the L_1-norm – see, e.g., Rudin [12; theorem 3.14]. Again, with a slight abuse in notation, the point values for V_x^a and V_y^a in (4.12) may be interpreted as point values of functions in $C_c([0,1])$ that are arbitrarily close to V_x^a and V_y^a with respect to the L_1-norm. The singularities in the functions $V_x^a(.,b)$, $V_y^a(.,b)$, $V_x^a(c,.)$, $V_y^a(c,.)$ will generally be jump discontinuities, i.e. the most common type of discontinuities of the first kind – see, e.g., [13; definition 4.26]. In principle, the approximation using the C_c functions may result in any value between the left and right limit of the jump.

From (4.11)-(4.13) we have

$$-\Delta P_x^a(b) = L_{11}(a,b)V_x^a(a,b) + L_{12}(a,b)V_y^a(a,b)$$

$$-\Delta P_y^a(c) = L_{21}(c,d)V_x^a(c,d) + L_{22}(c,d)V_y^a(c,d) \tag{4.15}$$

Equation (4.15) completes the transformation from the grid scale back to the sub-grid scale. We are now ready to define the effective permeability.

DEFINITION 4.2. The matrix of the effective permeability tensor that is assigned to the grid block S is given by

$$k^e = \begin{pmatrix} k_{11} & k_{12} \\ k_{21} & k_{22} \end{pmatrix} = \begin{pmatrix} L_{11}(a,b) & L_{12}(a,b) \\ L_{21}(c,d) & L_{22}(c,d) \end{pmatrix}^{-1} \tag{4.16}$$

if the indicated matrix inverse exists. ∎

Note that on the grid scale, the point $(a,b)\epsilon S$ is identified with the point $(c,d)\epsilon S$.

Definition 4.2 is useful only if there is some form of independency of the effective permeability from the boundary velocities V_i (i=1,..,4) given in definition 2.2. Denote by

$$BC = (V_1, V_2, V_3, V_4) \tag{4.17}$$

the four boundary conditions of definition 2.2, subject to the incompressibility condition (2.8). Within the class of Neumann boundary conditions considered in this paper, we distinguish the following sub-classes

$$M(f) = \{BC: \nabla P_h = \alpha.f + C; \alpha \varepsilon \Re, C = (C_1, C_2)^T, \tag{4.18}$$
$$C_1 \text{ and } C_2 \text{ constant functions on S}\},$$
$$f \varepsilon C^1(S) \times C^1(S),$$

where $C^1(S)$ is the space of continuously differentiable functions on S. Note that if

V_1, V_2, V_3, V_4 $\left.\begin{array}{l}\\ \\ \\ \end{array}\right\}$ $V_1 = V_2, V_3 = V_4 \rightarrow BC \varepsilon M(f)$; coordinate functions of the generic function f are zero functions (4.19)

constant functions on [0,1] $V_1 \neq V_2, V_3 \neq V_4 \rightarrow BC \varepsilon M(f)$; coordinate functions of the generic function f are linear functions. (4.20)

With respect to the existence and the independence from boundary conditions of the effective permeability, we have the following result, which can be proved using the functional representations given in (4.1) and (4.2):

PROPOSITION 4.3. For some sub-grid random permeability, given in assumption 2.1, and some $f \varepsilon C^1(S) \times C^1(S)$, k^e is constant for all $BC \varepsilon M(f)$. ∎

Proposition 4.3 implies that different effective permeabilities should be assigned to the same grid block dependent on the – sub-class of the – boundary value flow problem that is considered. This will be demonstrated in Section 5, in which the effective permeabilities for a grid block will be calculated for boundary conditions (4.19) and for boundary conditions (4.20). Furthermore, proposition 4.3 gives us a lead how to calculate the effective permeability. In view of equations (4.15), we can construct a set of linear equations in L_{11}, L_{12}, L_{21} and L_{22} by making different choices for the boundary conditions BC within a certain sub-class M(f). A simple linear least-squares estimation scheme will suffice to estimate L_{11}, L_{12}, L_{21} and L_{22} from the set of linear equations. To guarantee that k^e is symmetric, we impose the condition $L_{12} = L_{21}$.

Finally we note that assumption 4.1 plays an important role in the derivation of the effective permeability tensor. Assumption 4.1 is necessary because generally there is no averaged version of Darcy's law, in which the average pressure gradient on S is related to the average velocity field on S by a constant permeability tensor. If spatial averaging is used instead of assumption 4.1, the intuitively appealing idea that k^e can be associated with a transformation between the grid scale and the sub-grid scale is lost. For in this approach the sub-grid pressure gradient ∇P^a and the sub-grid velocity V^a are transformed to the grid scale by spatial averaging, and k^e follows from a Darcy-type relation between these grid-scale quantities.

In the next section we present some applications of the theoretical results.

5. APPLICATIONS

The results of the preceding sections can be illustrated using a simplified situation. In particular, we shall make a sort of weak stationarity assumption for the random permeability. We cannot employ the usual weak stationarity assumptions – see, e.g., Wong [17; pp. 88-89], since the random permeability is considered only on a finite domain.
Define

$$\mathcal{T}_{\Delta_x,\Delta_y}(k)(\omega)(x,y) = k(\omega)(x + \Delta_x, y + \Delta_y) \tag{5.1}$$

$\omega \varepsilon \Omega;\ (x,y),(x+\Delta_x,y+\Delta_y)\ \varepsilon S;\ -1 \leq \Delta_x, \Delta_y \leq 1$

where $\{\mathcal{T}_{\Delta_x,\Delta_y}\}$ is a two-parameter family of translation operators. We make the following weak stationarity assumption:

ASSUMPTION 5.1.

$$E\{k(\omega)(x,y)\} = a_k \tag{5.2}$$

$(x,y)\varepsilon\ S.$

where a_k is a real constant.

$$E\left\{\tilde{k}(\omega)(x_1,y_1) \cdot \tilde{k}(\omega)(x_2,y_2)\right\}$$

$$= E\{\mathcal{T}_{\Delta_x,\Delta_y}(\tilde{k})(\omega)(x_1,y_1) \cdot \mathcal{T}_{\Delta_x,\Delta_y}(\tilde{k})(\omega)(x_2,y_2)\} \tag{5.3}$$

$(x_1,y_1),\ (x_2,y_2)\ \varepsilon\ S$

where

$$\tilde{k}(\omega) = k(\omega) - a_k$$

$\omega\varepsilon\Omega.$ ■

The function

$$R(\Delta_x, \Delta_y) = E\left\{\tilde{k}(\omega)(x+\Delta_x, y+\Delta_y)\tilde{k}(\omega)(x,y)\right\} \tag{5.4}$$

is the covariance function of the random permeability. Furthermore, we choose

$$m_k = a_k \tag{5.5}$$

Equation (5.7) implies that in equation (4.2) $\hat{a} = 0$ and $\hat{R} = R$.

Realizations have been simulated of 2-D random permeabilities with constant mean m_k and two-point correlation function of $\tilde{k}$ given by

$$R(\Delta_x, \Delta_y) = \sigma^2 \exp\left(-\frac{|\Delta_x|}{\ell_y} - \frac{|\Delta_y|}{\ell_y}\right), \tag{5.6}$$

where σ^2 is the variance and ℓ_x and ℓ_y are the average correlation lengths in the x- and y-directions, respectively, using an algorithm developed by Poulisse and van Soldt [11]. This stationary situation is amply covered by the results of the preceding sections. In particular, the continuity of the correlation function R implies that the random permeability $k(\omega)(x,y)$ is a quadratic mean continuous process — see Wong [17; pp. 77].

All numerical results are based on second-order approximate solutions — see Section 3.

Effective permeability tensors, determined for boundary conditions of sub-class (4.19) are denoted by k_1^e and for boundary conditions of sub-class (4.20) by k_2^e.

The set of linear equations in L_{11}, L_{12}, L_{21} and L_{22}, which have to be constructed to obtain the effective permeability, consists of 10 equations.

Error estimates for the calculated effective permeability have been obtained by applying a second order statistical analysis to the simulated data.

The average correlation lengths ℓ_x and ℓ_y in (5.8) are given in percentages with respect to the lengths of the sides of the grid block S. The permeabilities are in milli-darcy.

EXAMPLE 1. $m_k = 500$, $\sigma_k^2 = 2.5 \cdot 10^5$, $\ell_x = \ell_y = 25\%$

Figure 1 shows a realization of the random permeability over S with the given parameters

$$k_1^e = \begin{pmatrix} 290 \pm 15 & 0.0 \\ 0.0 & 290 \pm 15 \end{pmatrix}$$

$$k_2^e = \begin{pmatrix} 220 \pm 10 & 90 \pm 10 \\ 90 \pm 10 & 220 \pm 10 \end{pmatrix}$$

The effective permeability tensor is diagonal for boundary conditions of sub-class (4.19), in accordance with the fact that this is a direct generalisation of the 1-D situation.

The following example shows the influence of the variance σ_k^2, i.e. the size of the permeability variations with respect to m_k.

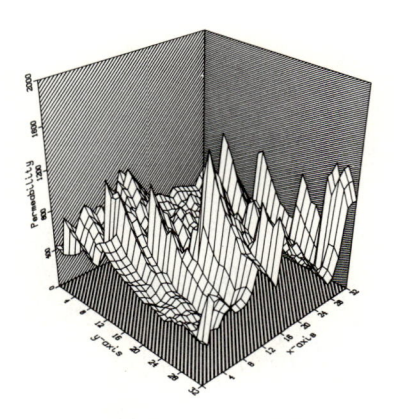

Fig. 1 Realization of Random Permeability with
$m_k = 500$, $\sigma_k^2 = 2.5*10^5$, $\ell_x = \ell_y = 25\%$

EXAMPLE 2. $m_k = 500$, $\sigma_k^2 = 5.0*10^5$, $\ell_x = \ell_y = 25\%$

$$k_2^e = \begin{pmatrix} 100 \pm 15 & 75 \pm 10 \\ 75 \pm 10 & 100 \pm 15 \end{pmatrix}$$

A larger variance means that smaller and larger values with respect to m_k appear in the realizations of the random permeability. The result shows that the smaller values dictate the effective permeability.

The next example shows what happens if $\ell_x \neq \ell_y$.

EXAMPLE 3. $m_k = 500$, $\sigma_k^2 = 2.5 \cdot 10^5$, $\ell_x = 25\%$, $\ell_y = 1000\%$. Figures 2 and 3 show two different realizations of the random permeability over S with the given parameters.

$$k_1^e = \begin{pmatrix} 50 \pm 5 & 0 \\ 0.0 & 270 \pm 15 \end{pmatrix}$$

$$k_2^e = \begin{pmatrix} 35 \pm 5 & 20 \pm 5 \\ 20 \pm 5 & 300 \pm 10 \end{pmatrix}$$

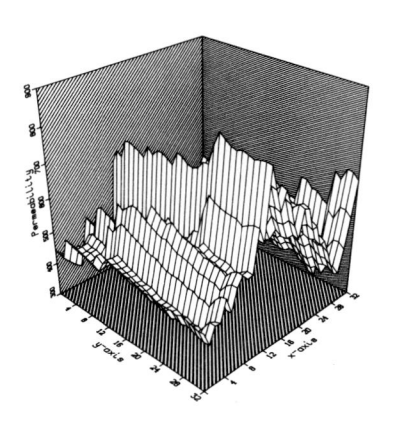

Fig. 2 Realization of Random Permeability with $m_k = 500$, $\sigma_k^2 = 2.5 \cdot 10^5$, $\ell_x = 25\%$, $\ell_y = 1000\%$

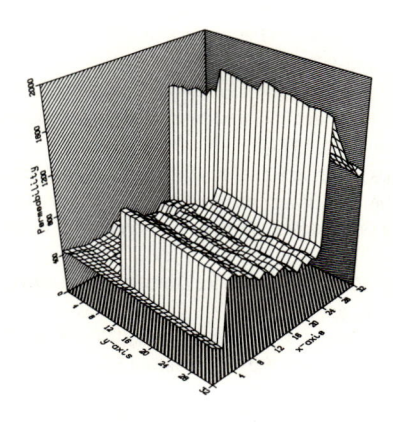

Fig. 3 Realization of Random Permeability with
same statistics as in Figure 2.

From a comparison of examples 1 and 3 it follows that a change in ℓ_y has a dramatic effect on k_{11} – see equation (4.16). This can be explained by the fact that the x-component of the velocity V_x encounters a low-permeability barrier in all realizations perpendicular to the x-direction, resulting from the long average correlation length of the random permeability in the y-direction.

Finally, we wish to emphasize that it can be misleading to interpret the calculated effective permeability in terms of one, or indeed a limited number of realizations of the random permeability, because the effective permeability is based on an ensemble of all possible realizations. Figures 2 and 3 demonstrate that the graphs of the individual realizations can be quite different.

REFERENCES

[1] G. Adomian, A New Approach to Nonlinear Partial Differential Equations, J. Math. Anal. Appl., 1984, **102**, pp. 420-434.
[2] J. Bear, "Dynamics of Fluids in Porous Media", American Elsevier, New York, 1972.
[3] M.J. Beran and J.J. McCoy, Mean Field Variation in Random Media, Quarterly Appl. Math., 1970, **XXVIII**, 2, pp. 245-258.
[4] A. Bharucha-Reid, "Random Integral Equations", Academic Press, New York, 1972.
[5] R.F. Curtain and A.J. Pritchard, "Functional Analysis in Modern Applied Mathematics", Academic Press, New York, 1977.
[6] U. Frisch, Wave Propagation in Random Media, In: "Probabilistic Methods in Applied Mathematics", Volume I, A. Bharucha-Reid (ed.), Academic Press, 1968, pp. 75-198.
[7] L.V. Kantorovich and G.P. Akilov, "Functional Analysis", Pergamon Press, Oxford, second ed., 1982.
[8] P.R. King, The Use of Renormalisation for Calculating Effective Permeability, Transport in Porous Media, 1989, 4, pp. 37-58.
[9] M. Loève, "Probability Theory, I", Springer Verlag, New York, 4th ed., 1977.
[10] P.C. Martin, E.D. Siggia and H.A. Rose, Statistical Dynamics of Classical Systems, Physical Review A, 1973, 8, 1, pp. 423-437.
[11] H.N.J. Poulisse and I.G. van Soldt, Generation of Realizations of 3-D Random Fields with Given Covariance Function, submitted to SIAM J., Sci. Stat. Comput.
[12] W. Rudin, "Real and Complex Analysis", TaTa McGraw-Hill, New Dehli, reprinted second ed., 1977.
[13] W. Rudin, "Principles of Mathematical Analysis", McGraw-Hill Kogakusha Ltd., Tokyo, third ed., 1976.
[14] L. Smith and R.A. Freeze, Stochastic Analysis of Steady-State Groundwater Flow in a Bounded Domain. 1. One-Dimensional Simulations, Water Res. Res., 1979, 15, 3, pp. 521-528.
[15] L. Smith and R.A. Freeze, Stochastic Analysis of Steady Groundwater Flow in a Bounded Domain. 2. Two-Dimensional Simulations, Water Res. Res., 1979, 15, 6, pp. 1543-1559.
[16] C.D. White and R.N. Horne, Computing Absolute Transmissibility in the Presence of Fine-Scale Heterogeneity, Soc. Pet. Eng., 1987, SPE 16011, pp. 209-220.
[17] E. Wong, "Stochastic Processes in Information and Dynamical Systems", Mc.Graw-Hill, New York, 1971.

RELATIVE PERMEABILITIES AND CAPILLARY PRESSURE ESTIMATION THROUGH LEAST SQUARE FITTING

C. Chardaire,
(Institut Français du Pétrole, Rueil-Malmaison, France),

G. Chavent,
(CEREMADE, Université Paris-Dauphine, and INRIA, Rocquencourt, France),

and
J. Jaffré, J. Liu,
(INRIA, Rocquencourt, France).

ABSTRACT
Relative permeabilities and capillary pressure are determined from two-phase immiscible flow laboratory experiments using numerical matching. An automatic method based on optimal control theory is proposed to identify simultaneously relative permeabilities and capillary pressure. These three parameters are determined by using a least square technique after discretizing the one-dimensional saturation equation by a higher order Godunov scheme associated with Van Leer's slope limiter. The Fortran codes for the numerical simulator and for the calculation of the gradient of the least square function are generated automatically by a Macsyma program that allows a fast and reliable programmation. Experimental data used for matching are production and pressure drop, and/or saturation profiles. New laboratory techniques (gammametry or tomodensidometry) provide saturation measurements in the core at different times and these extra data improve drastically the matching. The validity of the method is demonstrated on simulated data.

1. INTRODUCTION

Simulators of two-phase flow in porous media need relative permeabilities and capillary pressure curves among other data. They are estimated from laboratory experiments by matching the pressure and production history. Automatic adjustment methods have been described more often for relative permeabilities [1,2,3,4] than for capillary pressure [3,4]. In the past they had to be estimated from separate laboratory experiments due to insufficiency of available data : only pressure drop and production histories were measured.

Previous work[5] has shown that if saturations were also measured, the identification problem would be much easier. Such measurements are now available through gamma-ray or CT scanning. Therefore we shall describe below an automatic method for estimating simultaneously both relative permeabilities and capillary pressure from a single experiment.

Another feature of our method is the use of a numerical scheme with greater precision and smaller numerical diffusion when capillary effects are small, than tradional schemes that are commonly used. Consequently the results are less dependent on the mesh size. Moreover it is well-known that numerical diffusion in the scheme is usually compensated by increasing artificially the convexity in the convex part of the relative permeability curves. This defect is greatly reduced by using a numerical model which has less numerical diffusion.

Finally the program that performs the automatic adjustment has been generated automatically with a symbolic code generator (SCG) written in Macsyma[6]. Input for this code are the discretized equations and the least square error function. Output are the fortran codes which calculate this error function and its gradient with respect to the parameters and which will feed the optimizer. Such a procedure reduced significantly the development time and improved the reliability of the codes.

2. LABORATORY EXPERIMENT MODELS

Two-phase displacements in laboratory experiments are represented by the following one-dimensional incompressible model[7] :

$$\phi \frac{\partial S}{\partial t} - \frac{\partial}{\partial x} Ka(S) \frac{\partial S}{\partial x} + q_T(t) \frac{\partial b_T(S)}{\partial x} + \frac{\partial q_G b_G(S)}{\partial x} = 0 ,$$
$x \in]0,L[,$ (2.1)

$q_T = - Kd(S)(\nabla P - \rho(S)g\nabla z)$: total flow rate independent of x, (2.2)

$q_G = + K(\rho_w + \rho_{Nw})g \nabla z/2$: gravity field, (2.3)

where S is the saturation of the displacing gluid, P the global pressure[7]. The index w (resp. Nw) indicates the wetting fluid (resp. nonwetting fluid). The functions b_T, b_G, d and ρ depend only on relative permeabilities while a also depends on capillary pressure:

$b_T = \dfrac{k_W}{k_W + k_{NW}}$, $b_G = \dfrac{k_W k_{NW}}{k_W + k_{NW}} \dfrac{\rho_W - \rho_{NW}}{2(\rho_W + \rho_{NW})}$, $d = k_W + k_{NW}$,

$\rho = \dfrac{k_W \rho_W + k_{NW} \rho_{NW}}{2(\rho_W + \rho_{NW})}$, $a = \dfrac{k_W k_{NW}}{k_W + k_{NW}} p_c'$,

where $k_N, k_{NW}, \rho_N, \rho_{NW}$ are respectively the mobilities and the densities of the wetting and nonwetting phases and p_c, is the capillary pressure.

The following boundary conditions have been considered:

- <u>forced drainage</u> : with given injection rate or given pressure drop :
(i) no flow for the wetting phase at injection end : $Q_W = 0$,
(ii) pressure continuity at production end, i.e. unilateral boundary condition :
$(S_W - S_{WC})Q_{NW} = 0$, $S_W \geq S_{WC}$, $Q_{NW} \geq 0$. S_{WC} denote the saturation for which the capillary pressure vanishes.

- <u>gravity drainage</u> : where the flow is governed only by gravity and capillary effects :
(i) $Q_W = 0$ at injection end,
(iii) given wetting saturation at production end : $S_W = S_W$ initial.

- forced imbibition : with given injection rate or given pressure drop :
(iv) no flow for the nonwetting phase at injection end : $Q_{NW} = 0$,
(v) pressure continuity at production end, i.e. unilateral boundary condition [7] :
$(S_W - S_{WC})Q_W = 0$, $S_W \leq S_{WC}$, $Q_W \geq 0$.

- free imbibition : where the flow is governed only by gravity and capillary effects:
(vi) the wetting saturation is maximum at ends in contact with the wetting fluid :
$S_W = S_{Wmax}$,
(vii) no flow for the wetting fluid at ends in contact with the nonwetting fluid.

3. THE NUMERICAL MODEL

In this section we describe the numerical scheme that we use to calculate the solution of equation (2.1). The space approximation is second order for both convective terms and diffusion terms. Concerning time discretization the scheme is only first order. Diffusion terms are treated implicitly while convective terms are calculated explicitly.

The saturation is approximated in the space V of piecewise linear functions discontinuous at discretization points. Given a function $S^n \in V$, the function $S^{n+1} \in V$ is calculated in two steps. In the first step, we calculate the average values in discretizaion cells, and in the second step we derive slopes in the cells from these averages values by satisfying : slope limitation rules due to Van Leer [9].

Let us denote by $x_{i+1/2}$ the discretization points and by h the mesh size. Functions v of V are discontinuous at discretization points and we denote by $v^L_{i+1/2}$ and $v^R_{i+1/2}$ left-hand and right-hand limit values at $x_{i+1/2}$. Also $v_i = (v^L_{i-1/2} + v^R_{i+1/2}) / 2$ denotes the average value in the cell $(x_{i-1/2}, x_{i+1/2})$.

Step 1 : calculation of the average values

The average values are calculated by a conservative finite difference scheme :

$$\phi_i \frac{S_i^{n+1} - S_i^n}{\Delta t} + \frac{1}{h}\left[r_{i+1/2}^{n+\theta} + F_{i+1/2}^n - r_{i-1/2}^{n+\theta} + F_{i-1/2}^n\right] = 0$$

where $r_{i+1/2}^{n+\theta}$ denotes the capillary numerical flux and $F_{i+1/2}^n$ the convective numerical flux. The first one is given by standard finite differencing in space :

$$r_{i+1/2}^{n+\theta} = \frac{1}{h} \frac{2}{1/K_i + 1/K_{i+1}} (\alpha(S_{i+1}^{n+\theta}) - \alpha(S_i^{n+\theta})),$$

where θ satisfying $0.5 < \theta \leq 1$ is a parameter controlling the time stepping for the diffusion terms : $\theta = 0.5$ corresponds to Crank Nicolson discretization and $\theta = 1$ to forward Euler differencing. In practice θ is chosen so that diffusion terms are unconditionaly stable.

The convective numerical flux $F_{i+1/2}^n$ can be calculated by solving a Riemann problem with the two limit values at $x_{i+1/2}$ as the initial data, i.e. $S_{i+1/2}^{L,n}$, $S_{i+1/2}^{R,n}$. Therefore it is a function of these two values :

$$F_{i+1/2}^n = F_{i+1/2}^n (S_{i+1/2}^{L,n}, S_{i+1/2}^{R,n})$$

where the function $F_{i+1/2}^n$ can be defined for example by as the Godunov numerical flux :

$$F_{i+1/2}^n (u,v) = \begin{cases} \min_{w \in [u,v]} f^n(x_{i+1/2}, w) & \text{if } u \leq v, \\ \max_{w \in [u,v]} f^n(x_{i+1/2}, w) & \text{if } u > v. \end{cases}$$

Here the function $f^n(x_{i+1/2}, w)$ is the convective flow rate of the displacing fluid:

$$f^n(x_{i+1/2}, w) = q_T^n \, b_T(w) + q_{G,i+1/2} \, b_G(w).$$

Step 2 : calculation of the cell slopes

Once the average values of S^{n+1} have been calculated as in step 1, we are left with the calculation of the slopes p_i^{n+1} to complete the determination of $S^{n+1} \in V$. This is done as follows:

$p_i^{n+1} = 0$ if $S_i^{n+1} \leq \mathrm{Min}(S_{i+1}^{n+1}, S_{i-1}^{n+1})$ or if $S_i^{n+1} \geq \mathrm{Max}(S_{i+1}^{n+1}, S_{i-1}^{n+1})$,

$p_i^{n+1} = \mathrm{Min}[(S_i^{n+1} - S_{i-1}^{n+1})/(h_{i-1}+h_i)/2, (S_{i+1}^{n+1} - S_i^{n+1})/(h_i+h_{i+1})/2]$

$$\text{if } S_{i-1}^{n+1} \leq S_i^{n+1} \leq S_{i+1}^{n+1},$$

$p_i^{n+1} = \mathrm{Max}[(S_i^{n+1} - S_{i-1}^{n+1})/(h_{i-1}+h_i)/2, (S_{i+1}^{n+1} - S_i^{n+1})/(h_i+h_{i+1})/2]$

$$\text{if } S_{i-1}^{n+1} \geq S_i^{n+1} \geq S_{i+1}^{n+1}.$$

Figures 1 and 2 show comparisons of this scheme (EC2) with two standard schemes using first order space approximation of the convective terms, one using the same time discretization (EC1) and another one using a semi-implicit time discretization (SIC1). Fig.1 corresponds to case with capillary effects and Fig.2 to a case without capillary effects. Note on this second figure that the true solution is not a smooth as one would expect because the relative permeabilities used for this simulation were approximated by a continuous piecewise linear function.

These figures show that our numerical scheme is more precise and has less numerical diffusion than traditional schemes. Therefore results are less dependent on the mesh size and, when estimating relative permeability the convexity in the convex part of the curves will not be exaggerated to compensate a too large numerical diffusion.

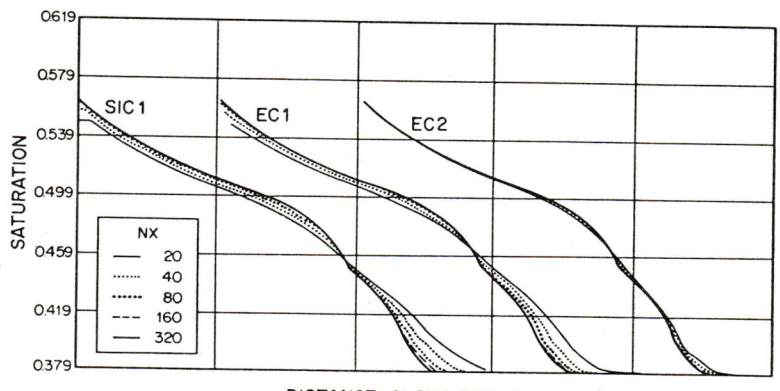

Figure 1 : Comparisons of convergence for three different schemes. Case with capillary pressure effects (NX = number of discretization points).
SIC1 = 1st order semi-implicit in time, 1st order in space for convective terms.
EC1 = 1st order explicit in time, 1st order in space for convective terms.
EC2 = 1st order explicit in time, 2nd order in space for convective terms.

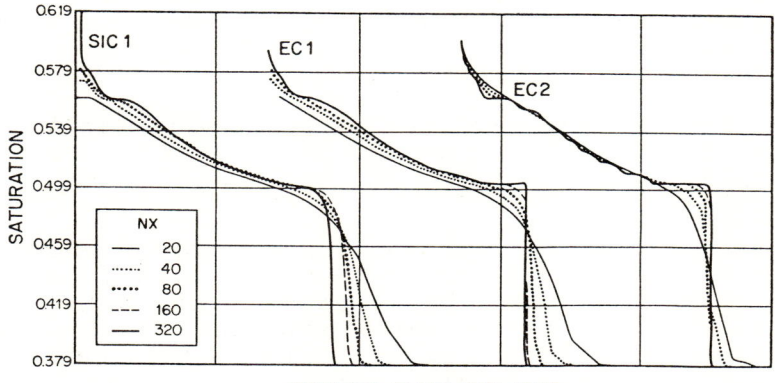

Figure 2 : Comparisons of convergence for three different schemes. Case without capillary pressure effects (NX = number of discretization points).
SIC1 = 1st order semi-implicit in time, 1st order in space for convective terms.
EC1 = 1st order explicit in time, 1st order in space for convective terms.
EC2 = 1st order explicit in time, 2nd order in space for convective terms.

4. PARAMETER ESTIMATION

Laboratory experiments provide the following data measurements:

- the cumulative production volumes Q_k^m measured at various times indexed by k,
- the pressure drop ΔP_k^m across the core measured at various times indexed by k (when injection rate is given),
- the saturation $S_{k,i}^m$ measured at various points indexed by i and at various times indexed by k.

Let us introduce an error function J of the vector a of the parameters defining the relative permeabilities and capillary pressure curves:

$$J(a) = W_Q \sum_k (Q_k^C - Q_k^m)^2 + W_P \sum_k (\Delta P_k^C - \Delta P_k^m)^2 + W_S \sum_k \sum_i (S_{k,i}^C - S_{k,i}^m)^2$$

where Q_k^C, ΔP_k^C and $S_{k,i}^C$ are calculated using the numerical model with relative permeabilities and capillary pressure defined by a and where W_Q, W_P, W_S are weights which are equal to zero when the corresponding measurement is missing.

The parameter estimation problem is set as to find a vector of parameters a for which the observed calculated quantities Q_k^C, ΔP_k^C, $S_{k,i}^C$ are as close as possible to the corresponding measured ones, i.e. which minimizes the function J. Minimizing J can be achieved by using an optimization routine which requires usually two subroutines, one calculating $J(a)$ and one calculating its gradient $\nabla J(a)$. Note that each calculation of $J(a)$ requires one numerical simulation of the laboratory experiment. As for $J(a)$ it can be calculated exactly using optimal control theory. This requires the solution of the adjoint equations[8,3,4].

The formulas giving $\nabla J(a)$ are obtained through complex and tedious calculations by hand but which follow an automatic procedure. A symbolic code generator (SCG)[6] has been used to decrease the amount of work necessary to implement the method and to improve the reliability of the produced program. Input for the SCG are the discretized equations as described in the following sections and the function J. To obtain the codes for the four models described in section II, we ran the SCG four times which required only a few days of manpower. The tests to check ∇J were successful the first time, demonstrating the reliability of this approach.

5. NUMERICAL EXPERIMENTS FOR PARAMETER ESTIMATION

In this section, we present experiments on simulated data. Experiments on realistic data are presented in Chavent et al [10]. Data are normalized : porosity ϕ, absolute permeability K and total velocity q_T are set to 1. The space and time intervals have unit length and gravity is neglected. Relative permeabilities and capillary pressure are :

$$k_W = 0.3 \, S^{3/2}, \quad k_{NW} = (1-S)^{5/2},$$
$$\frac{dp_C}{dS} = (0.3 - S + S^2) \quad \text{with } p_C(1) = 0.$$

First the program is run with these data to obtain observed data : saturation profiles, pressure drop and production. Then from these observed data, we try to recover the relative permeabilities and the capillary pressure by using our parameter estimation code. The optimization routine that we used is a standard one based on a quasi-Newton method and has been found in the IMSL library [12].

Let us first consider the case where the parameters are represented analytically :

$$k_W = a_W \, S^{b_W}, \quad k_{NW} = a_{NW} \, S^{b_{NW}}, \quad \frac{dp_C}{dS} = c_0 + c_1 S + c_2 S^2.$$

Therefore we have seven parameters to identify. In a first experiment the observed data were only 30 saturation profiles at equidistant times. After 30 iterations of optimization, the error function J is reduced by 10^4 and the following parameters are recovered with 3 digits : b_W, b_{NW}, a_{NW}/a_W, $a_W c_0$, $a_W c_1$, $a_W c_2$. It is easy to check that the solution of eq.(2.1) is depending only on these six parameters and not on the seven parameters defining relative permeabilities and capillary pressure. To estimate all the seven parameters, one has to take for observed data in addition to saturation profiles either the pressure drop or the production curve or both. When observing also the pressure drop up to twice the breakthrough time, the algorithm converged as easily but all the seven parameters were obtained. We also tried to estimate the parameters by observing only the pressure drop and the production curve. The convergence was then much slower. The error function J has to be reduced by 10^7. This explains why usually automatic adjustment does not succeed in this case for realistic experiments.

When representing relative permeabilities and capillary pressure curves with ten points, then 30 parameters have to be estimated. We used as observed data the same saturation profiles and pressure drop as before. The estimated parameters are shown on figures 3 and 4. Though the results are satisfactory, the optimization process was more difficult as the number of parameters is larger. It needed 150 iterations and the cost function has been reduced by 10^6. Also the estimated relative permeabilities have been smoothed three times during the optimization process and the estimated capillary pressure only once. The difficulty in obtaining the right values of relative permeabilities for high values of the saturation might be due to a too short length of time of observation (only up to twice the breakthrough time). This does not allow the core to be filled with enough of the injected fluid. Without saturation profiles, it is not possible at all to estimate the parameters.

RELATIVE PERMEABILITIES 731

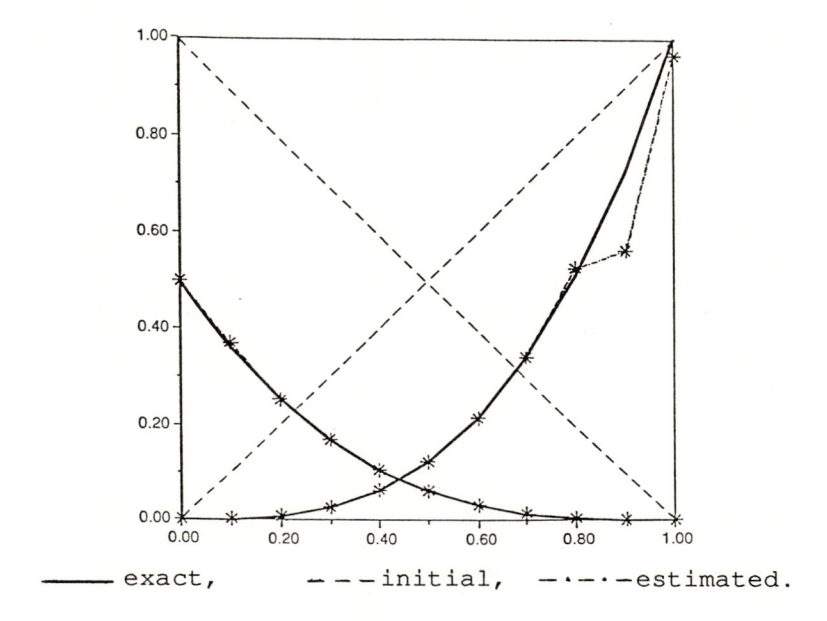

——— exact,   — — —initial,   —·—·—estimated.

Figure 3 : Relative permeabilities

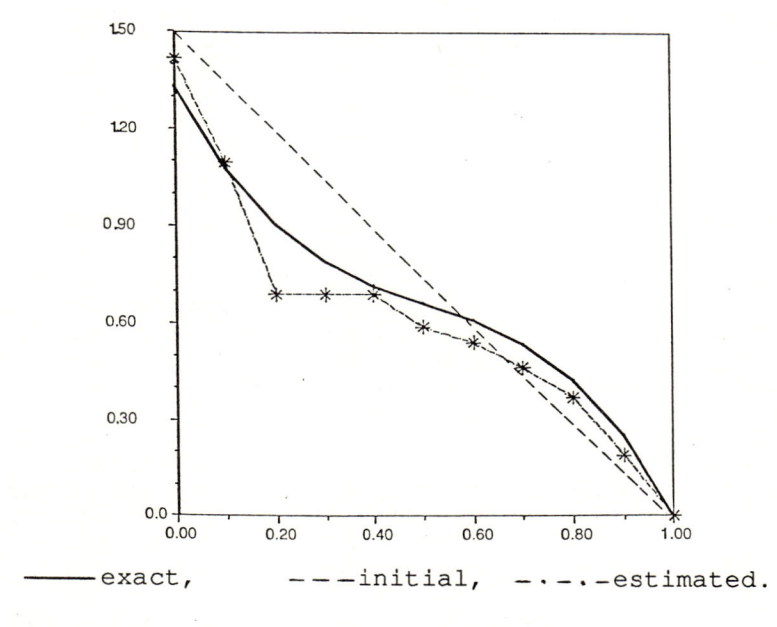

——— exact,   — — —initial,   —·—·—estimated.

Figure 4 : Capillary pressure

6. CONCLUSION

A method to estimate simultaneously relative permeabilities and capillary pressure from laboratory experiments has been presented. It minimizes an error function measuring the error between measured observations and observations calculated by the numerical model for a certain set of parameters. The gradient of this error function is calculated through techniques from optimal control theory. The two routines calculating the error function and its gradient are generated by a symbolic generator to decrease implementation time and to improve reliability. These two routines feed an optimization program.

Advantages of using more accurate methods for the numerical model have been demonstrated : results less dependent on the mesh size and on the numerical diffusion of the scheme. The method has been validated with simulated data. It is shown that the availability of measured saturation profiles is essential to improve the parameter estimation. The parameters are easier to estimate when represented analytically. When represented with discrete values, the parameters can be obtained but the optimization process is more difficult since more parameters have to be estimated. Research to improve the efficiency of the optimization process is under way[11].

REFERENCES

[1] Chavent G., Cohen G., "Numerical approximation and identification is a 1-D parabolic degenerated non linear diffusion and transport equations", Proceedings of the 8th IFIP Conference on Optimization Techniques, Wurzburg, Germany, Sept.5-9, 1977 (Ed. J. Stoer), Lecture Notes in Control and Information Sciences 6 (Springer-Verlag, 1978) 282-293.

[2] Chavent G., Cohen G., Espy M., "Determination of relative permeabilities and capillary pressures by an automatic adjustment method", SPE 9237, Proceedings of the 55th SPE Annual Fall Technical Conference, Dallas, Texas (Sept.21-24, 1980).

[3] Watson A.T., Seinfeld J.H., Gavalas G.R., Woo P.T., "History matching in two-phase petroleum

reservoirs", SPE 8250, 54th SPE Annual Fall Technical Conference, Las Vegas, Nevada, (Sept.23-26, 1979).

[4] Kerig P.D., Watson A.T., "A New Algorithm for Estimation Relative Permeabilities from Displacement Experiments", Soc. Pet. Eng. J., (Feb.1987) 103-112.

[5] Cohen G., "Resolution Numérique et Identification pour une Equation Quasi-Parabolique Non-Linéaire Dégénérée de Diffusion et Transport en Dimension Un", Thèse de l'Université Paris-Dauphine (1978).

[6] Liu J., "Gradpack : A symbolic system for automatic generation of numerical programs in parameter estimation", Proceedings of the 5th IFAC Symposium on Control of Distributed Parameter Systems, Perpignan, France (June 26-29, 1989).

[7] Chavent G., Jaffré J., "Mathematical Models and Mixed Finite Elements for Reservoir Simulation", (North-Holland, 1986).

[8] Chavent G., "Identification of functional parameters in P.D.E", in Identification of Parameters in Distributed Systems (Eds., R.E. Goodson, M. Polis), (ASME, New York, 1974).

[9] Van Leer B., "Towards the Ultimate Conservative Scheme : IV A New Approach to Numerical Convection", J comp. Phys. 23, (1977) 276-299.

[10] Chardaire C., Chavent G., Jaffré J., Liu J., Bourbiaux B., "Simultaneous estimation of relative permeabilities and capillary pressure", SPE 19680, SPE Annual Technical Conference and Exhibition, San Antonio, Texas, (Oct.8-11, 1989).

[11] Chavent G., Liu J., "Multiscale parametrization for the estimation of a diffusion coefficient in elliptic and parabolic problems", Proceedings of the 5th IFAC Symposium on Control of Distributed Parameter Systems, Perpignan, France (June 26-29, 1989).

[12] IMSL, 2500 City West Boulevard, Houston, Texas 77042-3020, U.S.A.

NOMENCLATURE

$k_j(S) = \dfrac{k_{rj}(S)}{\mu_j}$ = mobilities of fluid j, j=w, NW,

$p_C(S) = p_W - p_{NW}$ = capillary pressure,

$a = \dfrac{k_W k_{NW}}{k_W + k_{NW}} \, p_C'$,

$\alpha = \displaystyle\int_{S_C}^{S} a \, dS$,

$b_T = \dfrac{k_W}{k_W + k_{NW}}$,

$b_G = \dfrac{k_W k_{NW}}{k_W + k_{NW}} \dfrac{\rho_W - \rho_{NW}}{2(\rho_W + \rho_{NW})}$,

$d = k_W + k_{NW}$,

$\rho = \dfrac{k_W \rho_W + k_{NW} \rho_{NW}}{2(\rho_W + \rho_{NW})}$,

$\gamma = \displaystyle\int_{S_C}^{S} \left(\dfrac{k_W}{k_W + k_{NW}} - \dfrac{1}{2}\right) \dfrac{dp_C}{dS} \, dS$,

$P = (P_W + P_{NW})/2 + \gamma(S_W)$ = global pressure,

$\Delta P_W = \Delta P + \Delta\gamma + \Delta p_C/2$ = pressure drop for the wetting fluid,

$\Delta P_{NW} = \Delta P + \Delta\gamma - \Delta p_C/2$ = pressure drop for the nonwetting fluid.

CONING SIMULATION MORE ACCURATELY

Frans J.T. Floris
(TNO Institute of Applied Geoscience, Delft, Netherlands)

ABSTRACT

Calculations regarding water coning due to oil production with conventional (finite difference) simulators are considered to give poor results. A reliable and accurate method has been used to improve this. The method was originally developed for fresh and salt water, but has been generalized for water/oil calculations. Presently the method is applied to sharp interface situations. The method was selected because of its accurate representation and calculation of the interface. It uses an adaptive finite element grid, which incorporates the interface. The stream function is introduced in the equation describing the interface movement, which is then solved by the $S^{\alpha,\beta}$ method.

The method has been used to investigate the growth of the cone and the stability of the displacement process, as well as the behaviour of the cone near the well perforations, subject to different well conditions. The method has also been proved successful for modelling the problem of fingering of fluid-fluid interfaces.

1. INTRODUCTION

Oil production from an underground reservoir is often accompanied by coning, i.e. the rise of the water/oil contact near a well (fig 1). This can lead to undesirable water production. To control coning it is important to be able to predict the position of the water/oil interface during production. Conventional (finite difference) methods are considered to give results of limited accuracy [4].

Sharp interface models play an important role in improving our understanding of complicated multiphase flow phenomena, such as coning behaviour. When simplifications, such as the Dupuit assumption are not made, these models can also be used to test the performance of numerical simulators. The conventional

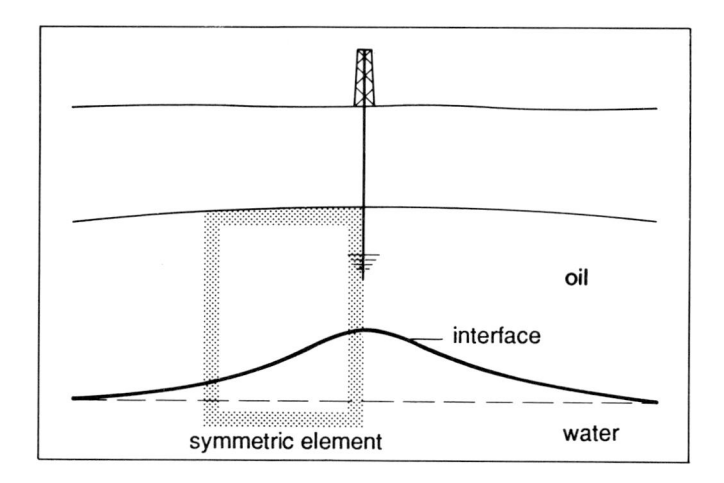

Fig 1. Water coning in an oil reservoir.

method of finite differences has disadvantages for sharp interface models. The main disadvantage is the inability to accurately represent the curvature of the interface [7]. The finite element method does not suffer from this disadvantage as it allows for an adaptive grid of triangles, which can be chosen along the interface (fig 2). This advantage still holds for future models, when steep saturation gradients are present.
A method developed for the calculation of the movement of the interface between fresh and salt water [2] has been extended for water/oil flow. It uses finite elements, the stream function and the $S^{\alpha,\beta}$ method to obtain an accurate calculation of the interface position.

The method is applied to study coning phenomena during oil recovery through horizontal and fractured wells. The method allows accurate calculation of breakthrough times and water cut after breakthrough. Other phenomena inherent to coning, such as fingering, are also studied.

2. PHYSICAL MODEL

The two dimensional model tested, assumes that only two fluids are flowing, e.g. water and oil. An (x,z) cross-section of the reservoir is used, but the method can be adapted to cylindrical symmetry too. The fluids are incompressible, satisfy Darcy's law and are separated by a sharp interface. The latter assumption can be justified if the capillary transition zone is small compared to reservoir dimensions and the mobility ratio is small enough so that the flow in terms of Buckley-Leverett displacement is dominated by shock behaviour. We shall assume that the reservoir is homogeneous, although the model allows for a heterogeneous description of the reservoir rock. The difference in oil and water flow properties only appears in the relative permeability end points.

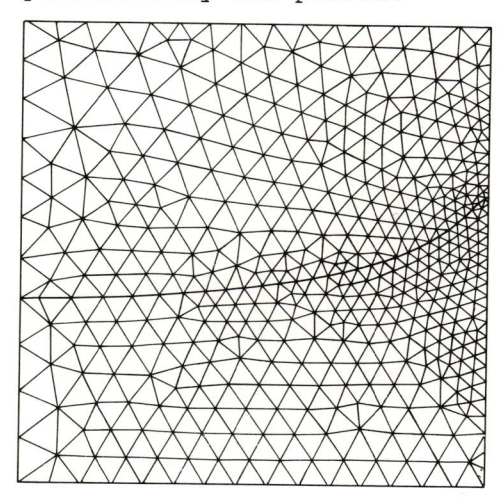

Fig 2. A finite element grid gives an accurate representation of the interface.

The top of the reservoir is sealed by a caprock. The outer and lower boundary can be considered either closed or open. In the latter case influx is considered to happen under hydrostatic conditions. Owing to symmetry only half of the reservoir needs to be simulated (fig 1). The well now becomes part of this inner boundary. The symmetry implies that there is no flow across the rest of the boundary. The well can either be a point well, representing a horizontal well in three dimensions, or a line well, modelling production from a fracture. The influx into the line well can be modelled as uniformly distributed (1), as distributed along the well according to fluid mobility (2) or as if the fracture is of infinite conductivity (3).

3. MATHEMATICAL MODEL AND NUMERICAL PROCEDURE

Since the fluids are incompressible, the Darcy velocity $\overline{q_f}$ is divergence free in both fluid domains, Ω_o and Ω_w. For the cross-sectional model

$$\overline{q_f} = (q_{fx}(x,z),\ 0,\ q_{fz}(x,z))^T\ ,\quad \text{div } \overline{q_f} = 0 \tag{1}$$

This is satisfied identically by the introduction of a stream function $\overline{\psi_f}$, as

$$\overline{\psi_f} = (0, \psi_f, 0)^T\ ,\quad \overline{q_f} = \text{curl } \overline{\psi_f} \tag{2}$$

Taking the curl of Darcy's Law eliminates the pressure gradient, and leaves only in the y-component a non-trivial equation

$$-\frac{\partial}{\partial x}\left(\frac{\mu_f}{k_f} q_{fz}\right) + \frac{\partial}{\partial z}\left(\frac{\mu_f}{k_f} q_{fx}\right) = \frac{\partial(\rho_f g)}{\partial x} \tag{3}$$

Substituting the stream function finally gives

$$-\frac{\partial}{\partial x}\left(\frac{\mu_f}{k_f}\frac{\partial \psi_f}{\partial x}\right) - \frac{\partial}{\partial z}\left(\frac{\mu_f}{k_f}\frac{\partial \psi_f}{\partial z}\right) = \frac{\partial(\rho_f g)}{\partial x} \quad \text{in } \Omega_f \tag{4}$$

Owing to the incompressibility of the fluids, the stream function is continuous across the interface, so one continuous stream function ψ can be defined over the whole reservoir domain Ω, which satisfies

$$-\text{div}\left(\frac{\tilde{\mu}}{\tilde{k}}\ \overline{\text{grad}}\ \psi\right) = g\frac{\partial \tilde{\rho}}{\partial x}\ ,\quad (x,z) \in \Omega \tag{5}$$

Here the permeability $\tilde{k}$, viscosity $\tilde{\mu}$ and density $\tilde{\rho}$

functions jump across the interface from water to oil values (fig 3). So by means of these parameters the interface position is included in the stream function equation. This elliptic equation is solved using finite elements and an adaptive grid every timestep.

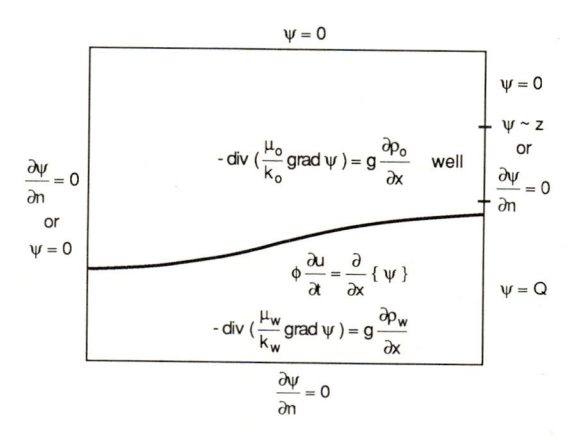

Fig 3. Mathematical model for the calculation of the time evolution of a sharp water/oil interface.

The displacement of the interface is described by the equation [3]

$$\frac{q_n}{\cos \theta} = \phi \frac{\partial u}{\partial t} \qquad (6)$$

Introduction of the stream function gives the interface equation

$$\phi \frac{\partial u}{\partial t} = \frac{\partial}{\partial x} \{ \psi(x,u(x,t),t) \}, \quad \text{on } \Gamma \qquad (7)$$

from which the time evolution of the interface can be calculated. This hyperbolic equation is solved using a two-step predictor-corrector method called the $S^{\alpha,\beta}$ method [6]. For particular choices of the parameters other well known schemes are obtained, e.g. $\alpha=\beta=1/2$ gives the two-step Lax-Wendroff scheme. Here, the parameters α,β have been chosen to obtain a second order accurate method in space, which optimize the shock behaviour of the method for the standard hyperbolic Burgers equation. For both predictor and corrector the stream function equation has to be solved. In the interface equation, values of the stream function are only required at nodal points on the interface.

No-flow boundaries are stream lines so here a Dirichlet boundary condition holds. Hydrostatic influx is modelled by a homogeneous Neumann boundary condition. The point well is modelled by a jump of the stream function, the line well by a linearly or piecewise linearly increasing stream function or a homogeneous Neumann condition.

4. RESULTS

4.1 Reference case

Using the method described above simulations were run to calculate the interface evolution, breakthrough times and water cut after breakthrough. The flow is characterized by two dimensionless parameters, viz. the mobility ratio $M = k_w\mu_o/k_o\mu_w$, and gravity number $G = \Delta\rho g h^2 k_o/Q\mu_o$. In the reference case M equals 1 and G equals 0.037. The finite element grid contains 15 elements along the upper and lower boundary. The number of elements along the other boundaries varies in time. Figure 4 shows the time evolution of the grid.

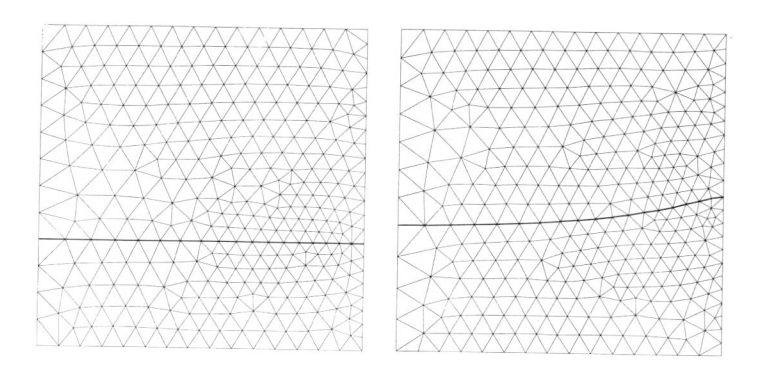

Fig 4. Time evolution of a finite element grid.

4.2 Influence of model parameters

Increasing the mobility ratio M, by increasing water mobility, accelerates coning. Increasing the gravity number G, e.g. by decreasing production rate or increasing gravity difference, mitigates this. To minimize simulation time, it is important to use the coarsest possible finite element grid. Figure 5 shows that apart from breakthrough time, the coarseness of the grid has little effect on the results. The reference grid has the optimal grid size without affecting the accuracy of the breakthrough calculation.

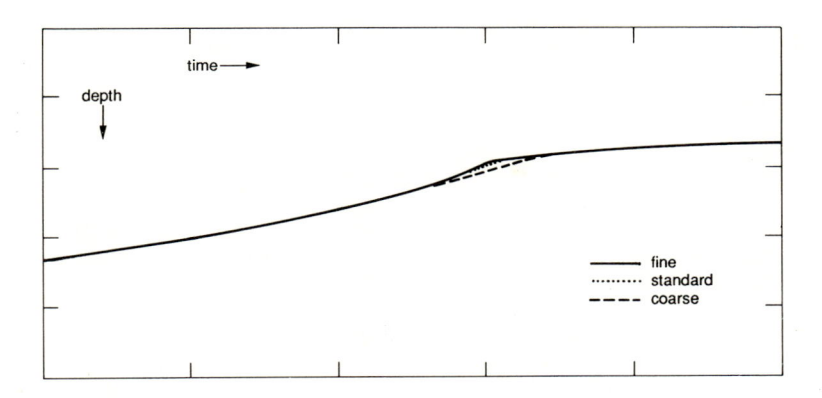

Fig 5. The effect of coarseness of the grid on the time evolution of the top of the cone. The number of elements on the boundary: 8, 15, 30.

To be able to investigate water cut performance the line well was used. The infinite conductivity model gave the fastest breakthrough. There was no difference in behaviour of the uniform and mobility dependent well model (section 2) before breakthrough, but after breakthrough the latter clearly showed a higher production rate for the more mobile fluid.

4.3 Approximation of stationary cones

In case the hydrostatic outer boundary condition is used, a stationary cone is formed for production rates below a certain critical rate [1]. Although a stationary cone was developing in the simulation, calculations finally led to wiggly solutions. This is probably because only a Courant-Friedrichs-Lewy stability condition [9] was used to determine the timestep. This condition involves the reciprocal of the velocity. As the interface approaches its stable position, the velocity declines, ultimately becoming zero. Very large timesteps are taken, and then the explicit calculation method leads to instabilities in the determination of the interface. Since all other tests in the article were run for moving interfaces, there are no consequences for further results.

4.4 Fingering

Perturbations imposed on the initial interface were shown to increase for mobility ratios above one: the higher the mobility ratio, the faster the increase. They remained stationary for unit mobility ratio and decreased for mobility ratios below one. This phenomenon is called viscous fingering [5,8]. Fingers are flattened by the effects of gravity. Further simulations revealed that clearly separated fingers of equal size remain separated, but small fingers are overtaken by a large central finger. Only a limited number of equal-sized fingers can grow simultaneously [5].

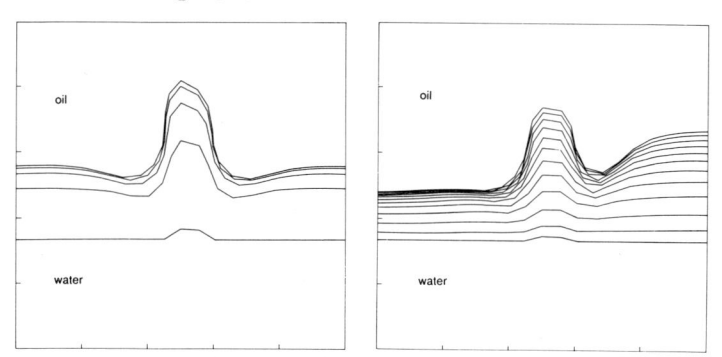

Fig 6. A small perturbation grows rapidly, despite shear flow from a cone.

These fingers can occur as small perturbations on top of a growing cone. Shear flow along the interface could flatten these perturbations, but simulation showed that a small perturbation on top of a cone behaves in the same way as a small perturbation without a cone (fig 6). Extremely small fingers still grew for very unfavourable mobility ratios, M=100, but if initial perturbations were not imposed fingering did not occur.

5. CONCLUSIONS

- For an accurate representation of the sharp interfaces, finite elements should be used.
- Combination of the stream function and the $S^{\alpha,\beta}$-method gives an accurate way of solving the interface equation.
- The algorithm described helps to understand many complicated coning phenomena.
- Coarse finite element grids can be used without sacrificing accuracy.
- Different well models lead to significantly different coning behaviour.
- The shear flow along a superimposed cone does not suppress fingering.

NOMENCLATURE

x	m	areal distance
z	m	upward vertical distance
$\mathbf{e}_z$	–	downward unit vector
Ω		reservoir
Γ		interface
$\mathbf{q}_f$	m.s^{-1}	Darcy velocity of fluid f
ψ	m^2.s^{-1}	stream function
Q	m^2.s^{-1}	production rate per m in symmetry direction
u	m	height of interface
θ	–	angle of interface with x-axis
h	m	height of reservoir
k	m^2	permeability
μ	Pa.s	viscosity
ϕ	–	porosity

ρ kg.m^{-3} density
g m.s^{-2} gravity constant

ACKNOWLEDGEMENT

I wish to acknowledge the contribution made by C.J. van Duyn of the Delft University of Technology in helping me gain insight into the method described and W. Scheulderman who used the algorithm for steam/oil interface calculations. I also thank J. Bruining and prof. H.J. de Haan of the Delft University of Technology for their constructive comments.

REFERENCES

[1] Bear, J.,(1975), "Dynamics of fluids in porous media", American Elsevier, New York, page 553
[2] Chan Hong, J.R., van Duyn, Hilhorst, van Kester,(1989), The interface between fresh and salt ground water: a numerical study, IMA J. of Appl. Math., 42, 209-240
[3] Chavent, G., Jaffre,(1986), "Mathematical models and finite elements for reservoir simulation", North-Holland, Amsterdam, page 23
[4] Ewing, R.E.,(1983), "The mathematics of reservoir simulation", SIAM, Philadelphia
[5] Homsy, G.M.,(1987), Viscous fingering in porous media, Ann. Rev. Fluid Mech, 19, 271-311
[6] Lerat, A., Peyret,(1973), Sur le choix de schéma aux différences du second ordre fournissannt des profils de choc sans oscillation, C.R. Acad. Sc. Paris, 277, 363-366
[7] Lynch D.R., Gray,(1980), Finite element simulation of flow in deforming regions, J. of Comp. Physics, 36, 135-153
[8] Matijevic E.,(1978), "Surface and colloid science", Vol 10, Plenum Press, New York, page 227-293
[9] Mitchell, A.R.,(1980), "The finite difference method in partial differential equations', John Wiley & Sons, New York, page 167

A finite difference scheme for a polymer flooding problem.

Aslak Tveito [*,†] Ragnar Winther[*,†]

Abstract

The Cauchy problem for a system of non-strictly hyperbolic conservation laws modelling polymer flooding is discussed. A finite difference scheme for the problem is developed. A subsequence of the family of approximate solutions generated by this scheme converges towards a weak solution of the Cauchy problem.

1 Introduction.

In order to enhance the oil recovery from a reservoir, water with a dissolved polymer is frequently used to displace the oil from the porous rock. This process can, under certain assumptions, be studied in terms of mathematical models. These models tend to be quite complicated and are generally not understood from a mathematical point of view. It is not known whether these models are well-posed, i.e. it is an open problem whether or not there exists a unique and stable solution of the non-linear partial differential equations modelling polymer flooding. The purpose of this paper is to discuss the existence part of this problem.

The non-linear partial differential equations modelling polymer flooding in a one dimensional homogeneous medium consist of a system of non-strictly hyperbolic conservation laws. A derivation and discussion of the model can be found in the paper by Pope [9]. The model is also discussed by Trangenstein [13], which also gives a general introduction to the theory of mathematical models for oil recovery processes.

Assume that the fluid in the medium consists of two incompressible and immiscible phases, an aqueous phase and an oleic phase. The aqueous phase consists of water and polymer. The polymer is assumed to be totally miscible

---

[*]This research has been supported by VISTA, a research cooperation between the Norwegian Academy of Science and Letters and Den norske stats oljeselskap a.s. (Statoil).

[†]Department of Informatics, P. O. Box 1080 Blindern, University of Oslo, Norway.

in water and we assume that the polymer remains in the aqueous phase. Furthermore, it assumed that the amount of polymer is so small that its volume is negligible. Then neglecting dispersive and gravitational effects, the following system of hyperbolic conservation laws can be derived,

$$s_t + f(s,c)_x = 0 \qquad (1.1)$$
$$(sc + a(c))_t + (cf(s,c))_x = 0.$$

Here s denotes the saturation of the aqueous phase and c denotes the concentration of the polymer in the aqueous phase. Both variables are in the unit interval. The saturation of the oleic phase is given by $1 - s$. The function f in (1.1) is called the fractional-flow function and is determined by the relative permeabilities and the viscosities of both phases. The adsorption of the polymer on the rock is modelled by the $a(c)$ term in the system. We refer to [6] for the detailed assumptions on the fractional flow function and the adsorption function.

In some cases several different polymer components are added to the aqueous phase. Such a process can be modelled by the following system of equations,

$$s_t + f(s, c_1, \ldots, c_n)_x = 0 \qquad (1.2)$$
$$(sc_i + a_i(c_i))_t + (c_i f(s, c_1, \ldots, c_n))_x = 0 \quad i = 1, \ldots, n.$$

Here c_i denotes the concentration of the i-th polymer component in the aqueous phase, and $a_i(c_i)$ models the adsorption of the i-th polymer component on the rock.

As we mentioned above, the Cauchy problem for these systems is not fully understood. However, the solution of the Riemann problem for the systems is known. Isaacson [2] solved the Riemann problem for the system (1.1) with $a \equiv 0$; a similar problem was solved independently by Keyfitz and Kranzer [8]. The solution of the Riemann problem for the system (1.1) was presented in [6]. This solution was generalized to the case of n different polymer components, i.e. the system (1.2), in the Paper [7].

The general Cauchy problem of (1.1) with $a \equiv 0$ was studied by Temple [12]. He established the existence of a weak solution to the system by using Glimms method [1].

The Cauchy problem of the system (1.2) with linear diffusion was studied in [14]. The problem was shown to be well posed in the sense that the existence of a unique and stable solution was established. The results obtained in [14] depend, however, strongly on the assumption that the diffusion coefficient ϵ satisfies $\epsilon > \epsilon_0 > 0$, they are therefore not applicable to the hyperbolic case.

As mentioned above, the solution of the Riemann problem for the system (1.2) is known. Hence it is possible to study the behaviour of classical numerical schemes simply by applying them to the some Riemann problems. Such numerical experiments with the Godunov method have been performed and reported in [4], [5] and [6]. All the experiments clearly indicate convergence of the approximate solutions towards the correct solution of the problem. We have also performed numerical experiments with other finite difference schemes, and they always seem to find the correct solution. This suggests that classical finite difference schemes are able to solve these problems.

For the system (1.1) with $a \equiv 0$, we are able to prove that a non-conservative finite difference scheme, under certain assumptions on the initial data, converges towards a weak solution of the system. We remark that, due to the consistency of the scheme (cf. [15]), convergence of the approximate solutions implies the existence of a weak solution of the system. To the authors knowledge, this is the first convergence result for a classical finite difference scheme applied to a non-strictly hyperbolic system of conservation laws.

2 A finite difference scheme.

In this section we will present a finite difference scheme for the system (1.1) when the adsorption term is neglected. Under certain assumptions on the initial-data, a subsequence of the family of approximate solutions will converge towards a weak solution of the problem.

We consider the following system of non-strictly hyperbolic conservation laws,

$$\begin{aligned} s_t + f(s,c)_x &= 0 \\ (sc)_t + (cf(s,c))_x &= 0. \end{aligned} \quad (2.1)$$

We assume that the fractional-flow function f is a smooth function with the following properties

$$\begin{aligned} f(0,c) &= 0, \quad \forall c \in [0,1] \\ f(1,c) &= 1, \quad \forall c \in [0,1] \\ \frac{\partial f}{\partial s} &\geq 0. \end{aligned} \quad (2.2)$$

We remark that these assumptions on the fractional flow function are reasonable from a physical point of view when gravitational effects are neglected.

We consider the system (2.1) with initial-data satisfying

$$\begin{aligned} (s^0(x), c^0(x)) &\in [0,1] \times [0,1] \quad \forall x \in \mathbf{R} \\ TV(s^0), TV(c^0) &\leq M < \infty \\ c^0 &\in Lip, \end{aligned} \quad (2.3)$$

where Lip denotes the space of Lipschitz continuous functions, and where $TV(s)$ denotes the total-variation of the function s, cf. Royden [10]. We remark that Isaacson and Temple [3] have shown that the system (2.1) is not well-posed for 'plug-data', that is an initial function consisting of three constant states. Such data are precluded from our analysis by the last condition of (2.3).

Let $g(s,c) = f(s,c)/s$. Then, for smooth solutions, the system (2.1) can be rewritten in following form,

$$s_t + f(s,c)_x = 0 \qquad (2.4)$$
$$c_t + g(s,c)c_x = 0.$$

Here the first equation is in conservative form, while the second equation is in non-conservative form.

We define a finite difference scheme for the problem by discretizing the system (2.4). Let Δx and Δt be the discretization parameters in the x and t direction respectively, and let (s_j^n, c_j^n) denote an approximation to

$$(s(j\Delta x, n\Delta t), c(j\Delta x, n\Delta t)).$$

Then the finite difference approximation is defined by

$$s_j^{n+1} = s_j^n - \mu(f_j^n - f_{j-1}^n) \qquad (2.5)$$
$$c_j^{n+1} = c_j^n - \mu g_j^n(c_j^n - c_{j-1}^n),$$

where $\mu = \Delta t/\Delta x$, $f_j^n = f(s_j^n, c_j^n)$ and $g_j^n = g(s_j^n, c_j^n)$. We assume that the mesh parameters satisfy the following CFL-condition

$$\frac{\Delta t}{\Delta x} \sup(\frac{\partial f}{\partial s}, g) \leq 1. \qquad (2.6)$$

The iteration is started by putting $s_j^0 = s^0(j\Delta x)$ and $c_j^0 = c^0(j\Delta x)$.

For this finite difference scheme, we have the following convergence theorem.

Theorem 1 Let (s_Δ, c_Δ) be the family of approximate solutions to the system (2.1) generated by the finite difference scheme (2.5). Then, assuming that the initial data satisfies (2.3) and that the fractional flow function satisfies (2.2), there is a subsequence $(s_{\Delta'}, c_{\Delta'})$ of approximate solutions converging in L^1_{loc} towards a weak solution of the system (2.1).

The weak solution $(s(x,t), c(x,t))$ generated as the limit of the subsequence of approximate solutions has the following properties:

1) $(s(x,t), c(x,t)) \in [0,1] \times [0,1] \quad \forall (x,t) \in \mathbf{R} \times \mathbf{R}^+$
2) $TV(s(\cdot,t)) < \infty \quad \text{and} \quad TV(c(\cdot,t)) \leq TV(c^0) < \infty \ , t < \infty$
3) $\dfrac{|c(x_2,t) - c(x_1,t)|}{|x_2 - x_1|} \leq K \quad \forall x_1, x_2 \in \mathbf{R} \ t \in \mathbf{R}^+,$

where K is a finite constant. □

Here L^1_{loc} denotes the space of functions which are locally in L^1, cf. Royden [10].

A proof of this theorem can found in [16]. The main steps in the proof are to establish the following properties of the approximate solutions: i) a maximum principle, ii) Lipschitz continuity of the polymer concentration, iii) total variation bounds for both variables, and finally iv) L^1-Lipschitz continuity of the approximate solutions in time. The estimates i), iii) and iv) imply convergence of a subsequence of the family of approximate solutions, cf. Smoller [11]. By using the estimates i), ii) and iii), one can prove that the subsequence converges towards a weak solution of the system (2.1), cf [15].

References

[1] J. Glimm, Solutions in the Large for Nonlinear Hyperbolic Systems, Comm. Pure Appl. Math. 18 (1965), pp. 697-715.

[2] E. Isaacson, Global solution of a Riemann problem for a non-strictly hyperbolic system of conservation laws arising in enhanced oil recovery, Rockefeller University preprints.

[3] E. Isaacson, B. Temple, The structure of asymptotic states in a singular system of conservation laws, preprint.

[4] T. Johansen, A. Tveito and R. Winther, Numerical and Analytical Solutions of a Model describing Polymer Flooding. in : Proceedings from the seminar on reservoir description and simulation with emphasis on EOR. Norway 1986.

[5] T. Johansen, A. Tveito and R. Winther, A Riemann solver for a two-phase multicomponent process, SIAM J. Sci. Stat. Comput. 10 (1989) pp. 846-879.

[6] T. Johansen and R. Winther, The Solution of the Riemann Problem for a Hyperbolic System of Conservation Laws Modelling Polymer Flooding, SIAM J. Math. Anal. 19 (1988) pp. 541-566.

[7] T. Johansen and R. Winther, The Riemann Problem for Multicomponent Polymer Flooding, SIAM J. Math. Anal. 20 (1989) pp. 908-929.

[8] B. Keyfitz, H. Kranzer, A system of non-strictly hyperbolic conservation laws arising in elasticity theory Arch. Rat. Mech. Anal. 72 (1980), 219-241.

[9] G. A. Pope, The Application of Fractional Flow Theory to Enhanced Oil Recovery, Soc. Pet. Eng. J. 20 (1980) pp. 191-205.

[10] H.L. Royden, Real Analysis, Collier Macmillan Publ.,1968.

[11] J. Smoller, Shock Waves and Reaction-Diffusion Equations, Springer Verlag, New York 1982.

[12] B. Temple, Global Solution of the Cauchy Problem for a Class of 2×2 Nonstrictly Hyperbolic Conservation Laws, Adv. in Appl. Math. 3 (1982) pp. 335-375.

[13] J. A. Trangenstein, Multi-phase flow in porous media: mechanics, mathematics and numerics, LLNL preprint 1987.

[14] A. Tveito, Convergence and stability of the Lax-Friedrichs scheme for a nonlinear parabolic polymer flooding problem, to appear in Adv. in Appl. Math.

[15] A. Tveito and R. Winther, Convergence of a non-conservative finite difference scheme for a system of hyperbolic conservation laws, To appear in Diff. and Integral Eq., 1990.

[16] A. Tveito and R. Winther, Existence, uniqueness and continuous dependence for a system of hyperbolic conservation laws modelling polymer flooding, Preprint, University of Oslo, Dep. of Informatics, 1990.

TWO-PHASE FLOW IN HETEROGENEOUS POROUS MEDIA: LARGE-SCALE CAPILLARY PRESSURE AND PERMEABILITY DETERMINATION

Michel Quintard, Henri Bertin and Stephen Whitaker[1*]
Laboratoire Energétique et Phénomènes de Transfert (UA CNRS 873), Ecole Nationale Supérieure d'Arts et Métiers - 33405 Talence Cedex, France

Abstract
In this paper we present an analysis of two-phase flow in heterogeneous porous media by the method of large-scale averaging, and a comparison of the theoretical results provided by this method with experimental data for flow in a stratified porous medium. We performed experiments of waterflooding on a stratified porous medium with a flow parallel to the strata. The saturation field was measured by gamma-ray absorption and the results were compared with a numerical solution of the large-scale averaged equations. The large-scale permeability tensors and the large-scale capillary pressure were determined theoretically by the resolution of the closure problem. The entire theoretical analysis was restricted by the quasi-static condition; however, we have recently extended the quasi-static analysis to obtain a theory that is capable of capturing many of the dynamic effects that influence two-phase flow in heterogeneous media (Quintard and Whitaker, 1990a and 1990b).

1. INTRODUCTION

Two-phase flow in heterogeneous porous media is a problem of central importance for oil recovery processes. In reservoir simulations pseudo-relative permeabilities and capillary pressures have been used for several years. The purposes of

---

[1]*Permanent address: Department of Chemical Engineering, University of California, Davis, CA 95616, U.S.A.

the different authors were to reduce the number of dimensions of the flow field or to save computer resources by decreasing the number of grid blocks (Hearn, 1970; Coats et al., 1971; Kyte and Berry, 1975; Killough and Foster, 1979; Yokoyama and Lake, 1981; Kortekaas, 1983; Thomas, 1983; Kossack et al., 1989). Pseudo are supposed to account for the heterogeneities effects within the grid blocks, they may also be used to deal with numerical dispersion, however, this latter aspect is beyond the scope of this paper.

The method of volume averaging produces transport equation and effective properties at a given length scale from the transport equations at a lower scale. This can be used from the pore-scale to the local scale to derive generalized Darcy's equations (Whitaker, 1986) for the flow of two phases in a porous medium. This method can be used for a subsequent change of scale from the local scale in a heterogeneous porous medium to the large-scale corresponding to a large-scale averaging volume including the heterogeneities. In the first analysis of this problem (Quintard and Whitaker, 1988) complete development was restricted by the quasi-static assumption. This assumption states that the fluid distribution at the closure level is determined by capillary equilibrium between the different regions.

In this paper theoretical developments are outlined, and the application of the quasi-static theory to the interpretation of experimental results is presented.

2. THEORY

The general problem under consideration is illustrated in Fig. 1 where we have shown three different length scales which are important in the study of two-phase flow in porous media. At Level I the flow is described by the Stokes equations and an appropriate set of boundary conditions. For reference we list the equations for the β-phase as

$$0 = -\nabla p_\beta + \rho_\beta g + \mu_\beta \nabla^2 v_\beta \quad , \quad \nabla \cdot v_\beta = 0 \qquad (1)$$

To develop the equations for Level II, the method of volume averaging is used to effect a change of scale which leads to the following variation of Eqs. 1 (Whitaker, 1986):

$$<v_\beta> = -\frac{1}{\mu_\beta} K_\beta \cdot (\nabla <p_\beta>^\beta - \rho_\beta g) \quad , \quad \frac{\partial \varepsilon_\beta}{\partial t} + \nabla \cdot <v_\beta> = 0 \qquad (2)$$

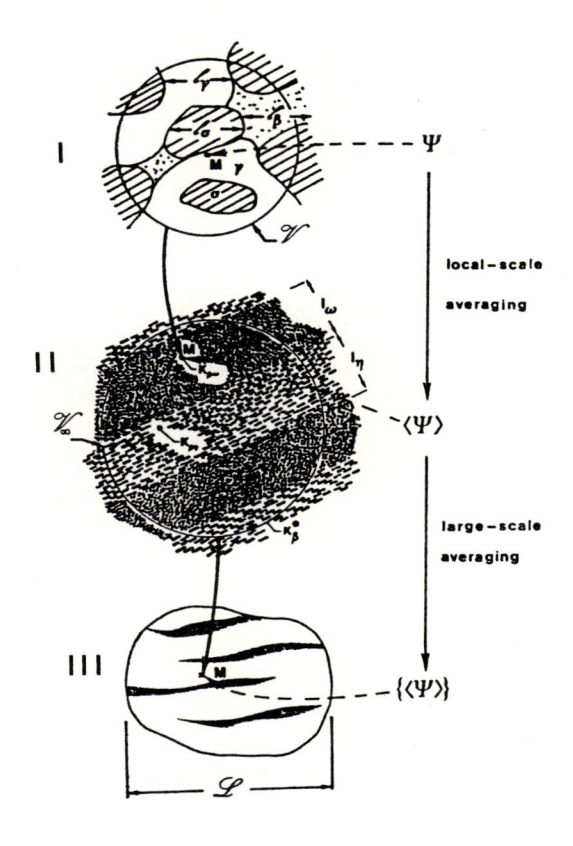

Figure 1: Multiple Length Scales for Two-Phase Flow in Heterogeneous Porous Media

In this process the point equations and boundary conditions are joined to produce Eqs. 2, and the closure problem associated with this theoretical development can be used to predict both $\underset{\approx}{K}_\beta$ and the capillary pressure-saturation relation. To capture the influence of the heterogeneities shown in Fig. 1, the method of large-scale averaging (Quintard and Whitaker, 1987) is used to derive the large-scale form of Eqs. 2. These equations are valid at Level III and the results obtained from Eqs. 2 can be expressed as (Quintard and Whitaker, 1988)

$$\{<\underset{\sim}{v}_\beta>\} = -\frac{1}{\mu_{*\beta}} \underset{\approx}{K}^*_\beta \cdot (\underset{\sim}{\nabla}\{<p_\beta>^\beta\}^\beta - \rho_\beta \underset{\sim}{g}) \qquad (3a)$$

$$\frac{\partial\{\varepsilon_\beta\}}{\partial t} + \underset{\sim}{\nabla}\cdot\{<\underset{\sim}{v}_\beta>\} = 0 \qquad (3b)$$

While these equations appear to be simple analogue of Eqs. 3, the theoretical analysis leading to the large-scale equations is quite complex and there are numerous restrictions that must be imposed in order to achieve these apparently simple extensions of Eqs. 2. These restrictions are presented below for two-phase flow excluding inactive zones within the large-scale averaging volume (i.e., zones corresponding to irreducible or residual saturations).

Large-scale averages

Large-scale properties are obtained in a general manner by defining three independent volumes over the large-scale averaging volume, V_∞: V_β is the volume of the active zone for the β-phase (i.e. zone where the relative permeability for the β-phase is different from zero), V_γ is the active zone for the γ-phase, and V_c is the capillary zone corresponding to $V_c = V_\beta \cap V_\gamma$. Large-scale averaged quantities are defined as follow

$$\{\psi_\beta\}^\beta = \frac{1}{V_\beta}\int_{V_\beta} \psi_\beta \, dV \quad \text{large-scale intrinsic phase average} \qquad (4)$$

$$\{\psi_\beta\} = \frac{1}{V_\infty}\int_{V_\beta} \psi_\beta \, dV \quad \text{large-scale phase average} \qquad (5)$$

$$\{\psi_\beta\}^* = \frac{1}{V_\infty}\int_{V_\infty} \psi_\beta \, dV \quad \text{large-scale spatial phase average} \qquad (6)$$

Similar averages exist for a function ψ_γ associated with the γ-phase. In addition, the large-scale capillary pressure is defined in terms of the volume of the __capillary region__ according to

$$\{p_c\}^c = \frac{1}{V_c} \int_{V_c} p_c \, dV \tag{7}$$

Our presentation of the large-scale averaging method is restricted in this paper to $V_\infty = V_\beta = V_\gamma = V_c$, but we retain the notation given by Eqs. 4 through 7 to be compatible with our previous publication where the influence of residual saturation effects is taken into account.

Continuity equation

The large-scale averaging of the local continuity equation produces in a straightforward manner the required large-scale continuity equation. For the β-phase we obtain

$$\frac{\partial \{\varepsilon_\beta\}^*}{\partial t} + \nabla \cdot \{\langle v_\beta \rangle\} = 0 \tag{8}$$

Deviations of the local values to the corresponding large-scale values are defined by

$$\langle p_\beta \rangle^\beta = \{\langle p_\beta \rangle^\beta\}^\beta + \hat{p}_\beta \quad ; \quad \varepsilon_\beta = \{\varepsilon_\beta\}^* + \hat{\varepsilon}_\beta$$

$$\langle v_\beta \rangle = \{\langle v_\beta \rangle\}^\beta + \hat{v}_\beta \tag{9}$$

The local continuity equation for the β-phase becomes

$$\frac{\partial \hat{\varepsilon}_\beta}{\partial t} + \nabla \cdot \hat{v}_\beta = 0 \tag{10}$$

At this point the quasi-static theory assumes that deviations depend upon the large-scale values in a quasi-steady manner, i.e., Eq. (10) reduces to

$$\nabla \cdot \hat{v}_\beta = 0 \tag{11}$$

Momentum equations

The large-scale average of the momentum equation for the β-phase can be expressed as

$$\{<v_\beta>\}^\beta = -\frac{1}{\mu_\beta}\left[\{K_\beta \cdot \nabla<p_\beta>^\beta\}^\beta - \{K_\beta\}^\beta \cdot \rho_\beta g\right] \tag{12}$$

and the presence of the average of the product, $K_\beta \cdot \nabla<p_\beta>^\beta$, requires the use of decompositions such as those illustrated by equations (9) and the development of a closure problem to predict the spatial deviations. At this point we only note that the spatial deviation form of equation (6) is given by (for the ω-region)

$$\hat{v}_{\beta\omega} + \{<v_\beta>\}^\beta = -\frac{1}{\mu_\beta}\left[\{K_\beta\}^\beta \cdot \Omega_\beta + \hat{K}_{\beta\omega} \cdot \Omega_\beta + K_{\beta\omega} \cdot \nabla\hat{p}_{\beta\omega}\right] \tag{13}$$

where $\Omega_\beta = \nabla<p_\beta>^\beta - \rho_\beta g$.

Capillary pressure

The capillary pressure relationship is expressed in terms of large-scale values and deviations, we have

$$p_c|_y = \left(\{<p_\gamma>^\gamma\}^\gamma - \{<p_\beta>^\beta\}^\beta\right)_x + \left(\Omega_\gamma - \Omega_\beta\right)_x \cdot y + (\rho_\gamma - \rho_\beta) g \cdot y +$$

$$+ (\hat{p}_\gamma - \hat{p}_\beta)_y + \frac{1}{2} yy : \left(\nabla\Omega_\gamma - \nabla\Omega_\beta\right)_x + \ldots \tag{14}$$

where $\underline{x}$ is the centroid of V_∞ and $\underline{y}$ is the position vector relative to the centroid. In this equation, the pressure effects have been simplified using a Taylor series expansion.

The large-scale form of this equation is

$$\{P_c\}_{\underline{x}}^c = \left(\{<p_\gamma>^\gamma\}^\gamma - \{<p_\beta>^\beta\}^\beta \right)_{\underline{x}} + \{\hat{p}_\gamma - \hat{p}_\beta\}_{\underline{x}}^c +$$

$$+ \frac{1}{2} \{\underline{yy}\}_{\underline{x}}^c : \{\nabla\underline{\Omega}_\gamma - \nabla\underline{\Omega}_\beta\}_{\underline{x}}^c + .. \quad (15)$$

Equation (14) can be written in the following compact form

$$P_c(\varepsilon_\beta)\Big|_{\underline{x}+\underline{y}} = \left(\{<p_\gamma>^\gamma\}^\gamma - \{<p_\beta>^\beta\}^\beta \right)\Big|_{\underline{x}} +$$

$$\left[\Delta\left(\{<p_\gamma>^\gamma\}^\gamma - \{<p_\beta>^\beta\}^\beta \right) + \hat{p}_\gamma - \hat{p}_\beta \right]\Big|_{\underline{x}+\underline{y}} \quad (16)$$

where $\Delta\psi = \psi(\underline{x}+\underline{y}) - \psi(\underline{x})$. Additional terms in the right handside of Eq. (16) reflect the dynamic behaviour of the general closure scheme. We note these additional terms, Δp, ε_β is given by

$$\varepsilon_\beta = P_c^{-1}\left(\left(\{<p_\gamma>^\gamma\}^\gamma - \{<p_\beta>^\beta\}^\beta \right)\Big|_{\underline{x}} + \Delta p\Big|_{\underline{x}+\underline{y}} \right)$$

$$\approx P_c^{-1}\left(\left(\{<p_\gamma>^\gamma\}^\gamma - \{<p_\beta>^\beta\}^\beta \right)\Big|_{\underline{x}} \right)$$

$$+ \left(\Delta p\Big|_{\underline{x}+\underline{y}} \right) \frac{\partial P_c^{-1}}{\partial p_c} \left(\left(\{<p_\gamma>^\gamma\}^\gamma - \{<p_\beta>^\beta\}^\beta \right)\Big|_{\underline{x}} \right) + \ldots \quad (17)$$

Under the condition expressed by

$$\left(\Delta p\Big|_{\underset{\sim}{x}+\underset{\sim}{y}}\right)\frac{\partial p_c^{-1}}{\partial p_c}\left(\left(\{<p_\gamma>^\gamma\}^\gamma-\{<p_\beta>^\beta\}^\beta\right)\Big|_{\underset{\sim}{x}}\right)/$$

$$p_c^{-1}\left(\left(\{<p_\gamma>^\gamma\}^\gamma-\{<p_\beta>^\beta\}^\beta\right)\Big|_{\underset{\sim}{x}}\right)\ll 1 \qquad (18)$$

we obtain the <u>quasi-static</u> equation

$$\varepsilon_\beta = p_c^{-1}\left(\left(\{<p_\gamma>^\gamma\}^\gamma-\{<p_\beta>^\beta\}^\beta\right)\Big|_{\underset{\sim}{x}}\right) \qquad (19)$$

Thus the large-scale capillary pressure relationship, in the quasi-static case, is

$$\{\varepsilon_\beta\}^* = \{p_c^{-1}\left(\left(\{<p_\gamma>^\gamma\}^\gamma-\{<p_\beta>^\beta\}^\beta\right)\Big|_{\underset{\sim}{x}}\right)\}^* \qquad (20)$$

where p_c is a function of the position within the averaging volume reflecting the existence of the heterogeneities.

When the quasi-static restriction is imposed on the representation for the large-scale capillary pressure, one can quickly calculate the large-scale capillary pressure for a two-region model such as we have shown in Fig. 1. In Fig. 2 we have illustrated $\{p_c\}^c$ for a system in which the ω-region occupies 36% of the space. This calculation requires that the local capillary pressure associated with Eqs. 2 be constant everywhere in the averaging volume. This allows us to determine the fluid distribution at any value of $\{p_c\}^c$, and with this known fluid distribution one can solve the large-scale closure problem to determine $\underset{\approx}{K}_\beta^*$.

Closure problem (quasi-static case)

Under the assumptions corresponding to the quasi-static case we can use the following approximations:
(i) expressions involving averaged values only are unchanged by an averaging operation,
(ii) as a consequence, the average of deviations is zero.

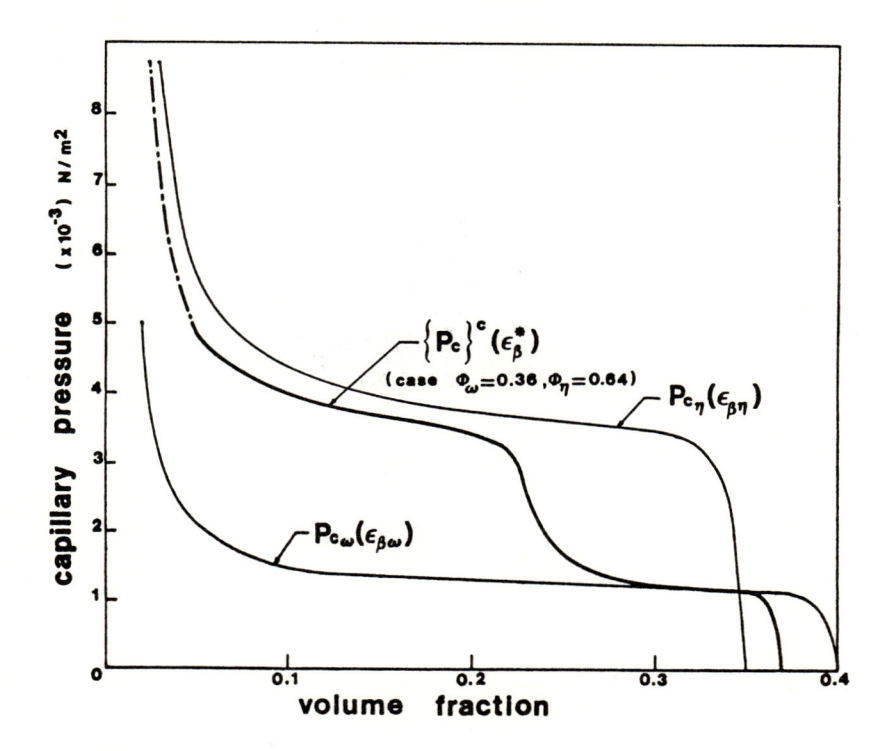

Figure 2: Capillary Pressure for a Two-Region Model of a Heterogeneous Porous Medium

The average of the momentum equation for the β-phase becomes

$$\{<\underline{v}_\beta>\} = -\frac{1}{\mu_\beta}\left[\{\underline{\underline{K}}_\beta\}\cdot\underline{\Omega}_\beta + \{\underline{\underline{\hat{K}}}_\beta\cdot\nabla\hat{P}_\beta\}\right] \quad (21)$$

The use of this last equation in Eq. (13) gives

$$\hat{\underline{v}}_{\beta\omega} = -\frac{1}{\mu_\beta}\left[\underline{\underline{\hat{K}}}_{\beta\omega}\cdot\underline{\Omega}_\beta + \underline{\underline{K}}_{\beta\omega}\cdot\nabla\hat{P}_{\beta\omega} - \{\underline{\underline{\hat{K}}}_\beta\cdot\nabla\hat{P}_\beta\}^\beta\right] \quad (22)$$

The problem is completed with similar equations for the η-region and the related boundary conditions.

Representations of the deviations are sought in the form

$$\hat{p}_{\beta\omega}=\underset{\sim}{b}_{\beta\omega}\cdot\underset{\sim}{\Omega}_\beta+\xi_{\beta\omega} \quad ; \quad \hat{\underset{\sim}{v}}_{\beta\omega}=\underset{\approx}{B}_{\beta\omega}\cdot\underset{\sim}{\Omega}_\beta+\underset{\sim}{\zeta}_{\beta\omega} \qquad (23)$$

and similar representations for the deviations in the η-region.

The tensors $\underset{\approx}{B}_\beta$ and the vectors $\underset{\sim}{b}_\beta$ are chosen in order to obtain a close approximation of Eqs. (22) and (11). They must obey the following boundary value problem for the representative unit cell.

$$\underset{\sim}{\nabla}\cdot\underset{\approx}{B}_{\beta\omega} = 0 \qquad (24)$$

$$\underset{\approx}{B}_{\beta\omega}=-\left[\hat{\underset{\approx}{K}}_{\beta\omega}+\underset{\approx}{K}_{\beta\omega}\cdot\underset{\sim\sim}{\nabla b}_{\beta\omega} - \{\hat{\underset{\approx}{K}}_\beta\cdot\underset{\sim\sim}{\nabla b}_\beta\}^\beta\right] \qquad (25)$$

B.C.1 $\quad \underset{\sim}{n}_{\omega\eta}\cdot\underset{\approx}{B}_{\beta\omega}=\underset{\sim}{n}_{\omega\eta}\cdot\underset{\approx}{B}_{\beta\eta} \quad$ at $A_{\omega\eta}$ $\qquad (26)$

B.C.2 $\quad \underset{\sim}{b}_{\beta\omega} = \underset{\sim}{b}_{\beta\eta} \quad$ at $A_{\omega\eta}$ $\qquad (27)$

$$\underset{\approx}{B}_{\beta\eta}=-\left[\hat{\underset{\approx}{K}}_{\beta\eta}+\underset{\approx}{K}_{\beta\eta}\cdot\underset{\sim\sim}{\nabla b}_{\beta\eta} - \{\hat{\underset{\approx}{K}}_\beta\cdot\underset{\sim\sim}{\nabla b}_\beta\}^\beta\right] \qquad (28)$$

$$\underset{\sim}{\nabla}\cdot\underset{\approx}{B}_{\beta\eta} = 0 \qquad (29)$$

$$\underset{\sim}{b}(\underset{\sim}{x}+\underset{\sim}{l}_i)=\underset{\sim}{b}(\underset{\sim}{x}) \quad ; \quad \{\underset{\sim}{b}\}^\beta=0 \quad ; \quad \{\underset{\approx}{B}\}^\beta=0 \qquad (30)$$

where $\underset{\sim}{l}_i$ are the lattice vectors that are needed to specify a unit cell.

With this choice it is demonstrated elsewhere, (Quintard and Whitaker, 1988), that the additional terms ξ_β and ζ_β in Eqs. (23) give negligible contributions to the deviation values. The representation explicited by Eqs. (23) gives the large-scale momentum equation Eq. (3a) while the large-scale β-phase permeability tensor is given by the following equation

$$\underset{\approx}{K}_\beta^* = \{\underset{\approx}{K}_\beta\} + \{\hat{\underset{\approx}{K}}_\beta\cdot\underset{\sim\sim}{\nabla b}\} \qquad (31)$$

Similar equations are used to calculate the γ-phase permeability tensor.

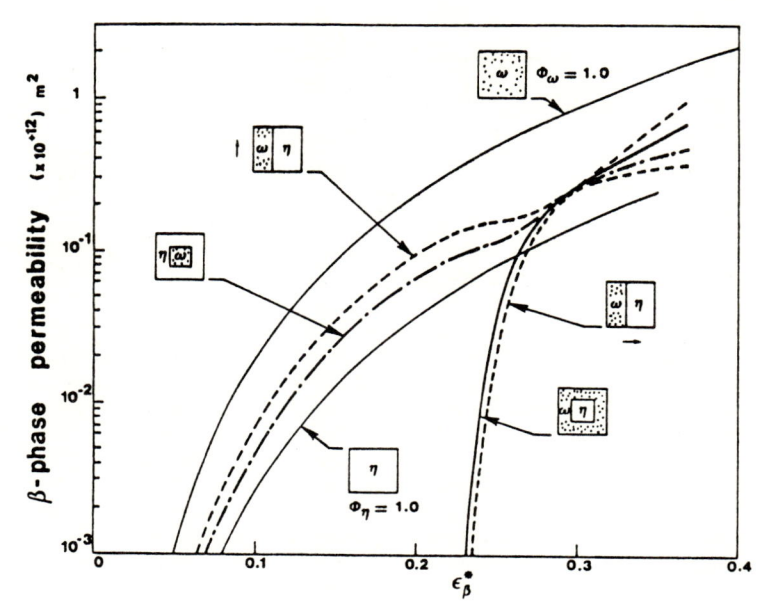

Figure 3: Large-Scale Permeabilities for the β-Phase

As the partial derivative equations are of elliptic type the boundary value problem for $\underset{\sim}{b}_\beta$ and $\underset{\approx}{B}_\beta$ is solved using a classical finite difference technique for the calculations presented in Fig. 3. Several calculated values of the components of $\underset{\approx}{K}_\beta$ are illustrated in Fig. 3 for $\Phi_\omega^* = 0.36$ (unless specified). Here it becomes very clear that $\underset{\approx}{K}_\beta$ is extremely sensitive to the manner in which the ω and η-regions are arranged, and it is not sufficient to know only the volume fraction of the ω-region, Φ_ω.

3. EXPERIMENT

The experiments (Quintard et al., 1989; Bertin et al., 1990) that we have performed correspond to the system illustrated in Fig. 4. The water-oil flow took place between the two ports denoted by A and B, while the twelve equally spaced ports denoted by C and D were used with water and oil flows in order to produce the initial water saturation for the water-flood experiments. The ω-region illustrated in Fig. 4 consisted of Aerolith-10, an artificial porous medium made by sintering a graded silica powder. This produces a porous medium of relatively high porosity and permeability. The

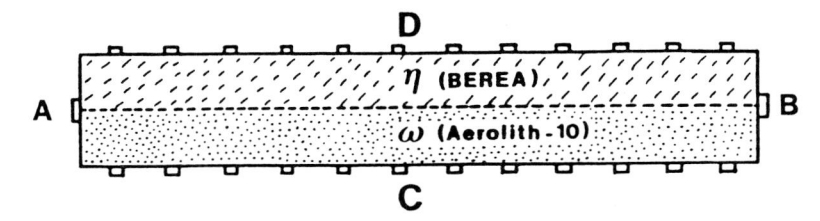

Figure 4: Experimental System for a Two-Layer, Heterogeneous Porous Medium

η-region was a Berea sandstone, a naturally porous media known to be relatively homogeneous; however, significant heterogeneities were found in both these materials. Variations in the porosity were determined by extensive gamma-ray absorption studies, while variations in the permeability were determined by removing cylindrical sections from both the Aerolith-10^{TM} and the Berea and performing individual flow experiments. The variations of the local permeability were on the order of ± 25% and this clearly indicates the difficulty of performing experiments that match the theoretical assumption of uniformity within each strata.

Experimental conditions

The experiments we performed were displacement of oil (μ=24Cpo) by water (with 50g/l of NaI) with the flow parallel to the strata of the medium shown Fig. 4. The water oil flow took place between ports A and B. Twelve equally spaced ports denoted C and D in Fig. 4 were used to initially saturate the system with water and then subsequently used to displace the water with oil to obtain the initial saturation $S_{\beta i}$. The saturation measurements were obtained by gamma-ray absorption over a 150 points grid measurements.

The waterflood was effected at a constant volumetric flow rate of 12 cc/h at the entry port A, this corresponds to an average filtration velocity, parallel to the strata, on the order of 30 cm/day. During the waterflooding, the entire saturation field was measured every two-hours.

Comparison with quasi-static large-scale results

Even though the ω and η-regions illustrated in Fig. 4 were not homogeneous, the theoretical analysis of the experiments was based on the assumption of homogeneous strata and the original quasi-static theory. The 1D large-scale

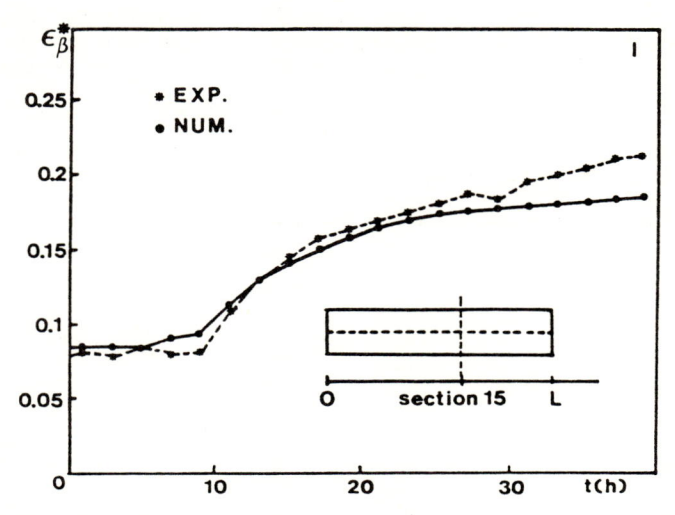

Figure 5: Experimental and Theoretical Results for the Large-Scale Volume Fraction of Water

problem describing the waterflooding experiments was simulated numerically. These computations provided large-scale saturation fields as functions of x and t. A comparison between theory and experiment for the large-scale volume fraction of water, $\{\varepsilon_\beta\}^*$, is shown in Fig. 5 and experimental and theoretical recovery curves are shown in Fig. 6. One uniform characteristic of the theory is that it underpredicts $\{\varepsilon_\beta\}^*$ for long times. This undoubtedly results from errors in the oil permeability, $\underset{\approx}{K}_\gamma^*$. At larges values of the water saturation, small errors in the absolute value of $\underset{\approx}{K}_\gamma^*$ result in large-percentage errors. If the oil permeability is too small in this domain, the volume fraction of $\{\varepsilon_\gamma\}^*$ will be too large and this leads to the long-time behaviour of the theory shown in Figs. 5 and 6. In addition to errors in K_γ^* it is shown elsewhere (Bertin et al., 1990) that local heterogeneities and dynamic effects can influence the crossflow between the strata. The influence of these effects will not be known until we obtain detailed solutions based on the dynamic theory.

4. CONCLUSION

The large-scale averaging method proved to simulate satisfactorily for two-phase flow experiments in heterogeneous porous systems when quasi-static conditions are achieved.

The general closure scheme features transient, dynamic, directional behaviour It can be used as a general approach of the problem of multiphase flow in heterogeneous porous media.

Acknowledgement: Financial support from ELF Petroleum Company is gratefully acknowledged.

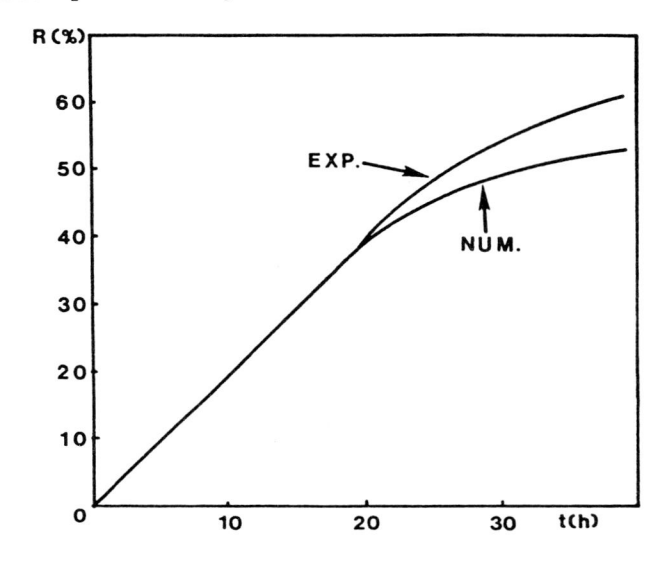

Figure 6: Experimental and Theoretical Recovery Curves

References

Bertin, H., Quintard, M., Corpel, Ph.V. and Whitaker, S., Two-phase flow in heterogeneous porous media III: Laboratory experiments for flow parallel to a stratified system, to be published in Transport in Porous Media, 1990.
Coats K.H., Dempsey, J.R. and Henderson, J.H., The use of vertical equilibrium in two-dimensional simulation of three-dimensional reservoir performance. Soc. Pet. Eng. J., March, 63-71, 1971.
Hearn, C.L., Simulation of stratified waterflooding by pseudo-relative permeability curves. Paper 2929 presented at the 45th Annual Fall Meeting of SPE, Houston, Oct.4-7, 1970.
Killough, J.E. and Foster, H.P. Jr., Reservoir simulation of the empire ABO field - the use of pseudos in a multilayered system. Soc. Pet. Eng. J., Oct., 279-291, 1979.
Kortekaas, T.F.M., Water/oil displacement characteristics in cross-bedded reservoir zones. SPE Conf., San Francisco, SPE 12112, 1983.

Kossack, C.A., Aasen, J.O. and Opdal, S.T., Scaling-up laboratory relative permeabilities and rock heterogeneities with pseudo functions for field simulations SPE symposium, Houston, SPE 18436, 1989.

Kyte, J.R. and Berry, D.W., New pseudos functions to control numerical dispersion.. Soc. Pet. Eng. J., August, 269-275, 1975.

Quintard, M., Bertin, H. and Whitaker, S., Two-phase flow in heterogeneous porous media: the method of large-scale averaging applied to laboratory experiments in a stratified system. SPE paper number 19682 presented at the 64th Annual Technical Conference and Exhibition. San Antonio, October, 1989.

Quintard, M. and Whitaker, S., Ecoulements monophasiques en milieu poreux: Effet des hétérogénéités locales, J. Méca. Théo. et Appli. $\underline{6}$, 691-726, 1987.

Quintard, M. and Whitaker, S., Two-phase flow in heterogeneous porous media: The method of large-scale averaging, Transport in Porous Media $\underline{3}$, 357-413, 1988.

Quintard, M. and Whitaker, S., Two-phase flow in heterogeneous porous media I: The influence of large spatial and temporal gradients, to be published in Transport in Porous Media, 1990.

Quintard, M. and Whitaker, S., Two-phase flow in heterogeneous porous media II: Numerical experiments for flow perpendicular to a stratified system, to be published in Transport in Porous Media, 1990.

Thomas, G.W., An extension of pseudofunction concepts. SPE paper 12274, 1983.

Whitaker, S., Flow in porous media I: a theoretical derivation of Darcy's law. Transport in Porous Media $\underline{1}$, 3-25, 1986.

Yokoyama, Y. and Lake L.W., The effects of capillary pressure on immiscible displacements in stratified porous media. Paper 10109 presented at the 56th Annual Fall Meeting of the SPE, San Antonio, Oct. 5-7, 1981.

A Matrix Injection Simulator

E. Touboul
Dowell Schlumberger

Abstract Matrix Acidizing is a complex process involving many physical and chemical parameters. Designing an adequate treatment can be greatly helped by simulation tools. A matrix injection simulator is described here. This simulator peforms a coupling between 2 phase flow in layered media, chemical reactions between acids and minerals for sandstone acidizing and particulate diversion. We first give the models for fluids displacement and chemical reactions. Then the numerical methods to solve the different coupled problems are presented. Finally the full interactivity of simulation and graphical display is demonstrated.

1 Introduction

1.1 Acidizing Treatment

A matrix acidizing treatment consists in injecting acids into the formation, in the near wellbore region, in order to dissolve damage or minerals, hence increase permeability and productivity. In sandstone formations, the damaging minerals to be dissolved are mainly clay particles coming from drilling muds or having migrated through the formation during a long time of production.

A "particulate diversion" process consisting in the injection of different slugs of fine oil soluble resin particles is often used to equilibrate the injectivity of the system of layers, in order to avoid the preferential penetration of acid in the high injectivity layers. The resin plugs at the sandface, in front of regions where the flow is originally higher.

1.2 Treatment Simulation

From inputs being the characteristics of the formation (geometry of the layers, permeability, mineral composition, well completion, injection flow rate ...), the simulator computes the placement of different slugs of fluid into the layered matrix, the state of dissolution of each mineral, the permeability evolution, the pressures into the matrix .... Thus the simulator is a tool to compute and understand the effectiveness of a given treatment. In this paper, we describe a Matrix Injection Simulator, which is a good compromise between realistic physics, short computing time and full interactivity.

2 Mathematical Modelling

All the computations are done in a 2 dimensional domain, either in a (R, Z) or (X, Y) co-ordinate system. The (R, Z) system represents a vertical plane including a vertical well. The medium is layered, each layer presenting different properties in terms of permeability, porosity, rock composition, fluid types and saturations, well completion ... etc.

2.1 Flow Module

The flow module describes the displacement of diphasic (water/oil, saturation S) slightly compressible fluids in a slightly compressible matrix. In each phase, several miscible fluids can be present. The presence of each fluid is characterized by a concentration function C (fraction of a given fluid in the total liquid). The basic unknowns are the pressure distribution and the concentration of each fluid.

The pressure equation is derived from generalized Darcy equation ([1]):

$$\sigma(x)\Phi_0(c_f + c_r)\frac{\partial P}{\partial t} - \nabla.(\sigma(x)B(x,P)\frac{Kr_1(S)}{\mu(C_{S1})} + \frac{Kr_2(S)}{\mu(C_{S2})}K(x)\nabla P) = 0 \qquad (2.1)$$

where σ is a geometrical factor, the generalized cross section, Φ_0 the reference porosity, c_f and c_r the fluid and rock compressibility, B the volume factor, Kr_1 and Kr_2 the relative permeability terms, μ the viscosity of each phase depending on the distribution of the concentration of the different miscible fluids in the phase, K the rock permeability tensor (assumed diagonal), and P the global pressure. The concentrations C_{Fi} of the different fluids enter in this equations in a non-linear way in both the relative permeability and the phase viscosity.

The "concentration equations" are mass balance equations relating the evolution of the concentration of each fluid according to the fractional flow rate of each fluid. There is one equation for each fluid considered:

$$\frac{\partial}{\partial t}(\sigma(x)\Phi(x,P)B(x,P)CF_i) + \nabla.(q_{Fi}(CF_i) - D\nabla CF_i) = 0 \qquad (2.2)$$

where C_{Fi} is the concentration of fluid i. The fractional flow i, q_{Fi} is computed from the fractional flow of the whole phase to which belongs the fluid i. For this computation, an approach close to Koval's one [3] was chosen. This "miscible" fractional flow favours the flow of less viscous fluid into more viscous one, and allows the modelling of fingering, with realistic results compared to laboratory experiments. The dispersion tensor D is neglected in this first version of the simulator. The dispersion of one fluid into another is due only to the viscosity contrast (and to numerical diffusion ...).

2.2 Diversion Module

The mass of diverter deposited on the perforations is calculated from the flow computations at each time step. The cake resistance at the well is a function of the total mass of diverter deposited. The relationship between mass deposited and cake resistance is derived from experimental results obtained earlier at STR. The eventual compressibility of the cake is taken into account through the experimental data [2].

2.3 Acidizing Module

The acidizing module is based on up-to date studies performed on sandstone mineral and formations [5]. Each layer of the matrix is composed of different minerals reacting with acid at different rates. The dissolution and removal of each mineral has a varying influence on the permeability increase, characterized by different exponents in the permeability/porosity law ([4]).

The reaction and transport of three chemicals are taken into account. Hydrofluoric acid, HF, dissolves the minerals, increasing both porosity and permeability:

$$\text{Mineral}_i + \alpha_i \, \text{HF} \longrightarrow \text{Product}_i \qquad (2.3)$$

where α_i is the stoicheiometric coefficient.

BF_4^-, is hydrolyzed into hydrofluoric acid giving a byproduct, BF_3OH^-:

$$BF_4^- + H_2O \rightleftharpoons HF + BF_3OH^- \qquad (2.4)$$

This hydrolysis is the basis of retarded acid system, allowing a deeper penetration of acid into the information.

The reactions are of first order in both ways, and the concentration in protons H^+ is assumed constant. We call C_{A1} the concentration of HF (moles/litre), C_{A2} the concentration BF_4^- and C_{A3} the concentration of BF_3OH^-. The mass balance equations of transport-reaction for the three chemicals can be written:

$$\frac{\partial}{\partial t}(\Phi(x,P)B(x,P)C_{A1}C_{FA}) + \frac{1.}{\sigma(x)}\nabla \cdot (q_{FA}C_{A1}) =$$
$$-\sum_{i=1,N_m} \alpha_i R_i A(M_i C_{A1} C_{FA} + \Phi C_{Fa}(H_1 C_{A2} - H_2 C_{A1} C_{A3}) \qquad (2.5)$$

$$\frac{\partial}{\partial t}(\Phi(x,P)B(x,P)C_{A2}C_{FA}) + \frac{1.}{\sigma(x)}\nabla \cdot (q_{FA}C_{A2}) = \Phi C_{FA}(H_1 C_{A2} H_2 C_{A1} C_{A3}) \qquad (2.6)$$

$$\frac{\partial}{\partial t}(\Phi(x,P)B(x,P)C_{A3}C_{FA}) + \frac{1.}{\sigma(x)}\nabla \cdot (q_{FA}C_{A3}) = \Phi C_{FA}(H_1 C_{A2} H_2 C_{A1} C_{A3}) \qquad (2.7)$$

where C_{FA} is the concentration of the fluid containing the chemicals in the total fluid phase, q_{FA} the fractional flow rate of this fluid, $A(M_i)$ the density of contact area of the ith mineral with acid depends on the concentration M_i of this mineral. This density of exposed mineral area is an important factor in such a reaction. H_1 and H_2 are the hydolysis rate constants (temperature dependent).

The mineral balance equation is for each mineral M_i:

$$\frac{\partial M_i}{\partial t} = -V_{Mi} R_i A(M_i) C_{A1} C_{FA} \qquad (2.8)$$

The acid reaction module is strongly coupled with the flow module: the acid convection is computed from the flow module, and the increase of porosity, and consequently of permeability, changes at each time the flow conditions.

2.4 Numerical Computations

The numerical computations are performed using finite differences to solve the partial differential equations, on a grid compressed in the near wellbore region.

A standard IMPES scheme is used for the pressure equation. The non-linearities of the fractional flow terms are treated explicitly. A radial harmonic weighting of the transmissibilities (factors $\frac{K_r}{\mu}$) has been adopted. The resulting linear system is solved using Cholesky method. The concentration equations for each fluid are solved explicitly, using an upwind scheme.

The diversion equation is solved explicitly, using a simple increase of the horizontal resistance of the blocks constituting the perforated well,

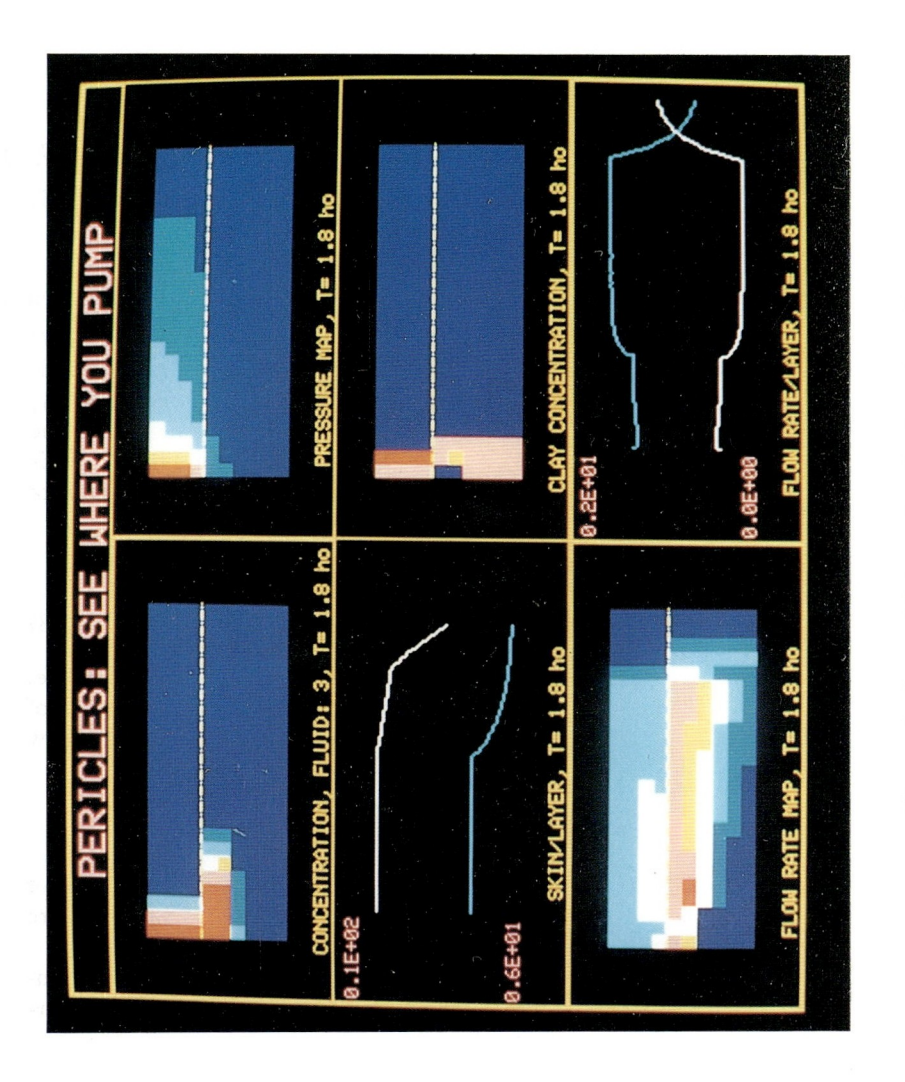

Figure 1: General interactive display of results.

according to the density and size of perforations, the cake concentration, the flow rate and experimental data relating the cake resistance to the deposited mass.

The chemicals convection is computed from the flow module. Then the reaction and hydrolysis equations are solved semi-analytically, using the exponential decrease of the acid strength with time. Each mineral concentration is then updated in each block according to the amount of moles of acid spent, and to the stoicheiometric coefficients. A new porosity is derived. The new permeability in each block is then computed with a law derived from Labrids formula, each mineral having a different contribution to the permeability increase.

The FORTRAN program is written in a modular way, favoured by the explicit numerical treatment of most of the equations. The computing time is very short, thanks to a numerical optimization of the computations. For a 25x12 grid on a Vax 11/785, a time step takes about one second. A total treatment can be simulated within 5 minutes, allowing a comfortable interactive use of the program.

3 Simulation and graphics Interactivity

While running the simulator, the user can act at any time on the injection conditions, as well as on the graphical display. The user can interactively change the injection flow rate or pressure; inject another fluid, an acid, a diverter; do a selective injection with either a packer or a coil tubing; shut down the injection, initiate a production phase ... This allows him a perfect monitoring in real time of the simulated treatment.

The screen is divided in up to ten windows. Curves or maps interactively chosen by the user are displayed at the same time in the windows, and updated automatically at each time step, creating an animation in each window, so the user can see on the screen the dynamic evolution of all the parameters he needs to monitor when designing a treatment. He can follow in a (R,Z) plane the displacement and penetration of each fluid as coloured saturation maps. He can follow the acids concentration evolution, the progressive dissolution of minerals, flow rate and pressure maps. Curves versus time can be displayed for the injection pressure, flow rate per layer and skin variations.

4 Examples of results

We give in figure 1 an example of the general display of the window system on a terminal screen. The case studied here is a two-layer matrix, with a low permeability upper layer and a not totally perforated lower layer. We see on the different windows various parameters of the problem on the form of curves or coloured maps. The efficiency of the diversion process to force the acid flow in the upper layer is clearly indicated by the slope variation in the skin curve.

In figure 2, some results of acidizing are given. A comparison between experimental results and simulations is first presented. Then a comparison between pure HF and retarded acid injected in a single layer matrix is given. The retarded acid allows a deeper penetration of live HF in the formation and leads to a deeper silicate (damage) removal.

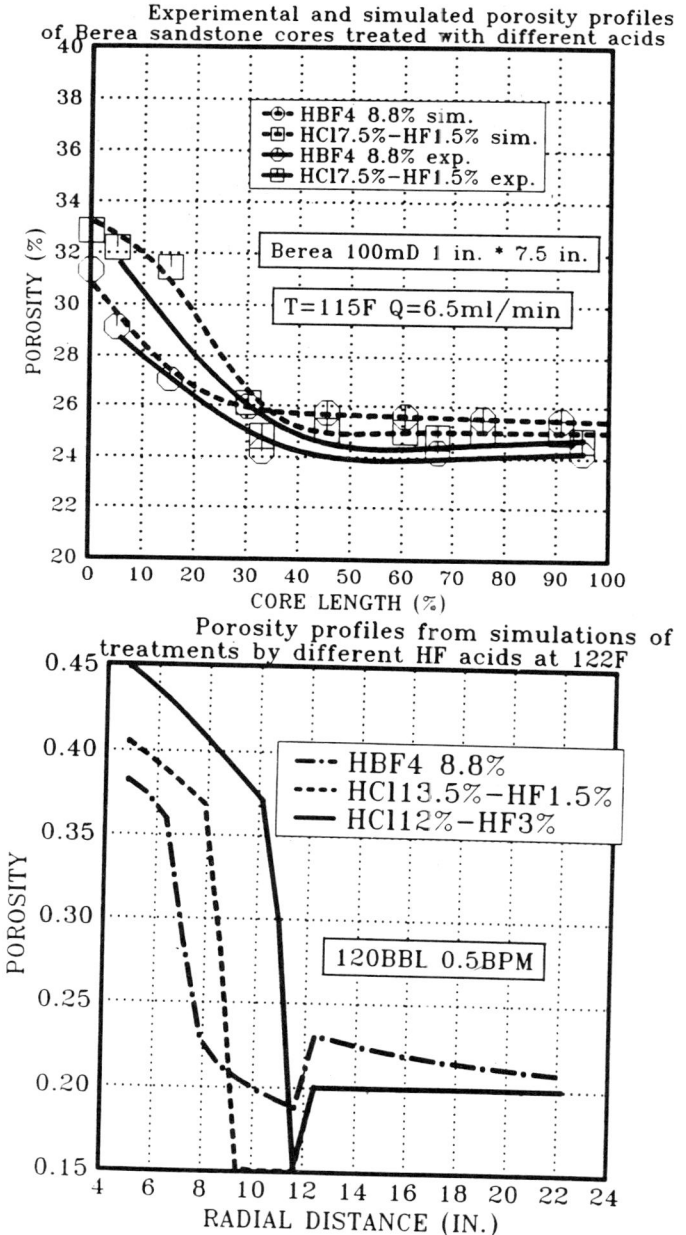

Figure 2: Effect of retarded acid (HBF4).

5 Conclusions

The Matrix Injection Simulator is only at its first experimental stage. But its performance already makes it an efficient tool for matrix treatment simulations. Further developments are planned. A physical model of foam flow, used as a diverter agent, is under study, as well as the possibility to treat carbonate formations. On the other hand, special subroutines will be implemented, allowing the user to perform simultaneously different simulations of a given part of the treatment, and to decide which is the best one using interactive graphic tools. In the future, the inverse problem for optimization of the design will also be tackled.

References

1. Chavent, G. and Jaffre, J., Mathematical models and finite elements for reservoir simulation (North Holland).

2. Doerler, N., Prouvost, L., (1987), SPE 16250 "Diverting agents: Laboratory Study and Modeling of Resultant Zone Injectivities".

3. Koval, E.J., (1963), A Method for Predicting the Performance of Unstable Miscible Displacement in Heterogeneous Media" (SPE).

4. Labrid, J.C., (1975), "Thermodynamic and Kinetic aspect of Argillaceous Sandstone Acidizing", SPEJ April, 117-128.

5. Piot, B.M., Perthuis, H.G., (1987), "Matrix Acidizing of Sandstones", in "Reservoir Simulation", Editors: M.J. Economides and K.G. Nolte, Schlumberger Educational Services.

EFFECT OF FLUID PROPERTIES ON CONVECTIVE DISPERSION:
COMPARISON OF ANALYTICAL MODEL WITH NUMERICAL SIMULATIONS

L.J.T.M. Kempers,
(Koninklijke/Shell Exploratie en Produktie Laboratorium,
Rijswijk, The Netherlands)

ABSTRACT

An analytical model for convective, longitudinal dispersion is outlined that includes the effect of fluid properties (viscosity contrast and density contrast). This model can be used to calculate efficiently the mixing between miscible fluid banks as a result of irregular spatial variations in permeability. The model displays good agreement with detailed numerical simulations of miscible displacements in a wide range of mobility ratios and gravity numbers.

1. INTRODUCTION

For the stable displacement of oil by a miscible solvent, mixing between oil and solvent can be modelled correctly only if the permeability variation (the main cause of mixing) is taken into account. However, reservoir simulators cannot cope with the rapid, irregular spatial variations in permeability as long as a realistic number of gridblocks are used. A correct way to describe mixing is by a random-walk process of fluid parcels superimposed on the average displacement velocity [7,3]. Addition of a diffusion-like term to the conservation-of-mass equation in the simulator is then sufficient to circumvent the large number of gridblocks needed. The coefficient of this diffusion-like term, the 'convective dispersion coefficient', is a function of the displacement velocity, reservoir properties and fluid properties. Gelhar et al. [4] have worked out the dependence of the dispersion coefficient on displacement velocity and reservoir properties (these are: correlation length of spatial permeability correlation function and standard deviation of permeability frequency distribution). In this paper, we present an outline of a model that includes the fluid properties. Special emphasis was placed on an extensive check of the model through a series of detailed numerical simulations of stable, miscible displacement in a heterogeneous reservoir.

2. THEORY OF GELHAR ET AL.

Gelhar et al. [4] have found the following relationship for the convective dispersion D in the longitudinal flow direction in a log-normal permeability distribution with standard deviation σ and a correlation function with exponential decay (correlation length ℓ):

$$\frac{D}{U} = \ell \, \sigma^2 , \qquad (1)$$

where U is the displacement velocity. (We have left out a correction factor in (1), which is minor for $\sigma < 1$.) Previously, Scheidegger [7], in his treatment of the displacement as a random-walk process, discovered the proportionality of D to U and to σ^2. According to Gelhar et al., the dispersivity, defined as D/U, is a material property, not dependent on the displacement process.

3. MODEL FOR THE EFFECT OF FLUID PROPERTIES

Here, we present only a short outline of the model to show the effect of fluid properties on the dispersion coefficient. As with Gelhar et al. we assume that the permeability is distributed randomly in all directions and has a log-normal frequency distribution and a small correlation length compared to the system length. Furthermore, we assume that displacement occurs in one of the three eigen directions of the permeability tensor, for example displacement parallel to the bedding. Another assumption is that at least one of the two correlation lengths perpendicular to the flow direction is small compared with the correlation length in the longitudinal flow direction. This assumption is satisfied in layered reservoirs with a layer thickness (= twice the across-dip correlation length) which is small compared with the correlation length in the flow direction.

Instead of calculating the movement of the fluid parcels in every part of the reservoir, we assume that the fluid parcels undergo a linear displacement with a random-walk process superimposed and we calculate the step size of the random-walk. To calculate the step size, we select one rectangular block of distinct permeability in the reservoir and consider this block as embedded in an infinite reservoir with effective properties (see Fig. 1). The block dimensions correspond to the permeability correlation lengths. The permeability of the selected block contrasts with the permeability of the surrounding medium (which has effective permeability k_0) by the standard deviation. Because we use a log-normal permeability distribution, the permeability k_1 of the selected block is equal to k_0/S (S: geometric standard deviation, $\sigma = \ln S$).

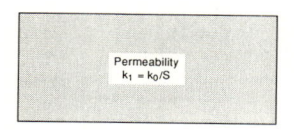

Fig. 1 Block of distinct permeability surrounded by an infinite medium with effective permeability

Figure 2 shows the distortion of the displacement front as it traverses the configuration of Fig. 1. For a log-normal permeability distribution, the difference in the logarithm of the distance travelled by the displacement front in the block, $\ln \ell_1$, and in the effective medium, $\ln \ell$, is a measure of the step size of the random-walk process and hence the dispersivity (for the definition of ℓ_1 see Fig. 2).

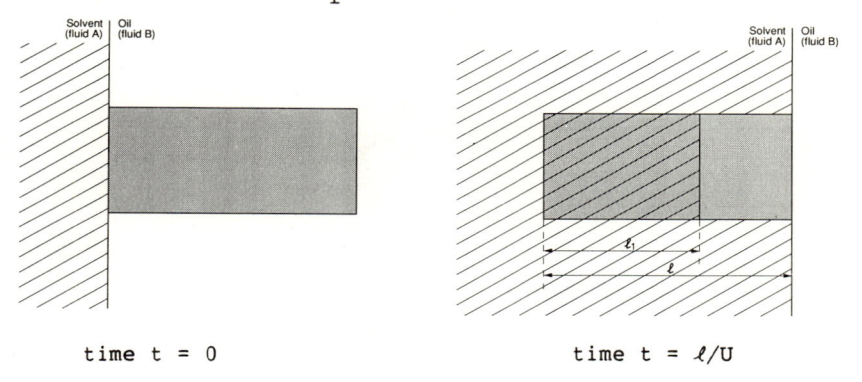

Fig. 2 Location of displacement front between oil and solvent

The dispersivity depends quadratically on this step size, according to Scheidegger, confirmed later by Gelhar et al. The dispersivity is thus given by:

$$\frac{D}{U} = \ell \ (\ln \ell/\ell_1)^2 \ . \qquad (2)$$

To calculate ℓ_1, we note that the pressure drop over the block is determined solely by the injection velocity, because
1) the model assumes the surrounding medium to be infinite
2) the block is thin compared with its length.

The equation of motion for the location of the displacement front includes Darcy and hydrostatic pressure terms. The appendix lists the equation of motion and the solution of ℓ_1. Here, we present the results graphically: in Fig. 3 the dispersivity ratio a (defined as the ratio of dispersivity taking fluid properties into account and dispersivity disregarding these), is plotted as a function of gravity number G, with mobility ratio M and geometric standard deviation S as parameters. (The appendix provides the definitions of M and G.) By definition a lies between 0 and 1. Substituting a in (2) gives

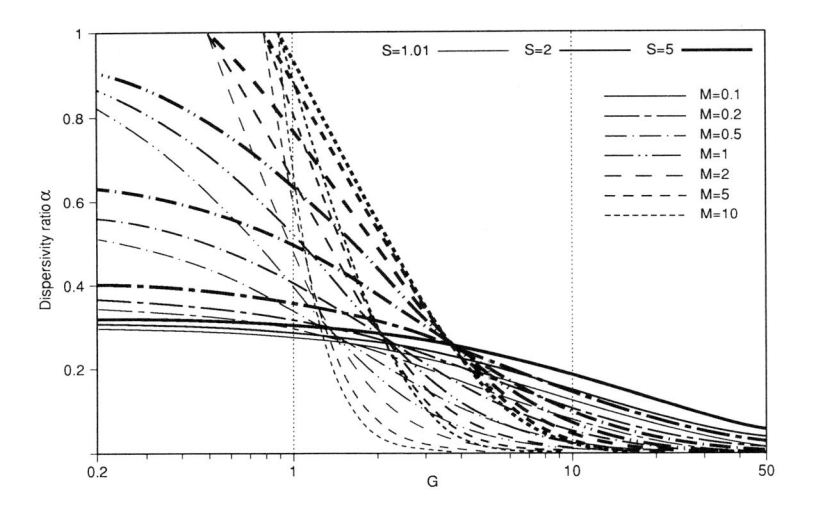

$$\frac{D}{U} = \ell\, \sigma^2\, a. \qquad (3)$$

Fig. 3 Dispersivity factor a as a function of gravity number G with parameters mobility ratio M and geometric standard deviation S

4. SET-UP OF SIMULATIONS

4.1 Simulator

The simulator we used is a fast miscible flood simulator for incompressible flow under gravity [2]. The TVD numerical scheme of the simulator has second-order truncation error in space and suppresses oscillations [5]. We applied this simulator to the linear displacement of oil by a miscible solvent.

4.2 Grid

To a grid of 20 squares by 20 squares, 400 random permeabilities from a log-normal distribution were assigned. Each square was subdivided into a fine mesh of 4 by 4 gridblocks. The result was a finely gridded configuration containing 6400 blocks. The length/width ratio of a square (and hence the whole configuration) was 6.8 to satisfy the model assumption that the correlation length of the permeability in the longitudinal direction is much larger than in the transverse direction. In test runs with M=1 and G=0 we found that dispersivity agreed with Gelhar's theory within the error limits of the simulation. We thereby adapted the spatial correlation function used by Gelhar to the triangular form of the correlation function of the grid, and also to the fact that the simulator uses harmonic weighting of permeabilities for adjacent blocks, and to the presence of grid boundaries.

We tested the analytical model with a grid with a geometric standard deviation S equal to 2.60, and in a similar grid using a smaller value (S = 1.57).

4.3 Evaluation

A special program calculated a dispersion coefficient for each injected pore volume for which a concentration pattern was output. This made it possible to monitor the development of the dispersion over time, which is a distinct advantage compared to the determination of dispersion from the production history. The calculation of dispersion is based on the following equation [1]:

$$C = \frac{1}{2} \text{ erfc } \left(\frac{\frac{x}{L} - I}{2\sqrt{I}} \sqrt{Pe} \right) \quad (4)$$

where L is the length of the total grid, x is the coordinate in longitudinal direction, I is injection in pore volumes, and Pe is the Peclet number, defined by Pe = UL/D. Equation (4) is a good approximation for high values of Pe times I. To match the concentration patterns with expression (4), the program determines the average concentration C of the solvent in an array of blocks in the transverse flow direction.

4.4 Example

An example of a concentration pattern generated by the simulator is shown in Fig. 4. The concentration averaged across the transverse flow direction is plotted in Fig. 5 showing an S-shaped profile. This profile was matched with expression (4) from which Pe was determined (see Fig. 5). The difference between the two profiles is at most two grid blocks.

Fig. 4 Concentration pattern of solvent at dimensionless time
I = 0.7 for simulation run with M = 0.2 and G = 0.
Figure is expanded by a factor of 6 in transverse
direction. Numbers represent 10 times injectant
concentration, X represents solvent concentration 1.0.

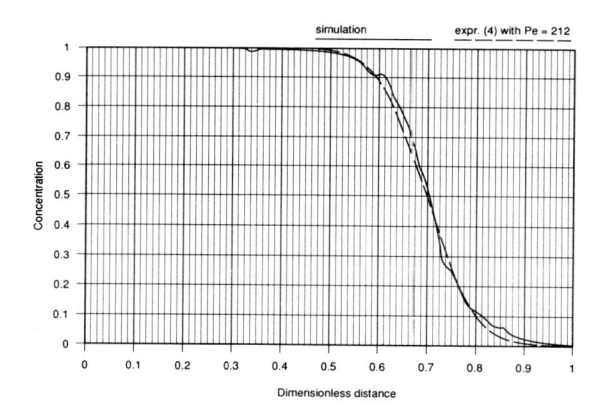

Fig. 5 Concentration profile at I = 0.7 of simulation run with
M = 0.2 and G = 0

The plot of Pe versus cumulative injection I, determined from a series of concentration patterns of this run, shown in Fig. 6, indicates that the value of Pe = UL/D does not fluctuate very much with I (apart from a transient below approximately I=0.35). For such a curve (without the transient) an average value of Pe can be calculated yielding a small standard deviation.

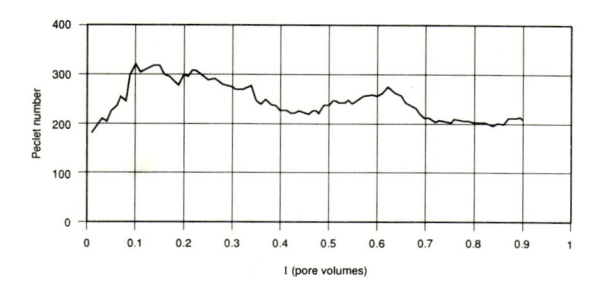

Fig. 6 Peclet number as a function of dimensionless time I for simulation run with M = 0.2 and G = 0

5. COMPARISON OF SIMULATIONS WITH MODEL

In addition to two reference runs for M = 1 and G = 0, we conducted fourteen simulation runs for a wide range of M and G values (M between 0.01 and 10; G between 0 and 14). We observed a dispersive behaviour of the transition zone in all runs. This is illustrated by the small standard deviation of the average Pe values (10% or smaller).

The dispersivity ratio a_s was calculated from the average Peclet value and the average Peclet value from the reference run. The table below lists the calculated a_s values (and their error due to the fluctuation of Pe as a function of I) and the a values predicted by our model (a_m). The table also lists the relative difference $\Delta a/a$.

S	M	G	a_m	a_s		$\Delta a/a$
2.60	0.2	0	0.390	0.392	±11%	0.5%
2.60	1	0.135	0.917	0.909	±11%	-0.9%
2.60	0.1	0	0.316	0.313	±11%	-1.0%
2.60	2	0.805	0.801	0.812	± 8%	1.4%
2.60	0.01	0	0.255	0.259	± 8%	1.5%
2.60	10	1.347	0.688	0.677	±13%	-1.6%
1.57	0.1	0	0.309	0.315	± 9%	1.9%
1.57	10	1.384	0.499	0.475	± 8%	5 %
2.60	0.5	0	0.631	0.588	±10%	-7 %
2.60	1	1.409	0.453	0.496	±11%	9 %
2.60	0.1	13.41	0.134	0.180	±12%	25 %
2.60	10	2.683	0.235	0.320	±16%	27 %
2.60	2	5.634	0.060	0.184	± 8%	67 %
2.60	10	5.365	0.026	0.212	± 7%	88 %

The table shows that of the fourteen runs, seven agree with the model prediction within 2%. Another three runs agree with the model prediction within 5 to 9%. For all these ten runs our model predicts an a value that agrees with a_s within the error bounds. The remaining four runs differ 25 to 88% from the model prediction. Although some of these four runs do not agree quantitatively with the model prediction, they still agree qualitatively with the model.

In Fig. 7 the a values predicted by the model and the a values determined from the simulation runs are plotted, together with the line $a_s = a_m$. Figure 7 shows that all but two data points are close to this line, illustrating the good quantitative agreement for these data points. Particularly at a small a_m (below 0.15), the data points are not close to the ideal line. This is due to a transverse flow effect that is not incorporated in the model.

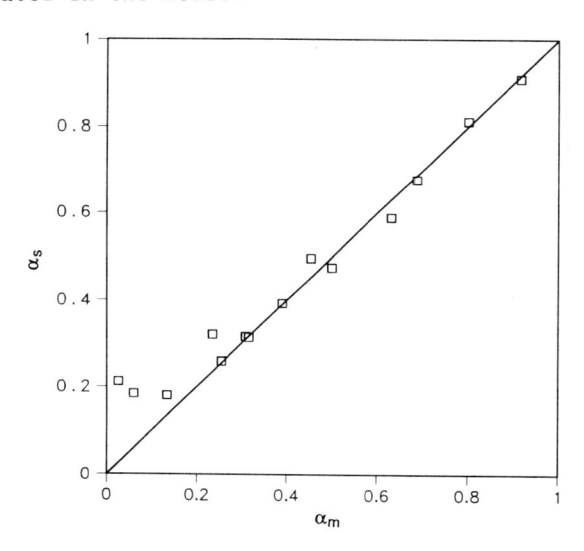

Fig. 7 Comparison of a derived from simulation (a_s) with a predicted by model (a_m)

6. CONCLUSIONS

All simulation runs that were carried out on the stable displacement of miscible fluid banks at mobility ratio not equal to 1 and gravity number not equal to 0 confirm the model assumption that the transition zone is dispersive. In practice this means that the transition zone in a stable miscible displacement grows proportionally with the square root of time.

There is good agreement between the model prediction and the simulation result for ten of the fourteen runs. The model prediction agrees with the dispersivity ratio determined from these simulations within the error limits of the simulation result (< 9%). The other four runs show that at small a agreement is only qualitative.

ACKNOWLEDGEMENT

The author is indebted to Shell Internationale Research Maatschappij B.V. for permission to publish this paper.

REFERENCES

1. Bear, J. & Verruijt, A. (1987), "Modelling groundwater flow and pollution", Reidel, Dordrecht.
2. Crump, J.G. (1988), Detailed simulations of the effects of process parameters on adverse mobility ratio displacements, SPE/DOE 17337.
3. Collins, R.E. (1976), "Flow of fluids through porous materials" Petr. Publ. Co., Tulsa.
4. Gelhar, L.W. and Axness, C.L. (1983), Three-dimensional stochastic analysis of macrodispersion in aquifers, Water Resources Research, 19, 161-180.
5. Harten, A. (1984), On a class of high-resolution total-variation-stable finite-difference schemes, SIAM J. Numer. Anal. 21, 1-23.
6. Jensen, J.L., Lake, L.W. & Hinkley, D.V. (1985), A statistical study of reservoir permeability: distributions, correlations and averages, SPE 14270.
7. Scheidegger, A.E. (1954), Statistical hydrodynamics in porous media, J. of Applied Physics, 25, 994-1001.

APPENDIX

The following equations of motion with Darcy pressure drops and hydrostatic pressures apply to the location of the displacement front inside and outside the block with permeability k_1 (see Fig. 2):

$$\Delta p = \frac{\mu_A}{k_0} \phi \frac{dx_0}{dt} x_0 + \frac{\mu_B}{k_0} \phi \frac{dx_0}{dt} (\ell - x_0) - \rho_A g x_0 \sin\gamma$$

$$- \rho_B g (\ell - x_0) \sin\gamma$$

$$= \frac{\mu_A}{k_1} \phi \frac{dx_1}{dt} x_1 + \frac{\mu_B}{k_1} \phi \frac{dx_1}{dt} (\ell - x_1) - \rho_A g x_1 \sin\gamma$$

$$- \rho_B g (\ell - x_1) \sin\gamma$$

(A.1)

with boundary conditions: at $t = 0$ is $x_1 = 0$
at $t = \ell/U$ is $x_1 = \ell_1$
$x_0 = U t$,

where Δp is the pressure drop over the block
μ_A is the viscosity of the displacing fluid
μ_B is the viscosity of the displaced fluid
ρ_A is the density of the displacing fluid
ρ_B is the density of the displaced fluid
ϕ is porosity
ℓ is the block length
ℓ_1 is the value of x_1 when $x_0 = \ell$
x_0 is the location of the displacement front in the surrounding medium
x_1 is the location of the displacement front in the block
t is time
g is the gravitational acceleration
γ is the dip angle

We define

$$M = \frac{\mu_B}{\mu_A}$$

$$G = \frac{M-1}{M} \frac{U_c}{U} = \frac{k_0 g (\rho_B - \rho_A)}{\mu_B \phi U} \sin\gamma$$

$$\omega = S \frac{U}{U_c} [(1-M) \ell_1/\ell + M] - \frac{1}{2}$$

$$\omega_0 = S \frac{U}{U_c} - \frac{1}{2}$$

$$a = S \frac{U}{U_c} (1 - \frac{U}{U_c}) - \frac{1}{4}$$

The solution to the differential equation (A.1) is the following set of equations depending on the sign of a:

if $a > 0$: $\quad 0 = -\ln M + \frac{1}{2} \ln(\frac{\omega^2 + a}{\omega_0^2 + a}) + \frac{1}{2\sqrt{a}} \arctan(\frac{\sqrt{a}(\omega-\omega_0)}{a+\omega\omega_0})$,

if $a = 0$: $\quad 0 = -\ln M + \ln \frac{|\omega|}{|\omega_0|} - \frac{1}{2\omega} + \frac{1}{2\omega_0}$, (A.2)

if $a < 0$: $\quad 0 = -\ln M + [\frac{1}{2}+\frac{1}{4\sqrt{-a}}] \ln(\frac{|\omega^2 + a|}{|\omega_0^2 + a|}) + \frac{1}{2\sqrt{-a}} \ln(\frac{\omega_0+\sqrt{-a}}{\omega +\sqrt{-a}})$

To find ℓ_1, the equations have to be solved numerically. The number of roots is 1 at maximum. If $a < 0$, a boundary extremum may be found at $\omega^2 = -a$.

FUNDAMENTAL MECHANISMS OF PARTICLE DEPOSITION ON A POROUS WALL : HYDRODYNAMICAL ASPECTS

P. Schmitz, C. Gouverneur, D. Houi.
Institut de Mécanique des Fluides de Toulouse
U.R.A. C.N.R.S.0005
av Camille Soula, 31400 Toulouse FRANCE

ABSTRACT

Investigation of 2D hydrodynamical field near a pore of a porous wall with suction enhances two classes of flow allowing or not particle deposition.
Trajectories of single particles moving in this wall region are determined by a balance of hydrodynamical and physico-chemical forces.
Then a statistical model is proposed to predetermine the surface deposit for different flow conditions and suspension types.

NOMENCLATURE

D	pore diameter
H	main channel height
x	dimensionless coordinate vector
X	dimensionless axial coordinate : X'/D
Y	dimensionless radial coordinate : Y'/D
t	dimensionless time : $t'.U_{max}/D$
Y_p	dimensionless distance between the centre of the particle and the wall : Y'_p/R_p
R_p	particle radius
Q	Hamacker's constant : 10^{-19}
u	dimensionless fluid velocity : $\mathbf{u'}/U_{max}$
v	dimensionless particle velocity : $\mathbf{v'}/U_{max}$
P	dimensionless pressure : $P'/\rho.U_{max}^2$
V_w	mean suction velocity
U_{max}	maximum axial velocity
T_s	dimensionless suction thickness : T'_s/D
Re	Reynolds number : $U_{max}.D/\nu$
Nst	Stokes number : $U_{max}.R_p/\nu$

1 INTRODUCTION

Particle deposition on a porous wall is considered as a

fundamental phenomena in crossflow microfiltration. As well in oil recovery processes, we are often concerned with suspensions flowing tangentially to porous surfaces. Such situations occur in well-fouling by muds or in well-reactivation when controlling permeability during acidizing treatments.
Our purpose is to analyse and describe phenomena inducing deposit formation.
At first we determine velocity and pressure fields in porous tube for laminar Newtonian flow with wall mass transfer by experimental and numerical models [8]. This preliminary study is needed to determine correct boundary conditions for investigation of local flows near a pore of the porous wall.
At microscopic scale we characterize by computation two different classes of flow as functions of suction velocity over axial velocity ratio. The first one seems to be able to carry away particles and the second one allows deposition of particles.
Then, in such flow conditions we calculate trajectories of single particles undergoing physico-chemical forces such as Van der Waals '.
Finally, injecting one by one many particles following those trajectories allows us to simulate a statistical deposit over a porous surface in order to determine its physical and geometrical characteristics.

2 FLOW NEAR A POROUS WALL

Typical suspensions are supposed to be at low concentration to assume that mean hydrodynamical field without particles is a representative model of velocity field of the suspension flow.
Let us consider a parallel laminar flow of a Newtonian fluid in the vicinity of a pore of a porous surface. In a 2D numerical approach this region is modelled by a rectangular cavity over a flat plate. Stokes ' flow is admitted because of the very low velocity magnitude near the wall and then the governing equations expressed in dimensionless form are :

$$-\nabla P + \frac{1}{Re} \nabla^2 u = 0 \tag{1}$$

$$\nabla \cdot u = 0 \tag{2}$$

with following boundary conditions (see figures 1 or 2) :
- uniform velocity gradient at the upstream edge,
- uniform axial velocity U_{max} at the upper fluid boundary,
- Fully developed Poiseuille flow of V_w mean velocity at the bottom of the pore,
- no slip condition at the walls.

2D Stokes flow near a cavity has been studied by O'Brien [5], Higdon [2] and Jasti [4] showing centred recirculating vortex in the cavity. If a suction flux is imposed at the bottom of the cavity, Tutty [9] has proved that the recirculation region occurs at the upstream edge of the pore and that the thickness of the disturbed main flow depends on the suction flux magnitude. We calculate local velocity field for a Reynolds number of 0.01 based on axial boundary velocity U_{max} and D diameter of the pore. Two different classes of flow characterize the importance of suction velocity V_w compared to boundary axial velocity U_{max}.

When $V_w/U_{max} \ll 1/100$ (small suction flux) only few streamlines near the downstream side of the pore go to the bottom of the pore due to the recirculating vortex at the upstream edge (figure 2).

When $V_w/U_{max} \gg 1/100$ recirculating vortex vanishes and all the streamlines under the dividing flow region line are deviated from axial flow to feed the pore (figure 1).

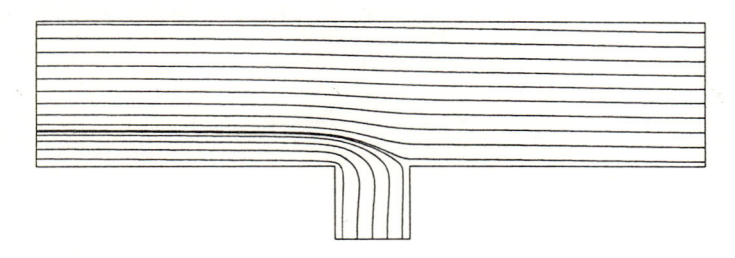

Fig. 1 : Streamlines near the pore for $V_w/U_{max} = 1/50$

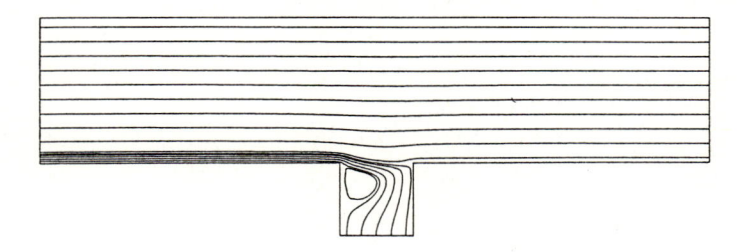

Fig. 2 : Streamlines near the pore for $V_w/U_{max} = 1/500$

General results are presented as variations of the suction thickness over V_w/U_{max} ratio. We define the suction thickness T_s as the initial Y coordinate of the streamline bounding the entrainment region, i.e. the upstream region over the flat plate occupied by the fluid which is eventually sucked by the pore. T_s is a square root function of the velocity ratio which can be derived from simple mass balance considerations as :

$$T_s = \left(\frac{V_w}{U_{max}} \cdot \frac{2H}{D}\right)^{1/2} \tag{3}$$

3 PARTICLE TRAJECTORIES NEAR A POROUS WALL

Trajectories of spherical solid buoyant monodisperse single particles flowing in the wall region are determined by balance of forces. We take into account hydrodynamical forces computed using 2D numerical model velocity field and physico-chemical forces expressed in the literature [6], [1], [8].
At first we solve the basic equations of mechanics (4) and (5) for a particle undergoing hydrodynamical and Van der Waals forces expressed respectively by Stokes formula and Spielman and Goren relation [8]. Nst denotes the Stokes number based on U_{max} axial velocity and R_p radius of the particle. We use Euler semi-implicit method to solve (4) and (5) for different V_w/U_{max} velocity ratio and R_p particle radius.

$$N_{st} \frac{dv}{dt} = -v + u - \frac{2}{3} \frac{Q}{6\pi\mu R_p U_{max}} \frac{1}{(Y_p^2-1)^2} \tag{4}$$

$$\frac{dx}{dt} = v \tag{5}$$

Results prove that particles can be captured by the wall under attractive forces (figure 3) contrary to expected effects of wall shear velocity carrying away these particles.

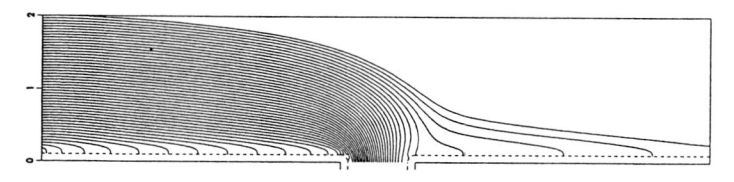

Fig. 3 : Particle trajectories, $R_p = 0.1$ μm, $V_w/U_{max} = 1/5$

4 SIMULATING A DEPOSIT

We build a deposit over porous wall surface by injecting particles from an initial random Y coordinate. Those particles follow prescribed trajectories until contact with the porous wall or an aggregated particle. Capture is imposed by behaviour rules defined by sticking angles [3]. Trajectory incidence represents suction intensity over intensity of entrainment flow; magnitude of sticking angles characterizes the importance of attractive forces compared with hydrodynamical and repulsive forces. These special quantities are estimated using values observed experimentally.

At first, linear trajectories are studied. We enhance deposit compactness of surface clusters for contact-capture and sticking angles assumptions (figures 4, 5 and 6).

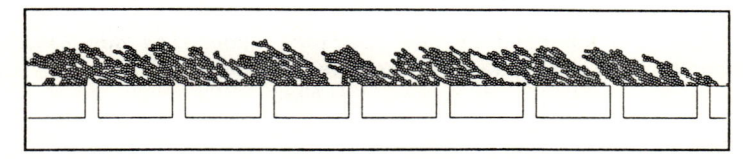

Fig. 4 : Contact-capture, incidence of 15°, $R_p = 0.1$ µm, $D = 1$ µm

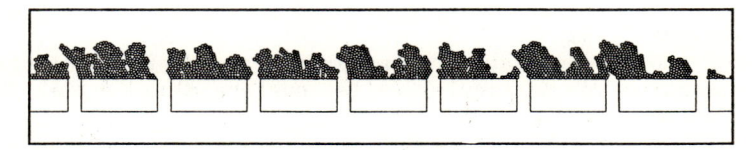

Fig. 5 : Low sticking angles without reentrainment, incidence of 15°, $R_p = 0.1$ µm, $D = 1$ µm

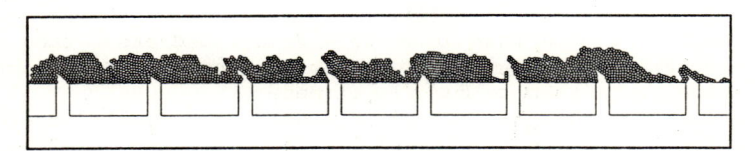

Fig. 6 : Low sticking angles with possibly reentrainment, incidence of 15°, $R_p = 0.1$ µm, $D = 1$ µm

In the case of sticking angles rules with possibly reentrainment, we have good agreement with experimental deposit visualisation of muds upon a filtrating ceramic membrane.

5 CONCLUSION

Hydrodynamical forces are not always strong enough to carry away particles from the porous wall with suction because of the importance of physico-chemical forces in the wall region. Our concern is to take into account these two effects on suspensions behaviour to simulate a statistical aggregation of particles on a porous wall in order to predetermine the deposits and know their physical and geometrical characteristics.

ACKNOWLEDGEMENTS

The authors wish to thank Mr B. LACROIX who helped with the numerical work.
This work is supported by the "Société Lyonnaise des Eaux" and the "Centre National de la Recherche Scientifique".

REFERENCES

[1] De Gennes, P.G., Dynamics of concentrated dispersions. A list of problems, Phys Chem Hydro, 2, 1, 31-44, 1981.
[2] Higdon, J.J.L., Stokes flow in arbitrary two-dimensionnal domains: shear flow over ridges and cavities, J Fluid Mech, 159, 195-226, 1988.
[3] Houi, D., Modélisation s expérimentales et théoriques de l'accumulation de particules à la surface d'un filtre, Thèse INPT, Toulouse, 1986.
[4] Jasti, J.K, and al, capacitance effects in porous media, SPE Reservoir Eng, 1207-1214, Nov, 1988.
[5] O'Brien, V., Closed streamlines associated with channel flow over a cavity, Phys of Fluids, 15, 12, 2089-2097, 1972.
[6] Overbeeck, J.T.G., Interparticle forces in colloid science, Powder Tech, 37, 195-208, 1984.
[7] Schmitz, P., Laminar flow in a dead ended prous tube with wall suction : application to hollow fiber crossflow filtration, submitted to Eur. J. of Mec.
[8] Spielman, L.A., and Goren, S.L, Capture of small particles by London forces from low speed liquids flow, Env Sci Tech, 4, 2, 135-140, 1970.
[9] Tutty, O.R., Flow in a tube with a small side branch, J. Fluid Mech, 191, 79-109, 1988.

WETTING PHENOMENA IN SQUARE-SECTIONAL CAPILLARIES

A. Winter

(Geological Survey of Denmark, Copenhagen, Denmark)

ABSTRACT

A computational strategy for the determination of thickness of the thin film of the aqueous phase in corners of a square-sectional capillary is described. The underlying mathematical model is given by the second order nonlinear differential equation derived from an augmented version of the Young-Laplace equation including the disjoining pressure term. The model is subsequently resolved as a two-point boundary value problem using the method of shooting.

1. INTRODUCTION

The phenomenon of wetting manifests itself by formation of the contact angle between a liquid and a solid substrate. In systems with a confined geometry, such as the square-sectional capillary shown in Figure 1 the contact angle is usually defined as the angle between the substrate and a continuation of the profile of the meniscus undisturbed by surface forces (cf. Figure 2).

Experimental investigations of oil droplets in water-filled square-sectional capillaries show that the shapes of the oil-water interfaces observed in the 45 and 90 degree planes with respect to the walls of a tube are not identical. In the case of the 90 degree plane, the oil-water interface becomes parallel to porewalls with no visible film of water in the plane of observation. On the other hand, in the 45 degree plane the oil-water interface is separated from the capillary corners by a thin film of the aqueous phase, cf. BIRDI et al. (1987) and ARRIOLA (1980).

This paper describes a computational strategy for the determination of the equilibrium shape of a liquid-liquid interface

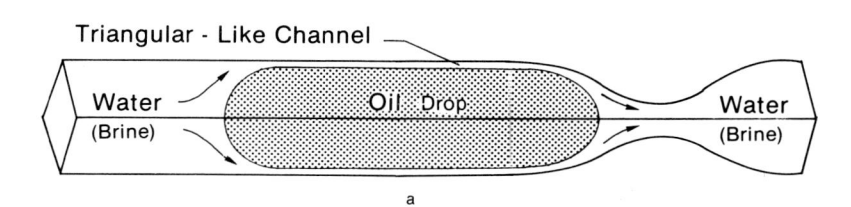

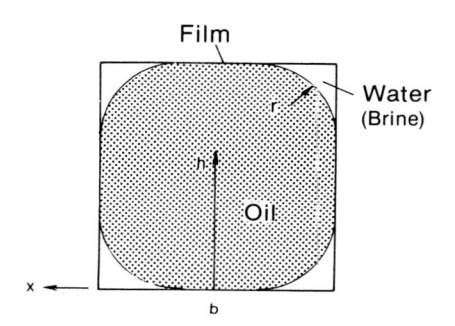

Fig. 1 Configuration of the oil-water interface in a square-sectional capillary tubing: (a) 45° plane; (b) 90° plane.

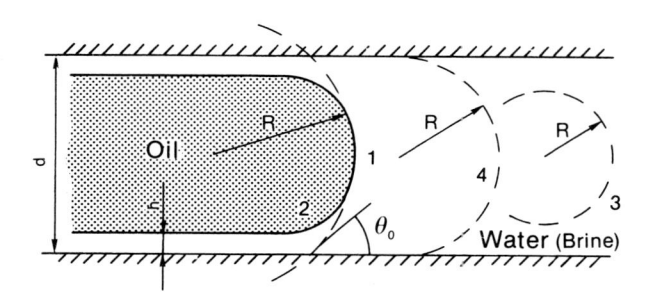

Fig. 2 Wetting angle in a square-sectional capillary tubing defined as the contact angle between the porewall and continuation of the oil-water interface undisturbed by surface forces (modified after CHURAEV (1987)).

in the 45° plane. It is organized as follows. The next section describes a mathematical model of the oil-water interface between an oil drop situated in a square-sectional capillary and the aqueous film confined by the tube walls. In the subsequent section the computational procedure used to solve the model equations is discussed. Finally, the results of numerical computations aimed at determination of the film thickness are described.

2. MATHEMATICAL MODEL OF THE OIL-WATER INTERFACE IN THE 45° PLANE

The equilibrium shape of the oil-water interface in a square-sectional capillary is described by an augmented version of the Young-Laplace equation

$$\Delta P_c = \sigma \Gamma + \Pi(h) \qquad (2.1)$$

where σ is the oil-water interfacial tension, h is the film thickness, $\Gamma = 1/R_1 + 1/R_2$ is the mean curvature of the interface (R_1 and R_2 are the orthogonal radii of curvature) and $\Pi(h)$ represents the disjoining pressure term.

Disjoining pressure is a macroscopic pressure correction accounting for long-range intermolecular interactions, cf. DERJAGIN et al. (1987). For nonpolar or slightly polar materials the dominating attractive forces between the individual molecules are London-van der Waals forces tending to destabilize thin films. On the other hand, presence of an electrolyte solution electrifies solid surfaces leading to the appearance of double ionic layers. The resulting ionic-electrostatic forces are repulsive in nature and tend to stabilize the aqueous films, cf. DERJAGIN et al. (1987).

In the case when the London-van der Waals and the ionic-electrostatic forces dominate the solid-liquid interactions, the disjoining pressure term is represented by the following expression, cf. DERJAGIN et al. (1987)

$$\Pi(h) = Ah^{-3} + Bh^{-2} \qquad (2.2)$$

where $A = A_H/6\pi$ (A_H is the Hamaker constant), B is another constant representing the ionic-electrostatic interactions and h is the film thickness.

The assumption that the interface is in the state of capillary equilibrium imposes the condition of an equal pressure drop across the oil-water interface. Consequently, the mean curvature, Γ, is the same at all points of the interface.

The easiest way of determination of the capillary pressure drop is to consider very long drops for which one of the radii of curvature (say, R_2) is very large and can be assumed as equal to ∞, cf. ARRIOLA (1980) and LENORMAND (1981). Elementary calculations show that the distance between the interface and the corner of the capillary is given by the following relation (cf. Figure 3)

$$H = R_1(\sqrt{2} - 1) \qquad (2.3)$$

where $R_1 = \sigma/\Delta P_c$.

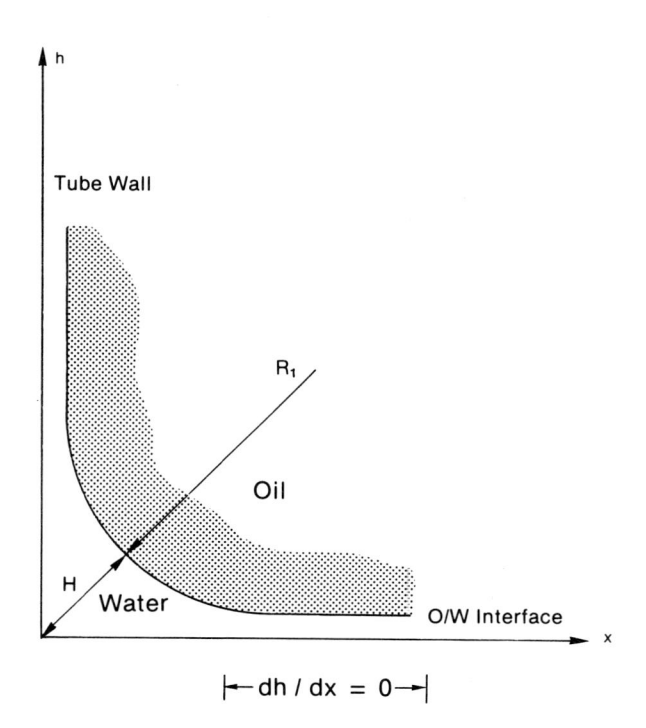

Fig. 3 Configuration of the oil-water interface confined by tube walls and an oil drop in a square-sectional capillary.

The capillary pressure term, ΔP_c, can be determined by assuming equality of the orthogonal radii of curvature observed at the 90° plane at the round ends of the drop. Experimental observations have shown (cf. BIRDI et al. (1987)) that these radii of curvature, R(90), are approximately equal to the half of the wall-to-wall distance, d (cf. Figure 2). Thus

$$\Delta P_c \simeq 4\sigma/d \qquad (2.4)$$

It should be noted that eq. (2.3) presupposes that in the corners of a square-sectional capillary the interfacial layers due to the porewall-wall- and water-oil interfaces do not overlap and, consequently, the disjoining pressure term can be neglected.

In the thin film zone thickness of the aqueous phase is given by the following form of the Young-Laplace equation obtained from eq. (2.1) after replacing R_1 by the well-known formula of differential geometry and assuming that $R_2 \simeq \infty$.

$$\frac{\Delta P_c - \Pi(h)}{\sigma} = \frac{d^2h/dx^2}{\{1 + (dh/dx)^2\}^{3/2}} \qquad (2.5)$$

where x denotes the horizontal coordinate (cf. Figure 3), $\Pi(h)$ is the disjoining pressure term defined by eq. (2.2) and all other terms are consistent with previously introduced notation.

The above equation has been solved for different ΔP_c's reflecting varying tube sizes. The initial values h_0 of the film thickness, h, have been determined from experimental investigations of the interface-to-corner distance, H (cf. Figure 3 and eq. (2.3)).

3. NUMERICAL SOLUTION OF THE AUGMENTED YOUNG-LAPLACE EQUATION

Thickness of the thin film surrounding a trapped oil ganglion can be found by solving the differential equation (2.5) subject to the following boundary conditions:

a. The film interface is parallel to the capillary wall in the centre of the capillary;

b. The approximate distance between the oil-water interface and the capillary corner, H, is assumed to be equal to a known experimental value.

Thus one has to resolve a two-point boundary value problem, e.g., by using the method of shooting, cf. ASCHER et al. (1988).

4. FINAL REMARKS

The results of the computations show that the film thickness is highly sensitive to changes of the oil-water interfacial tension. For example, for interfacial tension equal to 1 mN/m the film thickness is approximately 11 times larger than that appearing at the interfacial tension equal to 32.1 mN/m (cf. Figure 4).

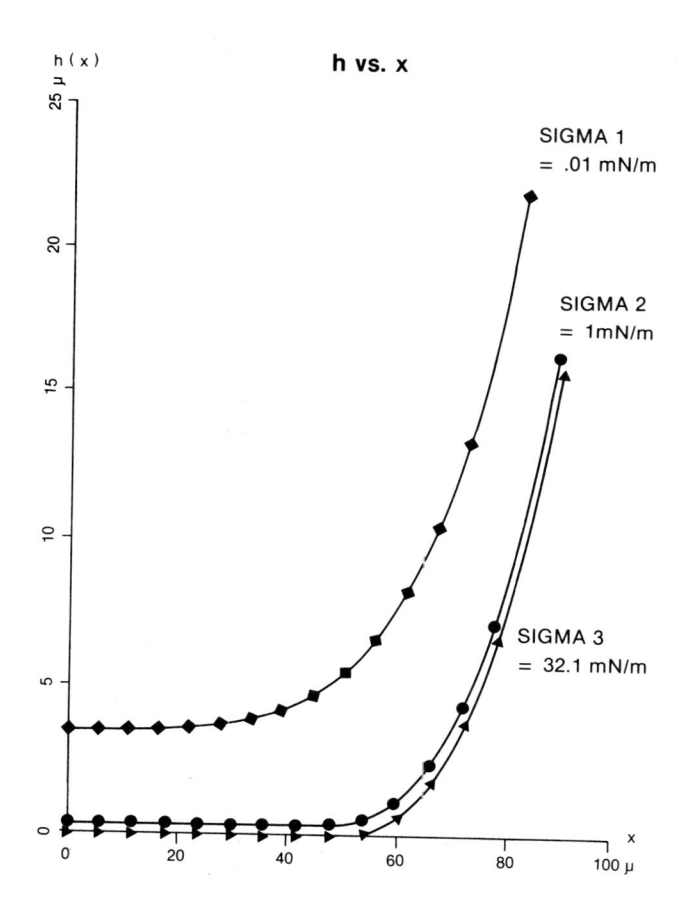

Fig. 4 Variation of the film thickness in a square-sectional capillary as one moves in the transversal plane from the centre of the drop towards the wall of the tubing. Three different interfacial regimes are shown ranging from high to ultra-low tensions.

The dependence of film thickness on the interfacial tension is important for enhanced oil recovery strategies in which the dominating mechanism is the presence of near-critical conditions in a multicomponent fluid mixture, cf. WINTER, (1987). More precisely, in the case where the mixture of pore fluids becomes near-critical, the capillary pressure component appearing in the eq. (2.1) decreases and the disjoining pressure becomes the decisive factor controlling the film thickness. In particular, it can be shown that above a certain value of the radius of interfacial curvature all the surface of the pore is wetted and below that value the wetting film appears only in corners of a capillary, cf. JOANNY (1985).

The determination of the limiting value of the aqueous film thickness in a square-sectional capillary tubing, corresponding to the transition between these two regimes, turns out to be an essential part of the wettability evaluation strategy, cf. WINTER (1990). According to that strategy the porous network can be represented by a self-similar (fractal) system consisting of square-sectional pores in a prespecified size range. The fractal porous networks allow determination of the fraction of partially wetted capillaries (i.e. those with the aqueous phase appearing only in the corners of capillary tubings) and, consequently, lead to evaluation of the global wettability preference.

5. ACKNOWLEDGEMENTS

The author is grateful to the anonymous reviewers for their critical remarks.

6. REFERENCES

1. Arriola, A. (1980), Ph.D. Thesis, University of Kansas.
2. Ascher, U.M., Mattheij, R.M. & Russell, R.D., (1988), Numerical Solution of Boundary Value Problems for Ordinary Differential Equations, Prentice Hall Inc., Englewood Cliffs, New Jersey.
3. Birdi, K.S., Vu, D.T. & Winter, A., (1987), Microscopic, Multiphase (Oil/Water) Displacement Phenomena in Rectangular Capillary Tubes. Capillary and Interfacial Forces in Improved Oil Recovery, in: Proceedings of the 4-th European Symposium on Enhanced Oil Recovery, Hamburg, October 27-29, 1987, Deutsche Wissenschaftliche Gesellschaft für Erdöl, Erdgas und Kohle E.V.
4. Churaev, N.V., (1987), Physical Mechanisms of Wetting of Solid Surfaces, Inzhenerno-Fizicheskii Zhurnal, 53, no. 5, 795-802 (in Russian).

5. Derjagin, B.V., Churaev, N.V. and Muller, V.M., (1987), Surface Forces, Consultants Bureau, New York and London.
6. Joanny, J.F., (1985), Thèse, Univ. de Paris VI.
7. Lenormand, R. (1981), Thèse, L'Institut National Polytechnique de Toulouse.
8. Winter, A., (1987), Griffiths Model of Tricritical Behavior and its Application to the Design of Enhanced Oil Recovery Processes, International Journal of Physicochemical Hydrodynamics, 9, no. 3-4, 589-603.
9. Winter, A., (1990), Estimation of Wettability in a Complex Porous Network, in: Phase Transitions in Soft Condensed Matter, Riste, T. and Sherrington, D. (eds.), to be published by Plenum Press.

A Numerical Method for Simulation of Centrifuge Drainage

Ivar Aavatsmark
Norsk Hydro, Bergen

Abstract. A finite-difference method for stable and fast solution of the saturation equation for centrifuge drainage is presented. The method is implicit, flux consistent, conservative and almost unconditionally monotonic. Unconditional monotonicity ensures that long time steps can be taken when the solution is approaching steady state.

The nonlinear equations are solved by Newton's method with a displacement test as the basic termination criterion. The Jacobian matrix is the sum of a tridiagonal M-matrix and a small rank-one matrix. The structure of the Jacobian matrix is utilized when solving the linear equations in the Newton iteration.

An example of a numerical simulation with six intermediate steady states is given.

1 Introduction

Numerical methods for simulation of centrifuge drainage have been given by O'Meara and Crump [5], Firoozabadi and Aziz [3], and Guo [4]. These methods are explicit, IMPES and implicit schemes. When approaching steady state it is desirable to be able to take large time steps. This can not be achieved by the explicit schemes. Also, implicitness is not sufficient for unconditional stability. A typical unstable behaviour is the appearance of a saturation "icicle" ahead of the wave foot, when steady state is approached. Such icicles contain negative saturations. The numerical method used by Aavatsmark [1] has a monotonicity property, which makes it appropriate for large time steps. The theory of this method applied to flooding, capillary imbibition and gravity drainage is given in Aavatsmark [2]. The implementation of this method for the equations of centrifuge drainage is presented here.

2 Mathematical description

Centrifuge drainage is the displacement of one phase by another in a porous core plug located in a centrifuge. The core is submerged in a bath of the displacing phase, and this phase has to be the nonwetting phase. The model equations are derived, assuming that the flow is one-dimensional, that the phases are incompressible, and that the porous medium is homogeneous. The displacement direction is assumed to be orthogonal to the rotational axis of the centrifuge. Gravity is neglected, and the only acting body force is the centrifugal force.

The two phases are termed phase 1 and phase 2. Let x be the spatial coordinate and t be the time coordinate. The saturation of phase 1 is termed u. For each phase i the quantities density ρ_i, volumetric flow density v_i, mobility λ_i, and pressure p_i are defined. The mobilities are functions of the saturation: $\lambda_i = \lambda_i(u)$. The largest saturation interval, in which both phases are mobile, is termed (u_{min}, u_{max}). Further quantities are porosity ϕ, angular velocity ω of the centrifuge, x-coordinate a of the rotational axis, and core length L. The angular velocity is assumed to be a nondecreasing function of time: $\omega = \omega(t)$ and $\omega'(t) \geq 0$.

The conservation of each phase is expressed by the equations

$$\phi \frac{\partial u}{\partial t} + \frac{\partial v_1}{\partial x} = 0, \qquad -\phi \frac{\partial u}{\partial t} + \frac{\partial v_2}{\partial x} = 0. \tag{1}$$

The motion of each phase is described by Darcy's law, equating the pressure force, the centrifugal force and the drag force:

$$\frac{\partial p_1}{\partial x} + \rho_1 \omega^2 (a - x) + \frac{v_1}{\lambda_1} = 0, \qquad \frac{\partial p_2}{\partial x} + \rho_2 \omega^2 (a - x) + \frac{v_2}{\lambda_2} = 0. \tag{2}$$

The quantities $\gamma(t) = (\rho_1 - \rho_2)\omega^2(t)$, capillary pressure $p = p_1 - p_2$, and total volumetric flow density $v = v_1 + v_2$ are introduced. The capillary pressure is an increasing function of saturation u: $p = p(u)$ and $p'(u) > 0$. Addition of the conservation equations (1) shows that the total volumetric flow density is a function of time only: $v = v(t)$. Subtraction between Darcy's laws (2) and subsequent multiplication with $\lambda_1 \lambda_2 / (\lambda_1 + \lambda_2)$ give

$$v_1 = \frac{\lambda_1}{\lambda_1 + \lambda_2} v(t) - \frac{\lambda_1 \lambda_2}{\lambda_1 + \lambda_2} \left(\gamma(t)(a - x) + \frac{\partial p}{\partial x} \right). \tag{3}$$

Substituting this equation into one of the equations (1) and using the functions

$$f_1(u) = \frac{1}{\phi} \frac{\lambda_1}{\lambda_1 + \lambda_2}, \qquad h(u) = \frac{1}{\phi} \frac{\lambda_1 \lambda_2}{\lambda_1 + \lambda_2}, \qquad g(u) = \int_{u_{min}}^{u} h(v) p'(v)\, dv, \tag{4}$$

the saturation equation

$$u_t + [v(t) f_1(u) - \gamma(t)(a - x) h(u) - g(u)_x]_x = 0 \tag{5}$$

emerges. The function $f_1(u)$ is a monotonically increasing S-shaped function, which vanishes in u_{min}. The function $h(u)$ is bell-shaped, and vanishes in u_{min} and u_{max}. The function $g'(u)$ is also usually bell-shaped, vanishing in both end points u_{min} and u_{max}.

The core, sealed on the side walls, but open at both ends, is submerged in a bath of the nonwetting phase. This phase is termed phase 1. It is assumed that the phases are not in equilibrium, such that phase 1 displaces phase 2. As argued in [1], [2], phase 2 will then flow out at one end, and phase 1 will flow in at the other end. The core is oriented such that $x = 0$ at the inlet boundary and $x = L$ at the outlet boundary. If $\rho_1 > \rho_2$, the displaced phase 2 migrates inwards, towards the rotational axis, and the radial distance $a - x$ becomes positive. If $\rho_1 < \rho_2$, the displaced phase 2 migrates outwards, towards the periphery, and the radial distance $a - x$ becomes negative. In both cases the product $\gamma(t)(x - a)$ is positive. If $\rho_1 = \rho_2$, no displacement takes place.

Since phase 1 is the nonwetting phase, the capillary pressure vanishes in u_{min} (or assumes its minimum value there).

At the outlet boundary ($x = L$) phase 2 must have the same pressure as phase 1 has in the bath outside the boundary. At the inlet boundary ($x = 0$) phase 1 must have the same pressure as it has in the bath outside the boundary. Outside the core (i.e. in the bath of phase 1) the pressure is hydrostatic (in the centrifugal force field). Hence

$$p_2(L,t) = p_1(0,t) - \rho_1 \omega^2(t) \int_0^L (a - x)\, dx. \tag{6}$$

Integration from 0 to L in Darcy's law for phase 2, equation (2), and substitution of expression (6) give

$$\int_0^L \frac{v_2}{\lambda_2}\, dx = p_2(0,t) - p_2(L,t) - \rho_2 \omega^2(t) \int_0^L (a-x)\, dx$$
$$= \gamma(t) \int_0^L (a-x)\, dx - p(u(0,t)). \tag{7}$$

Using equation (3) and the fact that $v_2 = v - v_1$, an expression for the total volumetric flow density

$$v(t) = \frac{\gamma(t) \int_0^L (a-x)\, dx - p(u(0,t)) - \phi \int_0^L f_1(u)(\gamma(t)(a-x) + p(u)_x)\, dx}{\int_0^L \frac{1}{\lambda_1(u)+\lambda_2(u)}\, dx} \tag{8}$$

emerges. The boundary conditions are easily stated. At the inlet boundary ($x = 0$) only phase 1 is flowing, i.e.

$$v(t) f_1(u) - \gamma(t)(a-x) h(u) - g(u)_x = v(t)/\phi. \tag{9}$$

The nonwetting phase 1 can never flow out of the core. Hence, the condition at the outlet boundary ($x = L$) is

$$v(t) f_1(u) - \gamma(t)(a-x) h(u) - g(u)_x = 0. \tag{10}$$

At the outlet boundary ($x = L$) the capillary pressure must vanish (or assume its minimum). Hence $u(L,t) = u_{min}$. This boundary condition has turned out to be numerically unfavourable compared to (10). However, expression (8) can be transformed using this condition. With

$$\Theta(u) = \int_{u_{min}}^{u} \left(1 - \phi f_1(v)\right) p'(v)\, dv, \tag{11}$$

there holds $-\Theta(u(0,t)) = \int_0^L (1-\phi f_1(u))p(u)_x\, dx = -\phi \int_0^L f_1(u)p(u)_x\, dx - p(u(0,t))$, and hence

$$v(t) = \frac{\gamma(t) \int_0^L \left(1 - \phi f_1(u)\right)(a - x)\, dx - \Theta(u(0,t))}{\int_0^L \frac{1}{\lambda_1(u)+\lambda_2(u)}\, dx}. \tag{12}$$

(If a capillary threshold pressure $p(u_{min}+) = p_0$ is used, then this must be accounted for in (11), say $\Theta(u) = \int_{u_{min}+}^{u} \left(1 - \phi f_1(v)\right) p'(v)\, dv + p_0$.)

3 Discrete solution

The initial-boundary-value problem (5), (9), (10), (12) is solved by the implicit, flux-consistent, conservative, and almost unconditionally monotonic method outlined by Aavatsmark [2] for other displacement types. Let $\{(x_i, t^n)\}$, $i = 0(1)N$, $n = 0, 1, \ldots$, be a grid in the xt-plane, such that $x_0 = 0$, $x_N = L$ and $t^0 = 0$. The grid spacings are termed $\Delta x_i = x_i - x_{i-1}$ and $\Delta t^n = t^n - t^{n-1}$. Further, let u_i^n be the approximation to $u(x, t^n)$ over Δx_i, and let $q_{i+1/2}$ approximate the flux $v(t)f_1(u) - \gamma(t)(a-x)h(u) - g(u)_x$ at $x = x_i$. The differential equation (5) with boundary conditions (9) and (10) is approximated by the implicit difference scheme on conservation form

$$\frac{u_i^n - u_i^{n-1}}{\Delta t^n} + \frac{q_{i+1/2}^n - q_{i-1/2}^n}{\Delta x_i} = 0, \quad 1 \leq i \leq N, \quad n \geq 1. \tag{13}$$

Splitting the function $h(u)$ in an increasing part $h_+(u)$ and a decreasing part $h_-(u)$,

$$h_+(u) = \int_{u_{min}}^{u} \max\{h'(v), 0\}\, dv, \qquad h_-(u) = \int_{u_{min}}^{u} \min\{h'(v), 0\}\, dv, \tag{14}$$

the flux can be approximated:

$$q_{i+1/2} = v f_1(u_i) - \gamma(a - x_i)\bigl(h_-(u_i) + h_+(u_{i+1})\bigr) - \frac{g(u_{i+1}) - g(u_i)}{\frac{1}{2}(\Delta x_i + \Delta x_{i+1})}, \tag{15}$$

$i = 1(1)N - 1$, together with $q_{1/2} = v/\phi$ and $q_{N+1/2} = 0$. The difference scheme (13), (15) is an implicit method of the form $F^n(u^n) = u^{n-1}$ with

$$F_i^n(u) = u_i + \lambda_i^n (q_{i+1/2} - q_{i-1/2}), \tag{16}$$

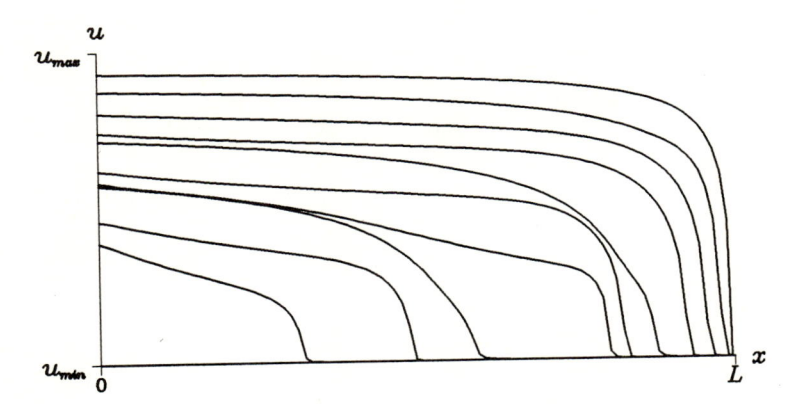

Fig. 1. Saturation profiles in centrifuge drainage.

where $\lambda_i^n = \Delta t^n / \Delta x_i$. The nonlinear equations (16) are solved by Newton's method. The Jacobian matrix has the form $(F^n)'(u) = A + cd^T$. A is a tridiagonal matrix, stemming from those terms of F^n (16), (15) that contain u_j explicitly. A is an M-matrix for all time-step lengths, and this fact makes the method almost unconditionally monotonic [2]. The vector d is the gradient of a discrete expression for v (12): $d = dv/du$.

Although the method can be used for any spatial grid, an equidistant grid (with grid spacing $\Delta x = L/N$) has been found most appropriate. In each iteration stage k on every time level n a linear system of equations of the form $(A + cd^T)x = b$ has to be solved. This is accomplished by solving the system of equations $Ay = b$ and $Az = c$ by Gaußian elimination, and then using the formula

$$x = y - \frac{d^T y}{1 - d^T z} z. \tag{17}$$

Although $\|cd^T\| \ll \|A\|$, it is important not to neglect the last term of equation (17). As termination criterion for the Newton iteration the displacement test $\|u^{(k)} - u^{(k-1)}\|_\infty \leq \epsilon$ is used. To avoid fluctuations in $v(t)$ the residual test with the weaker bound $\|u^{n-1} - F^n(u^{(k-1)})\|_\infty \leq N\epsilon$ should be used in addition. The time step Δt^n is controlled by demanding that the total number of iteration stages k_{max} satisfies the inequality $3 \leq k_{max} \leq 8$.

4 Simulation

Figures 1 and 2 show a simulation of centrifuge drainage. The angular velocity $\omega(t)$ was changed stepwise, and each stage was maintained until an intermediate steady state was reached. The time scale is changing by

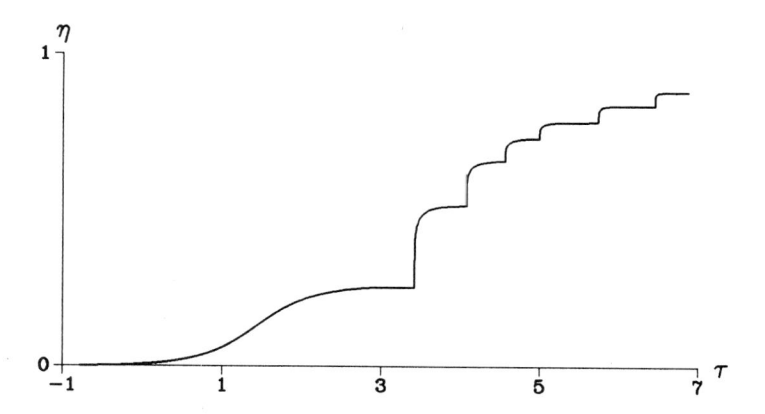

Fig. 2. Recovery efficiency versus logarithmic time.

several orders of magnitude during the experiment, making it very important to be able to take large time steps. This is possible because of the monotonicity property. Figure 1 shows the saturation u as a function of x for successive time levels. Figure 2 shows the recovery efficiency η versus $\tau = \lg(t/T)$, where T is some reference time.

5 References

1. Aavatsmark, I.: Simulator for endimensjonal sentrifugedrenering i labforsøk. Report R-033036, Norsk Hydro Research Center, Bergen, 1988.
2. Aavatsmark, I.: Modellierung von Verdrängungsvorgängen in Laborversuchen. Submitted to *Zeitschrift für angewandte Mathematik und Mechanik*.
3. Firoozabadi, A., and Aziz, K.: Relative permeability from centrifuge data. Paper SPE 15059 presented at the *SPE 56th California Regional Meeting, Oakland CA*, 1986.
4. Guo, Y.: Centrifuge experiment and relative permeabilities. Dissertation 1988:34, The Norwegian Institute of Technology, Trondheim, 1988.
5. O'Meara, D.J., and Crump, J.G.: Measuring capillary pressure and relative permeability in a single centrifuge experiment. Paper SPE 14419 presented at the *SPE 60th Annual Technical Conference and Exhibition, Las Vegas NV*, 1985.

GAS-CONING BY A HORIZONTAL WELL

Paul Papatzacos
(Høgskolesenteret i Rogaland, Norway)

ABSTRACT

The problem of gas-coning by a horizontal well is solved in the framework of a model which accounts for the movement of the gas-oil boundary. The main result is a plot of critical time versus rate.

The model assumes an infinitely long horizontal well in the oil-zone. Both fluids are incompressible and gas is at all times in static equilibrium. Flow of oil is then two-dimensional in the plane perpendicular to the axis of the well. In this plane, the gas-oil interface is a line which forms a moving boundary and the problem of calculating the flow of oil is a moving-boundary problem. The introduction of boundary-fitted orthogonal coordinates transforms the moving boundary problem into two classical boundary-value problems and one initial-value problem. The boundary-value problems are solved analytically. The initial-value problem, which controls the movement of the boundary, is in the form of a non-linear integro-differential equation. For small rates this equation is solved by transforming it into an infinite system of ordinary differential equations, and using acceleration of convergence as a method of closure. For large rates an iteration method is used.

1. INTRODUCTION

Horizontal wells are becoming a possible alternative to vertical wells at least for certain types of reservoirs, such as thin reservoirs or reservoirs with water and/or gas coning problems. See references [9], and [5].

This paper is a study of gas coning by a horizontal well. It is based on a model which has been developed for the study of coning by a horizontal well inside a thin oil zone sandwiched between a water zone and a gas cap [7]. The assumptions underlying the model are presented in the next section. The equations arising from these assumptions are given in section 3. Boundary-fitted normal coordinates are introduced in section 4. The coning problem is then solved as a moving boundary problem in section 5 where, in particular, critical time is plotted versus rate.

2. MODEL FOR GAS CONING BY A HORIZONTAL WELL

It is first assumed that capillary pressure is zero so that the gas-oil contact is at all times a well-defined surface. It is further assumed that the well has zero radius, that it is parallel to the original gas-oil contact and at a distance a below it, and that its length L is much larger than a. The last assumption implies that, to a good approximation, flow is two-dimensional and can be described in a vertical plane perpendicular to the axis of the well. In that plane the well is represented by a point and the original gas-oil contact by a line. The point representing the well is chosen as the origin of coordinates. (See figure 1. With reference to this figure, it is pointed out that most of this paper is written with dimensionless quantities, the relevant transformation formulas being given in the Appendix.)

The reservoir is supposed to have infinite extent areally, which means that there is no finite drainage radius and no lateral drive. It is then expected that no critical rate exists. In other words, no equilibrium position of the gas-oil interface exists. This interface will move towards the well and breakthrough will occur for any chosen rate of depletion, at a time called *critical time* (see [7] and the original work by Muskat [6]). The model is thus applicable to those situations where the reservoir is so large that the presence of its sealing boundaries is not felt before breakthrough.

The main goal of this paper is to find critical time as a function of rate through a calculation that accounts for the movement of the gas-oil interface during production. The coning problem is actually a free boundary or, more generally, a moving boundary problem, and methods have been developed these last decades for solving a variety of such problems. See for example [2] and [10]. To apply these methods it is necessary to introduce some simplifying assumptions in addition to the ones outlined above [7]. These additional assumptions are as follows.

1. Oil is incompressible.

2. Gas is in static equilibrium.

The first assumption above is probably not detrimental to the main purpose of the calculation, i.e., to obtain the critical time, since compressibility effects are important inside a usually very short time-interval after the well is opened to flow. Appreciable coning effects will take place at larger times, at least when the rate of depletion is chosen so as to postpone breakthrough as much as possible. The second assumption, namely that gas does not flow, has a considerable simplifying effect since it allows to work only with the one equation describing the flow of oil. The assumption is *a priori* reasonably good since one expects that the largest flow velocities will occur around the well and that for reasonably small rates of depletion the interface will move so slowly that the gas movement can be neglected.

3. FLOW-EQUATION AND BOUNDARY CONDITIONS

Oil-compressibility being zero the flow equation for oil reduces to a Poisson equation for the potential $\Phi(x, y, t)$:

$$(\partial^2/\partial x^2 + \partial^2/\partial y^2)\Phi = 2\pi Q \delta(x)\delta(y), \tag{1}$$

where $\delta(x)$ is the Dirac delta-function and it is reminded that all quantities are dimensionless (see the Appendix). The coordinates are shown in figure 1.

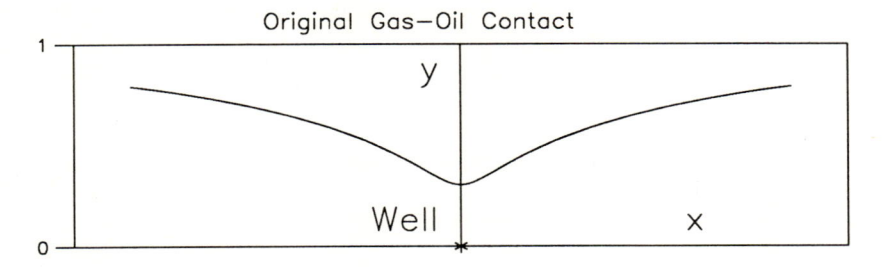

Figure 1: Well and gas-oil interface in a plane perpendicular to the well-axis

There are two types of boundary conditions, which one may call "static" and "dynamic" respectively. The dynamic boundary condition expresses the velocity of the fluid at a point of the gas-oil interface in terms of the potential Φ and it is convenient to postpone its mathematical formulation to the next section, where the boundary fitted orthogonal coordinates are defined. The static boundary condition expresses the equality of pressures across the gas-oil interface. Gas being in static equilibrium the latter boundary condition is found to be

$$\Phi(x_b, y_b, t) = y_b(t) - 1, \tag{2}$$

where (x_b, y_b) is a point on the interface. The fact that the potential is not constant on the boundary precludes the use of conformal mapping (see the discussion in [7] where references to work on the conformal mapping method are given). The method of solution presented here is based on the use of boundary-fitted orthogonal coordinates [10].

4. BOUNDARY-FITTED NORMAL COORDINATES. DYNAMIC BOUNDARY CONDITIONS

A new set of coordinates (ξ, η) is introduced, with the property that the $\xi = \xi_0$ curves are orthogonal to the $\eta = \eta_0$ curves. This can be achieved by demanding that [10,7]

$$\frac{\partial x}{\partial \xi} = \frac{\partial y}{\partial \eta}, \qquad \frac{\partial x}{\partial \eta} = -\frac{\partial y}{\partial \xi}. \tag{3}$$

The ranges of these new coordinates are

$$-\infty < \xi < +\infty, \tag{4}$$
$$-\infty < \eta \leq 1, \tag{5}$$

implying that lines $\xi = \xi_0 \to \infty \ (-\infty)$ are infinitely far to the right (left) and that, by symmetry, $\xi = 0$ is the y-axis. In the same way, the line $\eta = \eta_0 \to -\infty$ is at infinite depth, while the line $\eta = 1$ represents the gas-oil interface at all times.

A point at the interface is given by an arbitrary ξ and $\eta = 1$. Its velocity is given by Darcy's law,

$$\frac{\partial x}{\partial t} = -\frac{\partial \Phi}{\partial x}, \quad \frac{\partial y}{\partial t} = -\frac{\partial \Phi}{\partial y}, \tag{6}$$

where derivations are performed at constant ξ and $\eta = 1$. Eqs. 6 are the dynamic boundary conditions on Φ.

The change of variables is conveniently expressed as

$$x(\xi, \eta, t) = \xi + X(\xi, \eta, t), \tag{7}$$
$$y(\xi, \eta, t) = \eta + Y(\xi, \eta, t), \tag{8}$$

where, according to eqs. 3, functions X and Y satisfy

$$\frac{\partial X}{\partial \xi} = \frac{\partial Y}{\partial \eta}, \quad \frac{\partial X}{\partial \eta} = -\frac{\partial Y}{\partial \xi}. \tag{9}$$

This means that Y satisfies the Laplace equation

$$(\partial^2/\partial \xi^2 + \partial^2/\partial \eta^2)Y = 0. \tag{10}$$

X also satisfies the Laplace equation but if Y can be determined then the calculation of X follows from eqs. 9. Actually, we shall see that X need not be calculated.

Expressing eqs. 1 and 2 in the new coordinates one can set up the following boundary value problem for the potential:

$$\begin{cases} (\partial^2/\partial \xi^2 + \partial^2/\partial \eta^2)\Phi = 2\pi Q \delta(\xi)\delta(\eta - \eta_W), \\ \Phi(\xi, 1, t) = Y_g(\xi, t), \\ \Phi(\xi, -\infty, t) = 0, \\ \Phi(\pm\infty, \eta, t) = 0, \end{cases} \tag{11}$$

where Y_g is the vertical displacement of a point at the gas-oil interface,

$$Y_g(\xi, t) = Y(\xi, 1, t), \tag{12}$$

and η_W is the value of η at the well. This parameter is a function of time and a calculational procedure to determine it will be given below. The last two equations in the set of equations 11 express the fact that, very far from the well-axis in any direction, the pressure is in hydrostatic equilibrium (see the relation between pressure and potential in the Appendix).

The Y-function is also determined by a boundary value problem, namely:

$$\begin{cases} (\partial^2/\partial\xi^2 + \partial^2/\partial\eta^2)Y = 0, \\ Y(\xi,1,t) = Y_g(\xi,t), \\ Y(\xi,-\infty,t) = 0, \\ Y(\pm\infty,\eta,t) = 0, \end{cases} \quad (13)$$

The last two equations express the fact that, very far from the well-axis, the differences between coordinates (x,y) and (ξ,η) vanish (see equations 7 and 8).

Both boundary value problems above have the same Green function [11]:

$$G = \frac{1}{4\pi} \ln \frac{(\xi-\xi')^2 + (\eta-\eta')^2}{(\xi-\xi')^2 + (2-\eta-\eta')^2},$$

so that

$$Y(\xi,\eta,t) = \frac{1}{\pi} \int_{-\infty}^{+\infty} \frac{1-\eta}{(\xi-\xi')^2 + (1-\eta)^2} Y_g(\xi',t) \, d\xi', \quad (14)$$

$$\Phi(\xi,\eta,t) = Y(\xi,\eta,t) + \frac{Q}{2} \ln \frac{\xi^2 + (\eta-\eta_W)^2}{\xi^2 + (2-\eta-\eta_W)^2}. \quad (15)$$

The potential thus depends on Y_g and η_W.

4.1 An initial value problem for Y_g

An equation determining Y_g is found by making use of the dynamic boundary conditions, eqs. 6. The second of these equations yields, with eqs. 8 and 12,

$$\frac{\partial Y_g}{\partial t} = -\left.\frac{\partial \Phi}{\partial y}\right|_{\eta=1}.$$

The right-hand side of this equation can be expressed in terms of the partial derivatives of Φ with respect to ξ and η and one finds

$$\frac{\partial Y_g}{\partial t} = -\frac{\left(\frac{\partial Y_g}{\partial \xi}\right)^2 + \left(\frac{\partial \Phi}{\partial \eta}\right)_{\eta=1}\left[1+\left(\frac{\partial Y}{\partial \eta}\right)_{\eta=1}\right]}{\left(\frac{\partial Y_g}{\partial \xi}\right)^2 + \left[1+\left(\frac{\partial Y}{\partial \eta}\right)_{\eta=1}\right]^2}, \quad (16)$$

where the partial derivatives of Φ and Y on the right hand side can be expressed in terms of Y_g through eqs. 14 and 15:

$$\left(\frac{\partial Y}{\partial \eta}\right)_{\eta=1} = -\frac{1}{\pi}\int_{-\infty}^{+\infty} \frac{\partial Y_g}{\partial \xi'} \frac{d\xi'}{\xi'-\xi}, \quad (17)$$

$$\left(\frac{\partial \Phi}{\partial \eta}\right)_{\eta=1} = \left(\frac{\partial Y}{\partial \eta}\right)_{\eta=1} + 2Q\frac{1-\eta_W}{\xi^2 + (1-\eta_W)^2}. \quad (18)$$

Thus Y_g satisfies an initial value problem in the form of the nonlinear integro-differential equation 16 with initial condition

$$Y_g(\xi, 0) = 0. \tag{19}$$

The solution to equations 16–19 is the subject of section 5.

Note that the parametric equations of the gas-oil interface are obtained by setting $\eta = 1$ in eqs. 7 and 8. Thus the interface is known at a time t as soon as $Y(\xi, 1, t) = Y_g(\xi, t)$ and $X(\xi, 1, t) \equiv X_g(\xi, t)$ are known. Using eqs. 9, one finds

$$X(\xi, \eta, t) = -\frac{1}{\pi} \int_{-\infty}^{+\infty} Y_g \frac{\xi' - \xi}{(\xi' - \xi)^2 + (1 - \eta)^2} d\xi', \tag{20}$$

$$X_g(\xi, t) = -\frac{1}{\pi} \int_{-\infty}^{+\infty} Y_g \frac{d\xi'}{\xi' - \xi}. \tag{21}$$

4.2 An equation defining η_W

Since $\xi = 0, \eta = \eta_W$ at the well, one must have

$$x(0, \eta_W, t) = 0,$$
$$y(0, \eta_W, t) = 0.$$

Eq. 20 shows that the first of these equations is identically satisfied when account is taken of the fact that Y_g is a symmetric function of ξ. The second equation gives the equation defining η_W, i.e.

$$\eta_W + Y(0, \eta_W, t) = 0. \tag{22}$$

Eq. 14 gives the following functional relation between η_W and Y_g:

$$\eta_W + \frac{1 - \eta_W}{\pi} \int_{-\infty}^{+\infty} \frac{Y_g(\xi', t) d\xi'}{\xi'^2 + (1 - \eta_W)^2} = 0. \tag{23}$$

5. SOLUTION TO EQUATIONS 16–19

The non-linear integro-differential equation 16 will be solved by two methods, depending on the value of the dimensionless rate Q. For small Q (≤ 0.4), eq. 16 is transformed into a set of coupled, ordinary, first-order differential equations [1] which must be solved numerically. For large Q, iteration leads to a solution in closed form. The first method especially depends on transforming the integration in eqs. 17 and 18 from an infinite to a finite range. This is done in the next subsection.

5.1 Change of variable

Using the symmetry properties of the integrand (Y_g is symmetric, $\partial Y_g/\partial \xi$ is antisymmetric with respect to ξ) the integrations in equations 14, 17, 20, 21, and 23 can be transformed to integrations from 0 to ∞. Introducing a variable α and a function of time ϵ_W by

$$\xi = \tan\frac{\alpha}{2}, \qquad -\pi < \alpha < +\pi, \tag{24}$$

$$\eta_W = \frac{2\tan(\epsilon_W/2)}{1+\tan(\epsilon_W/2)}, \qquad 0 < \epsilon_W < \pi/2, \tag{25}$$

one finds, for the functions appearing in eq. 16:

$$\frac{\partial Y_g}{\partial \xi} = (1+\cos\alpha)\frac{\partial Y_g}{\partial \alpha}, \tag{26}$$

$$\left(\frac{\partial Y}{\partial \eta}\right)_{\eta=1} = \frac{1+\cos\alpha}{\pi}\int_0^\pi \frac{\partial Y_g}{\partial \alpha'}\frac{\sin\alpha'\,d\alpha'}{\cos\alpha'-\cos\alpha}, \tag{27}$$

$$\left(\frac{\partial \Phi}{\partial \eta}\right)_{\eta=1} = \left(\frac{\partial Y}{\partial \eta}\right)_{\eta=1} + Q\cos\epsilon_W \frac{1+\cos\alpha}{1-\sin\epsilon_W \cos\alpha}. \tag{28}$$

The equation defining η_W becomes, by using eqs. 24 and 25 in eq. 23,

$$\frac{2\tan(\epsilon_W/2)}{1+\tan(\epsilon_W/2)} + \frac{1}{\pi}\int_0^\pi \frac{Y_g(\alpha,t)\cos\epsilon_W}{1-\sin\epsilon_W \cos\alpha}\,d\alpha = 0. \tag{29}$$

Finally the equation for the horizontal displacement of a point at the gas-oil interface (eq. 21) becomes

$$X_g(\alpha,t) = \frac{\sin\alpha}{\pi}\int_0^\pi \frac{Y_g(\alpha',t)}{\cos\alpha'-\cos\alpha}\,d\alpha'. \tag{30}$$

5.2 Solution for small rates

$Y_g(\alpha,t)$ is expanded in a Fourier series with time-dependent coefficients:

$$Y_g(\alpha,t) = \tfrac{1}{2}A_0(t) + \sum_{n=1}^{\infty}\frac{A_n(t)}{n}\cos n\alpha, \tag{31}$$

and the problem is now to calculate the $A_n(t)$. Using this expansion in eqs. 26–28 one finds that eq. 16 can be written

$$\frac{\partial Y_g}{\partial t} = F(A_0,\ldots;\alpha), \tag{32}$$

where

$$F = -\frac{R_1^2 + (1+R_2)(R_0+R_2)}{R_1^2 + (1+R_2)^2}, \tag{33}$$

and
$$R_0 = Q \cos \epsilon_W (1 + \cos \alpha)/(1 - \sin \epsilon_W \cos \alpha), \tag{34}$$
$$R_1 = -(1 + \cos \alpha) \sum_{n=1}^{\infty} A_n(t) \sin n\alpha, \tag{35}$$
$$R_2 = (1 + \cos \alpha) \sum_{n=1}^{\infty} A_n(t) \cos n\alpha. \tag{36}$$

In eq. 34, ϵ_W is an implicit function of the A_n, found by using eqs. 24, 25, and 31 in eq. 29:
$$\frac{2\tan(\epsilon_W/2)}{1+\tan(\epsilon_W/2)} + \tfrac{1}{2} A_0 + \sum_{n=1}^{\infty} \frac{A_n}{n} \left(\tan \frac{\epsilon_W}{2} \right)^n = 0. \tag{37}$$
Using eq. 31 on the left-hand side of eq. 32 one obtains
$$\tfrac{1}{2} \dot{A}_0(t) + \sum_{n=1}^{\infty} \frac{\dot{A}_n}{n} \cos n\alpha = F(A_0, \ldots; \alpha),$$
so that
$$\dot{A}_n = \frac{2N_n}{\pi} \int_0^{\pi} F(A_0, \ldots; \alpha) \cos n\alpha \, d\alpha, \qquad (N_0 = 1; N_n = n, n > 0). \tag{38}$$
We thus have an infinite set of coupled, ordinary differential equations. The initial conditions, according to eqs. 19 and 31 are
$$A_n(0) = 0, \qquad n = 0, 1, \ldots. \tag{39}$$
The calculation of the right-hand side of eqs. 38 can be done very effectively by using a Fast Fourier Transform algorithm (FFT). The main problem is to limit the number of Fourier coefficients so as to reduce eqs. 38 to a finite set which can be solved by a standard solver. With acceleration of convergence [7,3,4] the number of equations can be reduced to 8 for Q less than about 0.2 but it turns out that it must be increased to 32 when Q is 0.4. For larger Q the method of this section breaks down. The next section describes an iteration method which is applicable to large Q.

When the appropriate number of Fourier coefficients has been calculated at time t then eq. 31 gives the vertical displacement of the interface by accelerating the sum. The horizontal displacement is given by eq. 30. Actually, it is more effective to substitute eq. 31 in eq. 30 and to perform the integration analytically. One finds
$$X_g(\alpha, t) = \sum_{n=1}^{\infty} \frac{A_n}{n} \sin n\alpha,$$
and acceleration of the sum gives the horizontal displacement. Setting $\eta = 1$ in eqs. 7 and 8 one gets a parametric representation for the curve representing the gas-oil interface. Figure 2 shows the movement of the interface for different values

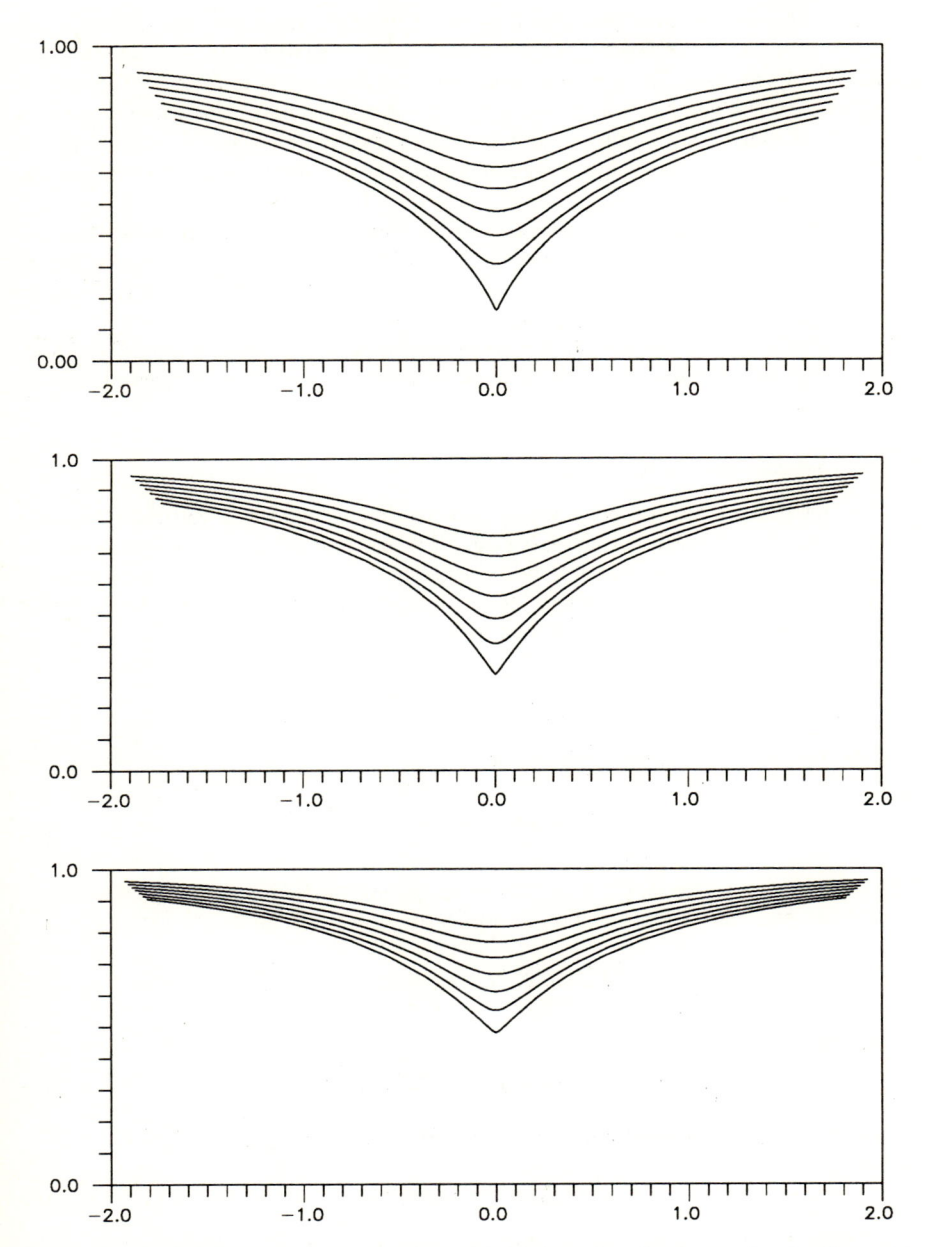

Figure 2: Displacement of the gas-oil interface for $Q = 0.2$ (top), $Q = 0.3$ (middle), and $Q = 0.4$ (bottom). The interface in each plot is shown at times $t_n = nt_c/10$, $n = 4, \ldots, 10$.

of Q. The figure shows that, for a given Q, the interface develops a cusp. This cusp appears at the time t_c, called the critical time, at which the denominator in eq. 16 vanishes. This means that the fluid velocity is infinite at t_c which is effectively the time at which breakthrough of gas occurs. The appearance of a cusp is a well known feature in moving boundary problems (see [7] and references given there). Note also that the cusp appears at a distance from the well, as was first noticed by Muskat [6].

It turns out that, as Q increases, higher and higher frequencies are needed in the Fourier expansion of eq. 31. For $Q > 0.4$ the high frequencies are so important that the acceleration procedure is not capable of guessing the correct behaviour of the function from a sample of low frequencies.

By solving eqs. 38 and 39 for given values of Q until the denominator of F (eq. 33) vanishes one can obtain a curve of t_c versus Q. See the full line on figure 3.

5.3 Critical time for large rates

The purpose of the calculations in this section is to obtain critical time as a function of rate, when the latter is large. The method is based on the observation that t_c becomes small when Q becomes large. An indication of the form of Y_g for small times can be obtained by setting $t = 0$ in eq. 16. Using the fact that $Y_g = 0$ and $\epsilon_W = 0$ when $t = 0$, one gets

$$\left.\frac{\partial Y_g}{\partial t}\right|_{t=0} = -Q(1+\cos\alpha).$$

One now assumes [7] that, when $t \ll 1$,

$$Y_g(\alpha, t) = Q\gamma(t)\cos\epsilon_W \frac{1+\cos\alpha}{1-\sin\epsilon_W \cos\alpha}, \qquad (40)$$

where $\epsilon_W = \epsilon_W(t)$, $\epsilon_W(0) = 0$. Substituting this formula in eq. 29 defining ϵ_W, one can perform the integration analytically:

$$\frac{1}{\pi}\int_0^\pi \frac{Y_g(\alpha,t)\cos\epsilon_W}{1-\sin\epsilon_W \cos\alpha}\,d\alpha = Q\gamma(t)\frac{1+\sin\epsilon_W}{\cos\epsilon_W}.$$

Eq. 40 gives now

$$Q\gamma(t) = Y_g(0,t)\frac{1-\sin\epsilon_W}{2}. \qquad (41)$$

Using these last two equations together with eq. 25 in eq. 29, one finds

$$\eta_W = -\tfrac{1}{2}Y_g(0,t). \qquad (42)$$

To obtain critical time one can now solve eq. 16 for the particular value $\alpha = 0$ which gives the point on the interface just above the well, i.e., the point where

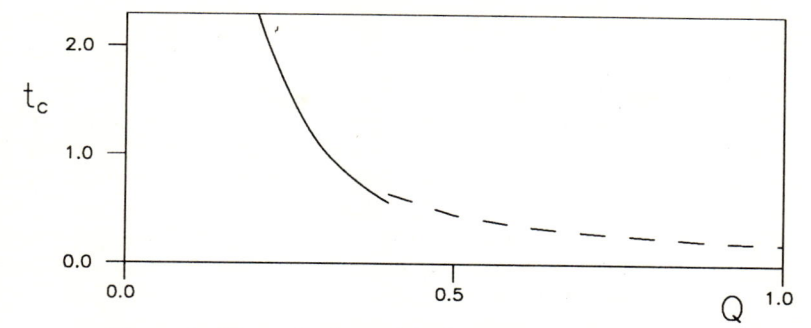

Figure 3: Dimensionless critical time versus dimensionless rate

the denominator of eq. 16 first vanishes. Since Y_g is symmetric in α one gets

$$\frac{d}{dt}Y_g(0,t) = -\frac{\left(\frac{\partial \Phi}{\partial \eta}\right)_{\substack{\eta=1 \\ \alpha=0}}}{1+\left(\frac{\partial Y}{\partial \eta}\right)_{\substack{\eta=1 \\ \alpha=0}}}, \qquad (43)$$

and the expressions on the right-hand side can be obtained by using eq. 40 in eqs. 27 and 28. The integrations can be performed analytically. Setting $\alpha = 0$ and using eq. 41 one then gets

$$\left(\frac{\partial Y}{\partial \eta}\right)_{\substack{\eta=1 \\ \alpha=0}} = Y_g(0,t)\frac{1+\sin\epsilon_W}{\cos\epsilon_W}, \qquad (44)$$

$$\left(\frac{\partial \Phi}{\partial \eta}\right)_{\substack{\eta=1 \\ \alpha=0}} = Y_g(0,t)\frac{1+\sin\epsilon_W}{\cos\epsilon_W} + 2Q\frac{\cos\epsilon_W}{1-\sin\epsilon_W}. \qquad (45)$$

Using eq. 25 and eq. 42 one gets the very simple differential equation:

$$\frac{d}{dt}Y_g(0,t) = -2\frac{Y_g+2Q}{3Y_g+2}, \qquad (46)$$

which shows that critical time is reached when $Y_g = -2/3$ that is when the lowest point of the gas-oil interface has covered two thirds of the distance to the well axis. Eq. 46, together with the initial condition $Y_g(0,0) = 0$ (see eq. 19) gives

$$\tfrac{3}{2}Y_g + (1-3Q)\ln\frac{Y_g+2Q}{2Q} = -t, \qquad (47)$$

so that critical time is given by

$$t_c = 1 - (3Q-1)\ln\frac{3Q}{3Q-1}. \qquad (48)$$

This curve is shown as the broken line on figure 3. There is unfortunately no overlap between this line and the full line which originates from the calculation with acceleration of convergence.

6. CONCLUSIONS

The main result of this work is the curve giving critical time versus rate, figure 3. It is reasonable to assume that the full line gives a good approximation to critical time because it corresponds to situations where the model is closest to reality. It will be recalled (see section 2) that one of the main assumptions in the model is that gas is at all times in static equilibrium. Now it is pointed out in section 5.2 that the velocity of the interface where it is closest to the well becomes infinite at critical time. With the dimensionless variables introduced in the Appendix, vertical velocity v' is related to dimensionless vertical velocity v through

$$v' = \frac{g(\rho_o - \rho_g)k_V}{\phi\mu} v.$$

A closer look at the numerical results shows that, for the small rates corresponding to the full line of figure 3, v at the interface is in absolute value less than one or of order one for most of the time-interval $(0, t_c)$ and that large values of v occur just before the formation of the cusp, inside a very short time-interval $(t_c - \epsilon, t_c)$ where ϵ is about one percent of t_c.

An evaluation of the model by comparison with simulations of two-phase flow will be published elsewhere [8].

APPENDIX: DIMENSIONLESS QUANTITIES

Let a be the distance from the well-axis to the original gas-oil contact, and let k_V, k_H, g, ρ_o, ρ_g, p, and μ be respectively the vertical and horizontal permeabilities, the acceleration due to gravity, the oil and gas densities, the oil pressure, and the oil viscosity. Further, let Q' be the volumetric rate of oil-production per unit length of perforated interval, let x', y', and t' be the space and time variables, and let Φ' be the potential defined by

$$\Phi'(x', y', t') = p(x', y', t') - p(\infty, 0, 0) + \rho_o g y'.$$

The dimensionless counterparts to the primed variables are defined as follows:

$$x = \sqrt{\frac{k_V}{k_H}} \frac{x'}{a}, \quad y = \frac{y'}{a}, \quad t = \frac{g(\rho_o - \rho_g)k_V t'}{a\phi\mu}$$

$$\Phi = \frac{\Phi'}{ag(\rho_o - \rho_g)}, \quad Q = \frac{\mu Q'}{2\pi\sqrt{k_V k_H} ag(\rho_o - \rho_g)},$$

References

[1] Bellman, R., and Adomian, G., "Partial Differential Equations" Reidel, Dordrecht, 1985.

[2] Crank, J., "Free and Moving Boundary Problems" Oxford University Press (Clarendon), Oxford, 1984.

[3] Gustafson, S.Å., Convergence Acceleration on a general class of Power Series, Computing, 21, 53–69, 1978.

[4] Gustafson, S.Å., Two Computer Codes for Convergence Acceleration, Computing, 21, 87–91, 1978.

[5] Joshi, S.D., A Review of Horizontal Well and Drainhole Technology, paper SPE 16868, presented at the 1987 Annual Meeting, Dallas, Sept. 27–30.

[6] M. Muskat: "Physical Principles of Oil Production" International Human Resources Development Corporation, Boston, 1981.

[7] Papatzacos, P., and Gustafson, S.Å., Incompressible Flow in Porous Media with Two Moving Boundaries, J. Comput. Phys., 78, 231–248, 1988.

[8] Papatzacos, P., Herring, T. R., Martinsen, R., and Skjæveland, S. M., Cone Breakthrough Time for Horizontal Wells,, paper SPE 19822, presented at the 1989 Annual Meeting, San Antonio, Oct. 8–11.

[9] Reiss, L.H., Horizontal Wells' Production After Five Years, paper SPE 14338, presented at the 1985 Annual Meeting, Las Vegas, Sept. 22–25.

[10] Ryskin, G., and Leal, L.G., Orthogonal Mapping, J. Comput. Phys., 50, 71–100, 1983.

[11] Williams, W.E., "Partial Differential Equations" Oxford University Press (Clarendon), Oxford, 1980.